Elements of Deep Learning

BIN TRAVERLER FORM

Cut By: [signature] Qty 14 Date 10/19/26

Scanned By: ______________ Qty ______ Date ______

Scanned Batch ID's

______________ ______________ ______________

Notes / Exceptions

____

Benyamin Ghojogh · Ali Ghodsi

Elements of Deep Learning

Springer

Benyamin Ghojogh
Department of Electrical and Computer Engineering
University of Waterloo
Waterloo, ON, Canada

Ali Ghodsi
Department of Statistics and Actuarial Science
University of Waterloo
Waterloo, ON, Canada

ISBN 978-3-032-10737-4 ISBN 978-3-032-10738-1 (eBook)
https://doi.org/10.1007/978-3-032-10738-1

This Springer imprint is published by the registered company Springer Nature Switzerland AG
The registered company address is: Gewerbestrasse 11, 6330 Cham, Switzerland

To Iran and its people—the land and the souls who nurtured us and gave without asking. This book is a small token of gratitude.

Benyamin Ghojogh:
To my lovely wife Golbahar Amanpour and my dear parents and parents-in-law, Shokuh Azam Zolfaghari, Bibiaysheh Rouhi, Yousef Ghojogh, and in memory of Siroos Amanpour.

Ali Ghodsi:
The final edits of this book were made beside my mother's hospital bed. I dedicate this book to her, whose love outlasted memory, and whose absence echoes where once her quiet grace lived.

To my father, whose silence still speaks, forty-five years after I lost him.

To Soroush and Maseeh, my pride, my joy—may you always carry the faith that I carry in you.

Preface

How This Book Can Be Useful

Motivation

Deep learning is reshaping the world—powering breakthroughs in natural language processing, healthcare, art, autonomous systems, and beyond. As its influence continues to grow, so too does the need for clear, up-to-date, and comprehensive educational resources.

This book aims to fill that need. Deep learning evolves at a rapid pace, and many existing textbooks struggle to keep up with its latest developments. Our goal is to provide a unified and coherent reference that not only explains the foundational theories of deep learning but also introduces emerging models and techniques that define the state of the art.

The textbook is structured to cover five main areas: Fundamentals of Deep Learning, Sequence Models, Generative Models, Emerging Topics in Deep Learning, and Deep Learning in Practice. It balances theory with practice—each major concept is explained with mathematical rigor, while the final chapters provide implementation guidance and sample deep learning codes to help readers gain hands-on experience.

Whether you are interested in feature extraction, image recognition, natural language processing, or signal analysis, the tools and insights presented here are widely applicable. Our intention is to offer readers both a strong theoretical foundation and a timely view of current trends in deep learning.

Targeted Readers

This book is designed to serve a broad audience interested in gaining a deep understanding of deep learning and neural networks—from foundational theory to modern practice. Whether in academia, research, or industry, readers will find value in both the structured explanations and practical guidance.

- Academic researchers and students—including undergraduate students, graduate students, post-graduate researchers, professors, lecturers, teachers, and librarians—can use this book as a textbook for deep learning and neural networks.

- Data scientists, machine learning scientists, computer vision scientists, computer scientists, statisticians, and mathematicians in the fields of applied mathematics and statistical learning can use this book as a reference for both technical and applied concepts.
- Industry professionals and practitioners—including applied engineers, data engineers, and engineers in various fields of science dealing with machine learning and deep learning—can use this book as a guidebook for learning the patterns of their data using neural networks.

This book is a valuable resource for instructors teaching advanced undergraduate or graduate-level courses in engineering, computer science, mathematics, statistics, and related disciplines. It is structured as a reference textbook that can support a wide range of instructional needs—from serving as the primary text in a deep learning course to acting as an in-depth supplementary resource. Researchers and practitioners seeking a deeper understanding of deep learning will also benefit from the comprehensive and accessible presentation.

Grounded in both theory and application, the book explains methods step by step, allowing readers to follow the reasoning and develop a solid understanding of the underlying concepts. No advanced background in machine learning is assumed; only undergraduate-level knowledge of linear algebra and calculus is required. Any additional mathematical tools are introduced within the text as needed.

Corresponding Courses

This book is suitable for a variety of advanced undergraduate or graduate-level courses, including deep learning, artificial neural networks, machine learning, artificial intelligence, computational intelligence, and data science. It supports both theoretical and applied learning outcomes and can be used as a primary textbook or a reference resource.

Beyond its pedagogical value, the book also serves as a historical guide to the development of deep learning as a distinct subfield of machine learning. Chapter 1 offers a chronological review of the evolution of key ideas in both statistical machine learning and deep learning.

Organization of the Book

Parts of the Book

This book is organized into five main parts, each focusing on a distinct area of deep learning: **Fundamentals of Deep Learning**, **Sequence Models**, **Generative Models**, **Emerging Topics in Deep Learning**, and **Deep Learning in Pratice**. Together, they provide a comprehensive and up-to-date roadmap for both newcomers and experienced practitioners.

Fundamentals of Deep Learning

We begin with the foundational principles that underpin deep learning. Chapter 1 introduces core machine learning concepts and their historical development, setting the stage for the chapters that follow. Chapter 2 covers the earliest neural models, from the Perceptron to feedforward neural networks, and includes important components such as activation functions, loss functions, ADALINE, logistic regression, radial basis function networks, and self-organizing maps. Chapter 3 presents the backpropagation algorithm and modern optimization techniques used in training deep networks. Chapter 4 explores convolutional neural networks, including architectures such as LeNet, AlexNet, VGG, Inception, GoogLeNet, U-Net, ResNet, and DenseNet. Chapter 5 discusses strategies for addressing overfitting, including ℓ_1-norm and ℓ_2-norm regularization, weight decay, noise injection, early stopping, and batch normalization. Chapter 6 discusses PAC learning and related theories of generalization. While PAC is not common in most deep learning textbooks, its inclusion provides readers with a formal view of generalization and error bounds. Finally, Chap. 7 introduces Boltzmann Machines, including restricted Boltzmann machines and deep belief networks.

Sequence Models

Part II of the book focuses on sequence modeling—an essential domain for applications such as language modeling and speech recognition. Chapter 8 introduces Recurrent Neural Networks (RNNs), backpropagation through time, Long Short-Term Memory (LSTM) networks, and bidirectional variants. Chapter 9 covers attention mechanisms, Transformers, and newer enhancements such as the Performer. Chapter 10 introduces State Space Models (SSMs), a promising alternative to Transformers. Notable examples such as S4 and Mamba are discussed in detail.

Chapter 11 presents Large Language Models (LLMs) including BERT, GPT, T5, and newer models like ChatGPT, RAG, LLaMa, Alpaca, OpenAssistant, Falcon, Claude, AlphaCode, PaLM, Gemini, Mistral, Qwen, DeepSeek, and Grok. This chapter also covers training limitations and key techniques like alignment, supervised fine-tuning, reward modeling, Reinforcement Learning from Human Feedback (RLHF), and Direct Preference Optimization (DPO). Although large language models are generative models (covered in Part 3 of the book), they are presented in Chap. 11—following the discussion of the attention mechanism (Chap. 9)—for better readability and flow.

Generative Models

This part of the book explores models that learn to generate data. Chapter 12 introduces variational inference and the Variational Autoencoder (VAE). Chapter 13 presents the Generative Moment Matching Network (GMMN) and the Maximum Mean Discrepancy (MMD) metric. Chapter 14 explains Generative Adversarial Networks (GANs) and their variants. Chapter 15 covers diffusion models, the current state-of-the-art in generative modeling.

Emerging Topics in Deep Learning

This part highlights innovative and rapidly evolving areas of deep learning. Chapter 16 introduces deep metric learning, Siamese networks, and self-supervised learning. Chapter 17 covers graph neural networks, including spectral GCNs, ChebNet, GCNs, GATs, and graph autoencoders. Chapter 18 focuses on deep reinforcement learning, featuring architectures like Deep Q-Networks (DQN). Chapter 19 discusses few-shot learning and meta-learning, both of which are critical for data-efficient learning. Chapter 20 presents compression techniques such as knowledge distillation, which are especially valuable in deploying large models efficiently.

Deep Learning in Practice

The final part of the book focuses on how to apply deep learning methods effectively. Chapter 21 offers practical guidance on real-world use of deep neural networks. It discusses training, validation, test, and dev sets, performance issues, data pipeline, architecture selection, optimizer choice, hyperparameter tuning (e.g., batch size and learning rate), and optimizing hardware and memory efficiency. Chapter 22 introduces essential coding and logging libraries for deep learning and provides PyTorch code samples to help readers translate theory into working implementations.

Exercises

To help reinforce key concepts and support active learning, each chapter concludes with a set of exercises. These are categorized by difficulty—*Easy*, *Moderate*, or *Difficult*—to guide readers through a progressively deeper understanding of the material. A detailed solutions manual is available and can be provided to instructors for use in grading and assessment.

Essential and Nonessential Sections

Some sections of the book are labeled as nonessential, denoted by , to help readers navigate the content based on their goals. Nonessential sections can be safely skipped on a first pass without losing the overall flow and may be revisited later for a more complete understanding. Readers seeking a fast understanding can first focus on the essential sections, and then optionally proceed to the nonessential sections for deeper insight and understanding.

Readers can approach this book in different ways: beginners may focus on Chaps. 1–5 and Chaps. 21 and 22, practitioners may emphasize Chaps. 21 and 22 alongside applied examples, and researchers may dive into the theoretical treatments in different chapters based on their field of research. This flexible structure makes the book suitable for a broad range of readers and goals.

Waterloo, Ontario, Canada
Summer 2025

Benyamin Ghojogh
Ali Ghodsi

Acknowledgments

The authors of this book thank Springer's editors for copy-editing the book. The authors acknowledge the efforts of Paul Drougas, Kate Lazaro, Kritheka Elango, Joseph Quatela, Pradeep Rajasekaran, Madanagopal D., Kalai Shahethya, and Rahul Sharma at Springer, during the publication process of the book. They are also thankful to their beloved families for supporting them during the write-up of this book. Last but not least, they thank all those who provided valuable feedback during the preparation of the book.

Some Other Related Books

While deep learning is a rapidly evolving field, several excellent books have already been written that readers may find useful:

- [4] is one of the first and most well-known books on deep learning; however, its materials are getting outdated because many new deep learning models have been proposed after publication of this book. Deep learning is improving very fast and many of the existing books in this field are getting outdated. For example, diffusion models, attention mechanism, Transformers, and meta-learning are not covered in this book. This book suitably covers the fundamental theories of deep learning.
- [6] is a comprehensive book covering shallow networks, backpropagation, regularization, convolutional networks, residual networks, Transformers, graph neural networks, GAN, normalizing flows, variational models, diffusion models, and reinforcement learning. It does not cover some topics such as PAC leaning, moment matching networks, and meta-learning.
- [2] is a comprehensive book explaining probabilistic deep learning, backpropagation, regularization, convolutional neural network, Transformers, graph neural networks, GAN, normalizing flows, autoencoders, and diffusion models.
- [1] is a comprehensive book covering fundamentals of deep learning, backpropagation, regularization, RBF networks, Boltzmann machines, recurrent neural network, convolutional networks, graph neural networks, reinforcement learning, and Transformers.
- [3] has a more mathematical approach than a machine learning approach. It is more suitable for mathematicians who want to analyze neural networks. It covers universal approximators, information capacity assessment, output manifolds, neuromanifolds, and some of the neural network architectures.
- [5] focuses on the fundamentals of deep learning such as backpropagation, regularization, universality of neural networks, and convolutional neural networks.
- [9] mainly covers the fundamentals of neural networks and the different architectures. It covers different convolutional neural networks, Boltzmann machines, autoencoders, GAN, and applications of deep learning in fingerprint recognition and character recognition.
- [7] introduces the research tools for interpretability of deep learning models, with a focus on neural network architectures. It covers backpropagation and neural network architectures including recurrent, convolutional,

autoencoder, GAN, and graph neural networks; however, its main focus is on interpretability in deep learning.

- [8] focuses on usage of PyTorch and Python for coding deep neural networks. It covers the PyTorch library, tensors, backpropagation using PyTorch, learning from images, coding convolutional networks, segmentation, and product deployment.

While the books above have each made valuable contributions to the field, this textbook takes a distinct approach. By organizing the material into five thematic parts—spanning foundational principles to practical implementation—it aims to offer both breadth and depth. The goal is to provide readers with a unified and accessible resource that combines theoretical rigor with real-world insight, bridging the gap between learning and application.

Competing Interests The authors have no competing interests to declare that are relevant to the content of this manuscript.

References

1. Charu C Aggarwal. *Neural networks and deep learning*. Vol. 10. 978. Springer, 2018.
2. Christopher M Bishop and Hugh Bishop. *Deep learning: Foundations and concepts*. Springer Nature, 2023.
3. Ovidiu Calin. *Deep learning architectures: a mathematical approach*. Springer, 2020.
4. Ian Goodfellow, Yoshua Bengio, and Aaron Courville. *Deep learning*. Vol. 1. 2. MIT press Cambridge, 2016.
5. Michael A Nielsen. *Neural networks and deep learning*. Vol. 25. Determination press San Francisco, CA, USA, 2015.
6. Simon JD Prince. *Understanding deep learning*. MIT press, 2023.
7. Ayush Somani, Alexander Horsch, and Dilip K Prasad. *Interpretability in deep learning*. Vol. 2. Springer, 2023.
8. Eli Stevens, Luca Antiga, and Thomas Viehmann. *Deep Learning with PyTorch: Build, train, and tune neural networks using Python tools*. Manning, 2020.
9. M Arif Wani, Farooq Ahmad Bhat, Saduf Afzal, and Asif Iqbal Khan. *Advances in deep learning*. Springer, 2020.

Contents

Part II Sequence Models

Part V Deep Learning in Practice

Introduction and History 1

Every model is a map, not the terrain—yet some maps help us find our way.

Abstract

This chapter first introduces the foundational philosophy and structure of machine learning through the lens of scientific modeling. It emphasizes that, like scientific theories, machine learning models are surrogates of reality, designed to both explain existing data and predict future outcomes. The machine learning process typically involves three core components: defining a hypothesis class (modeling assumption), choosing a scoring mechanism (loss function), and selecting a search strategy (optimization). These elements collectively form a general workflow for learning from data, reflecting the essential trade-off between model fit and generalization. The chapter continues by defining machine learning and its key fundamental tasks—classification, regression, dimensionality reduction, and reinforcement learning. It then surveys essential methods in statistical learning and concludes with a historical overview of neural networks and deep learning, situating them within the broader context of machine learning.

1 Introduction

When we try to understand the world around us, we inevitably construct models of it. None of these models is the world itself; rather, they are surrogates—some more useful than others. In science, we often have no way to determine if our model truly matches reality or diverges from it; the only criterion we can reliably check is whether the model is useful. A model is considered useful if (a) it explains current observations well and (b) it predicts future observations accurately. This principle lies at the heart of both empirical science and the field of machine learning.

As Karl Popper famously emphasized, scientific inquiry does not begin with pure observation but with conjectures—hypotheses about how the world might work. These conjectures are then tested against observations, and through a process of refinement and falsification, they gradually improve our understanding of the world [178]. Machine learning closely mirrors this philosophy: we start with a modeling assumption and test it against data, refining it as needed.

In machine learning, despite the diversity of algorithms, all methods share a common structure. We begin by making an assumption about how the data might be represented—a conjecture in the spirit

B. Ghojogh and A. Ghodsi, *Elements of Deep Learning*,
https://doi.org/10.1007/978-3-032-10738-1_1

of Popper's philosophy—and then refine this assumption using data. This process comprises three essential steps:

1. **Hypothesis Class (Conjecture):**
   First, we define a hypothesis class, which represents our modeling assumption about the data. For example, choosing a linear model implies that we assume a linear relationship between inputs and outputs. A tree-based model implicitly assumes that the data can be partitioned hierarchically based on feature splits. A neural network suggests we assume that the underlying relationship is compositional or layered, capturing complex patterns through learned feature hierarchies. This hypothesis class embodies our initial conjecture about the structure underlying the data.

2. **Scoring Mechanism (Loss Function):**
   Once we have a hypothesis class, we need a way to compare its members—that is, to determine which model performs better than others. This is accomplished using a scoring mechanism that assigns a value (often called a "loss") to each model, quantifying its performance on the data. Common examples of loss functions include the mean squared error (which measures the average squared difference between predicted and actual values) and the cross-entropy loss (often used for classification tasks). This step allows us to rank the models based on how well they fit the data and how effectively they capture the patterns we seek.

3. **Search Strategy (Optimization):**
   Finally, we need to select the best model—the one with the lowest loss—according to our chosen scoring mechanism. This is accomplished through optimization methods, which can vary depending on the structure of the hypothesis class and the loss function. If the optimization problem is convex, there may be efficient analytical or numerical methods to find the global optimum. If the problem is non-convex, as is often the case with neural networks, iterative techniques such as gradient descent are commonly used to search for good solutions, though they may only find local minima. Thus, the choice of optimization method depends critically on the nature of the problem and the properties of the loss landscape.

In summary, the general machine learning workflow can be expressed as follows:

1. Choose a hypothesis class, denoted by $\mathcal{F}$.
2. Define a scoring mechanism, denoted by $S(f)$ for $f \in \mathcal{F}$, where a lower score indicates a better model. In most cases, $S(f)$ represents the loss function.
3. Search for the best model by solving the optimization problem:

$$\widehat{f} := \arg\min_{f \in \mathcal{F}} S(f). \tag{1}$$

This process highlights the "No Free Lunch Theorem" in machine learning [249], which states that no learning algorithm can perform well across all possible problems without making assumptions. In other words, without assumptions, there can be no learning. By choosing a hypothesis class, we effectively impose a bias toward a certain kind of solution—this is known as the inductive bias of the model. The scoring mechanism and the optimization strategy define how we evaluate and search within this hypothesis class. Together, these choices—including the loss function, optimization landscape, and stopping criteria—collectively shape how the model generalizes from observed data to new, unseen situations, making them a fundamental part of the learning process.

Just as in science, where the aim is to build models that both explain and predict the world, the same goal applies in machine learning. A model is considered useful if (a) it explains the current data well, corresponding to a low training error, and (b) it effectively predicts future or unseen data, corresponding to a low test error. In this sense, the notion of a useful model in machine learning is a direct extension of the scientific principle of constructing models that both describe current phenomena and anticipate future ones. The balance between these two aspects—fit to known data and generalization to new data—defines the ultimate utility of the model.

This book begins by explaining what machine learning is and what its fundamental tasks are, and then overviews essential algorithms and ideas in statistical learning, concluding with a brief history of neural networks. This introduction helps the reader understand where deep learning fits within the broader landscape of machine learning.

Section 2 discusses the machine learning model as a specific instance within the hypothesis class introduced in this section. For example, a neural network with a particular set of weights is one member of the broader class of all possible neural networks with the same architecture. Section 3 introduces the fundamental and historical machine learning tasks. Section 4 reviews key algorithms and ideas in statistical learning. Finally, Sect. 5 provides an overview of the history of neural networks and deep learning.

2 Machine Learning Model

In machine learning, each model in the hypothesis class defines a function, or mapping, from inputs to outputs, parameterized by learnable parameters. Learning then involves using a chosen loss function and search strategy to find the best parameter values that yield the desired outputs for given inputs.

2.1 Dataset

Assume some quantity is measured. This quantity can be any data or signal, such as personal health data (e.g., blood pressure or blood sugar), multiple images, or a time series like weather temperature or stock market values. In machine learning, data is usually represented as vectors: if an object can be described by d features, then it can be expressed as a vector in a d-dimensional space. Let the dimensionality of the space, in which the data vectors exist, be d, meaning that every quantity is a d-dimensional vector in the d-dimensional Euclidean space $\mathbb{R}^d$. The set of d values for the quantity can be called the features of the quantity. This set of d-dimensional data points is referred to as a dataset. A set of images or the personal health data of a few people are two examples of datasets.

2.2 Learning Model

Consider a dataset of n data points, $\{\boldsymbol{x}_i \in \mathbb{R}^d\}_{i=1}^n$, each of which is a d-dimensional vector in the Euclidean space $\mathbb{R}^d$. These vectors can be arranged column-wise in a matrix $\boldsymbol{X} \in \mathbb{R}^{d \times n}$.

The learning model f is a function (or mapping) from the data space to an output space:

$$\begin{aligned} f_\theta &: \mathbb{R}^d \to \mathbb{R}^p, \\ \boldsymbol{x} &\mapsto f_\theta(\boldsymbol{x}), \end{aligned} \tag{2}$$

where θ denotes the parameters of the function f. These parameters are learned from the data. Typically, though not necessarily, $p \leq d$. Figure 1 illustrates the concept of the learning model.

Fig. 1 The learning model mapping inputs to the desired outputs

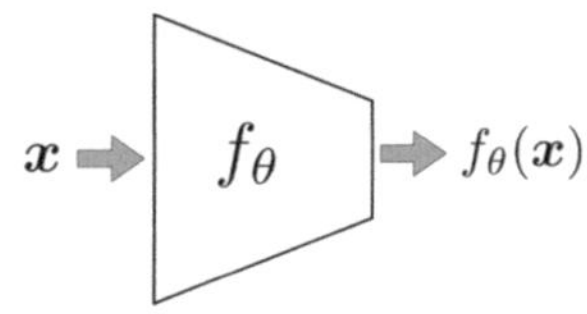

Machine learning algorithms search for the parameter values θ that yield outputs aligned with the desired task. The nature of these desired outputs depends on the specific task. For example, in classification, the desired output is the correct class label for each input data instance.

3 Machine Learning Tasks

Machine learning encompasses a wide range of tasks, which are commonly categorized into supervised, unsupervised, and reinforcement learning.

- **Supervised Learning** includes tasks like classification and regression, where the algorithm learns from labeled data to predict outputs for new, unseen inputs.
- **Unsupervised Learning** includes tasks like clustering and dimensionality reduction, where the goal is to uncover patterns or structure in unlabeled data. Although dimensionality reduction is typically unsupervised, there are also supervised variants that incorporate label information to guide the learning process.
- **Reinforcement Learning** focuses on learning from sequential interactions with an environment, where the model receives feedback in the form of rewards or penalties to improve its performance over time.

In addition to these traditional categories, modern machine learning has evolved to include specialized tasks that do not fit neatly into one category. Examples include semi-supervised learning (which combines labeled and unlabeled data), self-supervised learning (where the model generates its own labels from the data), few-shot learning, and zero-shot learning (where the model generalizes to new domains with little or no training data).

This section introduces these machine learning tasks from a historical and foundational perspective. To aid understanding, a metaphor is used in which the learning model is likened to a newborn baby that must be trained. Note that this metaphor serves only to illustrate the training process and does not imply that the learning model can acquire knowledge without assumptions or prior information.

3.1 Supervised Learning

3.1.1 Classification

Suppose you want to teach a young child to distinguish between apples and cucumbers. You might collect 5 examples of each, show them to the child one by one, and say “this is an apple” or “this is a cucumber.” By providing these labeled examples, the child gradually learns to recognize apples and cucumbers based on their features.

This scenario illustrates the fundamental task of **classification** in machine learning, where the goal is to assign each data instance to one of several predefined classes.

Technical Definition: Consider a dataset of n data points $\{(\boldsymbol{x}_i, y_i)\}_{i=1}^n$, where $\boldsymbol{x}_i \in \mathbb{R}^d$ is the input vector and y_i is the corresponding label. The label y_i takes values in a finite set of class labels

$\{\ell_1, \ell_2, \dots, \ell_m\}$. The task of classification is to learn a function f_θ such that for a new input $\boldsymbol{x}$, the model predicts the label $y = f_\theta(\boldsymbol{x})$:

$$f_\theta : \mathbb{R}^d \to \{\ell_1, \ell_2, \dots, \ell_m\}. \tag{3}$$

Machine learning algorithms for classification aim to find the parameter values θ that yield outputs matching the desired labels as closely as possible.

3.1.2 Regression

Regression is another fundamental task in machine learning, where the goal is to predict a continuous value based on input features.

Suppose you want to teach a child to estimate the diameter of apples. You might collect 5 apples, measure their diameters, and present them to the child while showing their corresponding sizes. With these labeled examples, the child learns to associate the visual features of an apple (such as color, texture, and shape) with its diameter.

Technical Definition: Consider a dataset of n data points $\{(\boldsymbol{x}_i, y_i)\}_{i=1}^n$, where $\boldsymbol{x}_i \in \mathbb{R}^d$ is the input vector and y_i is the corresponding label. Unlike classification, where y_i takes values in a finite set, in regression y_i takes values in $\mathbb{R}$. The task of regression is to learn a function f_θ such that for a new input $\boldsymbol{x}$, the model predicts a real-valued output $y = f_\theta(\boldsymbol{x})$:

$$f_\theta : \mathbb{R}^d \to \mathbb{R}. \tag{4}$$

Machine learning algorithms for regression aim to find the parameter values θ that yield outputs as close as possible to the true values for new, unseen data. This generalizes the training data, allowing the model to predict outputs for inputs that may not be identical to those seen during training.

3.2 Unsupervised Learning

3.2.1 Clustering

Clustering is another fundamental task in machine learning, where the goal is to group data points based on similarity without using any labels.

Suppose you want to teach a child to distinguish apples from cucumbers without telling them their names. You might collect 5 apples and 5 cucumbers, shuffle them, and ask the child to separate them into groups based on how they look. The child might notice that apples tend to be round while cucumbers are longer, and naturally group them accordingly. This grouping is based solely on the patterns of the objects, without any prior label information.

Technical Definition: Consider data $\{\boldsymbol{x}_i\}_{i=1}^n$ where $\boldsymbol{x}_i \in \mathbb{R}^d$. The task of clustering is to learn a function f_θ such that when we observe a new input $\boldsymbol{x}$, we assign it to one of the clusters:

$$f_\theta(\boldsymbol{x}) \in \{\ell_1, \ell_2, \dots, \ell_m\}. \tag{5}$$

Mathematically, the model is expressed as

$$f_\theta : \mathbb{R}^d \to \{\ell_1, \ell_2, \dots, \ell_m\}. \tag{6}$$

This formulation is similar to classification in that the model maps each input to one of a finite set of outputs. However, unlike classification, where we have labeled data $\{(\boldsymbol{x}_i, y_i)\}_{i=1}^n$ to train the model, in clustering we only have input data $\{\boldsymbol{x}_i\}_{i=1}^n$ without any labels. Thus, the learning process in clustering relies entirely on discovering patterns and structures within the input data.

Clustering highlights the inherent ambiguity of this task: one cluster can often be subdivided or merged with others, depending on the desired granularity. To manage this ambiguity, clustering algorithms typically require either specifying the number of clusters in advance or using hyperparameters to determine the clustering structure.

3.2.2 Dimensionality Reduction

Dimensionality reduction is a fundamental task in machine learning, where the goal is to map high-dimensional data into a lower-dimensional space while preserving the important structures or patterns.

Imagine teaching a child to understand the ripeness of fruits. Instead of asking them to memorize every detail—such as size, color, and texture—you show them how to combine these features into a single indicator that captures the overall ripeness. This new dimension represents a summary of the key features, simplifying the understanding of the fruit's condition while preserving important information.

Technical Definition: Consider a dataset of n data points $\{\boldsymbol{x}_i\}_{i=1}^n$ where each $\boldsymbol{x}_i \in \mathbb{R}^d$. The task of dimensionality reduction is to learn a function f_θ that maps each high-dimensional input vector $\boldsymbol{x}$ to a lower-dimensional representation $y = f_\theta(\boldsymbol{x})$, typically in $\mathbb{R}^p$ where $p < d$:

$$f_\theta : \mathbb{R}^d \to \mathbb{R}^p. \tag{7}$$

Here, the lower-dimensional data points are called the embedding data. Usually, $p \ll d$ to achieve significant reduction. When $p = 2$ or $p = 3$, dimensionality reduction can be used for visualization in 2D or 3D plots [156, 226].

Dimensionality reduction can be achieved using linear or nonlinear mappings. When nonlinear mappings are used, the technique is often referred to as manifold learning, since the data may lie on a lower-dimensional manifold within the high-dimensional space. It's worth noting that dimensionality reduction is sometimes called feature extraction in the literature [91].

3.3 Supervised Versus Unsupervised

In unsupervised dimensionality reduction, the goal is to embed similar data points close to each other while placing dissimilar points farther apart, relying solely on the patterns in the data $\{\boldsymbol{x}_i\}_{i=1}^n$. This means pushing similar points together and pulling dissimilar points away, exaggerating similarities and differences in the data structure. Techniques like Locally Linear Embedding (LLE) [78, 195] are among the early methods developed for this purpose.

Supervised dimensionality reduction extends the unsupervised case by incorporating *side information* (such as class labels, pairwise similarities, or other relevant metadata) in addition to the input data. The goal is to use this side information to improve the lower-dimensional embedding. For example, side information might help decrease intra-class variance (bringing similar points together) and increase inter-class variance (pushing different groups apart), improving class separability and interpretability in the reduced space. A classical method that exemplifies this approach is Fisher discriminant analysis [52, 76], which uses class labels as side information.

For example, imagine helping a color-blind and visually impaired person distinguish between red and blue balls on the ground. By pushing the red and blue balls apart (increasing inter-class variance) and then collapsing each group into a tight cluster (decreasing intra-class variance), we make the separation easier to perceive.

Dimensionality reduction—whether unsupervised or supervised—helps with visualization, noise reduction, and simplifying complex datasets while retaining meaningful information.

3.4 Reinforcement Learning

In addition to supervised and unsupervised learning, reinforcement learning is another major category of machine learning. Unlike supervised learning (which uses labeled data) or unsupervised learning (which finds structure in unlabeled data), reinforcement learning trains a model through interactions with its environment, using a system of rewards and penalties to guide learning.

Imagine teaching a young child to put their toys away after playing. Initially, the child may not know what to do when you ask them to clean up. If the child doesn't respond, you provide no reward. One day, by chance, the child starts putting the toys away, and you praise them or give them a treat. After this happens a few more times, the child begins to associate cleaning up with receiving a reward and becomes more likely to repeat the behavior. The more consistently you reward this behavior, the more the child learns that cleaning up leads to a positive outcome. If the child instead makes a mess, you might gently correct them or withhold praise, creating a negative reward that discourages the behavior. Through these positive and negative reinforcements, the child learns the desired behavior over time.

In machine learning, the learning model can be thought of as the child being trained through a system of rewards. In reinforcement learning [213], the model—called the agent—interacts with its environment and learns to take actions that maximize cumulative rewards.

Initially, the agent performs actions randomly. As it interacts with the environment and receives rewards or penalties based on its actions, it learns to adjust its behavior to maximize rewards. Over time, the agent learns which actions to take in which situations to achieve the best outcomes.

Technical Definition: Formally, in reinforcement learning, the learning model's output is an action a chosen from a set of possible actions denoted by $\mathcal{A}$:

$$f_\theta(\boldsymbol{x}) = a \in \mathcal{A}. \tag{8}$$

Researchers have successfully applied reinforcement learning to a wide range of challenging tasks beyond simple games. For example, reinforcement learning has been used to train models that can play the complex board game Go at superhuman levels [209], to optimize robotic control systems, and to develop agents that can navigate complex environments in simulation and real-world settings. These successes highlight the versatility and power of reinforcement learning in tackling tasks that require sequential decision-making under uncertainty.

4 Essential Algorithms and Ideas in Statistical Learning

In the long and rich history of machine learning, there have been many breakthroughs which have been impactful. It is a difficult job to list all of those methods. Here, some of them are listed and explained. Among these methods are neural networks which faced a winter and then gained success again. As this book is about deep learning, the neural network methods are explained in a separate section, i.e., Sect. 5. The current section overviews some of the essential algorithms and ideas in statistical learning.

4.1 Statistical Classification

4.1.1 Bayes Classifier

Thomas Bayes was an English statistician, who lived in 1702–1761. He studied how to compute the probability parameter in a binomial distribution. After he passed away, his friend Richard Price published his work in 1763 [11]. The Bayes' rule, proposed by Bayes, states the relation between the posterior, likelihood, and the prior:

$$\mathbb{P}(x \in \mathcal{C}_j \mid X = x) = \frac{\mathbb{P}(X = x \mid x \in \mathcal{C}_j)\,\mathbb{P}(x \in \mathcal{C}_j)}{\mathbb{P}(X = x)}, \tag{9}$$

where $\mathcal{C}_j$ denotes the jth class, $\mathbb{P}(x \in \mathcal{C}_j \mid X = x)$ denotes the posterior, $\mathbb{P}(X = x \mid x \in \mathcal{C}_j)$ is the likelihood (or class conditional), $\mathbb{P}(x \in \mathcal{C}_j)$ is the prior of the jth class, and $\mathbb{P}(X = x)$ denotes the prior of the data. This rule was inspired by philosophy and it models the a priori and a posteriori in epistemology of philosophy.[1] In epistemology, while a priori knowledge can be obtained without experience, a posteriori knowledge can only be obtained by experience [6]. These two types of knowledge can be combined with being analytic or synthetic which are about relationship between subject and predicate in statements. Therefore, four types of propositions are discussed in epistemology, especially by Immanuel Kant [123], which are analytic a priori, analytic a posteriori, synthetic a priori, and synthetic a posteriori. The prior and posterior in Eq. (9) also have similar meaning as in epistemology. The posterior refers to probability of belonging to a class after observing or experiencing the data x. The prior, however, is the probability of each class before experience of data.

The Bayes classifier maximizes the posteriors of the classes. It is an optimal classifier [162, Chap. 6, p. 175] because it is maximizing the posterior, which is relevant to the multiplication of likelihood and prior as shown in Eq. (9). That is why in the literature, it is also referred to as Bayes optimal classifier. However, the problem with Bayes classifier is that the posterior is usually not known or it is difficult to compute. Therefore, naive Bayes classifier was proposed which naively assume that, given the class label, the features of data are independent.

4.1.2 Linear Discriminant Analysis

While Bayes and naive Bayes classifiers do not make any assumption on the posterior or likelihood, Linear Discriminant Analysis (LDA) [72] assumes a unimodal Gaussian distribution for the likelihood of data in each class. Moreover, LDA makes an additional assumption which is assuming equal covariances for the classes. If this additional assumption is relaxed, Quadratic Discriminant Analysis (QDA) is obtained in which the class boundaries are quadratic while LDA results in linear class boundaries. It is possible to prove that LDA is equivalent to Fisher discriminant analysis (see [76]). Therefore, Ronald A. Fisher, a statistician, can be considered as the proposer of LDA in 1936 [52]. Fisher discriminant analysis will be discussed more in Sect. 4.3.1.

4.1.3 Logistic Regression

Logistic regression is a binary classifier where it assigns probability between zero and one for belonging to one of the classes. In contrast to Bayes classifier, LDA, and QDA—which work on likelihood and prior for dealing with the posterior—logistic regression works on the posterior directly. The logistic

[1] Note that, centuries ago, philosophy and mathematics were twisted together. Many researchers worked on both mathematics and philosophy at the same time. For example, Rene Descartes was a mathematician developing many theories such as Cartesian coordinate system, addition of vectors, and angular defect. He was also a philosopher. Another example is Immanuel Kant, the famous philosopher, who wrote an article about kinetic energy in his early twenties in 1747 [122] and his dissertation thesis in 1755 was about fire, heat, and physical and chemical principles [121].

function, used in logistic regression, was initially proposed in 1845 for modeling the population growth[2] [230]. It was further improved in the 20th century [233]. Refer to [39] for the history of logistic regression. Logistic regression considers the classification problem as a regression problem where it regresses (predicts) the probability of belonging to a class. It first considers a linear regression, but as there is no bound on this regression while the output is desired to be in the range [0, 1] to behave like a probability, it models the posterior using a logistic function, also called the sigmoid function, to make this regression between zero and one. Note that, as explained in Chap. 2, logistic regression can be considered as equivalent to a single neuron with a sigmoid activation function in a neural network.

4.1.4 Metric Learning and Nearest Neighbors Classifier

One of the simplest classifiers is to consider the centers or representatives of the classes in the space and calculate the distances of the test data instance from these centers. The smallest distance determines the class label of the test data. However, who says that Euclidean distance is the best distance for classification in the space? The best distance metric to be used for classification depends on the data. Metric learning is a family of algorithms which learn the best distance metric for classification, clustering, and other tasks [87].

Another simple classifier is k-nearest neighbors (KNN) classifier which classifies every point by a majority vote between the labels of its k-nearest neighbors in the training dataset. To improve the k-NN classification performance, the intra-class and inter-class variances of data can be decreased and increased, respectively [237]; an idea which later developed into the triplet loss for neural networks [206]. Note that KNN can also be used for regression where the label of a test data instance is determined by the summary statistics (e.g., mean) of the labels of its nearest neighbor training instances.

4.1.5 Tree, Random Forest, and Ensemble Learning

A simple-to-implement algorithm is the tree algorithm which partitions the space using trees or usually binary trees. At every node of the tree, it finds the best feature to split the space based on and then it calculates the best threshold for the split. The tree can be implemented and grown using recursion in programming.

Trees can be used for classification or regression. In the test phase of classification by trees, the test data instance falls in one of the end nodes of the tree and its label is determined by the majority vote of labels in that partition of the space. Regression trees, however, label a test instance using summary statistics (e.g., mean) of the partition it falls in.

The tree algorithm for classification is also called the decision tree because it decides the class label by tree. Some of important decision tree algorithms are CHAID (1980) [125], CART (1984) [23], ID3 (1986) [181], C4.5 (1993) [180], OC1 (1994) [167] which was the first method creating multivariate splits at every node, and MARS (1991) [61].

Random forest [24] uses multiple trees in aggregation. It improves the performance of tree because of model averaging [37, 114], also called ensemble learning [177] or bootstrap aggregating (bagging) [22, 28]. Bagging reduces the variance of estimation especially when the models in the ensemble are uncorrelated (see Chap. 5 for proof). Therefore, random forest selects from both data instances and the features for the trees to make the trees uncorrelated as much as possible. Some of the important random forest algorithms are regular random forest (1993) [105], randomized node optimization in random forest (2000) [45], and extremely random forest (2006) [70] for classification and regression.

[2] The value of the logistic function gradually goes from zero to one where it saturates at value one. It is suitable for modeling the population growth of a region because the population grows initially in that region. Then, after some growth, some of the population dye because of lack of hygiene, food, and resources. Therefore, the population saturates at some level.

Mondrian forest (2014–2016) was proposed for online data classification [140] and regression [139]. Isolation forest (2008) [148] was proposed for anomaly detection using trees. iMondrian forest [150] is a hybrid of isolation and Mondrian forests for online anomaly detection.

4.1.6 Support Vector Machine and Kernel Support Vector Machine

Perceptron was one of the earliest neural networks proposed for binary classification. It will be introduced with details in Sect. 5.1.3. If the two classes are linearly separable, Perceptron finds one of the many possible decision boundaries between the two classes, depending on its initialization. It was in 1974 that Vladimir Vapnik thought why we do not try to find the best decision boundary, among all options, between the two classes. In 1974, he proposed Support Vector Machine (SVM) [228] which, by optimization, finds the best decision boundary having equal distances from the two classes. This decision boundary has the best generalization of classification to unseen test data.

The optimization of SVM finally yields to quadratic programming. According to KKT conditions [85], it can be shown that SVM only cares about the data instances of classes which are closest to the decision boundary; these points are called support vectors. As SVM ignores many data instances except the support vectors, it is a sparse algorithm. Therefore, it is effective due to betting on sparsity principal [60, 219] which states that the sparse version of a dense algorithm mostly works better. SVM also simplifies the data by ignoring many of them; therefore, it is also justifiable by Occam's razor [46] stating that if two explanations justify a phenomenon, the simpler explanation is more valid.[3]

The first SVM, which is called the hard-margin SVM, can converge only when the two classes are completely linearly separable. Therefore, in 1995, he proposed soft-margin SVM [38] which tolerates if a few data instances of opposite class fall on the wrong side of decision boundary. In other words, soft margin SVM takes it easier on the data instances compared to hard-margin SVM.

Soft-margin SVM is effective when the two classes are roughly linearly separable but not when they are completely nonlinearly separable. Is such cases, kernel SVM is effective, proposed by Vapnik in 1992 [20, 227]. Kernel SVM maps data to a higher-dimensional space, namely Hilbert space, so that the classes hopefully become linearly separable in higher dimensions. The linear SVM is then applied in the Hilbert space, where hopefully the classes become linearly separable.

Vapnik borrowed the idea of kernels from functional analysis of mathematics and brought it to machine learning [86]. It was in 1904 when David Hilbert proposed his work on kernels and defined a definite kernel [107], and later in 1908, his and Erhard Schmidt's works proposed integral equations such as Fredholm integral equations [107, 203]. This was the introduction of a new space in mathematics which was named the Hilbert space. Reproducing Kernel Hilbert Space (RKHS) is a special case of Hilbert space with some properties having reproducing kernels [86]. Mercer kernels, which are used nowadays in machine learning, were proposed by Mercer in 1909 [158]. The first work on RKHS was in 1950 [5]. Later, the concepts of RKHS were improved further in 1964 [3]. In pure mathematics, eigenfunctions were also developed for eigenvalue problem applied on operators and functions [245]. By the appearance of quantum mechanics in 1920s, eigenfunctions were found to be useful in physics [138]. Kernels and the RKHS remained in pure mathematics and theoretical physics until they were used for the first time in machine learning by introduction of kernel SVM in 1992 [20, 227]. Eigenfunctions were also used in machine learning, for out-of-sample extension of spectral dimensionality reduction algorithms, by Yoshua Bengio in 2003 [16, 17].

[3] For better understanding of Occam's razor, consider the following exaggerated example. Assume you are hungry. It might have one of the following reasons: (1) either you have not eaten for hours or (2) an invisible magician has come and tricked you to become hungry. Both of these explain why you are hungry. Obviously, the first reason, which is simpler, is a better explanation because it has made less assumptions. Note that in logic, reasoning with least amount of assumptions is also called abductive reasoning.

4.2 Statistical Regression

4.2.1 Linear Regression

The simplest regression algorithm is linear regression which fits a hyperplane to the labels of data (e.g., it fits a line for one-dimensional data and labels). Linear regression, also called simple linear regression, uses least squares error between the target labels and the predicted labels to calculate the regression coefficients. Linear regression can also be seen as a linear projection onto the column space of its coefficients. Note that the predicted label is also called the response variable in the literature of regression. Simple linear regression is very old and its history goes back to centuries ago [212], probably to the works by Francis Galton in 1894 [67]; although, the least squared optimization has also been seen in the works of Legendre in 1805 and Gauss in 1809.

Linear regression does not have any control or regularization on the calculated coefficients. In order to penalize the large coefficients, regularization can be used (see Chap. 5). Ridge regression uses ℓ_2 norm regularization, also called Tikhonov regularization, proposed by Andrey Nikolayevich Tikhonov in 1943 [221] and David Phillips in 1962 [176]. Lasso (Least Absolute Shrinkage and Selection Operator) regression uses ℓ_1 norm regularization proposed by Robert Tibshirani at Stanford University in 1996 [218]. Using the unit ball of ℓ_1 norm, whose corners have many zeros in their coordinates, Tibshirani discussed that ℓ_1 norm regularization, or lasso regularization, makes the problem sparse [218] (see Chap. 5, for details). Therefore, the performance of lasso regression is justified by betting on sparsity principal [60, 219] and the Occam's razor [46].

Moreover, generalized linear regression was proposed 1972 [168]. This regression allows the linear regression to be relevant to the predicted labels through a link function. It also allows the variance of each data instance to be a function of its predicted label. It is also noteworthy that sufficient dimension reduction (first proposed in 1991 [145]), which will be introduced in Sect. 4.3.2, also deals with linear regression [83].

4.2.2 Quantile Regression

While linear regression minimizes the mean squared error between the target labels and the predictions, quantile regression minimizes the sum of absolute residuals weighted by a quantile-specific function. In other words, quantile regression considers the quantiles of the predictions rather than their means. According to [134], the history of quantile regression goes back to median regression proposed by Roger Joseph Boscovich in 1760. Later, Pierre-Simon Laplace and Francis Edgeworth continued the work until Roger Koenker combined the works and developed this methodology in the years 1998–2000 [133].

4.2.3 Nonlinear Regression

When the pattern of data is nonlinear, linear regression is not effective. In these cases, nonlinear regression should be used. Many nonlinear regression algorithms have been developed. One of these methods is polynomial regression which uses least squared optimization for fitting a polynomial function to the data. Its history goes back to Joseph Diez Gergonne in 1815 [69]. Some other nonlinear regression methods are regression using k-nearest neighbors, trees, and random forest. After development of artificial neural networks, they have also been used for nonlinear regression without a statistical approach. They are clearly nonlinear because of the nonlinear activation functions used in neural networks.

4.3 Statistical Dimensionality Reduction

Dimensionality reduction methods can be broken down into statistical dimensionality reduction and deep dimensionality reduction. The statistical dimensionality reduction is further divided into spectral dimensionality reduction and probabilistic dimensionality reduction, which will be discussed here. Some of the deep dimensionality reduction methods are generative autoencoders and self-supervised metric learning, which will be discussed in Sects. 5.7 and 5.9.3, respectively. For more information on dimensionality reduction, refer to our other textbook [74].

4.3.1 Spectral Dimensionality Reduction

Spectral dimensionality reduction methods are geometry-based and they usually end up to an eigenvalue problem [89]. It was in the year 1901 when the first unsupervised spectral dimensionality reduction method, named Principal Component Analysis (PCA), was proposed by Karl Pearson [174]. PCA was further developed by Harold Hotelling in 1933 [116]. PCA finds the most variant directions of data in a linear way. Consider a cloud of data points, for example. PCA finds the directions of spread of data in the space and projects data to a few of these directions to reduce the dimensionality of data. If all directions are used in PCA, this method can be used for canceling rotation of data to align to its most variant axis. Refer to [80] for theoretical details of PCA.

Then, the first supervised spectral dimensionality reduction method, named Fisher Discriminant Analysis (FDA), was proposed in 1936 [52] by Ronald Aylmer Fisher (1890–1962) who was a genius in statistics. Fisher proposed many important concepts in the modern statistics, such as variance [53] (can you imagine modern statistics without variance!?), FDA [52], Fisher information [58], and Analysis of Variance (ANOVA) [54]. The paper [52], which proposed FDA, was the first paper introducing the well-known Iris flower dataset. Fisher's work was mostly concentrating on the statistics in the area of genetics. Much of his work was about variance making no wonder for us why FDA is all about variance and scatters. FDA proposed the main idea behind supervised dimensionality reduction. It maximizes the variance between classes while reducing the variance in each class. It is possible to prove that LDA, introduced in Sect. 4.1.2, is equivalent to projecting data onto the FDA subspace and then using Euclidean distance from the means of classes for classification (see [76] for proof); that is why FDA is also called LDA or Fisher LDA in the literature.

Then, Multi-Dimensional Scaling (MDS) was proposed. The idea of MDS is to preserve the similarity [223] or dissimilarity/distances [12] of points in the low-dimensional embedding space. If two data points are similar (resp. dissimilar) in the input space, they should also be similar (resp. dissimilar) in the embedding space. It can be proved that one of the special cases of MDS is equivalent to PCA (see [79] for proof). In 1969, Sammon mapping was proposed by John Sammon [197]. Sammon mapping is a special case of MDS and it preserves the distances between the data points. While PCA, FDA, and many MDA variants were linear methods, Sammon mapping was the first nonlinear dimensionality reduction method.

In the years 1997 and 1998, kernel PCA was proposed by Bernhard Schölkopf et al. [204, 205]. As it was discussed in Sect. 4.1.6, kernels were first introduced to machine learning by Vapnik through kernel SVM. Ever since, many kernel methods in machine learning have been proposed especially by Schölkopf, focusing on kernels, at the Max Planck Institute. Note that for handling nonlinear data, two main approaches exist: using a nonlinear algorithm or transforming the data into a space where linear methods are applicable. Kernel PCA transforms data to another higher dimensional space and applies the linear PCA on the transformed data.

It was the year 2000 when two important nonlinear dimensionality reduction methods, named Isometric Mapping (Isomap) [217] and Locally Linear Embedding (LLE) [195], were proposed in the Science journal. Isomap uses geodesic distances, rather than Euclidean distances, between data points in the formulation of MDS. As calculation of geodesic distances between data instances is difficult, requiring differential geometry, Isomap approximates geodesic distances with piece-wise Euclidean distances. For illustration, consider two cities like Toronto and Tokyo on the Earth. Euclidean distance between these cities is if we dig the Earth from Toronto to Tokyo, while geometric distance between them is like going on the shortest curvy path on the Earth from Toronto to Tokyo. Approximation by piece-wise distances is like going from Toronto to Montreal, then to London, then Rome, then Istanbul, then Tehran, and many cities in between until we reach Tokyo. The piece-wise distances are each roughly Euclidean distances because of proximity of the cities in between the path. The curvy geodesic distance is approximated by the summation of these piece-wise distances.

LLE was proposed by Sam Roweis [195] who used to be one of the postdoctoral fellows of Hinton and became one of the effective researchers in machine learning and dimensionality reduction. While FDA had the main idea behind supervised dimensionality reduction, LLE proposed the main idea behind all unsupervised dimensionality reduction methods. It says embed the data points in a way that every embedded point is reconstructed by its embedded neighbors as in the input space. In other words, fit locally and think globally [198]; preserve the neighborhood information in the embedding space and the global structure will be hopefully preserved!

In the years 1999–2001, spectral clustering and Laplacian eigenmap were proposed [13, 169, 240]. Spectral clustering [169, 240] considers the dataset as a graph and it uses the Laplacian of the data graph for clustering. Laplacian eigenmap [13] adopted the idea of spectral clustering for dimensionality reduction. Refer to [77] for theoretical details of the Laplacian-based algorithms in machine learning.

In 2003, Yoshua Bengio et al. found out that most of the spectral dimensionality reduction methods, including LLE, Isomap, MDS, Laplacian eigenmap, and spectral clustering, are basically kernel PCA with different kernels [16, 17]. They all are learning eigenfunctions of their kernels (see Sect. 4.1.6 for more information on eigenfunctions). Now, a question arose: "what is the best kernel for a dataset?". This encouraged learning the best kernel, for each dataset, by Semidefinite Embedding (SDE), also called Maximum Variance Unfolding (MVU) proposed in 2004 [238, 239]. This method finds the best kernel using semidefinite programming optimization to unfold the underlying manifold of data. It is a slow method because of iteration solution of semidefinite programming.

In 2005, graph embedding was proposed [253, 254]. This method generalizes the Laplacian-based dimensionality reduction [77] discussed before. In the mean-time, many spectral metric learning methods—also discussed in Sect. 4.1.4—were proposed for learning the best distance metric, instead of regular Euclidean distance, for each dataset [87, 255].

Consider the following scenario. Assume whenever one person moves right, a second person also moves right, and whenever the first person moves left, the second person also moves left. In this case, they have positive linear dependence. If the second person always move in the opposite direction, they have negative linear dependence. Linear dependence is called correlation. Now, consider a scenario in which the second person moves in some seemingly arbitrary direction when the first person moves. It is not obvious whether the two people are dependent or their movements are just random. Therefore, calculation of dependence is much harder than calculation of correlation. In 2005, Arthur Gretton and his professor Bernhard Schölkopf et al. proposed Hilbert-Schmidt Independence Criterion (HSIC), which estimates the dependence of two random variables with the linear correlation of their kernels [96]. HSIC maps data to the Reproducing Kernel Hilbert Space (RKHS) using kernels, and then measures the dependence of variables by computing their correlation in the RKHS. It theoretically claims that correlation in the Hilbert space is a good estimate for the dependence in the input space! Using HSIC, Ali Ghodsi et al. proposed supervised PCA, in 2011 [10], which maximizes the dependence of the class labels and the projected data. It is proved that PCA is a special case of supervised PCA where labels are not used (see [80] for proof).

4.3.2 Probabilistic Dimensionality Reduction

One of the first probabilistic dimensionality reduction methods was factor analysis, proposed in 1950s [62]. Factor analysis assumes that there is a latent random variable, named latent factor, on which the data are conditioned. In other words, the data are the observed random variable and there is a latent factor for the observed data. It is assumed that data are obtained by a linear projection of the latent factor plus some noise. While the prior probability of the latent factor is assumed to be standard normal distribution, factor analysis learns the projection matrix and the covariance matrix of noise using variational inference. Variational inference makes use of the Expectation Maximization (EM) algorithm (note that EM will be discussed more in Sect. 4.4.2). Variational inference plays a game where, in the E-step of EM, it pushes the lower bound of likelihood of data to become a tight bound and, in the M-step of EM, it maximizes the likelihood again. It is like a game where a person runs after a second person, to reach to them, while the second person also runs away [75].

A special case of factor analysis is probabilistic PCA where the noise is assumed to be isotropic having a diagonal covariance matrix with equal variance in all dimensions. Two approaches for training probabilistic PCA have been proposed. One approach, proposed by Sam Roweis in 1997 [193], uses the iterative EM algorithm as in factor analysis. Another approach, proposed by Tipping and Bishop in 1999 [222], finds a closes form solution to the EM algorithm for probabilistic PCA. Note that the simplification of noise covariance, compared to factor analysis, makes the solution of probabilistic PCA simple and closed-form. Refer to [75] for theoretical details of factor analysis and probabilistic PCA.

People were developing different dimensionality reduction methods until, in 1984, Johnson-Lindenstrauss lemma [119] justified theoretically that even random projection onto a subspace roughly preserves the distances of the data points with some bound on the probability of error! That was a shocking news. Obviously, it is not as perfect as optimized projection matrices like PCA or FDA but it works up to some accuracy. Random projection explains why random initialization of neural networks is justified as a good initial solution to start with. Refer to [81] for the theory of Johnson-Lindenstrauss lemma and random projection.

Sufficient Dimension Reduction (SDR), first proposed in 1991 [145], is a family of methods that find a transformation of data to a lower dimensional space without changing the conditional of labels given data. Therefore, the prediction of labels or separation of classes does not change after projection onto the subspace by SDR. In fact, the subspace is sufficient for predicting the labels from the projected data onto the subspace. One of the important SDR algorithms is Kernel Dimension Reduction (KDR) which was proposed by Fukumizu, Bach, and Michael Jordan in 2003 [65] and it was further discussed by them in [63, 64]. Note that SDR methods have a statistical regression approach while KDR has both statistical and machine learning tastes. A summary of theory of SDR and KDR can be found in [83].

It 2003, Geoffrey Hinton and Sam Roweis proposed Stochastic Neighbor Embedding (SNE) [109]. SNE considers a Gaussian neighborhood distribution around each data point in the input space where each other data point has some probability of being a neighbor to that point. It tries to preserves the same neighborhood Gaussian distribution in the embedding space. In 2008, Hinton and his student Laurens van der Maaten improved SNE by proposing Student-t distributed SNE (t-SNE) in 2008 [151]. The improvement gets clear by the following example. Assume you want to fit one million people in a room. Obviously you cannot because the crowd is too big for the room; a problem referred to as the crowding problem. You can enlarge the room to be able to fit all those people. Likewise, t-SNE enlarges the tails of neighborhood in the embedding space by using Student-t distribution which has heaver tails than Gaussian distribution. SNE and t-SNE belong to a family of algorithms which deal with attractive force between similar points and repulsive force between dissimilar points. Another important algorithm in this family is Uniform Manifold Approximation and Projection (UMAP) proposed in 2018 [156]. Both t-SNE and UMAP are widely used for visualization of high-dimensional data, such as genes or deep learning features, in 2D or 3D plots. Refer to [82, 84] for theories of t-SNE and UMAP, respectively.

4.4 Some Other Statistical Learning Methods

4.4.1 Boosting

There are two approaches to have ensemble of machine learning models. One approach is to have multiple models and combine their outputs in some way such as model averaging [37, 114] or majority voting. This approach is referred to as ensemble learning [177] or bootstrap aggregating (bagging) [22, 28] (n.b. bagging was discussed in Sect. 4.1.5).

Another approach for having an ensemble of models is boosting [73]. In boosting, multiple models are boosted in concatenation or hierarchy where every model care more about the data instances incorrectly predicted by the previous model in the concatenation. In other words, if a machine learning model performs purely on some data instances, those instances are weighted more for the next model. In this way, the machine learning models used in boosting can be each weak models performing slightly better than random prediction [126, 127]. Finally, the overall prediction is a weighted average of the models in boosting. Different methods have been proposed for boosting, one of the most well-known ones is AdaBoost (Adaptive Boosting) proposed in 1996 [57].

AdaBoost showed its merit by its good empirical performance especially when using trees as models in boosting [179]. After the proposal of AdaBoost, researchers tried to theoretically analyze why it works so well. Many researchers, such as Hastie and Tibshirani at Stanford University, have analyzed boosting theoretically. For example, additive models [103] can be used to explain why boosting works [59, 189]. Moreover, it was found out that there is an upper bound on the generalization error of boosting [199]. Furthermore, both boosting and bagging reduce the variance of estimation [21, 199], especially for the models with high variance of estimation such as trees [179]. In another perspective, boosting can be seen as a method to maximize the margins of training error [199], as in SVM [228] (see Sect. 4.1.6), which results in a good generalization error. Refer to Chap. 5 and [73] for detailed theories of bagging and boosting.

4.4.2 Mixture Models

When the data have only one uni-modal probability distribution, then a probability distribution can be fit to data where the parameters of the distribution can be estimated using Maximum Likelihood Estimation (MLE) or Method Of Moments (MOM). For example, when the distribution is the normal (or Gaussian) distribution, the mean and covariance can be estimated by sample average and sample covariance matrix, respectively.

However, when the data have a complicated multi-modal probability distribution, a mixture model should be fit to data. Multiple techniques exist for fitting a mixture distribution to data. One of the methods is fitting a mixture distribution using Expectation Maximization (EM) [88, 102]. Expectation maximization, which was proposed in 1977 [43], is basically MLE but when some part of data is missing. The main idea of EM can be summarized in this short friendly conversation [88]:

"– What shall we do? Some part of data is missing! The likelihood is not known completely so MLE cannot be used (because MLE maximizes the likelihood of data).
– Mmm, probably we can replace the missing part of data with something...
– Aha! Let us replace it with its mean.
– You are right! We can take the mean of likelihood over the possible values of the missing data. Then everything in the likelihood will be known, and then...
– And then we can do MLE!"

EM has two stages which are calculating expectation and maximization (the name of algorithm comes from these two stages). The first stage calculates the expectation (or mean) of the missing data and then, when everything is known, the second stages maximizes the likelihood. However, as this maximization changes the expectation of missing data, this procedure should be repeated iteratively until convergence; hence, EM is iterative in contrast to MLE.

A mixture distribution is a weighted average of multiple probability distributions in order to have an overall multi-modal distribution. For example, Gaussian Mixture Model (GMM) is a multi-modal mixture of multiple Gaussian distributions with different means and covariances. Fitting a mixture distribution needs EM rather than MLE because the missing data are which of the distributions in the mixture every data instance belongs to. In other words, the weights in the weighted average of distributions is missing. Therefore, EM should be used to have expectation over the weights and then the likelihood is maximized to obtain the parameters of the distributions in the mixture. This is performed iteratively until convergence of parameters of distributions. Refer to [88] for theoretical details of the algorithm.

Another approach for fitting a mixture distribution is using Riemannian optimization [115]. One other approach is to use Kernel Density Estimation (KDE), first proposed independently by Rosenblatt in 1956 [192] and Parzen in 1962 [172]. KDE considers the histogram of values and then estimates the probability distribution by limiting the size of bins to zero in the histogram. When the size of bins in the histogram goes to zero, the histogram almost becomes continuous and like the probability distribution. Therefore, KDE can fit a distribution to data.

4.4.3 Markov Models

In machine learning, the data vectors are usually assumed to have the same length or the same number of dimensions. When the data vectors have different lengths, either the algorithm should be able to accept different lengths of input or the data vectors should be warped to have the same length. An example of the first approach is language models such as recurrent neural networks which will be discussed in Sect. 5.8. Two examples of the second approach are Dynamic Time Warping (DTW), proposed in 1959 [15], and IMGT numbering of antibody sequences in biotechnology, proposed in 1997 [144].

Another example for accepting different lengths of data without warping is using Markov models. A Markov model or Markov chain is a chain of random variables, called states, where each state goes to the next state or stays in the same state with some probability. It is called the Markov model because it posses the Markovian property. [The first order] Markovian property is an assumption which states that in a time series of random variables, every random variable is merely dependent on the latest previous random variable and not the others. The Markovian property makes sense because, for example, someone at age 30 is conditioned on that person on age 29. Their age 30 does not need to be conditioned on age 0 to 29 because almost all the information of age 0–28 is roughly encapsulated in their age 28.

Assume there is a time series of hidden random variables or states [71]. Every hidden random variable can generate or emit an observation symbol or output. Therefore, there is a set of states and a set of observation symbols. A Hidden Markov Model (HMM) has some hidden states and some emitted observation symbols [182]. HMM models the probability density of a sequence of observed symbols [194]. There are two main algorithms for training and evaluation in HMM. These two methods are using the EM algorithm [71] (see Sect. 4.4.2 for explanation of EM) and the forward-backward procedure [183]. One of the most well-known methods for estimation in HMM is the Viterbi algorithm proposed in 1967–1973 [56, 232]. Refer to [90] for theoretical details of these algorithms.

5 A Brief History of Neural Networks and Deep Learning

This section presents a concise history of neural networks and deep learning, highlighting key milestones and challenges in their development[4]

5.1 Early History of Neural Networks

5.1.1 McCulloch-Pitts Model (1943)

In 1943, Warren McCulloch, a neurophysiologist, and Walter Pitts, a logician, proposed the first model of a neuron [155]. Known as the McCulloch-Pitts model, it consisted of a simple unit with multiple binary inputs, each with equal and fixed weights. The unit produced a binary output that activated only when the sum of inputs reached a threshold. This model effectively functioned as a logical gate—such as AND, OR, or NOT—laying the foundation for representing logical operations with networks of interconnected units.

5.1.2 Hebbian Learning (1949)

Hebbian theory, proposed by Donald Hebb in his influential 1949 book *The Organization of Behavior* [106], is a foundational neuropsychological theory suggesting that "cells that fire together, wire together." It states that when a presynaptic neuron repeatedly and persistently activates a postsynaptic neuron, the connection between them strengthens. In the context of artificial neural networks, Hebbian learning inspired early algorithms for adjusting connection weights based on the correlation between neurons' activations.

5.1.3 Perceptron (1958–1969)

After the proposal of Hebbian learning, the Perceptron algorithm was introduced as a model of a single neuron that could learn from examples. Frank Rosenblatt implemented it in 1958 at Cornell Aeronautical Laboratory [191]. It was initially designed for binary classification, which aligned well with the binary logic needed for computer operations.

Rosenblatt was very optimistic about the Perceptron's potential. In a 1958 press conference, he gave an enthusiastic description of its abilities, leading the New York Times to declare it "the embryo of an electronic computer that expects it will be able to walk, talk, see, write, reproduce itself, and be conscious of its existence" [170]. At that time, some believed they were on the brink of achieving true artificial intelligence.

However, the single-layer Perceptron is fundamentally a linear classifier and cannot solve nonlinear problems like XOR. This limitation was highlighted in 1969 when Marvin Minsky and Seymour Papert published their influential book, Perceptrons [161], which demonstrated mathematically that the single-layer Perceptron could not represent XOR and similar nonlinear functions. This finding shattered the early optimism and led to a decline in interest in neural networks.

Rosenblatt had experimented with connecting Perceptrons in loosely structured ways and allowing their weights to be adjusted. However, these early experiments did not amount to a systematic multi-layer model with a clear training algorithm. While the idea of stacking Perceptrons in layers could, in theory, solve nonlinear problems like XOR, no effective learning algorithm was available at the time to train such a network. This technical hurdle prevented neural networks from progressing toward learning more complex tasks.

[4] For a different perspective on the history of neural networks, see [200, 202].

The Perceptron learning rule adjusts weights based on the difference between the predicted and actual outputs, gradually shifting the decision boundary to correctly classify data points.[5] Refer to Chap. 2 for the theory of the Perceptron.

5.1.4 ADALINE (1960–1990)

In 1960, Widrow and his student Hoff proposed a model named ADALINE (Adaptive Linear Neuron) for adjusting weights [241, 243]. This model was considered groundbreaking at the time, and some articles even claimed that "robots can think"! ADALINE differed from the Perceptron in its learning rule and error calculation: unlike the Perceptron, which updates weights based only on whether an instance is correctly classified, ADALINE minimizes the Least Mean Squares (LMS) error using gradient descent, which allows for continuous error correction even on correctly classified examples. This training method is known as the LMS algorithm or Widrow-Hoff learning rule (see Chap. 2 for details).

ADALINE had better generalization than the Perceptron because it aimed to minimize overall error rather than just fixing misclassifications. However, like the Perceptron, ADALINE is still a linear model.

To address the need for nonlinearity, Widrow and his students experimented with stacking several ADALINE neurons, resulting in a structure they called MADALINE (Many ADALINEs). Over time, they proposed different learning rules for MADALINE. MADALINE Rule 1 (MRI), proposed in 1962 [242], could not train the weights between hidden and output layers. MADALINE Rule 2 (MRII), proposed in 1988 [247], improved on this by allowing weight adaptation for the hidden-output layer. Finally, MADALINE Rule 3 (MRIII), introduced in 1990 [244], replaced the signum (binary) activation function with a sigmoid function to allow continuous outputs instead of merely binary outputs.

MADALINE and multi-layer Perceptrons were developed in parallel at Stanford University and Cornell University, respectively. Ultimately, multi-layer Perceptrons, especially after the backpropagation breakthrough, became the foundation of modern neural networks, overshadowing ADALINE-based models. For a more detailed history of early neural networks, refer to [50].

5.1.5 Backpropagation, SGD, Adam, and Optimization (1960–1986)

In 1960, Henry J. Kelley introduced the use of gradient-based methods for optimizing control systems, laying foundational concepts for backpropagation in neural networks [128]. Around the same time, in 1965, Alexey Ivakhnenko and Valentin Lapa developed the Group Method of Data Handling (GMDH), one of the earliest models with a multilayer structure and self-organizing capabilities [118].

In 1967, Shun'ichi Amari proposed applying Stochastic Gradient Descent (SGD) to train multilayer perceptrons, marking one of the first uses of SGD in neural network training [4]. In 1969, Arthur E. Bryson and Yu-Chi Ho described methods for optimizing dynamic systems, contributing to the theoretical underpinnings of backpropagation [26].

A significant step came in 1970 when Seppo Linnainmaa introduced reverse-mode automatic differentiation, providing an efficient way to compute gradients in computational graphs—a key component of backpropagation [147]. Although these developments laid the groundwork, it was not until 1986

[5] Formally, the weight update is: $\boldsymbol{w} \leftarrow \boldsymbol{w} + \eta(y - \hat{y})\boldsymbol{x}$, where y is the true label, $\hat{y}$ is the predicted label, and η is the learning rate. If the point is classified correctly, no update occurs; if misclassified, the update shifts the decision boundary toward correct classification.

that David E. Rumelhart, Geoffrey E. Hinton, and Ronald J. Williams successfully demonstrated the effectiveness of backpropagation in training multilayer neural networks[6] [196].

The use of SGD, which updates model parameters using small subsets of data, gradually became a standard optimization technique in machine learning. Although the concept of stochastic approximation was introduced by Robbins and Monro in 1951 [188], its application to neural networks gained prominence in the 1990s and later.

Meanwhile, as these foundational methods were being developed, research in other areas of neural networks, such as ADALINE and MADALINE at Stanford and the study of multilayer perceptrons at Cornell, was also advancing (see preceding sections).

For detailed discussions on backpropagation, SGD, and related optimization methods, refer to Chap. 3.

5.2 Biological-Inspired Neural Networks (1980s–Present)

As discussed earlier, artificial neural networks were originally inspired by the workings of the human brain. Over time, various training algorithms such as the Perceptron learning rule, ADALINE's LMS rule, and backpropagation were developed to adjust network weights. However, some researchers argued that these algorithms lacked sufficient biological plausibility, as they often treated neurons as simple, linear summing units.

In response, researchers proposed alternative architectures and learning rules that aimed to more closely mimic the dynamics of biological neurons. These are often called biologically inspired neural networks [42, 250]. One prominent family of such networks is spiking neural networks [92, 216], which model the discrete spikes or action potentials observed in real neurons, capturing their temporal dynamics.

Biologically inspired models are sometimes described as the third generation of neural networks, following the first generation (single-layer Perceptrons) and the second generation (multilayer networks trained with backpropagation). These third-generation models incorporate more realistic mechanisms of neural computation, offering insights into both neuroscience and machine learning.

Meanwhile, while these biologically inspired models were being explored, parallel efforts were underway to improve training algorithms for multilayer perceptrons, leading to breakthroughs in backpropagation and gradient-based optimization.

5.3 Recurrent Neural Networks and LSTMs (1990–1997)

While early artificial neural networks such as Perceptrons and ADALINE focused on processing static inputs, many real-world problems involve sequential or temporal data, such as language, speech, and time series. To address this, the concept of Recurrent Neural Networks (RNNs) was introduced.

The first widely recognized RNN architecture was proposed by Jeffrey Elman in 1990 [49]. Elman's network featured hidden states with feedback connections, allowing information to persist over time and enabling the network to capture temporal dependencies in data sequences. This opened the door for applications such as speech recognition, language modeling, and sequence prediction.

[6] Later developments introduced adaptive learning rate methods, such as AdaGrad (2011) [47], RMSProp (2012, unpublished slide deck) [220], and Adam (2014) [129], which built on these foundations. While these optimizers are widely used in modern deep learning, their development followed the earlier milestones laid in the 1960s-1980s. For recent techniques addressing adaptive learning rate methods and the sharpness of minima in the loss landscape—such as Sharpness-Aware Minimization (SAM, 2021) [55]—see Chap. 3.

However, Elman's simple RNN suffered from significant challenges in training, particularly the vanishing and exploding gradient problems that arise during backpropagation through time (BPTT). These issues made it difficult for RNNs to learn long-range dependencies.

To overcome these problems, Hochreiter and Schmidhuber introduced the Long Short-Term Memory (LSTM) network in 1997 [113]. LSTM introduced memory cells with gates (input, output, and forget gates) that could regulate the flow of information through the network, allowing it to maintain long-term dependencies and mitigating the vanishing gradient problem. This innovation made LSTMs the de facto standard for sequence modeling for many years.

Despite their advantages, LSTMs remained computationally expensive and somewhat challenging to train, and it was not until the resurgence of deep learning in the 2010s, with improved optimization techniques and hardware accelerators, that RNNs and LSTMs became widely used in practical applications. For further details on sequence modeling with RNNs and LSTMs, refer to Chap. 8.

5.4 Convolutional Neural Networks (1980–2017)

Convolutional Neural Networks (CNNs) emerged as a biologically inspired model for image recognition, motivated by the hierarchical structure of the visual cortex. In 1980, Kunihiko Fukushima in Japan introduced the Neocognitron [66], a neural network designed to process visual data by applying local connectivity and hierarchical feature extraction. His goal was to achieve translation invariance and to capture the way biological neurons respond to localized patterns.

This design contrasted with earlier fully connected networks, which required flattening an image into a long vector, losing the crucial two-dimensional spatial structure. By using local receptive fields and weight sharing, the Neocognitron and later CNNs were able to efficiently process images while preserving spatial relationships among pixels.

In the late 1980s, Yann LeCun and colleagues extended these ideas and applied them to digit recognition [143], combining convolutional layers with backpropagation to train the network end-to-end. This work culminated in LeNet-5 [142], which achieved significant success on tasks like handwritten digit classification (MNIST).

Over time, CNNs evolved and inspired a wave of architectures. Notable examples include AlexNet[7] (2012) [136], VGG (2014) [210], Inception (2015) [215], U-Net (2015) [190], ResNet (2016) [104], and DenseNet (2017) [117]. These later models, though part of modern deep learning practice, built on the foundational ideas of local connectivity and weight sharing that originated in the 1980s. (For details on modern CNN architectures, see Chap. 4.)

5.5 The Winter of Neural Networks and Competition with Kernel SVM (1980s–2006)

During the 1980s, neural networks enjoyed a period of rapid development and enthusiasm. However, this momentum gradually declined in the late 1980s and throughout the 1990s, leading to what is sometimes called the "AI Winter"—or more specifically, the "Winter of Neural Networks". This period lasted until around 2006.

Several factors contributed to this decline. First, training deep neural networks was practically infeasible due to challenges like the vanishing and exploding gradient problems, which caused gradients

[7] AlexNet, for example, was a key milestone in reviving deep learning by demonstrating impressive results on the ImageNet competition in 2012.

to either shrink toward zero or grow uncontrollably as they were propagated backward through many layers. These issues made it extremely difficult to train multilayer networks effectively. Additionally, computational hardware at the time lacked the power needed to handle the heavy demands of training large networks, and the size of available datasets was often too small to fully leverage the potential of neural networks.

Meanwhile, kernel-based methods—particularly the Support Vector Machine (SVM) and its extension via the kernel trick—gained prominence. Introduced in the early 1990s [20], kernel SVMs offered a mathematically robust and computationally efficient framework for classification tasks. By implicitly mapping inputs to a high-dimensional feature space, SVMs could effectively model nonlinearity without suffering from the gradient issues that hindered deep networks. Combined with their solid mathematical foundations and relative ease of use on small datasets, kernel SVMs became the preferred choice for many machine learning practitioners.

As a result, interest in neural networks waned during this period, and their potential remained largely unrealized. This "winter" persisted until 2006, when Geoffrey Hinton and his student, Ruslan Salakhutdinov, published a landmark paper on Deep Belief Networks (DBNs). They demonstrated that deep networks could be successfully trained by using unsupervised layer-wise pretraining to initialize weights before applying backpropagation [110]. This approach effectively mitigated the vanishing gradient problem and marked the beginning of a renewed wave of interest in deep learning.

Although later advances—such as the introduction of ReLU activations and the advent of GPU acceleration—would further enhance the training and scalability of deep networks, Hinton and Salakhutdinov's 2006 breakthrough was the key turning point that lifted neural networks out of the winter period and set the stage for the deep learning revolution.

This resurgence of deep learning laid the groundwork for remarkable advances in image processing and computer vision, which were soon to emerge with the development of convolutional neural networks and their success in tasks like object recognition and large-scale image classification.

5.6 Breakthrough in Image Recognition and the Resurgence of Deep Learning (2006–2012)

Following the 2006 success of Deep Belief Networks (DBNs) by Hinton and Salakhutdinov, which used RBMs for unsupervised pretraining, researchers began to explore ways to overcome the challenges of training deep networks without relying on RBM-based initialization. Developments such as the introduction of ReLU activation functions, which alleviated the vanishing gradient problem, and advances in stochastic gradient descent (SGD) optimization, paved the way for training deep networks directly with backpropagation.

These advances set the stage for a transformative breakthrough in 2012, when AlexNet—a deep convolutional neural network designed by Alex Krizhevsky under the supervision of Geoffrey Hinton—achieved a dramatic improvement in image classification performance at the ImageNet Large Scale Visual Recognition Challenge (ILSVRC) [137]. AlexNet's success, driven by the use of GPUs for large-scale training, ReLU activations, and improved regularization, signaled the start of a new era for deep learning.

The impressive performance of AlexNet sparked widespread interest in deep learning, leading to rapid adoption in image recognition, natural language processing, and speech processing [29, 111, 124, 141]. This period saw an explosion of deep learning frameworks and libraries, including TensorFlow by Google Brain [1] and PyTorch by Facebook [173] (see Chap. 22). These tools have since become foundational for modern deep learning research and applications.

5.7 Early Developments in Neural Network-Based Generative Models (2014–2018)

As neural networks advanced, they paved the way for the development of generative machine learning models that could learn to map latent noise to data instances in order to approximate the distribution of training data. This marked a significant shift from earlier statistical generative models, such as factor analysis [31, 36, 62, 75, 101].

At its core, a generative model aims to learn a transformation that maps a simple, well-understood distribution—such as a standard Gaussian or uniform distribution—into a complex, high-dimensional distribution like that of face images. For instance, if a face image is 1,000 by 1,000 pixels, then each image can be thought of as a single point in a one-million-dimensional space. Generating new, realistic face images therefore corresponds to sampling new points from this complicated distribution. Since we can easily sample from a standard Gaussian or uniform distribution, generative models define a function (usually a neural network) that transforms these simple samples into points that resemble real images, effectively learning the desired complex distribution.

One of the first neural network-based generative models was the *Variational Autoencoder (VAE)*, proposed in 2014 by Kingma and Max Welling [130]. VAEs employ variational inference in an autoencoder framework: the input data is fed to an encoder, which outputs parameters for a latent distribution (e.g., mean and variance of a Gaussian). A sample drawn from this latent space is then fed to the decoder, which reconstructs the data instance. In this way, VAEs learn to approximate the complex data distribution by transforming samples from a simple latent distribution. Refer to Chap. 12 for the theoretical details of variational inference and variational autoencoders.

Shortly after VAE,[8] the *Generative Adversarial Network (GAN)* was proposed by Ian Goodfellow and Yoshua Bengio et al. in 2014 [93]. GANs consist of two networks—a generator and a discriminator—trained in an adversarial framework: the generator aims to produce data that is indistinguishable from real data, while the discriminator attempts to distinguish between real and generated data. Training proceeds as a minimax game where the generator minimizes the discriminator's ability to differentiate real from fake, and the discriminator tries to maximize it. GANs generally produced higher-quality outputs than VAEs, but training them posed challenges, including instability and mode collapse.

Following GANs, adversarial autoencoder models were introduced by Alireza Makhzani and Brendan Frey, including the adversarial autoencoder (2015) [154], PixelGAN autoencoder (2017) [153], and Implicit autoencoder (2018) [152]. Since then, numerous GAN variants have emerged to improve training stability and output quality. Refer to Chap. 14 for details on GANs, adversarial learning, and related developments.

[8] For completeness, it is worth mentioning that around 2015, the Generative Moment Matching Network (GMMN) was proposed [48, 146]. It aimed to minimize the Maximum Mean Discrepancy (MMD) [94, 95] between the generated and dataset distributions. However, GMMN was less impactful in practice and is therefore noted here only for historical context.

5.8 From Sequential Models to Transformers and Large Language Models

5.8.1 The Attention Revolution (2014–2017)

Following the success of RNNs and LSTMs in sequential data processing (discussed in Sect. 5.3), the field of natural language processing was transformed by the introduction of the attention mechanism.[9] Attention mechanisms in neural networks were first introduced in 2014–2015 by Bahdanau et al. [7] and Luong et al. [149] to address limitations of recurrent neural networks in sequence-to-sequence tasks, particularly machine translation.

The attention mechanism addresses fundamental limitations of RNNs and LSTMs: their sequential nature prevents parallelization during training, and they struggle with long-range dependencies despite architectural improvements like gating mechanisms. Attention allows models to directly access any part of the input sequence, effectively creating shortcuts for information flow.

The core principle of attention can be understood through a simple analogy: when reading a complex sentence, humans naturally focus on the most relevant words rather than processing each word with equal weight. Similarly, attention mechanisms learn to assign different weights to different parts of the input, allowing the model to "attend" to the most relevant information for each prediction.

Formally, attention computes a weighted combination of input representations, where the weights are learned based on the relevance of each input element to the current context.

5.8.2 The Era of Large Language Models (2018–Present)

The Transformer architecture quickly became the foundation for two dominant paradigms in language modeling. In 2018, Google introduced BERT (Bidirectional Encoder Representations from Transformers) [44], which utilized the encoder component of the Transformer for bidirectional language understanding. Simultaneously, OpenAI developed GPT (Generative Pre-trained Transformer) [184], which employed the decoder component for autoregressive language generation.

BERT revolutionized natural language understanding tasks by learning bidirectional representations through masked language modeling. Its success in transfer learning established the paradigm of pre-training large models on vast text corpora and fine-tuning them for specific tasks.

GPT, designed as an autoregressive language model, demonstrated remarkable capabilities in text generation and few-shot learning. The successive iterations of GPT models—from GPT-1 to GPT-4 and beyond—showcased the benefits of scaling model size, data, and compute, leading to the emergence of large language models (LLMs) with increasingly sophisticated capabilities.

The development of LLMs has spawned numerous variants and architectural improvements [257], but the core Transformer architecture remains dominant. These models have demonstrated emergent capabilities that scale with model size, including in-context learning, chain-of-thought reasoning, and complex instruction following.

5.8.3 Model Alignment and Safety

As LLMs became more capable, the need for aligning their behavior with human values and intentions became critical. Two primary approaches have emerged for this alignment problem. Reinforcement

---

[9] During this period, embedding methods such as Word2Vec [159, 160] (2013) and GloVe [175] (2014) provided important foundations for semantic word representations. While these methods demonstrated that semantic relationships could be captured in vector spaces—famously showing that vector("king") - vector("man") + vector("woman") $\approx$ vector("queen")—they were limited by context-independent representations. The same word would have identical embeddings across different contexts, such as "bank" meaning both a financial institution and a river's edge. These methods, while influential for word-level semantics, were ultimately superseded by the contextual representations learned by Transformer-based (attention-based) models.

Learning with Human Feedback (RLHF), developed in 2022 [8, 171], uses human preferences to train a reward model that guides the language model's behavior through reinforcement learning.

More recently, Direct Preference Optimization (DPO), proposed in 2023 [185], provides a more direct approach to preference learning without requiring a separate reward model. These alignment techniques have become essential for deploying LLMs safely in real-world applications.

5.8.4 Alternative Architectures

While Transformers have dominated the field, researchers continue to explore alternative architectures to address computational limitations. State space models, inspired by control theory, offer promising alternatives with linear computational complexity. Notable examples include the Structured State Space Model (S4) [99] and Mamba [98], which attempt to combine the benefits of RNNs and Transformers. For detailed coverage of these alternatives, refer to Chap. 10.

5.9 Modern Developments in Deep Learning (2010s–Present)

Following the foundational breakthroughs discussed in previous sections, deep learning has continued to evolve with new paradigms, architectures, and applications. This section chronicles the major developments that have shaped modern deep learning, organized chronologically by their emergence and lasting impact on the field.

5.9.1 Deep Reinforcement Learning (2013–2016)

The integration of deep learning with reinforcement learning marked one of the most significant paradigm shifts in modern AI. While reinforcement learning theory dates back to Bellman's work in 1952 [14] and foundational algorithms like Q-learning [234] and REINFORCE [246] were established by the early 1990s, their combination with deep neural networks created unprecedented capabilities.

Deep Q Network (DQN), proposed by DeepMind in 2013 [164], was further developed to achieve superhuman performance in Atari games by 2015 [163]. This breakthrough sparked significant interest in deep reinforcement learning, leading to rapid developments across research institutions worldwide and establishing it as a major research direction.

The success continued with AlphaGo in 2016 [209], which combined deep neural networks with tree search to defeat world champion Go player Lee Sedol. This achievement demonstrated that deep RL could tackle complex strategic problems previously thought intractable for machine learning. Refer to Chap. 18 for detailed theories of reinforcement learning.

5.9.2 New Architectural Paradigms (2016–2017)

Graph Neural Networks: Traditional neural networks require vectorized input, limiting their applicability to graph-structured data ubiquitous in social networks, molecular structures, and knowledge graphs. Graph Neural Networks (GNNs) emerged to address this limitation by operating directly on graph structures.

The foundational work began in 2016 with Chebyshev Network (ChebNet) [41] and Graph Convolutional Network (GCN) [131]. ChebNet generalized convolutions to graphs using Chebyshev polynomials, while GCN simplified this approach with first-order approximations. Kipf and Welling also introduced Graph Autoencoder (GAE) [132] for unsupervised graph representation learning.

The field advanced with Graph Attention Network (GAT) in 2017 [229], which incorporated the attention mechanism (discussed in Sect. 5.8) to learn adaptive importance weights between connected vertices. Refer to Chap. 17 for comprehensive coverage of graph neural networks.

5.9.3 New Learning Paradigms (2016–Present)

Few-Shot Learning and Meta-Learning: Few-shot learning addresses one of deep learning's fundamental limitations: the need for large amounts of labeled data. Meta-learning, or "learning to learn," provides the theoretical framework for rapid adaptation to new tasks with minimal examples.

While meta-learning concepts trace back to Schmidhuber's 1987 thesis [201], practical deep learning applications emerged in 2016. Key methods include Matching Networks (2016) [231], Prototypical Networks (2017) [211], and Model-Agnostic Meta-Learning (MAML) (2017) [51], which became the most influential approach due to its broad applicability.

These methods typically use episodic training with support and query sets in an N-way, K-shot setup, where N classes are sampled with K examples each per episode [224]. Chapter 19 covers the theory of few-shot learning and meta-learning.

Self-Supervised Learning: Self-supervised learning emerged as a paradigm for learning representations without manual labels by generating supervisory signals from the data itself. While early work like Siamese networks (1993) [25] and contrastive loss (2006) [100] laid foundations, the modern self-supervised revolution began around 2018–2020.

Key breakthroughs include SimCLR (2020) [34], BYOL (2020) [97], and subsequent methods like SwAV (2020) [30], SimSiam (2021) [35], Barlow Twins (2021) [256], and VICReg (2022) [9]. These methods learn by pushing embeddings of augmented versions of the same data instance together while separating different instances, achieving remarkable success in computer vision tasks. Refer to Chap. 16 for comprehensive coverage.

5.9.4 Practical Advances and Applications (2015–Present)

Neural Network Compression: As deep learning models grew larger, deploying them on resource-constrained devices became critical. Neural network compression emerged as a practical necessity, particularly for mobile applications and edge computing.

The field encompasses multiple approaches: pruning removes less important weights [186, 236], quantization reduces precision [258], matrix decomposition factorizes weight matrices [252], and knowledge distillation transfers knowledge from large "teacher" networks to smaller "student" networks [27, 108]. These techniques have become essential for practical deployment of large models. Refer to Chap. 20 for detailed coverage.

Federated Learning: Federated learning addresses privacy and data distribution challenges by training models across decentralized data sources without centralizing the data itself. Early privacy-preserving work appeared in 2000–2008 [2, 225], but Google's formalization of federated learning in 2017 [157] established it as a major research direction.

The approach enables collaborative training while maintaining data privacy crucial for applications involving sensitive information like healthcare and finance. Both centralized and decentralized variants have been developed, each with distinct security and robustness trade-offs [120, 135].

5.9.5 Safety, Interpretability, and Robustness (2015–Present)

Adversarial Attacks and Defenses: The discovery that small, imperceptible perturbations could fool deep neural networks revealed critical vulnerabilities in these systems. While early concepts emerged from spam filtering discussions in 2004 [248], systematic study began with methods like DeepFool in 2016 [165].

This field addresses both attack methods that expose model vulnerabilities and defense strategies to improve robustness. The dual terminology of "adversarial" (used both for GANs and attacks) reflects distinct but important aspects of deep learning research. Refer to [32] for comprehensive surveys.

Explainable AI and Interpretability: As deep learning systems became more complex and were deployed in critical applications, understanding their decision-making processes became essential. While explainability research dates back to the 1970s–1980s [207, 214], deep learning's "black box" nature created new urgency.

Modern techniques focus on saliency maps and gradient-based explanations, including Layer-wise Relevance Propagation (LRP, 2016) [19], LIME (2016) [187], Grad-CAM (2017) [208], and improved variants [33, 166]. These methods help identify which input features most influence model decisions. Refer to [251] for historical context.

Shortcut Learning and Robustness: The phenomenon of shortcut learning, formally characterized in 2020 [68], revealed how neural networks can achieve good training performance by exploiting spurious correlations rather than learning intended patterns. This insight has important implications for model robustness and generalization, leading to improved training methodologies and regularization techniques (covered in Chap. 5).

5.9.6 Next-Generation Generative Models (2020–Present)

Diffusion Models: Diffusion models, introduced in 2020 [112], represent the latest breakthrough in generative modeling. These models learn to gradually add noise to data and then reverse this process to generate new samples from pure noise. They have demonstrated superior performance compared to GANs in both sample quality and training stability.

The success of diffusion models has led to widespread adoption across domains, from image generation to scientific applications like protein and antibody design [18, 235]. Their principled probabilistic framework and stable training dynamics have made them the preferred choice for many generative applications. Refer to Chap. 15 for theoretical details and [40] for comprehensive surveys of variants.

6 Conclusion

Machine learning can be broadly viewed as a three-step process: defining a hypothesis class (the modeling assumptions about data), specifying a scoring mechanism (loss function), and applying a search strategy (optimization). From another perspective, a machine learning model can be considered as a mapping from input data to predictions or outputs.

Traditionally, machine learning tasks have been categorized into supervised learning (encompassing classification and regression), unsupervised learning (including clustering and dimensionality reduction), and reinforcement learning.

Fundamental algorithms and ideas in statistical learning can be grouped into several areas: statistical classification (e.g., Bayes classifier, LDA, logistic regression, metric learning, random forests, and SVM), statistical regression (e.g., linear regression, quantile regression, and nonlinear regression), statistical dimensionality reduction (e.g., spectral and probabilistic methods), boosting, mixture models, and Markov models.

The history of neural networks begins with the McCulloch-Pitts model, Hebbian learning, the Perceptron, and ADALINE. This was followed by the development of backpropagation, stochastic gradient descent (SGD), and Adam. Biologically inspired neural networks also form an important category. Later, specialized architectures such as Recurrent Neural Networks (RNNs), Long Short-

Term Memory (LSTM) networks, and Convolutional Neural Networks (CNNs) were introduced for sequence and image-like data, respectively. After the so-called "AI winter", breakthroughs in image recognition sparked a resurgence of deep learning. Since then, a wide range of neural network models have emerged, spanning generative models, Transformers, state space models, large language models, deep reinforcement learning, graph neural networks, few-shot learning, meta-learning, self-supervised learning, and federated learning.

In addition, several important concepts have become central to modern deep learning, including neural network compression, adversarial robustness, explainable AI, and shortcut learning.

7 Exercises

1. (Level: Easy) Define what machine learning is and how it is different from classical statistics. Provide the fundamental machine learning tasks, briefly define each and give an example of where each of them can be used. Include some common algorithms for each of the machine learning tasks.
2. (Level: Easy) Categorize each of these machine learning scenarios into supervised learning, unsupervised learning, or reinforcement learning. Justify your reasoning for each case.

   (a) A neural network is trained to classify handwritten digits using the MNIST dataset, which contains 60,000 images of handwritten digits, along with the correct answer for each image.
   (b) A robot is programmed to learn how to play a video game. It does not have access to the game's rules, but it can observe its current score after each action. Over time, it learns to play better by maximizing its score.
   (c) A deep learning model is designed to segment medical images into different sections corresponding to specific organs. The training data consists of medical scans that have been annotated by experts to mark the boundaries of the organs.
   (d) A machine learning model is given 100,000 astronomical images of unknown stars and galaxies. Using dimensionality reduction techniques, it groups similar-looking objects based on their features, such as size and shape.

3. (Level: Moderate) Which of the following statements best describes the significance of the Structured State Space Model (S4) and Mamba architectures in the evolution of deep learning?

   (a) They replaced transformers in most NLP tasks by using self-attention mechanisms.
   (b) They are generative models that rely on contrastive learning to synthesize realistic data.
   (c) They offer alternatives to Transformers by combining the benefits of RNNs and state space models, achieving linear computational complexity.
   (d) They are graph-based architectures used primarily in processing molecular structures and social networks.

4. (Level: Moderate) What is the main advantage of self supervised learning methods like SimCLR, BYOL, and SwAV in deep learning?

   (a) They require large amounts of manually labeled data for training.
   (b) They learn useful data representations without needing manual labels by generating supervisory signals from the data itself.
   (c) They improve model compression by pruning unimportant weights.
   (d) They are primarily used for training reinforcement learning agents.

5. (Level: Moderate) Which of the following statements correctly describes the significance of the Deep Q Network (DQN) introduced by DeepMind in 2013?
   (a) It was the first neural network model to use attention mechanisms.
   (b) It introduced meta-learning concepts for few-shot learning.
   (c) It achieved superhuman performance in Atari games by combining deep learning with reinforcement learning.
   (d) It was the first model to use diffusion processes for generative modeling.

6. (Level: Difficult) Which of the following statements correctly describes an early foundational method for graph-structured data learning, as introduced in the mid-2010s?
   (a) Chebyshev Network (ChebNet) generalized graph convolutions using Chebyshev polynomials.
   (b) Graph Attention Network (GAT) was the first method to apply recurrent neural networks on graphs.
   (c) Graph Autoencoder (GAE) introduced attention mechanisms for supervised graph classification.
   (d) Prototypical Networks were the first graph-based few-shot learning approach.

References

1. Martín Abadi et al. "Tensorflow: Large-scale machine learning on heterogeneous distributed systems". In: *arXiv preprint* arXiv:1603.04467 (2016).
2. Rakesh Agrawal and Ramakrishnan Srikant. "Privacy-preserving data mining". In: *Proceedings of the 2000 ACM SIGMOD international conference on Management of data*. 2000, pp. 439–450.
3. Mark A Aizerman, E. M. Braverman, and L. I. Rozonoer. "Theoretical foundations of the potential function method in pattern recognition learning". In: *Automation and remote control* 25 (1964), pp. 821–837.
4. Shunichi Amari. "A theory of adaptive pattern classifiers". In: *IEEE Transactions on Electronic Computers* 3 (1967), pp. 299–307.
5. Nachman Aronszajn. "Theory of reproducing kernels". In: *Transactions of the American mathematical society* 68.3 (1950), pp. 337–404.
6. Robert Audi. *Epistemology: A contemporary introduction to the theory of knowledge*. Routledge, 2010.
7. Dzmitry Bahdanau, Kyunghyun Cho, and Yoshua Bengio. "Neural machine translation by jointly learning to align and translate". In: *arXiv preprint* arXiv:1409.0473 (2014).
8. Yuntao Bai et al. "Training a helpful and harmless assistant with reinforcement learning from human feedback". In: *arXiv preprint* arXiv:2204.05862 (2022).
9. Adrien Bardes, Jean Ponce, and Yann LeCun. "VICReg: Variance-invariance-covariance regularization for self-supervised learning". In: *International Conference on Learning Representations*. 2022.
10. Elnaz Barshan, Ali Ghodsi, Zohreh Azimifar, and Mansoor Zolghadri Jahromi. "Supervised principal component analysis: Visualization, classification and regression on subspaces and submanifolds". In: *Pattern Recognition* 44.7 (2011), pp. 1357–1371.
11. Thomas Bayes. "An essay towards solving a problem in the doctrine of chances". In: *Biometrika* 45.3-4 (1958), pp. 296–315.
12. Richard Beals, David H Krantz, and Amos Tversky. "Foundations of multidimensional scaling." In: *Psychological review* 75.2 (1968), p. 127.
13. Mikhail Belkin and Partha Niyogi. "Laplacian eigenmaps and spectral techniques for embedding and clustering". In: *Advances in neural information processing systems* 14 (2001), pp. 585–591.
14. Richard Bellman. "On the theory of dynamic programming". In: *Proceedings of the national Academy of Sciences* 38.8 (1952), pp. 716–719.
15. Richard Bellman and Robert Kalaba. "On adaptive control processes". In: *IRE Transactions on Automatic Control* 4.2 (1959), pp. 1–9.
16. Yoshua Bengio et al. "Out-of-sample extensions for LLE, Isomap, MDS, eigenmaps, and spectral clustering". In: *Advances in neural information processing systems* 16 (2003).

17. Yoshua Bengio et al. *Spectral clustering and kernel PCA are learning eigenfunctions*. Tech. rep. Departement d'Informatique et Recherche Operationnelle, Technical Report 1239, 2003.
18. Nathaniel R Bennett et al. "Atomically accurate de novo design of single-domain antibodies". In: *bioRxiv* (2024).
19. Alexander Binder, Grégoire Montavon, Sebastian Lapuschkin, Klaus-Robert Müller, and Wojciech Samek. "Layer-wise relevance propagation for neural networks with local renormalization layers". In: *Artificial Neural Networks and Machine Learning–ICANN 2016: 25th International Conference on Artificial Neural Networks, Barcelona, Spain, September 6-9, 2016, Proceedings, Part II 25*. Springer. 2016, pp. 63–71.
20. Bernhard E Boser, Isabelle M Guyon, and Vladimir N Vapnik. "A training algorithm for optimal margin classifiers". In: *Proceedings of the fifth annual workshop on Computational learning theory*. 1992, pp. 144–152.
21. Leo Breiman. "Arcing classifier (with discussion and a rejoinder by the author)". In: *The annals of statistics* 26.3 (1998), pp. 801–849.
22. Leo Breiman. "Bagging predictors". In: *Machine learning* 24.2 (1996), pp. 123–140.
23. Leo Breiman. *Classification and regression trees*. Routledge, 2017.
24. Leo Breiman. "Random forests". In: *Machine learning* 45 (2001), pp. 5–32.
25. Jane Bromley et al. "Signature verification using a "Siamese" time delay neural network". In: *International Journal of Pattern Recognition and Artificial Intelligence* 7.04 (1993), pp. 669–688.
26. Arthur Earl Bryson and Ho Yu-Chi. *Applied optimal control: optimization, estimation and control*. CRC Press, 1969.
27. Cristian BuciluǅŽ, Rich Caruana, and Alexandru Niculescu-Mizil. "Model compression". In: *Proceedings of the 12th ACM SIGKDD international conference on Knowledge discovery and data mining*. 2006, pp. 535–541.
28. Peter Lukas Bühlmann and Bin Yu. "Explaining bagging". In: *Research report/Seminar f ü r Statistik, Eidgen ö ssische Technische Hochschule Z ü rich*. Vol. 92. Seminar für Statistik, Eidgenössische Technische Hochschule (ETH). 2000.
29. Erik Cambria and Bebo White. "Jumping NLP curves: A review of natural language processing research". In: *IEEE Computational intelligence magazine* 9.2 (2014), pp. 48–57.
30. Mathilde Caron et al. "Unsupervised learning of visual features by contrasting cluster assignments". In: *Advances in neural information processing systems* 33 (2020), pp. 9912–9924.
31. Raymond B Cattell. "A biometrics invited paper. Factor analysis: An introduction to essentials I. The purpose and underlying models". In: *Biometrics* 21.1 (1965), pp. 190–215.
32. Anirban Chakraborty, Manaar Alam, Vishal Dey, Anupam Chattopadhyay, and Debdeep Mukhopadhyay. "Adversarial attacks and defences: A survey". In: *arXiv preprint* arXiv:1810.00069 (2018).
33. Aditya Chattopadhay, Anirban Sarkar, Prantik Howlader, and Vineeth N Balasubramanian. "Grad-CAM++: Generalized gradient-based visual explanations for deep convolutional networks". In: *2018 IEEE winter conference on applications of computer vision (WACV)*. IEEE. 2018, pp. 839–847.
34. Ting Chen, Simon Kornblith, Mohammad Norouzi, and Geoffrey Hinton. "A simple framework for contrastive learning of visual representations". In: *International conference on machine learning*. PMLR. 2020, pp. 1597–1607.
35. Xinlei Chen and Kaiming He. "Exploring simple siamese representation learning". In: *Proceedings of the IEEE/CVF conference on computer vision and pattern recognition*. 2021, pp. 15750–15758.
36. Dennis Child. *The essentials of factor analysis*. Cassell Educational, 1990.
37. Gerda Claeskens and Nils Lid Hjort. *Model selection and model averaging*. Cambridge Books, Cambridge University Press, 2008.
38. Corinna Cortes. "Support-Vector Networks". In: *Machine Learning* (1995).
39. Jan Salomon Cramer. *The origins of logistic regression*. Tech. rep. Tinbergen Institute Working Paper, 2002.
40. Florinel-Alin Croitoru, Vlad Hondru, Radu Tudor Ionescu, and Mubarak Shah. "Diffusion models in vision: A survey". In: *IEEE Transactions on Pattern Analysis and Machine Intelligence* 45.9 (2023), pp. 10850–10869.
41. Michaël Defferrard, Xavier Bresson, and Pierre Vandergheynst. "Convolutional neural networks on graphs with fast localized spectral filtering". In: *Advances in neural information processing systems*. Vol. 29. 2016, pp. 3844–3852.
42. Javier Del Ser et al. "Bio-inspired computation: Where we stand and what's next". In: *Swarm and Evolutionary Computation* 48 (2019), pp. 220–250.
43. Arthur P Dempster, Nan M Laird, and Donald B Rubin. "Maximum likelihood from incomplete data via the EM algorithm". In: *Journal of the royal statistical society: series B (methodological)* 39.1 (1977), pp. 1–22.
44. Jacob Devlin, Ming-Wei Chang, Kenton Lee, and Kristina Toutanova. "BERT: Pre-training of deep bidirectional transformers for language understanding". In: *arXiv preprint* arXiv:1810.04805 (2018).
45. Thomas G Dietterich. "An experimental comparison of three methods for constructing ensembles of decision trees: Bagging, boosting, and randomization". In: *Machine learning* 40 (2000), pp. 139–157.
46. Pedro Domingos. "The role of Occam's razor in knowledge discovery". In: *Data mining and knowledge discovery* 3.4 (1999), pp. 409–425.
47. John Duchi, Elad Hazan, and Yoram Singer. "Adaptive subgradient methods for online learning and stochastic optimization". In: *Journal of machine learning research* 12.7 (2011).

48. Gintare Karolina Dziugaite, Daniel M Roy, and Zoubin Ghahramani. "Training generative neural networks via maximum mean discrepancy optimization". In: *Proceedings of the Thirty-First Conference on Uncertainty in Artificial Intelligence*. 2015, pp. 258–267.
49. Jeffrey L Elman. "Finding structure in time". In: *Cognitive science* 14.2 (1990), pp. 179–211.
50. Laurene V Fausett. *Fundamentals of neural networks: architectures, algorithms and applications*. Pearson Education India, 2006.
51. Chelsea Finn, Pieter Abbeel, and Sergey Levine. "Model-agnostic meta-learning for fast adaptation of deep networks". In: *International conference on machine learning*. PMLR. 2017, pp. 1126–1135.
52. Ronald A Fisher. "The use of multiple measurements in taxonomic problems". In: *Annals of eugenics* 7.2 (1936), pp. 179–188.
53. Ronald A Fisher. "XV.—The correlation between relatives on the supposition of Mendelian inheritance." In: *Earth and Environmental Science Transactions of the Royal Society of Edinburgh* 52.2 (1919), pp. 399–433.
54. Ronald Aylmer Fisher. "Statistical methods for research workers". In: *Breakthroughs in statistics*. Springer, 1992, pp. 66–70.
55. Pierre Foret, Ariel Kleiner, Hossein Mobahi, and Behnam Neyshabur. "Sharpness-aware minimization for efficiently improving generalization". In: *International Conference on Learning Representations (ICLR)*. 2021.
56. G David Forney. "The Viterbi algorithm". In: *Proceedings of the IEEE* 61.3 (1973), pp. 268–278.
57. Yoav Freund and Robert E Schapire. "Experiments with a new boosting algorithm". In: *Proceedings of the Thirteenth International Conference on Machine Learning*. Morgan Kaufman, San Francisco. 1996, pp. 148–156.
58. B Roy Frieden. *Science from Fisher information: a unification*. Cambridge University Press, 2004.
59. Jerome Friedman, Trevor Hastie, and Robert Tibshirani. "Additive logistic regression: a statistical view of boosting". In: *The annals of statistics* 28.2 (2000), pp. 337–407.
60. Jerome Friedman, Trevor Hastie, and Robert Tibshirani. *The elements of statistical learning: Data Mining, Inference, and Prediction*. Vol. 2. Springer series in statistics, New York, NY, USA, 2009.
61. Jerome H Friedman. "Multivariate adaptive regression splines". In: *The annals of statistics* 19.1 (1991), pp. 1–67.
62. Benjamin Fruchter. *Introduction to factor analysis*. Van Nostrand, 1954.
63. Kenji Fukumizu, Francis R Bach, and Michael I Jordan. "Dimensionality reduction for supervised learning with reproducing kernel Hilbert spaces". In: *Journal of Machine Learning Research* 5.Jan (2004), pp. 73–99.
64. Kenji Fukumizu, Francis R Bach, and Michael I Jordan. "Kernel dimension reduction in regression". In: *The Annals of Statistics* 37.4 (2009), pp. 1871–1905.
65. Kenji Fukumizu, Francis R Bach, and Michael I Jordan. "Kernel dimensionality reduction for supervised learning". In: *Advances in neural information processing systems*. Vol. 16. 2003.
66. Kunihiko Fukushima. "Neocognitron: A self-organizing neural network model for a mechanism of pattern recognition unaffected by shift in position". In: *Biological cybernetics* 36.4 (1980), pp. 193–202.
67. Francis Galton. *Natural Inheritance*. (5th ed.), New York: Macmillan and Company, 1894.
68. Robert Geirhos et al. "Shortcut learning in deep neural networks". In: *Nature Machine Intelligence* 2.11 (2020), pp. 665–673.
69. Joseph Diaz Gergonne. "The application of the method of least squares to the interpolation of sequences". In: *Historia Mathematica* 1.4 (1974), pp. 439–447.
70. Pierre Geurts, Damien Ernst, and Louis Wehenkel. "Extremely randomized trees". In: *Machine learning* 63 (2006), pp. 3–42.
71. Zoubin Ghahramani. "An introduction to hidden Markov models and Bayesian networks". In: *Hidden Markov models: applications in computer vision*. World Scientific, 2001, pp. 9–41.
72. Benyamin Ghojogh and Mark Crowley. "Linear and quadratic discriminant analysis: Tutorial". In: *arXiv preprint* arXiv:1906.02590 (2019).
73. Benyamin Ghojogh and Mark Crowley. "The theory behind overfitting, cross validation, regularization, bagging, and boosting: tutorial". In: *arXiv preprint* arXiv:1905.12787 (2019).
74. Benyamin Ghojogh, Mark Crowley, Fakhri Karray, and Ali Ghodsi. *Elements of Dimensionality Reduction and Manifold Learning*. Springer Nature, 2023.
75. Benyamin Ghojogh, Mark Crowley, Fakhri Karray, and Ali Ghodsi. "Factor analysis and probabilistic principal component analysis". In: *Elements of Dimensionality Reduction and Manifold Learning*. Springer, 2023, pp. 355–374.
76. Benyamin Ghojogh, Mark Crowley, Fakhri Karray, and Ali Ghodsi. "Fisher Discriminant Analysis". In: *Elements of Dimensionality Reduction and Manifold Learning*. Springer, 2023, pp. 155–184.
77. Benyamin Ghojogh, Mark Crowley, Fakhri Karray, and Ali Ghodsi. "Laplacian-Based Dimensionality Reduction". In: *Elements of Dimensionality Reduction and Manifold Learning*. Springer, 2023, pp. 249–284.
78. Benyamin Ghojogh, Mark Crowley, Fakhri Karray, and Ali Ghodsi. "Locally linear embedding". In: *Elements of Dimensionality Reduction and Manifold Learning*. Springer, 2023, pp. 207–247.
79. Benyamin Ghojogh, Mark Crowley, Fakhri Karray, and Ali Ghodsi. "Multidimensional Scaling, Sammon Mapping, and Isomap". In: *Elements of Dimensionality Reduction and Manifold Learning*. Springer, 2023, pp. 185–205.

80. Benyamin Ghojogh, Mark Crowley, Fakhri Karray, and Ali Ghodsi. "Principal component analysis". In: *Elements of Dimensionality Reduction and Manifold Learning*. Springer, 2023, pp. 123–154.
81. Benyamin Ghojogh, Mark Crowley, Fakhri Karray, and Ali Ghodsi. "Random Projection". In: *Elements of Dimensionality Reduction and Manifold Learning*. Springer, 2023, pp. 395–425.
82. Benyamin Ghojogh, Mark Crowley, Fakhri Karray, and Ali Ghodsi. "Stochastic neighbour embedding". In: *Elements of Dimensionality Reduction and Manifold Learning*. Springer, 2023, pp. 455–477.
83. Benyamin Ghojogh, Mark Crowley, Fakhri Karray, and Ali Ghodsi. "Sufficient Dimension Reduction and Kernel Dimension Reduction". In: *Elements of Dimensionality Reduction and Manifold Learning*. Springer, 2023, pp. 427–453.
84. Benyamin Ghojogh, Mark Crowley, Fakhri Karray, and Ali Ghodsi. "Uniform manifold approximation and projection (UMAP)". In: *Elements of Dimensionality Reduction and Manifold Learning*. Springer, 2023, pp. 479–497.
85. Benyamin Ghojogh, Ali Ghodsi, Fakhri Karray, and Mark Crowley. "KKT conditions, first-order and second-order optimization, and distributed optimization: tutorial and survey". In: *arXiv preprint* arXiv:2110.01858 (2021).
86. Benyamin Ghojogh, Ali Ghodsi, Fakhri Karray, and Mark Crowley. "Reproducing Kernel Hilbert Space, Mercer's Theorem, Eigenfunctions, Nyström Method, and Use of Kernels in Machine Learning: Tutorial and Survey". In: *arXiv preprint* arXiv:2106.08443 (2021).
87. Benyamin Ghojogh, Ali Ghodsi, Fakhri Karray, and Mark Crowley. "Spectral, probabilistic, and deep metric learning: Tutorial and survey". In: *arXiv preprint* arXiv:2201.09267 (2022).
88. Benyamin Ghojogh, Aydin Ghojogh, Mark Crowley, and Fakhri Karray. "Fitting a mixture distribution to data: tutorial". In: *arXiv preprint* arXiv:1901.06708 (2019).
89. Benyamin Ghojogh, Fakhri Karray, and Mark Crowley. "Eigenvalue and generalized eigenvalue problems: Tutorial". In: *arXiv preprint* arXiv:1903.11240 (2019).
90. Benyamin Ghojogh, Fakhri Karray, and Mark Crowley. "Hidden Markov Model: Tutorial". In: *Engineering Archive* (2019).
91. Benyamin Ghojogh et al. "Feature selection and feature extraction in pattern analysis: A literature review". In: *arXiv preprint* arXiv:1905.02845 (2019).
92. Samanwoy Ghosh-Dastidar and Hojjat Adeli. "Spiking neural networks". In: *International journal of neural systems* 19.04 (2009), pp. 295–308.
93. Ian Goodfellow et al. "Generative adversarial nets". In: *Advances in neural information processing systems* 27 (2014).
94. Arthur Gretton, Karsten Borgwardt, Malte Rasch, Bernhard Schölkopf, and Alex Smola. "A kernel method for the two-sample-problem". In: *Advances in neural information processing systems* 19 (2006).
95. Arthur Gretton, Karsten M Borgwardt, Malte J Rasch, Bernhard Schölkopf, and Alexander Smola. "A kernel two-sample test". In: *The Journal of Machine Learning Research* 13.1 (2012), pp. 723–773.
96. Arthur Gretton, Olivier Bousquet, Alex Smola, and Bernhard Schölkopf. "Measuring statistical dependence with Hilbert-Schmidt norms". In: *International conference on algorithmic learning theory*. Springer. 2005, pp. 63–77.
97. Jean-Bastien Grill et al. "Bootstrap your own latent-a new approach to self-supervised learning". In: *Advances in neural information processing systems* 33 (2020), pp. 21271–21284.
98. Albert Gu and Tri Dao. "Mamba: Linear-Time Sequence Modeling with Selective State Spaces". In: *First Conference on Language Modeling*. 2024.
99. Albert Gu, Karan Goel, and Christopher Re. "Efficiently Modeling Long Sequences with Structured State Spaces". In: *International Conference on Learning Representations*. 2022.
100. Raia Hadsell, Sumit Chopra, and Yann LeCun. "Dimensionality reduction by learning an invariant mapping". In: *2006 IEEE Computer Society Conference on Computer Vision and Pattern Recognition (CVPR'06)*. Vol. 2. IEEE. 2006, pp. 1735–1742.
101. Harry H Harman. *Modern factor analysis*. University of Chicago press, 1976.
102. Trevor Hastie, Robert Tibshirani, and Jerome Friedman. *The elements of statistical learning: data mining, inference, and prediction*. Springer, 2009.
103. Trevor J Hastie and Robert Tibshirani. "Generalized Additive Models". In: *Statistical Science* 1.3 (1986), pp. 297–318.
104. Kaiming He, Xiangyu Zhang, Shaoqing Ren, and Jian Sun. "Deep residual learning for image recognition". In: *Proceedings of the IEEE conference on computer vision and pattern recognition*. 2016, pp. 770–778.
105. D Heath, S Kasif, and S Salzberg. "k-DT: A multi-tree learning method". In: *Proc. of the Second Int. Workshop on Multistrategy Learning*. 1993, pp. 138–149.
106. Donald Olding Hebb. *The Organization of Behavior*. New York: Wiley & Sons, 1949.
107. David Hilbert. "Grundzüge einer allgemeinen Theorie der linearen Integralrechnungen I". In: *Nachrichten von der Gesellschaft der Wissenschaften zu G ö ttingen, Mathematisch-Physikalische Klasse* (1904), pp. 49–91.
108. Geoffrey Hinton, Oriol Vinyals, and Jeff Dean. "Distilling the Knowledge in a Neural Network". In: *NIPS 2014 Deep Learning Workshop*. 2015.
109. Geoffrey E Hinton and Sam T Roweis. "Stochastic neighbor embedding". In: *Advances in neural information processing systems*. 2003, pp. 857–864.

110. Geoffrey E Hinton and Ruslan R Salakhutdinov. "Reducing the dimensionality of data with neural networks". In: *Science* 313.5786 (2006), pp. 504–507.
111. Julia Hirschberg and Christopher D Manning. "Advances in natural language processing". In: *Science* 349.6245 (2015), pp. 261–266.
112. Jonathan Ho, Ajay Jain, and Pieter Abbeel. "Denoising diffusion probabilistic models". In: *Advances in neural information processing systems* 33 (2020), pp. 6840–6851.
113. Sepp Hochreiter and Jürgen Schmidhuber. "Long short-term memory". In: *Neural computation* 9.8 (1997), pp. 1735–1780.
114. Jennifer A Hoeting, David Madigan, Adrian E Raftery, and Chris T Volinsky. "Bayesian model averaging: a tutorial". In: *Statistical science* (1999), pp. 382–401.
115. Reshad Hosseini and Suvrit Sra. "An alternative to EM for Gaussian mixture models: batch and stochastic Riemannian optimization". In: *Mathematical programming* 181.1 (2020), pp. 187–223.
116. Harold Hotelling. "Analysis of a complex of statistical variables into principal components." In: *Journal of educational psychology* 24.6 (1933), p. 417.
117. Gao Huang, Zhuang Liu, Laurens Van Der Maaten, and Kilian Q Weinberger. "Densely connected convolutional networks". In: *Proceedings of the IEEE conference on computer vision and pattern recognition*. 2017, pp. 4700–4708.
118. Alexey Grigorevich Ivakhnenko and Valentin Grigorévich Lapa. "Cybernetic predicting devices". In: *CCM Information Corporation* (1965).
119. William B Johnson and Joram Lindenstrauss. "Extensions of Lipschitz mappings into a Hilbert space". In: *Contemporary mathematics* 26 (1984).
120. Peter Kairouz et al. "Advances and open problems in federated learning". In: *Foundations and trends ® in machine learning* 14.1–2 (2021), pp. 1–210.
121. Immanuel Kant. "Meditationum quarundam de igne succinta delineatio". PhD thesis. University of Königsberg, 1755.
122. Immanuel Kant. *Thoughts on the true estimation of living forces*. Newcomb Livraria Press, 2012.
123. Immanuel Kant, Paul Guyer, and Allen W Wood. *Critique of pure reason*. Cambridge University Press, 1929.
124. Andrej Karpathy and Li Fei-Fei. "Deep visual-semantic alignments for generating image descriptions". In: *Proceedings of the IEEE conference on computer vision and pattern recognition*. 2015, pp. 3128–3137.
125. Gordon V Kass. "An exploratory technique for investigating large quantities of categorical data". In: *Journal of the Royal Statistical Society: Series C (Applied Statistics)* 29.2 (1980), pp. 119–127.
126. Michael Kearns. "Thoughts on hypothesis boosting". In: *Technical Report, Machine Learning class project* (1988), pp. 1–9.
127. Michael Kearns and Leslie Valiant. "Cryptographic limitations on learning Boolean formulae and finite automata". In: *Journal of the ACM (JACM)* 41.1 (1994), pp. 67–95.
128. Henry J Kelley. "Gradient theory of optimal flight paths". In: *Ars Journal* 30.10 (1960), pp. 947–954.
129. Diederik P Kingma and Jimmy Ba. "Adam: A method for stochastic optimization". In: *arXiv preprint* arXiv:1412.6980 (2014).
130. Diederik P Kingma and Max Welling. "Auto-encoding variational bayes". In: *International Conference on Learning Representations*. 2014.
131. Thomas N Kipf and Max Welling. "Semi-supervised classification with graph convolutional networks". In: *International Conference on Learning Representations*. 2017.
132. Thomas N Kipf and Max Welling. "Variational graph auto-encoders". In: *arXiv preprint* arXiv:1611.07308 (2016).
133. Roger Koenker. "Galton, Edgeworth, Frisch, and prospects for quantile regression in econometrics". In: *Journal of Econometrics* 95.2 (2000), pp. 347–374.
134. Roger Koenker. "Quantile regression". In: *Cambridge University Press* (2005).
135. Jakub Konečný. "Federated Learning: Strategies for Improving Communication Efficiency". In: *arXiv preprint* arXiv:1610.05492 (2016).
136. Alex Krizhevsky, Ilya Sutskever, and Geoffrey E Hinton. "ImageNet classification with deep convolutional neural networks". In: *Communications of the ACM* 60.6 (2017), pp. 84–90.
137. Alex Krizhevsky, Ilya Sutskever, and Geoffrey E Hinton. "Imagenet classification with deep convolutional neural networks". In: *Advances in neural information processing systems* 25 (2012).
138. Bruce R Kusse and Erik A Westwig. *Mathematical physics: applied mathematics for scientists and engineers*. 2nd ed. Wiley-VCH, 2006.
139. Balaji Lakshminarayanan, Daniel M Roy, and Yee Whye Teh. "Mondrian forests for large-scale regression when uncertainty matters". In: *Artificial Intelligence and Statistics*. PMLR. 2016, pp. 1478–1487.
140. Balaji Lakshminarayanan, Daniel M Roy, and Yee Whye Teh. "Mondrian forests: Efficient online random forests". In: *Advances in neural information processing systems* 27 (2014).
141. Yann LeCun, Yoshua Bengio, and Geoffrey Hinton. "Deep learning". In: *nature* 521.7553 (2015), pp. 436–444.
142. Yann LeCun, Léon Bottou, Yoshua Bengio, and Patrick Haffner. "Gradient-based learning applied to document recognition". In: *Proceedings of the IEEE* 86.11 (1998), pp. 2278–2324.

143. Yann LeCun et al. "Backpropagation applied to handwritten zip code recognition". In: *Neural computation* 1.4 (1989), pp. 541–551.
144. Marie-Paule Lefranc. "Unique database numbering system for immunogenetic analysis". In: *Immunology today* 11.18 (1997), p. 509.
145. Ker-Chau Li. "Sliced inverse regression for dimension reduction". In: *Journal of the American Statistical Association* 86.414 (1991), pp. 316–327.
146. Yujia Li, Kevin Swersky, and Rich Zemel. "Generative moment matching networks". In: *International conference on machine learning*. PMLR. 2015, pp. 1718–1727.
147. Seppo Linnainmaa. "The representation of the cumulative rounding error of an algorithm as a Taylor expansion of the local rounding errors". MA thesis. Master's Thesis (in Finnish), Univ. Helsinki, 1970.
148. Fei Tony Liu, Kai Ming Ting, and Zhi-Hua Zhou. "Isolation forest". In: *2008 eighth ieee international conference on data mining*. IEEE. 2008, pp. 413–422.
149. Minh-Thang Luong, Hieu Pham, and Christopher D Manning. "Effective Approaches to Attention-based Neural Machine Translation". In: *Proceedings of the 2015 Conference on Empirical Methods in Natural Language Processing*. 2015, pp. 1412–1421.
150. Haoran Ma, Benyamin Ghojogh, Maria N Samad, Dongyu Zheng, and Mark Crowley. "Isolation Mondrian forest for batch and online anomaly detection". In: *2020 IEEE International Conference on Systems, Man, and Cybernetics (SMC)*. IEEE. 2020, pp. 3051–3058.
151. Laurens van der Maaten and Geoffrey Hinton. "Visualizing data using t-SNE". In: *Journal of machine learning research* 9.Nov (2008), pp. 2579–2605.
152. Alireza Makhzani. "Implicit autoencoders". In: *arXiv preprint* arXiv:1805.09804 (2018).
153. Alireza Makhzani and Brendan Frey. "PixelGAN autoencoders". In: *Advances in neural information processing systems*. 2017.
154. Alireza Makhzani, Jonathon Shlens, Navdeep Jaitly, Ian Goodfellow, and Brendan Frey. "Adversarial autoencoders". In: *arXiv preprint* arXiv:1511.05644 (2015).
155. Warren S McCulloch and Walter Pitts. "A logical calculus of the ideas immanent in nervous activity". In: *The bulletin of mathematical biophysics* 5 (1943), pp. 115–133.
156. Leland McInnes, John Healy, and James Melville. "UMAP: Uniform manifold approximation and projection for dimension reduction". In: *arXiv preprint* arXiv:1802.03426 (2018).
157. Brendan McMahan, Eider Moore, Daniel Ramage, Seth Hampson, and Blaise Aguera y Arcas. "Communication-efficient learning of deep networks from decentralized data". In: *Artificial intelligence and statistics*. PMLR. 2017, pp. 1273–1282.
158. J Mercer. "Functions of positive and negative type and their connection with the theory of integral equations". In: *Philosophical Transactions of the Royal Society* A.209 (1909), pp. 415–446.
159. Tomas Mikolov, Kai Chen, Greg Corrado, and Jeffrey Dean. "Efficient estimation of word representations in vector space". In: *International Conference on Learning Representations*. 2013.
160. Tomas Mikolov, Ilya Sutskever, Kai Chen, Greg S Corrado, and Jeff Dean. "Distributed representations of words and phrases and their compositionality". In: *Advances in neural information processing systems*. 2013, pp. 3111–3119.
161. Marvin Minsky and Seymour A Papert. *Perceptrons*. MIT press, 1969.
162. Thomas Mitchell. *Machine learning*. McGraw Hill Higher Education, 1997.
163. Volodymyr Mnih et al. "Human-level control through deep reinforcement learning". In: *Nature* 518.7540 (2015), pp. 529–533.
164. Volodymyr Mnih et al. "Playing Atari with deep reinforcement learning". In: *arXiv preprint* arXiv:1312.5602 (2013).
165. Seyed-Mohsen Moosavi-Dezfooli, Alhussein Fawzi, and Pascal Frossard. "Deepfool: a simple and accurate method to fool deep neural networks". In: *Proceedings of the IEEE conference on computer vision and pattern recognition*. 2016, pp. 2574–2582.
166. T Nathan Mundhenk, Barry Y Chen, and Gerald Friedland. "Efficient saliency maps for explainable AI". In: *arXiv preprint* arXiv:1911.11293 (2019).
167. Sreerama K Murthy, Simon Kasif, and Steven Salzberg. "A system for induction of oblique decision trees". In: *Journal of artificial intelligence research* 2 (1994), pp. 1–32.
168. John Ashworth Nelder and Robert WM Wedderburn. "Generalized linear models". In: *Journal of the Royal Statistical Society Series A: Statistics in Society* 135.3 (1972), pp. 370–384.
169. Andrew Ng, Michael Jordan, and Yair Weiss. "On spectral clustering: Analysis and an algorithm". In: *Advances in neural information processing systems* 14 (2001), pp. 849–856.
170. Mikel Olazaran. "A sociological study of the official history of the perceptrons controversy". In: *Social Studies of Science* 26.3 (1996), pp. 611–659.
171. Long Ouyang et al. "Training language models to follow instructions with human feedback". In: *Advances in neural information processing systems* 35 (2022), pp. 27730–27744.
172. Emanuel Parzen. "On estimation of a probability density function and mode". In: *The annals of mathematical statistics* 33.3 (1962), pp. 1065–1076.

173. Adam Paszke et al. "PyTorch: An imperative style, high-performance deep learning library". In: *Advances in neural information processing systems* 32 (2019).
174. Karl Pearson. "LIII. On lines and planes of closest fit to systems of points in space". In: *The London, Edinburgh, and Dublin Philosophical Magazine and Journal of Science* 2.11 (1901), pp. 559–572.
175. Jeffrey Pennington, Richard Socher, and Christopher D Manning. "GloVe: Global vectors for word representation". In: *Proceedings of the 2014 conference on empirical methods in natural language processing*. 2014, pp. 1532–1543.
176. David L Phillips. "A technique for the numerical solution of certain integral equations of the first kind". In: *Journal of the ACM (JACM)* 9.1 (1962), pp. 84–97.
177. Robi Polikar. "Ensemble learning". In: *Ensemble machine learning*. Springer, 2012, pp. 1–34.
178. Karl R. Popper. *Conjectures and Refutations: The Growth of Scientific Knowledge*. London: Routledge & Kegan Paul, 1963.
179. J Ross Quinlan. "Bagging, boosting, and C4.5". In: *AAAI/IAAI Conference*. Vol. 1. 1996, pp. 725–730.
180. J Ross Quinlan. *C4.5: programs for machine learning*. Elsevier, 2014.
181. J. Ross Quinlan. "Induction of decision trees". In: *Machine learning* 1 (1986), pp. 81–106.
182. Lawrence R Rabiner. "A tutorial on hidden Markov models and selected applications in speech recognition". In: *Proceedings of the IEEE* 77.2 (1989), pp. 257–286.
183. Lawrence R Rabiner and Biing-Hwang Juang. "An introduction to hidden Markov models". In: *IEEE ASSP Magazine* 3.1 (1986), pp. 4–16.
184. Alec Radford, Karthik Narasimhan, Tim Salimans, and Ilya Sutskever. *Improving language understanding by generative pre-training*. Tech. rep. OpenAI, 2018.
185. Rafael Rafailov et al. "Direct preference optimization: Your language model is secretly a reward model". In: *Advances in Neural Information Processing Systems* 36 (2023), pp. 53728–53741.
186. Russell Reed. "Pruning algorithms-a survey". In: *IEEE transactions on Neural Networks* 4.5 (1993), pp. 740–747.
187. Marco Tulio Ribeiro, Sameer Singh, and Carlos Guestrin. " "Why should I trust you?" Explaining the predictions of any classifier". In: *Proceedings of the 22nd ACM SIGKDD international conference on knowledge discovery and data mining*. 2016, pp. 1135–1144.
188. Herbert Robbins and Sutton Monro. "A stochastic approximation method". In: *The annals of mathematical statistics* (1951), pp. 400–407.
189. Raúl Rojas. "AdaBoost and the super bowl of classifiers: a tutorial introduction to adaptive boosting". In: *Freie University, Berlin, Technical Report* (2009).
190. Olaf Ronneberger, Philipp Fischer, and Thomas Brox. "U-net: Convolutional networks for biomedical image segmentation". In: *Medical Image Computing and Computer-Assisted Intervention–MICCAI 2015: 18th International Conference, Munich, Germany, October 5-9, 2015, Proceedings, Part III 18*. Springer. 2015, pp. 234–241.
191. Frank Rosenblatt. *The Perceptron – a perceiving and recognizing automaton Project Para*. Tech. rep. Report 85-460-1, Cornell Aeronautical Laboratory., 1957.
192. Murray Rosenblatt. "Remarks on Some Nonparametric Estimates of a Density Function". In: *The Annals of Mathematical Statistics* (1956), pp. 832–837.
193. Sam Roweis. "EM algorithms for PCA and SPCA". In: *Advances in neural information processing systems* 10 (1997), pp. 626–632.
194. Sam Roweis. *Hidden Markov models*. Tech. rep. University of Toronto, 2003.
195. Sam T Roweis and Lawrence K Saul. "Nonlinear dimensionality reduction by locally linear embedding". In: *Science* 290.5500 (2000), pp. 2323–2326.
196. David E Rumelhart, Geoffrey E Hinton, and Ronald J Williams. "Learning representations by back-propagating errors". In: *nature* 323.6088 (1986), pp. 533–536.
197. John W Sammon. "A nonlinear mapping for data structure analysis". In: *IEEE Transactions on computers* 100.5 (1969), pp. 401–409.
198. Lawrence K Saul and Sam T Roweis. "Think globally, fit locally: unsupervised learning of low dimensional manifolds". In: *Journal of Machine Learning Research* 4 (2003), pp. 119–155.
199. Robert E Schapire, Yoav Freund, Peter Bartlett, and Wee Sun Lee. "Boosting the margin: A new explanation for the effectiveness of voting methods". In: *The annals of statistics* 26.5 (1998), pp. 1651–1686.
200. Jürgen Schmidhuber. "Deep learning in neural networks: An overview". In: *Neural networks* 61 (2015), pp. 85–117.
201. Jürgen Schmidhuber. "Evolutionary principles in self-referential learning, or on learning how to learn: the meta-meta-... hook". PhD thesis. Technische Universität München, 1987.
202. Jürgen Schmidhuber. *The most cited neural networks all build on work done in my labs*. AI Blog (accessed 2024), https://people.idsia.ch/~juergen/most-cited-neural-nets.html. 2021.
203. Erhard Schmidt. "Über die Auflösung linearer Gleichungen mit unendlich vielen Unbekannten". In: *Rendiconti del Circolo Matematico di Palermo (1884-1940)* 25.1 (1908), pp. 53–77.
204. Bernhard Schölkopf, Alexander Smola, and Klaus-Robert Müller. "Kernel principal component analysis". In: *International conference on artificial neural networks*. Springer. 1997, pp. 583–588.
205. Bernhard Schölkopf, Alexander Smola, and Klaus-Robert Müller. "Nonlinear component analysis as a kernel eigenvalue problem". In: *Neural computation* 10.5 (1998), pp. 1299–1319.

206. Florian Schroff, Dmitry Kalenichenko, and James Philbin. "Facenet: A unified embedding for face recognition and clustering". In: *Proceedings of the IEEE conference on computer vision and pattern recognition*. 2015, pp. 815–823.
207. A Carlisle Scott, William J Clancey, Randall Davis, and Edward H Shortliffe. "Explanation capabilities of production-based consultation systems". In: *American Journal of Computational Linguistics* (1977), pp. 1–50.
208. Ramprasaath R Selvaraju et al. "Grad-CAM: Visual explanations from deep networks via gradient-based localization". In: *Proceedings of the IEEE international conference on computer vision*. 2017, pp. 618–626.
209. David Silver et al. "Mastering the game of Go with deep neural networks and tree search". In: *Nature* 529.7587 (2016), pp. 484–489.
210. Karen Simonyan and Andrew Zisserman. "Very deep convolutional networks for large-scale image recognition". In: *arXiv preprint* arXiv:1409.1556 (2014).
211. Jake Snell, Kevin Swersky, and Richard Zemel. "Prototypical networks for few-shot learning". In: *Advances in neural information processing systems* 30 (2017).
212. Jeffrey M Stanton. "Galton, Pearson, and the peas: A brief history of linear regression for statistics instructors". In: *Journal of Statistics Education* 9.3 (2001).
213. Richard S Sutton. "Reinforcement learning: An introduction". In: *A Bradford Book* (2018).
214. William R Swartout. "Explaining and justifying expert consulting programs". In: *Computer-assisted medical decision making*. Springer, 1985, pp. 254–271.
215. Christian Szegedy et al. "Going deeper with convolutions". In: *Proceedings of the IEEE conference on computer vision and pattern recognition*. 2015, pp. 1–9.
216. Amirhossein Tavanaei, Masoud Ghodrati, Saeed Reza Kheradpisheh, Timothée Masquelier, and Anthony Maida. "Deep learning in spiking neural networks". In: *Neural networks* 111 (2019), pp. 47–63.
217. Joshua B Tenenbaum, Vin De Silva, and John C Langford. "A global geometric framework for nonlinear dimensionality reduction". In: *Science* 290.5500 (2000), pp. 2319–2323.
218. Robert Tibshirani. "Regression shrinkage and selection via the lasso". In: *Journal of the Royal Statistical Society: Series B (Methodological)* 58.1 (1996), pp. 267–288.
219. Robert Tibshirani, Martin Wainwright, and Trevor Hastie. *Statistical learning with sparsity: the lasso and generalizations*. Chapman and Hall/CRC, 2015.
220. Tijmen Tieleman and Geoffrey Hinton. "Lecture 6.5-rmsprop: Divide the gradient by a running average of its recent magnitude". In: *COURSERA: Neural networks for machine learning* 4.2 (2012), pp. 26–31.
221. Andrey Nikolayevich Tikhonov. "On the stability of inverse problems". In: *Dokl. akad. nauk sssr*. Vol. 39. 5. 1943, pp. 195–198.
222. Michael E Tipping and Christopher M Bishop. "Probabilistic principal component analysis". In: *Journal of the Royal Statistical Society: Series B (Statistical Methodology)* 61.3 (1999), pp. 611–622.
223. Warren S Torgerson. "Multidimensional scaling of similarity". In: *Psychometrika* 30.4 (1965), pp. 379–393.
224. Eleni Triantafillou et al. "Meta-dataset: A dataset of datasets for learning to learn from few examples". In: *International Conference on Learning Representations*. 2020.
225. Jaideep Vaidya, Hwanjo Yu, and Xiaoqian Jiang. "Privacy-preserving SVM classification". In: *Knowledge and Information Systems* 14 (2008), pp. 161–178.
226. Laurens Van der Maaten and Geoffrey Hinton. "Visualizing data using t-SNE". In: *Journal of machine learning research* 9.11 (2008).
227. Vladimir Vapnik. *The nature of statistical learning theory*. Springer science & business media, 2000.
228. Vladimir Vapnik and Alexey Chervonenkis. *Theory of pattern recognition*. Nauka, Moscow, 1974.
229. Petar VeliÄ?ković et al. "Graph attention networks". In: *International Conference on Learning Representations*. 2017.
230. Pierre François Verhulst. "Resherches mathematiques sur la loi d'accroissement de la population". In: *Nouveaux memoires de l'academie royale des sciences* 18 (1845), pp. 1–41.
231. Oriol Vinyals, Charles Blundell, Timothy Lillicrap, Daan Wierstra, et al. "Matching networks for one shot learning". In: *Advances in neural information processing systems* 29 (2016).
232. Andrew Viterbi. "Error bounds for convolutional codes and an asymptotically optimum decoding algorithm". In: *IEEE transactions on Information Theory* 13.2 (1967), pp. 260–269.
233. Strother H Walker and David B Duncan. "Estimation of the probability of an event as a function of several independent variables". In: *Biometrika* 54.1-2 (1967), pp. 167–179.
234. Christopher John Cornish Hellaby Watkins. "Learning from delayed rewards". PhD thesis. King's College, Cambridge United Kingdom, 1989.
235. Joseph L Watson et al. "De novo design of protein structure and function with RFdiffusion". In: *Nature* 620.7976 (2023), pp. 1089–1100.
236. Andreas Weigend, David Rumelhart, and Bernardo Huberman. "Generalization by weight-elimination with application to forecasting". In: *Advances in neural information processing systems* 3 (1990).
237. Kilian Q Weinberger, John Blitzer, and Lawrence K Saul. "Distance metric learning for large margin nearest neighbor classification". In: *Advances in Neural Information Processing Systems*. 2006, pp. 1473–1480.

238. Kilian Q Weinberger and Lawrence K Saul. "An introduction to nonlinear dimensionality reduction by maximum variance unfolding". In: *Proceedings of the AAAI Conference on Artificial Intelligence*. Vol. 6. 2006, pp. 1683–1686.
239. Kilian Q Weinberger, Fei Sha, and Lawrence K Saul. "Learning a kernel matrix for nonlinear dimensionality reduction". In: *Proceedings of the twenty-first international conference on Machine learning*. 2004, p. 106.
240. Yair Weiss. "Segmentation using eigenvectors: a unifying view". In: *Proceedings of the seventh IEEE international conference on computer vision*. Vol. 2. IEEE. 1999, pp. 975–982.
241. Bernard Widrow. *An adaptive "ADALINE" neuron using chemical "memistors"*. Tech. rep. No. 1553-2, Stanford Electronics Laboratories, 1960.
242. Bernard Widrow. "Generalization and information storage in networks of adaline neurons". In: *Self-organizing systems* (1962), pp. 435–461.
243. Bernard Widrow and Marcian E Hoff. *Adaptive switching circuits*. Tech. rep. Stanford University, California, Stanford Electronics Labs, 1960.
244. Bernard Widrow and Michael A Lehr. "30 years of adaptive neural networks: perceptron, madaline, and backpropagation". In: *Proceedings of the IEEE* 78.9 (1990), pp. 1415–1442.
245. Christopher Williams and Matthias Seeger. "The effect of the input density distribution on kernel-based classifiers". In: *Proceedings of the 17th international conference on machine learning*. 2000.
246. Ronald J Williams. "Simple statistical gradient-following algorithms for connectionist reinforcement learning". In: *Machine learning* 8 (1992), pp. 229–256.
247. Capt Rodney Winter and B Widrow. "MADALINE RULE II: A training algorithm for neural networks". In: *Second Annual International Conference on Neural Networks*. 1988, pp. 1–401.
248. Gregory L Wittel and Shyhtsun Felix Wu. "On Attacking Statistical Spam Filters". In: *CEAS*. 2004.
249. David H Wolpert. "The lack of a priori distinctions between learning algorithms". In: *Neural computation* 8.7 (1996), pp. 1341–1390.
250. Stanisław Woźniak, Angeliki Pantazi, Thomas Bohnstingl, and Evangelos Eleftheriou. "Deep learning incorporating biologically inspired neural dynamics and in-memory computing". In: *Nature Machine Intelligence* 2.6 (2020), pp. 325–336.
251. Feiyu Xu et al. "Explainable AI: A brief survey on history, research areas, approaches and challenges". In: *Natural language processing and Chinese computing: 8th cCF international conference, NLPCC 2019, dunhuang, China, October 9–14, 2019, proceedings, part II 8*. Springer. 2019, pp. 563–574.
252. Jian Xue, Jinyu Li, and Yifan Gong. "Restructuring of deep neural network acoustic models with singular value decomposition". In: *Interspeech*. 2013, pp. 2365–2369.
253. Shuicheng Yan, Dong Xu, Benyu Zhang, and Hong-Jiang Zhang. "Graph embedding: A general framework for dimensionality reduction". In: *2005 IEEE Computer Society Conference on Computer Vision and Pattern Recognition (CVPR'05)*. Vol. 2. IEEE. 2005, pp. 830–837.
254. Shuicheng Yan et al. "Graph embedding and extensions: A general framework for dimensionality reduction". In: *IEEE transactions on pattern analysis and machine intelligence* 29.1 (2006), pp. 40–51.
255. Liu Yang and Rong Jin. "Distance metric learning: A comprehensive survey". In: *Michigan State Universiy* 2.2 (2006), p. 4.
256. Jure Zbontar, Li Jing, Ishan Misra, Yann LeCun, and Stéphane Deny. "Barlow Twins: Self-supervised learning via redundancy reduction". In: *International conference on machine learning*. PMLR. 2021, pp. 12310–12320.
257. Wayne Xin Zhao et al. "A survey of large language models". In: *arXiv preprint* arXiv:2303.18223 (2023).
258. Shuchang Zhou et al. "DoReFa-Net: Training low bitwidth convolutional neural networks with low bitwidth gradients". In: *arXiv preprint* arXiv:1606.06160 (2016).

Part I
Fundamentals of Deep Learning

2 From Perceptron to Feedforward Neural Networks: Foundations and Early Models

Abstract

This chapter introduces the foundational concepts of artificial neural networks by focusing on the Perceptron, its limitations, and the development of Multi-Layer Perceptrons (MLPs) or feedforward neural networks. Beginning with early neuron models inspired by biological systems, it presents the Perceptron as the first trainable neural network and explores its capacity and limitations, such as the inability to solve nonlinearly separable problems. These challenges motivate the transition to multi-layer architectures, which form the basis of deep learning. The chapter explains the structure of MLPs, including neurons, layers, activation functions, and common loss functions. The chapter also offers an overview of related models—including ADALINE, logistic regression, radial basis function networks, and self-organizing map—that provide additional historical and conceptual context but are not central to the main developmental trajectory of deep learning. The universal approximation theorem, in neural networks, is also discussed in this chapter.

1 Introduction

Artificial neural networks are among the most powerful tools in modern machine learning, enabling state-of-the-art performance across a wide range of tasks, from computer vision to natural language processing. At the core of these models lies a simple computational unit inspired by the biological neuron. Understanding this foundational building block and its evolution is essential for grasping the structure and function of deep learning systems.

This chapter begins with the Perceptron, one of the earliest and most influential models in the history of neural networks. Despite its simplicity, the Perceptron introduces key ideas such as weighted inputs, activation functions, and weight updates based on feedback. However, the Perceptron has important limitations—most notably, its inability to solve nonlinear classification problems such as XOR. These limitations prompted the development of more expressive models, most notably Multi-Layer Perceptrons (MLPs) or feedforward neural networks, which are capable of representing complex decision boundaries.

Section 2 focuses first on the Perceptron and its limitations, and then builds toward the architecture of MLPs. Then, Sect. 3 introduces several related models and historical extensions that complement the main narrative, including ADALINE, logistic regression, radial basis function networks, and self-organizing map. Finally, Sect. 4 discusses the universal approximation theorem in neural networks.

B. Ghojogh and A. Ghodsi, *Elements of Deep Learning*,
https://doi.org/10.1007/978-3-032-10738-1_2

2 From Biology to Abstraction

2.1 The Biological Neuron

Artificial neural networks are composed of units called artificial neurons, which are loosely inspired by biological neurons. A biological neuron consists of several key components, illustrated in Fig. 1:

- **Soma:** The cell body of the neuron, responsible for integrating incoming signals.
- **Nucleus:** Located within the soma, it contains the neuron's genetic material and regulates cellular activity.
- **Dendrites:** Branch-like extensions that receive electrochemical signals from other neurons and transmit them toward the soma.
- **Synapse:** A small gap between the axon terminal of one neuron and the dendrites of another, across which signals are transmitted chemically.

When a biological neuron receives signals from neighboring neurons via its dendrites and synapses, it accumulates these signals at the soma. If the total incoming signal exceeds a certain threshold, the neuron "fires" by generating an action potential that travels down its axon to communicate with other neurons.

This threshold-based firing mechanism inspired early mathematical models of artificial neurons, which aim to mimic this behavior using simplified abstractions. The first of these models, introduced in the mid-twentieth century, is discussed in the next subsection.

2.2 The McCulloch–Pitts Model

In 1943, Warren McCulloch, a physiologist, and Walter Pitts, a logician, proposed the first mathematical model of a neuron inspired by its biological counterpart [20]. This model, later known as the McCulloch–Pitts neuron, represents one of the earliest attempts to formalize neural computation.

The McCulloch–Pitts model consists of two or more binary inputs and a single binary output, as illustrated in Fig. 2. A key feature of the model is that it only activates when the sum of its inputs exceeds a certain threshold; otherwise, it remains inactive. All inputs are assigned equal, fixed weights, and no learning takes place—the behavior is entirely predefined.

This threshold-based mechanism allows the neuron to simulate basic logical operations such as AND, OR, and NOT gates, depending on the configuration of inputs and the threshold.

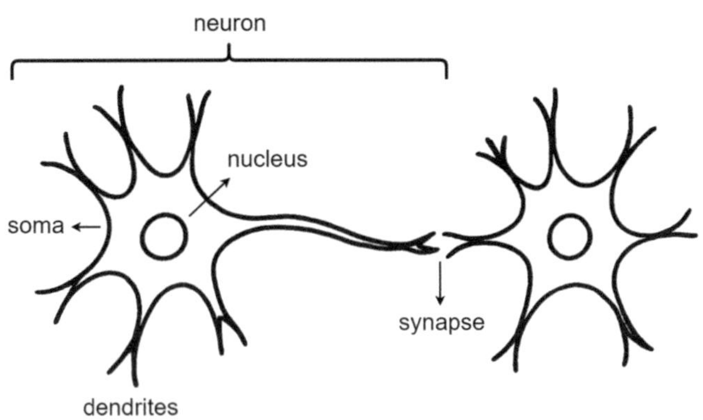

Fig. 1 The main components of a biological neuron

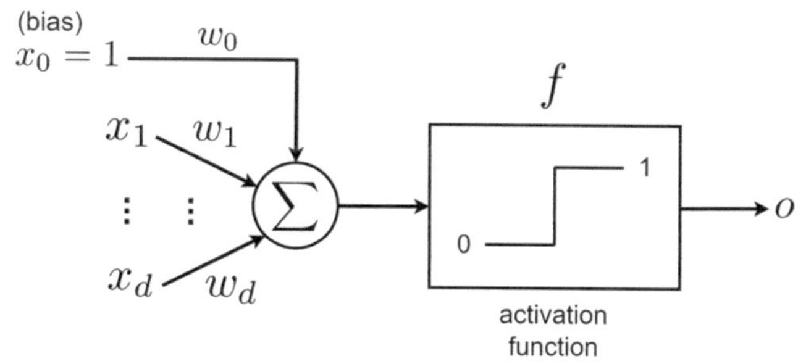

Fig. 2 The McCulloch–Pitts model of a neuron

The mathematical formulation of the model is

$$o = f\Big(\sum_{j=1}^{d} w_j x_j + w_0\Big) = f\Big(\sum_{j=0}^{d} w_j x_j\Big), \tag{1}$$

where $\{x_i \in \mathbb{R}\}_{i=1}^d$ are the input signals, $x_0 = 1$ is for the bias (intercept), $\{w_i \in \mathbb{R}\}_{i=0}^d$ are the learnable weights (parameters), and $f(\cdot)$ is the activation function. The neuron computes a weighted sum of the inputs and applies the activation function f to produce the output o.

In the McCulloch–Pitts model, the activation function is a binary step function, so the output o takes values in $\{0, 1\}$.

2.3 The Perceptron

Following the McCulloch–Pitts model, the next major step in neural network development was the introduction of the **Perceptron**, a model that not only mimicked the structure of a neuron but also introduced a supervised learning rule to adjust weights. The Perceptron was proposed and implemented in 1958 by physiologist Frank Rosenblatt at the Cornell Aeronautical Laboratory [25]. Rosenblatt's goal was to develop a system that could emulate aspects of human visual perception [1]. Today, the Perceptron is recognized as a foundational building block of modern neural networks.

The Perceptron was primarily designed for **binary classification**, which was both practical and conceptually aligned with the binary logic used in digital computers. This made it relevant both for general pattern recognition and for computer science applications, which were emerging rapidly during that era. Even for problems involving multiple classes, the task could be decomposed into binary classification subproblems.

In the initial design, Rosenblatt emphasized random connectivity between input and output neurons, based on his belief that the human retina was randomly connected to the visual cortex [1]. While early inspirations came from Hebbian learning [14], which emphasized strengthening synaptic weights through repeated co-activation, the Perceptron ultimately adopted a distinct supervised learning rule, driven by labeled data and classification error.

The model garnered significant public attention. At a 1958 press conference organized by the U.S. Navy, Rosenblatt described the Perceptron as a potential foundation for intelligent machines. Following the event, the *New York Times* published an article declaring the Perceptron to be "the embryo of an electronic computer that [the Navy] expects will be able to walk, talk, see, write, reproduce itself, and be conscious of its existence" [23]. These sensational claims contributed to both public excitement and eventual skepticism.

2.4 Perceptron with Binary Activation Function

The structure of the Perceptron is shown in Fig. 3. In this model, a linear combination of the inputs $\{x_j \in \mathbb{R}\}_{j=1}^d$ and the bias x_0 is computed using learnable weights $\{w_i \in \mathbb{R}\}_{i=0}^d$, and the result is passed through an activation function f. There are n sets of inputs $\{x_j \in \mathbb{R}\}_{j=1}^d$, where each set of input is a vector $\boldsymbol{x}_i = [x_{i1}, \dots, x_{id}]^\top \in \mathbb{R}^d$. Let $\boldsymbol{x}_i \in \mathbb{R}^d$ be the ith input data vector (out of n vectors) and $x_{ij} \in \mathbb{R}$ denote the jth element (out of d elements) in the vector $\boldsymbol{x}_i$. Each input data vector $\boldsymbol{x}_i$ has a ground-truth label y_i.

For each data vector $\boldsymbol{x}_i$ as input to the Perceptron, the output o_i is compared with the ground-truth label y_i, and the weights are updated based on the difference. While the Perceptron learning rule is often associated with Hebbian principles, it is technically a supervised rule distinct from classical Hebbian learning [14]. The update rule is given by

$$\begin{aligned} \boldsymbol{w} &:= \boldsymbol{w} + \eta(y_i - o_i)\boldsymbol{x}_i, \quad \forall i \in \{1, \dots, n\}, \\ w_0 &:= w_0 + \eta(y_i - o_i), \quad \forall i \in \{1, \dots, n\}, \end{aligned} \tag{2}$$

where $\eta > 0$ is the learning rate, $\boldsymbol{w}_i = [w_1, \dots, w_d]^\top \in \mathbb{R}^d$ is the vector of weights, $\boldsymbol{x}_i = [x_{i1}, \dots, x_{id}]^\top \in \mathbb{R}^d$ is the ith input data, $y_i \in \{0, 1\}$ is its target label, and o_i is the Perceptron's predicted output for that input. When $o_i = y_i$, no update is performed, as expected, since $y_i - o_i = 0$. Otherwise, the weights are adjusted in proportion to the classification error.

These two equations can be written more compactly as

$$\boldsymbol{w} := \boldsymbol{w} + \eta(y_i - o_i)\boldsymbol{x}_i, \quad \forall i \in \{1, \dots, n\}, \tag{3}$$

where the bias input is incorporated using $x_{i0} := 1$.

Perceptron training can also be performed in batch mode rather than one data point at a time. In this scenario, each batch contains b input vectors, denoted $\{\boldsymbol{x}_i\}_{i=1}^b$, where b is the batch size. The batch-wise update rule becomes

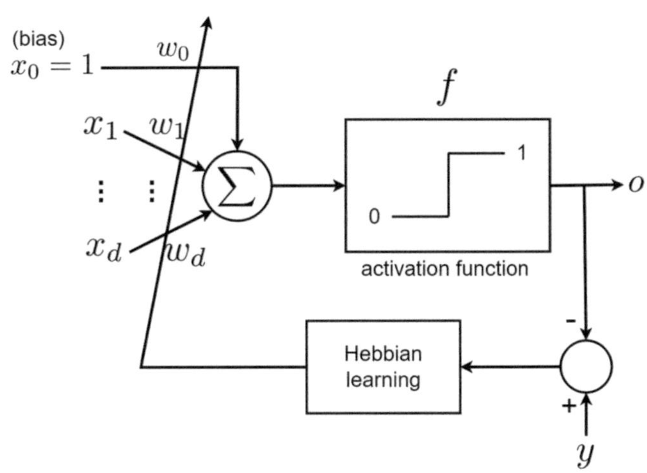

Fig. 3 The Perceptron model of a neuron with binary activation function

$$\begin{aligned} \boldsymbol{w} &:= \boldsymbol{w} + \eta \sum_{i=1}^{b} (y_i - o_i)\boldsymbol{x}_i, \\ w_0 &:= w_0 + \eta \sum_{i=1}^{b} (y_i - o_i). \end{aligned} \tag{4}$$

For interpretation, one can consider the rule without summation:

$$\boldsymbol{w} := \boldsymbol{w} + \eta(y_i - o_i)\boldsymbol{x}_i, \quad \forall i \in \{1, \ldots, n\}. \tag{5}$$

If the predicted output o_i matches the true label y_i, the update is zero:

$$o_i = y_i \implies \eta(y_i - o_i)\boldsymbol{x}_i = \mathbf{0}.$$

2.5 Perceptron with Signum Activation Function

As shown in Fig. 4, the activation function can be the sign function, also known as the signum function, instead of the binary step function. In this case, the weight update rule is defined as

$$\boldsymbol{w} := \begin{cases} \boldsymbol{w} + 2\eta y_i \boldsymbol{x}_i, & \text{if } o_i \neq y_i, \\ \boldsymbol{w}, & \text{if } o_i = y_i, \end{cases} \tag{6}$$

for all $i \in \{1, \ldots, n\}$, and the bias is updated as

$$w_0 := \begin{cases} w_0 + 2\eta y_i, & \text{if } o_i \neq y_i, \\ w_0, & \text{if } o_i = y_i. \end{cases} \tag{7}$$

When training in batch mode, the weight updates are aggregated over a batch of size b. The corresponding batch-wise update rules become

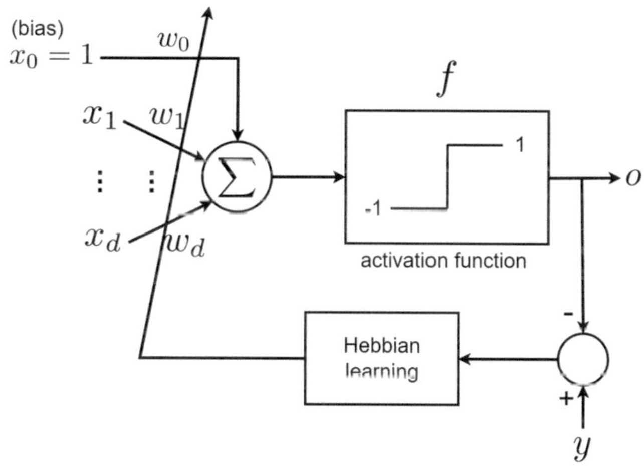

Fig. 4 The Perceptron model of a neuron with signum activation function

$$\boldsymbol{w} := \boldsymbol{w} + 2\eta \sum_{i=1}^{b} y_i\, \boldsymbol{x}_i\, \mathbb{I}(o_i \neq y_i), \tag{8}$$

and the bias update is

$$w_0 := w_0 + 2\eta \sum_{i=1}^{b} y_i\, \mathbb{I}(o_i \neq y_i). \tag{9}$$

We now provide two complementary justifications for the Perceptron weight update rule when using the signum activation function. The first is an algebraic argument based on label values, and the second is a geometric interpretation using gradient ascent/descent with respect to a linear decision boundary.

We begin with the first justification, which is algebraic. Recall Eq. (5):

$$\boldsymbol{w} := \boldsymbol{w} + \eta(y_i - o_i)\boldsymbol{x}_i, \quad \forall i \in \{1, \ldots, n\}.$$

When the output o_i and the target label y_i have levels ± 1:

$$\begin{aligned}
& y_i, o_i \in \{-1, 1\}, \\
& y_i = 1, o_i = 1 \implies \eta(y_i - o_i)\boldsymbol{x}_i = \boldsymbol{0}, \\
& y_i = 1, o_i = -1 \implies \eta(y_i - o_i)\boldsymbol{x}_i = \eta(1 - (-1))\boldsymbol{x}_i = 2\eta\boldsymbol{x}_i = 2\eta y_i \boldsymbol{x}_i, \\
& y_i = -1, o_i = 1 \implies \eta(y_i - o_i)\boldsymbol{x}_i = \eta(-1 - 1)\boldsymbol{x}_i = -2\eta\boldsymbol{x}_i = 2\eta y_i \boldsymbol{x}_i, \\
& y_i = -1, o_i = -1 \implies \eta(y_i - o_i)\boldsymbol{x}_i = \eta(-1 - (-1))\boldsymbol{x}_i = \boldsymbol{0}.
\end{aligned}$$

This provides Eqs. (8) and (9).

An alternative justification is based on gradient descent optimization. The goal is to identify a linear decision boundary that separates the two classes such that each class lies on a different side of the boundary. The equation defining such a linear decision boundary is given by

$$\boldsymbol{w}^\top \boldsymbol{x} + w_0 = 0, \tag{10}$$

where $\boldsymbol{x} \in \mathbb{R}^d$ is the data point, $\boldsymbol{w} \in \mathbb{R}^d$ is the normal vector of the linear line, and w_0 is the bias (intercept) of the line. Consider any two points $\boldsymbol{x}_1$ and $\boldsymbol{x}_2$ on the decision boundary, as shown in Fig. 5. As the line passes through each of them, they both satisfy Eq. (10):

$$\begin{aligned}
& \boldsymbol{w}^\top \boldsymbol{x}_1 + w_0 = 0, \quad \boldsymbol{w}^\top \boldsymbol{x}_2 + w_0 = 0 \implies \boldsymbol{w}^\top \boldsymbol{x}_1 + w_0 = \boldsymbol{w}^\top \boldsymbol{x}_2 + w_0, \\
& \implies \boldsymbol{w}^\top (\boldsymbol{x}_1 - \boldsymbol{x}_2) = 0 \implies \boldsymbol{w} \perp (\boldsymbol{x}_1 - \boldsymbol{x}_2),
\end{aligned}$$

which verifies that $\boldsymbol{w}$ is the normal vector of the decision boundary.

As depicted in Fig. 6, consider a point $\boldsymbol{x}_0$ on the decision boundary and a point $\boldsymbol{x}$ on one of the sides of the decision boundary. Therefore:

$$\boldsymbol{w}^\top \boldsymbol{x}_0 + w_0 = 0 \implies w_0 = -\boldsymbol{w}^\top \boldsymbol{x}_0. \tag{11}$$

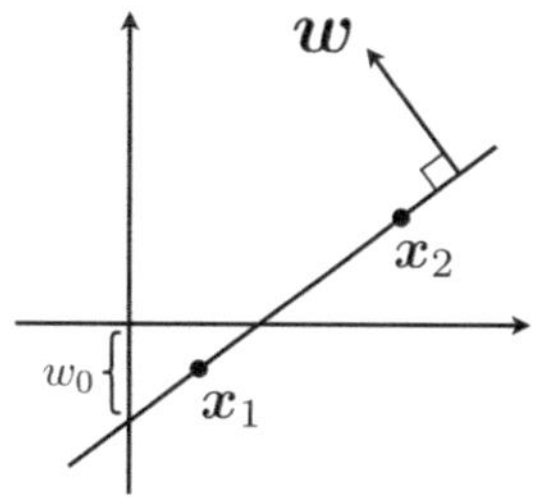

Fig. 5 Two points $\boldsymbol{x}_1$ and $\boldsymbol{x}_2$ on the linear decision boundary in $\mathbb{R}^2$

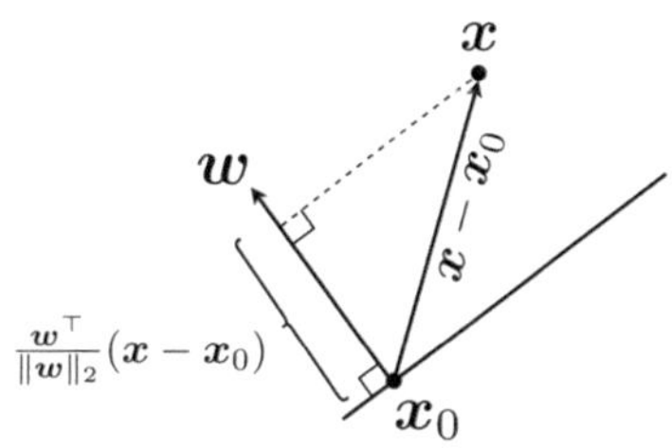

Fig. 6 A point $\boldsymbol{x}_0$ on the decision boundary and a point $\boldsymbol{x}$ on one side of the boundary

Assume the normal vector $\boldsymbol{w}$ is normalized, i.e., it has unit length. The distance of point $\boldsymbol{x}$ from the decision boundary is

$$d = \boldsymbol{w}^\top(\boldsymbol{x} - \boldsymbol{x}_0) = \boldsymbol{w}^\top\boldsymbol{x} - \boldsymbol{w}^\top\boldsymbol{x}_0 \overset{(11)}{=} \boldsymbol{w}^\top\boldsymbol{x} + w_0. \tag{12}$$

The distance should be non-negative so there is

$$d = |\boldsymbol{w}^\top\boldsymbol{x} + w_0|.$$

However, the absolute value function is non-smooth and non-differentiable. This non-smooth function can be bypassed by reparameterizing the distance in a way that is always non-negative:

$$y = +1 \implies \boldsymbol{w}^\top\boldsymbol{x} + w_0 > 0, \quad y = -1 \implies \boldsymbol{w}^\top\boldsymbol{x} + w_0 < 0, \tag{13}$$

$$\implies d := y(\boldsymbol{w}^\top\boldsymbol{x} + w_0). \tag{14}$$

A possible cost function is the number of misclassified data points to be minimized. But it is discrete and hard to optimize. Therefore, we can optimize the distance as the cost function. The objective is to maximize the distance of points from the decision boundary, so gradient ascent is used for maximizing the distances of points from the decision boundary[1]:

$$\text{distance} = \sum_{i=1}^{b} y_i(\boldsymbol{w}^\top\boldsymbol{x}_i + w_0),$$

$$\frac{\partial\text{distance}}{\partial\boldsymbol{w}} = \sum_{i=1}^{b} y_i\boldsymbol{x}_i, \quad \frac{\partial\text{distance}}{\partial w_0} = \sum_{i=1}^{b} y_i.$$

[1] Unlike support vector machines, which maximize the margin only for support vectors (the closest points to the decision boundary), this approach considers the distances of all data points from the decision boundary.

Gradient ascent should be used to maximize the distances:

$$\begin{aligned} \boldsymbol{w} &:= \boldsymbol{w} + \eta \sum_{i=1}^{b} y_i \boldsymbol{x}_i, \\ w_0 &= w_0 + \eta \sum_{i=1}^{b} y_i, \end{aligned} \tag{15}$$

where updates can be performed only for $o_i \neq y_i$ cases to have Eqs. (8) and (9).

In another approach, it is possible to take the distances of misclassified data points as the cost function. According to Eq. (13), the distance in Eq. (14) is for correctly classified data points. So, the distance should be multiplied by -1 to have the distances of misclassified points:

$$\text{error} = \sum_{i=1}^{b} -y_i(\boldsymbol{w}^\top \boldsymbol{x}_i + w_0) = -\sum_{i=1}^{b} y_i(\boldsymbol{w}^\top \boldsymbol{x}_i + w_0),$$

$$\frac{\partial \text{error}}{\partial \boldsymbol{w}} = -\sum_{i=1}^{b} y_i \boldsymbol{x}_i, \quad \frac{\partial \text{error}}{\partial w_0} = -\sum_{i=1}^{b} y_i.$$

Gradient descent should be used to minimize the error:

$$\begin{aligned} \boldsymbol{w} &:= \boldsymbol{w} + \eta \sum_{i=1}^{b} y_i \boldsymbol{x}_i, \\ w_0 &= w_0 + \eta \sum_{i=1}^{b} y_i, \end{aligned} \tag{16}$$

where updates can be performed only for $o_i \neq y_i$ cases to have Eqs. (8) and (9).

2.6 Limitations of the Perceptron Algorithm

While the Perceptron algorithm marked a significant milestone in the development of neural networks, it is inherently restricted to binary classification tasks involving linearly separable data. This limitation was rigorously analyzed by Minsky and Papert in their influential 1969 book *Perceptrons* [21], where they demonstrated that the Perceptron cannot solve problems requiring nonlinear decision boundaries, such as the XOR classification problem.

The recognition of this shortcoming led to a sharp decline in research interest in neural networks for many years. Although it was hypothesized that stacking multiple Perceptrons into a multi-layer architecture could overcome the nonlinearity barrier, the lack of efficient training algorithms at the time posed a significant challenge.

Even when classes are linearly separable, the Perceptron also suffers from limited generalization capacity. Depending on the initial random weights, it may converge to different decision boundaries. Some of these may lie dangerously close to one of the classes, as illustrated in Fig. 7. While such solutions may yield perfect training accuracy, they often generalize poorly to unseen data.

To address these limitations—especially poor generalization and the inability to handle nonlinearly separable data—the ADALINE (Adaptive Linear Neuron) algorithm was introduced by Widrow and

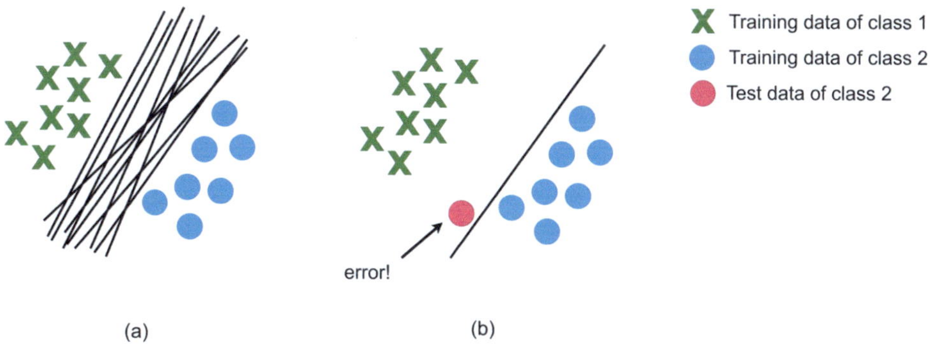

Fig. 7 **a** Various decision boundaries that the Perceptron algorithm can produce depending on weight initialization, and **b** an example of a poor decision boundary with low generalization

Hoff in 1960. Unlike the Perceptron, ADALINE minimizes a continuous loss function using the Least Mean Squares (LMS) algorithm, leading to more stable convergence and improved generalization. A detailed discussion of ADALINE can be found in Sect. 3.

Despite these advantages, ADALINE did not become the foundational unit in deep learning. Its linear output is incompatible with the hierarchical and nonlinear feature representations required by deep architectures. In contrast, the Perceptron's nonlinear activation makes it a more suitable primitive for composing multi-layer networks. As a result, the Perceptron remains the conceptual building block underlying modern neural networks trained via backpropagation.[2]

2.7 Multi-Layer Perceptron

While the Perceptron model was a groundbreaking development, one of its major limitations is its inability to model even simple functions such as XOR due to its linear decision boundary. This shortcoming was recognized early on, and researchers—including Rosenblatt himself—proposed that a hierarchy of Perceptrons arranged in multiple layers could mitigate this issue. However, in the early days of neural networks, the lack of an effective training algorithm for such multi-layer architectures posed a significant challenge and slowed progress.

Eventually, the backpropagation algorithm emerged as a practical solution for training multi-layer networks, revitalizing interest in neural networks and laying the foundation for deep learning. The core building block in these modern networks remains the Perceptron neuron—extended in depth and empowered by nonlinear activations.

2.7.1 The Structure of Neural Network

A feedforward neural network is a stack of layers of neural network where in every layer, all neurons of the previous layer are connected to all neurons of the next layer. A feedforward neural network is also referred to as a fully-connected neural network. Every layer of the feedforward neural network is called a fully connected layer or a dense layer. Each neuron in the fully connected neural network is a

[2] Several decades later, Support Vector Machines (SVMs) emerged as an elegant and theoretically grounded solution to the generalization problem and the inability to handle nonlinearly separable data discussed here. SVMs select the decision boundary that maximizes the margin between classes, thereby improving robustness to unseen data. While highly effective for classification tasks, SVMs are not based on neural network architectures and thus fall outside the scope of this book.

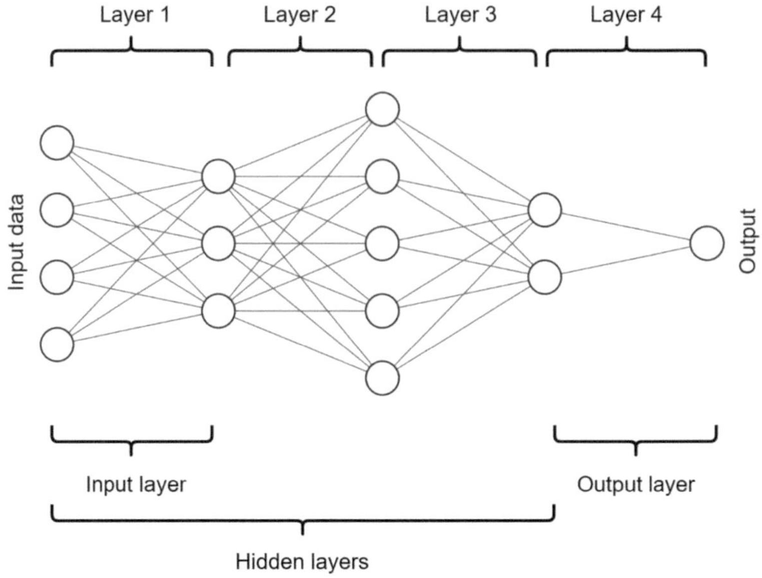

Fig. 8 A feedforward (fully connected or MLP) neural network with four layers

Perceptron neuron. That is why this network is also called the Multi-Layer Perceptron[3] (MLP). MLP was proposed by Rosenblatt in 1958, in the same paper as Perceptron [26]. In that paper, he proposed an MLP with three layers.

Figure 8 illustrates a feedforward network with four layers. The layer accepting the input data is called the input layer and the layer giving the output of network is named the output layer. All the layers except the output layers are called the hidden layers. Note that, the input layer is usually considered to be one of the hidden layers in the terminology of the literature. For a sample PyTorch code of feedforward neural network, refer to Chap. 22.

2.7.2 What Happens in a Neuron

Each neuron in the feedforward neural network is a Perceptron neuron (see Fig. 9). Therefore, it has summation of the outputs of previous layer multiplied by the weights of previous layer:

$$a_i = \sum_{\ell=1}^{m} w_{i\ell} z_\ell, \tag{17}$$

where $\{w_{i\ell}\}_{\ell=1}^{m}$ are the weights connecting to the ith neuron in the layer and m is the number of neurons in the previous layer.

The ith neuron has an activation function $\sigma_i(\cdot)$, which is usually a nonlinear function, where the output of Eq. (17) passes through it (see Fig. 9):

$$z_i := \sigma_i(a_i). \tag{18}$$

[3] The phrases feedforward neural network, fully connected neural network, and multi-layer Perceptron (MLP) are different names for the same type of neural network. These names are used interchangeably in the literature.

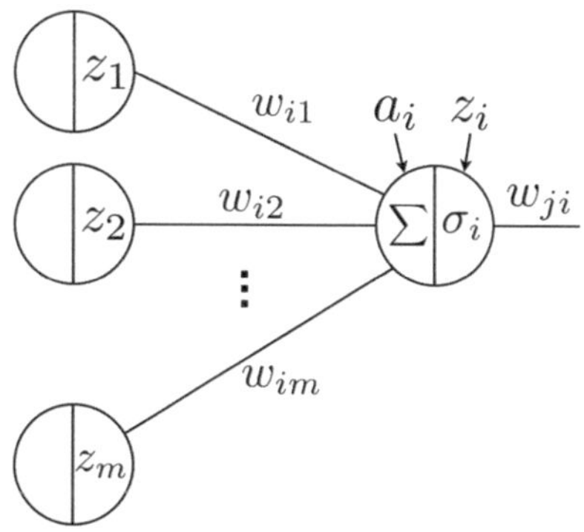

Fig. 9 A Perceptron neuron in the fully-connected neural network

2.7.3 The Necessity of Nonlinear Activation Functions

As Fig. 10 depicts, every layer in the feedforward network can be seen as a linear projection followed by an activation function:

$$\boldsymbol{y} = \sigma_3\Big(\boldsymbol{W}_3^\top \sigma_2\big(\boldsymbol{W}_2^\top \sigma_1(\boldsymbol{W}_1^\top \boldsymbol{x})\big)\Big). \tag{19}$$

The activation function is usually a nonlinear function because if all activation functions are linear in the network, the entire network is collapsed to be one linear projection. This behavior is explained in the following. It is known that $\boldsymbol{W}_i^\top \boldsymbol{x}$ is the linear projection of $\boldsymbol{x}$ onto the column space of $\boldsymbol{W}_i$. Assume that the activation functions are linear functions and let $\boldsymbol{V}_i := \sigma_i \boldsymbol{W}_i^\top$ be a new linear projection because both σ_i and $\boldsymbol{W}^\top$ are linear operators. Then, in this case, Eq. (19) simplifies to:

$$\boldsymbol{y} = \boldsymbol{V}_3^\top \boldsymbol{V}_2^\top \boldsymbol{V}_1^\top \boldsymbol{x} = \boldsymbol{V}^\top \boldsymbol{x}, \tag{20}$$

where

$$\boldsymbol{V} := \boldsymbol{V}_1 \boldsymbol{V}_2 \boldsymbol{V}_3, \tag{21}$$

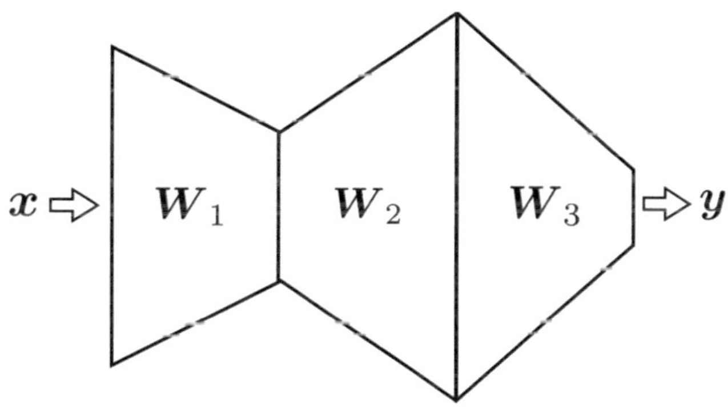

Fig. 10 Viewing the weights of layers as linear projection matrices between the activation functions

and Eq. (20) is a linear projection of the input $\boldsymbol{x}$ onto the column space of $\boldsymbol{V}$. In other words, the nonlinearity of neural networks is because of the nonlinear activation functions. Thus, for a network to be able to handle nonlinear patterns of data, it should have nonlinear activation functions.

2.8 Activation Functions

As discussed earlier, the expressive power of neural networks relies crucially on the nonlinearity introduced by activation functions. Without them, even deep networks would reduce to a single linear transformation. Activation functions introduce the nonlinearity necessary for modeling complex patterns in data, enabling networks to learn decision boundaries that are not limited to linear separability.

Over the years, a wide variety of activation functions have been developed, each with distinct characteristics that make them suitable for different purposes and architectures. Below, we review the most commonly used activation functions. For each, we provide the formula, its range, its derivative, and a plot in Fig. 11.

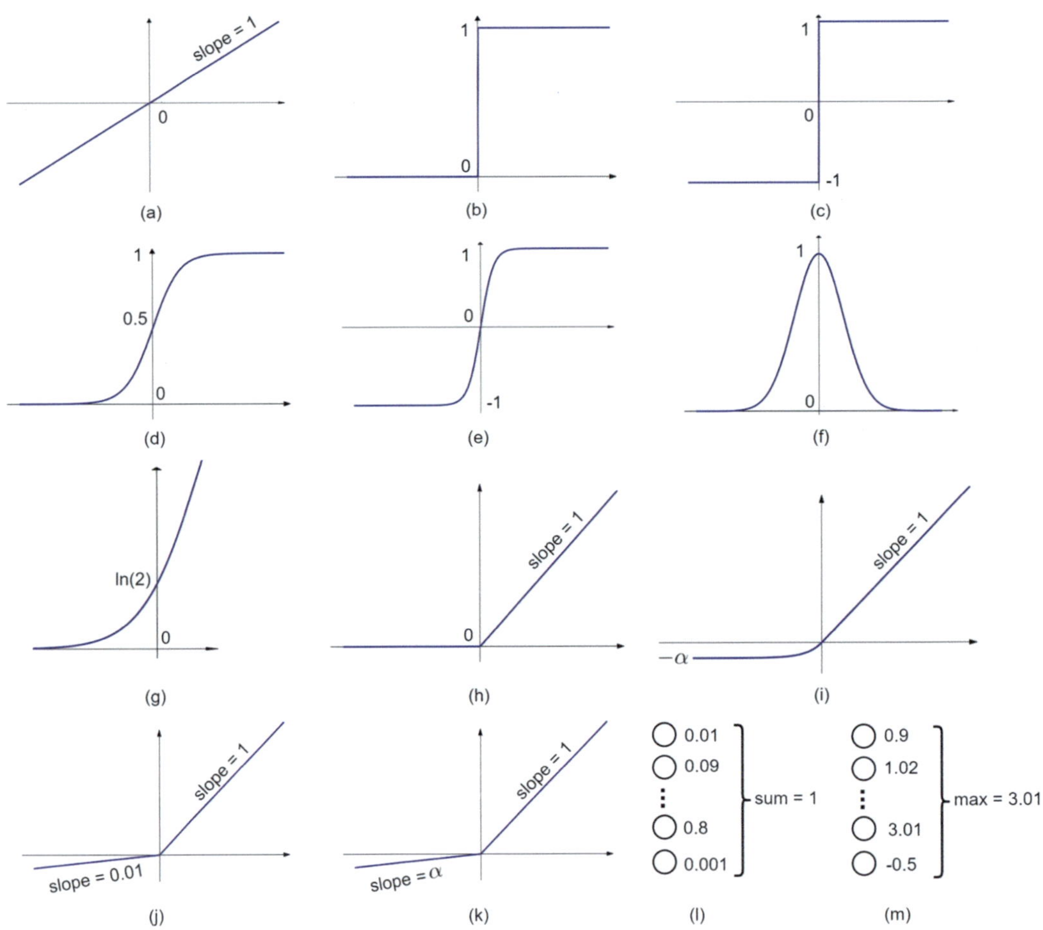

Fig. 11 Well-known activation functions: **a** linear, **b** binary, **c** sign (signum), **d** logistic (sigmoid), **e** hyperbolic tangent (tanh), **f** Gaussian (RBF), **g** softplus, **h** ReLU, **i** ELU, **j** Leaky ReLU, **k** PReLU, **l** softmax, and **m** maxout activation functions

- **Linear (identity) Function:** This function applies no transformation and is equivalent to omitting the activation. It is rarely used in hidden layers but may be applied in output layers for regression tasks.

$$\sigma(a) = a, \quad \sigma(a) \in (-\infty, \infty), \quad \sigma'(a) = 1. \tag{22}$$

- **Binary Step:** A threshold function with values 0 and 1. Its non-differentiability at $a = 0$ limits its use in modern deep learning frameworks.

$$\sigma(a) = \begin{cases} 0 \text{ if } a < 0 \\ 1 \text{ if } a \geq 0 \end{cases}, \quad \sigma(a) \in [0, 1], \quad \sigma'(a) = \begin{cases} 0 & \text{if } a \neq 0 \\ \text{undefined} & \text{if } a = 0. \end{cases} \tag{23}$$

- **Sign (signum) Function:** This function maps inputs to ± 1. Unlike binary step, it preserves the sign of inputs and is occasionally used for quantized networks.

$$\sigma(a) = \begin{cases} -1 & \text{if } a < 0 \\ 1 & \text{if } a \geq 0 \end{cases}, \quad \sigma(a) \in [-1, 1], \quad \sigma'(a) = \begin{cases} 0 & \text{if } a \neq 0 \\ \text{undefined} & \text{if } a = 0. \end{cases} \tag{24}$$

- **Logistic (sigmoid):** A widely used nonlinear activation, particularly in the output layer of binary classifiers using cross-entropy loss. It maps inputs to the range [0, 1] and can be interpreted as a probability.

$$\sigma(a) = \frac{1}{1 + e^{-a}}, \quad \sigma(a) \in [0, 1], \quad \sigma'(a) = \sigma(a)(1 - \sigma(a)). \tag{25}$$

- **Hyperbolic Tangent (tanh):** Similar to the sigmoid function, but its output range is $[-1, 1]$, which can lead to faster convergence in practice due to its zero-centered output.

$$\sigma(a) = \frac{e^{a} - e^{-a}}{e^{a} + e^{-a}}, \quad \sigma(a) \in [-1, 1], \quad \sigma'(a) = 1 - \sigma(a)^2. \tag{26}$$

- **Gaussian (RBF):** A bell-shaped function used in radial basis function networks. While uncommon in deep networks, it remains of historical and theoretical interest.

$$\sigma(a) = e^{-a^2}, \quad \sigma(a) \in (0, 1], \quad \sigma'(a) = -2ae^{-a^2}. \tag{27}$$

- **Softplus** [10]**:** A smooth approximation to ReLU with always-positive output, offering better theoretical properties but often slower convergence.

$$\sigma(a) = \ln(1 + e^{a}), \quad \sigma(a) \in [0, \infty), \quad \sigma'(a) = \frac{1}{1 + e^{-a}}. \tag{28}$$

- **ReLU (Rectified Linear Unit)** [22]**:** A milestone in deep learning. It enabled training deeper networks by avoiding vanishing gradients. ReLU enforces sparsity by zeroing out negative activations, which helps generalization.

$$\sigma(a) = \begin{cases} 0 \text{ if } a < 0 \\ a \text{ if } a \geq 0 \end{cases} = \max(0, a), \quad \sigma(a) \in [0, \infty), \quad \sigma'(a) = \begin{cases} 0 & \text{if } a < 0 \\ \text{undefined} & \text{if } a = 0 \\ 1 & \text{if } a > 0. \end{cases} \tag{29}$$

- **ELU (Exponential Linear Unit)** [5]**:** A smooth alternative to ReLU that allows for negative outputs, which can improve learning dynamics.

$$\sigma(a) = \begin{cases} \alpha(e^a - 1) & \text{if } a < 0 \\ a & \text{if } a \geq 0 \end{cases}, \quad \sigma(a) \in (-\alpha, \infty), \quad \sigma'(a) = \begin{cases} \alpha e^a & \text{if } a < 0 \\ 1 & \text{if } a = 0, \alpha = 1 \\ 1 & \text{if } a > 0. \end{cases} \tag{30}$$

- **Leaky ReLU** [19]**:** Addresses the "dying ReLU" problem by allowing a small, non-zero gradient for negative inputs.

$$\sigma(a) = \begin{cases} 0.01a & \text{if } a < 0 \\ a & \text{if } a \geq 0 \end{cases}, \quad \sigma(a) \in (-\infty, \infty), \quad \sigma'(a) = \begin{cases} 0.01 & \text{if } a < 0 \\ \text{undefined} & \text{if } a = 0 \\ 1 & \text{if } a > 0. \end{cases} \tag{31}$$

- **Parametric ReLU (PReLU)** [13]**:** A generalization of Leaky ReLU where the negative slope is a learnable parameter α.

$$\sigma(a) = \begin{cases} \alpha a & \text{if } a < 0 \\ a & \text{if } a \geq 0 \end{cases}, \quad \sigma(a) \in (-\infty, \infty), \quad \sigma'(a) = \begin{cases} \alpha & \text{if } a < 0 \\ \text{undefined} & \text{if } a = 0 \\ 1 & \text{if } a > 0. \end{cases} \tag{32}$$

- **Softmax:** A generalization of sigmoid for multi-class classification. It transforms a vector of scores into a probability distribution over classes. Unlike pointwise activations, it operates across an entire layer.

$$\sigma_i(\boldsymbol{a}) = \frac{e^{a_i}}{\sum_{i=1}^{m} e^{a_i}}, \ \forall i \in \{1, \ldots, m\}, \quad \sigma_i(\boldsymbol{a}) \in (0, 1), \quad \sigma'(a) = \sigma_i(\boldsymbol{a})\big(\delta_{ij} - \sigma_j(\boldsymbol{a})\big), \tag{33}$$

  where δ_{ij} is the Kronecker delta:

$$\delta_{ij} := \begin{cases} 1 & \text{if } i = j \\ 0 & \text{if } i \neq j. \end{cases} \tag{34}$$

- **Maxout** [11]**:** Operates across multiple inputs and returns the maximum, offering universal approximation properties and robustness to dropout regularization.

$$\sigma_i(\boldsymbol{a}) = \max_j(\sigma_j), \quad \sigma_i(\boldsymbol{a}) \in (-\infty, \infty), \quad \sigma'(a) = \begin{cases} 1 & \text{if } i = \arg\max_j(\sigma_j) \\ 0 & \text{if } i \neq \arg\max_j(\sigma_j). \end{cases} \tag{35}$$

2.9 Common Loss Functions

Loss functions, also known as cost functions, play a central role in training neural networks by guiding the optimization of the model parameters. A variety of loss functions have been developed to suit different tasks, such as regression, classification, and deep metric learning. In this section, we introduce several commonly used loss functions in supervised learning.

Let θ denote the set of network weights, $\boldsymbol{x}_i$ the input data, $\mathbf{f}_o(\boldsymbol{x}_i)$ the network output for input $\boldsymbol{x}_i$, and y_i the corresponding target label. Loss functions specifically designed for deep metric learning are discussed separately in Chap. 16.

2.9.1 Mean Squared Error and Mean Absolute Error

The Mean Squared Error (MSE) is a widely used loss function for regression. It penalizes the squared difference between predicted and true values:

$$\underset{\theta}{\text{minimize}} \quad \sum_{i=1}^{b} (\mathbf{f}_o(\boldsymbol{x}_i) - y_i)^2. \tag{36}$$

While MSE is differentiable and thus suitable for gradient-based optimization, it can be sensitive to outliers due to the squaring operation.

The Mean Absolute Error (MAE), in contrast, uses the absolute difference between predictions and targets:

$$\underset{\theta}{\text{minimize}} \quad \sum_{i=1}^{b} |\mathbf{f}_o(\boldsymbol{x}_i) - y_i|. \tag{37}$$

Also known as the Manhattan distance, MAE is more robust to outliers and can encourage sparsity in learned representations. Although it is not differentiable at zero, modern optimization frameworks handle this through subgradients or numerical differentiation.

2.9.2 Huber Loss

The Huber loss offers a compromise between MSE and MAE, combining their respective advantages:

$$\underset{\theta}{\text{minimize}} \quad \sum_{i=1}^{b} \begin{cases} 0.5(\mathbf{f}_o(\boldsymbol{x}_i) - y_i)^2 & \text{if } |\mathbf{f}_o(\boldsymbol{x}_i) - y_i| \leq \delta \\ \delta(|\mathbf{f}_o(\boldsymbol{x}_i) - y_i| - 0.5\delta) & \text{otherwise,} \end{cases} \tag{38}$$

where $\delta > 0$ is a tunable hyperparameter that determines the threshold between the quadratic and linear regimes.

2.9.3 KL-Divergence Loss

The Kullback–Leibler (KL) divergence [18] measures the dissimilarity between two probability distributions. In supervised learning, it is used to align the predicted distribution $\mathbb{P}(\mathbf{f}(\boldsymbol{x}))$ with the target distribution $\mathbb{P}(y)$. When both the predicted outputs $\mathbf{f}(\boldsymbol{x}_i)$ and the target labels y_i represent probability distributions (e.g., soft labels or one-hot vectors), the KL-divergence loss for a batch of size b is defined as:

$$\underset{\theta}{\text{minimize}} \quad \sum_{i=1}^{b} \sum_{l=1}^{c} (y_i)_l \log\left(\frac{(y_i)_l}{\mathbf{f}(\boldsymbol{x}_i)_l}\right), \tag{39}$$

where $(y_i)_l$ and $\mathbf{f}(\boldsymbol{x}_i)_l$ denote the lth component of the target and predicted distributions, respectively.

In practice, when y_i is one hot encoded, this loss simplifies to the cross entropy loss. KL divergence becomes especially useful when y_i represents a soft target distribution, such as in knowledge distillation.

2.9.4 Hinge Loss

In binary classification settings where labels are encoded as $y_i \in \{-1, 1\}$, the hinge loss is often used:

$$\underset{\theta}{\text{minimize}} \sum_{i=1}^{b} \left[m - y_i \, \mathbf{f}_o(\boldsymbol{x}_i)\right]_+, \tag{40}$$

where $[\cdot]_+ := \max(\cdot, 0)$ and $m > 0$ is a margin hyperparameter.

This loss penalizes incorrect predictions and also correct predictions that fall within the margin. When y_i and $\mathbf{f}_o(\boldsymbol{x}_i)$ share the same sign and the prediction exceeds the margin ($|\mathbf{f}_o(\boldsymbol{x}_i)| \geq m$), the loss is zero. Otherwise, the model is penalized proportionally to how far it is from satisfying the margin constraint.

Note that $[\,\cdot\,]_+$ is equivalent to the ReLU function introduced earlier in Eq. (29). Hinge loss also underlies more complex loss formulations such as contrastive loss [12] and triplet loss [27], which are discussed in Sect. 4.4 of Chap. 16.

2.9.5 Cross-Entropy Loss

Cross-entropy loss is the most commonly used loss function for classification tasks. Unlike previous losses that use linear output activations, cross-entropy requires a softmax activation (for $c > 2$ classes) or a sigmoid activation (for binary classification) in the output layer to ensure that predictions lie in $[0, 1]$ and sum to one.

Given c classes and one-hot encoded target vectors $\boldsymbol{y}_i \in \{0, 1\}^c$, and predictions $\mathbf{f}_o(\boldsymbol{x}_i) \in \mathbb{R}^c$, the cross-entropy loss is defined as:

$$\underset{\theta}{\text{minimize}} \quad -\sum_{i=1}^{b} \sum_{l=1}^{c} (\boldsymbol{y}_i)_l \log\left(\mathbf{f}_o(\boldsymbol{x}_i)_l\right), \tag{41}$$

where $(\boldsymbol{y}_i)_l$ and $\mathbf{f}_o(\boldsymbol{x}_i)_l$ denote the lth components of the target and prediction vectors, respectively.

Minimizing the cross-entropy encourages the network to assign high probabilities to the correct classes. Moreover, it leads to discriminative features in the penultimate layer [2, 28], enhancing the embedding's utility for downstream tasks such as retrieval or clustering.

3 Complementary Models and Historical Perspectives

While the core ideas of this chapter have focused on the Perceptron and its foundational role in modern neural networks, several other models emerged historically or conceptually that offer valuable complementary insights. Although not essential for understanding the main learning framework developed so far, these models help situate neural networks in a broader context of early machine learning methods.

This section presents several such models, including ADALINE, logistic regression, radial basis function (RBF) networks, and self-organizing maps. Some of these, like ADALINE, emerged from early neural network research. Others, such as logistic regression, were not originally part of the neural network lineage but can be interpreted as special cases of simple neural networks. For example, logistic regression can be viewed as a single-neuron network with a sigmoid activation function, providing useful insight into how probabilistic outputs and gradient-based training relate to neural computation.

Readers interested in alternative training mechanisms, probabilistic interpretations, or historical neural architectures may find this section enriching.

3.1 ADALINE

The limitations of the Perceptron, particularly its reliance on a binary activation function and its poor generalization, prompted researchers to seek more stable and analytically tractable learning methods. In 1960, Bernard Widrow and Ted Hoff introduced one such alternative: the **ADALINE** (Adaptive Linear Neuron) [32, 34]. Their approach offered a new perspective on training neural units using continuous error signals, and it attracted considerable public and academic attention—some media even went so far as to proclaim that "robots can think."

ADALINE differs fundamentally from the Perceptron in how it computes and uses the prediction error. Rather than applying a hard threshold and then comparing the binary output with the target, ADALINE compares the ground-truth label directly with the weighted sum of inputs—i.e., the neuron's output before any activation is applied. This continuous-valued output during training enables smoother gradient-based updates, often resulting in better generalization.

As illustrated in Fig. 12, the ADALINE neuron computes the output as a weighted sum of the input vector $\boldsymbol{x}_i = [x_{i1}, \ldots, x_{id}]^\top \in \mathbb{R}^d$, and uses this linear combination to define the prediction error. The training objective is to minimize the *least mean squares* (LMS) error, a quadratic loss that penalizes the squared difference between the predicted and true labels:

$$e = \frac{1}{2}\sum_{i=1}^{b}\left(y_i - \left(\sum_{j=1}^{d} w_j x_{ij} + w_0\right)\right)^2, \tag{42}$$

where b is the batch size, y_i is the ground-truth label, and w_0 is the bias term. This learning procedure, known as the *LMS algorithm* or the *Widrow-Hoff rule*, is an early form of gradient descent applied to a linear model.

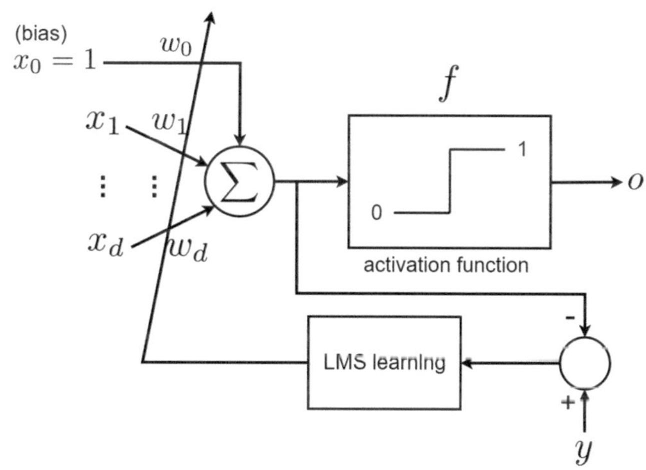

Fig. 12 The ADALINE model of neuron

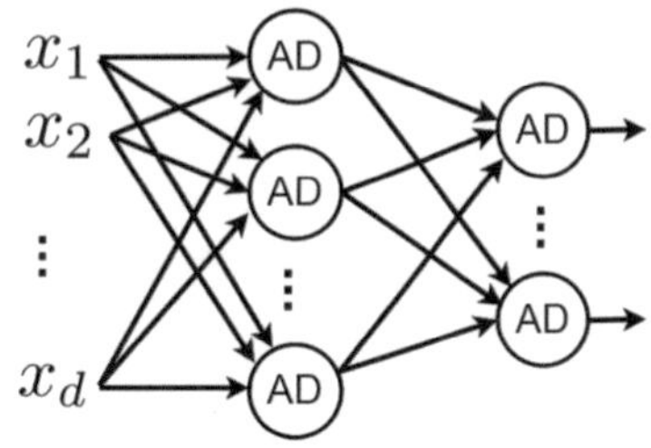

Fig. 13 The MADALINE neural network. Every neuron denoted by AD is an ADALINE neuron

The gradients of the error with respect to the weights and bias are

$$\begin{aligned}\frac{\partial e}{\partial w_j} &= -\sum_{i=1}^{b}\left(y_i - \left(\sum_{j=1}^{d} w_j x_{ij} + w_0\right)\right) x_{ij}, \quad \forall j \in \{1, \dots, d\}, \\ \frac{\partial e}{\partial w_0} &= -\sum_{i=1}^{b}\left(y_i - \left(\sum_{j=1}^{d} w_j x_{ij} + w_0\right)\right).\end{aligned} \tag{43}$$

Accordingly, the weights and bias are updated using gradient descent:

$$\begin{aligned} w_j &:= w_j - \eta \frac{\partial e}{\partial w_j}, \quad \forall j \in \{1, \dots, d\}, \\ w_0 &:= w_0 - \eta \frac{\partial e}{\partial w_0}, \end{aligned} \tag{44}$$

where $\eta > 0$ is the learning rate.

To extend ADALINE to more complex tasks, Widrow and collaborators proposed stacking multiple ADALINE units to form a layered architecture called **MADALINE** (Many ADALINEs), as shown in Fig. 13. In 1962, they introduced *MADALINE Rule I (MRI)* for training this network [33], although this early rule could only update the weights between input and hidden layers.

Later improvements addressed this limitation. In 1988, *MADALINE Rule II (MRII)* was proposed [35], enabling updates to both input-to-hidden and hidden-to-output connections. This was followed by *MADALINE Rule III (MRIII)* in 1990, which offered further refinements for multi-layer training. These efforts represented some of the earliest attempts to build and train deep neural architectures, even before the widespread adoption of backpropagation.

3.2 Logistic Regression

Although not part of the historical development of neural networks, **logistic regression** remains a foundational model in statistics and machine learning—especially in fields like bio-statistics. It predates the modern neural network framework, yet closely resembles a one-neuron network with a sigmoid (logistic) activation function.

We include logistic regression in this chapter not for its historical relevance, but because it provides useful mathematical insight into the behavior of simple neural models. In particular, it helps illuminate how a single neuron with a nonlinear activation can be used for binary classification in a probabilistic framework. This connection can help the reader better understand the foundations of neural network training, especially when it comes to interpreting outputs probabilistically and using loss functions derived from likelihood principles.

Let $\boldsymbol{x} \in \mathbb{R}^d$ be input data and $y \in \mathbb{R}$ be the binary class label. According to Bayes' rule, we have

$$\mathbb{P}(y|\boldsymbol{x}) = \frac{\mathbb{P}(\boldsymbol{x}|y)\mathbb{P}(y)}{\mathbb{P}(\boldsymbol{x})}, \tag{45}$$

where $\mathbb{P}(y|\boldsymbol{x})$ and $\mathbb{P}(\boldsymbol{x}|y)$ are the posterior and likelihood, respectively, and $\mathbb{P}(\boldsymbol{x})$ and $\mathbb{P}(y)$ are the priors. In contrast to Linear Discriminant Analysis (LDA), logistic regression works on the posterior $\mathbb{P}(y|\boldsymbol{x})$ directly rather than working on likelihood $\mathbb{P}(\boldsymbol{x}|y)$ and prior $\mathbb{P}(y)$.

Logistic regression is a binary classifier where it assigns probability between zero and one for belonging to one of the classes. The logistic function, used in logistic regression, was initially proposed in 1845 for modeling the population growth [30]. It was further improved in the twentieth century [31]. See [6] for the history of logistic regression.

Logistic regression considers the classification problem as a regression problem where it regresses, or predicts, the probability of belonging to a class. It first considers a linear regression $\boldsymbol{\beta}^\top \boldsymbol{x} + \beta_0$. However, in order to not have the bias, it assumes that $\boldsymbol{x}$ is $(d+1)$ dimensional with an additional element of 1 for bias, i.e., $\boldsymbol{x} = [1, x_1, \ldots, x_d]^\top$. The $\boldsymbol{\beta} \in \mathbb{R}^{d+1}$ is the learnable parameter of the logistic regression model. As a result, the linear regression becomes $\boldsymbol{\beta}^\top \boldsymbol{x}$. However, there is no bound on this regression while logistic regression desires the output to be in the range [0, 1] to behave like a probability. Therefore, logistic regression models the posterior using a logistic function, also called the sigmoid function, to make this regression between zero and one.

Consider a binary classification problem where there are two classes, $y \in \{0, 1\}$. Logistic regression models the posterior using a logistic function, also called the sigmoid function:

$$\mathbb{P}(y = 1|X = x) = \frac{e^{\boldsymbol{\beta}^\top \boldsymbol{x}}}{1 + e^{\boldsymbol{\beta}^\top \boldsymbol{x}}}, \tag{46}$$

$$\mathbb{P}(y = 0|X = x) = 1 - \mathbb{P}(y = 1|X = x) = \frac{1}{1 + e^{\boldsymbol{\beta}^\top \boldsymbol{x}}}, \tag{47}$$

where $\boldsymbol{\beta} \in \mathbb{R}^d$ is the learnable parameter of the logistic regression model (see Fig. 14).

As Fig. 15 depicts, logistic regression can be seen as a neural network with one neuron where the activation function is the nonlinear sigmoid (logistic) function.

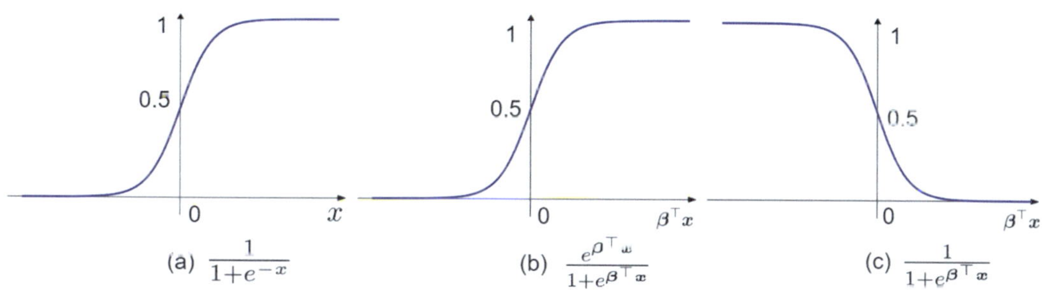

Fig. 14 **a** The logistic (sigmoid) function, **b** the logistic function with parameter $\boldsymbol{\beta}^\top \boldsymbol{x}$, and **c** one minus the logistic function with parameter $\boldsymbol{\beta}^\top \boldsymbol{x}$

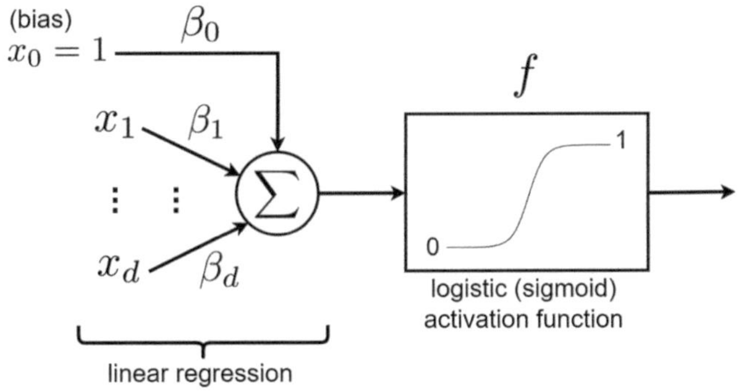

Fig. 15 Logistic regression interpreted as a neural network with one neuron

Consider n data points $\{(\boldsymbol{x}_i, y_i)\}_{i=1}^n$ in the dataset. Assuming that they are independent and identically distributed (i.i.d), the posterior over all data points is

$$\mathbb{P}(y|X) = \prod_{i=1}^{n} \Big(\mathbb{P}(y_i = 1|X = x_i)\mathbb{I}(y_i = 1) + \mathbb{P}(y_i = 0|X = x_i)\mathbb{I}(y_i = 0)\Big), \tag{48}$$

where $\mathbb{I}(.)$ is the indicator function which is one if its condition is satisfied and is zero otherwise. As the labels are either zero or one, i.e., $y_i \in \{0, 1\}$, this equation can be restated as:

$$\mathbb{P}(y|X) = \prod_{i=1}^{n} \big(\mathbb{P}(y_i = 1|X = x_i)\big)^{y_i} \big(\mathbb{P}(y_i = 0|X = x_i)\big)^{1-y_i}. \tag{49}$$

Substituting Eqs. (46) and (47) in this equation gives

$$\mathbb{P}(y|X) = \prod_{i=1}^{n} \big(\frac{e^{\boldsymbol{\beta}^\top \boldsymbol{x}_i}}{1 + e^{\boldsymbol{\beta}^\top \boldsymbol{x}_i}}\big)^{y_i} \big(\frac{1}{1 + e^{\boldsymbol{\beta}^\top \boldsymbol{x}_i}}\big)^{1-y_i}. \tag{50}$$

The log posterior is

$$\begin{aligned}
\ell(\boldsymbol{\beta}) := \mathbb{P}(y|X = x) &= \log \prod_{i=1}^{n} \big(\frac{e^{\boldsymbol{\beta}^\top \boldsymbol{x}_i}}{1 + e^{\boldsymbol{\beta}^\top \boldsymbol{x}_i}}\big)^{y_i} \big(\frac{1}{1 + e^{\boldsymbol{\beta}^\top \boldsymbol{x}_i}}\big)^{1-y_i} \\
&= \sum_{i=1}^{n} \Big(\log \big(\frac{e^{\boldsymbol{\beta}^\top \boldsymbol{x}_i}}{1 + e^{\boldsymbol{\beta}^\top \boldsymbol{x}_i}}\big)^{y_i} + \log \big(\frac{1}{1 + e^{\boldsymbol{\beta}^\top \boldsymbol{x}_i}}\big)^{1-y_i}\Big) \\
&= \sum_{i=1}^{n} \Big(y_i \log(e^{\boldsymbol{\beta}^\top \boldsymbol{x}_i}) - y_i \log(1 + e^{\boldsymbol{\beta}^\top \boldsymbol{x}_i}) - (1 - y_i) \log(1 + e^{\boldsymbol{\beta}^\top \boldsymbol{x}_i})\Big) \\
&= \sum_{i=1}^{n} \Big(y_i \boldsymbol{\beta}^\top \boldsymbol{x}_i - y_i \log(1 + e^{\boldsymbol{\beta}^\top \boldsymbol{x}_i}) - \log(1 + e^{\boldsymbol{\beta}^\top \boldsymbol{x}_i}) + y_i \log(1 + e^{\boldsymbol{\beta}^\top \boldsymbol{x}_i})\Big) \\
&= \sum_{i=1}^{n} \Big(y_i \boldsymbol{\beta}^\top \boldsymbol{x}_i - \log(1 + e^{\boldsymbol{\beta}^\top \boldsymbol{x}_i})\Big).
\end{aligned}$$

Newton's method can be used to find the optimum $\boldsymbol{\beta}$. The first derivative, or the gradient, it:

$$\frac{\partial \ell(\boldsymbol{\beta})}{\partial \boldsymbol{\beta}} = \sum_{i=1}^{n} \Big(y_i \boldsymbol{x}_i - \frac{1}{1 + e^{\boldsymbol{\beta}^\top \boldsymbol{x}_i}} e^{\boldsymbol{\beta}^\top \boldsymbol{x}_i} \boldsymbol{x}_i \Big) = \sum_{i=1}^{n} (y_i - \frac{e^{\boldsymbol{\beta}^\top \boldsymbol{x}_i}}{1 + e^{\boldsymbol{\beta}^\top \boldsymbol{x}_i}}) \boldsymbol{x}_i. \tag{51}$$

Its transpose is

$$\frac{\partial \ell(\boldsymbol{\beta})}{\partial \boldsymbol{\beta}^\top} = \sum_{i=1}^{n} (y_i - \frac{e^{\boldsymbol{\beta}^\top \boldsymbol{x}_i}}{1 + e^{\boldsymbol{\beta}^\top \boldsymbol{x}_i}}) \boldsymbol{x}_i^\top.$$

The second derivative is

$$\begin{aligned}
\frac{\partial^2 \ell(\boldsymbol{\beta})}{\partial \boldsymbol{\beta} \partial \boldsymbol{\beta}^\top} &= \frac{\partial}{\partial \boldsymbol{\beta}} (\frac{\partial \ell(\boldsymbol{\beta})}{\partial \boldsymbol{\beta}^\top}) = \frac{\partial}{\partial \boldsymbol{\beta}} \Big(\sum_{i=1}^{n} (y_i - \frac{e^{\boldsymbol{\beta}^\top \boldsymbol{x}_i}}{1 + e^{\boldsymbol{\beta}^\top \boldsymbol{x}_i}}) \boldsymbol{x}_i^\top \Big) \\
&= \sum_{i=1}^{n} \Big(- \frac{\partial}{\partial \boldsymbol{\beta}} (\frac{e^{\boldsymbol{\beta}^\top \boldsymbol{x}_i}}{1 + e^{\boldsymbol{\beta}^\top \boldsymbol{x}_i}}) \Big) \boldsymbol{x}_i^\top.
\end{aligned}$$

We define

$$\mathbb{P}(\boldsymbol{x}_i | \boldsymbol{\beta}) := \frac{e^{\boldsymbol{\beta}^\top \boldsymbol{x}_i}}{1 + e^{\boldsymbol{\beta}^\top \boldsymbol{x}_i}}. \tag{52}$$

Therefore:

$$\frac{\partial^2 \ell(\boldsymbol{\beta})}{\partial \boldsymbol{\beta} \partial \boldsymbol{\beta}^\top} = - \sum_{i=1}^{n} \Big(\frac{\partial}{\partial \boldsymbol{\beta}} (\mathbb{P}(\boldsymbol{x}_i | \boldsymbol{\beta})) \Big) \boldsymbol{x}_i^\top. \tag{53}$$

We have

$$\begin{aligned}
\frac{\partial}{\partial \boldsymbol{\beta}} (\mathbb{P}(\boldsymbol{x}_i | \boldsymbol{\beta})) &= \frac{\partial}{\partial \boldsymbol{\beta}} (\frac{e^{\boldsymbol{\beta}^\top \boldsymbol{x}_i}}{1 + e^{\boldsymbol{\beta}^\top \boldsymbol{x}_i}}) \\
&= \frac{1}{(1 + e^{\boldsymbol{\beta}^\top \boldsymbol{x}_i})^2} \Big(e^{\boldsymbol{\beta}^\top \boldsymbol{x}_i} \boldsymbol{x}_i (1 + e^{\boldsymbol{\beta}^\top \boldsymbol{x}_i}) - e^{\boldsymbol{\beta}^\top \boldsymbol{x}_i} (e^{\boldsymbol{\beta}^\top \boldsymbol{x}_i} \boldsymbol{x}_i) \Big) \\
&= \frac{e^{\boldsymbol{\beta}^\top \boldsymbol{x}_i}}{(1 + e^{\boldsymbol{\beta}^\top \boldsymbol{x}_i})^2} \Big(1 + e^{\boldsymbol{\beta}^\top \boldsymbol{x}_i} - e^{\boldsymbol{\beta}^\top \boldsymbol{x}_i} \Big) \boldsymbol{x}_i = \frac{e^{\boldsymbol{\beta}^\top \boldsymbol{x}_i}}{(1 + e^{\boldsymbol{\beta}^\top \boldsymbol{x}_i})^2} \boldsymbol{x}_i \\
&= \frac{e^{\boldsymbol{\beta}^\top \boldsymbol{x}_i}}{(1 + e^{\boldsymbol{\beta}^\top \boldsymbol{x}_i})} \frac{1}{(1 + e^{\boldsymbol{\beta}^\top \boldsymbol{x}_i})} \boldsymbol{x}_i \overset{(52)}{=} \mathbb{P}(\boldsymbol{x}_i | \boldsymbol{\beta}) \big(1 - \mathbb{P}(\boldsymbol{x}_i | \boldsymbol{\beta}) \big) \boldsymbol{x}_i
\end{aligned}$$

Substituting it in Eq. (53) gives the second derivative, i.e., the Hessian matrix:

$$\frac{\partial^2 \ell(\boldsymbol{\beta})}{\partial \boldsymbol{\beta} \partial \boldsymbol{\beta}^\top} = - \sum_{i=1}^{n} \Big(\mathbb{P}(\boldsymbol{x}_i | \boldsymbol{\beta}) \big(1 - \mathbb{P}(\boldsymbol{x}_i | \boldsymbol{\beta}) \big) \boldsymbol{x}_i \Big) \boldsymbol{x}_i^\top. \tag{54}$$

It is possible to write the Newton's method in matrix form. We define

$$\mathbb{R}^{(d+1)\times n} \ni \boldsymbol{X} := \begin{bmatrix} 1 & 1 & \dots & 1 \\ \boldsymbol{x}_1 & \boldsymbol{x}_2 & \dots & \boldsymbol{x}_n \end{bmatrix},$$

$$\mathbb{R}^{n\times n} \ni \boldsymbol{W} := \textbf{diag}\Big(\mathbb{P}(\boldsymbol{x}_i|\boldsymbol{\beta})\big(1 - \mathbb{P}(\boldsymbol{x}_i|\boldsymbol{\beta})\big)\Big),$$

$$\mathbb{R}^{n} \ni \boldsymbol{y} := [y_1, \dots, y_n]^\top,$$

$$\mathbb{R}^{n} \ni \boldsymbol{p} := [\frac{e^{\boldsymbol{\beta}^\top \boldsymbol{x}_1}}{1 + e^{\boldsymbol{\beta}^\top \boldsymbol{x}_1}}, \dots, \frac{e^{\boldsymbol{\beta}^\top \boldsymbol{x}_n}}{1 + e^{\boldsymbol{\beta}^\top \boldsymbol{x}_n}}]^\top.$$

Equations (51) and (54) can be restated as

$$\mathbb{R}^{(d+1)} \ni \frac{\partial \ell(\boldsymbol{\beta})}{\partial \boldsymbol{\beta}} = \boldsymbol{X}(\boldsymbol{y} - \boldsymbol{p}), \tag{55}$$

$$\mathbb{R}^{(d+1)\times(d+1)} \ni \frac{\partial^2 \ell(\boldsymbol{\beta})}{\partial \boldsymbol{\beta} \partial \boldsymbol{\beta}^\top} = -\boldsymbol{X}\boldsymbol{W}\boldsymbol{X}^\top. \tag{56}$$

Using Newton's method for maximization of the log posterior is

$$\begin{aligned} \boldsymbol{\beta}^{(\tau+1)} &:= \boldsymbol{\beta}^{(\tau)} + (\frac{\partial^2 \ell(\boldsymbol{\beta})}{\partial \boldsymbol{\beta} \partial \boldsymbol{\beta}^\top})^{-1} \frac{\partial \ell(\boldsymbol{\beta})}{\partial \boldsymbol{\beta}} \implies \\ \boldsymbol{\beta}^{(\tau+1)} &:= \boldsymbol{\beta}^{(\tau)} - (\boldsymbol{X}\boldsymbol{W}\boldsymbol{X}^\top)^{-1}\boldsymbol{X}(\boldsymbol{y} - \boldsymbol{p}), \end{aligned} \tag{57}$$

where τ is the iteration index. It is repeated until convergence of $\boldsymbol{\beta}$.

In the test phase of logistic regression, the class of a point $\boldsymbol{x}$ is determined as

$$y = \begin{cases} 1 \text{ if } \frac{e^{\boldsymbol{\beta}^\top \boldsymbol{x}}}{1+e^{\boldsymbol{\beta}^\top \boldsymbol{x}}} \geq 0.5, \\ 0 \text{ Otherwise.} \end{cases} \tag{58}$$

3.3 Radial Basis Function Network

While Radial Basis Function (RBF) networks are not commonly used in modern deep learning practice, they represent an important conceptual and historical development in the study of neural architectures. Introduced in 1988 [3, 4], RBF networks offer an alternative to the multi-layer Perceptron by using localized, kernel-like activation functions. Their structure also provides a different geometric intuition for how data can be transformed and separated in high-dimensional spaces.

We include RBF networks in this chapter to provide a broader perspective on early attempts to address nonlinear decision boundaries using neural architectures. Understanding RBF networks can deepen one's insight into the variety of strategies explored before the dominance of deep multi-layer networks trained via backpropagation.

As shown in Fig. 16, the RBF network consists of two layers. The first layer connects the input data $\boldsymbol{x} \in \mathbb{R}^d$ to a set of m fixed basis (kernel) functions $\phi_i(\boldsymbol{x})i = 1^m$. The second layer consists of trainable weights that map these basis function outputs to the final predictions $\widehat{y}jj = 1^k$. Each weight wij connects the ith basis function to the jth output neuron.

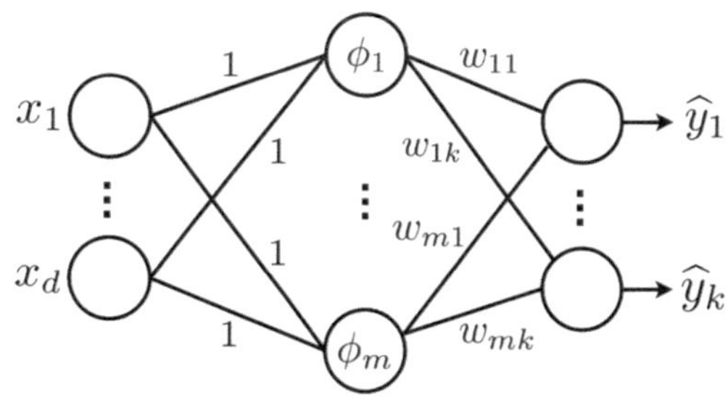

Fig. 16 The RBF neural network

The basis functions can be various kernel functions such as:

$$\text{Gaussian distribution:} \quad \phi_i(\boldsymbol{x}) = e^{-\frac{\|\boldsymbol{x}_i - \boldsymbol{\mu}_i\|_2^2}{2\sigma_i^2}}, \tag{59}$$

$$\text{Logistic (sigmoid) function:} \quad \phi_i(\boldsymbol{x}) = \frac{1}{1 + e^{-\frac{\|\boldsymbol{x}_i - \boldsymbol{\mu}_i\|_2^2}{2\sigma_i^2}}}, \tag{60}$$

where $\boldsymbol{\mu}_i \in \mathbb{R}^d$ and $\sigma_i^2 \in \mathbb{R}$ are the mean and variance for $\phi_i(\boldsymbol{x})$. At the first step, the means $\{\boldsymbol{\mu}_i\}_{i=1}^m$ are found by applying a clustering method, such as K-means, on the training data with m clusters. The variances of clusters determine the variances $\{\sigma_i\}_{i=1}^m$.

An RBF network can be considered as an additive model. An additive model, first proposed in [9], maps data to a space with several basis functions and then tries to learn a weighted average of those bases. In the RBF network, the output is obtained as

$$\widehat{y}_j = w_{1j}\,\phi_1(\boldsymbol{x}) + \cdots + w_{mj}\,\phi_m(\boldsymbol{x}) = \sum_{i=1}^{m} w_{ij}\,\phi_i(\boldsymbol{x}). \tag{61}$$

Consider n data points together in a matrix $\boldsymbol{X} \in \mathbb{R}^{k \times n}$. Equation (61) can be stated in a matrix form:

$$\widehat{\boldsymbol{Y}} = \boldsymbol{W}^\top \boldsymbol{\Phi}, \tag{62}$$

where

$$\mathbb{R}^{k\times n} \ni \widehat{\boldsymbol{Y}} = \begin{bmatrix} \widehat{y}_{11} & \dots & \widehat{y}_{1n} \\ \vdots & \ddots & \vdots \\ \widehat{y}_{k1} & \dots & \widehat{y}_{kn} \end{bmatrix}, \quad \mathbb{R}^{m\times k} \ni \boldsymbol{W} = \begin{bmatrix} w_{11} & \dots & w_{1k} \\ \vdots & \ddots & \vdots \\ w_{m1} & \dots & w_{mk} \end{bmatrix},$$

$$\mathbb{R}^{m\times n} \ni \boldsymbol{\Phi} = \begin{bmatrix} \phi_1(x_1) & \dots & \phi_1(x_n) \\ \vdots & \ddots & \vdots \\ \phi_m(x_1) & \dots & \phi_m(x_n) \end{bmatrix}.$$

Least squares error minimization between the label of data $\boldsymbol{Y} \in \mathbb{R}^{k\times n}$ and the output of network $\widehat{\boldsymbol{Y}} \subset \mathbb{R}^{k\times n}$ is

$$\underset{\boldsymbol{W}}{\text{minimize}} \quad \|\boldsymbol{Y} - \widehat{\boldsymbol{Y}}\|_F^2 = \|\boldsymbol{Y} - \boldsymbol{W}^\top \boldsymbol{\Phi}\|_F^2. \tag{63}$$

The cost function can be simplified:

$$\begin{aligned}\|\boldsymbol{Y} - \boldsymbol{W}^\top \boldsymbol{\Phi}\|_F^2 &= \textbf{tr}\big((\boldsymbol{Y} - \boldsymbol{W}^\top \boldsymbol{\Phi})^\top (\boldsymbol{Y} - \boldsymbol{W}^\top \boldsymbol{\Phi})\big) = \textbf{tr}\big((\boldsymbol{Y}^\top - \boldsymbol{\Phi}^\top \boldsymbol{W})(\boldsymbol{Y} - \boldsymbol{W}^\top \boldsymbol{\Phi})\big) \\ &= \textbf{tr}(\boldsymbol{Y}^\top \boldsymbol{Y} - \boldsymbol{Y}^\top \boldsymbol{W}^\top \boldsymbol{\Phi} - \boldsymbol{\Phi}^\top \boldsymbol{W} \boldsymbol{Y} + \boldsymbol{\Phi}^\top \boldsymbol{W} \boldsymbol{W}^\top \boldsymbol{\Phi}) \\ &\overset{(a)}{=} \textbf{tr}(\boldsymbol{Y}^\top \boldsymbol{Y}) - \textbf{tr}(\boldsymbol{W}^\top \boldsymbol{\Phi} \boldsymbol{Y}^\top) - \textbf{tr}(\boldsymbol{W} \boldsymbol{Y} \boldsymbol{\Phi}^\top) + \textbf{tr}(\boldsymbol{W}^\top \boldsymbol{\Phi} \boldsymbol{\Phi}^\top \boldsymbol{W}),\end{aligned}$$

where (a) is because of the linearity and cyclic property of the trace operator. Solving this optimization problem gives

$$\begin{aligned}&\frac{\partial}{\partial \boldsymbol{W}} \|\boldsymbol{Y} - \boldsymbol{W}^\top \boldsymbol{\Phi}\|_F^2 = -\boldsymbol{\Phi} \boldsymbol{Y}^\top - \boldsymbol{\Phi} \boldsymbol{Y}^\top + 2\boldsymbol{\Phi}\boldsymbol{\Phi}^\top \boldsymbol{W} \overset{\text{set}}{=} \boldsymbol{0} \implies \boldsymbol{\Phi}\boldsymbol{\Phi}^\top \boldsymbol{W} = \boldsymbol{\Phi} \boldsymbol{Y}^\top \\ &\implies \boldsymbol{W} = (\boldsymbol{\Phi}\boldsymbol{\Phi}^\top)^{-1} \boldsymbol{\Phi} \boldsymbol{Y}^\top. \end{aligned} \tag{64}$$

If $\boldsymbol{\Phi}\boldsymbol{\Phi}^\top$ is a singular matrix, the pseudo-inverse can be used:

$$\boldsymbol{W} = (\boldsymbol{\Phi}\boldsymbol{\Phi}^\top)^\dagger \boldsymbol{\Phi} \boldsymbol{Y}^\top, \tag{65}$$

where † denotes the pseudo-inverse of matrix. The output of the RBF network, for either the training or test data, is

$$\widehat{\boldsymbol{Y}} = \boldsymbol{W}^\top \boldsymbol{\Phi} = \big((\boldsymbol{\Phi}\boldsymbol{\Phi}^\top)^{-1} \boldsymbol{\Phi} \boldsymbol{Y}^\top\big)^\top \boldsymbol{\Phi} = \boldsymbol{Y} \boldsymbol{\Phi}^\top (\boldsymbol{\Phi}\boldsymbol{\Phi}^\top)^{-\top} \boldsymbol{\Phi} = \boldsymbol{Y} \boldsymbol{\Phi}^\top (\boldsymbol{\Phi}\boldsymbol{\Phi}^\top)^{-1} \boldsymbol{\Phi}. \tag{66}$$

3.4 Self-Organizing Map

The Self-Organizing Map (SOM), also known as the Kohonen network [16, 17], represents another historically significant yet conceptually distinct neural model. Unlike most networks discussed in this chapter, SOMs are designed for unsupervised learning and clustering, not supervised prediction. Their unique training method—driven by self-organization principles—offered early insight into how neural networks might be used to reveal structure in high-dimensional data without labeled examples.

Although SOMs are not part of the standard toolkit in modern deep learning, they remain influential in data visualization, topology preservation, and prototype-based clustering. We include SOMs here to give readers a broader appreciation of the diversity of early neural architectures and how some of their ideas echo in more recent models such as topological autoencoders.

SOM is a single-layer neural network that connects input vectors $\boldsymbol{x} = [x_1, \ldots, x_d]^\top \in \mathbb{R}^d$ to k neurons, each of which corresponds to a cluster prototype. Let w_{ij} denote the weight connecting input component x_i to the jth neuron. During training, the network adapts these weights so that each neuron becomes representative of a region in the input space. The neurons themselves are arranged in a predefined topological layout—typically 1D or 2D—so that similar input vectors activate neighboring neurons. This layout is illustrated in Fig. 17.

Every neuron has a neighborhood around it in the 1D or 2D structure of neurons. This neighborhood is decreased gradually during the training phase. Let $\mathcal{N}_j^{(\tau)}$ denote the neighborhood of the jth neuron at iteration τ. The training algorithm of SOM is as follows:

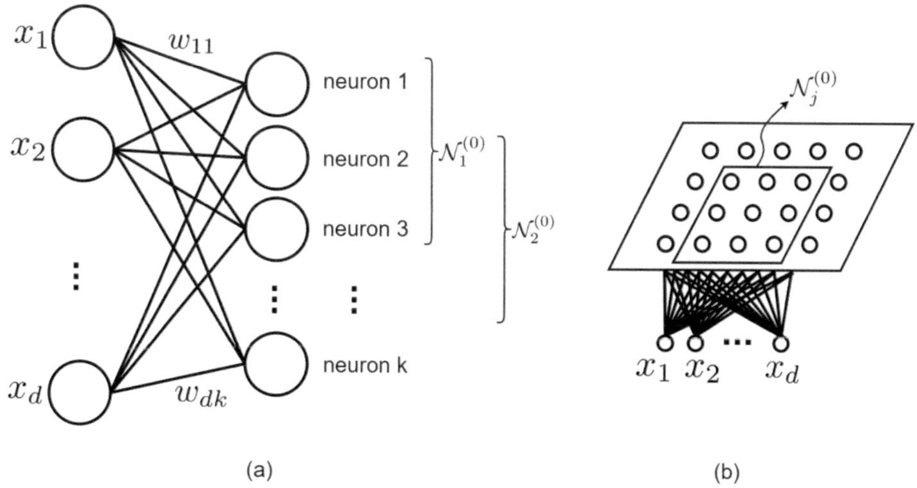

Fig. 17 The SOM with **a** 1D and **b** 2D structure of neurons

- Step 1: Initialize all weights to small random values. Set all neighborhoods $\{\mathcal{N}_j^{(0)}\}_{j=1}^k$ to half of the neuron structure grid. Set the initial learning rate $\eta^{(0)}$ to a number in range (0, 1].
- Step 2: Select one of the input data points, $\boldsymbol{x} = [x_1, \ldots, x_d]^\top$, and feed it to the network. Select the winning neuron z by:

$$z := \arg\min_j \sum_{i=1}^d \|x_i - w_{ij}\|_2. \tag{67}$$

- Step 3: Update the weights of the neurons in the neighborhood of the winning neuron z:

$$w_{ij}^{(\tau+1)} := \begin{cases} w_{ij}^{(\tau)} + \eta^{(\tau)}(x_i - w_{ij}^{(\tau)}) & \text{if } j \in \mathcal{N}_z^{(\tau)} \\ w_{ij}^{(\tau)} & \text{Otherwise,} \end{cases} \tag{68}$$

  for all $i \in \{1, \ldots, d\}$.
- Step 4: Decrease the learning rate and the neighborhoods:

$$\eta^{(\tau+1)} := \eta^{(0)}(1 - \frac{\tau}{t}), \tag{69}$$

$$\mathcal{N}_j^{(\tau+1)} := \mathcal{N}_j^{(\tau+1)}/2, \quad \forall j \in \{1, \ldots, k\}, \tag{70}$$

  where t denotes the total number of training iterations.
- Step 5: Increase τ and go to step 2.

This training algorithm clusters the data points into k clusters. In the test phase, when a data instance x_i is fed to the network, its winning neuron z, calculated by Eq. (67), determines the cluster it belongs to.

4 Universal Approximation Theorem

The Universal Approximation Theorem (UAT) is one of the foundational results in the theory of neural networks and sits squarely in the category of representation theorems [8, 29]. Just like Fourier or wavelet representations approximate arbitrary functions via a particular basis, UAT says that neural networks can approximate a wide class of functions to arbitrary precision under certain conditions.

Theorem 1 (Universal Approximation Theorem [7]) *Let $\sigma : \mathbb{R} \to \mathbb{R}$ be any continuous, bounded, and non-constant activation function. Then, for any continuous function $f : [0, 1]^n \to \mathbb{R}$ and any $\varepsilon > 0$, there exists a function:*

$$F(x) = \sum_{i=1}^{N} \alpha_i \, \sigma(w_i^T x + \theta_i), \tag{71}$$

such that:

$$\sup_{x \in [0,1]^n} |f(x) - F(x)| < \varepsilon. \tag{72}$$

Proof (sketch) The sketch of proof is provided below in four steps.

Step 1: Define the Hypothesis Space
Let $C([0, 1]^n)$ denote the Banach space of continuous functions on the unit cube with the sup norm. Define the hypothesis space:

$$\mathcal{H}_\sigma := \operatorname{span} \left\{ x \mapsto \sigma(w^T x + \theta) \mid w \in \mathbb{R}^n, \theta \in \mathbb{R} \right\}.$$

We aim to show that $\mathcal{H}_\sigma$ is dense in $C([0, 1]^n)$.

Step 2: Suppose the Contrary
Assume that $\mathcal{H}_\sigma$ is not dense. Then, by the Hahn-Banach theorem, there exists a nonzero continuous linear functional $L \in C([0, 1]^n)^*$ that annihilates $\mathcal{H}_\sigma$, i.e.,

$$L(h) = 0, \quad \forall h \in \mathcal{H}_\sigma.$$

Step 3: Apply the Riesz Representation Theorem
By the Riesz Representation Theorem, there exists a signed Radon measure μ such that for all $f \in C([0, 1]^n)$:

$$L(f) = \int_{[0,1]^n} f(x) \, d\mu(x).$$

Hence:

$$\int_{[0,1]^n} \sigma(w^T x + \theta) \, d\mu(x) = 0, \quad \forall w, \theta.$$

Step 4: Use Separation Property
Let $g_{w,\theta}(x) = \sigma(w^T x + \theta)$. Then, the assumption gives:

$$\int \sigma(w^T x + \theta)\, d\mu(x) = 0, \quad \forall w, \theta.$$

If σ is continuous, bounded, and non-constant, the set of all such $g_{w,\theta}$ is rich enough to separate measures. Therefore, this implies $\mu \equiv 0$, contradicting $L \neq 0$. Hence, $\mathcal{H}_\sigma$ must be dense in $C([0, 1]^n)$. □

This theorem shows that functions of the form:

$$x \mapsto \sum_{i=1}^{N} \alpha_i \, \sigma(w_i^T x + \theta_i), \tag{73}$$

can approximate any continuous function on a compact domain arbitrarily well. Neural networks are in the form of this function. Therefore, neural networks form a *basis* for approximating functions, analogous to trigonometric functions in Fourier series. The UAT guarantees the expressive power of neural networks but does not provide a constructive method for finding the parameters w_i, θ_i, α_i. Refer to [15, 24] for further reading about UAT.

5 Conclusion

The early development of neural networks included the McCulloch-Pitts model, the Perceptron, ADALINE, the multi-layer Perceptron, the radial basis function (RBF) network, and the self-organizing map. Activation functions and loss functions are also fundamental building blocks of deep learning. Logistic regression itself can be viewed as a single neuron in a neural network. Moreover, the universal approximation theorem establishes that neural networks are capable of representing any continuous function. Together, these concepts form the foundation for understanding modern neural networks.

6 Exercises

1. (Level: Moderate) We have four data points in the dataset:

$$(\boldsymbol{x}_1 = [1, 2, 3]^\top, y_1 = 1), \quad (\boldsymbol{x}_2 = [1, 1, 1]^\top, y_2 = 1),$$
$$(\boldsymbol{x}_3 = [-1, -1, -1]^\top, y_3 = -1), \quad (\boldsymbol{x}_4 = [-2, -2, -2]^\top, y_4 = -1).$$

Consider a Perceptron having a bias and with a signum function as its activation function. Do two iterations of the mini-batch training of Perceptron with batches:

- mini-batch for iteration 1: data points 1 and 4
- mini-batch for iteration 2: data points 2 and 3

Use the learning rate $\eta = 0.01$. Let the initial weights be $w^{(0)} = [0, 0.1, 0.2, -0.3]^\top$.

2. (Level: Moderate) Consider a toy classification dataset $D = \{(x_i, y_i)\}_{i=1}^{8}$, where $x_i \in \mathbb{R}^3$ represents the input features and $y_i \in \{-1, +1\}$ represents the corresponding class label. The dataset is given as follows:

x_i	y_i
$(0, 0, 0)$	$+1$
$(1, 0, 0)$	-1
$(0, 1, 0)$	-1
$(0, 0, 1)$	-1
$(0, 1, 1)$	$+1$
$(1, 0, 0)$	-1
$(1, 1, 0)$	-1
$(1, 1, 1)$	$+1$

   Determine whether a single Perceptron can learn this mapping. If such a Perceptron exists, find the weights $w \in \mathbb{R}^3$ and bias $b \in \mathbb{R}$ of the separating hyperplane. Otherwise, explain why it is not possible for a single Perceptron to learn this mapping.
3. (Level: Difficult) Prove that the Perceptron learning algorithm converges in a finite number of steps if the dataset is linearly separable.
   Hint: Assume that the dataset $\{(\boldsymbol{x}_i, y_i)\}_{i=1}^{n}$ is linearly separable, where $\boldsymbol{x}_i \in \mathbb{R}^d$ are the input vectors, and $y_i \in \{-1, 1\}$ are their corresponding labels. Show that there exists a weight vector $\boldsymbol{w}^*$ and a bias b^* such that $y_i(\boldsymbol{w}^* \cdot \boldsymbol{x}_i + b^*) > 0$ for all i, and use this assumption to bound the number of updates made by the algorithm.
4. (Level: Moderate) Prove that a Perceptron cannot converge for the XOR problem.
5. (Level: Moderate) This problem is about using Perceptrons to implement logic functions. Assume a dataset of the form $x_1, x_2 \in \{0, 1\}$, and a Perceptron defined as: $y = H(\beta_0 + \beta_1 x_1 + \beta_2 x_2)$, where H is the Heaviside step function, defined as:

$$H(z) = \begin{cases} 1, & \text{if } z \geq 0, \\ 0, & \text{if } z < 0. \end{cases}$$

   (a) Find weights β_1, β_2 and bias β_0 for a single Perceptron that implements the AND function.
   (b) Find the weights β_1, β_2 and bias β_0 for a single Perceptron that implements the OR function.
   (c) Given the truth table for the XOR function:

x_1	x_2	$x_1 \oplus x_2$
0	0	0
0	1	1
1	0	1
1	1	0

   show that it cannot be learned by a single Perceptron. Find a small neural network of multiple Perceptrons that can implement the XOR function. (Hint: a hidden layer with 2 Perceptrons).

6. (Level: Easy) Answer the following questions:

   (a) Consider a deep neural network with a single input, a single output, and k hidden layers, each containing d hidden units. How many parameters does this network have in total?
   (b) Now, generalize the problem: if the number of inputs is n and the number of outputs is m, how many parameters does this network have in total?

7. (Level: Easy) Thinking about feedforward model with sigmoid activation, compute the output of a single-layer neural network with 3 inputs and 1 output. Assume the input vector is $x = (0.1, 0.4, 0.6)$, the weights are $w = (0.2, 0.3, 0.5)$, the bias is $b = 0.1$, and the activation function is

   (a) Sigmoid activation function: $f(z) = \frac{1}{1+e^{-z}}$
   (b) ReLU activation function: $f(z) = \max(0, z)$
   (c) Tanh activation function: $f(z) = \tanh(z) = \frac{e^{z}-e^{-z}}{e^{z}+e^{-z}}$

8. (Level: Moderate) We have two data points in the first batch of dataset:

$$(\boldsymbol{x}_1 = [2, 1, 3, 2]^\top, y_1 = -1), \quad (\boldsymbol{x}_2 = [-2, -1, -1, -3]^\top, y_2 = +1).$$

   Consider an ADALINE having a bias and with a binary function as its activation function. Let the initial weights be $\boldsymbol{w}^{(0)} = [1, 0.5, -1, 1, -0.5]^\top$ whose first element corresponds to the bias. Do one iteration of the mini-batch training of ADALINE with the given first batch. Use the learning rate $\eta = 0.01$.
   Hint: it is easier if you calculate $\left(\sum_{j=1}^{d} w_j x_{ij} + w_0\right)$ once for every $i \in \{1, 2\}$ and then use it multiple times in the formulas.

References

1. James A Anderson and Edward Rosenfeld. *Talking nets: An oral history of neural networks*. MIT Press, 2000.
2. Malik Boudiaf et al. "A unifying mutual information view of metric learning: cross-entropy vs. pairwise losses". In: *European Conference on Computer Vision*. Springer. 2020, pp. 548–564.
3. David Broomhead and David Lowe. "Multivariable functional interpolation and adaptive networks, complex systems, vol. 2". In: *Complex Systems* 2 (1988), pp. 321–355.
4. David S Broomhead and David Lowe. *Radial basis functions, multi-variable functional interpolation and adaptive networks*. Tech. rep. Royal Signals and Radar Establishment Malvern (United Kingdom), 1988.
5. Djork-Arné Clevert, Thomas Unterthiner, and Sepp Hochreiter. "Fast and accurate deep network learning by exponential linear units (elus)". In: *arXiv preprint* arXiv:1511.07289 (2015).
6. Jan Salomon Cramer. "The origins of logistic regression". In: (2002).
7. George Cybenko. "Approximation by superpositions of a sigmoidal function". In: *Mathematics of control, signals and systems* 2.4 (1989), pp. 303–314.
8. Ronald A DeVore and George G Lorentz. *Constructive Approximation*. Grundlehren der mathematischen Wissenschaften. Springer, 1993.
9. Jerome H Friedman and Werner Stuetzle. "Projection pursuit regression". In: *Journal of the American statistical Association* 76.376 (1981), pp. 817–823.
10. Xavier Glorot, Antoine Bordes, and Yoshua Bengio. "Deep sparse rectifier neural networks". In: *Proceedings of the fourteenth international conference on artificial intelligence and statistics*. JMLR Workshop and Conference Proceedings. 2011, pp. 315–323.
11. Ian Goodfellow, David Warde-Farley, Mehdi Mirza, Aaron Courville, and Yoshua Bengio. "Maxout networks". In: *International conference on machine learning*. PMLR. 2013, pp. 1319–1327.
12. Raia Hadsell, Sumit Chopra, and Yann LeCun. "Dimensionality reduction by learning an invariant mapping". In: *2006 IEEE Computer Society Conference on Computer Vision and Pattern Recognition (CVPR'06)*. Vol. 2. IEEE. 2006, pp. 1735–1742.

13. Kaiming He, Xiangyu Zhang, Shaoqing Ren, and Jian Sun. "Delving deep into rectifiers: Surpassing human-level performance on imagenet classification". In: *Proceedings of the IEEE international conference on computer vision*. 2015, pp. 1026–1034.
14. Donald Olding Hebb. *The Organization of Behavior*. New York: Wiley & Sons, 1949.
15. Kurt Hornik, Maxwell Stinchcombe, and Halbert White. "Multilayer feedforward networks are universal approximators". In: *Neural networks* 2.5 (1989), pp. 359–366.
16. Teuvo Kohonen. "Self-organized formation of topologically correct feature maps". In: *Biological cybernetics* 43.1 (1982), pp. 59–69.
17. Teuvo Kohonen and Timo Honkela. "Kohonen network". In: *Scholarpedia* 2.1 (2007), p. 1568.
18. Solomon Kullback and Richard A Leibler. "On information and sufficiency". In: *The annals of mathematical statistics* 22.1 (1951), pp. 79–86.
19. Andrew L Maas, Awni Y Hannun, Andrew Y Ng, et al. "Rectifier nonlinearities improve neural network acoustic models". In: *Proc. icml*. Vol. 30. 1. Atlanta, Georgia, USA. 2013, p. 3.
20. Warren S McCulloch and Walter Pitts. "A logical calculus of the ideas immanent in nervous activity". In: *The bulletin of mathematical biophysics* 5 (1943), pp. 115–133.
21. Marvin Minsky and Seymour A Papert. *Perceptrons*. MIT press, 1969.
22. Vinod Nair and Geoffrey E Hinton. "Rectified linear units improve restricted boltzmann machines". In: *Proceedings of the 27th international conference on machine learning (ICML-10)*. 2010, pp. 807–814.
23. Mikel Olazaran. "A sociological study of the official history of the perceptrons controversy". In: *Social Studies of Science* 26.3 (1996), pp. 611–659.
24. Allan Pinkus. "Approximation theory of the MLP model in neural networks". In: *Acta numerica* 8 (1999), pp. 143–195.
25. Frank Rosenblatt. *The Perceptron – a perceiving and recognizing automaton Project Para*. Tech. rep. Report 85-460-1, Cornell Aeronautical Laboratory., 1957.
26. Frank Rosenblatt. "The perceptron: a probabilistic model for information storage and organization in the brain." In: *Psychological review* 65.6 (1958), p. 386.
27. Florian Schroff, Dmitry Kalenichenko, and James Philbin. "Facenet: A unified embedding for face recognition and clustering". In: *Proceedings of the IEEE conference on computer vision and pattern recognition*. 2015, pp. 815–823.
28. Milad Sikaroudi et al. "Supervision and source domain impact on representation learning: A histopathology case study". In: *2020 42nd Annual International Conference of the IEEE Engineering in Medicine & Biology Society (EMBC)*. IEEE. 2020, pp. 1400–1403.
29. Marshall H Stone. "The generalized Weierstrass approximation theorem". In: *Mathematics Magazine* 21.5 (1948), pp. 237–254.
30. Pierre François Verhulst. "Resherches mathematiques sur la loi d'accroissement de la population". In: *Nouveaux memoires de l'academie royale des sciences* 18 (1845), pp. 1–41.
31. Strother H Walker and David B Duncan. "Estimation of the probability of an event as a function of several independent variables". In: *Biometrika* 54.1-2 (1967), pp. 167–179.
32. Bernard Widrow. *An adaptive "ADALINE" neuron using chemical "memistors"*. Tech. rep. No. 1553-2, Stanford Electronics Laboratories, 1960.
33. Bernard Widrow. "Generalization and information storage in networks of adaline neurons". In: *Self-organizing systems* (1962), pp. 435–461.
34. Bernard Widrow and Marcian E Hoff. *Adaptive switching circuits*. Tech. rep. Stanford University, California, Stanford Electronics Labs, 1960.
35. Capt Rodney Winter and B Widrow. "MADALINE RULE II: A training algorithm for neural networks". In: *Secondf Annual International Conference on Neural Networks*. 1988, pp. 1–401.

Backpropagation and Optimization in Deep Neural Networks

3

All life can be seen as optimization,
With ups and downs that its cost brings.
Everyone tries to minimize the cost in a descent direction,
Although variables of life are dependent like chain rings.

Abstract

This chapter focuses on optimization as the core mechanism behind deep learning. At its heart, training a neural network involves minimizing a loss function with respect to the model's parameters. The essential material presented in this chapter includes the foundational concept of gradient descent and its application via the backpropagation algorithm. These form the backbone of neural network training and are indispensable for understanding deep learning. Beyond this core material, the chapter offers several complementary sections intended to deepen understanding or provide historical and practical context. These include classical optimization techniques such as line search, momentum, and steepest descent, as well as advanced methods like stochastic and mini-batch gradient descent, adaptive learning rate algorithms (AdaGrad, RMSProp, and Adam), and Sharpness-Aware Minimization (SAM). While these sections are not essential for grasping the fundamentals of neural network training, they enrich the reader's perspective on optimization strategies used in practice. The chapter concludes with theoretical insights into the convergence behavior of optimization algorithms, especially in the over-parameterized regime typical of deep learning. These discussions provide further context and are particularly relevant for readers interested in the theory behind the practice.

1 Introduction

As discussed in Chap. 1, any machine learning algorithm can be understood through three fundamental steps: (1) formulating a *conjecture* or hypothesis class, (2) defining a *ranking mechanism* via a loss function, and (3) *searching* for the best hypothesis that minimizes this loss. This chapter focuses on the third step—the optimization or search component.

B. Ghojogh and A. Ghodsi, *Elements of Deep Learning*,
https://doi.org/10.1007/978-3-032-10738-1_3

In deep learning, this search process involves navigating a high-dimensional, often non-convex loss landscape induced by multi-layer neural networks. While the optimization goal is conceptually straightforward—minimize a loss function—the practical challenges are significant, especially when models involve many layers and millions of parameters.

The breakthrough came with the backpropagation algorithm [37], which efficiently computes the gradients needed to update parameters in deep networks. Although gradient descent as a concept was well understood, it was not immediately obvious how to apply it to multi-layer architectures. The challenge lay in computing how a parameter in an intermediate layer influences the final loss. Backpropagation addressed this difficulty by systematically applying the chain rule in a recursive manner, enabling efficient gradient computation throughout the network.

While second-order methods like Newton's method [22] offer appealing theoretical properties such as faster local convergence, they are computationally prohibitive in deep learning due to the need to calculate and invert the Hessian matrix. In addition to this computational burden, empirical evidence suggests that first-order methods—especially stochastic ones—tend to find flatter local minima that generalize better to unseen data. In contrast, second-order methods may converge to sharper minima, which are often associated with poorer generalization. The reasons behind this behavior are still not fully understood and remain an active area of research. Nevertheless, first-order methods such as gradient descent have proven remarkably effective for training large-scale neural networks in practice.

This chapter begins with the foundational optimization techniques needed to understand backpropagation. Section 2 first introduces gradient descent and then Sect. 3 presents the backpropagation algorithm in detail. These constitute the essential core of the chapter. Following this, we explore additional techniques and refinements that, while not strictly necessary for implementing backpropagation, provide useful insight and practical enhancements. These include line-search methods, momentum, and steepest descent (Sect. 2); stochastic and mini-batch gradient descent and their convergence properties (Sects. 4 and 4.1); adaptive learning rate methods such as AdaGrad, RMSProp, and Adam (Sect. 5); and sharpness-aware minimization (Sect. 6). Finally, Sect. 7 concludes the chapter with theoretical perspectives on the convergence of optimization algorithms in over-parameterized networks.

2 Gradient Descent

Backpropagation and many optimization algorithms in neural networks rely on first-order optimization methods. Among these, the most well-known and widely used is **gradient descent**. Since backpropagation combines gradient descent with the chain rule in derivatives, we begin this chapter by exploring gradient descent as a foundational tool.

Gradient descent is one of the fundamental methods in numerical optimization. Historically, it was first suggested by Cauchy in 1874 [28] and later refined by Hadamard in 1908 [24]. Its convergence properties were further analyzed by Curry in 1944 [12].

Consider the problem of minimizing a differentiable function $f(\cdot)$ with respect to the variable $\boldsymbol{w}$:

$$\underset{\boldsymbol{w}}{\text{minimize}} \quad f(\boldsymbol{w}). \tag{1}$$

Numerical optimization begins with a randomly chosen initial point and iteratively updates it using a step $\Delta\boldsymbol{w}$:

$$\boldsymbol{w}^{(k+1)} := \boldsymbol{w}^{(k)} + \Delta\boldsymbol{w}. \tag{2}$$

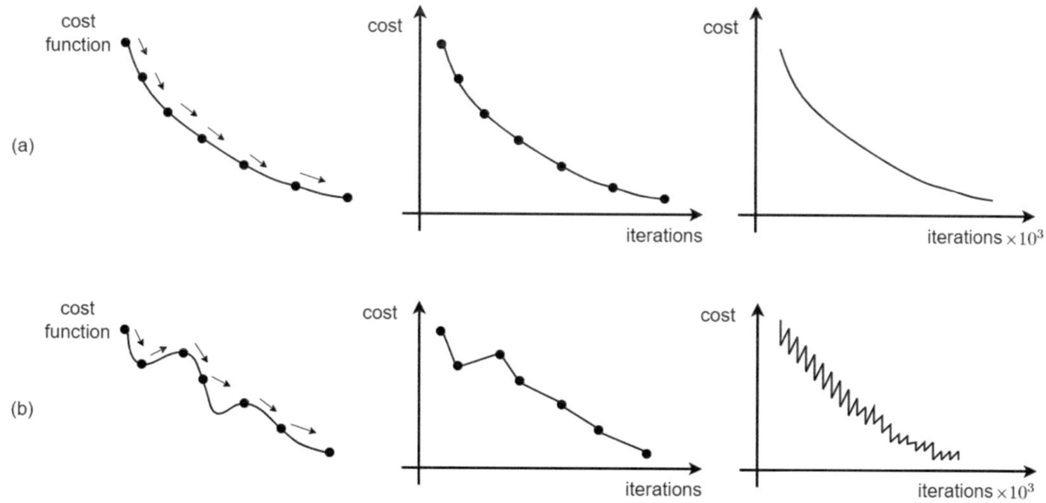

Fig. 1 The gradient descent steps for **a** a convex cost function and **b** a non-convex cost function. The left to right figures depict the optimization steps on cost function, the cost value at the iterations of optimization, and the cost value in large number of iterations

This process continues until the algorithm converges (or gets sufficiently close) to the desired optimum $\boldsymbol{w}^*$.

Assume the gradient of $f(\boldsymbol{w})$ is L-smooth, where L is the Lipschitz constant.[1] In this setting, the update rule is (see [22, 23] for proof):

$$\Delta \boldsymbol{w} = -\frac{1}{L}\nabla f(\boldsymbol{w}^{(k)}) \implies \boldsymbol{w}^{(k+1)} := \boldsymbol{w}^{(k)} - \frac{1}{L}\nabla f(\boldsymbol{w}^{(k)}). \tag{3}$$

In practice, the Lipschitz constant L is often unknown or difficult to compute. As a result, a fixed step size $\eta > 0$—known as the **learning rate**—is typically used:

$$\Delta \boldsymbol{w} = -\eta \nabla f(\boldsymbol{w}^{(k)}), \text{ i.e., } \boldsymbol{w}^{(k+1)} := \boldsymbol{w}^{(k)} - \eta \nabla f(\boldsymbol{w}^{(k)}). \tag{4}$$

If the problem is a maximization rather than minimization, the step becomes $\Delta \boldsymbol{w} = \eta \nabla f(\boldsymbol{w}^{(k)})$, in which case the algorithm is known as **gradient ascent**.

The learning rate η can be chosen using **line search**, a technique more commonly used in traditional optimization than in deep learning. Line search will be explained in Sect. 2.2.

For a convex objective, the sequence $\{\boldsymbol{w}^{(k)}\}$ converges to the global minimum $\boldsymbol{w}^*$, and the objective values decrease monotonically:

$$\{\boldsymbol{w}^{(0)}, \boldsymbol{w}^{(1)}, \boldsymbol{w}^{(2)}, \dots\} \to \boldsymbol{w}^*,$$
$$f(\boldsymbol{w}^{(0)}) \geq f(\boldsymbol{w}^{(1)}) \geq f(\boldsymbol{w}^{(2)}) \geq \dots \geq f(\boldsymbol{w}^*).$$

If $f(\cdot)$ is non-convex, the solution is a local minimum. As illustrated in Fig. 1, gradient descent still performs reasonably well for non-convex objectives: although it may oscillate, the overall trend of cost values tends to decrease.

[1] The function $f(\cdot)$ is Lipschitz with Lipschitz constant L if $|f(\boldsymbol{w}_1) - f(\boldsymbol{w}_2)| \leq L\,\|\boldsymbol{w}_1 - \boldsymbol{w}_2\|_2$ for all $\boldsymbol{w}_1, \boldsymbol{w}_2 \in \mathcal{D}$, where $\mathcal{D}$ is the domain. Here, we assume the gradient is Lipschitz smooth, i.e., $|\nabla f(\boldsymbol{w}_1) - \nabla f(\boldsymbol{w}_2)| \leq L\,\|\boldsymbol{w}_1 - \boldsymbol{w}_2\|_2$.

2.1 Convergence Criteria

For all numerical optimization methods including gradient descent, there exist several convergence criteria to terminate the optimization process:

- **Small norm of the gradient:**

$$\|\nabla f(\boldsymbol{w}^{(k+1)})\|_2 \leq \epsilon,$$

where ϵ is a small positive number. This is motivated by the first-order optimality condition, which states that $\|\nabla f(\boldsymbol{w}^*)\|_2 = 0$ at a local optimum. Note that in non-convex settings, this criterion may result in convergence to saddle points.

- **Small change in the cost function:**

$$|f(\boldsymbol{w}^{(k+1)}) - f(\boldsymbol{w}^{(k)})| \leq \epsilon.$$

- **Small change in the gradient:**

$$|\nabla f(\boldsymbol{w}^{(k+1)}) - \nabla f(\boldsymbol{w}^{(k)})| \leq \epsilon.$$

- **Maximum number of iterations:** A predefined maximum iteration count is reached.

2.2 Line Search

The step size η in gradient descent is ideally chosen based on the Lipschitz constant L of the gradient. However, L is often unknown or difficult to compute in practice. An alternative is to determine an appropriate step size adaptively at each iteration using a procedure called *line search*.

Line search begins each iteration with an initial step size $\eta = 1$. If the proposed update $\Delta\boldsymbol{w} = -\eta\nabla f(\boldsymbol{w}^{(k)})$ does not reduce the cost function,

$$f(\boldsymbol{w}^{(k)} - \eta\nabla f(\boldsymbol{w}^{(k)})) < f(\boldsymbol{w}^{(k)}), \tag{5}$$

then the step size is halved, i.e., $\eta \leftarrow \eta/2$.

This process is repeated until Eq. (5) is satisfied—i.e., until a sufficient decrease in the objective function is observed. This will eventually happen once the step size becomes small enough to satisfy

$$\eta < \frac{1}{L}. \tag{6}$$

See [22, 23] for proof.

More advanced line-search strategies include the *Armijo rule* [3] (also known as *backtracking line search*) and *Wolfe conditions* [44]. See [22, 23] for detailed discussion.

The algorithm of gradient descent with line search is given in Algorithm 1. Note that line search introduces an inner loop of step-size adjustment within each gradient descent iteration.

```
Initialize w^(0)
for iteration k = 0, 1, ... do
    Initialize η := 1
    for iteration τ = 1, 2, ... do
        Check line-search condition, i.e., Eq. (5)
        if not satisfied then
            η ← 1/2 × η
        else
            w^(k+1) := w^(k) − η∇f(w^(k))
            break the loop
    Check convergence criterion
    if converged then
        return w^(k+1)
```

Algorithm 1: Gradient descent with line search

2.3 Steepest Descent

Steepest descent is closely related to gradient descent, but with one key difference: rather than using a fixed or predefined step size, it computes the step size at each iteration to minimize the objective function as much as possible in the direction of the negative gradient.

At iteration k, the step size $\eta^{(k)}$ is chosen by solving the following one-dimensional optimization problem [10]:

$$\eta^{(k)} := \arg\min_{\eta} f\big(\boldsymbol{w}^{(k)} - \eta \nabla f(\boldsymbol{w}^{(k)})\big). \tag{7}$$

Once this optimal step size is found, the solution is updated in the same direction as in gradient descent:

$$\boldsymbol{w}^{(k+1)} := \boldsymbol{w}^{(k)} - \eta^{(k)} \nabla f(\boldsymbol{w}^{(k)}).$$

2.4 Momentum

Gradient descent and other first-order methods can be enhanced by introducing a **momentum** term. Proposed by Rumelhart et al. [37], momentum helps accelerate convergence and reduces oscillations, particularly in ravines and narrow valleys of the loss surface.

Instead of relying solely on the current gradient, momentum incorporates a fraction of the previous update. This modifies the standard update rule in Eq. (4) to:

$$(\Delta \boldsymbol{w})^{(k)} := \alpha (\Delta \boldsymbol{w})^{(k-1)} - \eta^{(k)} \nabla f(\boldsymbol{w}^{(k)}), \tag{8}$$

where $\alpha > 0$ is the momentum coefficient, controlling how much of the previous update is retained. The new iterate is then updated as follows:

$$\boldsymbol{w}^{(k+1)} := \boldsymbol{w}^{(k)} + (\Delta \boldsymbol{w})^{(k)}.$$

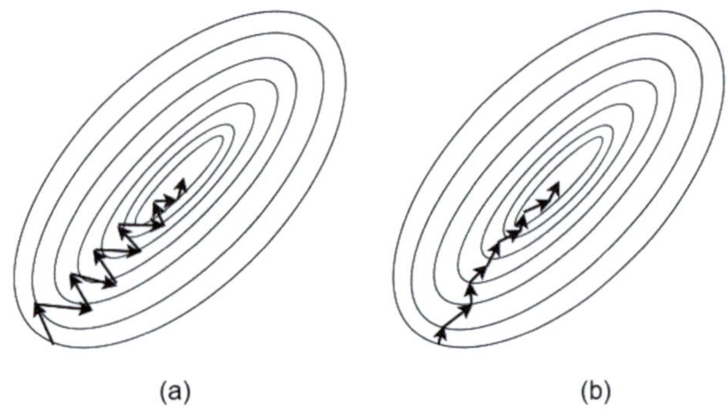

Fig. 2 Gradient descent **a** without and **b** with momentum. Each contour shows the same cost value in the optimization. As the figure shows, momentum reduces oscillation of optimization

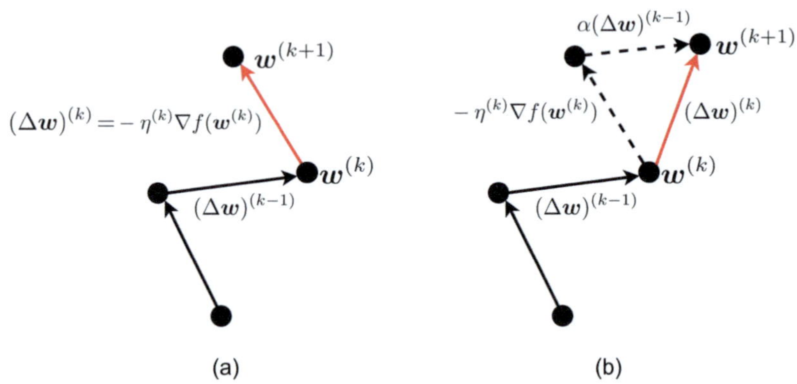

Fig. 3 Update of solution in gradient descent **a** without and **b** with momentum according to Eq. (8). According to the addition of terms in this equation, oscillation is reduced

The key idea is that the update becomes smoother by preserving some of the direction of the previous steps. This helps the algorithm move more consistently in valleys of the loss surface and reduces oscillatory behavior, as illustrated in Fig. 2. The vectorial addition of the gradient and history term is shown in Fig. 3, highlighting the cumulative effect of momentum in directing the optimization path. This vector illustration clarifies why momentum helps in reducing oscillation.

3 Backpropagation

Backpropagation is the primary algorithm used to train feedforward neural networks through gradient-based optimization. It efficiently computes the gradient of the loss function with respect to the network's weights by applying the chain rule layer by layer in reverse order—from the output layer back to the input.

To understand how backpropagation works, consider three layers of a feedforward neural network as shown in Fig. 4. This figure depicts three consecutive layers of the network, which we denote by ℓ, i, and j. Each of these symbols refers to an entire layer of neurons.

Each neuron in a given layer has two associated quantities: a_i and z_i. For a neuron in layer i, the value a_i is the weighted sum of all inputs it receives from the neurons in the previous layer (layer ℓ)—that is, the sum over all incoming edges. The value z_i is obtained by applying the neuron's activation function to a_i.

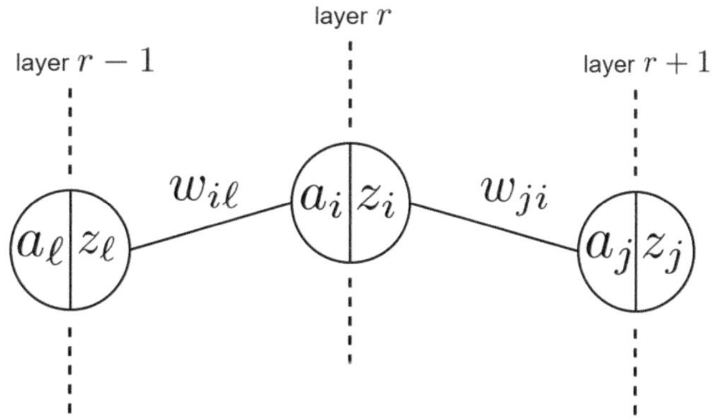

Fig. 4 Three neurons in three successive layers of a network

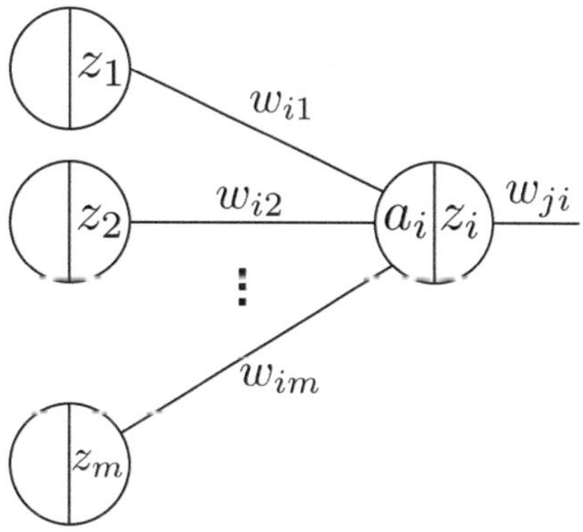

Fig. 5 Neuron i in the neural network

Consider a neuron i in the neural network, as illustrated in Fig. 5. Let $w_{i\ell}$ denote the weight connecting neuron ℓ in the previous layer to neuron i in the current layer. Then:

$$a_i = \sum_{\ell=1}^{m} w_{i\ell} z_\ell, \tag{9}$$

$$z_i := \sigma_i(a_i), \tag{10}$$

where a_i is the pre-activation input, z_i is the post-activation output of neuron i, and $\sigma_i(\cdot)$ is the activation function.

To compute the gradient of the error e with respect to any weight $w_{i\ell}$ (which connects neuron ℓ in one layer to neuron i in the next), we apply the chain rule:

$$\frac{\partial e}{\partial w_{i\ell}} = \frac{\partial e}{\partial a_i} \times \frac{\partial a_i}{\partial w_{i\ell}} \overset{(a)}{=} \delta_i \times z_\ell, \tag{11}$$

where δ_i is defined as follows:

$$\delta_i := \frac{\partial e}{\partial a_i}. \tag{12}$$

The equality (a) uses the fact that $a_i = \sum_\ell w_{i\ell} z_\ell$, hence $\partial a_i / \partial w_{i\ell} = z_\ell$.

Now, consider three consecutive layers in the network. If neuron i is in a hidden layer (not the output layer), its δ_i must be computed based on the downstream gradients. By another application of the chain rule:

$$\delta_i = \frac{\partial e}{\partial a_i} = \sum_j \left(\frac{\partial e}{\partial a_j} \times \frac{\partial a_j}{\partial a_i} \right) = \sum_j (\delta_j \times \frac{\partial a_j}{\partial a_i}). \tag{13}$$

The term $\partial a_j / \partial a_i$ can again be computed using the chain rule:

$$\frac{\partial a_j}{\partial a_i} = \frac{\partial a_j}{\partial z_i} \times \frac{\partial z_i}{\partial a_i} \overset{(a)}{=} w_{ji}\, \sigma'(a_i), \tag{14}$$

where $a_j = \sum_i w_{ji} z_i$ and $z_i = \sigma(a_i)$. Substituting Eq. (14) into Eq. (13) gives

$$\delta_i = \sigma'(a_i) \sum_j (\delta_j\, w_{ji}). \tag{15}$$

Putting this into Eq. (11) yields

$$\frac{\partial e}{\partial w_{i\ell}} = z_\ell\, \sigma'(a_i) \sum_j (\delta_j\, w_{ji}). \tag{16}$$

Backpropagation uses this gradient to update the weights via gradient descent:

$$w_{i\ell}^{(k+1)} := w_{i\ell}^{(k)} - \eta^{(k)} \frac{\partial e}{\partial w_{i\ell}}, \quad \forall i, \ell. \tag{17}$$

In particular: For weights leading into the output layer (i.e., when i is in the final layer):

$$w_{i\ell}^{(k+1)} := w_{i\ell}^{(k)} - \eta^{(k)} z_\ell \frac{\partial e}{\partial a_i}, \quad \forall i, \ell, \tag{18}$$

by Eqs. (11) and (12). For weights in hidden layers:

$$w_{i\ell}^{(k+1)} := w_{i\ell}^{(k)} - \eta^{(k)} z_\ell\, \sigma'(a_i) \sum_j (\delta_j\, w_{ji}), \quad \forall i, \ell, \tag{19}$$

by Eq. (16).

The full backpropagation procedure is summarized in Algorithm 2. At each iteration of training, gradients are computed layer by layer from the output back to the input, and weights are updated accordingly. First introduced by Rumelhart et al. [37], backpropagation is conceptually gradient descent applied in layered systems where derivatives are computed recursively using the chain rule.

Initialize $\boldsymbol{w}^{(0)}$
for *iteration* $k = 0, 1, \ldots$ **do**
  Initialize the learning rate $\eta^{(0)}$
  for *layer r from the last layer to the first layer* **do**
    for *neuron i in the layer r* **do**
      for *neuron* ℓ *in the layer* $(r-1)$ **do**
        if *layer r is the last layer* **then**
          $w_{i\ell}^{(k+1)} := w_{i\ell}^{(k)} - \eta^{(k)} z_\ell \frac{\partial e}{\partial a_i}$
        else
          $w_{i\ell}^{(k+1)} := w_{i\ell}^{(k)} - \eta^{(k)} z_\ell\, \sigma'(a_i) \sum_j (\delta_j\, w_{ji})$
  Check the convergence criterion
  if *converged* **then**
    return all weights of neural network
  Adapt (update) the learning rate $\eta^{(k)}$.

Algorithm 2: Backpropagation algorithm

4 Stochastic Gradient Descent

Backpropagation provides an efficient way to compute the gradient of the loss function with respect to the network parameters. However, computing the full gradient over the entire training set at every iteration can be computationally expensive, especially for large datasets. This motivates the use of *Stochastic Gradient Descent (SGD)*, which approximates the gradient using a small subset—or even a single example—from the dataset, allowing for faster and more scalable updates during training.

Assume there is a dataset of n data instances, $\{\boldsymbol{x}_i \in \mathbb{R}^d\}_{i=1}^n$, with corresponding labels $\{l_i \in \mathbb{R}\}_{i=1}^n$. Let the total cost function $f(\cdot)$ be written as the average of n individual loss terms, where $f_i(\boldsymbol{w})$ is the loss corresponding to the data point $\boldsymbol{x}_i$:

$$f(\boldsymbol{w}) = \frac{1}{n} \sum_{i=1}^{n} f_i(\boldsymbol{w}).$$

This formulation is common in supervised learning problems, where the overall loss is the average loss over the dataset. The optimization problem becomes

$$\underset{\boldsymbol{w}}{\text{minimize}} \quad \frac{1}{n} \sum_{i=1}^{n} f_i(\boldsymbol{w}). \tag{20}$$

In this case, the full gradient is the average of the individual gradients:

$$\nabla f(\boldsymbol{w}) = \frac{1}{n} \sum_{i=1}^{n} \nabla f_i(\boldsymbol{w}), \tag{21}$$

so the update in gradient descent, i.e., Eq. (3) becomes

$$\Delta \boldsymbol{w} = -\eta \nabla f(\boldsymbol{w}^{(k)}) = -\frac{\eta}{n} \sum_{i=1}^{n} \nabla f_i(\boldsymbol{w}^{(k)}).$$

This is the expression used in full-batch gradient descent to update the parameters at each iteration.

However, calculating the full gradient is computationally expensive and inefficient when n is large, especially since it must be recomputed at every iteration. Stochastic Gradient Descent (SGD), also known as the stochastic gradient method, addresses this by approximating the full gradient using a single randomly sampled data point at each iteration. It updates the parameters using:

$$\boldsymbol{w}^{(k+1)} := \boldsymbol{w}^{(k)} - \eta^{(k)} \nabla f_i(\boldsymbol{w}^{(k)}), \tag{22}$$

rather than the full gradient update:

$$\boldsymbol{w}^{(k+1)} := \boldsymbol{w}^{(k)} - \eta \nabla f(\boldsymbol{w}^{(k)}).$$

The idea of stochastic approximation was first introduced in 1951 [36] and was adopted for machine learning applications in 1998 [4].

As shown in Eq. (22), SGD typically uses an adaptive step size $\eta^{(k)}$ that changes over iterations. Larger step sizes are used in the early stages of optimization—when the iterate is far from the optimum—while smaller steps are used in later stages to fine-tune convergence near the minimum. Some common choices for adapting the step size include

$$\eta^{(k)} := \frac{1}{k},$$
$$\eta^{(k)} := \frac{1}{\sqrt{k}},$$
$$\eta^{(k)} := \eta.$$

In practice, SGD and its variants are the default optimization methods for training deep neural networks due to their computational efficiency and scalability.

4.1 Mini-Batch Stochastic Gradient Descent

As discussed in the previous section, Stochastic Gradient Descent (SGD) approximates gradient descent by updating the parameters using a single randomly sampled data point at each iteration. While this makes SGD computationally efficient for very large datasets, it also introduces high variance in the updates and can lead to unstable convergence.

A practical and widely used compromise is to compute the update using a small batch of b randomly sampled points instead of a single point or the full dataset. This method is known as *mini-batch stochastic gradient descent* (or simply mini-batch SGD). It offers a trade-off between the efficiency of SGD and the stability and accuracy of full gradient descent. Mini-batch SGD is especially popular in deep learning, where large datasets and models are common.

Typically, before the optimization begins, the n data points are randomly divided into $\lfloor n/b \rfloor$ disjoint batches of size b. This procedure corresponds to simple random sampling without replacement. Let the dataset be denoted by $\mathcal{D}$ with $|\mathcal{D}| = n$, and the ith batch be $\mathcal{B}_i$ with $|\mathcal{B}_i| = b$. The batches satisfy

$$\bigcup_{i=1}^{\lfloor n/b \rfloor} \mathcal{B}_i = \mathcal{D}, \tag{23}$$

$$\mathcal{B}_i \cap \mathcal{B}_j = \varnothing, \quad \forall i, j \in \{1, \dots, \lfloor n/b \rfloor\},\ i \neq j. \tag{24}$$

Alternatively, one may sample a new batch during optimization using bootstrapping, i.e., sampling with replacement. In this case, batches are no longer disjoint, and the above equalities do not hold.

Definition 1 (*Epoch*) In mini-batch SGD, one *epoch* is completed when all $\lfloor n/b \rfloor$ batches have been used once for optimization. After an epoch, the process typically continues for several more epochs until convergence.

Let $\mathcal{B}_{k'}$ denote the batch used in the kth iteration of optimization. The parameter update rule in mini-batch SGD is

$$\boldsymbol{w}^{(k+1)} := \boldsymbol{w}^{(k)} - \eta^{(k)} \frac{1}{b} \sum_{i \in \mathcal{B}_{k'}} \nabla f_i(\boldsymbol{w}^{(k)}). \tag{25}$$

Note that the factor $1/b$ is sometimes omitted in practice. The literature and implementations in deep learning often refer to mini-batch SGD simply as "SGD," so one should not confuse this with the single-sample update discussed in Sect. 4.

Mini-batch SGD is highly efficient for large-scale optimization, especially when implemented in parallel on GPUs or distributed systems. By leveraging batches, it balances the low computation cost of SGD with the stability and accuracy of full-batch gradient descent.

Proposition 1 (Convergence rates for mini-batch SGD) *Consider a function $f(\boldsymbol{w}) = \sum_{i=1}^{n} f_i(\boldsymbol{w})$ that is bounded below, with each f_i differentiable. Let $f(\cdot)$ be defined on domain $\mathcal{D}$ and assume its gradient is L-smooth. Suppose the learning rate is fixed, $\eta^{(k)} = \eta$. The mini-batch gradient is an approximation of the full gradient with an error term e_t at iteration t:*

$$\frac{1}{b} \sum_{i \in \mathcal{B}_{t'}} \nabla f_i(\boldsymbol{w}^{(t)}) = \nabla f(\boldsymbol{w}^{(t)}) + e_t. \tag{26}$$

Then the convergence rates are as follows:

For non-convex or convex $f(\cdot)$:

$$\mathcal{O}\Big(\frac{1}{t} + \|e_t\|_2^2\Big). \tag{27}$$

If the functions f_i are μ-strongly convex:

$$\mathcal{O}\Big(\Big(1 - \frac{\mu}{L}\Big)^t + \|e_t\|_2^2\Big). \tag{28}$$

As the batch size b increases, the mini-batch gradient approaches the full gradient, and the approximation error $\|e_t\|_2^2$ in Eq. (26) decreases. Consequently, the convergence rate in Eq. (27) improves and can become similar to that of full-batch gradient descent (see Eq. (31)).

The expected error $\mathbb{E}[\|e_t\|_2^2]$ differs depending on whether sampling is done with or without replacement:

$$\mathbb{E}[\|e_t\|_2^2] = \left(1 - \frac{b}{n}\right)\frac{\sigma^2}{b}, \tag{29}$$

$$\mathbb{E}[\|e_t\|_2^2] = \frac{\sigma^2}{b}, \tag{30}$$

where σ^2 is the variance of the dataset. In both cases, increasing the batch size (i.e., $b \to n$ or $b \to \infty$) reduces the gradient noise but also slows down each iteration. Thus, mini-batch SGD embodies a fundamental trade-off between convergence speed and per-iteration computational cost.

We now turn to the theoretical understanding of these optimization methods and compare their long-term behavior using convergence analysis.

4.2 Convergence Analysis of Stochastic Gradient Descent

Having introduced the basic algorithm of Stochastic Gradient Descent (SGD) and its mini-batch variant, we now examine the convergence properties of these methods in comparison to full-batch gradient descent. The convergence behavior depends on factors such as the function's smoothness, convexity, and the choice of step size.

Proposition 2 (Convergence rate of gradient descent with full gradient) *Consider a convex and differentiable function $f(\cdot)$, with domain $\mathcal{D}$, whose gradient is L-smooth. Let f^* be the minimum of the cost function and $\boldsymbol{w}^*$ be the minimizer. Starting from the initial solution $\boldsymbol{w}^{(0)}$, after t iterations of the optimization algorithm, the convergence rate of gradient descent is*

$$f(\boldsymbol{w}^{(t+1)}) - f^* \leq \frac{2L\|\boldsymbol{w}^{(0)} - \boldsymbol{w}^*\|_2^2}{t+1} = \mathcal{O}\left(\frac{1}{t}\right). \tag{31}$$

Proposition 3 (Convergence rate of stochastic gradient descent) *Consider a function $f(\boldsymbol{w}) = \sum_{i=1}^{n} f_i(\boldsymbol{w})$ that is bounded below, where each f_i is differentiable. Let the domain of $f(\cdot)$ be $\mathcal{D}$ and assume its gradient is L-smooth. Suppose that for some constant β, the gradient of each individual term satisfies:*

$$\mathbb{E}[\|\nabla f_i(\boldsymbol{w}_k)\|_2^2 \mid \boldsymbol{w}_k] \leq \beta^2.$$

Depending on the step size, the convergence rate of SGD is

$$f(\boldsymbol{w}^{(t+1)}) - f^* \leq \mathcal{O}\left(\frac{1}{\log t}\right) \quad \textit{if} \;\; \eta^{(\tau)} = \frac{1}{\tau}, \tag{32}$$

$$f(\boldsymbol{w}^{(t+1)}) - f^* \leq \mathcal{O}\left(\frac{\log t}{\sqrt{t}}\right) \quad \textit{if} \;\; \eta^{(\tau)} = \frac{1}{\sqrt{\tau}}, \tag{33}$$

$$f(\boldsymbol{w}^{(t+1)}) - f^* \leq \mathcal{O}\left(\frac{1}{t} + \eta\right) \quad \textit{if} \;\; \eta^{(\tau)} = \eta. \tag{34}$$

Here, τ denotes the iteration index.

If the functions f_i are additionally μ-strongly convex, then the convergence rate of SGD is

$$f(\boldsymbol{w}^{(t+1)}) - f^* \leq \mathcal{O}\left(\frac{1}{t}\right) \quad \textit{if} \quad \eta^{(\tau)} = \frac{1}{\mu\tau}, \tag{35}$$

$$f(\boldsymbol{w}^{(t+1)}) - f^* \leq \mathcal{O}\left(\left(1 - \frac{\mu}{L}\right)^t + \eta\right) \quad \textit{if} \quad \eta^{(\tau)} = \eta. \tag{36}$$

Equations (34) and (36) show that with a fixed step size η, SGD converges sublinearly for a non-convex function and exponentially in the early stages for a strongly convex function. However, as $t \to \infty$, the algorithm stagnates in a neighborhood of $\mathcal{O}(\eta)$ around the optimal point and never exactly reaches it. For example, from Eq. (34):

$$\lim_{t\to\infty} f(\boldsymbol{w}^{(t+1)}) - f^* \leq \lim_{t\to\infty} \mathcal{O}\left(\frac{1}{t} + \eta\right) = \mathcal{O}(\eta)$$
$$\Rightarrow f(\boldsymbol{w}^{(t+1)}) = f^* + \mathcal{O}(\eta).$$

In contrast, full gradient descent achieves exact convergence:

$$\lim_{t\to\infty} f(\boldsymbol{w}^{(t+1)}) - f^* \leq \lim_{t\to\infty} \mathcal{O}\left(\frac{1}{t}\right) = 0$$
$$\Rightarrow f(\boldsymbol{w}^{(t+1)}) = f^*.$$

Thus, while SGD is less accurate than full gradient descent, its main advantage is the lower per-iteration cost, especially when the number of data points n is very large.

It is also important to note that in SGD, the full gradient is not available for use in convergence checks. One may instead monitor the norm of the gradient on the sampled data point or use alternative stopping criteria. Additionally, SGD can be combined with momentum and line-search techniques to improve performance.

5 Adaptive Learning Rate

Recall from Sect. 2.2 that a suitable learning rate can be determined using line search. However, in deep learning, the learning rate is usually initialized manually and then adapted over iterations. Adaptive learning rate methods dynamically adjust the learning rate during training to improve convergence and stability.

This section introduces three widely used methods for adaptive learning rates: AdaGrad, RMSProp, and Adam.

5.1 Adaptive Gradient (AdaGrad)

Adaptive Gradient (AdaGrad) [18] updates the parameters iteratively as follows:

$$\boldsymbol{w}^{(k+1)} := \boldsymbol{w}^{(k)} - \eta^{(k)} \boldsymbol{G}^{-1} \nabla f_i(\boldsymbol{w}^{(k)}), \tag{37}$$

where $\boldsymbol{G}$ is a $(d \times d)$ diagonal matrix with (j, j)th entry:

$$\boldsymbol{G}(j, j) := \sqrt{\varepsilon + \sum_{\tau=0}^{k} \left(\nabla_j f_{i_\tau}(\boldsymbol{w}^{(\tau)})\right)^2}. \tag{38}$$

Here, $\varepsilon \geq 0$ ensures numerical stability, i_τ is the sampled data index at iteration τ, and $\nabla_j f_{i_\tau}(.)$ denotes the jth component of the gradient.

Substituting Eq. (38) into Eq. (37) yields a simplified form:

$$\boldsymbol{w}_j^{(k+1)} := \boldsymbol{w}_j^{(k)} - \frac{\eta^{(k)}}{\sqrt{\varepsilon + \sum_{\tau=0}^{k} \left(\nabla_j f_{i_\tau}(\boldsymbol{w}^{(\tau)})\right)^2}} \nabla f_j(\boldsymbol{w}_j^{(k)}). \tag{39}$$

AdaGrad accumulates historical gradient information. For dimensions where gradients have been large over time, it dampens the update step, effectively giving more attention to less active dimensions to ensure balanced parameter updates.

5.2 Root Mean Square Propagation (RMSProp)

Root Mean Square Propagation (RMSProp), proposed in [42], improves upon AdaGrad by preventing the learning rate from decaying too aggressively.

Inspired by momentum (see Eq. (8)):

$$(\Delta \boldsymbol{w})^{(k)} := \alpha(\Delta \boldsymbol{w})^{(k-1)} - \eta^{(k)} \nabla f(\boldsymbol{w}^{(k)}),$$

RMSProp maintains a moving average of squared gradients:

$$v^{(k+1)} := \gamma v^{(k)} + (1 - \gamma)\|\nabla f_i(\boldsymbol{w}^{(k)})\|_2^2, \tag{40}$$

where $\gamma \in [0, 1]$ is the forgetting factor (e.g., $\gamma = 0.9$).

The parameter update is then:

$$\boldsymbol{w}^{(k+1)} := \boldsymbol{w}^{(k)} - \frac{\eta^{(k)}}{\sqrt{\varepsilon + v^{(k+1)}}} \nabla f_j(\boldsymbol{w}_j^{(k)}), \tag{41}$$

where ε ensures numerical stability. Equation (41) shows that RMSProp behaves similar to AdaGrad but with an exponentially decaying average instead of a cumulative sum.

5.3 Adam Optimizer

Adam (Adaptive Moment Estimation) [25] combines ideas from momentum and RMSProp. It maintains both the first and second moment estimates of the gradient.

It computes

$$\boldsymbol{m}^{(k+1)} := \gamma_1 \boldsymbol{m}^{(k)} + (1 - \gamma_1)\nabla f_i(\boldsymbol{w}^{(k)}), \tag{42}$$

$$v^{(k+1)} := \gamma_2 v^{(k)} + (1 - \gamma_2)\|\nabla f_i(\boldsymbol{w}^{(k)})\|_2^2, \tag{43}$$

where $\gamma_1, \gamma_2 \in [0, 1]$. These are bias-corrected as follows:

$$\widehat{\boldsymbol{m}}^{(k+1)} := \frac{1}{1-\gamma_1^k}\boldsymbol{m}^{(k+1)}, \tag{44}$$

$$\widehat{v}^{(k+1)} := \frac{1}{1-\gamma_2^k}v^{(k+1)}. \tag{45}$$

The final update rule is

$$\boldsymbol{w}^{(k+1)} := \boldsymbol{w}^{(k)} - \frac{\eta^{(k)}}{\sqrt{\varepsilon + \widehat{v}^{(k+1)}}}\widehat{\boldsymbol{m}}^{(k+1)}, \tag{46}$$

which performs SGD with momentum and adaptive scaling based on past gradients.

Adam is one of the most widely used optimizers in modern deep learning libraries. Most deep learning frameworks offer both SGD (interpreted as mini-batch SGD) and Adam as standard options.

6 Sharpness-Aware Minimization (SAM)

While backpropagation is capable of locating a local minimum of the loss function, not all local minima are equally desirable. Empirical studies have shown [20] that smoother (flatter) minima tend to result in better generalization than sharp ones. In this context, a "smooth" minimum refers to a region where the loss remains nearly constant in a neighborhood around the solution, rather than spiking sharply at a single point (see Fig. 6).

This observation suggests a possible link between the smoothness of the converged solution and the neural network's ability to generalize to unseen test data [20]. However, this claim is still debated in the literature. Some works argue that flatness is not the only contributor to generalization performance, and that other factors may also play a significant role [43].

6.1 Vanilla SAM

Motivated by the desire to converge to flatter solutions that may generalize better, the *Sharpness-Aware Minimization (SAM)* algorithm was introduced [20]. SAM modifies the optimization objective by

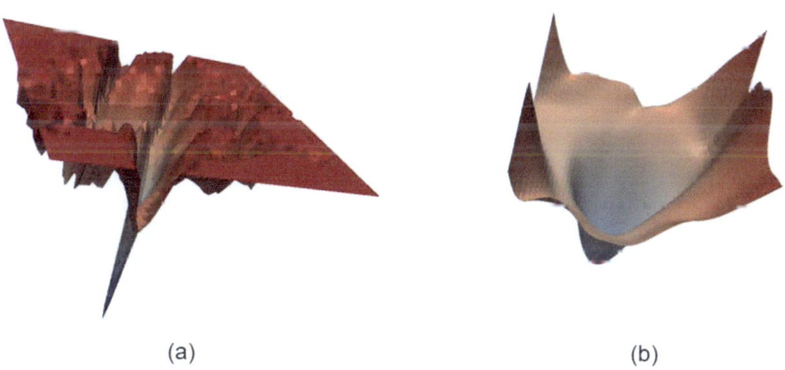

Fig. 6 Visual comparison of **a** a sharp local minimum and **b** an almost flat (smooth) local minimum. The credit of image is for [20]. See [29] for visualization of loss functions in neural networks

incorporating a min-max formulation that simultaneously minimizes the loss and penalizes sharpness, thereby encouraging smoother minima:

$$\underset{\boldsymbol{w}}{\text{minimize}}\ \mathcal{L}_{\text{SAM}}(\boldsymbol{w}) + \lambda \|\boldsymbol{w}\|_2^2, \tag{47}$$

$$\mathcal{L}_{\text{SAM}}(\boldsymbol{w}) := \underset{\epsilon:\|\epsilon\|_p \leq \rho}{\text{maximize}}\ \mathcal{L}(\boldsymbol{w} + \epsilon), \tag{48}$$

where $\boldsymbol{w}$ is the weights of neural network, $\mathcal{L}(.)$ is the loss function of neural network, $\mathcal{L}_{\text{SAM}}(.)$ is the SAM loss function, $\lambda \geq 0$ is the regularization parameter, $\rho \geq 0$ is a hyperparameter, and $p \in [1, \infty]$ where $p = 2$ is recommended [20]. It is possible to drop the regularization term and write the loss as a min-max optimization problem:

$$\underset{\boldsymbol{w}}{\text{minimize}}\ \underset{\epsilon:\|\epsilon\|_p \leq \rho}{\text{maximize}}\ \mathcal{L}(\boldsymbol{w} + \epsilon). \tag{49}$$

Equation (49) is first finding the maximum loss in a neighborhood of radius ϵ around the solution $\boldsymbol{w}$ and then minimizes that. This forces all the local neighborhood of the solution to be small and hence flat or smooth.

Note that Eq. (48) can be restated as [41] follows:

$$\begin{aligned}
\mathcal{L}_{\text{SAM}}(\boldsymbol{w}) &:= \underset{\epsilon:\|\epsilon\|_p \leq \rho}{\text{maximize}}\ \mathcal{L}(\boldsymbol{w} + \epsilon) \\
&= \mathcal{L}(\boldsymbol{w}) - \mathcal{L}(\boldsymbol{w}) + \underset{\epsilon:\|\epsilon\|_p \leq \rho}{\text{maximize}}\ \mathcal{L}(\boldsymbol{w} + \epsilon) \\
&= \underbrace{\mathcal{L}(\boldsymbol{w})}_{\text{empirical loss}} + \underbrace{\underset{\epsilon:\|\epsilon\|_p \leq \rho}{\text{maximize}} \Big(\mathcal{L}(\boldsymbol{w} + \epsilon) - \mathcal{L}(\boldsymbol{w})\Big)}_{\text{sharpness}}.
\end{aligned} \tag{50}$$

The first term in this equation is the empirical loss and the second term is the sharpness measure. The sharpness term can be generalized into a class of sharpness measurements [41].

Consider the inner maximization in Eq. (49):

$$\begin{aligned}
\epsilon^*(\boldsymbol{w}) &:= \arg \max_{\|\epsilon\|_p \leq \rho} \mathcal{L}(\boldsymbol{w} + \epsilon) \\
&\overset{(a)}{\approx} \arg \max_{\|\epsilon\|_p \leq \rho} \mathcal{L}(\boldsymbol{w}) + \epsilon^\top \nabla_{\boldsymbol{w}} \mathcal{L}(\boldsymbol{w}) \\
&\overset{(b)}{=} \arg \max_{\|\epsilon\|_p \leq \rho} \epsilon^\top \nabla_{\boldsymbol{w}} \mathcal{L}(\boldsymbol{w}),
\end{aligned}$$

where $\nabla_{\boldsymbol{w}}$ denotes derivative with respect to $\boldsymbol{w}$, (a) is because of the first-order Taylor series expansion of $\mathcal{L}(\boldsymbol{w} + \epsilon)$ around $\mathbf{0}$ and (b) is because $\mathcal{L}(\boldsymbol{w})$ is not a function of ϵ. This maximization is a classical dual norm problem whose solution is [20]:

$$\epsilon^*(\boldsymbol{w}) = \rho\, \text{sign}\big(\nabla_{\boldsymbol{w}} \mathcal{L}(\boldsymbol{w})\big) \frac{|\nabla_{\boldsymbol{w}} \mathcal{L}(\boldsymbol{w})|^{q-1}}{(\|\nabla_{\boldsymbol{w}} \mathcal{L}(\boldsymbol{w})\|_q^q)^{(1/p)}}, \tag{51}$$

where $(1/p) + (1/q) = 1$. With $p = 2$, it simplifies to:

$$\epsilon^*(\boldsymbol{w}) = \rho \frac{\nabla_{\boldsymbol{w}}\mathcal{L}(\boldsymbol{w})}{\|\nabla_{\boldsymbol{w}}\mathcal{L}(\boldsymbol{w})\|_2}. \tag{52}$$

It is noteworthy that some discussions exist in the literature that Eq. (52) is an upper bound, and not an exact bound, on the classification error. We refer interested readers to [45] for those discussions.

Let us put the found ϵ, i.e., Eq. (52), in Eq. (49) and calculate the gradient of the loss function:

$$\begin{aligned}
\nabla_{\boldsymbol{w}}\mathcal{L}(\boldsymbol{w} + \epsilon^*(\boldsymbol{w})) &\overset{(a)}{=} \frac{d(\boldsymbol{w} + \epsilon^*(\boldsymbol{w}))}{d\boldsymbol{w}} \nabla_{\boldsymbol{w}}\mathcal{L}(\boldsymbol{w})\Big|_{\boldsymbol{w}+\epsilon^*(\boldsymbol{w})} \\
&= \nabla_{\boldsymbol{w}}\mathcal{L}(\boldsymbol{w})\Big|_{\boldsymbol{w}+\epsilon^*(\boldsymbol{w})} + \frac{d\epsilon^*(\boldsymbol{w})}{d\boldsymbol{w}} \nabla_{\boldsymbol{w}}\mathcal{L}(\boldsymbol{w})\Big|_{\boldsymbol{w}+\epsilon^*(\boldsymbol{w})} \\
&\overset{(b)}{\approx} \nabla_{\boldsymbol{w}}\mathcal{L}(\boldsymbol{w})\Big|_{\boldsymbol{w}+\epsilon^*(\boldsymbol{w})},
\end{aligned} \tag{53}$$

where (a) is because of chain rule in derivatives and (b) is because the second term contains multiplication of two derivatives which is small compared to the first term and hence can be ignored. Equation (53) is the gradient of SAM loss function and it can be used in backpropagation. This gradient can be numerically approximated by the deep learning libraries such as PyTorch [35].

6.2 Efficient SAM

The gradient computation in vanilla SAM is computationally intensive and inefficient. To address this, *Efficient SAM (ESAM)* was proposed [14] to improve the computational efficiency of SAM. ESAM introduces two key techniques: *Stochastic Weight Perturbation (SWP)* and *Sharpness-sensitive Data Selection (SDS)*.

6.2.1 Stochastic Weight Perturbation (SWP)

SWP [14] approximates $\epsilon^*(\boldsymbol{w})$ more efficiently by perturbing only a random subset of weights instead of the entire weight vector. Let n denote the total number of weights in the neural network, and let the weights be $\{w_1, \ldots, w_n\}$.

In each iteration of backpropagation, SWP constructs a random binary gradient mask $\boldsymbol{m} = [m_1, \ldots, m_n]^\top$, where each m_i is independently sampled from a Bernoulli distribution: $m_i \overset{\text{i.i.d.}}{\sim} \text{Bernoulli}(\beta)$, with β as the probability parameter.

Using this mask, the solution to the inner maximization, $\epsilon^*(\boldsymbol{w})$, is approximated by [14]:

$$\widehat{\epsilon}(\boldsymbol{w}) \approx \frac{1}{\beta}\boldsymbol{m}^\top \epsilon^*(\boldsymbol{w}), \tag{54}$$

which is used in the gradient of loss in Eq. (53). In other words, SWP does not use the entire weights and, instead, uses a subset of weights for $\epsilon^*(\boldsymbol{w})$. This simplification does not affect the expectation of $\epsilon^*(\boldsymbol{w})$:

$$\mathbb{E}[\widehat{\epsilon}(\boldsymbol{w})_i] \overset{(54)}{=} \frac{1}{\beta}\mathbb{E}[m_i\, \boldsymbol{\epsilon}^*(\boldsymbol{w})_i] \overset{(a)}{=} \frac{1}{\beta}\mathbb{E}[m_i]\,\mathbb{E}[\boldsymbol{\epsilon}^*(\boldsymbol{w})_i]$$
$$\overset{(b)}{=} \frac{1}{\beta}\beta\mathbb{E}[\boldsymbol{\epsilon}^*(\boldsymbol{w})_i] = \mathbb{E}[\boldsymbol{\epsilon}^*(\boldsymbol{w})_i],$$

where $\boldsymbol{\epsilon}(\boldsymbol{w})_i$ denotes the ith element of $\boldsymbol{\epsilon}(\boldsymbol{w})$, (a) is because m_i and $\boldsymbol{\epsilon}^*(\boldsymbol{w})_i$ are independent, and (b) is because expectation of the Bernoulli distribution of m_i.

6.2.2 Sharpness-Sensitive Data Selection (SDS)

While SWP reduces computation by perturbing a subset of weights, further efficiency can be gained by reducing the number of training examples used in each iteration. *Sharpness-sensitive Data Selection (SDS)* [14] addresses this by selectively applying SAM only to the most influential samples in the mini-batch.

Instead of using the entire mini-batch $\mathcal{B}$, SDS identifies and selects a subset of examples that are more sensitive to sharpness. The mini-batch is thus partitioned as follows [14]:

$$\begin{aligned} \mathcal{B}^+ &:= \{(x_i, y_i) \in \mathcal{B} \,|\, \mathcal{L}(\boldsymbol{w} + \boldsymbol{\epsilon}^*(\boldsymbol{w})) - \mathcal{L}(\boldsymbol{w}) > \alpha\}, \\ \mathcal{B}^- &:= \{(x_i, y_i) \in \mathcal{B} \,|\, \mathcal{L}(\boldsymbol{w} + \boldsymbol{\epsilon}^*(\boldsymbol{w})) - \mathcal{L}(\boldsymbol{w}) < \alpha\}, \end{aligned} \tag{55}$$

where $\mathcal{B}^+$ is the sharpness-sensitive subset of mini-batch and $\alpha > 0$ is a hyperparameter. SDS uses the sharpness-sensitive $\mathcal{B}^+$ rather than the entire mini-batch $\mathcal{B}$ in the loss function of SAM. Therefore, calculation is faster with less samples in the mini-batch.

6.3 Adaptive SAM

A limitation of the original SAM formulation is that it uses a fixed perturbation radius ρ in the constraint $\|\epsilon\|_p \leq \rho$. As a result, SAM is sensitive to re-scaling the neural network's weights—for example, scaling by a matrix $\boldsymbol{A}$ can change the perceived sharpness of the solution [26]. This sensitivity can negatively affect optimization stability and generalization.

To address this issue, *Adaptive SAM (ASAM)* was proposed [26], which introduces a scale-invariant sharpness measure by adapting the perturbation radius to the scale of the weights.

$$\underset{\epsilon:\|\epsilon\|_p \leq \rho}{\text{maximize}}\ \mathcal{L}(\boldsymbol{w} + \epsilon) \neq \underset{\epsilon:\|\epsilon\|_p \leq \rho}{\text{maximize}}\ \mathcal{L}(\boldsymbol{A}\boldsymbol{w} + \epsilon).$$

In other words, neural networks with different weight scales have different sharpness values.

Adaptive SAM (ASAM) [26] makes SAM robust to weight re-scaling. We define a normalization operator of $\boldsymbol{w}$, denoted by $T_{\boldsymbol{w}}^{-1}$, where

$$T_{\boldsymbol{A}\boldsymbol{w}}^{-1}\boldsymbol{A} = T_{\boldsymbol{w}}^{-1}, \tag{56}$$

for any invertible scaling matrix $\boldsymbol{A}$. ASAM defines and optimizes the adaptive sharpness of $\boldsymbol{w}$ [26]:

$$\underset{\epsilon:\|T_{\boldsymbol{w}}^{-1}\epsilon\|_p \leq \rho}{\text{maximize}}\ \mathcal{L}(\boldsymbol{w} + \epsilon) - \mathcal{L}(\boldsymbol{w}), \tag{57}$$

while the regular sharpness, sensitive to scale, of $\boldsymbol{w}$ is defined as in Eq. (50).

$$\underset{\epsilon:\|\epsilon\|_p\leq\rho}{\text{maximize}}\ \mathcal{L}(\boldsymbol{w}+\epsilon)-\mathcal{L}(\boldsymbol{w}).$$

The adaptive sharpness, defined in Eq. (57), is scale-invariant because for the scaled weights $\boldsymbol{A}\boldsymbol{w}$, there is

$$\begin{aligned}
\underset{\epsilon:\|T_{\boldsymbol{A}\boldsymbol{w}}^{-1}\epsilon\|_p\leq\rho}{\text{maximize}}\ \mathcal{L}(\boldsymbol{A}\boldsymbol{w}+\epsilon) &\overset{(a)}{=} \underset{\epsilon:\|T_{\boldsymbol{A}\boldsymbol{w}}^{-1}\epsilon\|_p\leq\rho}{\text{maximize}}\ \mathcal{L}(\boldsymbol{w}+\boldsymbol{A}^{-1}\epsilon) \\
&\overset{(b)}{=} \underset{\epsilon':\|T_{\boldsymbol{A}\boldsymbol{w}}^{-1}\boldsymbol{A}\epsilon'\|_p\leq\rho}{\text{maximize}}\ \mathcal{L}(\boldsymbol{w}+\epsilon') \\
&\overset{(56)}{=} \underset{\epsilon':\|T_{\boldsymbol{w}}^{-1}\epsilon'\|_p\leq\rho}{\text{maximize}}\ \mathcal{L}(\boldsymbol{w}+\epsilon') \\
&\overset{(c)}{=} \underset{\epsilon:\|T_{\boldsymbol{w}}^{-1}\epsilon\|_p\leq\rho}{\text{maximize}}\ \mathcal{L}(\boldsymbol{w}+\epsilon),
\end{aligned}$$

where (a) is because of left-multiplying the input of loss with $\boldsymbol{A}$, (b) is because of change of variable $\epsilon' := \boldsymbol{A}^{-1}\epsilon$ (so that $\epsilon := \boldsymbol{A}\epsilon'$), and (c) is because of renaming the dummy variable ϵ' to ϵ.

7 Convergence Guarantees for Optimization in Over-parameterized Neural Networks

The previous sections focused on practical optimization methods used in training neural networks. In this section, we shift our attention to theoretical results that explain the convergence behavior of these methods under specific assumptions.

One of the most surprising phenomena in deep learning is that heavily over-parameterized neural networks—i.e., networks with significantly more learnable parameters than training data instances—often generalize well to unseen data. Classical statistical learning theory suggests that such models should overfit. However, empirical results show otherwise, leading to a fundamental puzzle in modern machine learning.

Over-parameterization typically refers to neural networks where the number of parameters exceeds the number of training samples. This can occur due to large layer widths (many neurons per layer), great depth (many layers), or both. The over-parameterized setting ensures the network has sufficient capacity to memorize the training data, but it raises questions about optimization and generalization.

While the generalization mystery remains unresolved, several theoretical works have shown that gradient based methods (e.g., gradient descent and SGD) converge to global or near-global minima in over-parameterized networks—despite the non convexity of the loss landscape. These results provide insight into why optimization is successful, at least in terms of training loss minimization, for both shallow and deep architectures [1, 2, 40].

7.1 Convergence Guarantees for Optimization in Shallow Networks

Consider a shallow network with one hidden layer and an output layer. Let d be the dimensionality of the input data, k be the number of hidden neurons, and suppose the output layer consists of a single neuron for predicting the label of data. This network maps a data point $\boldsymbol{x} \in \mathbb{R}^d$ to a scalar output as follows:

$$\boldsymbol{x} \mapsto \boldsymbol{v}^\top \sigma(\boldsymbol{W}\boldsymbol{x}),$$

where $\boldsymbol{W} \in \mathbb{R}^{k \times d}$ is the weight matrix from the input to the hidden layer, $\boldsymbol{v} \in \mathbb{R}^k$ is the weight vector of the output layer, and $\sigma(\cdot)$ is the activation function.

Assume that the loss function for a data instance $\boldsymbol{x}_i$ is the mean squared error between the network's output and the target label y_i:

$$\mathcal{L}(\mathbf{W}) = \frac{1}{2n} \sum_{i=1}^{n} \left(y_i - \boldsymbol{v}^\top \sigma(\mathbf{W}\boldsymbol{x}_i)\right)^2,$$

where n is the number of training instances.

Proposition 4 ([40, Theorem 1]) *In the above-mentioned shallow network, suppose the activation function is a quadratic function, i.e., $\sigma(z) = z^2$. If $k \geq 2d$ and the weights of the output layer $\boldsymbol{v}$ contain at least d positive values and d negative values, then:*

1. *There are no spurious local minima, i.e., all local minima are global minima.*
2. *All saddle points have a direction of strictly negative curvature. In other words, at any saddle point $\boldsymbol{W}_s$ of the loss function, there exists a direction $\boldsymbol{U} \in \mathbb{R}^{k \times d}$ such that:*

$$\boldsymbol{vec}(\boldsymbol{U})^\top \nabla^2 \mathcal{L}(\boldsymbol{W}_s)\, \boldsymbol{vec}(\boldsymbol{U}) < 0,$$

   where $\nabla^2 \mathcal{L}(\boldsymbol{W}_s)$ is the Hessian (second-order derivative) of the loss at the saddle point, and $\mathbf{vec}(\cdot)$ is the vectorization operator that flattens a matrix into a vector.
3. *If $d \leq n \leq cd^2$, where $c > 0$ is a constant, the global optimum of the loss function $\mathcal{L}(\mathbf{W})$ achieves zero loss.*

Other variants of Proposition 4, involving more relaxed assumptions and broader activation functions, can be found in [40].

Item 1 in Proposition 4 implies that (almost) all local minima of the neural network are as good as the global minimum. In other words, the situation illustrated in Fig. 7 occurs. This means that, regardless of the initialization of the network's weights, the optimization typically converges to a global minimum—even though the loss function is non-convex [8, 9, 16, 17, 19]. Different local solutions tend to yield similar performance on training data [11]. This is partially attributed to the fact that neural networks implicitly map data into a reproducing kernel Hilbert space [21], in which the loss landscape is much smoother and most local minima are globally optimal [13].

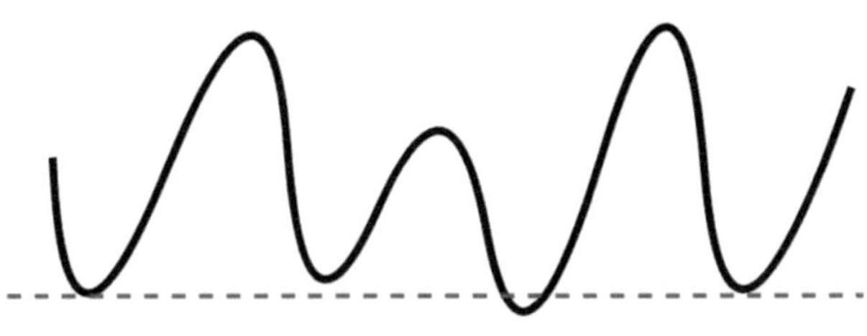

Fig. 7 In a neural network, (almost) all local solutions are good enough to be the global solution

Item 2 in Proposition 4 states that neural networks exhibit escape directions at saddle points—directions with strictly negative curvature. This guarantees that first-order methods such as gradient descent, commonly used in backpropagation, can typically avoid saddle points given random initializations [8, 13, 27, 34].

Finally, item 3 asserts that the network can perfectly fit the training data under the stated conditions, i.e., it reaches zero training loss. However, whether this perfect fit also generalizes well to unseen test data is a separate question and is not addressed by this proposition.

7.2 Convergence Guarantees for Optimization in Deep Networks

In addition to shallow networks, theoretical convergence guarantees have also been established for deep neural networks [1, 2]. These results show that even for deep, highly over-parameterized networks with non-convex objectives, gradient-based optimization can provably converge under certain conditions.

Proposition 5 ([2, Theorems 1 and 2]) *Consider a fully connected l-layer neural network with mean squared error loss and ReLU activation function. Without loss of generality,*[2] *assume the data instances are normalized to have unit length and the last dimension of each instance is $1/\sqrt{2}$.*

Let d be the dimensionality of the input data, and let δ denote a lower bound on the Euclidean distance between any two training instances:

$$\|\boldsymbol{x}_i - \boldsymbol{x}_j\|_2 \geq \delta, \quad \forall i, j \subset \{1, \ldots, n\},$$

where n is the number of training samples. Assume the network weights are randomly initialized. Then:

- *Suppose the network width satisfies $m \geq \Omega(\mathbf{poly}(n, l, \delta^{-1})\, d)$, where $\Omega(\cdot)$ denotes a lower bound and $\mathbf{poly}(\cdot)$ denotes a polynomial function of its arguments. Using a learning rate $\eta = \Theta(\frac{d\delta}{\mathbf{poly}(n,l)\, m})$, gradient descent converges to a loss value less than ϵ after:*

$$T = \Theta\Big(\frac{\mathbf{poly}(n, l)\log(\epsilon^{-1})}{\delta^2}\Big), \tag{58}$$

iterations, with high probability at least $1 - e^{-\Omega(\log^2(m))}$.
- *If instead mini-batch SGD is used with batch size $b \in \{1, \ldots, n\}$, and network width satisfies $m \geq \Omega\left(\frac{\mathbf{poly}(n,l,\delta^{-1})\, d}{b}\right)$, then with learning rate $\eta = \Theta\left(\frac{bd\delta}{\mathbf{poly}(n,l)\, m \log^2(m)}\right)$, the mini-batch SGD converges to a loss less than ϵ after:*

$$T = \Theta\Big(\frac{\mathbf{poly}(n, l)\log(\epsilon^{-1})\log^2(m)}{\delta^2 b}\Big), \tag{59}$$

iterations, with high probability at least $1 - e^{-\Omega(\log^2(m))}$.

Proposition 5 shows that, under appropriate over-parameterization and data separation assumptions, deep networks trained with gradient-based methods converge to solutions with arbitrarily small training loss, and do so in polynomial time with high probability.

[2] It is always possible to normalize data and add an auxiliary dimension with value $1/\sqrt{2}$.

7.3 Other Works on Convergence Guarantees for Optimization in Neural Networks

Beyond the results discussed so far, several other works have investigated convergence guarantees for neural networks under various settings. These include analyses involving different activation functions, architectures, and data assumptions.

For example, convergence guarantees have been established for networks using quadratic activations [15], ReLU activations [7, 39, 46], and leaky ReLU [5].

In addition to standard fully connected networks, convergence analyses have been extended to more complex architectures such as ResNets [17, 38], as well as settings involving structured data [30].

Other theoretical studies focus on the characterization of critical points—locations where the gradient of the loss function vanishes or changes sign—within the loss landscape of neural networks [33, 47]. Further results include convergence guarantees for binary classification tasks [31], loss surface analysis using tools from algebraic geometry [32], and error bounds on gradient descent methods in neural networks [6].

8 Conclusion

The optimization of neural networks relies on backpropagation, which is built upon gradient descent and its stochastic variant. Several adaptive learning rate methods, including AdaGrad, RMSProp, and Adam, have been developed to improve training efficiency and stability. More recent approaches, such as Sharpness-Aware Minimization (SAM), suggest that converging to smoother minima leads to better generalization. Finally, this chapter also discussed convergence guarantees for optimization in neural networks, providing theoretical insights into their training dynamics.

9 Exercises

1. (Level: Moderate) Consider a neural network with a single real-valued input $x \in \mathbb{R}$ and a single output $y \in \mathbb{R}$. The network architecture is defined as follows:

$$y = v_1 \cdot \tanh(w_1 x) + v_2 \cdot \tanh(w_2 x),$$

where $w = [w_1, w_2] \in \mathbb{R}^2$ are the weights of the first layer, $v = [v_1, v_2] \in \mathbb{R}^2$ are the weights of the output layer, and $\phi(z) = \tanh(z)$ is the activation function used in the hidden layer.
Given a training sample $(x^{(1)}, y_{\text{true}})$, we use the mean squared error (MSE) loss function:

$$L = \frac{1}{2}(y - y_{\text{true}})^2.$$

Compute the following quantities:

(a) Derive expressions for the gradients $\frac{\partial L}{\partial v_1}$ and $\frac{\partial L}{\partial v_2}$.
(b) Derive expressions for the gradients $\frac{\partial L}{\partial w_1}$ and $\frac{\partial L}{\partial w_2}$.
(c) Given $x^{(1)} = 1$, $y_{\text{true}} = 0$, $w_1 = w_2 = 1$, and $v_1 = v_2 = 1$, compute the numerical values of all the gradients obtained in the backward pass.

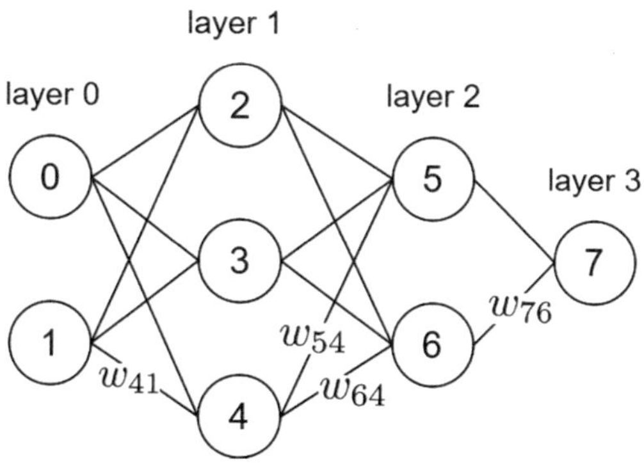

Fig. 8 The neural network of Exercise 2

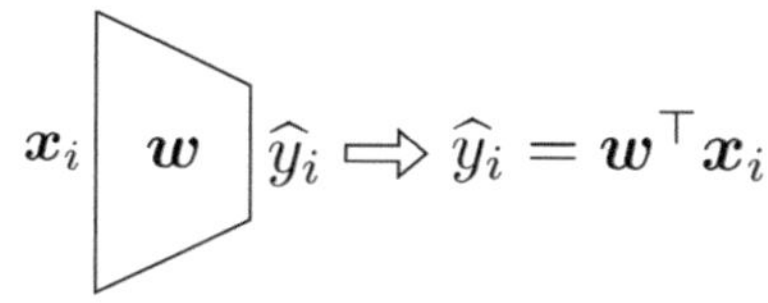

Fig. 9 The neural network of Exercise 3

2. (Level: Moderate) Consider the fully connected neural network depicted in Fig. 8. Consider the data point $\boldsymbol{x} = [2, 1]^\top$ with label $\ell = 1$. Let the initial weights be $w_{ij} = 1, \forall i, j$. Assume all neurons including the last layer have sigmoid (logistic) activation functions:

$$\sigma(a) = \frac{1}{1 + e^{-a}}.$$

The loss function for training the network is

$$e = (z_7 - \ell)^2.$$

(a) Feed the data to the network and get the output of network for one feedforward pass.
(b) Do one iteration of backpropagation to update the weights w_{76} and w_{64}. Let the learning rate be fixed to $\eta = 0.1$.

3. (Level: Moderate) This question is about stochastic gradient descent. Consider a one-layer neural network, with weights $\boldsymbol{w} \in \mathbb{R}^d$ and without activation function, as illustrated in Fig. 9 ($d = 2$ here). Let the loss function be

$$e = \frac{1}{n} \sum_{i=1}^{n} f_i(\boldsymbol{w}), \tag{60}$$

where

$$f_i(\boldsymbol{w}) := (y_i - \widehat{y}_i)^2, \tag{61}$$

where $\widehat{y}_i$ is the output of network. We have four data points in the dataset:

$$(\boldsymbol{x}_1 = [1, 2]^\top, y_1 = 0.1), \quad (\boldsymbol{x}_2 = [0, 0.1]^\top, y_2 = 0), \quad (\boldsymbol{x}_3 = [2, 2]^\top, y_3 = 0.3), \quad (\boldsymbol{x}_4 = [3, 2]^\top, y_4 = 0.5).$$

Note that the optimization variable is the vector $\boldsymbol{w}$ which is the weights of neural network.

(a) Do two iterations of the gradient descent with full gradient. Use a fixed learning rate $\eta = 0.01$. Start with the initial solution $\boldsymbol{w}^{(0)} = [0.5, 0.5]^\top$ which is a uniform solution.
(b) Do two iterations of the mini-batch gradient descent with batches:

- mini-batch for iteration 1: data points 1 and 3.
- mini-batch for iteration 2: data points 2 and 4.

4. (Level: Difficult) Let $f : \mathbb{R}^d \to \mathbb{R}$ be a twice-differentiable non-convex function with multiple local minima. We make two additional definitions:
Definition 1: f is said to satisfy the modified Polyak–Łojasiewicz (PL) condition if there exists some $\mu > 0$ such that for all x in a neighborhood U around a local minimum x^* of f, we have:

$$\langle \nabla f(x), x - x^* \rangle \geq \mu \|x - x^*\|^2,$$

where $\langle ., . \rangle$ denotes inner product and $\|.\|$ denotes the norm operator.
Definition 2: The gradient of f, denoted ∇f, is said to be L-Lipschitz in U if there exists some $L > 0$ such that for any $x, y \in U$,

$$\|\nabla f(x) - \nabla f(y)\| \leq L\|x - y\|, \quad \forall x, y \in U.$$

Consider the gradient descent update step:

$$x_t = x_{t-1} - \eta \nabla f(x_{t-1}),$$

which generates a sequence of iterates $\{x_t\}_{t \geq 1}$.
Show that if the learning rate satisfies $0 < \eta < \frac{2\mu}{L^2}$ and f satisfies the modified PL condition and L-Lipschitz conditions, respectively, in a local neighborhood U around a local minimum x^* of f, then there exists some $\rho \in (0, 1)$ such that:

$$\|x_{t+1} - x^*\| \leq \rho \|x_t - x^*\|,$$

when the sequence $\{x_t\}_{t \geq 1}$ is initialized in U.

References

1. Zeyuan Allen-Zhu, Yuanzhi Li, and Yingyu Liang. "Learning and generalization in overparameterized neural networks, going beyond two layers". In: *Advances in neural information processing systems* 32 (2019).
2. Zeyuan Allen-Zhu, Yuanzhi Li, and Zhao Song. "A convergence theory for deep learning via over-parameterization". In: *International conference on machine learning*. PMLR. 2019, pp. 242–252.
3. Larry Armijo. "Minimization of functions having Lipschitz continuous first partial derivatives". In: *Pacific Journal of mathematics* 16.1 (1966), pp. 1–3.
4. Léon Bottou et al. "Online learning and stochastic approximations". In: *On-line learning in neural networks* 17.9 (1998), p. 142.
5. Alon Brutzkus, Amir Globerson, Eran Malach, and Shai Shalev-Shwartz. "SGD learns over-parameterized networks that provably generalize on linearly separable data". In: *arXiv preprint* arXiv:1710.10174 (2017).

6. Yuan Cao and Quanquan Gu. "Generalization bounds of stochastic gradient descent for wide and deep neural networks". In: *Advances in neural information processing systems* 32 (2019).
7. Yuan Cao and Quanquan Gu. "Generalization error bounds of gradient descent for learning over-parameterized deep relu networks". In: *Proceedings of the AAAI Conference on Artificial Intelligence*. Vol. 34. 04. 2020, pp. 3349–3356.
8. Yuxin Chen, Yuejie Chi, Jianqing Fan, and Cong Ma. "Gradient descent with random initialization: Fast global convergence for nonconvex phase retrieval". In: *Mathematical Programming* 176 (2019), pp. 5–37.
9. Lenaic Chizat and Francis Bach. "On the global convergence of gradient descent for over-parameterized models using optimal transport". In: *Advances in neural information processing systems* 31 (2018).
10. Edwin KP Chong and Stanislaw H Zak. *An introduction to optimization*. John Wiley & Sons, 2004.
11. Anna Choromanska, Mikael Henaff, Michael Mathieu, Gérard Ben Arous, and Yann LeCun. "The loss surfaces of multilayer networks". In: *Artificial intelligence and statistics*. PMLR. 2015, pp. 192–204.
12. Haskell B Curry. "The method of steepest descent for non-linear minimization problems". In: *Quarterly of Applied Mathematics* 2.3 (1944), pp. 258–261.
13. Yann N Dauphin et al. "Identifying and attacking the saddle point problem in high-dimensional non-convex optimization". In: *Advances in neural information processing systems* 27 (2014).
14. Jiawei Du et al. "Efficient sharpness-aware minimization for improved training of neural networks". In: *International Conference on Learning Representations (ICLR)*. 2022.
15. Simon Du and Jason Lee. "On the power of over-parametrization in neural networks with quadratic activation". In: *International conference on machine learning*. PMLR. 2018, pp. 1329–1338.
16. Simon Du, Jason Lee, Haochuan Li, Liwei Wang, and Xiyu Zhai. "Gradient descent finds global minima of deep neural networks". In: *International conference on machine learning*. PMLR. 2019, pp. 1675–1685.
17. Simon S Du, Xiyu Zhai, Barnabas Poczos, and Aarti Singh. "Gradient descent provably optimizes over-parameterized neural networks". In: *arXiv preprint* arXiv:1810.02054 (2018).
18. John Duchi, Elad Hazan, and Yoram Singer. "Adaptive subgradient methods for online learning and stochastic optimization". In: *Journal of machine learning research* 12.7 (2011).
19. Soheil Feizi, Hamid Javadi, Jesse Zhang, and David Tse. "Porcupine neural networks:(almost) all local optima are global". In: *arXiv preprint* arXiv:1710.02196 (2017).
20. Pierre Foret, Ariel Kleiner, Hossein Mobahi, and Behnam Neyshabur. "Sharpness-aware minimization for efficiently improving generalization". In: *International Conference on Learning Representations (ICLR)*. 2021.
21. Benyamin Ghojogh, Mark Crowley, Fakhri Karray, and Ali Ghodsi. "Background on Kernels". In: *Elements of Dimensionality Reduction and Manifold Learning* (2023), pp. 43–73.
22. Benyamin Ghojogh, Mark Crowley, Fakhri Karray, and Ali Ghodsi. "Background on Optimization". In: *Elements of Dimensionality Reduction and Manifold Learning*. Springer, 2023, pp. 75–120.
23. Benyamin Ghojogh, Ali Ghodsi, Fakhri Karray, and Mark Crowley. "KKT Conditions, First-Order and Second-Order Optimization, and Distributed Optimization: Tutorial and Survey". In: *arXiv preprint* arXiv:2110.01858 (2021).
24. Jacques Hadamard. *M é moire sur le probl è me d'analyse relatif à l' é quilibre des plaques é lastiques encastr é es*. Vol. 33. Imprimerie nationale, 1908.
25. Diederik P Kingma and Jimmy Ba. "Adam: A method for stochastic optimization". In: *arXiv preprint* arXiv:1412.6980 (2014).
26. Jungmin Kwon, Jeongseop Kim, Hyunseo Park, and In Kwon Choi. "ASAM: Adaptive sharpness-aware minimization for scale-invariant learning of deep neural networks". In: *International Conference on Machine Learning*. PMLR. 2021, pp. 5905–5914.
27. Jason D Lee et al. "First-order methods almost always avoid strict saddle points". In: *Mathematical programming* 176 (2019), pp. 311–337.
28. Claude Lemaréchal. "Cauchy and the gradient method". In: *Doc Math Extra* 251.254 (2012), p. 10.
29. Hao Li, Zheng Xu, Gavin Taylor, Christoph Studer, and Tom Goldstein. "Visualizing the loss landscape of neural nets". In: *Advances in neural information processing systems* 31 (2018).
30. Yuanzhi Li and Yingyu Liang. "Learning overparameterized neural networks via stochastic gradient descent on structured data". In: *Advances in neural information processing systems* 31 (2018).
31. Shiyu Liang, Ruoyu Sun, Yixuan Li, and Rayadurgam Srikant. "Understanding the loss surface of neural networks for binary classification". In: *International Conference on Machine Learning*. PMLR. 2018, pp. 2835–2843.
32. Dhagash Mehta, Tianran Chen, Tingting Tang, and Jonathan D Hauenstein. "The loss surface of deep linear networks viewed through the algebraic geometry lens". In: *IEEE Transactions on Pattern Analysis and Machine Intelligence* 44.9 (2021), pp. 5664–5680.
33. Maher Nouiehed and Meisam Razaviyayn. "Learning deep models: Critical points and local openness". In: *INFORMS Journal on Optimization* 4.2 (2022), pp. 148–173.
34. Ioannis Panageas, Georgios Piliouras, and Xiao Wang. "First-order methods almost always avoid saddle points: The case of vanishing step-sizes". In: *Advances in Neural Information Processing Systems* 32 (2019).
35. Adam Paszke et al. "PyTorch: An imperative style, high-performance deep learning library". In: *Advances in neural information processing systems* 32 (2019).

36. Herbert Robbins and Sutton Monro. "A stochastic approximation method". In: *The annals of mathematical statistics* (1951), pp. 400–407.
37. David E Rumelhart, Geoffrey E Hinton, and Ronald J Williams. "Learning representations by back-propagating errors". In: *nature* 323.6088 (1986), pp. 533–536.
38. Ohad Shamir. "Are ResNets provably better than linear predictors?" In: *Advances in neural information processing systems* 31 (2018).
39. Arsalan Sharifnassab, Saber Salehkaleybar, and S Jamaloddin Golestani. "Bounds on over-parameterization for guaranteed existence of descent paths in shallow ReLU networks". In: *International conference on learning representations*. 2020.
40. Mahdi Soltanolkotabi, Adel Javanmard, and Jason D Lee. "Theoretical insights into the optimization landscape of over-parameterized shallow neural networks". In: *IEEE Transactions on Information Theory* 65.2 (2018), pp. 742–769.
41. Behrooz Tahmasebi, Ashkan Soleymani, Dara Bahri, Stefanie Jegelka, and Patrick Jaillet. "A Universal Class of Sharpness-Aware Minimization Algorithms". In: *International Conference on Machine Learning (ICML)*. 2024.
42. Tijmen Tieleman and Geoffrey Hinton. "Lecture 6.5-rmsprop: Divide the gradient by a running average of its recent magnitude". In: *COURSERA: Neural networks for machine learning* 4.2 (2012), pp. 26–31.
43. Kaiyue Wen, Zhiyuan Li, and Tengyu Ma. "Sharpness minimization algorithms do not only minimize sharpness to achieve better generalization". In: *Advances in Neural Information Processing Systems* 36 (2024).
44. Philip Wolfe. "Convergence conditions for ascent methods". In: *SIAM review* 11.2 (1969), pp. 226–235.
45. Wanyun Xie, Fabian Latorre, Kimon Antonakopoulos, Thomas Pethick, and Volkan Cevher. "Improving SAM Requires Rethinking its Optimization Formulation". In: *International Conference on Machine Learning (ICML)*. 2024.
46. Xiao Zhang, Yaodong Yu, Lingxiao Wang, and Quanquan Gu. "Learning one-hidden-layer ReLU networks via gradient descent". In: *The 22nd international conference on artificial intelligence and statistics*. PMLR. 2019, pp. 1524–1534.
47. Yi Zhou and Yingbin Liang. "Critical points of neural networks: Analytical forms and landscape properties". In: *arXiv preprint* arXiv:1710.11205 (2017).

4 Convolutional Neural Networks

Abstract

This chapter introduces Convolutional Neural Networks (CNNs), a class of neural networks specifically designed to process data with a grid-like structure, such as images. Unlike fully connected networks—which require input images to be flattened into vectors, thereby discarding spatial relationships—CNNs preserve the two-dimensional structure of images and effectively capture local pixel correlations. They achieve this by replacing standard matrix multiplications with convolution operations in one or more of their layers. The chapter begins by presenting the general architecture of CNNs and then details the components of a convolutional layer, including the convolution, activation (detector), and pooling stages. It also discusses strategies for initializing network weights. Several landmark CNN architectures, including LeNet, AlexNet, VGG, Inception, GoogLeNet, U-Net, ResNet, and DenseNet, are introduced. Collectively, these topics provide a solid foundation for understanding the design and functionality of modern convolutional neural networks.

1 Introduction

Images are high-dimensional data with strong spatial correlations—nearby pixels tend to be related. When using a fully connected neural network (i.e., a multi-layer perceptron) to process an image, the image must first be reshaped into a long vector. This reshaping discards the two-dimensional spatial structure, and the fully connected layers treat all pixels as equally related, regardless of their position. As a result, fully connected networks struggle to capture local features, such as edges, textures, or object parts, which are critical for image understanding.

To address this limitation, Convolutional Neural Networks (CNNs) were developed. A CNN operates directly on the original two-dimensional structure of the image, enabling it to effectively capture local patterns and spatial hierarchies. Intuitively, CNNs learn small filters that slide over the image to detect local features.

More formally, a CNN can be viewed as a special case of a fully connected network with two key structural modifications: it is *sparse* and employs *weight sharing*. To illustrate this, consider a standard fully connected layer where every input unit connects to every output unit with a distinct weight. Now imagine a situation where

B. Ghojogh and A. Ghodsi, *Elements of Deep Learning*,
https://doi.org/10.1007/978-3-032-10738-1_4

- Each output unit is connected only to a small local region of the input (sparsity).
- All such local regions use the same set of weights (shared weights).

This pattern of sparse and shared connectivity defines a convolutional layer.

This architectural design drastically reduces the number of parameters and incorporates the prior assumption that local patterns—such as edges or corners—are useful features that may appear in multiple locations of the image. It also induces *translation equivariance*, meaning that if the input shifts, the output shifts in a corresponding manner.

This chapter presents the foundational components and architecture of convolutional neural networks. Section 2 introduces LeNet and the basic structure of a CNN. The internal stages of a convolutional layer—including convolution, activation (detector), and pooling—are described in Sect. 3. Weight initialization strategies are discussed in Sect. 4. Section 5 reviews several influential CNN architectures, including AlexNet, VGG, Inception, GoogLeNet, U-Net, ResNet, and DenseNet. For a practical PyTorch implementation of a CNN, refer to Chap. 22.

2 LeNet and Regular CNN

Convolutional Neural Networks (CNNs) were initially proposed by Kunihiko Fukushima in 1980 [7], under the name *neocognitron*. The concept was later adapted and refined for the task of handwritten digit recognition by a team led by Yann LeCun around 1989 [22]. In 1998, it was further improved by Yann LeCun, Yoshua Bengio, and collaborators, and the resulting model was named *LeNet* after LeCun [20]. The same paper also introduced the Modified NIST (MNIST) dataset, a widely used benchmark dataset for handwritten digit recognition in machine learning.

The structure of LeNet is illustrated in Fig. 1. As shown in the figure, convolutional layers are applied to the input image, producing feature maps of size 28×28 with six channels. Next, subsampling or pooling—explained later in this chapter—is used to reduce the feature map size to 14×14 while maintaining six channels. Further convolution and subsampling operations yield feature maps with 16 channels and sizes of 10×10 and 5×5, respectively. These feature maps are then reshaped into vectors and passed through multiple fully connected layers with 120, 84, and 10 neurons. The final layer contains ten neurons, corresponding to classification into the digit classes 0 through 9.

This architecture—which begins with convolution and pooling layers and concludes with fully connected layers—has served as a foundational model that has inspired the design of modern convolutional neural networks.

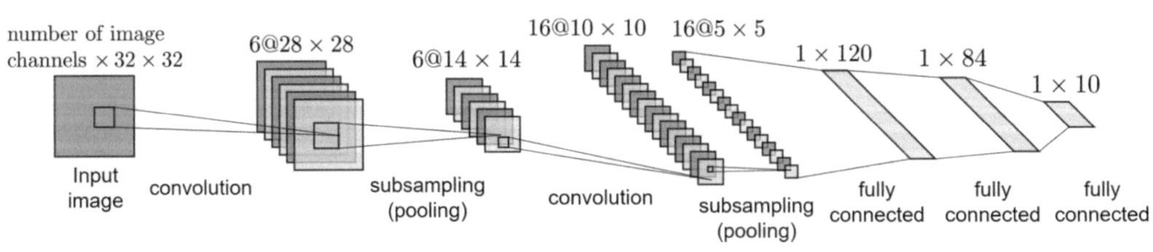

Fig. 1 The LeNet convolutional neural network, which has inspired regular convolutional neural networks

3 The Stages in a Convolutional Layer

Every convolutional layer consists of three main stages: the convolution stage, the detector stage (activation function), and the pooling stage. These stages are introduced in the following subsections.

3.1 Convolution Stage

Recall filtering in image processing. Various kernel filters are used for different tasks such as edge detection, corner detection, smoothing, and sharpening. For example, the following two kernels in the Prewitt filter [27] detect horizontal and vertical edges, respectively:

$$\begin{bmatrix} 1 & 1 & 1 \\ 0 & 0 & 0 \\ -1 & -1 & -1 \end{bmatrix}, \quad \begin{bmatrix} 1 & 0 & -1 \\ 1 & 0 & -1 \\ 1 & 0 & -1 \end{bmatrix}. \tag{1}$$

These specific kernels perform edge detection—that is, they extract edge-related features. In general, different tasks require different filters for effective feature extraction. The optimal filter kernel may vary not only across tasks (e.g., classification versus regression) but also across datasets, even within the same task.

Consequently, rather than manually designing filter values, convolutional layers treat these values as learnable weights that can be optimized for the specific task and dataset. This learning is performed using the backpropagation algorithm [29].

3.1.1 The Discrete Convolution

The discrete convolution is defined as follows:

$$a[t] = (x * k)[t] = \sum_{a=-\infty}^{\infty} x[a]w[t-a]. \tag{2}$$

In the case of a 2D image I, convolution is often applied across two spatial dimensions. Let the 2D filter kernel be denoted by K. The two-dimensional convolution is then given by

$$s[i, j] = (I * K)[i, j] = \sum_m \sum_n I[m, n]K[i-m, j-n], \tag{3}$$

where the summation is taken over the dimensions of the kernel. In practice, convolution is not performed over the entire infinite range $(-\infty, \infty)$; rather, it is restricted to the size of the kernel. One may assume a conceptual infinite range, but values outside the kernel's support are treated as zero.

In theoretical convolution, one of the signals (typically the kernel) is flipped. However, in practical implementations—especially in image processing and CNNs—the kernel is usually not flipped. This makes no difference in training because the network will simply learn either the flipped or non-flipped version of the kernel, depending on the convention used.

The non-flipped variant, known as cross-correlation, is formulated as follows:

$$s[i, j] = (I * K)[i, j] = \sum_{m=1}^{k} \sum_{n=1}^{k} I[m, n]K[m - i, n - j], \tag{4}$$

where k is the kernel size.

3.1.2 Sparsity of the Convolutional Layer

A comparison between a dense (fully connected) layer and a convolutional layer is shown in Fig. 2. As illustrated, the convolutional layer exhibits sparse connectivity, in contrast to the dense connectivity of a fully connected layer. In a fully connected layer, each output neuron interacts with every input neuron. In contrast, a convolutional layer connects each output neuron only to a local region of the input, resulting in sparse learnable weights.

This sparsity is advantageous for several reasons. It aligns with the "betting on sparsity" principle [6, 34], supports the philosophy of Occam's razor [4], and often leads to lower-rank weight matrices compared to those in fully connected layers [19].

As shown in the figure, this sparsity arises because the kernel is typically smaller than the input size. If the kernel size equals the input size, the layer effectively becomes fully connected.

Let m and n denote the number of input and output neurons, respectively. In a fully connected layer, the runtime, memory usage, and number of parameters are all $\mathcal{O}(m \times n)$. In contrast, for a convolutional layer with kernel size k (where $k < m$), these metrics are reduced to $\mathcal{O}(k \times n)$. Therefore, convolutional layers offer improved time and space complexity compared to fully connected layers.

3.1.3 Parameter Sharing in the Convolutional Layer

Another key property of convolutional layers is parameter sharing. The parameters—i.e., the weights—of the kernel are shared across all spatial positions. In other words, the same kernel is applied repeatedly as it slides over the input neurons, as illustrated in Fig. 3.

In a fully connected layer, each element of the weight matrix is used only once—each weight connects a specific input neuron to a specific output neuron. In contrast, in a convolutional layer, each element of the kernel is reused at every spatial position of the input. Rather than learning separate parameters for every position, the network learns a single set of parameters that are shared across all positions.

3.1.4 Translation Equivalence of the Convolution Operator

Another key property of convolutional layers is translation equivalence: if the input is shifted, the output feature map shifts in the same way. This means convolution preserves the relative spatial arrangement of features. However, it is not invariant to transformations like rotation or scaling.

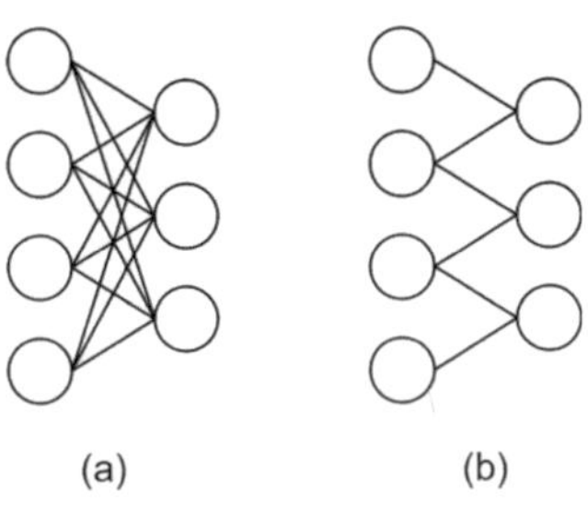

Fig. 2 Comparison of **a** a dense layer and **b** sparsity in a convolutional layer

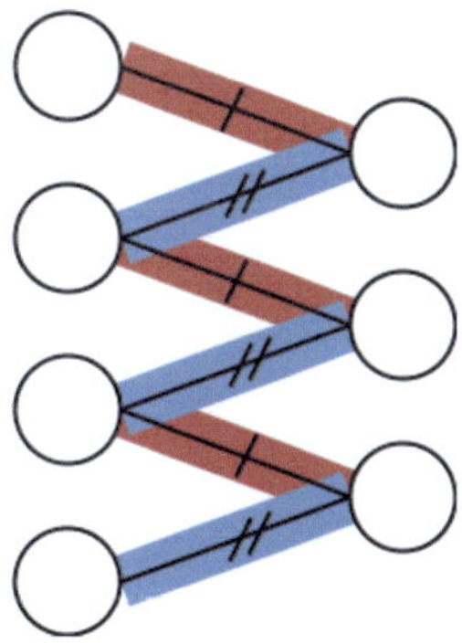

Fig. 3 Parameter sharing in the convolutional layer

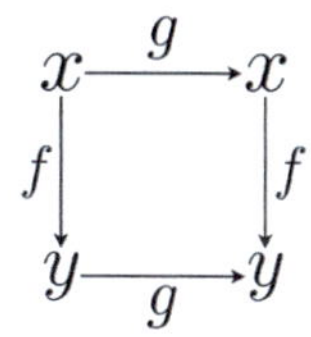

Fig. 4 The equivalence property

To be more specific, a function $f(x)$ is said to be equivalent to a function $g(x)$ if:

$$f(g(x)) = g(f(x)). \quad (5)$$

This concept is illustrated in Fig. 4. The convolution operator satisfies this property with respect to translation. In other words, the following two operations yield equivalent results:

- Translate the input and then apply convolution.
- Apply convolution and then translate the result.

However, convolution does not exhibit equivalence with respect to other transformations such as rotation or scaling.

3.1.5 Stride, Width, Number of Channels, and Padding in Convolution

Several important parameters govern the behavior of the convolutional stage, including stride, kernel width, number of channels, and padding.

The *stride* refers to the number of steps the kernel moves when sliding across the input. The *width* typically refers to both the width and height of the kernel, as kernels are usually square in shape. Smaller kernels emphasize the central values more, whereas larger kernels aggregate information from a broader neighborhood.

The *number of channels* of a kernel refers to its depth and should not be confused with the number of output channels (feature maps) produced after convolution. The kernel depth must match the number of channels in the input data it is applied to.

Padding is used to expand the spatial dimensions of the input. Common padding methods include the following.

- **Zero padding**. Pads the input with zeros.
- **Constant padding**: Pads with a specified constant value.
- **Replicate padding**: Replicates the edge values.
- **Reflect padding**: Mirrors the input values along the border.

- **Circular padding**: Wraps the input values circularly.

When no padding is applied, the case is referred to as *valid padding*.

3.1.6 Output Dimensions of Feature Maps

The spatial dimensions of the output feature map after applying a convolution are given by

$$\text{Output Dimensions} = \left\lfloor \frac{d - f + 2p}{s} + 1 \right\rfloor \times \left\lfloor \frac{d - f + 2p}{s} + 1 \right\rfloor, \tag{6}$$

where d is the width (or height) of the input, f is the width of the kernel, p is the padding size, s is the stride, and $\lfloor \cdot \rfloor$ denotes the floor operation (rounding down to the nearest integer).

3.1.7 Examples of Convolution in the Convolutional Layer

Figure 5 shows an example of convolution applied to an input with a single channel. A more complex example, involving three input channels and two filter kernel channels, is illustrated in Fig. 6.

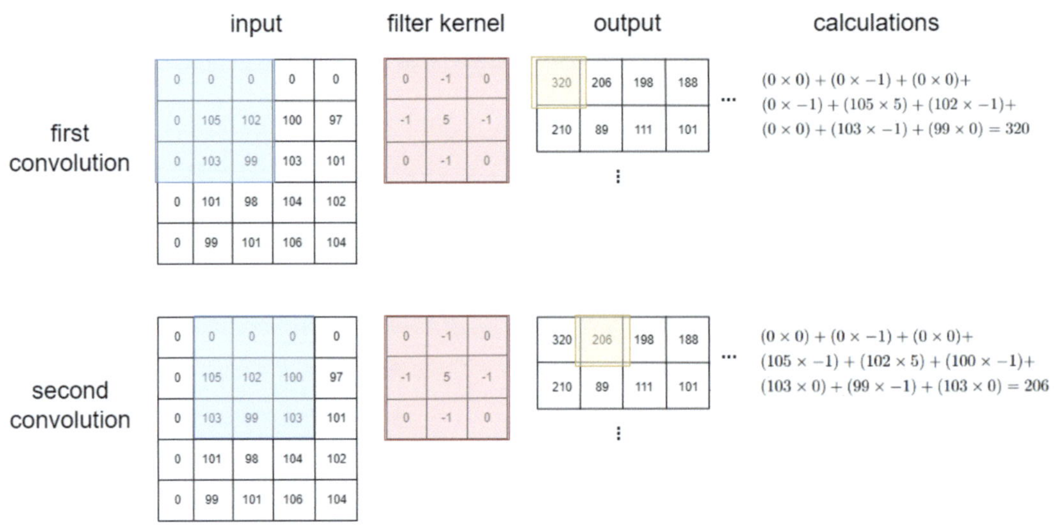

Fig. 5 An example of convolution with one channel of input

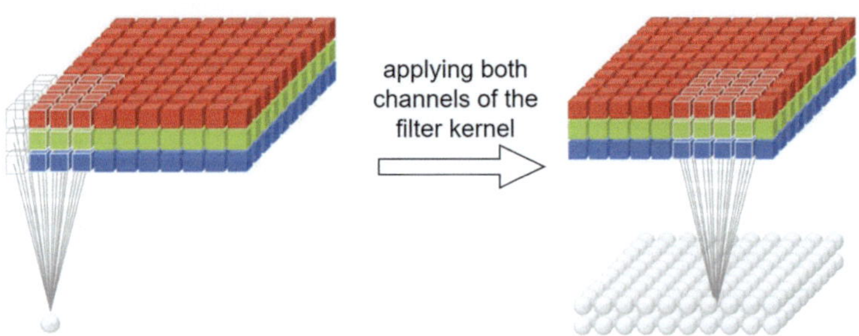

Fig. 6 An example of convolution with three channels of input and two channels of filter kernels. This figure uses an image from https://codelabs.developers.google.com/codelabs/keras-flowers-convnets/#4, accessed in 2024

3.2 Detector and Pooling Stages

The *detector stage* refers to the application of an activation function. These functions are sometimes described as detectors because they activate in response to input signals exceeding certain thresholds. Activation functions were introduced in Chap. 2 and the same functions can be used in convolutional neural networks.

3.2.1 Pooling Methods

In the pooling stage, pooling operations are applied to compute summary statistics over local regions of the input signal. These operations involve a kernel that slides over the input and computes a statistic within each local region.

Common pooling methods include the following:

- **Max pooling**: Returns the maximum value.
- **Average pooling**: Returns the mean value.
- **Weighted average pooling**: Returns a weighted mean.
- **Median pooling**: Returns the median value.
- **ℓ_2-norm pooling**: Returns the ℓ_2 norm of values in the region.

3.2.2 Stride, Width, Number of Channels, and Padding in Pooling

The key parameters of the pooling stage include stride, kernel width, number of channels, and padding:

- **Stride**: The number of steps the pooling kernel moves over the input. If the stride is smaller than the kernel width, pooling regions will overlap. Otherwise, pooling is non-overlapping.
- **Width**: The spatial size of the pooling kernel. Pooling kernels are typically square, having equal width and height. Larger widths incorporate more neighboring values, while smaller widths focus more on local values.
- **Number of channels**: The depth of the pooling kernel, which matches the number of channels in the input.
- **Padding**: Used to expand the spatial dimensions of the input before pooling. Padding methods are the same as those described in Sect. 3.1.5, including zero padding, constant padding, replicate padding, reflect padding, and circular padding.

3.2.3 Examples of Pooling

Consider the following input signal:

$$\begin{bmatrix} 2 & 1 & 7 & 5 \\ 2 & 3 & 2 & 4 \\ 8 & -1 & -5 & 1 \\ 7 & 3 & 2 & 6 \end{bmatrix}.$$

Max pooling (stride 2, width 2):

$$\begin{bmatrix} 3 & 7 \\ 8 & 6 \end{bmatrix}.$$

Median pooling (stride 2, width 2):

$$\begin{bmatrix} 2 & 5.5 \\ 5 & 1.5 \end{bmatrix}.$$

Average pooling (stride 2, width 2):

$$\begin{bmatrix} 2 & 4.5 \\ 4.25 & 1 \end{bmatrix}.$$

Max pooling (stride 1, width 2):

$$\begin{bmatrix} 3 & 7 & 7 \\ 8 & 3 & 4 \\ 8 & 3 & 6 \end{bmatrix}.$$

Max pooling (stride 1, width 3):

$$\begin{bmatrix} 8 & 7 \\ 8 & 6 \end{bmatrix}.$$

Max pooling with zero padding (stride 2, width 2): Applying zero padding to the original signal

$$\begin{bmatrix} 0 & 0 & 0 & 0 & 0 & 0 \\ 0 & 2 & 1 & 7 & 5 & 0 \\ 0 & 2 & 3 & 2 & 4 & 0 \\ 0 & 8 & -1 & -5 & 1 & 0 \\ 0 & 7 & 3 & 2 & 6 & 0 \\ 0 & 0 & 0 & 0 & 0 & 0 \end{bmatrix}$$

yields the following result after max pooling:

$$\begin{bmatrix} 3 & 7 & 5 \\ 8 & 6 & 6 \\ 7 & 6 & 6 \end{bmatrix}.$$

3.2.4 Invariance to Local Translations by Pooling

Pooling makes features approximately invariant to local translations. For example, in Fig. 7, all four values in the input to the pooling stage change as a result of a small shift (translation). However, after applying max pooling, only two of the resulting features change.

In most applications, invariance to local translations is desirable because the presence of a specific feature is often more important than its exact location. For instance, in Fig. 8, detecting the digit "5" is more important than knowing its precise position in the input.

However, in some tasks, spatial position is critical. In such cases, the invariance introduced by pooling may be detrimental, as it can obscure important positional information.

3.2.5 Downsampling by Pooling

Pooling also facilitates downsampling of the data, reducing both the spatial dimensions of the representation and the computational burden on subsequent layers. Downsampling occurs when the stride is greater than one. For example, a stride of two reduces the feature size by half.

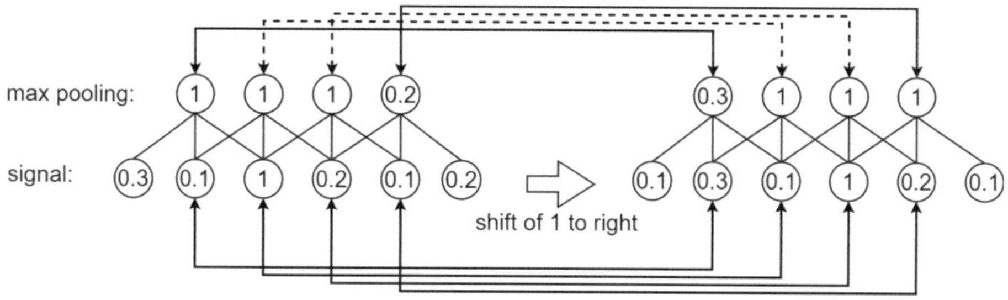

Fig. 7 Illustration of why pooling is robust to local translations. In this example, a shift of one to the right is applied to the signal. As a result, all four values of the signal change. However, only two features change when max pooling is applied. The dashed lines highlight unchanged values

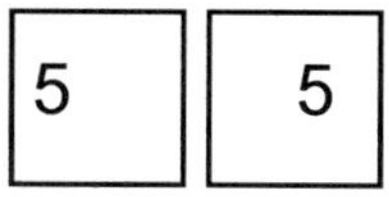

Fig. 8 An example in which local translation is not important; recognizing the digit "5" is the main objective

This reduction in dimensionality is beneficial for managing computational complexity and improving generalization by reducing the number of learnable parameters. The role of pooling in downsampling is also illustrated in the LeNet architecture shown in Fig. 1. An additional example is provided in Fig. 9.

3.2.6 Infinitely Strong Prior

Both convolution and pooling stages can be interpreted as imposing an *infinitely strong prior*. As illustrated in Fig. 10, such a prior effectively sets certain weights to zero. This is due to the inherent sparsity of convolution and pooling operations, as discussed earlier in Fig. 2 and Sect. 3.1.2.

This built-in sparsity supports the "betting on sparsity" principle—"use a procedure that performs well under sparsity, since no procedure performs well under high density" [6, 34]. It also aligns with Occam's razor [4], which asserts that simpler solutions are generally more likely to be correct than complex ones.

3.2.7 Dynamic Pooling

Pooling need not be limited to fixed, local neighborhoods. *Dynamic pooling* strategies allow features to be pooled based on their content or structure rather than just spatial proximity.

For instance, a clustering algorithm such as K-means can be applied to the locations of salient features. Features within the same cluster are then pooled together [2]. In this approach, different pooling regions are dynamically determined for each input.

Fig. 9 An example of downsampling by pooling

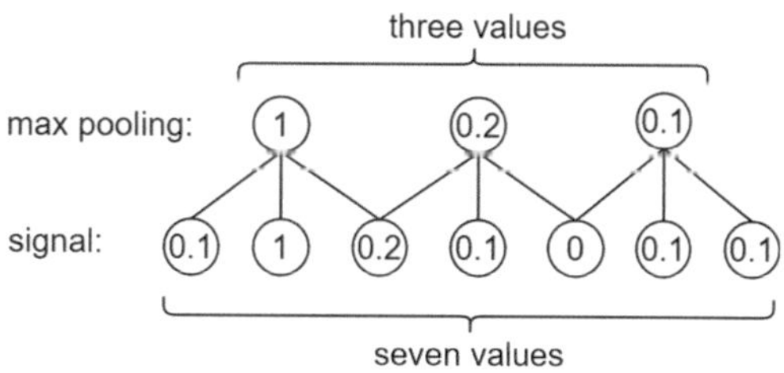

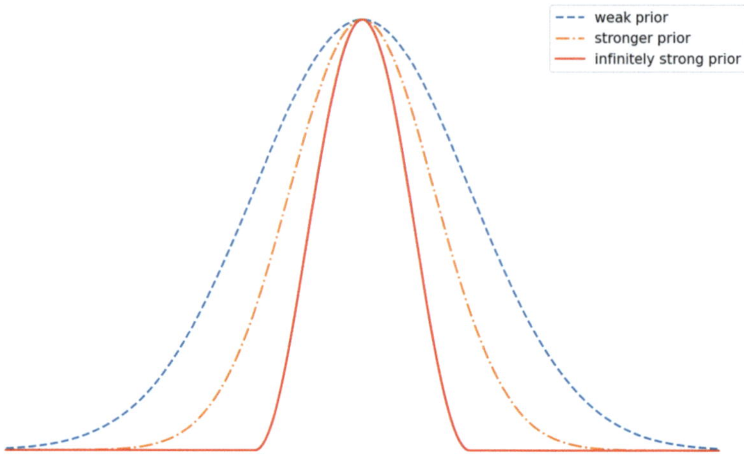

Fig. 10 Weak prior (weak constraint), stronger prior, and infinitely strong prior

Alternatively, a fixed pooling structure can be learned during training and then applied consistently across all data inputs [16].

3.2.8 Using Pooling for Handling Different Input Sizes

Standard CNN architectures require input images to have fixed dimensions (i.e., consistent width and height). However, in practice, inputs may vary in size across different samples.

One solution is to resize all images to a fixed shape using cropping or zero padding, although this can result in the loss of valuable information.

A more flexible approach is to use pooling to accommodate varying input sizes. For example, as shown in Fig. 11, the kernel width of pooling can be adjusted to match the input dimensions—larger for larger inputs and smaller for smaller ones.

Another widely used method is *spatial pyramid pooling* [13], which replaces the final pooling layer with a set of fixed-size bins that adapt to the size of the input. This allows CNNs to accept images of arbitrary size without requiring resizing or cropping.

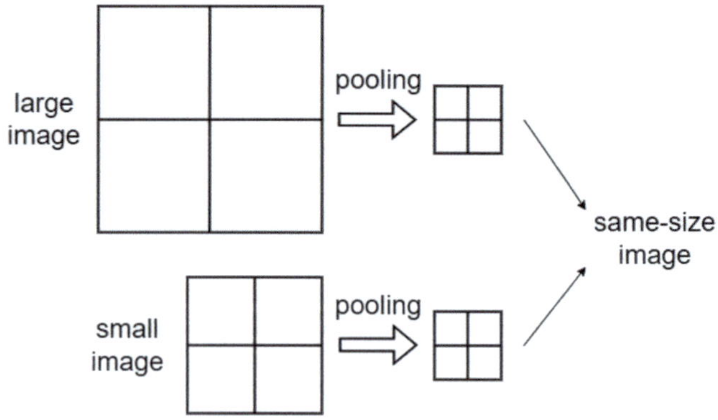

Fig. 11 Using different kernel widths in pooling to handle varying image sizes

3.2.9 Global Average Pooling in Convolutional Neural Networks

In modern CNNs, global average pooling is often used instead of fully connected layers at the end of the network [23]. This reduces the number of parameters and enforces a direct link between feature maps and output classes.

4 Initialization of the Weights of Neural Network

In a CNN, the weights of the network correspond to the values of the convolutional kernel filters. These weights must be optimized to minimize the network's loss function.

The most widely used optimization method for neural network weights is backpropagation, introduced in Chap. 3. As explained in that chapter, backpropagation combines gradient descent with the chain rule of calculus. Because it is an iterative optimization algorithm, the weights (i.e., optimization variables) must be initialized before training begins.

There are various methods for initializing weights. One of the initialization methods is *random initialization*, which is also related to the concept of random projection and the Johnson–Lindenstrauss lemma [9]. Another possible approach involves learning weights through an unsupervised method such as Principal Component Analysis (PCA) [8]. However, in recent practice, popular initialization methods include Xavier/Glorot initialization [10] (suited for tanh and sigmoid activations) and He initialization [12] (suited for ReLU activations). These balance the variance of activations and gradients across layers, reducing vanishing/exploding gradient problems. Random Gaussian or uniform initialization [1, 21] is less common in modern deep learning. In the following, random initialization and unsupervised initialization are introduced.

4.1 Random Initialization of Weights

Weights in neural networks can be initialized randomly using values drawn from stochastic distributions such as the Gaussian or uniform distribution. These initial values must be chosen carefully: if they are too large, gradient explosion may occur; if they are too small, gradient vanishing can become a problem.

Random initialization is applicable to both feedforward and convolutional neural networks. It is closely related to the concepts of random projection and the Johnson–Lindenstrauss lemma [9].

As discussed in Sect. 2.7.3 of Chap. 2, the weights of network layers can be interpreted as projection matrices in a linear transformation. When these weights are randomly initialized, the transformation can be viewed as a random projection.

The Johnson–Lindenstrauss lemma [9] states that random projections approximately preserve pairwise distances with high probability, within a small error bound. This theoretical result helps justify why random initialization of weights is effective in neural networks.

For more details on random projection and the Johnson–Lindenstrauss lemma, refer to [9].

4.2 Unsupervised Initialization of Weights

It is also possible to initialize the weights of a CNN using unsupervised learning methods. In this approach, the weights of a convolutional layer are divided into patches, where the patch size corresponds to the kernel stride s.

To make the extracted features robust to small shifts in the input data, each patch is slightly shifted (e.g., by a few pixels), and both the original and shifted patches are collected. These patches are then reshaped into column vectors and used as input to an unsupervised feature extraction method such as Principal Component Analysis (PCA) [8].

By applying PCA to both original and shifted patches, the resulting features become invariant to small translations. PCA produces $s \times s$ eigenvector matrices, which can then be used to replace the patch weights in the convolutional layer. These eigenvectors serve as the initial kernel values for that layer.

This process can be repeated across all layers in the network to initialize the convolutional kernels in an unsupervised manner.

5 Important Convolutional Neural Networks

Numerous convolutional neural networks have been developed over the years. Some of the most influential and widely recognized models include the following:

- **AlexNet** (2012) [18]—Introduced by Alex Krizhevsky, Ilya Sutskever, and Geoffrey Hinton; it won the ImageNet Large-Scale Visual Recognition Challenge (ILSVRC) in 2012 and is often credited with kickstarting the deep learning revolution in computer vision. The original paper appeared in NeurIPS 2012 and was later republished in Communications of the ACM in 2017.
- **VGG** (2014) [31].
- **Inception and GoogLeNet** (2015) [32].
- **U-Net** (2015) [28].
- **ResNet** (2016) [11].
- **DenseNet** (2017) [14].

Many of these architectures were developed by major technology companies and research institutions and frequently won or placed highly in the ImageNet Large-Scale Visual Recognition Challenge (ILSVRC) [30], an annual benchmark competition in machine learning and computer vision.

An important point is that these networks have progressively built upon one another. Each new architecture typically incorporates and extends ideas from its predecessors, contributing to the cumulative advancement of convolutional network design.

5.1 About ImageNet and ILSVRC

ImageNet is a large-scale image dataset containing approximately 15 million labeled high-resolution images across around 22,000 object categories. The images were collected from the Internet and manually annotated by human workers using Amazon's Mechanical Turk platform [3].

Since 2010, the ImageNet Large-Scale Visual Recognition Challenge (ILSVRC) [30] has been organized annually as part of the PASCAL Visual Object Challenge [5]. ILSVRC is a prominent benchmark for evaluating image classification and object detection algorithms.

The ILSVRC dataset is a curated subset of ImageNet, consisting of approximately 1.2 million high-resolution images across 1,000 classes. Each class contains about 1,000 images. All images are RGB with a resolution of 224×224 pixels.

5.2 AlexNet

AlexNet was proposed by Alex Krizhevsky during his time as a student under the supervision of Geoffrey Hinton [18]. The network is named after Krizhevsky and is widely considered one of the first highly successful deep neural networks for image classification. Its breakthrough results helped catalyze the widespread adoption of deep learning in the field of image processing.

The architecture of AlexNet is shown in Fig. 12. It consists of eight layers: five convolutional layers followed by three fully connected layers. This structure is inspired by LeNet, which also stacks convolutional layers before transitioning into fully connected layers.

Due to hardware limitations at the time, the network was distributed across two GPUs to fit within the 3GB memory constraint of each device. This setup also accelerated training. The configuration of the layers is as follows:

- **Layer 1:** Filters $224 \times 224 \times 3$ input images using 96 kernels of size $11 \times 11 \times 3$, with a stride of 4.
- **Layer 2:** Filters the output feature maps using 256 kernels of size $5 \times 5 \times 48$.
- **Layer 3:** Applies 384 kernels of size $3 \times 3 \times 256$.
- **Layer 4:** Uses 384 kernels of size $3 \times 3 \times 192$.
- **Layer 5:** Uses 256 kernels of size $3 \times 3 \times 192$.
- **Layers 6 and 7:** Fully connected layers with 4,096 neurons each.
- **Layer 8:** A fully connected output layer with 1,000 neurons (one per class).

All layers, except the final one, use the Rectified Linear Unit (ReLU) activation function [26], which empirically improves training speed and convergence. The final layer employs the softmax activation function to perform classification over 1,000 classes.

In Layers 1 and 2, local response normalization is applied after the ReLU activation. This normalization was found to improve generalization in AlexNet. Denoting the ReLU-activated output of kernel i at position (x, y) as $a^i_{x,y}$, the normalization is defined as follows:

$$b^i_{x,y} = \frac{a^i_{x,y}}{\left(k + \alpha \sum_{j=\max(0,i-n/2)}^{\min(N-1,i+n/2)} (a^j_{x,y})^2\right)^{\beta}}, \tag{7}$$

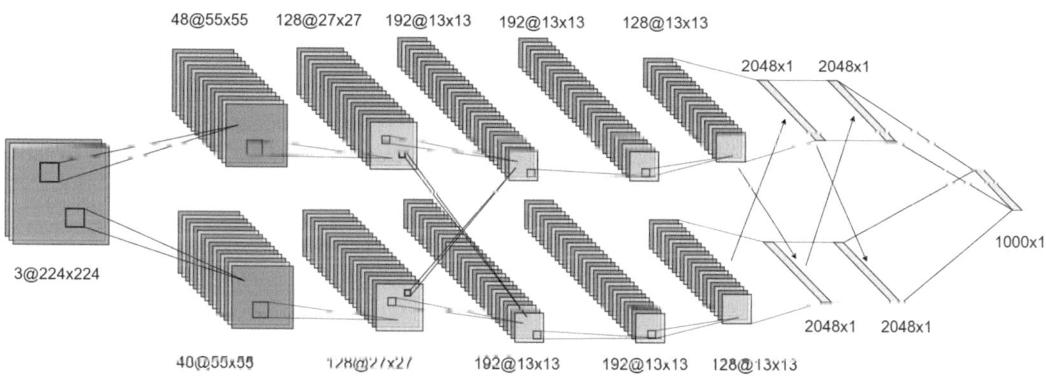

Fig. 12 The architecture of the AlexNet neural network. Arrows indicate layer connections

where the summation runs over n neighboring feature maps at the same spatial location, and N is the total number of kernels in the layer. The hyperparameters were set via cross-validation to $k = 2$, $n = 5$, $\alpha = 10^{-4}$, and $\beta = 0.75$.

Max pooling is applied after Layers 1, 2, and 5. In Layers 1 and 2, pooling follows the local response normalization step. The pooling kernel has a width of 3 and a stride of 2, resulting in overlapping pooling.

AlexNet contains approximately 60 million parameters and 650,000 neurons. In the ILSVRC 2010 competition, it achieved top-1 and top-5 error rates of 37.5% and 17.0%, respectively.

The loss function used in AlexNet is based on multinomial logistic regression. For a data instance $\boldsymbol{x}_i$ with target label ℓ, the network maximizes the output of the ℓth softmax neuron:

$$\max_{\theta} \quad \mathbb{P}(y_i = \ell) = \frac{e^{o_\ell}}{\sum_{j=1}^{1000} e^{o_j}} \quad \overset{(a)}{\Longrightarrow} \quad \min_{\theta} \quad -\mathbb{P}(y_i = \ell) = -\frac{e^{o_\ell}}{\sum_{j=1}^{1000} e^{o_j}}, \tag{8}$$

where o_j denotes the jth output of the network before applying softmax. Step (a) reflects the fact that neural network training is formulated as a minimization problem.

5.3 VGG

The VGG neural network was proposed by the Visual Geometry Group (VGG) at the University of Oxford [31]. VGG addressed the role of network depth by significantly increasing the number of layers, thereby allowing the model to learn more complex and hierarchical features. To counter the potential risk of overfitting due to a large number of learnable parameters, a larger training dataset was used.

VGG can be regarded as a deep convolutional neural network composed of many convolutional layers. It primarily uses very small convolutional filter kernels of size 3×3. This design choice helps to control the number of parameters and reduce overfitting, which is especially important when dealing with very deep networks.

As a preprocessing step, the mean RGB value of the $224 \times 224 \times 3$ input images is subtracted to zero-center the data. Most convolutional kernels in VGG are of size 3×3, with a few 1×1 kernels used to apply linear transformations to the features. All convolutions use a stride of 1, and the 3×3 convolutions use a padding of size 1 to preserve spatial dimensions.

Max pooling is employed with a window of width 2 and a stride of 2, resulting in non-overlapping pooling. Depending on the version, VGG contains a large number of convolutional layers—typically much deeper than earlier architectures. These are followed by three fully connected layers: the first two have 4,096 neurons, and the final layer has 1,000 neurons for classification into 1,000 categories.

All hidden layers use the Rectified Linear Unit (ReLU) activation function [26], while the output layer uses the softmax activation function. Unlike AlexNet, most versions of VGG do not use local response normalization, as it was empirically found not to improve performance on the ILSVRC dataset.

Training was conducted using mini-batch stochastic gradient descent with momentum. The batch size was set to 256, which is a common practice as it is a power of 2. The momentum coefficient was 0.9, a value chosen to provide a high level of momentum without causing excessive oscillation during training.

To regularize the model and prevent overfitting, weight decay using the ℓ_2 norm was applied, with a regularization parameter of 5×10^{-4}. This parameter is typically small so that it does not overshadow the main loss function. The initial learning rate was set to 10^{-2} and was reduced by a factor of 10 whenever the validation accuracy plateaued. A learning rate must strike a balance: it should not be too high to avoid instability, nor too low to prevent slow convergence.

5.4 Inception and GoogLeNet

The Inception module was introduced in [32], which also proposed the GoogLeNet architecture built from stacked Inception modules. The name "GoogLeNet" is inspired by two sources: it was developed by researchers at Google, and it pays homage to the earlier LeNet architecture [20], one of the foundational convolutional networks.

GoogLeNet won first place in the ILSVRC 2014 competition. In general, there are two main ways to increase the capacity of a neural network: by increasing the depth (number of layers) or by increasing the width (number of neurons per layer), or both. However, increasing the number of learnable parameters has two major drawbacks: (1) it increases the risk of overfitting, especially when labeled training data are limited, and (2) it significantly raises computational demands.

To address both issues, the Inception module was introduced. It promotes sparsity in the network, which aligns with the "betting on sparsity principle" [6, 34] and the principle of Occam's razor [4].

An early design of the Inception module is illustrated in Fig. 13. Each neuron in a given layer can be thought of as corresponding to a localized region of the input image. In earlier layers of the network, correlated neurons tend to be spatially clustered, forming dense local regions. These local clusters can be efficiently processed using 1×1 convolutions, which act as linear transformations of features.

In contrast, neurons with broader spatial correlations require larger convolutional kernels. However, larger kernels (e.g., 5×5) are more computationally expensive and introduce more parameters. To maintain efficiency and reduce redundancy, the Inception module restricts itself to small kernels, such as 1×1, 3×3, and 5×5. Additionally, 3×3 max pooling is used, given its known benefits discussed earlier in the chapter.

GoogLeNet constructs its architecture by stacking Inception modules. The proportions of 1×1, 3×3, and 5×5 convolutions vary across layers. In deeper layers near the output, more abstract features are extracted, requiring larger receptive fields. Consequently, these layers include more 3×3 and 5×5 convolutions.

However, deeper stacks of large convolutions lead to computational inefficiency. To mitigate this, the final design of the Inception module incorporates dimensionality reduction via 1×1 convolutions. These are applied both before the expensive 3×3 and 5×5 convolutions and after the max pooling operation, as shown in Fig. 14.

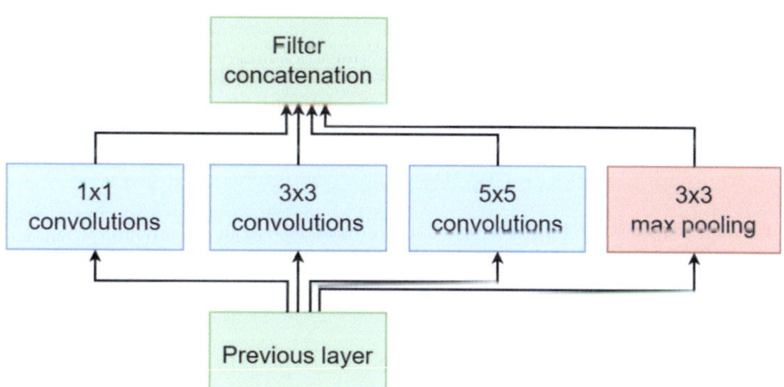

Fig. 13 The initial design of an Inception module used in GoogLeNet

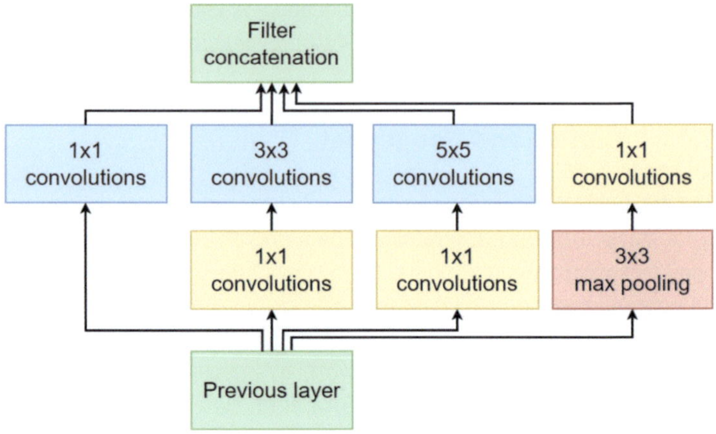

Fig. 14 The final design of an Inception module used in GoogLeNet

The overall architecture of GoogLeNet is presented in Fig. 15. As a preprocessing step, the mean RGB values are subtracted from the $224 \times 224 \times 3$ input images to zero-center the data. GoogLeNet comprises stacked Inception modules as its core building blocks. All convolutional operations, including those within Inception modules, use the Rectified Linear Unit (ReLU) activation function [26].

5.5 U-Net

U-Net was introduced in [28] for biomedical image segmentation and was the winning submission for the ISBI challenge on segmentation of neural structures in electron microscopy stacks. The architecture is named "U-Net" due to its symmetric U-shaped layout.

U-Net consists of a contracting path (corresponding to the left side of the U) and an expansive path (corresponding to the right side of the U). The contracting path captures contextual information, while the expansive path enables precise localization. The overall architecture of U-Net is illustrated in Fig. 16.

The contracting path follows the pattern of a typical convolutional network. Each step consists of two 3×3 convolutions, each followed by a ReLU activation function [26]. Downsampling is performed via a 2×2 max pooling operation with stride 2. At each downsampling step, the number of feature channels is doubled to increase the network's capacity.

The expansive path mirrors the contracting path. Each step again consists of two 3×3 convolutions followed by ReLU activations. However, instead of pooling, upsampling is performed using 2×2 deconvolutions—also known as transpose convolutions or up-convolutions (see Fig. 17 for an illustration). At each upsampling stage, the number of feature channels is halved. After each upsampling, the resulting feature map is concatenated with the cropped corresponding feature map from the contracting path. Cropping is required because the original U-Net used unpadded convolutions, which reduced feature map size at each step. Padding can also be used to avoid this issue in modern implementations.

U-Net was originally implemented using the Caffe deep learning framework [17]. Due to the large size of biomedical image tiles, the batch size was set to one during training. Similar to VGG, U-Net employed a momentum parameter of 0.9 to promote stable convergence and reduce oscillations during training.

The final output of U-Net has the same spatial dimensions as the input image. The last feature map consists of c channels—one per segmentation class—and uses the softmax activation function. The

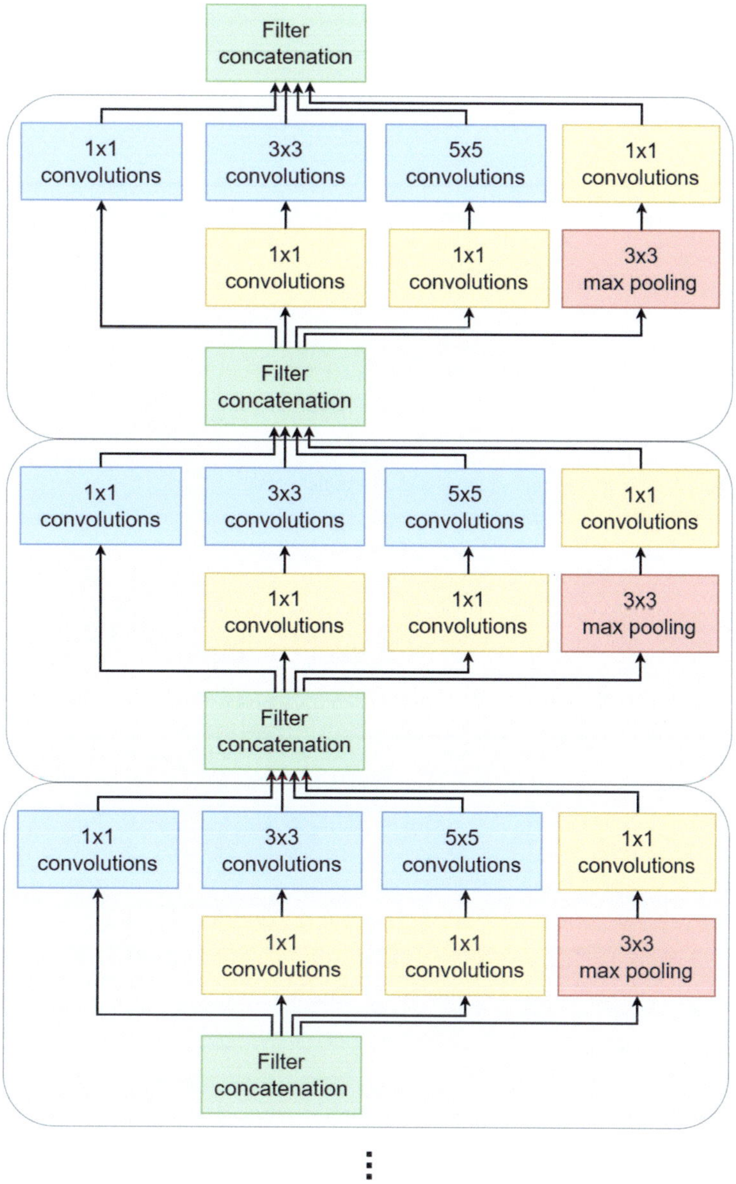

Fig. 15 The architecture of the GoogLeNet neural network

loss function used is the cross-entropy loss, appropriate for pixel-wise classification in segmentation tasks.

Notably, U-Net has inspired several successor architectures. One such model is Y-Net [25], which features a Y-shaped structure instead of the U-shaped layout of U-Net.

5.6 ResNet

Empirical evidence shows that in very deep neural networks—typically those with more than 20 layers—increasing the network depth can cause the training accuracy to saturate and then degrade.

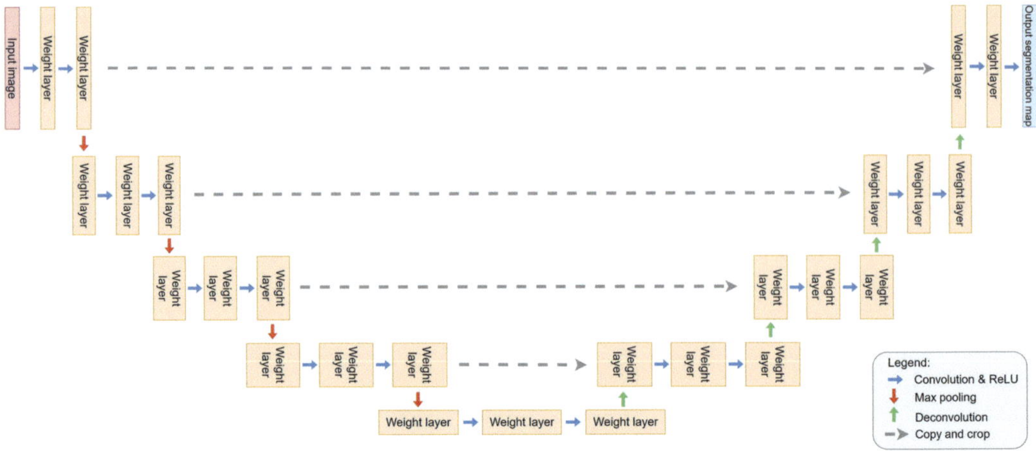

Fig. 16 The architecture of the U-Net neural network

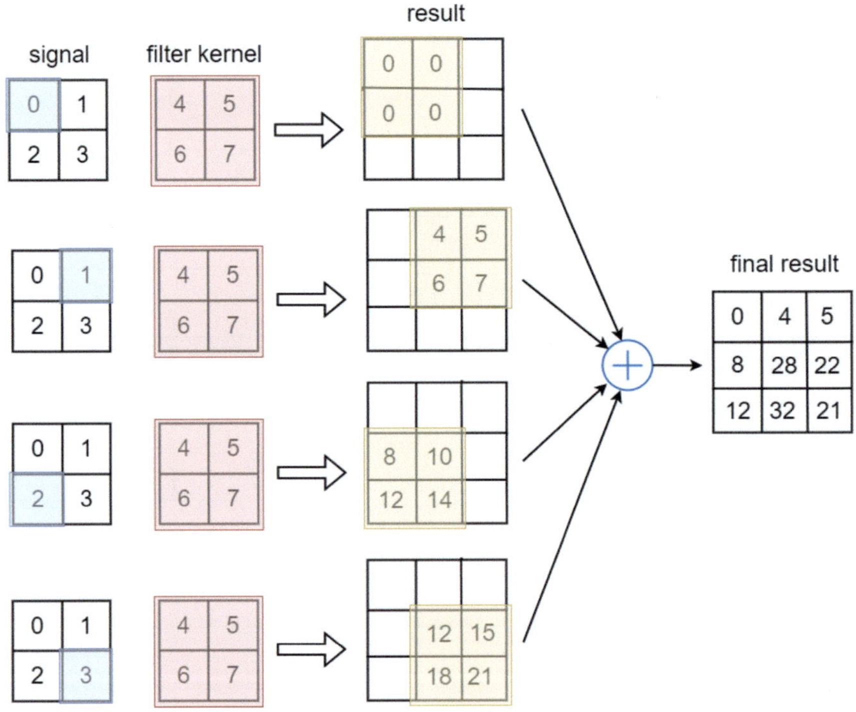

Fig. 17 An example of deconvolution, also called transpose convolution or up-convolution

This phenomenon, known as the *degradation problem*, is counterintuitive. Theoretically, a deeper network should perform at least as well as a shallower one, since the added layers could simply learn identity mappings, thereby preserving the performance of the shallower model. However, in practice, the addition of layers can increase the training error.

To address this problem, the Residual Network (ResNet) was introduced in [11], proposing a strategy known as *deep residual learning*. In standard neural networks, it is expected that stacked layers learn a direct mapping to the target function. In contrast, ResNet enables each group of layers to learn a *residual function* with respect to the input. Let $\mathcal{H}(x)$ denote the desired underlying mapping for input x. Instead of learning $\mathcal{H}(x)$ directly, ResNet learns

$$\mathcal{F}(x) := \mathcal{H}(x) - x, \tag{9}$$

which implies

$$\mathcal{H}(x) = \mathcal{F}(x) + x. \tag{10}$$

This computation is implemented through a structure called a *residual block*, shown in Fig. 18. A complete ResNet architecture consists of multiple stacked residual blocks.

Residual learning provides multiple benefits. One key advantage is improved gradient flow during backpropagation, reducing the risk of gradient vanishing. The shortcut connection (identity mapping) allows gradients to propagate more directly from deeper to shallower layers. Additionally, it has been empirically validated that optimizing the residual mapping $\mathcal{F}(x)$ is easier than optimizing the original mapping $\mathcal{H}(x)$, which lacks a reference.

Today, ResNet remains one of the most widely used CNN architectures. While earlier research focused heavily on designing new network architectures, contemporary research often assumes standard backbones (e.g., ResNet or standard CNNs like LeNet) and focuses on developing new loss functions.

There are several well-known ResNet variants, such as ResNet-18, ResNet-34, ResNet-50, and ResNet-101, where the numbers refer to the total number of layers (residual blocks) in the network. The choice among these variants depends on available computational resources, memory, and the complexity of the application and dataset. For most laptops and standard personal computers, ResNet-18 is typically sufficient.

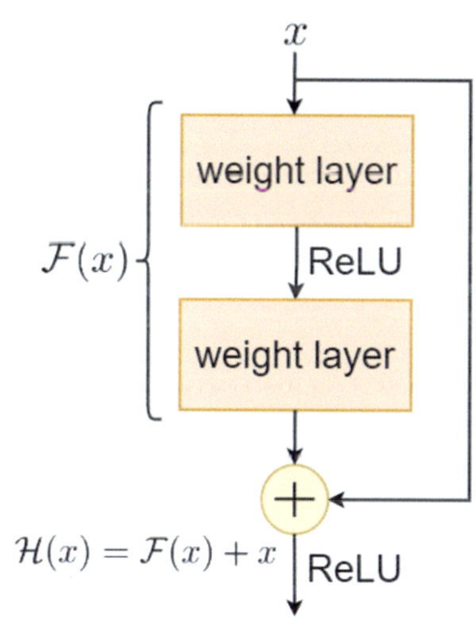

Fig. 18 A residual block used in ResNet

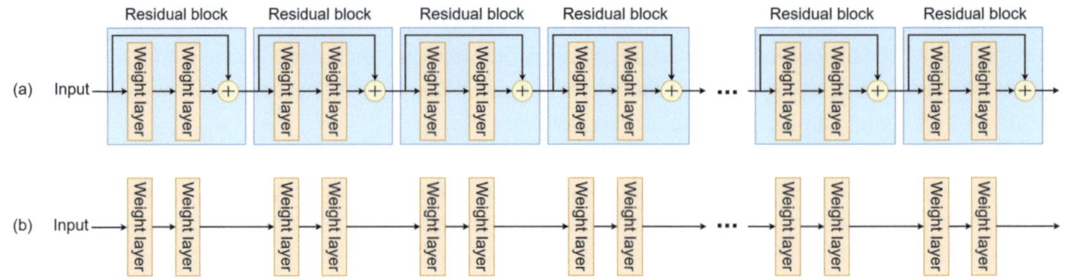

Fig. 19 Comparison of the architectures of **a** ResNet and **b** plain CNN or VGG neural networks

Figure 19 compares the architecture of ResNet with that of a plain CNN and the VGG network.

5.7 DenseNet

Empirical studies have shown that convolutional neural networks can benefit significantly from introducing shorter connections between earlier and later layers. In particular, such connections can enable deeper networks that are both more accurate and easier to train. The Dense Convolutional Network (DenseNet) [14] embodies this principle by incorporating *dense blocks*, within which each layer is connected to every other layer in a feedforward fashion, as illustrated in Fig. 20.

DenseNet may be viewed as an extreme extension of the shortcut connections introduced in ResNet. Whereas ResNet connects each layer to its immediate successor using residual links, DenseNet connects each layer to all subsequent layers in a dense manner. In a standard network with l layers, there are l weight matrices, each corresponding to a connection between adjacent layers. In contrast, a DenseNet with l layers includes $l(l + 1)/2$ weight matrices, as every layer is directly connected to all its successors.

Fig. 20 The architecture of a dense block used in DenseNet

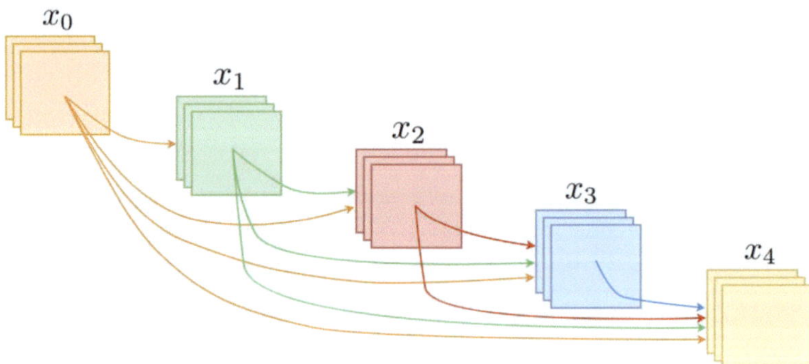

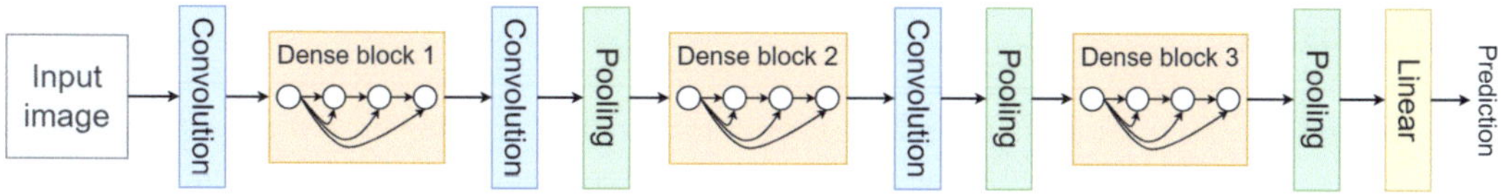

Fig. 21 The architecture of a DenseNet containing dense blocks

DenseNet offers several advantages:

- It mitigates the gradient vanishing problem by facilitating improved gradient flow during backpropagation, as gradients can directly propagate through the multiple skip connections.
- It enhances feature propagation and encourages feature reuse.
- It reduces the number of learnable parameters by requiring fewer layers, while still maintaining expressive capacity through the dense connections.

Let $\mathcal{F}_\ell(x)$ denote the composite function at layer ℓ, consisting of batch normalization [15], convolution, ReLU activation [26], and possibly pooling. While ResNet adopts the form:

$$x_\ell = \mathcal{F}_\ell(x_{\ell-1}) + x_{\ell-1}, \tag{11}$$

DenseNet modifies this by concatenating the outputs of all preceding layers:

$$x_\ell = \mathcal{F}_\ell\big([x_0, x_1, \ldots, x_{\ell-1}]^\top\big), \tag{12}$$

where $[x_0, x_1, \ldots, x_{\ell-1}]$ denotes the concatenation of the outputs from all earlier layers.

As depicted in Fig. 21, a DenseNet consists of multiple dense blocks, each containing layers that are fully connected to all subsequent layers within the block. Between dense blocks, *transition layers* are used to reduce dimensionality and control model complexity. Each transition layer performs batch normalization, followed by a 1×1 convolution and 2×2 average pooling. The final transition layer includes average pooling and a fully connected output layer with softmax activation for classification.

Within each dense block, the function $\mathcal{F}_\ell(x)$ specifically comprises batch normalization, a 3×3 convolution, and ReLU activation. The number of output feature maps from each convolution layer is referred to as the *growth rate* k. If k_0 denotes the number of channels in the input to the dense block, then the ℓth layer will have $k_0 + k(\ell - 1)$ input feature maps. Empirical results show that even a relatively small growth rate (e.g., $k = 12$) can be effective in DenseNet [14].

5.8 Other Convolutional Networks: EfficientNet and ConvNeXt

After DenseNet [14], architectures such as EfficientNet (2019) [33] introduced compound scaling of depth, width, and resolution, achieving state-of-the-art accuracy with fewer parameters. More recently, ConvNeXt (2022) [24] modernized CNN design to compete with Transformers [35] (see Chap. 9), showing that CNNs remain competitive in the era of attention-based models.

6 Conclusion

Convolutional Neural Networks (CNNs) have been central to deep learning breakthroughs in computer vision. Starting from LeNet and AlexNet, successive architectures like VGG, Inception, ResNet, and DenseNet progressively improved depth, efficiency, and feature reuse. While attention-based models such as Transformers are now dominant in many areas, CNNs remain highly relevant in computer vision, medical imaging, and edge applications. They continue to inspire new architectures and remain an indispensable foundation for understanding modern deep learning.

7 Exercises

1. (Level: Moderate) Consider a simple convolutional neural network (CNN) applied to a grayscale image I of size 6×6 pixels. The CNN consists of a single convolutional layer followed by a max pooling layer, configured as follows:

   - Convolutional Layer:
     - Employs a single filter (kernel) K of size 3×3.
     - The filter weights are defined by the matrix:

$$K = \begin{bmatrix} 1 & 0 & -1 \\ 1 & 0 & -1 \\ 1 & 0 & -1 \end{bmatrix}$$

     - Utilizes valid padding (no padding).
     - The stride is set to 1.
     - No bias is added.
     - The activation function is the identity function (no activation).

   - Pooling Layer:
     - Applies max pooling.
     - The pooling size is 2×2.
     - The stride is set to 2.

   Answer the following questions:

   (a) Calculate the output dimensions of the feature map after the convolutional layer.
   (b) Explain how the convolution operation is performed. Then, compute the output feature map given the input image:

$$I = \begin{bmatrix} 3 & 0 & 1 & 2 & 7 & 4 \\ 1 & 5 & 8 & 9 & 3 & 1 \\ 2 & 7 & 2 & 5 & 1 & 3 \\ 0 & 1 & 3 & 1 & 7 & 8 \\ 4 & 2 & 1 & 6 & 2 & 8 \\ 2 & 4 & 5 & 2 & 3 & 9 \end{bmatrix}.$$

(c) Calculate the output dimensions of the feature map after the pooling layer.
(d) Perform the max pooling operation on the feature map obtained from the convolutional layer.

2. (Level: Easy) Consider the following image pixels:

$$I = \begin{bmatrix} 2 & 1 & -1 & 3 \\ 4 & 3 & 2 & 1 \\ 1 & -1 & 0.5 & 2 \\ -0.5 & 1 & 2 & 1 \end{bmatrix}.$$

Consider the following filter kernel:

$$\frac{1}{3} \times \begin{bmatrix} 0 & 0.5 & 0 \\ 0.5 & 1 & 0.5 \\ 0 & 0.5 & 0 \end{bmatrix}.$$

Do the following steps:

- Zero-pad the image with one pixel.
- Convolve the filter with the image with stride 1. Do the convolution without flipping.
- Apply max pooling with size 2×2 and stride 1.
- Apply the sigmoid activation function:

$$\sigma(a) = \frac{1}{1 + e^{-a}}.$$

3. (Level: Easy) Sobel filter is a popular edge-detection filter. Sobel filter can be seen as combining a 1D "derivative filter," $[1, 0, -1]$, in one direction and a 1D "blur" filter, $[1, 2, 1]^\top$, in the other direction.

(a) Horizontal Sobel Filter: it is often stated as follows:

$$\begin{bmatrix} 1 & 0 & -1 \\ 2 & 0 & -2 \\ 1 & 0 & -1 \end{bmatrix}.$$

See how this comes from multiplying $\begin{bmatrix} 1 \\ 2 \\ 1 \end{bmatrix}$ by $\begin{bmatrix} 1 & 0 & -1 \end{bmatrix}$.

(b) Vertical Sobel Filter: Construct the 3×3 matrix for the vertical Sobel filter by swapping the roles of "blur" and "derivative," and write it out explicitly.

(c) Application: Given the 3×3 image patch $\begin{bmatrix} 2 & 2 & 2 \\ 2 & 5 & 7 \\ 1 & 6 & 9 \end{bmatrix}$, apply your vertical Sobel filter to the center pixel (i.e., do a 3×3 convolution) and calculate the filter's response.
Hint: For the vertical Sobel, think of smoothing left-to-right, then taking the derivative top-to-bottom.

4. (Level: Easy) Consider a 3×3 kernel $K = \begin{bmatrix} 1 & 2 & 1 \\ 0 & 0 & 0 \\ -1 & -2 & -1 \end{bmatrix}$
   (a) What is the effect of this convolution kernel?
   (b) What is the effect of the transpose kernel $K^\top$?
5. (Level: Moderate) This problem shows that padding and stride affect the behavior of a convolutional layer.
   (a) Suppose a CNN applies a 3×3 filter to an input of size 32×32 with zero padding P and stride S. Compute the minimum and maximum values of P and S such that the output feature map remains at least 16×16.
   (b) Show how "same" padding and "valid" padding affect the output size differently.
   (c) Given the following scenarios, compute the output size:
   - 3×3 filter, $P = 1$, $S = 1$ on 64×64 input.
   - 5×5 filter, $P = 2$, $S = 2$ on 64×64 input.
6. (Level: Moderate) Consider a convolutional neural network (CNN) used for image classification. The network consists of a convolutional layer followed by a max pooling layer. The convolutional layer uses a 3×3 kernel with a stride of 1, and the max pooling layer uses a 2×2 kernel with a stride of 2. Given an input image of size 32×32 with 3 color channels (RGB), calculate the dimensions of the output feature map after passing through both the convolutional and max pooling layers.
7. (Level: Moderate) Let $f : \mathbb{R}^{H \times W \times C} \to \mathbb{R}^{H' \times W' \times C'}$ represent a convolutional layer in a CNN with kernel size $k \times k$, stride s, and zero padding p. Assume the layer has C' output channels and uses a standard linear convolution operation without depthwise separability or dilation. Then, increasing k while keeping s, p, and C' fixed always results in a strictly larger receptive field and an increase in the number of learnable parameters. Is this true?
8. (Level: Difficult) In convolutional neural networks, convolution operations exhibit equivalence to input transformations, while pooling operations enhance invariance to such transformations.
   (a) Prove that the convolution operation is equivalent to translation. That is, given an input $x(i)$ and a convolutional kernel $w(i)$, if the input is translated by k, the output feature map is also translated accordingly:

$$(x * w)(i) = \sum_j x(j)w(i - j).$$

   Show that:

$$(T_k x * w)(i) = T_k(x * w)(i),$$

   where T_k denotes a translation by k units.
   (b) Explain how pooling operations disrupt equivalence and enhance model invariance.
9. (Level: Moderate) Consider training a deep neural network where the gradients of the loss with respect to parameters sometimes explode to very large values. To mitigate this, gradient clipping is

applied with a threshold $\tau = 5$. Let $g = (g_1, g_2)$ be a gradient vector with norm $|g| = \sqrt{g_1^2 + g_2^2}$. The clipped gradient is defined as follows:

$$g_{\text{clipped}} = g \cdot \min\left(1, \frac{\tau}{\|g\|}\right).$$

Suppose at an iteration, we have $g_1 = 6$ and $g_2 = 8$.

(a) Compute the clipped gradient g_{clipped}.
(b) Explain why gradient clipping is necessary in deep networks and its effect on training dynamics.
(c) Assume we are training a deep neural network with gradient clipping and a learning rate of 0.01. The norm of the unmodified gradient at time step t is $|g_t| = 20$, and the updated gradient after clipping is $|g_{\text{clipped},t}| = 5$. Compute the weight update for a parameter w given that the gradient component in that direction is $g_t^w = 10$. If gradient clipping was not applied, how would the update change, and what potential issue could occur?
(d) Consider the same scenario, but now the threshold for gradient clipping is increased from $\tau = 5$ to $\tau = 10$. Recompute the clipped gradient g_{clipped} using this new threshold and compare it to the previously computed result. How does increasing the threshold affect the gradient update?

References

1. Yoshua Bengio, Patrice Simard, and Paolo Frasconi. "Learning long-term dependencies with gradient descent is difficult". In: *IEEE Transactions on Neural Networks* 5.2 (1994), pp. 157–166.
2. Y-Lan Boureau, Nicolas Le Roux, Francis Bach, Jean Ponce, and Yann LeCun. "Ask the locals: multi-way local pooling for image recognition". In: *2011 International Conference on Computer Vision*. IEEE. 2011, pp. 2651–2658.
3. Kevin Crowston. "Amazon mechanical turk: A research tool for organizations and information systems scholars". In: *Shaping the Future of ICT Research. Methods and Approaches: IFIP WG 8.2, Working Conference, Tampa, FL, USA, December 13-14, 2012. Proceedings*. Springer. 2012, pp. 210–221.
4. Pedro Domingos. "The role of Occam's razor in knowledge discovery". In: *Data mining and knowledge discovery* 3.4 (1999), pp. 409–425.
5. Mark Everingham, Luc Van Gool, Christopher KI Williams, John Winn, and Andrew Zisserman. "The Pascal visual object classes (VOC) challenge". In: *International journal of computer vision* 88 (2010), pp. 303–338.
6. Jerome Friedman, Trevor Hastie, and Robert Tibshirani. *The elements of statistical learning: Data Mining, Inference, and Prediction*. Vol. 2. Springer series in statistics, New York, NY, USA, 2009.
7. Kunihiko Fukushima. "Neocognitron: A self-organizing neural network model for a mechanism of pattern recognition unaffected by shift in position". In: *Biological cybernetics* 36.4 (1980), pp. 193–202.
8. Benyamin Ghojogh, Mark Crowley, Fakhri Karray, and Ali Ghodsi. "Principal component analysis". In: *Elements of Dimensionality Reduction and Manifold Learning*. Springer, 2023, pp. 123–154.
9. Benyamin Ghojogh, Mark Crowley, Fakhri Karray, and Ali Ghodsi. "Random Projection". In: *Elements of Dimensionality Reduction and Manifold Learning*. Springer, 2023, pp. 395–425.
10. Xavier Glorot and Yoshua Bengio. "Understanding the difficulty of training deep feedforward neural networks". In: *Proceedings of the Thirteenth International Conference on Artificial Intelligence and Statistics (AISTATS)*. JMLR Workshop and Conference Proceedings. 2010, pp. 249–256.
11. Kaiming He, Xiangyu Zhang, Shaoqing Ren, and Jian Sun. "Deep residual learning for image recognition". In: *Proceedings of the IEEE conference on computer vision and pattern recognition*. 2016, pp. 770–778.
12. Kaiming He, Xiangyu Zhang, Shaoqing Ren, and Jian Sun. "Delving deep into rectifiers: Surpassing human-level performance on ImageNet classification". In: *Proceedings of the IEEE International Conference on Computer Vision (ICCV)*. 2015, pp. 1026–1034.
13. Kaiming He, Xiangyu Zhang, Shaoqing Ren, and Jian Sun. "Spatial pyramid pooling in deep convolutional networks for visual recognition". In: *IEEE transactions on pattern analysis and machine intelligence* 37.9 (2015), pp. 1904–1916.

14. Gao Huang, Zhuang Liu, Laurens Van Der Maaten, and Kilian Q Weinberger. "Densely connected convolutional networks". In: *Proceedings of the IEEE conference on computer vision and pattern recognition*. 2017, pp. 4700–4708.
15. Sergey Ioffe and Christian Szegedy. "Batch normalization: Accelerating deep network training by reducing internal covariate shift". In: *International conference on machine learning*. pmlr. 2015, pp. 448–456.
16. Yangqing Jia, Chang Huang, and Trevor Darrell. "Beyond spatial pyramids: Receptive field learning for pooled image features". In: *2012 IEEE Conference on Computer Vision and Pattern Recognition*. IEEE. 2012, pp. 3370–3377.
17. Yangqing Jia et al. "Caffe: Convolutional architecture for fast feature embedding". In: *Proceedings of the 22nd ACM international conference on Multimedia*. 2014, pp. 675–678.
18. Alex Krizhevsky, Ilya Sutskever, and Geoffrey E Hinton. "ImageNet classification with deep convolutional neural networks". In: *Communications of the ACM* 60.6 (2017), pp. 84–90.
19. Brian Kulis, Mátyás Sustik, and Inderjit Dhillon. "Learning low-rank kernel matrices". In: *Proceedings of the 23rd international conference on Machine learning*. 2006, pp. 505–512.
20. Yann LeCun, Léon Bottou, Yoshua Bengio, and Patrick Haffner. "Gradient-based learning applied to document recognition". In: *Proceedings of the IEEE* 86.11 (1998), pp. 2278–2324.
21. Yann LeCun, Léon Bottou, Genevieve B Orr, and Klaus-Robert Müller. "Efficient backprop". In: *Neural Networks: Tricks of the Trade* (1998), pp. 9–50.
22. Yann LeCun et al. "Backpropagation applied to handwritten zip code recognition". In: *Neural computation* 1.4 (1989), pp. 541–551.
23. Min Lin, Qiang Chen, and Shuicheng Yan. "Network in network". In: *International Conference on Learning Representations (ICLR)*. 2014.
24. Zhuang Liu et al. "A ConvNet for the 2020s". In: *Proceedings of the IEEE/CVF Conference on Computer Vision and Pattern Recognition (CVPR)*. 2022.
25. Sachin Mehta et al. "Y-Net: joint segmentation and classification for diagnosis of breast biopsy images". In: *International conference on medical image computing and computer-assisted intervention*. Springer. 2018, pp. 893–901.
26. Vinod Nair and Geoffrey E Hinton. "Rectified linear units improve restricted boltzmann machines". In: *Proceedings of the 27th international conference on machine learning (ICML-10)*. 2010, pp. 807–814.
27. Judith MS Prewitt et al. "Object enhancement and extraction". In: *Picture processing and Psychopictorics* 10.1 (1970), pp. 15–19.
28. Olaf Ronneberger, Philipp Fischer, and Thomas Brox. "U-net: Convolutional networks for biomedical image segmentation". In: *Medical Image Computing and Computer-Assisted Intervention–MICCAI 2015: 18th International Conference, Munich, Germany, October 5-9, 2015, Proceedings, Part III 18*. Springer. 2015, pp. 234–241.
29. David E Rumelhart, Geoffrey E Hinton, and Ronald J Williams. "Learning representations by back-propagating errors". In: *nature* 323.6088 (1986), pp. 533–536.
30. Olga Russakovsky et al. "ImageNet large scale visual recognition challenge". In: *International journal of computer vision* 115 (2015), pp. 211–252.
31. Karen Simonyan and Andrew Zisserman. "Very deep convolutional networks for large-scale image recognition". In: *arXiv preprint* arXiv:1409.1556 (2014).
32. Christian Szegedy et al. "Going deeper with convolutions". In: *Proceedings of the IEEE conference on computer vision and pattern recognition*. 2015, pp. 1–9.
33. Mingxing Tan and Quoc V. Le. "EfficientNet: Rethinking Model Scaling for Convolutional Neural Networks". In: *Proceedings of the 36th International Conference on Machine Learning (ICML)*. 2019.
34. Robert Tibshirani, Martin Wainwright, and Trevor Hastie. *Statistical learning with sparsity: the lasso and generalizations*. Chapman and Hall/CRC, 2015.
35. Ashish Vaswani et al. "Attention is all you need". In: *Advances in neural information processing systems*. 2017, pp. 5998–6008.

Regularization and Overfitting

5

Abstract

This chapter presents foundational techniques for improving model generalization in supervised learning by addressing challenges such as overfitting and underfitting. It begins with a formal introduction to datasets, including training, test, and validation splits, and their respective roles in model evaluation. The chapter then explores key concepts such as Mean Squared Error (MSE), bias, and variance, which are essential for understanding model behavior. Core strategies for mitigating overfitting—such as cross-validation and regularization using ℓ_1 and ℓ_2 norms—are introduced, followed by advanced regularization methods commonly employed in deep learning, including weight decay, noise injection, early stopping, dropout, and batch normalization. The chapter also discusses ensemble methods like bagging, offering theoretical insights into why these approaches improve generalization performance.

1 Introduction

Assume there is a dataset of instances $\{(\boldsymbol{x}_i, y_i)\}_{i=1}^n$ with sample size n, where each input $\boldsymbol{x}_i \in \mathbb{R}^d$ and each corresponding observation (label) $y_i \in \mathbb{R}$. The set $\{\boldsymbol{x}_i\}_{i=1}^n$ represents the input data, and $\{y_i\}_{i=1}^n$ represents the observed outputs. Let the dataset be denoted by $\mathcal{D}$, such that $n := |\mathcal{D}|$. This dataset is typically partitioned into two disjoint subsets: the training set $\mathcal{T}$ and the test set $\mathcal{R}$, i.e.,

$$\mathcal{D} = \mathcal{T} \cup \mathcal{R}, \tag{1}$$

$$\mathcal{T} \cap \mathcal{R} = \varnothing. \tag{2}$$

In the training set, the labels y_i are available and used for training the model. For the test set, although the labels may also be available, they are not utilized during training. In classification tasks, the observations belong to a finite discrete set, while in regression tasks, they are continuous.

Let the sizes of the training and test sets be $n_{tr} := |\mathcal{T}|$ and $n_{te} := n - n_{tr}$, respectively. Thus, the training set can be denoted by $\{(\boldsymbol{x}_i, y_i)\}_{i=1}^{n_{tr}}$. In some scenarios, a validation set $\mathcal{V}$ is also included, resulting in three disjoint subsets:

B. Ghojogh and A. Ghodsi, *Elements of Deep Learning*,
https://doi.org/10.1007/978-3-032-10738-1_5

$$\mathcal{D} = \mathcal{T} \cup \mathcal{R} \cup \mathcal{V}, \tag{3}$$

$$\mathcal{T} \cap \mathcal{R} = \varnothing, \quad \mathcal{T} \cap \mathcal{V} = \varnothing, \quad \mathcal{V} \cap \mathcal{R} = \varnothing. \tag{4}$$

The purposes of the training, test, and validation sets will be discussed in more detail later in this chapter.

This chapter introduces overfitting, cross-validation, and regularization from a theoretical perspective. The concepts of mean squared error, variance, and bias are reviewed in Sect. 2. The mean squared error in the estimation of observations is addressed in Sect. 3. Overfitting, underfitting, and generalization are discussed in Sect. 4. Section 5 introduces cross-validation techniques. Regularization using ℓ_1 and ℓ_2 norms is covered in Sect. 6. Finally, advanced regularization techniques used in deep learning—namely weight decay, input noise injection, early stopping, dropout, and batch normalization—are presented in Sects. 7, 8, 9, 10, and 11, respectively.

2 Mean Squared Error, Variance, and Bias

2.1 Measures for a Random Variable

Assume there is a variable X to be estimated. Let the random variable $\widehat{X}$ denote the estimate of X. The *variance* of estimating this random variable is defined as

$$\mathbb{V}\mathrm{ar}(\widehat{X}) := \mathbb{E}\big((\widehat{X} - \mathbb{E}(\widehat{X}))^2\big), \tag{5}$$

which represents the average squared deviation of $\widehat{X}$ from its expected value, $\mathbb{E}(\widehat{X})$. The squaring ensures symmetry in the measure of deviation.

This variance can also be expressed as

$$\begin{aligned}\mathbb{V}\mathrm{ar}(\widehat{X}) &= \mathbb{E}\big(\widehat{X}^2 + (\mathbb{E}(\widehat{X}))^2 - 2\widehat{X}\mathbb{E}(\widehat{X})\big) \\ &\overset{(a)}{=} \mathbb{E}(\widehat{X}^2) + (\mathbb{E}(\widehat{X}))^2 - 2\mathbb{E}(\widehat{X})\mathbb{E}(\widehat{X}) \\ &= \mathbb{E}(\widehat{X}^2) - (\mathbb{E}(\widehat{X}))^2, \end{aligned} \tag{6}$$

where (a) follows from the linearity of expectation and the fact that $\mathbb{E}(\widehat{X})$ is not random.

Estimates may also exhibit bias. The *bias* of an estimate is defined as

$$\mathbb{B}\mathrm{ias}(\widehat{X}) := \mathbb{E}(\widehat{X}) - X, \tag{7}$$

which quantifies the deviation of the expected estimate from the true value X.

The *Mean Squared Error (MSE)* of the estimate $\widehat{X}$ is given by:

$$\mathrm{MSE}(\widehat{X}) := \mathbb{E}\big((\widehat{X} - X)^2\big), \tag{8}$$

which measures the expected squared deviation between the estimate and the actual value.

The intuition behind bias, variance, and MSE is illustrated in Fig. 1, where estimation is visualized as a dart game. The figure presents four cases representing different combinations of high/low bias and high/low variance.

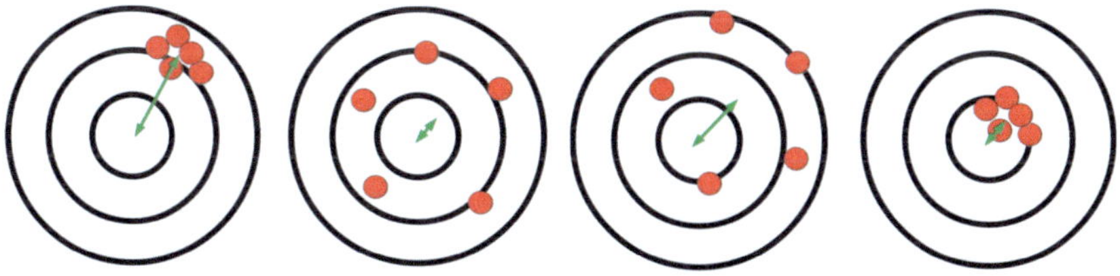

Fig. 1 The dart analogy for **a** high bias and low variance, **b** low bias and high variance, **c** high bias and high variance, and **d** low bias and low variance. The worst and best cases are (**c**) and (**d**), respectively. The center of the circles represents the true value of the variable

The relationship between MSE, variance, and bias is as follows:

$$\begin{aligned}
\text{MSE}(\widehat{X}) &= \mathbb{E}\big((\widehat{X} - X)^2\big) \\
&= \mathbb{E}\big((\widehat{X} - \mathbb{E}(\widehat{X}) + \mathbb{E}(\widehat{X}) - X)^2\big) \\
&= \mathbb{E}\big((\widehat{X} - \mathbb{E}(\widehat{X}))^2 + (\mathbb{E}(\widehat{X}) - X)^2 + 2(\widehat{X} - \mathbb{E}(\widehat{X}))(\mathbb{E}(\widehat{X}) - X)\big) \\
&\overset{(a)}{=} \mathbb{E}\big((\widehat{X} - \mathbb{E}(\widehat{X}))^2\big) + (\mathbb{E}(\widehat{X}) - X)^2 + 2\,\underbrace{(\mathbb{E}(\widehat{X}) - \mathbb{E}(\widehat{X}))}_{0}(\mathbb{E}(\widehat{X}) - X) \\
&\overset{(b)}{=} \mathbb{V}\text{ar}(\widehat{X}) + (\mathbb{B}\text{ias}(\widehat{X}))^2,
\end{aligned} \tag{9}$$

where (a) is due to the linearity of expectation and the constancy of X and $\mathbb{E}(\widehat{X})$, and (b) uses Eqs. (5) and (7).

For two random variables $\widehat{X}$ and $\widehat{Y}$, the variance of their linear combination is

$$\begin{aligned}
\mathbb{V}\text{ar}(a\widehat{X} + b\widehat{Y}) &\overset{(6)}{=} \mathbb{E}\big((a\widehat{X} + b\widehat{Y})^2\big) - \big(\mathbb{E}(a\widehat{X} + b\widehat{Y})\big)^2 \\
&\overset{(a)}{=} a^2\,\mathbb{E}(\widehat{X}^2) + b^2\,\mathbb{E}(\widehat{Y}^2) + 2ab\,\mathbb{E}(\widehat{X}\widehat{Y}) \\
&\quad - a^2\,(\mathbb{E}(\widehat{X}))^2 - b^2\,(\mathbb{E}(\widehat{Y}))^2 - 2ab\,\mathbb{E}(\widehat{Y})\mathbb{E}(\widehat{Y}) \\
&\overset{(6)}{=} a^2\,\mathbb{V}\text{ar}(\widehat{X}) + b^2\,\mathbb{V}\text{ar}(\widehat{Y}) + 2ab\,\mathbb{C}\text{ov}(\widehat{X}, \widehat{Y}),
\end{aligned} \tag{10}$$

where (a) follows from the linearity of expectation. The *covariance* is defined as:

$$\mathbb{C}\text{ov}(\widehat{X}, \widehat{Y}) := \mathbb{E}(\widehat{X}\widehat{Y}) - \mathbb{E}(\widehat{X})\,\mathbb{E}(\widehat{Y}). \tag{11}$$

If the two random variables are independent, i.e., $X \perp\!\!\!\perp Y$, then:

$$\begin{aligned}
\mathbb{E}(\widehat{X}\widehat{Y}) &\overset{(a)}{=} \iint \widehat{x}\widehat{y} f(\widehat{x}, \widehat{y}) d\widehat{x} d\widehat{y} \overset{\perp\!\!\!\perp}{=} \iint \widehat{x}\widehat{y} f(\widehat{x}) f(\widehat{y}) d\widehat{x} d\widehat{y} \\
&= \int \widehat{y} f(\widehat{y}) \underbrace{\int \widehat{x} f(\widehat{x}) d\widehat{x}}_{\mathbb{E}(\widehat{X})} d\widehat{y} = \mathbb{E}(\widehat{X}) \underbrace{\int \widehat{y} f(\widehat{y}) d\widehat{y}}_{\mathbb{E}(\widehat{Y})} \\
&= \mathbb{E}(\widehat{X})\,\mathbb{E}(\widehat{Y}) \implies \mathbb{C}\text{ov}(\widehat{X}, \widehat{Y}) = 0,
\end{aligned} \tag{12}$$

where (a) follows from the definition of expectation. Note, however, that the converse of Eq. (12) is not generally true and can be disproved by counterexample.

Equations (10) and (11) extend naturally to multiple random variables:

$$\mathbb{V}\text{ar}\Big(\sum_{i=1}^{k} a_i X_i\Big) = \sum_{i=1}^{k} a_i^2\, \mathbb{V}\text{ar}(X_i) + \sum_{i=1}^{k} \sum_{j=1, j\neq i}^{k} a_i a_j \mathbb{C}\text{ov}(X_i, X_j), \tag{13}$$

$$\mathbb{C}\text{ov}\Big(\sum_{i=1}^{k_1} a_i X_i, \sum_{j=1}^{k_2} b_j Y_j\Big) = \sum_{i=1}^{k_1}\sum_{j=1}^{k_2} a_i\, b_j\, \mathbb{C}\text{ov}(X_i, Y_j), \tag{14}$$

where a_i and b_j are constant (non-random) coefficients.

2.2 Measures for a Model

Assume there is a *function* f that receives the ith input $\boldsymbol{x}_i$ and produces an output $f_i = f(\boldsymbol{x}_i)$. Figure 2 illustrates this function along with its inputs and outputs.

The goal is to learn this function, referred to as the *true model*, even though it is inaccessible in practice. Furthermore, the true outputs f_i are not directly observable, as the actual observations are corrupted by additive noise:

$$y_i = f_i + \varepsilon_i, \tag{15}$$

where the noise term follows a Gaussian distribution, $\varepsilon_i \sim \mathcal{N}(0, \sigma^2)$. Therefore,

$$\mathbb{E}(\varepsilon_i) = 0, \quad \mathbb{E}(\varepsilon_i^2) \overset{(6)}{=} \mathbb{V}\text{ar}(\varepsilon_i) + (\mathbb{E}(\varepsilon_i))^2 = \sigma^2. \tag{16}$$

Since f_i is deterministic (i.e., not random), its expectation is

$$\mathbb{E}(f_i) = f_i. \tag{17}$$

Given the input training data $\{\boldsymbol{x}_i\}_{i=1}^{n_{tr}}$ and their corresponding noisy observations $\{y_i\}_{i=1}^{n_{tr}}$, the task is to approximate the true model f using a learned model $\widehat{f}$. The resulting estimates of the observations, denoted $\{\widehat{y}_i\}_{i=1}^{n_{tr}}$, should be as close as possible to the observed values $\{y_i\}_{i=1}^{n_{tr}}$ for the given training inputs. The model is trained using this data in an attempt to approximate the unknown true model. After training, the model can then be applied to both the training inputs $\{\boldsymbol{x}_i\}_{i=1}^{n_{tr}}$ and previously unseen test inputs $\{\boldsymbol{x}_i\}_{i=1}^{n_{te}}$, yielding estimated outputs $\{\widehat{y}_i\}_{i=1}^{n_{tr}}$ and $\{\widehat{y}_i\}_{i=1}^{n_{te}}$, respectively. This process is illustrated in Fig. 2.

In this text, the estimate of the observation for the ith instance is denoted by either $\widehat{y}_i$ or $\widehat{f}_i$. The model $\widehat{f}$ may represent either a *regression* (prediction) model, where outputs are continuous, or a *classification* model, where outputs belong to a discrete set of possible classes.

The definitions of variance, bias, and mean squared error (MSE) given in Eqs. (5), (7), and (8) also apply to the model estimate $\widehat{f}_i$ of the true value f_i. The relationship in Eq. (9) similarly holds and can be visualized for the model setting using Fig. 3, which illustrates this decomposition based on the Pythagorean theorem.

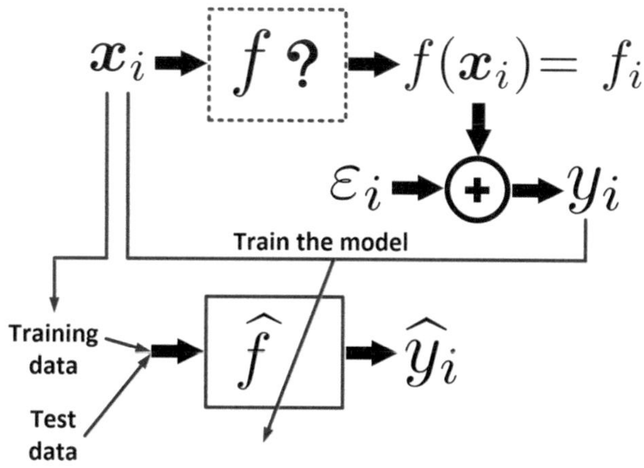

Fig. 2 The true model and the estimated model trained using the input training data and their observations. The observations are derived from the true model outputs corrupted by noise. Once trained, the estimated model can be used to predict outputs for both training and test input data

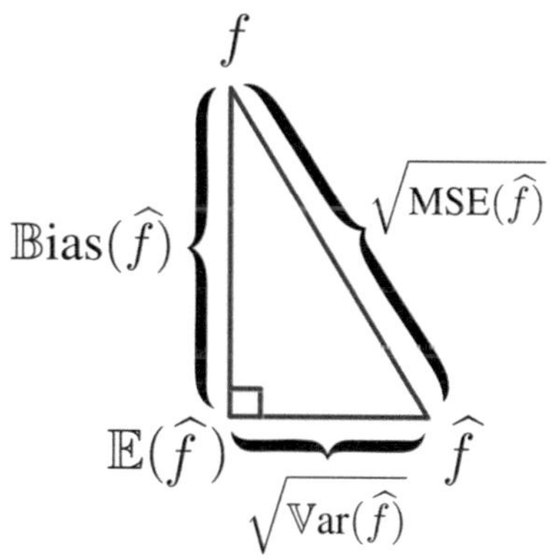

Fig. 3 The triangle of variance, bias, and MSE. Here, f and $\widehat{f}$ represent the true and estimated models, respectively

3 Mean Squared Error of the Estimation of Observations

Suppose there is an instance $(\boldsymbol{x}_0, y_0)$, which may belong to either the training set or the test/validation set. Both scenarios will be discussed in the following.

According to Eq. (15), the observation y_0 is expressed as:

$$y_0 = f_0 + \varepsilon_0. \tag{18}$$

Assume the model's estimate of y_0 is denoted by $\widehat{f}_0$. Then, using Eq. (8), the Mean Squared Error (MSE) of the estimation is

$$\begin{aligned}
\mathbb{E}\big((\widehat{f}_0 - y_0)^2\big) &\overset{(18)}{=} \mathbb{E}\big((\widehat{f}_0 - f_0 - \varepsilon_0)^2\big) \\
&= \mathbb{E}\big((\widehat{f}_0 - f_0)^2 + \varepsilon_0^2 - 2\,\varepsilon_0(\widehat{f}_0 - f_0)\big) \\
&= \mathbb{E}\big((\widehat{f}_0 - f_0)^2\big) + \mathbb{E}(\varepsilon_0^2) - 2\,\mathbb{E}\big(\varepsilon_0(\widehat{f}_0 - f_0)\big) \\
&\overset{(16)}{=} \mathbb{E}\big((\widehat{f}_0 - f_0)^2\big) + \sigma^2 - 2\,\mathbb{E}\big(\varepsilon_0(\widehat{f}_0 - f_0)\big).
\end{aligned} \tag{19}$$

The final term in Eq. (19) is

$$\mathbb{E}\big(\varepsilon_0(\widehat{f_0} - f_0)\big) \overset{(18)}{=} \mathbb{E}\big((y_0 - f_0)(\widehat{f_0} - f_0)\big). \tag{20}$$

To evaluate this term, two cases must be considered: (I) the instance $(\boldsymbol{x}_0, y_0)$ is *not* in the training set, or (II) the instance *is* in the training set. In other words, whether the instance was used to train the model (estimator) or not.

3.1 Case I: Instance Not In the Training Set

Assume the instance $(\boldsymbol{x}_0, y_0)$ was not included in the training set, i.e., it was not used to train the model. In other words, $y_0 \notin \mathcal{T}$. The instance may belong to either the validation set or the test set.

Because y_0 was not used during training, the estimate $\widehat{f_0}$ is independent of the observation y_0. Therefore:

$$\begin{aligned}
&\therefore \quad y_0 \perp\!\!\!\perp \widehat{f_0} \implies (y_0 - f_0) \perp\!\!\!\perp (\widehat{f_0} - f_0) \\
&\implies \mathbb{E}\big((y_0 - f_0)(\widehat{f_0} - f_0)\big) \overset{(a)}{=} \mathbb{E}(y_0 - f_0)\,\mathbb{E}(\widehat{f_0} - f_0) \overset{(b)}{=} 0 \times \mathbb{E}(\widehat{f_0} - f_0) = 0,
\end{aligned}$$

where (a) follows from the independence of $(y_0 - f_0)$ and $(\widehat{f_0} - f_0)$, and (b) follows from:

$$\mathbb{E}(y_0 - f_0) = \mathbb{E}(y_0) - \mathbb{E}(f_0) \overset{(c)}{=} f_0 - f_0 = 0,$$

where (c) uses Eq. (17) and:

$$\mathbb{E}(y_0) \overset{(18)}{=} \mathbb{E}(f_0) + \mathbb{E}(\varepsilon_0) = f_0 + 0 = f_0.$$

Therefore, the last term in Eq. (19) is zero, and the MSE simplifies to:

$$\mathbb{E}\big((\widehat{f_0} - y_0)^2\big) = \mathbb{E}\big((\widehat{f_0} - f_0)^2\big) + \sigma^2. \tag{21}$$

Now suppose the number of instances not in the training set is n_{te}. Using Monte Carlo approximation for the expectation terms [18], we have

$$\frac{1}{n_{te}}\sum_{i=1}^{n_{te}}(\widehat{f_i} - y_i)^2 = \frac{1}{n_{te}}\sum_{i=1}^{n_{te}}(\widehat{f_i} - f_i)^2 + \sigma^2 \implies \sum_{i=1}^{n_{te}}(\widehat{f_i} - y_i)^2 = \sum_{i=1}^{n_{te}}(\widehat{f_i} - f_i)^2 + n_{te}\sigma^2. \tag{22}$$

The term $\sum_{i=1}^{n_{te}}(\widehat{f_i} - y_i)^2$ is the squared error between the predicted outputs and the observed labels. This is referred to as the *empirical error*, denoted by **err**. The term $\sum_{i=1}^{n_{te}}(\widehat{f_i} - f_i)^2$ represents the squared error between the predictions and the true (unknown) outputs, referred to as the *true error*, denoted by **Err**. Thus:

$$\mathbf{err} = \mathbf{Err} + n_{te}\,\sigma^2 \implies \mathbf{Err} = \mathbf{err} - n_{te}\,\sigma^2. \tag{23}$$

Since $n_{te}\sigma^2$ is a constant, it can be ignored during optimization. Therefore, in this case, the empirical error is a reliable proxy for the true error, and minimizing the empirical error effectively minimizes the true error.

3.2 Case II: Instance In the Training Set

In this case, the instance is part of the training set. To analyze this scenario, we utilize a mathematical result known as Stein's Lemma, which underlies the formulation of *Stein's Unbiased Risk Estimate (SURE)*.

Consider a multivariate random variable $\boldsymbol{z} = [z_1, \ldots, z_d]^\top \in \mathbb{R}^d$ whose components are independent and normally distributed, i.e., $z_i \sim \mathcal{N}(\mu_i, \sigma)$. Define $\boldsymbol{\mu} = [\mu_1, \ldots, \mu_d]^\top \in \mathbb{R}^d$ and let $\boldsymbol{g}(z) = [g_1, \ldots, g_d]^\top$ be a function such that $\boldsymbol{g} : \mathbb{R}^d \to \mathbb{R}^d$. Stein's Lemma states:

$$\mathbb{E}\big((\boldsymbol{z} - \boldsymbol{\mu})^\top \boldsymbol{g}(\boldsymbol{z})\big) = \sigma^2 \sum_{i=1}^{d} \mathbb{E}\Big(\frac{\partial g_i}{\partial z_i}\Big), \tag{24}$$

which is used to derive SURE [33]. See Appendix 13 for the proof of Eq. (24).

In the univariate case, Stein's Lemma reduces to:

$$\mathbb{E}\big((z - \mu)\, g(z)\big) = \sigma^2\, \mathbb{E}\Big(\frac{\partial g(z)}{\partial z}\Big). \tag{25}$$

Now, consider setting $z = \varepsilon_0$, $\mu = 0$, and $g(z) = \widehat{f}_0 - f_0$ in the scalar form of Stein's Lemma. Then, using Eq. (25), the last term in Eq. (19) becomes

$$\begin{aligned}
\mathbb{E}\big((\varepsilon_0 - 0)(\widehat{f}_0 - f_0)\big) &= \sigma^2\, \mathbb{E}\Big(\frac{\partial(\widehat{f}_0 - f_0)}{\partial \varepsilon_0}\Big) \\
&= \sigma^2\, \mathbb{E}\Big(\frac{\partial \widehat{f}_0}{\partial \varepsilon_0} - \frac{\partial f_0}{\partial \varepsilon_0}\Big) \overset{(a)}{=} \sigma^2\, \mathbb{E}\Big(\frac{\partial \widehat{f}_0}{\partial \varepsilon_0}\Big) \\
&\overset{(b)}{=} \sigma^2\, \mathbb{E}\Big(\frac{\partial \widehat{f}_0}{\partial y_0} \cdot \frac{\partial y_0}{\partial \varepsilon_0}\Big) \overset{(c)}{=} \sigma^2\, \mathbb{E}\Big(\frac{\partial \widehat{f}_0}{\partial y_0}\Big),
\end{aligned}$$

where (a) follows from the fact that the true model f_0 is not a function of the noise ε_0, (b) applies the chain rule, and (c) uses the fact that:

$$y_0 \overset{(18)}{=} f_0 + \varepsilon_0 \quad \Rightarrow \quad \frac{\partial y_0}{\partial \varepsilon_0} = 1.$$

Substituting this into Eq. (19) yields

$$\mathbb{E}\big((\widehat{f}_0 - y_0)^2\big) = \mathbb{E}\big((\widehat{f}_0 - f_0)^2\big) + \sigma^2 - 2\sigma^2\, \mathbb{E}\Big(\frac{\partial \widehat{f}_0}{\partial y_0}\Big). \tag{26}$$

Now suppose the number of training instances is n_{tr}. Applying Monte Carlo approximation to the expectations [18], we obtain:

$$\frac{1}{n_{tr}}\sum_{i=1}^{n_{tr}}(\widehat{f_i}-y_i)^2 = \frac{1}{n_{tr}}\sum_{i=1}^{n_{tr}}(\widehat{f_i}-f_i)^2 + \sigma^2 - \frac{2\sigma^2}{n_{tr}}\sum_{i=1}^{n_{tr}}\frac{\partial \widehat{f_i}}{\partial y_i}$$

$$\implies \sum_{i=1}^{n_{tr}}(\widehat{f_i}-y_i)^2 = \sum_{i=1}^{n_{tr}}(\widehat{f_i}-f_i)^2 + n_{tr}\sigma^2 - 2\sigma^2\sum_{i=1}^{n_{tr}}\frac{\partial \widehat{f_i}}{\partial y_i}. \tag{27}$$

The term $\sum_{i=1}^{n_{tr}}(\widehat{f_i}-y_i)^2$ represents the squared error between the model predictions and the observed labels, and is referred to as the *empirical error*, denoted by **err**. The term $\sum_{i=1}^{n_{tr}}(\widehat{f_i}-f_i)^2$ measures the error relative to the true (but unknown) targets, and is known as the *true error*, denoted by **Err**. Therefore:

$$\mathbf{Err} = \mathbf{err} - n_{tr}\,\sigma^2 + 2\sigma^2\sum_{i=1}^{n_{tr}}\frac{\partial \widehat{f_i}}{\partial y_i}. \tag{28}$$

The final term in Eq. (28) serves as a measure of the model's *complexity* or degree of *overfitting*. Specifically, the derivative $\partial\widehat{f_i}/\partial y_i$ indicates how sensitive the model's estimate of the ith training instance is to perturbations in that same instance. Higher sensitivity suggests a more complex (and possibly overfitted) model.

To build intuition, consider a simple linear regression model trained via least squares. If a training point is perturbed, the fitted line does not change much—indicating a simple, underfitted model. In contrast, a highly flexible regression model that passes through all training points will change significantly if even one point is modified, which reflects a high degree of overfitting. See Fig. 4 for an illustration of this behavior.

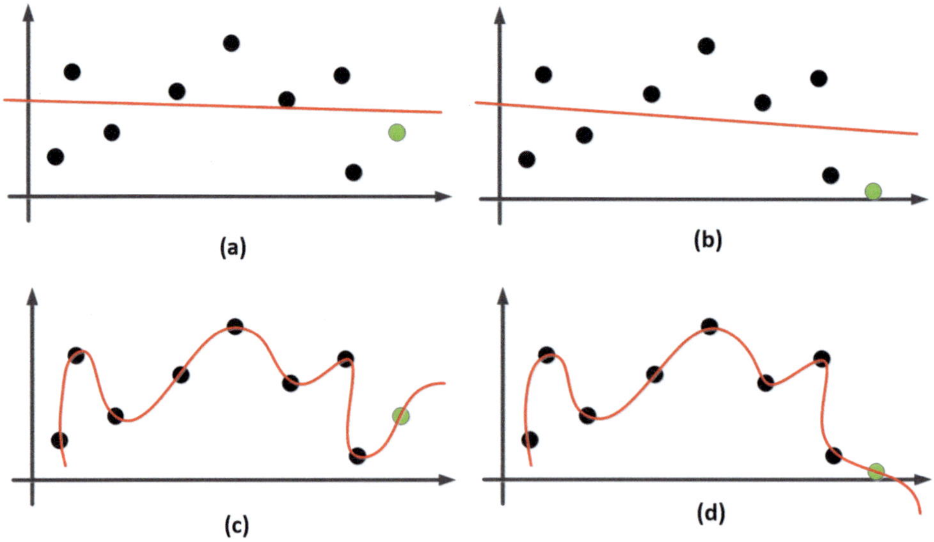

Fig. 4 An example of a simple and a complex model: **a** a simple model, **b** the modified simple model after moving a training instance (shown in green), **c** a complex model, and **d** the modified complex model after moving a training instance (shown in green). The complex model is highly sensitive to perturbations in the training data, while the simple model is more stable

According to Eq. (28), when the instance is part of the training set, the empirical error is not a reliable estimator of the true error. Minimizing **err** alone tends to increase model complexity, which can eventually offset or even reverse gains in true error reduction beyond a certain point in training.

3.3 Estimation of Variance of Noise in MSE of the Model

In both of the previously discussed cases, the MSE of the model involves σ, which denotes the standard deviation of the noise. An unbiased estimate of the noise variance can be computed as

$$\sigma^2 \approx \frac{1}{n_{tr} - 1} \sum_{i=1}^{n_{tr}} (y_i - \widehat{f}_i), \quad (29)$$

where the residuals between the training observations and their model-based estimates are used.

However, the model's predictions $\widehat{f}_i$ depend on the model's complexity. As such, using these predictions directly in Eq. (29) may introduce bias into the estimation of σ^2, especially if the model is complex or overfitted.

To address this, a practical approach is to estimate σ^2 using a model with high bias and low variance—typically a simple model. For instance, a linear model fitted to the training data via least squares (linear regression) can be used to obtain the estimates $\widehat{f}_i$ for Eq. (29). This ensures that the estimated σ^2 is not heavily influenced by model complexity.

Once estimated, the value of σ is typically held fixed for simplicity, and not updated as the model complexity changes.

4 Overfitting, Underfitting, and Generalization

When a model is trained in an overly simplistic manner, such that its predictions exhibit low variance but high bias, the model is said to suffer from *underfitting*. Underfitting is sometimes also referred to as *over-generalization*. Conversely, when a model is excessively complex—leading to predictions with high variance but low bias—the result is *overfitting*.

To summarize:

- **Underfitting**: low variance, high bias, low complexity.
- **Overfitting**: high variance, low bias, high complexity.

The desirable scenario lies between these two extremes and is referred to as a *good fit*.

Figure 5 illustrates examples of underfitting, good fit, and overfitting. As shown, both underfitting and overfitting can result in poor performance on a test instance. In contrast, a well-fitted model is capable of estimating the output for a test instance—which was not seen during training—with relatively low error.

The ability of a model to perform well on previously unseen test (i.e., out-of-sample) data is referred to as *generalization*. Poor generalization is the primary reason why both underfitting and especially overfitting are undesirable. In overfitting, the training error (denoted by **err**) may be very small, yet the test error (denoted by **Err**) is typically much worse.

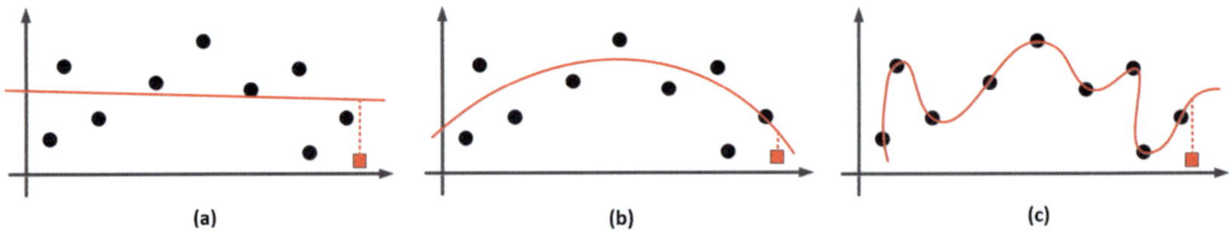

Fig. 5 An example of (a) underfitting, (b) good fit, and (c) overfitting. The black circles and red square represent training and test instances, respectively. The red curve is the fitted model

5 Cross Validation

5.1 Definition

Cross validation should be used to either (I) determine how much complexity the model should have, or (II) tune the model's hyper-parameters [1].

In cross validation, the dataset $\mathcal{D}$ is divided into two disjoint subsets: the training set $\mathcal{T}$ and the test set $\mathcal{R}$, such that:

$$\mathcal{T} \cup \mathcal{R} = \mathcal{D}, \tag{30}$$

$$\mathcal{T} \cap \mathcal{R} = \varnothing. \tag{31}$$

The model is trained on $\mathcal{T}$ and evaluated on $\mathcal{R}$ to assess its generalization performance.

There are several strategies for performing cross validation. Two of the most commonly used methods are K-fold cross validation and Leave-One-Out Cross Validation (LOOCV).

In K-fold cross validation, the dataset $\mathcal{D}$ is partitioned into K subsets $\{\mathcal{D}_1, \ldots, \mathcal{D}_K\}$ such that:

$$|\mathcal{D}_1| \approx |\mathcal{D}_2| \approx \cdots \approx |\mathcal{D}_K|, \tag{32}$$

$$\bigcup_{i=1}^{K} \mathcal{D}_i = \mathcal{D}, \tag{33}$$

$$\mathcal{D}_i \cap \mathcal{D}_j = \varnothing, \quad \forall i, j \in \{1, \ldots, K\},\ i \neq j, \tag{34}$$

where $|.|$ denotes the cardinality of a set. Optionally, the dataset can be shuffled prior to splitting to improve randomness. Either simple random sampling without replacement or stratified sampling [4, 18] may be used.

The K-fold procedure consists of K iterations. In each iteration, one partition is used as the test set and the remaining $K - 1$ partitions are used for training. The final estimation error is the average of the K test errors. In practice, $K \in \{2, 5, 10\}$, with $K = 10$ being the most common. The algorithm is presented in Algorithm 1.

Randomly split $\mathcal{D}$ into K partitions with approximately equal sizes.
for *k from* 1 *to* K **do**
  $\mathcal{R} \leftarrow$ Partition k from $\mathcal{D}$.
  $\mathcal{T} \leftarrow \mathcal{D} \setminus \mathcal{R}$.
  Use $\mathcal{T}$ to train the model.
  $\mathbf{Err}_k \leftarrow$ Use the trained model to predict $\mathcal{R}$.
$\mathbf{Err} \leftarrow \frac{1}{K} \sum_{k=1}^{K} \mathbf{Err}_k$

Algorithm 1: K-fold Cross Validation

In LOOCV, the dataset size is $|\mathcal{D}| = n$. Each iteration selects one instance as the test set (i.e., $|\mathcal{R}| = 1$), with the remaining $n-1$ instances used for training. The final error is the average of all n test errors. LOOCV is particularly useful when the dataset is small. The algorithm is shown in Algorithm 2.

for *k from* 1 *to* $|\mathcal{D}| = n$ **do**
  $\mathcal{R} \leftarrow$ Take the k-th instance from $\mathcal{D}$.
  $\mathcal{T} \leftarrow \mathcal{D} \setminus \mathcal{R}$.
  Use $\mathcal{T}$ to train the model.
  $\mathbf{Err}_k \leftarrow$ Use the trained model to predict $\mathcal{R}$.
$\mathbf{Err} \leftarrow \frac{1}{|\mathcal{D}|} \sum_{k=1}^{|\mathcal{D}|} \mathbf{Err}_k$

Algorithm 2: Leave-One-Out Cross Validation

If the goal is to train the model and then evaluate it, the training and test sets must be disjoint, i.e., $\mathcal{T} \cap \mathcal{R} = \varnothing$. Otherwise, the model may have access to part or all of the test data during training. This results in better test performance, but it violates the assumption that test data should be unseen during training. In real-world applications, test data is not available at training time. Therefore, if $\mathcal{T} \cap \mathcal{R} \neq \varnothing$ occurs, it is referred to as *cheating in machine learning* (we call it *cheating #1* here).

In some situations, the model has hyper-parameters that need to be selected.[1] In such cases, the dataset $\mathcal{D}$ is split into three subsets: the training set $\mathcal{T}$, the validation set $\mathcal{V}$, and the test set $\mathcal{R}$. Usually, $|\mathcal{T}| > |\mathcal{R}|$ and $|\mathcal{T}| > |\mathcal{V}|$.

To select the best hyper-parameters:

- The model is trained on $\mathcal{T}$ using different hyper-parameter values.
- Each trained model is evaluated on the validation set $\mathcal{V}$.
- The hyper-parameter configuration that yields the best performance on $\mathcal{V}$ is selected.
- The model is then retrained on $\mathcal{T}$ using the selected configuration and finally tested on $\mathcal{R}$.

In this setup, the sets must remain disjoint:

$$\mathcal{T} \cap \mathcal{R} = \varnothing, \quad \mathcal{T} \cap \mathcal{V} = \varnothing, \quad \mathcal{V} \cap \mathcal{R} = \varnothing. \tag{35}$$

The validation and test sets must not overlap, because hyper-parameters should be selected based only on validation performance. In practice, the test set should remain completely unseen during both

[1] Note that there is a difference between parameters and hyper-parameters of the model. Parameters are learnable and their optimal values are obtained using optimization. Hyper-parameters are set by the user. Some examples include learning rate, number of epochs, and the number of layers in a neural network.

training and validation. If $\mathcal{V} \cap \mathcal{R} \neq \varnothing$ occurs, it is referred to as *cheating in machine learning* (we call it *cheating #2* here). Unfortunately, this mistake is common in the literature, where some works optimize hyper-parameters using the test set rather than using a separate validation set.

Likewise, $\mathcal{T} \cap \mathcal{R} \neq \varnothing$ is again *cheating #1*. Although $\mathcal{T} \cap \mathcal{V} \neq \varnothing$ is not considered cheating, it is still discouraged and will be addressed in Sect. 5.2.

To incorporate a validation set in cross validation, the dataset is typically first split into a temporary training set $\mathcal{T}'$ and a test set $\mathcal{R}$ such that $\mathcal{T}' \cup \mathcal{R} = \mathcal{D}$ and $\mathcal{T}' \cap \mathcal{R} = \varnothing$. Then, $\mathcal{T}'$ is further divided into the training and validation sets: $\mathcal{T} \cup \mathcal{V} = \mathcal{T}'$ and $\mathcal{T} \cap \mathcal{V} = \varnothing$, typically with $|\mathcal{T}| > |\mathcal{V}|$.

The K-fold and LOOCV algorithms can be extended to include validation sets. In LOOCV, it is common to set $|\mathcal{V}| = 1$.

5.2 Theory

Recall Eqs. (23) and (28), which relate the true error to the empirical error for test instances (not in the training set) and training instances, respectively.

When an instance belongs to the training set, the true error **Err** and the empirical error **err** behave differently, as illustrated in Fig. 6a. In the early stages of training, both **err** and **Err** decrease. However, as training progresses and the model becomes increasingly complex, overfitting can occur—at which point **Err** begins to increase, even though **err** continues to decrease. Training should therefore be stopped when **Err** begins to rise, which marks the point of a good fit.

To identify this point, models are often evaluated on a validation set at regular intervals (e.g., after each iteration or epoch). This is standard practice when training neural networks, where the validation error **Err** is tracked over time. A record of **Err** values is typically maintained for recent epochs, and if a sustained increase is observed, training is terminated. The best model—the one corresponding to the lowest **Err**—is retained. Since complex models like neural networks may exhibit small fluctuations in **Err**, the stopping criterion must be robust to minor variations. This strategy is known as *early stopping* in neural networks [29], and it will be discussed in more detail in Sect. 9.

The reason **Err** increases during continued training can be understood from Eq. (28). Omitting the constant term $n_{tr}\sigma^2$ yields

$$\mathbf{Err} \approx \mathbf{err} + 2\,\sigma^2 \sum_{i=1}^{n_{tr}} \frac{\partial \widehat{f_i}}{\partial y_i}, \tag{36}$$

where the term $2\,\sigma^2 \sum_{i=1}^{n_{tr}} \partial \widehat{f_i} / \partial y_i$ reflects the model's complexity.

Figure 6b shows both **err** and the model complexity as functions of the training stage (e.g., iteration count). According to Eq. (36), **Err** is the sum of these two components, which explains its rise during overfitting. This justifies why a model should not be trained excessively: beyond a certain point, additional training reduces generalization performance.

Figure 6a and Eq. (28) together emphasize the importance of having $\mathcal{T} \cap \mathcal{V} = \varnothing$. If, for example, $\mathcal{T} = \mathcal{V}$, then the validation error **Err** becomes equivalent to the training error **err**, and will continue to decrease—even during overfitting. This prevents overfitting from being detected. Equation (23) also shows that validation or test error serves as a good estimator of the true error, which is why either can be used to determine an appropriate stopping point during training.

Finally, it is important to summarize the implications of various intersections among the training, validation, and test sets, based on the above discussion and Sect. 5.1.

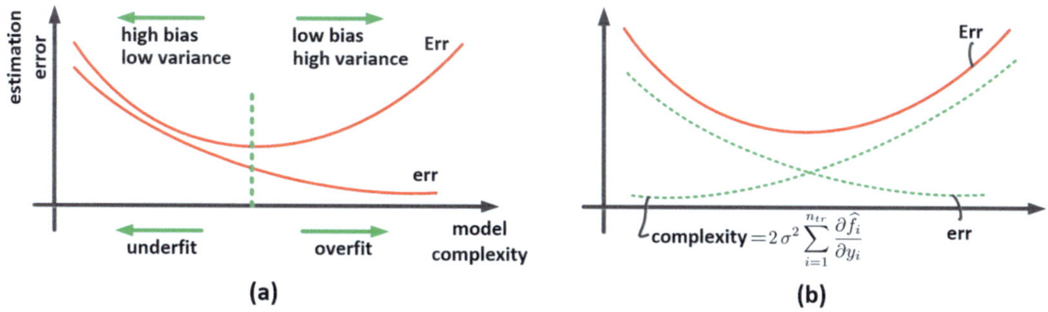

Fig. 6 The overfitting of a model: **a** comparison of training error and true error over training stages, **b** illustration of Eq. (28)

If only training and test sets are used (no validation set):

- $\mathcal{T} \cap \mathcal{R} \neq \varnothing \quad \Rightarrow$ cheating #1

If training, validation, and test sets are all present:

- $\mathcal{T} \cap \mathcal{R} \neq \varnothing \quad \Rightarrow$ cheating #1
- $\mathcal{V} \cap \mathcal{R} \neq \varnothing \quad \Rightarrow$ cheating #2
- $\mathcal{T} \cap \mathcal{V} \neq \varnothing \quad \Rightarrow$ harmful to training (prevents proper detection of overfitting)

The first two cases—$\mathcal{T} \cap \mathcal{R} \neq \varnothing$ and $\mathcal{V} \cap \mathcal{R} \neq \varnothing$—may improve apparent performance on the test data, but they constitute cheating. Moreover, they may degrade performance on future, truly unseen test data. The third case, $\mathcal{T} \cap \mathcal{V} \neq \varnothing$, is not cheating but is nonetheless detrimental: it hinders the detection of overfitting and can lead to poorer generalization. Therefore, it should also be avoided.

6 Regularization

6.1 Definition

The true error, **Err**, can be minimized through optimization. According to Eq. (28), the following objective arises

$$\text{minimize} \quad \mathbf{err} - n_{tr}\,\sigma^2 + 2\,\sigma^2 \sum_{i=1}^{n_{tr}} \frac{\partial \widehat{f}_i}{\partial y_i}. \tag{37}$$

The term $n_{tr}\,\sigma^2$ is constant and can be dropped. Moreover, computing the derivatives $\partial \widehat{f}_i/\partial y_i$ is typically difficult. Therefore, a penalty function is introduced to approximate this behavior. The penalty increases with model complexity and serves to imitate the effect of the derivative term.

This leads to the formulation of a *regularized optimization* problem:

$$\underset{\boldsymbol{x}}{\text{minimize}} \quad \widetilde{J}(\boldsymbol{x};\theta) := J(\boldsymbol{x};\theta) + \alpha\,\Omega(\boldsymbol{x}), \tag{38}$$

where θ denotes the parameters of the cost function, $J(\cdot)$ represents the empirical error **err**, $\Omega(\cdot)$ is a penalty function that captures model complexity, $\alpha > 0$ is the regularization coefficient, and $\widetilde{J}(\cdot)$ is the regularized objective function.

Various forms of the penalty function Ω can be used, such as the ℓ_2 norm [17], ℓ_1 norm [31, 34], and $\ell_{2,1}$ norm [11]. The ℓ_1 and $\ell_{2,1}$ norms are particularly useful for promoting sparsity [2, 3].

Sparsity is highly effective due to the *"bet on sparsity"* principle: "Use a procedure that does well in sparse problems, since no procedure does well in dense problems" [17, 35]. The value of sparsity can also be justified by Occam's razor [15], which asserts that "simpler solutions are more likely to be correct than complex ones" or that "simplicity is a goal in itself."

It is important to emphasize that in Eqs. (37) and (38), the optimization seeks to minimize the true error **Err** (i.e., $\widetilde{J}(\boldsymbol{x};\theta)$), not merely the empirical error **err** (i.e., $J(\boldsymbol{x};\theta)$). As discussed in Sects. 4 and 5, minimizing **err** alone leads to overfitting. Regularization helps to avoid this by penalizing overly complex models.

6.2 Theory for L2 Norm Regularization

This section briefly explains the theory behind ℓ_2 norm regularization [17]. The ℓ_2 norm regularization is also known as *ridge regression* or *Tikhonov regularization* [19]. It is defined as:

$$\underset{\boldsymbol{x}}{\text{minimize}} \quad \widetilde{J}(\boldsymbol{x};\theta) := J(\boldsymbol{x};\theta) + \frac{\alpha}{2}\,||\boldsymbol{x}||_2^2. \tag{39}$$

Suppose $\boldsymbol{x}^*$ is the minimizer of $J(\boldsymbol{x};\theta)$, i.e.,

$$\nabla J(\boldsymbol{x}^*;\theta) = 0. \tag{40}$$

A Taylor series expansion of $J(\boldsymbol{x};\theta)$ up to the second-order term at $\boldsymbol{x}^*$ gives

$$\begin{aligned}\widehat{J}(\boldsymbol{x};\theta) &\approx J(\boldsymbol{x}^*;\theta) + \nabla J(\boldsymbol{x}^*;\theta) + \frac{1}{2}(\boldsymbol{x}-\boldsymbol{x}^*)^\top \boldsymbol{H}(\boldsymbol{x}-\boldsymbol{x}^*)\\ &= J(\boldsymbol{x}^*;\theta) + \frac{1}{2}(\boldsymbol{x}-\boldsymbol{x}^*)^\top \boldsymbol{H}(\boldsymbol{x}-\boldsymbol{x}^*),\end{aligned} \tag{41}$$

where $\boldsymbol{H} \in \mathbb{R}^{d\times d}$ is the Hessian matrix.

Substituting this approximation into the regularized objective [19]:

$$\begin{aligned}\widetilde{J}(\boldsymbol{x};\theta) &= \widehat{J}(\boldsymbol{x};\theta) + \frac{\alpha}{2}\,||\boldsymbol{x}||_2^2\\ &= J(\boldsymbol{x}^*;\theta) + \frac{1}{2}(\boldsymbol{x}-\boldsymbol{x}^*)^\top \boldsymbol{H}(\boldsymbol{x}-\boldsymbol{x}^*) + \frac{\alpha}{2}\,||\boldsymbol{x}||_2^2,\\ \frac{\partial \widetilde{J}(\boldsymbol{x};\theta)}{\partial \boldsymbol{x}} &= \boldsymbol{H}(\boldsymbol{x}^\dagger - \boldsymbol{x}^*) + \alpha\,\boldsymbol{x}^\dagger \overset{\text{set}}{=} \boldsymbol{0},\\ &\implies (\boldsymbol{H}+\alpha\boldsymbol{I})\,\boldsymbol{x}^\dagger = \boldsymbol{H}\boldsymbol{x}^*\\ &\implies \boldsymbol{x}^\dagger = (\boldsymbol{H}+\alpha\boldsymbol{I})^{-1}\boldsymbol{H}\boldsymbol{x}^*,\end{aligned} \tag{42}$$

where $\boldsymbol{x}^\dagger$ is the minimizer of $\widetilde{J}(\boldsymbol{x};\theta)$. Note that $\partial J(\boldsymbol{x}^*;\theta)/\partial \boldsymbol{x} = \boldsymbol{0}$ since $J(\boldsymbol{x}^*;\theta)$ is constant with respect to $\boldsymbol{x}$.

Equation (42) is consistent: when $\alpha = 0$, the regularizer vanishes and we recover $\boldsymbol{x}^\dagger = \boldsymbol{x}^*$, as expected.

Now, suppose eigenvalue decomposition is applied to the Hessian:

$$\boldsymbol{H} = \boldsymbol{U}\boldsymbol{\Lambda}\boldsymbol{U}^\top, \tag{43}$$

where $\boldsymbol{U}$ contains the eigenvectors and $\boldsymbol{\Lambda}$ is a diagonal matrix of eigenvalues. Substituting into Eq. (42) yields:

$$\begin{aligned}\boldsymbol{x}^\dagger &= (\boldsymbol{U}\boldsymbol{\Lambda}\boldsymbol{U}^\top + \alpha\boldsymbol{I})^{-1}\boldsymbol{U}\boldsymbol{\Lambda}\boldsymbol{U}^\top\boldsymbol{x}^* \overset{(a)}{=} (\boldsymbol{U}\boldsymbol{\Lambda}\boldsymbol{U}^\top + \boldsymbol{U}\boldsymbol{U}^\top\alpha\boldsymbol{I})^{-1}\boldsymbol{U}\boldsymbol{\Lambda}\boldsymbol{U}^\top\boldsymbol{x}^* \\ &\overset{(b)}{=} (\boldsymbol{U}\boldsymbol{\Lambda}\boldsymbol{U}^\top + \boldsymbol{U}\alpha\boldsymbol{I}\boldsymbol{U}^\top)^{-1}\boldsymbol{U}\boldsymbol{\Lambda}\boldsymbol{U}^\top\boldsymbol{x}^* = \Big(\boldsymbol{U}(\boldsymbol{\Lambda} + \alpha\boldsymbol{I})\boldsymbol{U}^\top\Big)^{-1}\boldsymbol{U}\boldsymbol{\Lambda}\boldsymbol{U}^\top\boldsymbol{x}^* \\ &\overset{(c)}{=} \boldsymbol{U}(\boldsymbol{\Lambda} + \alpha\boldsymbol{I})^{-1}\underbrace{\boldsymbol{U}^{-1}\boldsymbol{U}}_{\boldsymbol{I}}\boldsymbol{\Lambda}\boldsymbol{U}^\top\boldsymbol{x}^* = \boldsymbol{U}(\boldsymbol{\Lambda} + \alpha\boldsymbol{I})^{-1}\boldsymbol{\Lambda}\boldsymbol{U}^\top\boldsymbol{x}^*,\end{aligned} \tag{44}$$

where (a) and (c) rely on the orthogonality of $\boldsymbol{U}$ (i.e., $\boldsymbol{U}^{-1} = \boldsymbol{U}^\top$, $\boldsymbol{U}^\top\boldsymbol{U} = \boldsymbol{I}$, and $\boldsymbol{U}\boldsymbol{U}^\top = \boldsymbol{I}$), and (b) uses the scalar nature of α.

Equation (44) shows that $\boldsymbol{x}^*$ is projected into the eigenspace of $\boldsymbol{H}$ using $\boldsymbol{U}^\top$, scaled via the shrinkage matrix $(\boldsymbol{\Lambda} + \alpha\boldsymbol{I})^{-1}\boldsymbol{\Lambda}$, and mapped back via $\boldsymbol{U}$.

Interpretations from Eq. (44) are as follows:

- If $\alpha = 0$:

$$\boldsymbol{x}^\dagger = \boldsymbol{U}\underbrace{\boldsymbol{\Lambda}^{-1}\boldsymbol{\Lambda}}_{\boldsymbol{I}}\boldsymbol{U}^\top\boldsymbol{x}^* = \boldsymbol{U}\boldsymbol{U}^\top\boldsymbol{x}^* \overset{(a)}{=} \underbrace{\boldsymbol{U}\boldsymbol{U}^{-1}}_{\boldsymbol{I}}\boldsymbol{x}^* = \boldsymbol{x}^*,$$

  where (a) and (b) use the orthogonality and completeness of $\boldsymbol{U}$. This confirms that without regularization, $\boldsymbol{x}^\dagger$ recovers the original minimizer $\boldsymbol{x}^*$.
- If $\alpha \neq 0$, the shrinkage matrix becomes

$$(\boldsymbol{\Lambda} + \alpha\boldsymbol{I})^{-1}\boldsymbol{\Lambda} = \begin{bmatrix} \frac{\lambda_1}{\lambda_1+\alpha} & 0 & \cdots & 0 \\ 0 & \frac{\lambda_2}{\lambda_2+\alpha} & \cdots & 0 \\ \vdots & \vdots & \ddots & \vdots \\ 0 & 0 & \cdots & \frac{\lambda_d}{\lambda_d+\alpha} \end{bmatrix},$$

  where $\boldsymbol{\Lambda} = \text{diag}([\lambda_1, \ldots, \lambda_d]^\top)$. The following cases arise:

  – If $\lambda_j \gg \alpha$, then $\lambda_j/(\lambda_j + \alpha) \approx 1$. The jth direction is retained, i.e., $\boldsymbol{x}^\dagger \approx \boldsymbol{x}^*$.
  – If $\lambda_j \ll \alpha$, then $\lambda_j/(\lambda_j + \alpha) \approx 0$. The jth direction is suppressed, i.e., $\boldsymbol{x}^\dagger \approx \boldsymbol{0}$ in that direction.

  Thus, ℓ_2 regularization preserves important directions and shrinks less relevant ones toward zero.

The following quantity is known as the *effective number of parameters* or *degrees of freedom* [17]:

$$\sum_{j=1}^{d} \frac{\lambda_j}{\lambda_j + \alpha}, \tag{45}$$

as it quantifies the number of directions retained after shrinkage. The term $\lambda_j/(\lambda_j + \alpha)$, or equivalently $(\boldsymbol{\Lambda} + \alpha\boldsymbol{I})^{-1}\boldsymbol{\Lambda}$, is referred to as the *shrinkage factor* because it modulates how much each direction is preserved under regularization.

6.3 Theory for L1 Norm Regularization

As previously discussed, sparsity is both useful and effective. If $\boldsymbol{x} = [x_1, \dots, x_d]^\top$, one way to encourage sparsity is through *subset selection*, implemented as:

$$\underset{\boldsymbol{x}}{\text{minimize}} \quad \widetilde{J}(\boldsymbol{x}; \theta) := J(\boldsymbol{x}; \theta) + \alpha \, ||\boldsymbol{x}||_0, \tag{46}$$

where:

$$||\boldsymbol{x}||_0 := \sum_{j=1}^{d} \mathbb{I}(x_j \neq 0) = \begin{cases} 0 \text{ if } x_j = 0, \\ 1 \text{ if } x_j \neq 0, \end{cases} \tag{47}$$

is the "ℓ_0" norm. Although widely used in terminology, it is not a true norm (hence the quotation marks) since it does not satisfy the properties of a norm [5]. The "ℓ_0" norm counts the number of non-zero elements in $\boldsymbol{x}$, so minimizing it promotes sparse solutions.

According to [16], the convex relaxation of the "ℓ_0" norm is the ℓ_1 norm. Consequently, the regularized optimization becomes:

$$\underset{\boldsymbol{x}}{\text{minimize}} \quad \widetilde{J}(\boldsymbol{x}; \theta) := J(\boldsymbol{x}; \theta) + \alpha \, ||\boldsymbol{x}||_1. \tag{48}$$

The ℓ_1 norm regularization is also referred to as *lasso* (least absolute shrinkage and selection operator) regularization [34].

Several methods exist for solving optimization problems involving ℓ_1 norm regularization, including the proximal algorithm with soft thresholding [27] and coordinate descent [38, 39]. Here, we explain the coordinate descent method.

The coordinate descent algorithm operates similarly to Gibbs sampling [10], updating one coordinate at a time. Following the same approach as used in deriving Eq. (42), we write:

$$\begin{aligned} \widetilde{J}(\boldsymbol{x}; \theta) &= \widehat{J}(\boldsymbol{x}; \theta) + \alpha \, ||\boldsymbol{x}||_1 \\ &= J(\boldsymbol{x}^*; \theta) + \frac{1}{2}(\boldsymbol{x} - \boldsymbol{x}^*)^\top \boldsymbol{H} (\boldsymbol{x} - \boldsymbol{x}^*) + \alpha \, ||\boldsymbol{x}||_1. \end{aligned}$$

To derive an interpretable form, assume that the Hessian matrix is diagonal [19]. Then, for the jth coordinate:

$$\begin{aligned} \widetilde{J}(x_j; \theta) &= \widehat{J}(x_j; \theta) + \alpha \, |x_j| \\ &= J(x_j^*; \theta) + \frac{1}{2}(x_j - x_j^*)^2 h_j + \alpha \, |x_j| + c, \end{aligned}$$

where $\boldsymbol{x} = [x_1, \dots, x_d]^\top$, $\boldsymbol{x}^* = [x_1^*, \dots, x_d^*]^\top$, h_j is the (j, j)th element of the Hessian (assumed diagonal), and c is constant with respect to x_j. Taking the derivative with respect to x_j yields:

$$\frac{\partial \widetilde{J}(x_j; \theta)}{\partial x_j} = 0 + (x_j - x_j^*) \, h_j + \alpha \, \textbf{sign}(x_j) \overset{\text{set}}{=} 0 \implies$$

$$x_j^\dagger = x_j^* - \frac{\alpha}{h_j} \, \textbf{sign}(x_j^\dagger) = \begin{cases} x_j^* - \frac{\alpha}{h_j} \text{ if } x_j^\dagger > 0, \\ x_j^* + \frac{\alpha}{h_j} \text{ if } x_j^\dagger < 0, \end{cases} \tag{49}$$

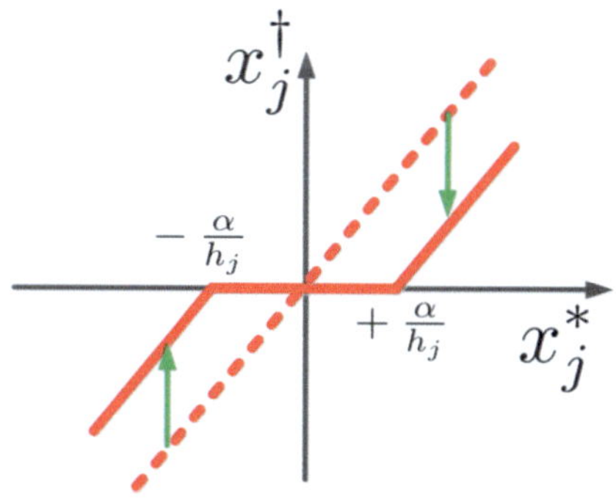

Fig. 7 The soft thresholding function

which defines the soft thresholding function. This is visualized in Fig. 7. As shown, if $|x_j^*| < \alpha/h_j$, the solution $x_j^\dagger$ becomes exactly zero. In contrast to ℓ_2 norm regularization, which shrinks weak entries toward zero, ℓ_1 norm regularization sets them exactly to zero—yielding sparser solutions.

Note also that even strong coordinates are slightly shrunk from the identity line $x_j^\dagger = x_j^*$, as was also the case for ℓ_2 regularization.

Another intuition for why ℓ_1 regularization yields sparse solutions is illustrated in Fig. 8 [34]. In this geometric view, the contours of the objective $J(\boldsymbol{x}; \theta)$ form elliptical (bowl-shaped) levels, while the regularization constraint defines a norm ball. For the ℓ_2 norm, this ball is circular; for ℓ_1, it is diamond-shaped [5].

As shown in Fig. 8, for ℓ_2 regularization, the minimum of the objective and penalty may occur at a point with small but nonzero coordinates. In contrast, for ℓ_1 regularization, the contact point often lies on an axis, where some coordinates are exactly zero. This provides further geometric intuition for the sparsity induced by ℓ_1 norm regularization.

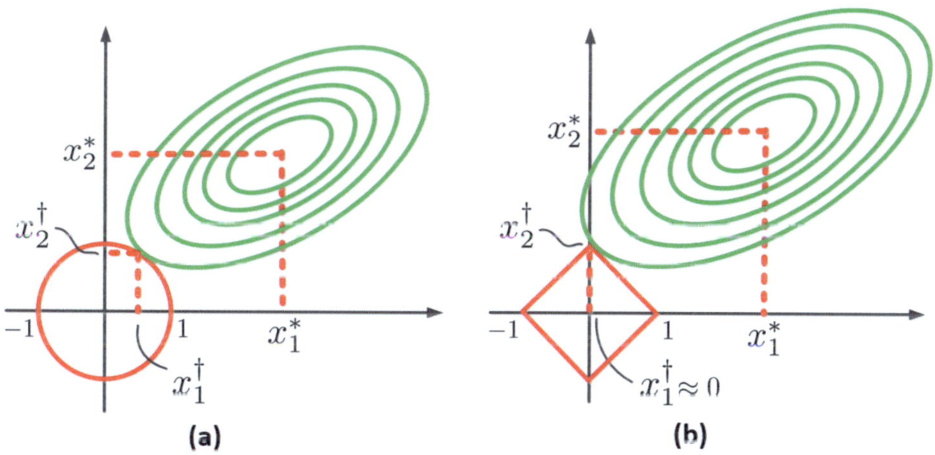

Fig. 8 The unit balls for ℓ_1 and ℓ_2 norm regularizations: (a) ℓ_2 norm regularization and (b) ℓ_1 norm regularization. The green curves represent contour levels of the unregularized objective function. The red balls show the unit balls of the norm penalties. The data are assumed to be two-dimensional; a third dimension (vertical) represents the value of the cost function. This illustration was first proposed in [34]

7 Weight Decay

Recall Eq. (39). If the objective variable $\boldsymbol{x}$ is replaced with the vector of neural network weights $\boldsymbol{w}$, we obtain:

$$\underset{\boldsymbol{w}}{\text{minimize}} \quad \widetilde{J}(\boldsymbol{w}; \theta) := J(\boldsymbol{w}; \theta) + \frac{\alpha}{2} ||\boldsymbol{w}||_2^2, \tag{50}$$

which can be used as the loss function to optimize in a neural network [19]. Penalizing the weights through regularization is referred to as *weight decay* [12, 24].

This penalty discourages the neural network from becoming excessively nonlinear (i.e., overly complex), which would result in overfitting. The reasoning is that in nonlinear activation functions such as the hyperbolic tangent, very large weights—either highly positive or highly negative—push the activations into their highly nonlinear regions. While a neural network should be sufficiently nonlinear to capture complex patterns, it should not be too nonlinear, which risks memorizing the training data. Weight decay encourages relatively smaller weights, which keep activations closer to their linear regions, thereby striking a balance between linearity and nonlinearity.

According to Eq. (44), the result of Eq. (50) is:

$$\boldsymbol{w}^{\dagger} = \boldsymbol{U}(\boldsymbol{\Lambda} + \alpha \boldsymbol{I})^{-1} \boldsymbol{\Lambda} \boldsymbol{U}^{\top} \boldsymbol{w}^{*}, \tag{51}$$

which shares the same interpretation discussed previously.

8 Noise Injection to Input

8.1 Justification of Noise Injection to Input As Data Augmentation

In training neural networks, it is beneficial to inject noise into the input data [25]. One interpretation of why noise injection improves training is through the lens of *data augmentation* [14, 36]. Data augmentation is especially valuable for deep networks, which contain a large number of learnable parameters (weights). Without sufficient training data, these models are prone to overfitting. By varying the training data, data augmentation promotes better generalization to unseen test data.

8.2 Theoretical Justification of Noise Injection to Input

Another perspective on noise injection is to interpret it as a form of regularization [19, 20]. Consider the following optimization objective for a neural network:

$$\underset{\boldsymbol{w}}{\text{minimize}} \quad J := \mathbb{E}((\widehat{y}(\boldsymbol{x}) - y)^2), \tag{52}$$

where $\boldsymbol{x}$ is the input, $\widehat{y}(\boldsymbol{x})$ is the network's prediction, and y is the corresponding label.

Assume Gaussian noise $\varepsilon \sim \mathcal{N}(\boldsymbol{0}, \sigma^2 \boldsymbol{I})$ is added to the input. The modified objective becomes

$$\begin{aligned}
\widetilde{J} &:= \mathbb{E}((\widehat{y}(\boldsymbol{x} + \varepsilon) - y)^2) \\
&= \mathbb{E}(\widehat{y}^2(\boldsymbol{x} + \varepsilon) - 2y\widehat{y}(\boldsymbol{x} + \varepsilon) + y^2) \\
&= \mathbb{E}(\widehat{y}^2(\boldsymbol{x} + \varepsilon)) - 2\mathbb{E}(y\widehat{y}(\boldsymbol{x} + \varepsilon)) + \mathbb{E}(y^2).
\end{aligned}$$

Assuming the noise variance is small, we can expand $\widehat{y}(\boldsymbol{x}+\varepsilon)$ using Taylor series:

$$\widehat{y}(\boldsymbol{x}+\varepsilon)=\widehat{y}(\boldsymbol{x})+\varepsilon^{\top}\nabla_{\boldsymbol{x}}\widehat{y}(\boldsymbol{x})+\frac{1}{2}\varepsilon^{\top}\nabla_{\boldsymbol{x}}^{2}\widehat{y}(\boldsymbol{x})\,\varepsilon+o(\varepsilon^{3}).$$

Therefore, the modified objective becomes

$$\begin{aligned}\widetilde{J}&\approx\mathbb{E}\Big(\big(\widehat{y}(\boldsymbol{x})+\varepsilon^{\top}\nabla_{\boldsymbol{x}}\widehat{y}(\boldsymbol{x})+\frac{1}{2}\varepsilon^{\top}\nabla_{\boldsymbol{x}}^{2}\widehat{y}(\boldsymbol{x})\,\varepsilon\big)^{2}\Big)\\&\quad-2\mathbb{E}\Big(y\widehat{y}(\boldsymbol{x})+y\varepsilon^{\top}\nabla_{\boldsymbol{x}}\widehat{y}(\boldsymbol{x})+\frac{1}{2}y\varepsilon^{\top}\nabla_{\boldsymbol{x}}^{2}\widehat{y}(\boldsymbol{x})\,\varepsilon\Big)+\mathbb{E}(y^{2})\\&=\mathbb{E}\Big(\widehat{y}(\boldsymbol{x})^{2}+y^{2}-2y\widehat{y}(\boldsymbol{x})\Big)-2\mathbb{E}\Big(\frac{1}{2}y\varepsilon^{\top}\nabla_{\boldsymbol{x}}^{2}\widehat{y}(\boldsymbol{x})\,\varepsilon\Big)\\&\quad+\mathbb{E}\Big(\widehat{y}(\boldsymbol{x})\varepsilon^{\top}\nabla_{\boldsymbol{x}}^{2}\widehat{y}(\boldsymbol{x})\varepsilon+(\varepsilon^{\top}\nabla_{\boldsymbol{x}}\widehat{y}(\boldsymbol{x}))^{2}+o(\varepsilon^{3})\Big).\end{aligned}$$

The first term,

$$\mathbb{E}(\widehat{y}(\boldsymbol{x})^{2}+y^{2}-2y\widehat{y}(\boldsymbol{x}))=\mathbb{E}((\widehat{y}(\boldsymbol{x})-y)^{2}),$$

is simply the original loss function defined in Eq. (52).

Since $\varepsilon\sim\mathcal{N}(\mathbf{0},\sigma^{2}\boldsymbol{I})$, we have $\mathbb{E}(\varepsilon^{\top}\varepsilon)=\sigma^{2}$. Given that the noise is independent of the input, i.e., $\varepsilon\perp\!\!\!\perp\boldsymbol{x}$, the following simplifies:

$$\mathbb{E}\Big((\varepsilon^{\top}\nabla_{\boldsymbol{x}}\widehat{y}(\boldsymbol{x}))^{2}\Big)=\mathbb{E}(\varepsilon^{\top}\varepsilon)\,\mathbb{E}(||\nabla_{\boldsymbol{x}}\widehat{y}(\boldsymbol{x})||_{2}^{2})=\sigma^{2}\,\mathbb{E}(||\nabla_{\boldsymbol{x}}\widehat{y}(\boldsymbol{x})||_{2}^{2}),$$

and the remaining terms become

$$\begin{aligned}&\mathbb{E}\Big(\widehat{y}(\boldsymbol{x})\varepsilon^{\top}\nabla_{\boldsymbol{x}}^{2}\widehat{y}(\boldsymbol{x})\varepsilon\Big)-2\mathbb{E}\Big(\frac{1}{2}y\varepsilon^{\top}\nabla_{\boldsymbol{x}}^{2}\widehat{y}(\boldsymbol{x})\,\varepsilon\Big)\\&=\mathbb{E}(\varepsilon^{\top}\varepsilon)\,\mathbb{E}\big(\widehat{y}(\boldsymbol{x})\nabla_{\boldsymbol{x}}^{2}\widehat{y}(\boldsymbol{x})\big)-\mathbb{E}(\varepsilon^{\top}\varepsilon)\,\mathbb{E}\big(y\nabla_{\boldsymbol{x}}^{2}\widehat{y}(\boldsymbol{x})\big)\\&=\sigma^{2}\,\mathbb{E}\Big(\big(\widehat{y}(\boldsymbol{x})-y\big)\nabla_{\boldsymbol{x}}^{2}\widehat{y}(\boldsymbol{x})\Big).\end{aligned}$$

Thus, the full regularized loss function becomes

$$\widetilde{J}\approx J+\sigma^{2}\,\mathbb{E}\Big(\big(\widehat{y}(\boldsymbol{x})-y\big)\nabla_{\boldsymbol{x}}^{2}\widehat{y}(\boldsymbol{x})\Big)+\sigma^{2}\,\mathbb{E}(||\nabla_{\boldsymbol{x}}\widehat{y}(\boldsymbol{x})||_{2}^{2}),\tag{53}$$

which corresponds to a regularized optimization problem with an ℓ_2 norm penalty (see Eq. (39)). The penalty acts on the second derivatives of the network's output, discouraging overly sharp changes in output values—thereby mitigating overfitting.

It is also worth noting that injecting noise into the weights of a neural network [19, 21] can be interpreted analogously to input noise. Thus, weight noise injection can also be viewed as regularization, where the penalty term takes the form $\sigma^{2}\,\mathbb{E}(||\nabla_{\boldsymbol{w}}\widehat{y}(\boldsymbol{x})||_{2}^{2})$, with $\boldsymbol{w}$ being the vector of weights [19].

8.3 Noise Injection to Input in Denoising Autoencoders

The technique of noise injection is also employed in denoising autoencoders [37]. Additionally, for overcomplete autoencoders with a single hidden layer [19]—where the number of hidden units exceeds the dimensionality of the input—noise is necessary; otherwise, the network may simply learn to copy the input to the output without learning a meaningful latent representation.

9 Early Stopping

9.1 Definition of Early Stopping

As discussed in the explanation of Fig. 6a, a neural network should be trained only up to the point where overfitting begins. This approach is known as *early stopping* [29, 40], which serves as a strategy to prevent overfitting [9].

It is noteworthy that empirical studies [41] suggest that noise injection and weight decay are often more effective than early stopping at mitigating overfitting, though early stopping still offers practical benefits.

9.2 Theoretical Justification of Early Stopping

According to Eq. (41), we have

$$\nabla_{\boldsymbol{w}} \widehat{J}(\boldsymbol{w}) \approx \nabla_{\boldsymbol{w}} J(\boldsymbol{w}^*) + \boldsymbol{H}(\boldsymbol{w} - \boldsymbol{w}^*) \overset{(40)}{=} \boldsymbol{H}(\boldsymbol{w} - \boldsymbol{w}^*).$$

The gradient descent algorithm, with learning rate η, used in backpropagation for neural networks is given by [5]:

$$\begin{aligned}
\boldsymbol{w}^{(t)} &:= \boldsymbol{w}^{(t-1)} - \eta \nabla_{\boldsymbol{w}} \widehat{J}(\boldsymbol{w}^{(t)}) \\
&= \boldsymbol{w}^{(t-1)} - \eta \boldsymbol{H}(\boldsymbol{w}^{(t-1)} - \boldsymbol{w}^*) \\
\implies \boldsymbol{w}^{(t)} - \boldsymbol{w}^* &= (\boldsymbol{I} - \eta \boldsymbol{H})(\boldsymbol{w}^{(t-1)} - \boldsymbol{w}^*),
\end{aligned}$$

where t denotes the iteration index.

Using Eq. (43), we obtain

$$\boldsymbol{w}^{(t)} - \boldsymbol{w}^* = (\boldsymbol{I} - \eta \boldsymbol{U}\boldsymbol{\Lambda}\boldsymbol{U}^\top)(\boldsymbol{w}^{(t-1)} - \boldsymbol{w}^*).$$

Assuming the initial weights are $\boldsymbol{w}^{(0)} = 0$, we get

$$\begin{aligned}
\boldsymbol{w}^{(1)} - \boldsymbol{w}^* &= -(\boldsymbol{I} - \eta \boldsymbol{U}\boldsymbol{\Lambda}\boldsymbol{U}^\top)\boldsymbol{w}^* \\
\implies \boldsymbol{w}^{(1)} &= \big(\boldsymbol{I} - (\boldsymbol{I} - \eta \boldsymbol{U}\boldsymbol{\Lambda}\boldsymbol{U}^\top)\big)\boldsymbol{w}^* \\
\overset{(a)}{\implies} \boldsymbol{w}^{(1)} &= \big(\boldsymbol{U}\boldsymbol{U}^\top - (\boldsymbol{U}\boldsymbol{U}^\top - \eta \boldsymbol{U}\boldsymbol{\Lambda}\boldsymbol{U}^\top)\big)\boldsymbol{w}^* \\
\implies \boldsymbol{w}^{(1)} &= \boldsymbol{U}\big(\boldsymbol{I} - (\boldsymbol{I} - \eta \boldsymbol{\Lambda})\big)\boldsymbol{U}^\top \boldsymbol{w}^*,
\end{aligned}$$

where (a) follows from the orthogonality of $\boldsymbol{U}$, implying $\boldsymbol{U}\boldsymbol{U}^\top = \boldsymbol{I}$.

By induction, we arrive at

$$
\begin{aligned}
\boldsymbol{w}^{(t)} &= \boldsymbol{U}\big(\boldsymbol{I} - (\boldsymbol{I} - \eta\,\boldsymbol{\Lambda})^t\big)\boldsymbol{U}^\top \boldsymbol{w}^*, \\
\implies \boldsymbol{U}^\top \boldsymbol{w}^{(t)} &= \boldsymbol{U}^\top \boldsymbol{U}\big(\boldsymbol{I} - (\boldsymbol{I} - \eta\,\boldsymbol{\Lambda})^t\big)\boldsymbol{U}^\top \boldsymbol{w}^*, \\
\overset{(a)}{\implies} \boldsymbol{U}^\top \boldsymbol{w}^{(t)} &= \big(\boldsymbol{I} - (\boldsymbol{I} - \eta\,\boldsymbol{\Lambda})^t\big)\boldsymbol{U}^\top \boldsymbol{w}^*,
\end{aligned}
\tag{54}
$$

where (a) again uses the orthogonality of $\boldsymbol{U}$.

Now recall Eq. (51):

$$
\begin{aligned}
\boldsymbol{w}^\dagger &= \boldsymbol{U}(\boldsymbol{\Lambda} + \alpha \boldsymbol{I})^{-1}\boldsymbol{\Lambda}\boldsymbol{U}^\top \boldsymbol{w}^*, \\
\implies \boldsymbol{U}^\top \boldsymbol{w}^\dagger &= (\boldsymbol{\Lambda} + \alpha \boldsymbol{I})^{-1}\boldsymbol{\Lambda}\boldsymbol{U}^\top \boldsymbol{w}^*, \\
\overset{(a)}{\implies} \boldsymbol{U}^\top \boldsymbol{w}^\dagger &= \big(\boldsymbol{I} - (\boldsymbol{\Lambda} + \alpha \boldsymbol{I})^{-1}\alpha\big)\boldsymbol{U}^\top \boldsymbol{w}^*,
\end{aligned}
\tag{55}
$$

where (a) results from algebraic manipulation as outlined in [19].

By comparing Eqs. (54) and (55), we observe that early stopping can be viewed as a form of ℓ_2 norm regularization, or weight decay [19].

The two expressions are equivalent when:

$$
(\boldsymbol{I} - \eta\,\boldsymbol{\Lambda})^t = (\boldsymbol{\Lambda} + \alpha \boldsymbol{I})\ ^1\alpha, \tag{56}
$$

for some values of η, t, and α.

Taking logarithms and applying the Taylor series expansion of $\log(1 + x)$ yields

$$
\begin{aligned}
\log(\boldsymbol{I} - \eta\,\boldsymbol{\Lambda})^t &= t \log(\boldsymbol{I} - \eta\,\boldsymbol{\Lambda}) \\
&\approx -t\,(\eta\boldsymbol{\Lambda} + \frac{1}{2}\eta^2\boldsymbol{\Lambda}^2 + \frac{1}{3}\eta^3\boldsymbol{\Lambda}^3 + \cdots),
\end{aligned}
\tag{57}
$$

$$
\begin{aligned}
\log(\boldsymbol{\Lambda} + \alpha \boldsymbol{I})^{-1}\alpha &= -\log(\boldsymbol{\Lambda} + \alpha \boldsymbol{I}) + \log \alpha \\
&= -\log(\alpha(\boldsymbol{I} + \frac{1}{\alpha}\boldsymbol{\Lambda})) + \log \alpha \\
&= -\log \alpha - \log(\boldsymbol{I} + \frac{1}{\alpha}\boldsymbol{\Lambda}) + \log \alpha \\
&\approx \frac{-1}{\alpha}\boldsymbol{\Lambda} + \frac{1}{2\alpha^2}\boldsymbol{\Lambda}^2 - \frac{1}{3\alpha^3}\boldsymbol{\Lambda}^3 + \cdots.
\end{aligned}
\tag{58}
$$

Equating Eqs. (57) and (58) as per Eq. (56) gives

$$
\alpha \approx \frac{1}{t\,\eta}, \quad t \approx \frac{1}{\alpha\,\eta}, \tag{59}
$$

which indicates that the inverse of the number of training iterations is proportional to the weight decay regularization parameter α. In other words, longer training leads to weaker regularization and potentially more overfitting. This relationship provides theoretical justification for early stopping, which halts training before the network overfits.

10 Dropout

10.1 Definition of Bagging

Bagging, short for Bootstrap AGGregatING, was first introduced in 1996 [7]. It is a meta-algorithm that can be applied to any type of model, including classifiers, regressors, and others.

10.1.1 Bootstrapping

The concept of *bootstrapping* is defined as follows. Suppose we have a sample $\{\boldsymbol{x}_i\}_{i=1}^n$ of size n, drawn from an unknown distribution $f(\boldsymbol{x})$, i.e., $\boldsymbol{x}_i \overset{iid}{\sim} f(\boldsymbol{x})$. When $f(\boldsymbol{x})$ is unknown, bootstrapping approximates sampling from it by randomly drawing from the existing sample. This is achieved through simple random sampling with replacement. The resulting sample is called a *bootstrap sample*.

10.1.2 Bagging

In *bagging*, k bootstrap samples are drawn, each of a certain sample size. A separate model h_j is then trained on the jth bootstrap sample for every $j \in \{1, \ldots, k\}$. This results in k trained models instead of just one. Finally, the predictions of the k models for an input instance $\boldsymbol{x}$ are aggregated:

$$\widehat{f}(\boldsymbol{x}) = \frac{1}{k} \sum_{j=1}^{k} h_j(\boldsymbol{x}). \tag{60}$$

If the model is a classifier, the sign function may be applied:

$$\widehat{f}(\boldsymbol{x}) = \text{sign}\Big(\frac{1}{k} \sum_{j=1}^{k} h_j(\boldsymbol{x})\Big). \tag{61}$$

10.2 Theory of Bagging

Let e_j denote the estimation error of the jth model for a given instance. Assume this error follows a normal distribution with zero mean: $e_j \overset{iid}{\sim} \mathcal{N}(0, s)$, where $s := \sigma^2$. Let c denote the covariance between the errors of any two models trained on different bootstrap samples. Then:

$$\begin{aligned}
&\mathbb{E}(e_j^2) = s \implies \mathbb{V}\text{ar}(e_j) = \mathbb{E}(e_j^2) - (\mathbb{E}(e_j))^2 = s - 0 = s \\
&\quad \implies \mathbb{V}\text{ar}(h_j(\boldsymbol{x})) = s, && (62)\\
&\mathbb{E}(e_j\, e_\ell) = c \implies \mathbb{C}\text{ov}(e_j, e_\ell) = \mathbb{E}(e_j\, e_\ell) - \mathbb{E}(e_j)\mathbb{E}(e_\ell) = c - (0 \times 0) = c \\
&\quad \implies \mathbb{C}\text{ov}(h_j(\boldsymbol{x}), h_\ell(\boldsymbol{x})) = c, && (63)
\end{aligned}$$

for all $j, \ell \in \{1, \ldots, k\}$, $j \neq \ell$.

Using Eqs. (60), (62), and (63), we have

$$\begin{aligned}\mathbb{V}\mathrm{ar}\big(\widehat{f}(\boldsymbol{x})\big) &= \frac{1}{k^2}\mathbb{V}\mathrm{ar}\big(\sum_{j=1}^{k} h_j(\boldsymbol{x})\big) \\ &\overset{(13)}{=} \frac{1}{k^2}\sum_{j=1}^{k}\mathbb{V}\mathrm{ar}(h_j(\boldsymbol{x})) + \frac{1}{k^2}\sum_{j=1}^{k}\sum_{\ell=1,\ell\neq j}^{k}\mathbb{C}\mathrm{ov}(h_j(\boldsymbol{x}), h_\ell(\boldsymbol{x})) \\ &= \frac{1}{k^2}ks + \frac{1}{k^2}k(k-1)c = \frac{1}{k}s + \frac{k-1}{k}c. \end{aligned} \tag{64}$$

This expression has the following interpretation. If the trained models are highly correlated (i.e., $c \approx s$), then:

$$\lim_{c\to s}\mathbb{V}\mathrm{ar}\big(\widehat{f}(\boldsymbol{x})\big) = \frac{1}{k}s + \frac{k-1}{k}s = s, \tag{65}$$

and if the models are uncorrelated (i.e., $c \approx 0$), then:

$$\lim_{c\to 0}\mathbb{V}\mathrm{ar}\big(\widehat{f}(\boldsymbol{x})\big) = \frac{1}{k}s + \frac{k-1}{k}0 = \frac{1}{k}s. \tag{66}$$

Thus, if the models are highly correlated, bagging offers little to no advantage over a single model. However, if the models are diverse, bagging significantly reduces variance by a factor of k.

This analysis shows that bagging is never harmful—it either has no effect or improves variance reduction [7, 8].

As illustrated in Fig. 5, more complex models typically exhibit higher variance and lower bias. This trade-off is further depicted in Fig. 6. Since increased variance is associated with overfitting, and bagging reduces variance, bagging can help mitigate overfitting. Therefore, it is an effective meta-algorithm for variance reduction and improved generalization [6].

Furthermore, bagging is a form of *ensemble learning* [28], which benefits from *model averaging* [13, 22].

10.3 Dropout

Dropout [32] is a regularization technique for neural networks, whose effectiveness can be explained through the lens of bagging. During each training iteration, dropout randomly deactivates neurons with probability $p = 0.5$, i.e., sampling from a Bernoulli distribution. At test time, all neurons are used, but their outputs are scaled by p.

This process of randomly omitting neurons during training is analogous to training different models, as in bagging. Scaling neuron outputs at test time simulates model averaging, consistent with Eq. (60). This provides a theoretical justification for why dropout helps prevent overfitting.

Another intuition behind dropout is that it induces sparsity in the network by randomly deactivating units. Sparsity is beneficial due to the *principle of sparsity* [17, 35] and Occam's razor [15], as discussed previously.

11 Batch Normalization

Batch normalization is a regularization technique introduced in 2015 [23]. It is applicable to various network architectures, including both fully connected and convolutional layers. Batch normalization stabilizes training by normalizing intermediate activations and improving gradient flow, thereby accelerating the training process. Although batch normalization was originally believed to mitigate covariate shift (distributional shift across mini-batches) [23], later studies demonstrated that this explanation was incorrect [30].

Assume the input to a layer is $\boldsymbol{x} = [x_1, x_2, \ldots, x_d]^\top$, where the layer contains d neurons. For a mini-batch, let the input be the set $\{\boldsymbol{x}_i\}_{i=1}^b$, where b denotes the batch size. Let x_{ij} represent the jth element of $\boldsymbol{x}_i$, i.e., $\boldsymbol{x}_i = [x_{i1}, \ldots, x_{id}]^\top$. Batch normalization performs standardization (Z-score normalization) on each feature (each neuron's input) across the mini-batch:

$$\mathbb{R} \ni \mu_j := \frac{1}{b} \sum_{i=1}^{b} x_{ij}, \tag{67}$$

$$\mathbb{R} \ni \sigma_j^2 := \frac{1}{b} \sum_{i=1}^{b} (x_{ij} - \mu_j)^2, \tag{68}$$

$$x_{ij} \leftarrow \frac{x_{ij} - \mu_j}{\sqrt{\sigma_j^2 + \varepsilon}}, \tag{69}$$

where $\varepsilon > 0$ is a small constant added for numerical stability and to avoid division by zero.

12 Conclusion

Overfitting and generalization in deep learning can be studied through tools such as Stein's Unbiased Risk Estimate (SURE), as discussed in this chapter. Cross-validation provides a principled way to split data for model selection and evaluation. Regularization plays a central role in mitigating overfitting, with classical approaches including ℓ_2 and ℓ_1 penalties. In deep learning, additional techniques such as weight decay, noise injection, early stopping, dropout, and batch normalization are widely used to improve generalization performance.

13 Proof of Stein's Lemma

The proof of Stein's lemma, i.e., Eq. (24), is presented as follows. Since the components of $\boldsymbol{z} = [z_1, \ldots, z_d]^\top \in \mathbb{R}^d$ are independent random variables with normal distributions, i.e., $z_i \sim \mathcal{N}(\mu_i, \sigma)$, we have

$$
\begin{aligned}
f(z) = f(z_1, \ldots, z_d) &\overset{(a)}{=} f(z_1) \times \cdots \times f(z_d) \\
&= \prod_{i=1}^{d} \frac{1}{\sqrt{2\pi\sigma^2}} \exp(-\frac{(z_i - \mu_i)^2}{2\sigma^2}) \\
&= \frac{1}{\sqrt{(2\pi\sigma^2)^d}} \exp(-\frac{\sum_{i=1}^{d}(z_i - \mu_i)^2}{2\sigma^2}) \\
&= \frac{1}{\sqrt{(2\pi\sigma^2)^d}} \exp(-\frac{||z - \mu||_2^2}{2\sigma^2}),
\end{aligned}
$$

where (a) follows from the independence of $z_1, \ldots, z_d$, i.e., $z_1 \perp\!\!\!\perp \ldots \perp\!\!\!\perp z_d$.

In addition, the following vector-form expression can be rewritten in scalar form:

$$
(z - \mu)^\top g(z) = \sum_{i=1}^{d} (z_i - \mu_i)\, g_i.
$$

According to the definition of expectation, we have

$$
\begin{aligned}
\mathbb{E}\big((z - \mu)^\top g(z)\big) &= \int_{\mathbb{R}^d} f(z)(z - \mu)^\top g(z)\, dz \\
&= \int_{\mathbb{R}^d} \frac{1}{\sqrt{(2\pi\sigma^2)^d}} \exp\big(-\frac{||z - \mu||_2^2}{2\sigma^2}\big) \sum_{i=1}^{d}(z_i - \mu_i)\, g_i\, dz_1 \ldots dz_d \\
&\overset{(a)}{=} \sigma^2 \sum_{i=1}^{d} \int_{\mathbb{R}^d} \frac{1}{\sqrt{(2\pi\sigma^2)^d}} \exp\big(-\frac{||z - \mu||_2^2}{2\sigma^2}\big) \frac{\partial g_i}{\partial z_i}\, dz_1 \ldots dz_d \\
&\overset{(b)}{=} \sigma^2 \sum_{i=1}^{d} \mathbb{E}(\frac{\partial g_i}{\partial z_i}),
\end{aligned}
$$

where (a) uses integration by parts, and (b) follows from the definition of expectation.

14 Exercises

1. (Level: Easy) Suppose $y \sim N(\theta, \sigma^2 I)$, where $\theta \in \mathbb{R}^n$ is the true parameter, y is the observed data, and σ^2 is the variance of the noise. Consider the following estimator for θ: $\widehat{\theta} = ay$, where $a \in \mathbb{R}$ is a scalar constant. Derive the Stein's Unbiased Risk Estimator (SURE) for the estimator $\widehat{\theta}$.
2. (Level: Moderate) Stein's Unbiased Risk Estimator (SURE) is a method used for estimating the expected prediction error of a model without relying on a validation set. The SURE formula is given by

$$
\text{SURE} = \frac{1}{n} \sum_{i=1}^{n} (y_i - \widehat{f}(x_i))^2 + \frac{2\sigma^2}{n} \sum_{i=1}^{n} \frac{\partial \widehat{f}(x_i)}{\partial x_i},
$$

where y_i's are the observed data points, $\widehat{f}(x_i)$ is the model prediction, σ^2 is the noise variance, $\partial \widehat{f}(x_i)/\partial x_i$ is the model sensitivity to inputs.

Consider a linear regression model $\widehat{f}(x) = wx + b$. The dataset consists of five points:

x_i	y_i
1	2.2
2	2.8
3	3.5
4	3.9
5	5.1

The estimated model parameters are $w = 0.9$, $b = 1.1$. The noise variance is $\sigma^2 = 0.1$.

(a) Compute the empirical error $(1/n)\sum_{i=1}^{n}(y_i - \widehat{f}(x_i))^2$.
(b) Compute the SURE correction term $(2\sigma^2/n)\sum_{i=1}^{n}(\partial\widehat{f}(x_i)/\partial x_i)$.
(c) Compute SURE for this dataset.

3. (Level: Moderate) Critically evaluate the application of Stein's Unbiased Risk Estimator (SURE) for image denoising in scenarios where ground truth data is unavailable. Discuss the implementation of SURE within convolutional neural networks (CNNs) for image recovery, focusing on the use of the divergence term:

$$\operatorname{div}\mathbf{J} = \operatorname{trace}(\nabla\mathbf{f}(\mathbf{x})),$$

from the trace of the Jacobian matrix of the estimator with respect to observed data. Explain how this approach, as explored in [26], assists in parameter optimization and mitigates the challenges posed by the absence of clean reference images.

4. (Level: Moderate) Given the loss function of a ridge regression:

$$f(\theta) = \sum_i (y_i - \widehat{y}_i)^2 + \lambda||\theta||_2^2.$$

(a) Isolate the weights θ for ridge regression and compare the ridge regression weights against that of an ordinary least squares (OLS) regression $f(\theta) = \sum_i (y_i - \widehat{y}_i)^2$.
(b) Show how the weights of the ridge regression are updated using the Newton-Raphson method.
(c) Discuss the benefits of ridge regression over the OLS regression.

5. (Level: Easy) Regularization is a key technique in deep learning to prevent overfitting by adding constraints to the model. One common form of regularization is ℓ_2-norm weight decay, which penalizes large weight values by adding a regularization term to the loss function:

$$L_{\text{reg}} = L + \lambda\sum_i w_i^2,$$

where L_{reg} is the regularized loss, L is the original loss function, λ is the regularization strength, and w_i's are the model weights.

Consider a linear regression model $\widehat{y} = wx + b$. Given a dataset:

x_i	y_i
1	2.1
2	2.9
3	3.7
4	4.4
5	5.2

The estimated model parameters are $w = 1.0$, $b = 1.0$. The original loss function is Mean Squared Error (MSE):

$$L = \frac{1}{n} \sum_{i=1}^{n} (y_i - \widehat{y}_i)^2.$$

The regularization parameter is $\lambda = 0.1$.

(a) Compute the original loss L.
(b) Compute the ℓ_2-norm regularization term $\lambda \sum w_i^2$.
(c) Compute the regularized loss L_{reg}.

6. (Level: Easy) Suppose you are given the following mini-batch of inputs: $X = [-10, -8, 2, 10]$.

(a) Apply the sigmoid activation function directly to X.
(b) Now, normalize the inputs X using batch normalization. Calculate the output of the layer $Y = \gamma \hat{X} + \beta$ where $\hat{X}$ is the batch normalized X and γ and β are the weights of layer. Assume that the layer has learned parameters $\gamma = 2$ and $\beta = 0.5$. Then, apply the sigmoid activation to the normalized inputs.
(c) Why can batch normalization be helpful in models that use activations such as the sigmoid function?

7. (Level: Difficult) Recall that the least-squares regression model can be obtained by minimizing the squared error:

$$\min_{\beta \in \mathbb{R}^d} \sum_{i=1}^{n} \left(y_i - \beta^T x_i \right)^2,$$

where $\{(x_i, y_i)\}_{i=1}^{n} \subset \mathbb{R}^d \times \mathbb{R}$ is the training set, and $\beta \in \mathbb{R}^d$ is the coefficient vector (including the bias term).
Suppose we perturb the input features to obtain an augmented collection of features $\tilde{x}_i = g(x_i, \epsilon_i)$, where g is a data augmentation function that applies independent random noise ϵ_i to each x_i. We can obtain a new objective function by taking the expectation over the randomness induced by ϵ_i:

$$\min_{\beta \in \mathbb{R}^d} \sum_{i=1}^{n} \mathbb{E} \left[\left(y_i - \beta^T \tilde{x}_i \right)^2 \right]. \tag{70}$$

For the following choices of g and ϵ_i, show that Eq. (70) can be expressed as a penalized least squares problem with respect to the original dataset, specifically:

$$\min_{\beta \in \mathbb{R}^d} \sum_{i=1}^{n} \left(y_i - \beta^T x_i\right)^2 + P(\beta).$$

(a) Define $g(x_i, \epsilon_i) = x_i + \epsilon_i$, where $\epsilon_i \overset{iid}{\sim} \mathcal{N}(0, \lambda \boldsymbol{I})$ is sampled independently from the multivariate Normal distribution with mean 0 and covariance matrix $\lambda \boldsymbol{I}$ where $\boldsymbol{I}$ is the identity matrix.

(b) Define $g(x_i, \epsilon_i) = x_i \odot \epsilon_i$, where $\epsilon_{ij} \overset{iid}{\sim} \text{Bernoulli}(p)$ for each component j in the noise vector ϵ_i, and $\odot$ represents Hadamard (element-wise) multiplication.

References

1. Sylvain Arlot and Alain Celisse. "A survey of cross-validation procedures for model selection". In: *Statistics surveys* 4 (2010), pp. 40–79.
2. Francis Bach, Rodolphe Jenatton, Julien Mairal, and Guillaume Obozinski. "Convex optimization with sparsity-inducing norms". In: *Optimization for Machine Learning* 5 (2011), pp. 19–53.
3. Francis Bach, Rodolphe Jenatton, Julien Mairal, and Guillaume Obozinski. "Optimization with sparsity-inducing penalties". In: *Foundations and Trends ® in Machine Learning* 4.1 (2012), pp. 1–106.
4. Vick Barnett. *Elements of sampling theory*. English Universities Press, London, 1974.
5. Stephen Boyd and Lieven Vandenberghe. *Convex optimization*. Cambridge university press, 2004.
6. Leo Breiman. "Arcing classifier (with discussion and a rejoinder by the author)". In: *The annals of statistics* 26.3 (1998), pp. 801–849.
7. Leo Breiman. "Bagging predictors". In: *Machine learning* 24.2 (1996), pp. 123–140.
8. Peter Lukas Bühlmann and Bin Yu. "Explaining bagging". In: *Research report/Seminar f ü r Statistik, Eidgen ö ssische Technische Hochschule Z ü rich*. Vol. 92. Seminar für Statistik, Eidgenössische Technische Hochschule (ETH). 2000.
9. Rich Caruana, Steve Lawrence, and C Lee Giles. "Overfitting in neural nets: Backpropagation, conjugate gradient, and early stopping". In: *Advances in neural information processing systems*. 2001, pp. 402–408.
10. George Casella and Edward I George. "Explaining the Gibbs sampler". In: *The American Statistician* 46.3 (1992), pp. 167–174.
11. Yale Chang. "$L_{2,1}$ Norm and its Applications". In: *Technical Report, University of Central Florida* ().
12. Ching-Tai Chiu, Kishan Mehrotra, Chilukuri K Mohan, and Sanjay Ranka. "Modifying training algorithms for improved fault tolerance". In: *Proceedings of 1994 IEEE International Conference on Neural Networks (ICNN'94)*. Vol. 1. IEEE. 1994, pp. 333–338.
13. Gerda Claeskens and Nils Lid Hjort. *Model selection and model averaging*. Cambridge Books, Cambridge University Press, 2008.
14. Terrance DeVries and Graham W Taylor. "Dataset augmentation in feature space". In: *arXiv preprint* arXiv:1702.05538 (2017).
15. Pedro Domingos. "The role of Occam's razor in knowledge discovery". In: *Data mining and knowledge discovery* 3.4 (1999), pp. 409–425.
16. David L Donoho. "For most large underdetermined systems of linear equations the minimal ℓ_1-norm solution is also the sparsest solution". In: *Communications on Pure and Applied Mathematics: A Journal Issued by the Courant Institute of Mathematical Sciences* 59.6 (2006), pp. 797–829.
17. Jerome Friedman, Trevor Hastie, and Robert Tibshirani. *The elements of statistical learning: Data Mining, Inference, and Prediction*. Vol. 2. Springer series in statistics, New York, NY, USA, 2009.
18. Benyamin Ghojogh, Hadi Nekoei, Aydin Ghojogh, Fakhri Karray, and Mark Crowley. "Sampling algorithms, from survey sampling to Monte Carlo methods: Tutorial and literature review". In: *arXiv preprint* arXiv:2011.00901 (2020).
19. Ian Goodfellow, Yoshua Bengio, and Aaron Courville. *Deep learning*. MIT press, 2016.
20. Yves Grandvalet, Stéphane Canu, and Stéphane Boucheron. "Noise injection: Theoretical prospects". In: *Neural Computation* 9.5 (1997), pp. 1093–1108.
21. Kevin Ho, Chi-sing Leung, and John Sum. "On weight-noise-injection training". In: *International Conference on Neural Information Processing*. Springer. 2008, pp. 919–926.
22. Jennifer A Hoeting, David Madigan, Adrian E Raftery, and Chris T Volinsky. "Bayesian model averaging: a tutorial". In: *Statistical science* (1999), pp. 382–401.

23. Sergey Ioffe and Christian Szegedy. "Batch normalization: Accelerating deep network training by reducing internal covariate shift". In: *International conference on machine learning*. pmlr. 2015, pp. 448–456.
24. Anders Krogh and John A Hertz. "A simple weight decay can improve generalization". In: *Advances in neural information processing systems*. 1992, pp. 950–957.
25. Kiyotoshi Matsuoka. "Noise injection into inputs in back-propagation learning". In: *IEEE Transactions on Systems, Man, and Cybernetics* 22.3 (1992), pp. 436–440.
26. Christopher A Metzler, Ali Mousavi, Reinhard Heckel, and Richard G Baraniuk. "Unsupervised learning with Stein's unbiased risk estimator". In: *arXiv preprint* arXiv:1805.10531 (2018).
27. Neal Parikh and Stephen Boyd. "Proximal algorithms". In: *Foundations and Trends ® in Optimization* 1.3 (2014), pp. 127–239.
28. Robi Polikar. "Ensemble learning". In: *Ensemble machine learning*. Springer, 2012, pp. 1–34.
29. Lutz Prechelt. "Early stopping-but when?" In: *Neural Networks: Tricks of the trade*. Springer, 1998, pp. 55–69.
30. Shibani Santurkar, Dimitris Tsipras, Andrew Ilyas, and Aleksander Madry. "How does batch normalization help optimization?" In: *Advances in neural information processing systems* 31 (2018).
31. Mark Schmidt. "Least squares optimization with L1-norm regularization". In: *CS542B Project Report* 504 (2005), pp. 195–221.
32. Nitish Srivastava, Geoffrey Hinton, Alex Krizhevsky, Ilya Sutskever, and Ruslan Salakhutdinov. "Dropout: a simple way to prevent neural networks from overfitting". In: *The Journal of Machine Learning Research* 15.1 (2014), pp. 1929–1958.
33. Charles M Stein. "Estimation of the mean of a multivariate normal distribution". In: *The annals of Statistics* (1981), pp. 1135–1151.
34. Robert Tibshirani. "Regression shrinkage and selection via the lasso". In: *Journal of the Royal Statistical Society: Series B (Methodological)* 58.1 (1996), pp. 267–288.
35. Robert Tibshirani, Martin Wainwright, and Trevor Hastie. *Statistical learning with sparsity: the lasso and generalizations*. Chapman and Hall/CRC, 2015.
36. David A Van Dyk and Xiao-Li Meng. "The art of data augmentation". In: *Journal of Computational and Graphical Statistics* 10.1 (2001), pp. 1–50.
37. Pascal Vincent, Hugo Larochelle, Yoshua Bengio, and Pierre-Antoine Manzagol. "Extracting and composing robust features with denoising autoencoders". In: *Proceedings of the 25th international conference on Machine learning*. ACM. 2008, pp. 1096–1103.
38. Stephen J Wright. "Coordinate descent algorithms". In: *Mathematical Programming* 151.1 (2015), pp. 3–34.
39. Tong Tong Wu and Kenneth Lange. "Coordinate descent algorithms for lasso penalized regression". In: *The Annals of Applied Statistics* 2.1 (2008), pp. 224–244.
40. Yuan Yao, Lorenzo Rosasco, and Andrea Caponnetto. "On early stopping in gradient descent learning". In: *Constructive Approximation* 26.2 (2007), pp. 289–315.
41. Richard M Zur, Yulei Jiang, Lorenzo L Pesce, and Karen Drukker. "Noise injection for training artificial neural networks: A comparison with weight decay and early stopping". In: *Medical physics* 36.10 (2009), pp. 4810–4818.

Theoretical Foundations of Generalization in Deep Learning 6

Abstract

This chapter explores prominent theoretical frameworks used to understand learning and generalization in deep learning. It covers Probably Approximately Correct (PAC) learning, error bounds and generalization in neural networks, the information bottleneck principle, double descent, grokking, benign overfitting, implicit bias of stochastic gradient descent, neural network kernel theory, lottery ticket hypothesis, and scaling laws. First, it introduces PAC learning theory as a statistical framework that aims to bound the true (test) error in terms of the empirical (training) error. Concepts such as realizable and agnostic PAC learnability, VC dimension, and complexity-dependent error bounds are discussed in the context of their applicability to neural networks. However, the chapter emphasizes a fundamental disconnect between PAC learning theory and the empirical success of modern over-parameterized deep networks, demonstrating that PAC theory does not fully account for their generalization behavior. The second part presents the information-theoretic perspective, wherein a neural network is viewed as an information bottleneck that compresses input data while preserving only the most task-relevant information. This viewpoint offers an alternative explanation for generalization in deep learning, especially in highly over-parameterized settings. Collectively, these two theoretical approaches provide complementary insights into the challenges and open questions surrounding the effectiveness of deep learning. Finally, the third part of chapter briefly introduces a few theoretical advances in deep learning generalization, including double descent, grokking, benign overfitting, implicit bias of stochastic gradient descent, neural network kernel theory, lottery ticket hypothesis, and scaling laws. These concepts explain why deep neural networks often exhibit good generalization to unseen data.

1 Introduction

Probably Approximately Correct (PAC) learning [42] is a theoretical framework for understanding statistical learning, including deep learning. The goal of PAC learning is to provide an upper bound on the true (test) error based on the empirical (training) error. In other words, PAC learning offers guarantees for good generalization—i.e., achieving low true error alongside low empirical error [35]. Typically, PAC learning establishes lower bounds on the number of training instances required to identify the best hypothesis (i.e., machine learning model) for prediction or classification tasks. It also

B. Ghojogh and A. Ghodsi, *Elements of Deep Learning*,
https://doi.org/10.1007/978-3-032-10738-1_6

addresses whether machine learning can be theoretically proven to work in principle [6]. Some studies, such as [12], attempt to explain deep learning through the lens of PAC learning. It is important to note that PAC learnability does not fully explain why neural networks generalize. Instead, it serves more as a conceptual baseline than a practical account of generalization in deep networks.

Alternatively, machine learning and deep learning can be analyzed using information theory [7]. A neural network can be interpreted as an information bottleneck [40], where data is compressed to retain only the most task-relevant information. The hidden layers of a neural network can be viewed as this bottleneck [41]. There is evidence of a relationship between the information bottleneck principle and generalization in neural networks [21].

Moreover, there are some other theoretical advances in deep learning generalization. Some of them are double descent [5, 27], grokking [31], benign overfitting [1, 2], implicit bias of stochastic gradient descent [16, 38], neural network kernel theory [19, 24], lottery ticket hypothesis [10], and scaling laws [20].

Section 2 introduces realizable and agnostic PAC learnability and explores PAC learnability in deep learning. Section 3 discussed error bounds and generalization in neural networks and it demonstrates the misalignment between PAC learning theory and the empirical success of deep learning. Specifically, it shows that PAC theory does not adequately explain the generalization observed in over-parameterized neural networks. Section 4 analyzes deep learning through the lens of the information bottleneck principle. Finally, Sect. 5 briefly introduces a few theoretical advances in deep learning generalization, including double descent, grokking, benign overfitting, implicit bias of stochastic gradient descent, and neural network kernel theory.

It is recommended that readers review Chaps. 3 and 5 prior to studying this chapter. Chapter 3 analyzes the generalization of over-parameterized neural networks from an optimization theory perspective. Chapter 5 discusses generalization in deep learning in the context of Stein's Unbiased Risk Estimate (SURE) [39].

2 PAC Learnability in Deep Learning

This section discusses PAC learnability in machine learning and deep learning. PAC learnability is a foundational framework for reasoning about learning and sample complexity. However, for modern overparameterized neural networks, the resulting bounds are usually vacuous, so PAC theory is more of a conceptual baseline than a practical explanation of why deep networks generalize.

Consider a hypothesis class $\mathcal{H}$ consisting of hypotheses that predict or classify the labels of a dataset. Each hypothesis is a model or function that receives input data and outputs the predicted label. Every model in the hypothesis class aims to minimize the empirical risk, which is the loss evaluated on the training dataset. Let $0 \lesssim \epsilon < 1$, typically a small positive number, represent the desired maximum error rate of the hypothesis obtained through Empirical Risk Minimization (ERM). Additionally, let $0 \lesssim \delta < 1$, also typically small, denote the tolerated probability of failure—i.e., the probability that the ERM-selected hypothesis exceeds the error rate ϵ. These variables are required for defining PAC learnibility in the following.

2.1 Realizable PAC Learnability

Assume the data are generated i.i.d. (independently and identically distributed) by distribution $\mathcal{D}$. The error of a hypothesis $h \in \mathcal{H}$ on distribution $\mathcal{D}$ is defined as [35]:

$$\mathrm{err}_{\mathcal{D}}(h) := \mathbb{P}_{(\boldsymbol{x},y)\sim\mathcal{D}}[h(\boldsymbol{x}) \neq y], \tag{1}$$

where $\mathbb{P}$ denotes probability, $\boldsymbol{x}$ denotes data instances, and y denotes the labels. In the realizable case where labels are generated by some target labeling function $f \in \mathcal{H}$, this error reduces to:

$$\mathrm{err}_{\mathcal{D}}(h) = \mathbb{P}_{\boldsymbol{x}\sim\mathcal{D}}[h(\boldsymbol{x}) \neq f(\boldsymbol{x})]. \tag{2}$$

This is the true error (generalization error) of a hypothesis h which is the probability that h misclassifies an input drawn from distribution $\mathcal{D}$:

Definition 1 (*(Realizable) PAC learnability* [35]) For accuracy ϵ and confidence δ, an algorithm is (realizable) PAC-learnable if with probability $\geq 1 - \delta$, it outputs $h \in \mathcal{H}$ such that:

$$\mathrm{err}_{\mathcal{D}}(h) \leq \epsilon, \tag{3}$$

assuming the labels are generated by some target function $f \in \mathcal{H}$.

Equation (3) states that the probability of the true (generalization) error $\mathrm{err}_{\mathcal{D}}(h)$ exceeding ϵ is less than δ. In other words, with high probability, the learner can find a hypothesis whose true error is small. This regular PAC learnability is also referred to as realizable PAC learnability.

2.2 Agnostic PAC Learnability

Thus far, it has been assumed that the true function to be learned lies within the hypothesis class $\mathcal{H}$. In other words, the hypotheses are assumed to contain the exact solution. This assumption, which often does not hold in practice, is known as the realizability assumption (realizable PAC model). A more realistic scenario involves a lack of a priori knowledge—meaning that the learner does not know whether the true classifier or predictor is in $\mathcal{H}$.

Another realistic consideration is that the labels may not be fully determined by the input features. In other words, some randomness may exist in the labels due to factors such as noise in the data, unobserved variables, or inherently stochastic processes. Under such circumstances, even the best possible model in $\mathcal{H}$ may not perfectly predict the labels for all data instances. Learning thus becomes more challenging: the learner must aim to find the best possible hypothesis within $\mathcal{H}$, even if it is not perfect.

These realistic considerations are known as agnostic PAC learning. In other words, the agnostic PAC learning is the non-realizable PAC learning. In the realizable PAC model, we assume there exists a hypothesis in $\mathcal{H}$ that labels all data correctly. In the agnostic PAC model, this assumption is removed. The agnostic PAC learning is defined formally in the following.

Definition 2 (*Agnostic PAC learnability* [35]) For accuracy ϵ and confidence δ, an algorithm is agnostic PAC-learnable if with probability $\geq 1 - \delta$, it outputs $h \in \mathcal{H}$ such that:

$$\mathrm{err}_{\mathcal{D}}(h) \leq \inf_{h' \in \mathcal{H}} (\mathrm{err}_{\mathcal{D}}(h')) + \epsilon, \tag{4}$$

Equation (4) states that the probability of the true (generalization) error $\mathrm{err}_{\mathcal{D}}(h)$ exceeding the error of the best hypothesis $\inf_{h' \in \mathcal{H}}(\mathrm{err}_{\mathcal{D}}(h'))$ by more than ϵ is less than δ. In other words, with high probability, the learner can find a hypothesis whose true error is close to that of the best possible hypothesis in $\mathcal{H}$, provided that enough training instances are available.

There are several key differences between PAC learnability, given in Eq. (3), and agnostic PAC learnability, given in Eq. (4).

The first difference lies in their assumptions. PAC learnability relies on the realizability assumption—i.e., it assumes that the perfect hypothesis exists within the hypothesis class $\mathcal{H}$. In contrast, agnostic PAC learnability does not make this assumption. Instead, it seeks the best hypothesis available in $\mathcal{H}$, even if none is perfect.

The second difference lies in the interpretation of Eqs. (3) and (4). In Eq. (3), PAC learnability guarantees that the algorithm can identify the true model with small error ϵ, with high probability, given sufficient training data. However, in Eq. (4), agnostic PAC learnability does not guarantee identification of the best model. Instead, it ensures that, with high probability, the learned model performs nearly as well as the best model in $\mathcal{H}$, within a margin of ϵ, given sufficient training data.

For intuitive understanding, consider the following analogy. PAC learnability is like promising to achieve the highest possible return on a stock investment after one year, assuming perfect information about influencing factors. Agnostic PAC learnability, on the other hand, does not guarantee the highest return, as real-world uncertainties such as wars or natural disasters may intervene. Instead, it ensures performance close to the best investor under the same conditions: if others gain significantly, it does too; if others gain modestly, it gains modestly. The key promise is that its performance will not be much worse than the best.

2.3 The VC Dimension

A natural question that arises is whether PAC learnability can be guaranteed. A crucial factor in establishing such a guarantee is the Vapnik-Chervonenkis (VC) dimension of the hypothesis class $\mathcal{H}$, introduced below.

Definition 3 (*Dichotomy, shattering, and VC dimension* [42]) The concepts of dichotomy, shattering, and VC dimension are defined as follows:

- A **dichotomy** of a set is a partition of the set into disjoint subsets.
- A hypothesis **shatters** a set if, for every possible dichotomy of the set, there exists a hypothesis that correctly classifies all data instances according to that dichotomy. In other words, a hypothesis can shatter a set if it can perfectly represent every possible labeling of that set.
- The **VC dimension** of $\mathcal{H}$ is the size of the largest finite subset of data that can be shattered by $\mathcal{H}$. That is, it is the maximum number of data instances for which hypotheses in $\mathcal{H}$ can realize all possible labelings.

For example, the VC dimension of a line in two-dimensional space is three. As shown in Fig. 1, any three points can be correctly classified by a line in 2D space, but this is not always true for four points.

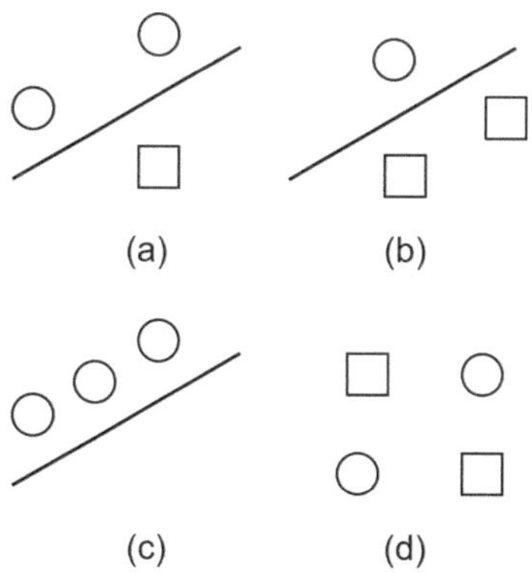

Fig. 1 The VC dimension of a line in 2D is three, as illustrated by cases (**a**), (**b**), and (**c**). However, configurations like **d** show that a line cannot shatter four points in 2D. Note that it is assumed the points do not all lie on a single line

Theorem 1 (Fundamental theorem of statistical learning [35]) *A hypothesis class $\mathcal{H}$ is PAC learnable if and only if its VC dimension is finite.*

Recent theoretical results establish nearly-tight bounds on the VC dimension of neural networks with ReLU (or more generally, piecewise-linear) activation functions. In particular, Harvey, Liaw, and Mehrabian [17], later extended by Bartlett et al. [3], proved that for a network with w parameters (weights and biases) and l layers:

$$\Omega\Big(wl \log \frac{w}{l}\Big) \;\leq\; \text{VCdim} \;\leq\; \mathcal{O}(wl \log w), \tag{5}$$

where $\Omega(.)$ and $\mathcal{O}(.)$ denote the big-Ω and big-O notations in complexity. Thus, the VC dimension grows essentially linearly with the number of parameters (up to logarithmic factors). In many practical regimes, where the depth l is much smaller than the total number of parameters w, these bounds simplify to:

$$\text{VCdim} = \Theta(w \log w), \tag{6}$$

where $\Theta(.)$ denotes the big-Θ notation in complexity.

Equation (6) shows that the VC dimension of neural networks grows on the order of $w \log w$ where w is the number of parameters, with mild additional dependence on depth. While these bounds show very high expressive capacity, they do not align with the fact that neural networks often generalize well even when their VC dimension far exceeds the sample size.

2.4 Sample Complexity Bounds

Definition 4 (*Sample complexity of a hypothesis class* [35]) Let $\mathcal{H}$ be a hypothesis class. The sample complexity function:

$$n_{\mathcal{H}} : (0, 1)^2 \to \mathbb{N}, \tag{7}$$

is defined such that for every $\epsilon, \delta \in (0, 1)$, for every distribution $\mathcal{D}$ on $\mathcal{X}$, and for every target function $f \in \mathcal{H}$, if the sample size satisfies $n \geq n_{\mathcal{H}}(\epsilon, \delta)$, then with probability at least $1 - \delta$, a PAC-learnable (or agnostic PAC-learnable) algorithm A outputs $h = A(S)$ that satisfies Eq. (3) (or Eq. (4)).

Theorem 2 (Sample complexity bounds) *For a hypothesis class with VC dimension d, the sample complexities are as follows:*

- *Realizable case:*

$$n_{\mathcal{H}} = \mathcal{O}\big(\tfrac{1}{\epsilon}\big(d \log \tfrac{1}{\epsilon} + \log \tfrac{1}{\delta}\big)\big) . \tag{8}$$

- *Agnostic case:*

$$n_{\mathcal{H}} = \mathcal{O}\Big(\tfrac{1}{\epsilon^2}\big(d + \log \tfrac{1}{\delta}\big)\Big) . \tag{9}$$

Theorem 2 provides a lower bound on the number of training data instances required to find the best hypothesis in the hypothesis class. It states that for sufficiently large $n_{\mathcal{H}}$, it is possible to find a hypothesis with true error satisfying Eq. (3) or Eq. (4).

2.5 Complexity-Dependant Error Bound

There exists a bound—dependent on the complexity of the model—on the true (test) error of the hypothesis. This bound is presented below.

Theorem 3 (Complexity-dependant error bound) *The true error of a hypothesis h is bounded by a function of its complexity c(h) and the number of training data instances n:*

$$err_{\mathcal{D}}(h) \leq \inf_{h' \in \mathcal{H}} (err_{\mathcal{D}}(h')) + f\big(c(h), n\big), \tag{10}$$

where $f\big(c(h), n\big)$ is an increasing function dependent on the complexity measure $c(h)$ of the hypothesis h and the training sample size n. This function captures the relationship between hypothesis complexity, sample size, and achievable error.

Theorem 3 incorporates the notion of hypothesis complexity into the error bound. Equation (10) indicates that as the complexity of the model increases, the gap between the true loss $\mathrm{err}_{\mathcal{D}}(h)$ and the best possible loss $\inf_{h' \in \mathcal{H}}(\mathrm{err}_{\mathcal{D}}(h'))$ also increases. In other words, the upper bound on the true error becomes less tight.

Classical complexity-based bounds, such as Eq. (10), suggest that models with far more parameters than data should overfit. Yet in practice, modern deep networks often generalize in this regime, highlighting a major mismatch between theory and empirical behavior.

3 Error Bounds and Generalization in Neural Networks

3.1 Expressive Power of Neural Networks

A machine learning model is said to be *expressible* if it has sufficient complexity to fit the training data. For example, in regression, a polynomial function of degree $n - 1$ is sufficient to fit n distinct points. The following Theorem 4 formalizes the remarkable expressive power of neural networks.

Theorem 4 (Expressive power of neural networks) *Let d denote the dimensionality of the data, and let $s(d)$ be an exponential function of d. There exists a neural network with $s(d)$ neurons that can approximate any function f:*

$$f : [0, 1]^d \rightarrow [0, 1],$$

with error at most $\epsilon \gtrsim 0$. Note that if the data features are not in the range $[0, 1]$, *they can be normalized accordingly.*

Theorem 4 states that neural networks can approximate any continuous function and depth of neural network crucially provides exponential efficiency compared to shallow networks. A deep network with modest width can represent functions that would require exponentially more neurons in a single hidden layer.

3.2 Complexity-Dependant Error Bound for Neural Networks

Theorem 3 can be adapted for neural networks, leading to the following bound:

$$\mathrm{err}_{\mathcal{D}}(h) \leq \inf_{h' \in \mathcal{H}} (\mathrm{err}_{\mathcal{D}}(h')) + \sqrt{\frac{|\mathcal{E}| + \log(\frac{1}{\delta})}{n}}, \tag{11}$$

where n is the number of training data instances and $|\mathcal{E}|$ is the number of edges (weights) in the neural network, i.e., the number of learnable parameters in the hypothesis h.

Equation (11) shows that the true error $\mathrm{err}_{\mathcal{D}}(h)$ of a neural network exceeds the best error in the hypothesis class $\inf_{h' \in \mathcal{H}}(\mathrm{err}_{\mathcal{D}}(h'))$ by at most $\sqrt{\frac{|\mathcal{E}|+\log(1/\delta)}{n}}$. Several important interpretations follow:

- If the neural network becomes large—whether in terms of the number of neurons or layers—then $|\mathcal{E}|$ increases. Consequently, the term $\sqrt{\frac{|\mathcal{E}|+\log(1/\delta)}{n}}$ increases, making the bound looser and the true error $\mathrm{err}_{\mathcal{D}}(h)$ larger. In simpler terms, larger networks tend to generalize worse unless properly managed.
- If the number of training instances n increases, the term $\sqrt{\frac{|\mathcal{E}|+\log(1/\delta)}{n}}$ decreases. This tightens the bound, reducing the true error $\mathrm{err}_{\mathcal{D}}(h)$. Thus, more training data typically improves generalization.
- If the training set size n is much larger than the number of parameters $|\mathcal{E}|$, then $\sqrt{\frac{|\mathcal{E}|+\log(1/\delta)}{n}}$ becomes small, leading to a tighter bound and improved generalization.

This analysis explains why a large neural network requires a correspondingly large training dataset in order to generalize well to unseen data. Otherwise, increasing the network size without increasing the amount of training data can lead to overfitting, resulting in poor generalization.

3.3 Theoretical Generalization Bounds Do Not Work in Practice for Neural Networks!

In Eq. (11), if the number of training instances n is approximately equal to the number of weights in the neural network $|\mathcal{E}|$, i.e., if $n \approx |\mathcal{E}| + \log(1/\delta)$, then the term $\sqrt{\frac{|\mathcal{E}|+\log(1/\delta)}{n}}$ becomes one, and the error bound remains loose. This implies that to achieve good generalization, the number of training instances must be much larger than the number of weights in the neural network, i.e., $n \gg |\mathcal{E}|$.

However, this theoretical result does not align with empirical observations in neural networks. In practice, large neural networks have been shown to generalize well, even when n is not much greater than $|\mathcal{E}|$. For example, large language models with millions of learnable parameters are trained on large corpora of Internet text. Although the condition $n \gg |\mathcal{E}|$ does not necessarily hold, these models still generalize effectively to unseen data.

This discrepancy clearly illustrates that theoretical generalization bounds do not accurately describe the practical behavior of deep learning. Several works in the literature have discussed this phenomenon, as summarized below.

3.3.1 Training Neural Networks with Random Labels or Data

The works [44, 45] highlight the paradox of generalization in deep learning—namely, that large neural networks are capable of generalizing well despite their size. Traditionally, small generalization errors are attributed to regularization techniques used during training (see Chap. 5). However, empirical evidence in [44, 45] suggests that generalization in deep learning cannot be fully explained by regularization alone.

Their experiments show that state-of-the-art Convolutional Neural Networks (CNNs) for image classification, when trained using stochastic gradient descent (see Chap. 3), can perfectly fit randomly labeled training data. That is, even when random labels were assigned to images, the networks could still achieve zero training error. As expected, test performance was poor in such cases because the training data lacked any meaningful structure. Notably, this capacity to memorize random labels persisted even in the presence of explicit regularization.

In another experiment, the networks were trained using correct labels but with inputs replaced by unstructured random noise. Again, the training error was driven to zero, despite the absence of useful signal in the input. Predictably, the test performance remained poor, due to the lack of structure in the input data.

These results indicate that neural networks with a sufficient number of parameters can perfectly fit any finite training set, regardless of label or input quality. This finding contradicts the expectations derived from PAC learning theory, as discussed in Sect. 3.3.

Some studies have proposed that the apparent lack of overfitting can be explained by properties of the loss landscape or implicit regularization arising from the optimization process itself [30].

3.3.2 Uniform Convergence Analysis in Deep Learning

Uniform convergence analysis seeks to determine whether a learning algorithm's performance on the training data converges uniformly to its performance on the underlying distribution, thereby enabling generalization to unseen data.

The work [26] questions the sufficiency of uniform convergence as a framework for explaining generalization in over-parameterized neural networks. Surprisingly, they show that uniform convergence-based bounds may increase—rather than decrease—with more training data, despite decreasing test error.

Their results demonstrate that even when neural networks achieve low test errors (i.e., generalize well), the generalization bounds obtained via uniform convergence remain ineffective. Consequently, uniform convergence alone cannot explain generalization in deep learning.

The authors argue that new or complementary theoretical frameworks are needed to adequately account for the generalization behavior of neural networks.

In addition to the aforementioned studies, other works have also explored the question of generalization in deep learning [22, 28].

4 Information Bottleneck in Deep Learning

When information is passed through a bottleneck, it undergoes compression such that only the most informative and relevant features are retained [40] (see Fig. 2a). In the information bottleneck framework, irrelevant information is minimized while task-relevant information is maximized [21].

A neural network can be interpreted as an information bottleneck. For instance, in an autoencoder, the compressed code layer between the encoder and decoder functions as the bottleneck [14] (see Fig. 2b). Similarly, in a standard multi-layer neural network, the hidden layers can be viewed as forming an information bottleneck between the input and output layers [41] (see Fig. 2c).

The information bottleneck provides one influential lens for understanding representation learning, emphasizing compression of inputs while preserving label-relevant information. However, this view is debated: empirical studies show that many networks generalize without clear compression, so information bottleneck should be seen as a possible but not universal explanation.

4.1 Deep Learning as Rate-Distortion Optimization

4.1.1 Rate-Distortion Optimization for Compression

Rate-distortion optimization is a central concept in information theory and is widely applied in data compression tasks [7]. Compression algorithms such as JPEG (Joint Photographic Experts Group) [25, 43] are based on this principle.

Rate-distortion optimization involves a trade-off between minimizing the *rate*, which corresponds to the amount of information retained (compression), and minimizing the *distortion*, which refers to the loss of essential information. Since reducing the rate typically increases distortion, both cannot be minimized simultaneously. As a result, achieving an effective compression requires a balanced trade-off.

Interestingly, many machine learning algorithms can also be interpreted through the lens of rate-distortion optimization [40]. Examples include quantized Fisher discriminant analysis [15] and neural networks [41].

The information bottleneck problem can itself be formulated as a rate-distortion optimization task. Regardless of whether the neural network is an autoencoder or a regular architecture, let x denote the input, $\widetilde{x}$ the compressed representation (e.g., hidden layers or code layer), $\widehat{y}$ the network output, and y the target labels. These can be considered as random variables X, $\widetilde{X}$, $\widehat{Y}$, and Y, respectively.

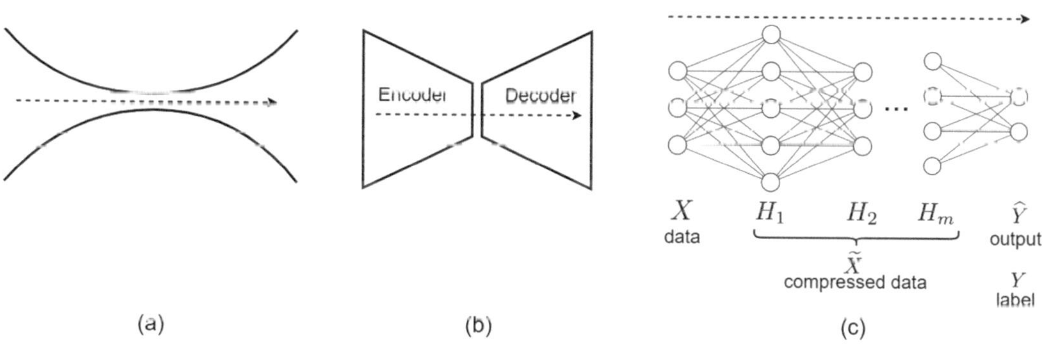

Fig. 2 Illustration of **a** information bottleneck, **b** autoencoder as an information bottleneck, and **c** a regular multi-layer neural network as an information bottleneck

By the law of total probability, we have

$$\mathbb{P}(\widetilde{x}) = \sum_{x} \mathbb{P}(\widetilde{x}|x)\mathbb{P}(x),$$

and by Bayes' rule and the chain rule of probability:

$$\mathbb{P}(y|\widetilde{x}) = \frac{\mathbb{P}(y, \widetilde{x})}{\mathbb{P}(\widetilde{x})} \overset{(a)}{=} \frac{\sum_{x} \mathbb{P}(y|x)\mathbb{P}(\widetilde{x}|x)\mathbb{P}(x)}{\mathbb{P}(\widetilde{x})},$$

where $\mathbb{P}(.)$ denotes probability and step (a) follows from the law of total probability.

The information bottleneck seeks to determine the optimal conditional distribution $\mathbb{P}(\widetilde{x}|x)$ by solving the following optimization problem:

$$\underset{\mathbb{P}(\widetilde{x}|x)}{\text{minimize}} \quad I(X; \widetilde{X}) - \beta I(\widetilde{X}; Y), \tag{12}$$

where $I(X_1; X_2)$ denotes the mutual information between random variables X_1 and X_2, and $\beta > 0$ is a regularization parameter that controls the trade-off between compression and task relevance.

According to the properties of probability distributions, the conditional distribution $\mathbb{P}(\widetilde{x}|x)$ must sum to one over all possible values of x and $\widetilde{x}$. Thus, the optimization in Eq. (12) can be reformulated as a constrained optimization problem:

$$\begin{aligned} &\underset{\mathbb{P}(\widetilde{x}|x)}{\text{minimize}} \quad && I(X; \widetilde{X}) - \beta I(\widetilde{X}; Y) \\ &\text{subject to} \quad && \sum_{x,\widetilde{x}} \mathbb{P}(\widetilde{x}|x) = 1. \end{aligned} \tag{13}$$

In either Eq. (12) or Eq. (13), the first term, $I(X; \widetilde{X})$, represents the rate to be minimized, thereby encouraging compression of X into $\widetilde{X}$. The second term, $-I(\widetilde{X}; Y)$, is minimized by maximizing the mutual information between the compressed representation $\widetilde{X}$ and the target Y, which corresponds to reducing distortion.

If $\beta = 0$, the solution collapses to a minimal representation—potentially as small as a single neuron—maximizing compression regardless of task relevance. Conversely, as $\beta \to \infty$, the optimization prioritizes preserving all task-relevant information, approaching a hard clustering interpretation of the information bottleneck [37].

4.1.2 Mutual Information of the Data, Hidden Layers, Labels, and Output of Neural Network

According to the definition of mutual information [7], we have

$$\begin{aligned} I(X; \widetilde{X}) &= \sum_{x,\widetilde{x}} \mathbb{P}(\widetilde{x}, x) \log\Big(\frac{\mathbb{P}(\widetilde{x}, x)}{\mathbb{P}(\widetilde{x})\mathbb{P}(x)}\Big) \\ &= \sum_{x,\widetilde{x}} \mathbb{P}(\widetilde{x}|x)\mathbb{P}(x) \log\Big(\frac{\mathbb{P}(\widetilde{x}|x)\mathbb{P}(x)}{\mathbb{P}(\widetilde{x})\mathbb{P}(x)}\Big) \\ &= \sum_{x,\widetilde{x}} \mathbb{P}(\widetilde{x}|x)\mathbb{P}(x) \log\Big(\frac{\mathbb{P}(\widetilde{x}|x)}{\mathbb{P}(\widetilde{x})}\Big), \end{aligned} \tag{14}$$

$$
\begin{aligned}
I(\widetilde{X}; Y) &= \sum_{\widetilde{x}, y} \mathbb{P}(\widetilde{x}, y) \log \Big(\frac{\mathbb{P}(\widetilde{x}, y)}{\mathbb{P}(\widetilde{x})\mathbb{P}(y)} \Big) \\
&= \sum_{\widetilde{x}, y} \mathbb{P}(\widetilde{x}, y) \log \Big(\frac{\mathbb{P}(\widetilde{x}|y)\mathbb{P}(y)}{\mathbb{P}(\widetilde{x})\mathbb{P}(y)} \Big) \\
&= \sum_{\widetilde{x}, y} \mathbb{P}(\widetilde{x}, y) \log \Big(\frac{\mathbb{P}(\widetilde{x}|y)}{\mathbb{P}(\widetilde{x})} \Big). \qquad (15)
\end{aligned}
$$

It is important to note that the mutual information between the compressed data $\widetilde{X}$ and the labels Y is always less than or equal to the mutual information between the original data X and the labels Y [40]:

$$I(\widetilde{X}; Y) \leq I(X; Y). \qquad (16)$$

This inequality reflects the fact that some information is inevitably lost during compression.

Let H_i denote the output of the ith hidden layer of a neural network (see Fig. 2c). Then, the mutual information between the network output $\widehat{Y}$ and the labels Y is bounded above by the mutual information between the hidden layers H_i and the labels Y. Moreover, these are also upper-bounded by the mutual information between the input X and the labels Y [41]:

$$I(\widehat{Y}; Y) \leq I(H_i; Y) \leq I(H_j; Y) \leq I(X; Y), \qquad (17)$$

for $i \geq j$. This follows from the fact that the input and labels are highly correlated, and although the network aims to replicate the labels at the output, the match is not perfect. Additionally, deeper hidden layers tend to lose more information due to increased compression. Therefore, the mutual information between the labels and shallower layers is typically greater than that of deeper layers.

When the joint distribution $\mathbb{P}(x, y)$ is estimated using n training instances, where n is the size of the dataset, let $|\mathcal{X}|$ and $|\mathcal{Y}|$ denote the cardinalities of the input and label spaces, respectively. Let $|\widetilde{\mathcal{X}}|$ denote the number of distinct values the compressed representation $\widetilde{x}$ can take. Let $\widehat{I}(\cdot\,;\,\cdot)$ denote the empirical mutual information estimated from the finite dataset. The following upper bounds apply [36]:

$$I(\widetilde{X}; Y) \leq \widehat{I}(\widetilde{X}; Y) + \mathcal{O}\Big(\frac{|\widetilde{\mathcal{X}}||\mathcal{Y}|}{\sqrt{n}}\Big), \qquad (18)$$

$$I(X; \widetilde{X}) \leq \widehat{I}(X; \widetilde{X}) + \mathcal{O}\Big(\frac{|\widetilde{\mathcal{X}}|}{\sqrt{n}}\Big), \qquad (19)$$

where $\mathcal{O}(\cdot)$ denotes big O complexity.

These bounds become looser as $|\widetilde{\mathcal{X}}|$ increases. Interestingly, they do not depend on the cardinality $|\mathcal{X}|$ of the original input space. This suggests that the information bottleneck is more reliably estimated when there is significant compression (i.e., when $|\widetilde{\mathcal{X}}|$ is small), and becomes less reliable under complex compressions with large $|\widetilde{\mathcal{X}}|$.

4.1.3 Solution of the Rate-Distortion Optimization for Compression

Theorem 5 ([40]) *The optimal solution to the information bottleneck optimization (13) is*

$$\mathbb{P}(\widetilde{x}|x) = \frac{\mathbb{P}(\widetilde{x})}{Z(x, \beta)} \exp\Big(-\beta KL\big(\mathbb{P}(y|x)\|\mathbb{P}(y|\widetilde{x})\big)\Big), \tag{20}$$

where $KL(.\|.)$ denotes the Kullback-Leibler (KL) divergence [23] and:

$$Z(x, \beta) := \sum_{\widetilde{x}} \mathbb{P}(\widetilde{x}) \exp\Big(-\beta KL\big(\mathbb{P}(y|x)\|\mathbb{P}(y|\widetilde{x})\big)\Big), \tag{21}$$

is the normalization factor, also called the partition function (see Chap. 7).

Equation (20) gives the distribution of the perfect compression $\widetilde{x}$ for the data x.

Proof The Lagrangian of the optimization problem (13), denoted by $\mathcal{L}$, is [13]:

$$\begin{aligned}
\mathcal{L} &= I(X; \widetilde{X}) - \beta I(\widetilde{X}; Y) - \sum_{x,\widetilde{x}} \lambda(x)(\mathbb{P}(\widetilde{x}|x) - 1) \\
&\overset{(a)}{=} \sum_{x,\widetilde{x}} \mathbb{P}(\widetilde{x}|x)\mathbb{P}(x) \log\Big(\frac{\mathbb{P}(\widetilde{x}|x)}{\mathbb{P}(\widetilde{x})}\Big) \\
&\quad - \beta \sum_{\widetilde{x},y} \mathbb{P}(\widetilde{x}, y) \log\Big(\frac{\mathbb{P}(\widetilde{x}|y)}{\mathbb{P}(\widetilde{x})}\Big) - \sum_{x,\widetilde{x}} \lambda(x)(\mathbb{P}(\widetilde{x}|x) - 1),
\end{aligned}$$

where $\lambda(x)$ is the Lagrange multiplier (dual variable) and (a) is because of Eqs. (14) and (15).

According to the law of total probability, there are

$$\mathbb{P}(\widetilde{x}) = \sum_{x} \mathbb{P}(\widetilde{x}|x)\mathbb{P}(x) \implies \frac{\partial \mathbb{P}(\widetilde{x})}{\partial \mathbb{P}(\widetilde{x}|x)} = \mathbb{P}(x), \tag{22}$$

$$\mathbb{P}(\widetilde{x}|y) = \sum_{x} \mathbb{P}(\widetilde{x}|x)\mathbb{P}(x|y) \implies \frac{\partial \mathbb{P}(\widetilde{x}|y)}{\partial \mathbb{P}(\widetilde{x}|x)} = \mathbb{P}(x|y), \tag{23}$$

where the summation goes away because the derivative is for a specific x and not the dummy variable x.

The derivative of the first term in the Lagrangian with respect to the primal optimization:

$$\begin{aligned}
&\frac{\partial}{\partial \mathbb{P}(\widetilde{x}|x)}\Big(\sum_{x,\widetilde{x}} \mathbb{P}(\widetilde{x}|x)\mathbb{P}(x) \log\Big(\frac{\mathbb{P}(\widetilde{x}|x)}{\mathbb{P}(\widetilde{x})}\Big)\Big) = \\
&\frac{\partial}{\partial \mathbb{P}(\widetilde{x}|x)}\Big(\sum_{x,\widetilde{x}} \mathbb{P}(\widetilde{x}|x)\mathbb{P}(x) \log\big(\mathbb{P}(\widetilde{x}|x)\big) - \sum_{x,\widetilde{x}} \mathbb{P}(\widetilde{x}|x)\mathbb{P}(x) \log\big(\mathbb{P}(\widetilde{x})\big)\Big) \\
&\overset{(a)}{=} \Big[\mathbb{P}(x) \log\big(\mathbb{P}(\widetilde{x}|x)\big) + \cancel{\mathbb{P}(\widetilde{x}|x)}\mathbb{P}(x)\frac{1}{\cancel{\mathbb{P}(\widetilde{x}|x)}}\Big] \\
&\quad - \Big[\mathbb{P}(x) \log\big(\mathbb{P}(\widetilde{x})\big) + \mathbb{P}(\widetilde{x}|x)\mathbb{P}(x)\frac{1}{\mathbb{P}(\widetilde{x})}\frac{\partial \mathbb{P}(\widetilde{x})}{\partial \mathbb{P}(\widetilde{x}|x)}\Big]
\end{aligned}$$

$$
\begin{aligned}
&\overset{(b)}{=} \mathbb{P}(x)\Big[1 + \log\big(\mathbb{P}(\widetilde{x}|x)\big)\Big] - \Big[\mathbb{P}(x)\log\big(\mathbb{P}(\widetilde{x})\big) + \cancel{\mathbb{P}(\widetilde{x})}\frac{1}{\cancel{\mathbb{P}(\widetilde{x})}}\mathbb{P}(x)\Big] \\
&= \mathbb{P}(x)\Big[1 + \log\big(\mathbb{P}(\widetilde{x}|x)\big)\Big] - \mathbb{P}(x)\Big[1 + \log\big(\mathbb{P}(\widetilde{x})\big)\Big] \\
&= \mathbb{P}(x)\Big[\log\big(\mathbb{P}(\widetilde{x}|x)\big) - \log\big(\mathbb{P}(\widetilde{x})\big)\Big] = \mathbb{P}(x)\log\Big(\frac{\mathbb{P}(\widetilde{x}|x)}{\mathbb{P}(\widetilde{x})}\Big),
\end{aligned}
$$

where (a) is noticing that the summations go away because the derivative is for a specific x and $\widetilde{x}$. The step (b) is because of Eq. (22) and $\sum_{x,\widetilde{x}} \mathbb{P}(\widetilde{x}|x)\mathbb{P}(x) = \sum_{\widetilde{x}} \mathbb{P}(\widetilde{x})$ by law of total probability, where the summation over $\widetilde{x}$ is also dropped because of derivative with respect to a specific $\widetilde{x}$.

The derivative of the second term in the Lagrangian with respect to the primal optimization:

$$
\begin{aligned}
&\frac{\partial}{\partial \mathbb{P}(\widetilde{x}|x)}\Big(-\beta \sum_{\widetilde{x},y} \mathbb{P}(\widetilde{x}, y)\log\Big(\frac{\mathbb{P}(\widetilde{x}|y)}{\mathbb{P}(\widetilde{x})}\Big)\Big) \overset{(a)}{=} \\
&\frac{\partial}{\partial \mathbb{P}(\widetilde{x}|x)}\Big(-\beta \sum_{y} \mathbb{P}(\widetilde{x}, y)\log\big(\mathbb{P}(\widetilde{x}|y)\big) + \beta \sum_{y} \mathbb{P}(\widetilde{x}, y)\log\big(\mathbb{P}(\widetilde{x})\big)\Big) \\
&\overset{(b)}{=} \frac{\partial}{\partial \mathbb{P}(\widetilde{x}|x)}\Big(-\beta \sum_{y} \mathbb{P}(\widetilde{x}|y)\mathbb{P}(y)\log\big(\mathbb{P}(\widetilde{x}|y)\big) + \beta\mathbb{P}(\widetilde{x})\log\big(\mathbb{P}(\widetilde{x})\big)\Big) \\
&= -\beta\Big[\sum_{y} \mathbb{P}(y)\frac{\partial \mathbb{P}(\widetilde{x}|y)}{\partial \mathbb{P}(\widetilde{x}|x)}\log\big(\mathbb{P}(\widetilde{x}|y)\big) \\
&\quad + \sum_{y} \mathbb{P}(y)\cancel{\mathbb{P}(\widetilde{x}|y)}\frac{1}{\cancel{\mathbb{P}(\widetilde{x}|y)}}\frac{\partial \mathbb{P}(\widetilde{x}|y)}{\partial \mathbb{P}(\widetilde{x}|x)}\Big] \\
&\quad + \beta\Big[\frac{\partial \mathbb{P}(\widetilde{x})}{\partial \mathbb{P}(\widetilde{x}|x)}\log\big(\mathbb{P}(\widetilde{x})\big) + \cancel{\mathbb{P}(\widetilde{x})}\frac{1}{\cancel{\mathbb{P}(\widetilde{x})}}\frac{\partial \mathbb{P}(\widetilde{x})}{\partial \mathbb{P}(\widetilde{x}|x)}\Big] \\
&= -\beta \sum_{y} \frac{\partial \mathbb{P}(\widetilde{x}|y)}{\partial \mathbb{P}(\widetilde{x}|x)}\mathbb{P}(y)\Big(1 + \log\big(\mathbb{P}(\widetilde{x}|y)\big)\Big) \\
&\quad + \beta\frac{\partial \mathbb{P}(\widetilde{x})}{\partial \mathbb{P}(\widetilde{x}|x)}\Big(1 + \log\big(\mathbb{P}(\widetilde{x})\big)\Big) \\
&\overset{(c)}{=} -\beta \sum_{y} \mathbb{P}(x|y)\mathbb{P}(y)\Big(1 + \log\big(\mathbb{P}(\widetilde{x}|y)\big)\Big) + \beta\mathbb{P}(x)\Big(1 + \log\big(\mathbb{P}(\widetilde{x})\big)\Big) \\
&\overset{(d)}{=} -\beta \sum_{y} \mathbb{P}(x)\Big(1 + \log\big(\mathbb{P}(\widetilde{x}|y)\big)\Big) + \beta\mathbb{P}(x)\Big(1 + \log\big(\mathbb{P}(\widetilde{x})\big)\Big),
\end{aligned}
$$

where (a) is noticing that the summation over $\widetilde{x}$ goes away because the derivative is for a specific $\widetilde{x}$. The step (b) is because of $\mathbb{P}(\widetilde{x}, y) = \mathbb{P}(\widetilde{x}|y)\mathbb{P}(y)$ and $\sum_y \mathbb{P}(\widetilde{x}, y) = \mathbb{P}(\widetilde{x})$ by law of total probability. Finally, the step (c) is because of Eqs. (22) and (23), and the step (d) is because of $\mathbb{P}(x|y)\mathbb{P}(y) = \mathbb{P}(x)$.

Putting the terms together, the derivative of the Lagrangian with respect to the primal optimization variable is

$$
\begin{aligned}
\frac{\partial \mathcal{L}}{\partial \mathbb{P}(\widetilde{x}|x)} &= \mathbb{P}(x)\log\Big(\frac{\mathbb{P}(\widetilde{x}|x)}{\mathbb{P}(\widetilde{x})}\Big) - \beta\sum_{y}\mathbb{P}(x)\Big(1+\log\big(\mathbb{P}(\widetilde{x}|y)\big)\Big) \\
&\quad + \beta\mathbb{P}(x)\Big(1+\log\big(\mathbb{P}(\widetilde{x})\big)\Big) - \lambda(x) \\
&\overset{(a)}{=} \mathbb{P}(x)\log\Big(\frac{\mathbb{P}(\widetilde{x}|x)}{\mathbb{P}(\widetilde{x})}\Big) - \beta\sum_{y}\mathbb{P}(x)\Big(1+\log\big(\mathbb{P}(\widetilde{x}|y)\big)\Big) \\
&\quad + \beta\mathbb{P}(x)\Big(\sum_{y}\mathbb{P}(y)+\sum_{y}\log\big(\mathbb{P}(\widetilde{x},y)\big)\Big) - \lambda(x) \\
&\overset{(b)}{=} \mathbb{P}(x)\Big[\log\Big(\frac{\mathbb{P}(\widetilde{x}|x)}{\mathbb{P}(\widetilde{x})}\Big) - \beta\sum_{y}\mathbb{P}(y|x)\log\Big(\frac{\mathbb{P}(y|\widetilde{x})}{\mathbb{P}(y)}\Big) - \frac{\lambda(x)}{\mathbb{P}(x)}\Big],
\end{aligned}
$$

where (a) is because $\sum_y \mathbb{P}(y) = 1$ and $\sum_y \log\big(\mathbb{P}(\widetilde{x}, y)\big) = \log\big(\mathbb{P}(\widetilde{x})\big)$, and (b) is because of the rearrangement mentioned in [40].

The derivative can be restated as

$$
\begin{aligned}
\frac{\partial \mathcal{L}}{\partial \mathbb{P}(\widetilde{x}|x)} &= \mathbb{P}(x)\Big[\log\Big(\frac{\mathbb{P}(\widetilde{x}|x)}{\mathbb{P}(\widetilde{x})}\Big) + \beta\sum_{y}\mathbb{P}(y|x)\log\Big(\frac{\mathbb{P}(y|x)}{\mathbb{P}(y|\widetilde{x})}\Big) \\
&\quad + \beta\sum_{y}\mathbb{P}(y|x)\log\Big(\frac{\mathbb{P}(y|x)}{\mathbb{P}(y|\widetilde{x})}\Big) - \frac{\lambda(x)}{\mathbb{P}(x)}\Big] \\
&\overset{(a)}{=} \mathbb{P}(x)\Big[\log\Big(\frac{\mathbb{P}(\widetilde{x}|x)}{\mathbb{P}(\widetilde{x})}\Big) + \beta\sum_{y}\mathbb{P}(y|x)\log\Big(\frac{\mathbb{P}(y|x)}{\mathbb{P}(y|\widetilde{x})}\Big) - \widetilde{\lambda}(x)\Big]
\end{aligned}
$$

where (a) is because we define

$$
\widetilde{\lambda}(x) := -\beta\sum_{y}\mathbb{P}(y|x)\log\Big(\frac{\mathbb{P}(y|x)}{\mathbb{P}(y|\widetilde{x})}\Big) + \frac{\lambda(x)}{\mathbb{P}(x)}.
$$

Setting the derivative of Lagrangian to zero results in:

$$
\mathbb{P}(\widetilde{x}|x) = \frac{\mathbb{P}(\widetilde{x})}{Z(x,\beta)}\exp\Big(-\beta\sum_{y}\mathbb{P}(y|x)\log\Big(\frac{\mathbb{P}(y|x)}{\mathbb{P}(y|\widetilde{x})}\Big)\Big),
$$

which is equivalent to Eq. (20), where:

$$
Z(x,\beta) = \exp\big(\beta\widetilde{\lambda}(x)\big) = \sum_{\widetilde{x}}\mathbb{P}(\widetilde{x})\exp\Big(-\beta\mathrm{KL}\big(\mathbb{P}(y|x)\|\mathbb{P}(y|\widetilde{x})\big)\Big).
$$

□

4.2 Upper-Bound on the Generalization Gap

It has been observed and suggested in some works that the generalization error of a neural network correlates with the amount of unnecessary information present in its hidden layers [21]. In other words, a network's ability to generalize is related to the extent of information bottlenecking.

However, other studies, such as [33, 34], argue that there is no such relationship between compression and generalization. They provide examples of neural networks that generalize well without compressing data in their hidden layers.

Here, we examine the theoretical relationship between generalization and compression under the assumption that such a correlation exists.

The generalization gap is defined as the difference between the expected loss and the training loss [21]:

$$\Delta(\mathcal{D}) := \mathbb{E}_{(X,Y)\sim\mathbb{P}(X,Y)}[\ell(f(X), Y)] - \frac{1}{n}\sum_{i=1}^{n}\ell(f(x_i) - y_i), \tag{24}$$

where ℓ is the neural network's loss function, $\mathcal{D}$ is the training dataset, $\mathbb{P}(X, Y)$ is the underlying data distribution, n is the number of training instances, and (x_i, y_i) denotes the ith training sample and its label.

Theorem 6 ([21, Theorem 2]) *For any $\delta > 0$, with high probability at least $1 - \delta$, there exists an upper bound on the generalization of the neural network:*

$$\lim_{n\to\infty} \mathcal{O}(\Delta(\mathcal{D})) \leq \min_{l\in\{1,\ldots,L\}} \sqrt{\frac{I(X; H_l|Y) + 1}{n}}, \tag{25}$$

where L is the number of layers in the neural network, n is the number of training instances, and $I(X; H_l|Y)$ denotes the mutual information between the input X and the lth hidden layer H_l, conditioned on the label Y.

Theorem 6 establishes an upper bound on the generalization gap in neural networks. This bound depends on the mutual information between the data and the representations at different hidden layers. Therefore, it supports the idea that generalization in neural networks is related to the information bottleneck.

It is also worth noting that some works, such as [18], utilize the information bottleneck not just for theoretical analysis but also as a practical tool for improving generalization during training.

4.3 Critiques of the Information Bottleneck Theory in Deep Learning

The information bottleneck theory has been proposed as a comprehensive framework for understanding generalization in deep neural networks. However, recent studies have highlighted several limitations:

1. Intractability of estimating mutual information: Estimating the mutual information $I(X; \widetilde{X})$ between the input X and the hidden representation $\widetilde{X}$ in high dimensional spaces is computationally challenging and intractable. Variational methods, such as Mutual Information Neural Estimation (MINE) and InfoNCE, have been developed to approximate these quantities, but they come with their own set of assumptions and limitations [4, 29].
2. Lack of compression in deep networks: Empirical observations suggest that deep neural networks do not always exhibit the compression phase predicted by the information bottleneck theory. Specifically, the mutual information between the input and the hidden layers does not consistently decrease during training, challenging the notion that compression is a universal mechanism for generalization [11, 32].

3. Alternative explanations for generalization: While the information bottleneck theory posits a direct link between compression and generalization, other factors may contribute to a model's ability to generalize. For example, the geometry of the loss landscape, the role of stochastic gradient descent, and the capacity of the model have also been implicated in influencing generalization performance [28, 46].

5 Theoretical Advances in Deep Learning Generalization

This section provides a brief overview of several theoretical advances in deep learning generalization, including double descent, grokking, benign overfitting, the implicit bias of stochastic gradient descent, neural network kernel theory, lottery ticket hypothesis, and scaling laws. These concepts offer insights into why deep neural networks frequently achieve strong generalization on unseen data.

5.1 Double Descent

Traditionally, as discussed in Chap. 5, test/validation error follows a U-shaped curve: it decreases as model complexity grows, then increases once the model begins to overfit. This U-shaped curve of test/validation error, while the training error goes down, is depicted on the left side of Fig. 3. However, empirical evidence shows that when models become extremely large—beyond the interpolation threshold, where they can perfectly fit the training data—the test/validation error unexpectedly decreases again (see Fig. 3). This phenomenon is known as double descent [5, 27]. According to this perspective, very large neural networks, such as large language models (see Chap. 11), exhibit test error that continues to decline, deviating from the traditionally observed U-shaped behavior.

The behavior of double descent is illustrated in Fig. 3 where the horizontal axis is the number of parameters (weights) in neural network. As this figure shows, in medium-sized neural networks, the

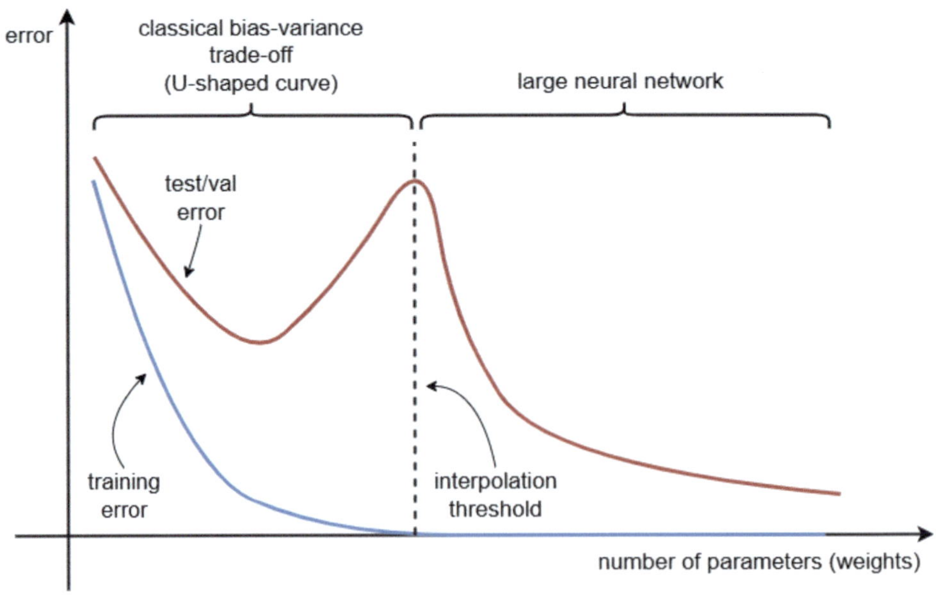

Fig. 3 The training and test/validation errors in double descent scenario.

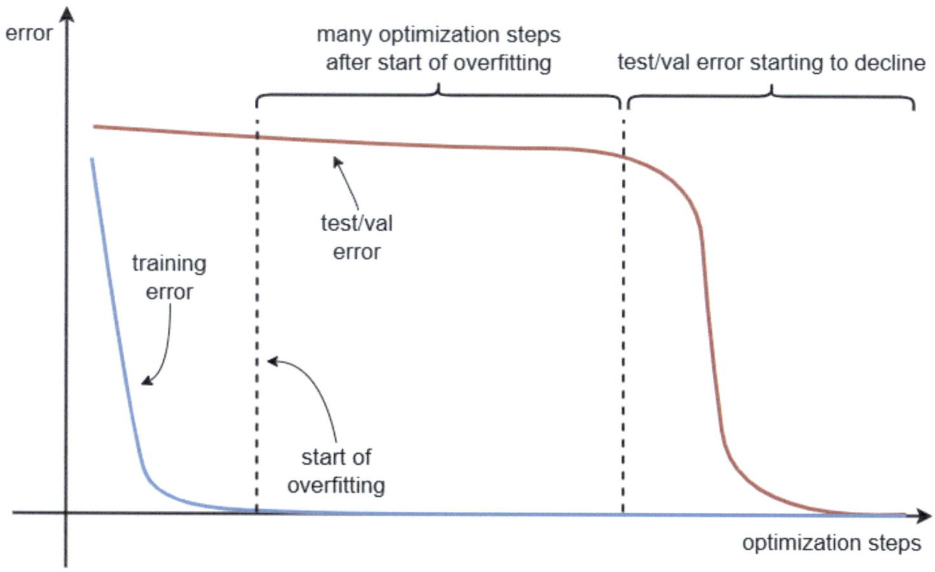

Fig. 4 The training and test/validation errors in grokking scenario.

test/validation error goes up after overfitting following the classical bias-variance trade-off. However, in very large neural networks, the test/validation error goes down even after overfitting.

5.2 Grokking

On small algorithmic datasets, a neural network may quickly achieve perfect training accuracy (by overfitting and memorizing the training data) while failing to generalize, resulting in poor validation or test performance. Surprisingly, if training is continued for orders of magnitude longer after overfitting, test accuracy can suddenly rise from chance level to near-perfect. This phenomenon, known as grokking [31], describes delayed generalization beyond the overfitting regime, where a sudden transition to strong generalization occurs well after the model has memorized the training data.

The behavior of grokking is depicted in Fig. 4. The horizontal axis of this figure is the optimization steps. As this figure shows, the training error quickly goes down, the neural network overfits to the training data, and the test/validation error is high because the dataset is small and algorithmic. The figure demonstrates that, after many optimization steps beyond start of overfitting, the test/validation error suddenly starts to decline.

5.3 Benign Overfitting

Consider data with an underlying true structure (the signal) alongside some noise in the labels or inputs, where the noise does not overwhelm the signal. Suppose a neural network is sufficiently overparameterized to fit both the signal and the noise perfectly. In this setting, even though the model interpolates the training data (including noise), it can still achieve low test error—provided the signal dominates the noise and the data distribution satisfies certain favorable properties (e.g., feature covariance structure in high-dimensional settings). This phenomenon is known as benign overfitting [1, 2].

The intuition is that in high-dimensional spaces, the geometry of overparameterized neural networks disperses the effect of noise across many parameters, diluting its negative impact on predictions. As a result, the dominant signal directions remain influential, allowing the model to generalize well despite technically memorizing noise. In other words, in overparameterized regimes, models can interpolate noisy training data and still generalize effectively, as long as the underlying data distribution has sufficient structure.

5.4 Implicit Bias of Stochastic Gradient Descent

The implicit bias of an optimization algorithm refers to its tendency to prefer certain solutions among the many that fit the training data equally well (i.e., achieve zero training error). This preference is not explicitly encoded in the loss function but emerges from the optimization dynamics [16, 38].

In deep learning, loss functions are often underdetermined, admitting many possible minima. Stochastic Gradient Descent (SGD) does not converge to an arbitrary solution; rather, its trajectory is influenced by factors such as step size, mini-batch noise, and initialization. This helps explain why neural networks frequently generalize well despite being heavily overparameterized.

Recall from Chap. 3 that SGD introduces noise through mini-batch sampling, which acts as an implicit regularizer by biasing the optimization process toward flatter minima—regions of the loss landscape characterized by small eigenvalues of the Hessian. Such flatter minima have been empirically associated with better generalization. Note that this implicit bias of SGD is closely related to the principles underlying Sharpness-Aware Minimization (SAM) [9] (see Chap. 3).

5.5 Neural Tangent Kernel Theory

The Neural Tangent Kernel (NTK) [19, 24] provides a framework for analyzing neural networks in the infinite-width regime, where each layer contains infinitely many neurons. In this limit, training a network with gradient descent becomes mathematically equivalent to performing kernel regression with a specific kernel—the NTK. This kernel, determined by the network's architecture and initialization, remains fixed during training in the infinite-width setting. As a result, instead of studying a highly nonlinear network, one can analyze it as a linear model in a high-dimensional feature space. This perspective helps explain why very wide neural networks often exhibit simple and predictable training dynamics.

The NTK further shows that wide networks do not evolve their effective features during training: the features are essentially "frozen" at initialization, with only the linear coefficients being updated. This behavior accounts for their resemblance to kernel methods and their stable optimization properties. By contrast, in finite-width networks, the NTK evolves during training, enabling feature learning and allowing such models to go beyond the limitations of pure kernel methods.

The NTK perspective complements the Universal Approximation Theorem [8] (see Chap. 2). While the universal approximation result shows that sufficiently wide networks can represent any continuous function, the NTK highlights how such wide networks are trained—revealing that, in the infinite-width limit, they behave like kernel machines rather than adaptive feature learners.

5.6 Lottery Ticket Hypothesis and Scaling Laws

Other important theoretical perspectives include the lottery ticket hypothesis [10], which shows that large overparameterized networks often contain smaller sub-networks—so-called "winning tickets"—that, when trained in isolation, can achieve performance comparable to the full model. This idea suggests that redundancy in large networks is not merely wasted capacity but may serve as a mechanism that enables the discovery of trainable subnetworks with favorable initialization and inductive biases. In practice, this line of work motivates research on network pruning, efficiency, and the implicit role of initialization in deep learning.

A complementary perspective is provided by scaling laws [20], which demonstrate that model performance follows predictable power-law relationships with respect to model size, dataset size, and computational budget. This implies that, given sufficient resources, performance improvements can be systematically achieved by scaling up along these axes, and that deviations from the power-law trend often signal limitations such as data quality or model architecture. Scaling laws therefore provide not only a descriptive account of empirical behavior in large models but also a practical tool for forecasting performance and guiding the allocation of computational resources.

Together, the lottery ticket hypothesis and scaling laws highlight both the efficiency hidden within large networks and the predictable benefits of scaling, making them central to understanding modern large-scale deep learning.

6 Conclusion

This chapter covered theoretical foundations of generalization in deep learning. Concepts like PAC learning, error bounds, expressive power, information bottleneck, double descent, grokking, benign overfitting, implicit bias of SGD, neural network kernel theory, lottery ticket hypothesis, and scaling laws were discussed.

In practice, classical theories like PAC learning and VC dimension are not predictive of neural network behavior. What matters more for practitioners are phenomena such as double descent, implicit bias from optimization, scaling laws, and architectural choices. These capture how deep learning models actually behave and generalize in real-world use.

7 Exercises

1. (Level: Easy) Suppose you are training a neural network classifier on a binary classification task using a large dataset. Under the PAC learning framework, which of the following statements is most accurate?
   (a) A model is PAC-learnable if it has 100% accuracy on the training data.
   (b) A hypothesis class is PAC-learnable if, for any small ε (error) and δ (confidence), there exists a sample size large enough such that the trained model has error less than ε with probability at least $1 - \delta$.
   (c) PAC learning guarantees zero generalization error for deep learning models.
   (d) PAC learning only applies to decision trees, not to neural networks.
2. (Level: Moderate) In the PAC learning framework, assume a hypothesis class $\mathcal{H}$ has a finite VC dimension d. You are training a neural network model (considered as a function in $\mathcal{H}$). Which of

the following best describes the relationship between the sample size n, the VC dimension d, the error ε, and the confidence δ?

(a) To achieve generalization error $\leq \varepsilon$ with confidence $\geq 1-\delta$, it suffices to have $n = \mathcal{O}(1/\varepsilon)$.
(b) The number of training samples required grows exponentially with the VC dimension.
(c) To ensure probably approximately correct learning, the required number of samples is $n = \mathcal{O}(\frac{d \log(1/\varepsilon)+\log(1/\delta)}{\varepsilon})$.
(d) The required number of training samples is independent of the VC dimension as long as ε and δ are small.

3. (Level: Difficult) Which of the following statements is true regarding the PAC-learnability of neural networks, particularly deep networks, under the standard assumptions of the PAC learning framework?

(a) Every neural network is PAC-learnable in polynomial time, regardless of its depth or activation function.
(b) Neural networks with non-polynomial activation functions (e.g., ReLU) are not PAC-learnable under any distribution.
(c) The class of Boolean functions represented by a feedforward neural network with a fixed number of layers and polynomially many weights is PAC-learnable, but it may not be efficiently PAC-learnable (i.e., in polynomial time), unless certain complexity-theoretic assumptions (like P=NP) hold.
(d) Deep neural networks are always PAC-learnable because they have universal approximation capabilities.

4. (Level: Easy) Consider datasets with binary labels in $\{0, 1\}$. Find the VC dimension for each of the following classification models:

(a) Let f_0 be a constant classifier that classifies everything as 0.
(b) Let f_a be a threshold classifier for $x \in \mathbb{R}$, with a single parameter, a. If $x > a$, f_a outputs 1. Otherwise, it outputs 0.
(c) Let f_3 be a decision tree classifier with 3 splits, for a one-dimensional input $x \in \mathbb{R}$.

5. (Level: Moderate) Let $\mathcal{H}$ be a hypothesis class with VC dimension d. In the PAC learning framework, to learn $\mathcal{H}$ with error at most ϵ and confidence at least $1-\delta$, it suffices to have:

$$n \geq \frac{1}{\epsilon}\left(d \log\left(\frac{1}{\epsilon}\right) + \log\left(\frac{1}{\delta}\right)\right).$$

Given a hypothesis class $\mathcal{H}$ with VC dimension $d = 10$, compute the minimum number of samples m required to achieve $\epsilon = 0.05$ and $\delta = 0.05$.

6. (Level: Difficult) Answer the following questions about PAC learning and ERM (Empirical Risk Minimization):

(a) What is the difference in settings between an agnostic PAC learning and a non-agnostic PAC learning?
(b) Derive the lower bound for training data size for ERM.
Hint: Use the Hoeffding's inequality and the union bound.

7. (Level: Medium) Answer the following questions:

(a) Define the key conditions under which a hypothesis class $\mathcal{H}$ is agnostically PAC learnable.
(b) Explain why it is difficult to prove that deep neural networks are agnostically PAC learnable.

(c) What role do regularization techniques, such as dropout or weight decay, play in addressing these theoretical limitations?

8. (Level: Difficult) Let a neural network have $|\mathcal{E}|$ learnable parameters and suppose the following complexity-dependent generalization bound holds with probability at least $1-\delta$ over the training set of size n:

$$\mathrm{err}_{\mathcal{D}}(h) \leq \inf_{h' \in \mathcal{H}} \mathrm{err}_{\mathcal{D}}(h') + \sqrt{\frac{|\mathcal{E}| + \log(1/\delta)}{n}}.$$

Answer the following mathematically (show all steps).

(a) For a prescribed tolerance $\varepsilon > 0$, derive an explicit necessary condition on n (as an inequality) that guarantees the excess term:

$$\sqrt{\frac{|\mathcal{E}| + \log(1/\delta)}{n}} \leq \varepsilon.$$

(b) Solve the inequality in part (a) for $|\mathcal{E}|$ to obtain the largest number of parameters the model may have while still guaranteeing the excess term is at most ε, given n and δ. Provide the result in a closed form.

(c) Suppose n and δ are fixed. Show that to halve the excess tolerance ε (i.e. replace ε by $\varepsilon/2$), how must $|\mathcal{E}|$ change asymptotically to keep the inequality in (a) satisfied? Give a brief mathematical justification.

9. (Level: Easy) Which of the following best distinguishes benign overfitting from the other generalization phenomena (double descent, grokking, and implicit bias of SGD)?

(a) It describes the situation where test error decreases again after the interpolation threshold is crossed, deviating from the classical U-shaped curve.
(b) It refers to delayed generalization, where test accuracy suddenly improves long after the model has memorized the training data.
(c) It highlights how overparameterized models can interpolate both signal and noise in training data, yet still generalize well when the signal dominates the noise.
(d) It emphasizes that optimization dynamics, rather than the explicit loss function, bias the solution towards flatter minima with better generalization.

References

1. Peter L Bartlett, Philip M Long, Gábor Lugosi, and Alexander Tsigler. "Benign overfitting in linear regression". In: *Proceedings of the National Academy of Sciences* 117.48 (2020), pp. 30063–30070.
2. Peter L Bartlett, Andrea Montanari, and Alexander Rakhlin. "Deep learning: a statistical viewpoint". In: *Acta numerica* 30 (2021), pp. 87–201.
3. Peter L. Bartlett, Nick Harvey, Christopher Liaw, and Abbas Mehrabian. "Nearly-tight VC dimension and pseudodimension bounds for piecewise linear neural networks". In: *Journal of Machine Learning Research* 20.1 (2019), pp. 1–17.
4. Ilyas Belghazi et al. "Mutual information neural estimation". In: *arXiv preprint* arXiv:1801.04062 (2018).
5. Mikhail Belkin, Daniel Hsu, Siyuan Ma, and Soumik Mandal. "Reconciling modern machine-learning practice and the classical bias–variance trade-off". In: *Proceedings of the National Academy of Sciences* 116.32 (2019), pp. 15849–15854.

6. Shai Ben-David, Pavel Hrubeš, Shay Moran, Amir Shpilka, and Amir Yehudayoff. "Learnability can be undecidable". In: *Nature Machine Intelligence* 1.1 (2019), pp. 44–48.
7. Thomas M Cover and Joy A Thomas. *Elements of information theory*. John Wiley & Sons, 2012.
8. George Cybenko. "Approximation by superpositions of a sigmoidal function". In: *Mathematics of control, signals and systems* 2.4 (1989), pp. 303–314.
9. Pierre Foret, Ariel Kleiner, Hossein Mobahi, and Behnam Neyshabur. "Sharpness-aware minimization for efficiently improving generalization". In: *International Conference on Learning Representations (ICLR)*. 2021.
10. Jonathan Frankle and Michael Carbin. "The lottery ticket hypothesis: Finding sparse, trainable neural networks". In: *International Conference on Learning Representations (ICLR)*. 2019.
11. Bernhard C. Geiger. "On information plane analyses of neural network classifiers—A review". In: *IEEE Transactions on Neural Networks and Learning Systems* 31.9 (2020), pp. 3485–3500.
12. Cédric Gerbelot et al. "Applying statistical learning theory to deep learning". In: *Journal of Statistical Mechanics: Theory and Experiment* 2024.10 (2024), p. 104003.
13. Benyamin Ghojogh, Mark Crowley, Fakhri Karray, and Ali Ghodsi. "Background on Optimization". In: *Elements of Dimensionality Reduction and Manifold Learning*. Springer, 2023, pp. 75–120.
14. Benyamin Ghojogh, Mark Crowley, Fakhri Karray, and Ali Ghodsi. "Introduction". In: *Elements of Dimensionality Reduction and Manifold Learning*. Springer, 2023, pp. 1–14.
15. Benyamin Ghojogh, Ali Saheb Pasand, Fakhri Karray, and Mark Crowley. "Quantized Fisher discriminant analysis". In: *arXiv preprint* arXiv:1909.03037 (2019).
16. Suriya Gunasekar, Jason D Lee, Daniel Soudry, and Nati Srebro. "Implicit bias of gradient descent on linear convolutional networks". In: *Advances in neural information processing systems* 31 (2018).
17. Nick Harvey, Christopher Liaw, and Abbas Mehrabian. "Nearly-tight VC-dimension bounds for piecewise linear neural networks". In: *Proceedings of the 2017 Conference on Learning Theory (COLT)*. 2017, pp. 1064–1068.
18. Zhanxuan Hu, Danyang Wu, Feiping Nie, and Rong Wang. "Generalization bottleneck in deep metric learning". In: *Information Sciences* 581 (2021), pp. 249–261.
19. Arthur Jacot, Franck Gabriel, and Clément Hongler. "Neural tangent kernel: Convergence and generalization in neural networks". In: *Advances in neural information processing systems* 31 (2018).
20. Jared Kaplan et al. "Scaling laws for neural language models". In: *arXiv preprint* arXiv:2001.08361 (2020).
21. Kenji Kawaguchi, Zhun Deng, Xu Ji, and Jiaoyang Huang. "How does information bottleneck help deep learning?" In: *International Conference on Machine Learning*. PMLR. 2023, pp. 16049–16096.
22. Kenji Kawaguchi, Leslie Pack Kaelbling, and Yoshua Bengio. "Generalization in Deep Learning". In: *Mathematical Aspects of Deep Learning*. Cambridge University Press, 2022. DOI:https://doi.org/10.1017/9781009025096.003.
23. Solomon Kullback and Richard A Leibler. "On information and sufficiency". In: *The annals of mathematical statistics* 22.1 (1951), pp. 79–86.
24. Jaehoon Lee et al. "Wide neural networks of any depth evolve as linear models under gradient descent". In: *Advances in neural information processing systems* 32 (2019).
25. J Mitchell. "Digital compression and coding of continuous-tone still images: Requirements and guidelines". In: *ITU-T Recommendation T* 81 (1992).
26. Vaishnavh Nagarajan and J Zico Kolter. "Uniform convergence may be unable to explain generalization in deep learning". In: *Advances in Neural Information Processing Systems* 32 (2019).
27. Preetum Nakkiran et al. "Deep double descent: Where bigger models and more data hurt". In: *Journal of Statistical Mechanics: Theory and Experiment* 2021.12 (2021), p. 124003.
28. Behnam Neyshabur, Srinadh Bhojanapalli, David McAllester, and Nati Srebro. "Exploring generalization in deep learning". In: *Advances in neural information processing systems* 30 (2017).
29. Aaron van den Oord, Oriol Vinyals, Dai He, and et al. "Representation learning: A review and new perspectives". In: *IEEE Transactions on Pattern Analysis and Machine Intelligence* 41.8 (2018), pp. 1797–1818.
30. Tomaso Poggio et al. "Theory of deep learning III: explaining the non-overfitting puzzle". In: *arXiv preprint* arXiv:1801.00173 (2017).
31. Alethea Power, Yuri Burda, Harri Edwards, Igor Babuschkin, and Vedant Misra. "Grokking: Generalization beyond overfitting on small algorithmic datasets". In: *arXiv preprint* arXiv:2201.02177 (2022).
32. Nuno Sa and Naftali Tishby. "Information bottleneck analysis of deep learning: A review". In: *Entropy* 20.10 (2018), p. 754.
33. Andrew M Saxe et al. "On the information bottleneck theory of deep learning". In: *Journal of Statistical Mechanics: Theory and Experiment* 2019.12 (2019), p. 124020.
34. Andrew Michael Saxe et al. "On the Information Bottleneck Theory of Deep Learning". In: *International Conference on Learning Representations*. 2018.
35. Shai Shalev-Shwartz and Shai Ben-David. *Understanding machine learning: From theory to algorithms*. Cambridge university press, 2014.
36. Ohad Shamir, Sivan Sabato, and Naftali Tishby. "Learning and generalization with the information bottleneck". In: *Theoretical Computer Science* 411.29-30 (2010), pp. 2696–2711.

37. Noam Slonim and Naftali Tishby. "Agglomerative information bottleneck". In: *Advances in neural information processing systems* 12 (1999).
38. Daniel Soudry, Elad Hoffer, Mor Shpigel Nacson, Suriya Gunasekar, and Nathan Srebro. "The implicit bias of gradient descent on separable data". In: *Journal of Machine Learning Research* 19.70 (2018), pp. 1–57.
39. Charles M Stein. "Estimation of the mean of a multivariate normal distribution". In: *The annals of Statistics* (1981), pp. 1135–1151.
40. Naftali Tishby, Fernando C Pereira, and William Bialek. "The information bottleneck method". In: *arXiv preprint physics/0004057* (2000).
41. Naftali Tishby and Noga Zaslavsky. "Deep learning and the information bottleneck principle". In: *2015 IEEE information theory workshop (ITW)*. IEEE. 2015, pp. 1–5.
42. Vladimir Vapnik. *The nature of statistical learning theory*. Springer science & business media, 2000.
43. Gregory K Wallace. "The JPEG still picture compression standard". In: *IEEE transactions on consumer electronics* 38.1 (1992), pp. xviii–xxxiv.
44. Chiyuan Zhang, Samy Bengio, Moritz Hardt, Benjamin Recht, and Oriol Vinyals. "Understanding deep learning (still) requires rethinking generalization". In: *Communications of the ACM* 64.3 (2021), pp. 107–115.
45. Chiyuan Zhang, Samy Bengio, Moritz Hardt, Benjamin Recht, and Oriol Vinyals. "Understanding deep learning requires rethinking generalization". In: *International Conference on Learning Representations (ICLR)* (2017).
46. Chiyuan Zhang, Yoshua Bengio, Moritz Hardt, Benjamin Recht, and Oriol Vinyals. "Understanding deep learning requires rethinking generalization". In: *arXiv preprint* arXiv:1611.03530 (2016).

Boltzmann Machines 7

Abstract

This chapter explores the foundations, evolution, and significance of Boltzmann Machines (BMs) and their variants within the context of neural networks and deep learning. Rooted in the Boltzmann distribution from statistical physics, these energy-based models draw inspiration from the Ising model and the Hopfield network, offering a probabilistic framework for learning data representations. The chapter traces the development of BMs and Restricted Boltzmann Machines (RBMs), emphasizing their architectures, learning algorithms, and their role in unsupervised feature learning through maximum likelihood estimation and Gibbs sampling. The chapter also introduces Deep Belief Networks (DBNs), which are formed by stacking multiple RBMs into deep generative models. This greedy layer-wise training approach addressed the vanishing gradient problem and marked a significant milestone in the resurgence of deep learning. However, the emergence of Rectified Linear Units (ReLU) and dropout regularization subsequently enhanced the effectiveness of backpropagation, reducing the need for unsupervised pretraining. The chapter begins with essential background concepts, including graphical models, Gibbs sampling, and the Ising and Hopfield models. It then presents the RBM, its conditional variant, and concludes with the structure and training of DBNs. Collectively, these models represent a pivotal historical phase in the evolution of deep learning. Today, their main importance is historical and conceptual, as modern neural networks typically rely on backpropagation-based training without RBM pretraining.

1 Introduction

The Boltzmann distribution [6], also known as the Gibbs distribution [15], was introduced many years ago. This energy-based probability distribution proved useful for modeling physical systems using statistical methods [29]. One such system was the Ising model, which described how particles with binary spins interact [30, 40]. Later, researchers recognized that the Ising model could be interpreted as a type of neural network [41]. This realization led to the development of the Hopfield network, which employed the Ising model structure to model memory in a network [27].

Inspired by the Hopfield network [27, 41], which itself was based on the Ising model [30, 40], Hinton and his colleagues introduced the Boltzmann Machine (BM) and the Restricted Boltzmann Machine (RBM) [1, 26]. These models are classified as energy-based models [39], named after the

B. Ghojogh and A. Ghodsi, *Elements of Deep Learning*,
https://doi.org/10.1007/978-3-032-10738-1_7

Boltzmann distribution [6, 15] used in their formulation. A BM includes weighted connections both between and within layers, whereas an RBM restricts the model by removing intra-layer connections.

In both BMs and RBMs, one layer represents the input data, while the other layer represents learned features or embeddings of the data. These models can be viewed as special cases of the Ising model, where the interaction strengths (weights) are learned. They are also related to the Hopfield network but differ in that they use maximum likelihood estimation to learn weights, rather than the Hebbian learning method [18] employed by the Hopfield network.

The Hebbian method struggled with generalization to new, unseen data. To address this limitation, the backpropagation algorithm was introduced [52], based on gradient descent and the chain rule. However, early deep networks suffered from a critical issue: as networks grew deeper, gradients often became very small, making learning difficult. This issue, known as the vanishing gradient problem, coincided with the rise in popularity of support vector machines using kernel methods [7], leading to a decline in interest in neural networks. This period is often referred to as the "winter of neural networks."

Despite this downturn, Hinton continued his research on neural networks. Along with his collaborators, including Max Welling, he revisited RBMs and proposed a method for training them using maximum likelihood estimation [22, 65]. BMs and RBMs are generative models, capable of producing new data samples via Gibbs sampling [13]. Hinton observed that the weights between any two layers of a neural network could define an RBM.

In 2006, he introduced a layer-by-layer greedy training approach[1] [4], where each layer is trained as an RBM [23, 24]. Stacking these RBMs led to the formulation of Deep Belief Networks (DBNs) [21, 23]. This method provided a strong initialization for deep networks and helped mitigate the vanishing gradient problem.

This breakthrough enabled successful training of deeper networks, which became more effective at learning from complex data. DBNs were applied successfully in various domains, including speech recognition [43–45] and action recognition [63]. Hinton expressed strong enthusiasm for RBMs and believed DBNs could shape the future of neural networks.

However, two new techniques soon became prominent: the Rectified Linear Unit (ReLU) activation function [16] and the dropout regularization method [58]. These techniques addressed overfitting (see Chap. 5) and also helped resolve the vanishing gradient problem, even without RBM-based pretraining. As a result, backpropagation alone became sufficiently powerful when combined with these methods. Neural networks began achieving success in many domains [38], including image recognition [31].

This chapter introduces Boltzmann Machines. Section 2 reviews the necessary background on probabilistic graphical models, Markov random fields, Gibbs sampling, statistical physics, the Ising model, and the Hopfield network. Restricted Boltzmann Machines are introduced in Sect. 3. Distributions of visible and hidden variables are covered in Sect. 4. Section 5 discusses the Conditional Restricted Boltzmann Machine. Finally, Sect. 6 presents the Deep Belief Network.

[1] A greedy algorithm makes each decision based on the best choice at the current step, without looking ahead to future steps. It hopes that many good small steps will lead to a good final result.

2 Background

2.1 Probabilistic Graphical Models and Markov Random Fields

Probabilistic Graphical Models (PGMs) are a class of statistical models that represent the structure of high-dimensional probability distributions using graphs [33]. In these models, nodes correspond to random variables, while edges encode (in)dependencies among those variables. PGMs provide a compact and efficient way to model conditional (in)dependence relationships.

PGMs can be broadly categorized into two types: Bayesian networks and Markov networks [33]. Bayesian networks use directed acyclic graphs to represent dependencies, whereas Markov networks—also known as Markov Random Fields (MRFs)—employ undirected graphs. Boltzmann Machines (BMs) and Restricted Boltzmann Machines (RBMs) are examples of MRFs due to their undirected graph structure [20].

2.2 Gibbs Sampling

Gibbs sampling is a Markov Chain Monte Carlo (MCMC) method introduced by [13]. It enables sampling from a multivariate distribution $\mathbb{P}(X)$ by drawing from its lower-dimensional conditional distributions. Gibbs sampling is especially useful when direct sampling from $\mathbb{P}(X)$ is intractable, but sampling from the conditionals $\mathbb{P}(x_j \mid x_1, x_2, \ldots, x_{j-1}, x_{j+1}, \ldots, x_d)$ is feasible [5].

Let $X \in \mathbb{R}^d$ and $\boldsymbol{x} = [x_1, x_2, \ldots, x_d]^\top$. Gibbs sampling begins with an arbitrary initial point and updates one dimension at a time, conditioned on the others [14]:

$$x_j \sim \mathbb{P}(x_j \mid x_1, \ldots, x_{j-1}, x_{j+1}, \ldots, x_d). \tag{1}$$

Each iteration cycles through all dimensions and repeats until convergence. Early samples are typically discarded during a burn-in phase. In most applications, the burn-in phase completes quickly, and only a few iterations are typically needed to obtain a valid sample.

Gibbs sampling is a special case of the Metropolis-Hastings algorithm in which the acceptance rate is always one [14]. In BMs and RBMs, it is commonly used to alternately sample between the hidden and visible layers.

2.3 Statistical Physics and the Ising Model

2.3.1 Boltzmann (Gibbs) Distribution

Consider a physical system of particles $\{x_i\}_{i=1}^d$, where each particle can occupy a discrete state (e.g., an electron with spin $+1$ or -1). The probability of the system being in a configuration x [29] is given by the *Boltzmann distribution* [6], also known as the *Gibbs distribution* [15]:

$$\mathbb{P}(x) = \frac{e^{-\beta E(x)}}{Z}, \tag{2}$$

where $E(x)$ is the energy of configuration x, and Z is the *partition function*:

$$Z := \sum_{\boldsymbol{x} \in \mathbb{R}^d} e^{-\beta E(x)}. \tag{3}$$

The parameter β is the inverse temperature:

$$\beta := \frac{1}{k_\beta T}, \tag{4}$$

where k_β is the Boltzmann constant and T is the absolute thermodynamic temperature.

The free energy of the system is defined as:

$$F(\beta) := \frac{-1}{\beta} \ln(Z), \tag{5}$$

and the internal energy is

$$\begin{aligned} U(\beta) &:= \frac{\partial}{\partial \beta}(\beta F(\beta)) = \frac{\partial}{\partial \beta}(-\ln(Z)) = \frac{-1}{Z}\frac{\partial Z}{\partial \beta} \\ &\overset{(3)}{=} \sum_{x \in \mathbb{R}^d} E(x) \frac{e^{-\beta E(x)}}{Z} \overset{(2)}{=} \sum_{x \in \mathbb{R}^d} \mathbb{P}(x) E(x). \end{aligned} \tag{6}$$

The entropy is

$$\begin{aligned} H(\beta) &:= -\sum_{x \in \mathbb{R}^d} \mathbb{P}(x) \ln\big(\mathbb{P}(x)\big) \overset{(2)}{=} -\sum_{x \in \mathbb{R}^d} \mathbb{P}(x)\big(-\beta E(x) - \ln(Z)\big) \\ &= \beta \sum_{x \in \mathbb{R}^d} \mathbb{P}(x) E(x) + \ln(Z) \underbrace{\sum_{x \in \mathbb{R}^d} \mathbb{P}(x)}_{=1} \overset{(a)}{=} -\beta F(\beta) + \beta U(\beta), \end{aligned} \tag{7}$$

where (a) follows from Eqs. (6) and (3).

Lemma 1 *Physical systems tend to occupy states of lower energy; as a result, they continuously dissipate energy over time.*

Proof According to the second law of thermodynamics, the entropy of an isolated physical system increases over time [9]. Entropy reflects the level of randomness or disorder in the system. Additionally, when a system releases energy to its environment, it transitions to a more disordered state. Consequently, the system's energy decreases over time, leading to an increase in entropy. □

Corollary 1 *Based on Eq. (2), in equilibrium, states with lower energy have higher probability. The system does not monotonically increase the probability of a state over time; instead, it explores states stochastically and converges to a stationary distribution given by the Boltzmann distribution.*

This corollary aligns with the intuition that physical systems evolve toward states of higher probability. A similar principle underlies simulated annealing[2], which was introduced in [32]. The concept of annealing—particularly temperature scheduling—has also been applied in training Boltzmann Machine (BM) models [2, 3, 50].

[2] Simulated annealing is a metaheuristic optimization method where a temperature parameter modulates the balance between global and local search. As the temperature is gradually lowered, the algorithm transitions from exploration to exploitation.

2.3.2 Ising Model

The Ising model [8, 30, 40] describes a physical system in which variables (spins) $x_i \in \{-1, +1\}$ interact through a graph. The joint probability is given by the Boltzmann distribution, Eq. (2), with the following energy function:

$$E(x) := \mathcal{H}(x) = -\sum_{(i,j)} J_{ij}\, x_i\, x_j, \tag{8}$$

where $\mathcal{H}(x)$ denotes the Hamiltonian, and J_{ij} represents the interaction (coupling) strength between particles i and j.

The underlying graph structure can take various forms (e.g., grid, torus), and interactions are typically local. The sign of J_{ij} determines the type of interaction:

- $J_{ij} \geq 0$: ferromagnetic (neighbors prefer to align),
- $J_{ij} < 0$: anti-ferromagnetic (neighbors prefer to disagree),
- mixed signs: spin glass,
- constant J_{ij}: homogeneous.

As $J_{ij} \to \infty$, a ferromagnetic Ising model converges to either the all-$+1$ or all--1 configuration with equal probability. The Ising model is equivalent to a normal factor graph [46, 47]. In BMs and RBMs, the weights can be interpreted as the coupling terms J_{ij} and are typically learned via maximum likelihood estimation [20]. These models fall under the broader category of energy-based models [39].

2.4 Hopfield Network

Inspired by the Ising model, Little proposed a neural variant in [41], which was later formalized by Hopfield as a model for associative memory [27]. A Hopfield network is a fully connected recurrent neural network composed of binary units $x_i \in \{-1, +1\}$. The weights are learned using Hebbian learning [18]:

$$w_{ij} := \begin{cases} x_i x_j & \text{if } i \neq j, \\ 0 & \text{if } i = j. \end{cases} \tag{9}$$

The network updates its state according to:

$$x_i := \begin{cases} +1 & \text{if } \sum_{j=1}^{d} w_{ij} x_j \geq \theta, \\ -1 & \text{otherwise}, \end{cases} \tag{10}$$

where θ is a threshold. For binary units $x_i \in \{0, 1\}$, Hopfield employed a modified rule $w_{ij} := (2x_i - 1)(2x_j - 1)$ for $i \neq j$ [27].

In the original Hopfield network formulation [27], the binary states are $x_i \in \{0, 1\}$ for all i, and the Hebbian learning rule becomes $w_{ij} := (2x_i - 1)(2x_j - 1)$ for all $i \neq j$. The Hopfield network can be regarded as a specific instance of the Ising model, as it employs the same energy function given in Eq. (8). This energy function also underlies the Boltzmann distribution in Eq. (2).

In addition to binary-state versions, Hopfield networks have also been developed to support continuous-valued states [28]. Modern developments, such as the model proposed in [51], are typically grounded in the framework of dense associative memories [37]. Further advancements in associative memory models are presented in [35, 36]. Notably, Boltzmann Machines (BM) and Restricted Boltzmann Machines (RBM) can also be considered Hopfield-type networks. However, unlike the original Hopfield model, their weights are learned through maximum likelihood estimation rather than Hebbian learning.

3 Restricted Boltzmann Machine

3.1 Structure of Restricted Boltzmann Machine

A Boltzmann Machine (BM) is a generative model and a type of Probabilistic Graphical Model (PGM) [5], which serves as a foundational component for many probabilistic models. The model derives its name from the Boltzmann distribution [6, 15], which underlies its formulation. BMs were first introduced for use in machine learning in [1, 26], and later extended in [22, 65].

A BM consists of a visible (or observation) layer $\boldsymbol{v} = [v_1, \ldots, v_d] \in \mathbb{R}^d$ and a hidden layer $\boldsymbol{h} = [h_1, \ldots, h_p] \in \mathbb{R}^p$. The visible layer corresponds to observed data, while the hidden layer consists of latent variables that capture meaningful features or embeddings of the visible data. Although the dimensionalities of these layers may differ (i.e., $d \neq p$), the two are meaningfully connected.

In the graphical model of a BM, connections (links) exist between units in $\boldsymbol{v}$ and units in $\boldsymbol{h}$. Additionally, each unit in $\boldsymbol{v}$ and $\boldsymbol{h}$ has an associated bias, and links may also exist between units within the same layer [55]. Let w_{ij} denote the link between v_i and h_j, l_{ij} the link between v_i and v_j, and j_{ij} the link between h_i and h_j. Let b_i and c_i denote the biases for v_i and h_i, respectively. The corresponding matrices and vectors are $\boldsymbol{W} = [w_{ij}] \in \mathbb{R}^{d \times p}$, $\boldsymbol{L} = [l_{ij}] \in \mathbb{R}^{d \times d}$, $\boldsymbol{J} = [j_{ij}] \in \mathbb{R}^{p \times p}$, $\boldsymbol{b} = [b_1, \ldots, b_d] \in \mathbb{R}^d$, and $\boldsymbol{c} = [c_1, \ldots, c_p] \in \mathbb{R}^p$. Note that $\boldsymbol{W}$ is symmetric, i.e., $w_{ij} = w_{ji}$. Since there are no self-connections, the diagonal elements of $\boldsymbol{L}$ and $\boldsymbol{J}$ are zero: $l_{ii} = j_{ii} = 0, \forall i$.

A Restricted Boltzmann Machine (RBM) is a special case of a BM in which there are no intra-layer connections—i.e., there are no links between units within the visible layer or within the hidden layer. This constraint is mathematically represented as $\boldsymbol{L} = \boldsymbol{J} = \boldsymbol{0}$. Figure 1 illustrates the structures of both a BM and an RBM.

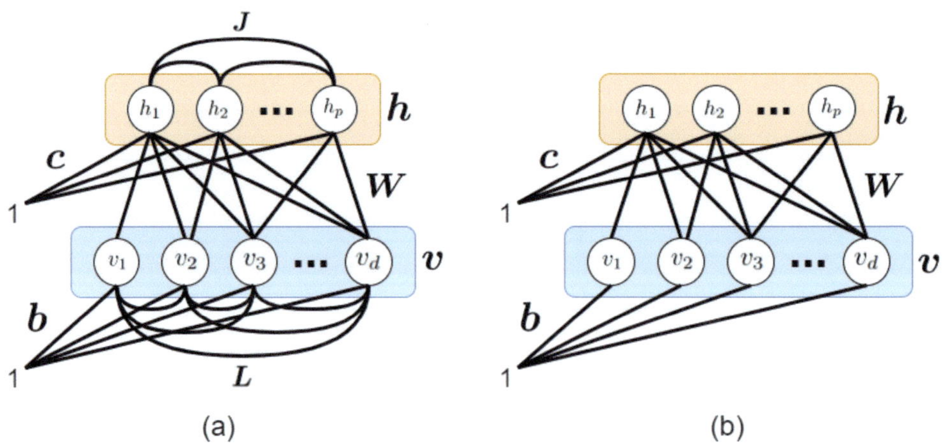

Fig. 1 The structures of **a** a Boltzmann machine and **b** a restricted Boltzmann machine

This section focuses on the RBM. Recall that an RBM is a type of Ising model. As seen in Eq. (8), the energy of an Ising model can be written as [1, 26]:

$$\mathbb{R} \ni E(\boldsymbol{v}, \boldsymbol{h}) := -\boldsymbol{b}^\top \boldsymbol{v} - \boldsymbol{c}^\top \boldsymbol{h} - \boldsymbol{v}^\top \boldsymbol{W}\boldsymbol{h}, \tag{11}$$

which is defined based on the interactions between connected units.

As introduced in Eq. (2), the joint probability distribution over visible and hidden units follows the Boltzmann distribution [19]:

$$\begin{aligned}\mathbb{P}(\boldsymbol{v}, \boldsymbol{h}) &= \frac{1}{Z}\exp(-E(\boldsymbol{v}, \boldsymbol{h})) \\ &\overset{(11)}{=} \frac{1}{Z}\exp(\boldsymbol{b}^\top \boldsymbol{v} + \boldsymbol{c}^\top \boldsymbol{h} + \boldsymbol{v}^\top \boldsymbol{W}\boldsymbol{h}),\end{aligned} \tag{12}$$

where Z is the partition function:

$$Z := \sum_{\boldsymbol{v}\in\mathbb{R}^d} \sum_{\boldsymbol{h}\in\mathbb{R}^p} \exp(-E(\boldsymbol{v}, \boldsymbol{h})). \tag{13}$$

According to Lemma 1, both BMs and RBMs aim to reduce their energy. Training a BM or RBM involves maximizing the log-likelihood of the observed data, which corresponds to assigning lower energy to data configurations while raising the energy of non-data configurations [1, 26].

3.2 Conditional Distributions

Proposition 1 (Conditional Independence of Variables) *In an RBM, given the visible variables, the hidden variables are conditionally independent. Likewise, given the hidden variables, the visible variables are conditionally independent. This property does not hold in a general BM due to the presence of intra-layer connections.*

Proof According to the Bayes' rule, there is

$$\begin{aligned}\mathbb{P}(\boldsymbol{h}|\boldsymbol{v}) &= \frac{\mathbb{P}(\boldsymbol{h}, \boldsymbol{v})}{\mathbb{P}(\boldsymbol{v})} = \frac{\mathbb{P}(\boldsymbol{v}, \boldsymbol{h})}{\sum_{\boldsymbol{h}\in\mathbb{R}^p} \mathbb{P}(\boldsymbol{v}, \boldsymbol{h})} \\ &\overset{(12)}{=} \frac{\frac{1}{Z}\exp(\boldsymbol{b}^\top \boldsymbol{v} + \boldsymbol{c}^\top \boldsymbol{h} + \boldsymbol{v}^\top \boldsymbol{W}\boldsymbol{h})}{\sum_{\boldsymbol{h}\in\mathbb{R}^p} \frac{1}{Z}\exp(\boldsymbol{b}^\top \boldsymbol{v} + \boldsymbol{c}^\top \boldsymbol{h} + \boldsymbol{v}^\top \boldsymbol{W}\boldsymbol{h})} \\ &= \frac{\frac{1}{Z}\exp(\boldsymbol{b}^\top \boldsymbol{v})\exp(\boldsymbol{c}^\top \boldsymbol{h})\exp(\boldsymbol{v}^\top \boldsymbol{W}\boldsymbol{h})}{\frac{1}{Z}\sum_{\boldsymbol{h}\in\mathbb{R}^p} \exp(\boldsymbol{b}^\top \boldsymbol{v})\exp(\boldsymbol{c}^\top \boldsymbol{h})\exp(\boldsymbol{v}^\top \boldsymbol{W}\boldsymbol{h})} \\ &\overset{(a)}{=} \frac{\exp(\boldsymbol{b}^\top \boldsymbol{v})\exp(\boldsymbol{c}^\top \boldsymbol{h})\exp(\boldsymbol{v}^\top \boldsymbol{W}\boldsymbol{h})}{\exp(\boldsymbol{b}^\top \boldsymbol{v})\sum_{\boldsymbol{h}\in\mathbb{R}^p} \exp(\boldsymbol{c}^\top \boldsymbol{h})\exp(\boldsymbol{v}^\top \boldsymbol{W}\boldsymbol{h})} \\ &= \frac{\exp(\boldsymbol{c}^\top \boldsymbol{h})\exp(\boldsymbol{v}^\top \boldsymbol{W}\boldsymbol{h})}{\sum_{\boldsymbol{h}\in\mathbb{R}^p} \exp(\boldsymbol{c}^\top \boldsymbol{h})\exp(\boldsymbol{v}^\top \boldsymbol{W}\boldsymbol{h})},\end{aligned}$$

where (a) is because the term $\exp(\boldsymbol{b}^\top \boldsymbol{v})$ does not have $\boldsymbol{h}$ in it. Note that $\sum_{\boldsymbol{h}\in\mathbb{R}^p}$ denotes summation over all possible p-dimensional hidden variables for the sake of marginalization. Let $Z' := \sum_{\boldsymbol{h}\in\mathbb{R}^p} \exp(\boldsymbol{c}^\top \boldsymbol{h}) \exp(\boldsymbol{v}^\top \boldsymbol{W}\boldsymbol{h})$. Therefore:

$$\begin{aligned}\mathbb{P}(\boldsymbol{h}|\boldsymbol{v}) &= \frac{1}{Z'} \exp(\boldsymbol{c}^\top \boldsymbol{h} + \boldsymbol{v}^\top \boldsymbol{W}\boldsymbol{h}) \\ &= \frac{1}{Z'} \exp\Big(\sum_{j=1}^{p} c_j h_j + \sum_{j=1}^{p} \boldsymbol{v}^\top \boldsymbol{W}_{:j} h_j\Big) \\ &= \frac{1}{Z'} \prod_{j=1}^{p} \exp(c_j h_j + \boldsymbol{v}^\top \boldsymbol{W}_{:j} h_j), \end{aligned} \tag{14}$$

where $\boldsymbol{W}_{:j} \in \mathbb{R}^d$ denotes the jth column of matrix $\boldsymbol{W}$. Equation (14) shows that given the visible variables, the hidden variables are conditionally independent because their joint distribution is the product of every distribution. Similar expressions can be written for the probability $\mathbb{P}(\boldsymbol{v}|\boldsymbol{h})$:

$$\begin{aligned}\mathbb{P}(\boldsymbol{v}|\boldsymbol{h}) &= \frac{1}{Z''} \exp(\boldsymbol{b}^\top \boldsymbol{v} + \boldsymbol{v}^\top \boldsymbol{W}\boldsymbol{h}) \\ &= \frac{1}{Z''} \exp\Big(\sum_{i=1}^{d} b_i v_i + \sum_{i=1}^{d} v_i \boldsymbol{W}_{i:} \boldsymbol{h}\Big) \\ &= \frac{1}{Z''} \prod_{i=1}^{d} \exp(b_i v_i + v_i \boldsymbol{W}_{i:} \boldsymbol{h}), \end{aligned} \tag{15}$$

where $\boldsymbol{W}_{i:} \in \mathbb{R}^p$ denotes the ith row of matrix $\boldsymbol{W}$ and $Z'' := \sum_{\boldsymbol{v}\in\mathbb{R}^d} \exp(\boldsymbol{b}^\top \boldsymbol{v}) \exp(\boldsymbol{v}^\top \boldsymbol{W}\boldsymbol{h})$. This equation shows that given the hidden variables, the visible variables are conditionally independent. □

According to Eq. (14) and considering the rule $\mathbb{P}(\boldsymbol{h}|\boldsymbol{v}) = \mathbb{P}(\boldsymbol{h}, \boldsymbol{v})/\mathbb{P}(\boldsymbol{v})$, there is

$$\begin{aligned}\mathbb{P}(\boldsymbol{h}|\boldsymbol{v}) &= \frac{1}{Z'} \prod_{j=1}^{p} \exp(c_j h_j + \boldsymbol{v}^\top \boldsymbol{W}_{:j} h_j) \\ &= \frac{1}{Z'} \prod_{j=1}^{p} \mathbb{P}(h_j, \boldsymbol{v}) \end{aligned}$$

$$\begin{aligned}\implies \mathbb{P}(h_j, \boldsymbol{v}) &= \exp(c_j h_j + \boldsymbol{v}^\top \boldsymbol{W}_{:j} h_j) \\ &= \exp(c_j h_j + \sum_{i=1}^{d} v_i w_{ij} h_j). \end{aligned} \tag{16}$$

Similarly, according to Eq. (15) and considering the rule $\mathbb{P}(\boldsymbol{v}|\boldsymbol{h}) = \mathbb{P}(\boldsymbol{h}, \boldsymbol{v})/\mathbb{P}(\boldsymbol{h})$, there is

$$
\begin{aligned}
\mathbb{P}(\boldsymbol{h}|\boldsymbol{v}) &= \frac{1}{Z''} \prod_{i=1}^{d} \exp(b_i v_i + v_i \boldsymbol{W}_{i:} \boldsymbol{h}) \\
&= \frac{1}{Z''} \prod_{i=1}^{d} \mathbb{P}(\boldsymbol{h}, v_i) \\
\implies \mathbb{P}(\boldsymbol{h}, v_i) &= \exp(b_i v_i + v_i \boldsymbol{W}_{i:} \boldsymbol{h}) \\
&= \exp(b_i v_i + \sum_{j=1}^{p} v_i w_{ij} h_j).
\end{aligned} \tag{17}
$$

These equations will be used later in this chapter.

3.3 Sampling Hidden and Visible Variables

3.3.1 Gibbs Sampling

Gibbs sampling can be used to sample and generate the hidden and visible units in an RBM. Let ν denote the iteration index of Gibbs sampling. Samples are drawn iteratively as follows:

$$
\boldsymbol{h}^{(\nu)} \sim \mathbb{P}(\boldsymbol{h} \mid \boldsymbol{v}^{(\nu)}), \tag{18}
$$

$$
\boldsymbol{v}^{(\nu+1)} \sim \mathbb{P}(\boldsymbol{v} \mid \boldsymbol{h}^{(\nu)}), \tag{19}
$$

until the burn-in phase converges. As discussed in Sect. 2.2, only a few iterations of Gibbs sampling are typically sufficient. After the burn-in, the samples are approximate samples from the joint distribution $\mathbb{P}(\boldsymbol{v}, \boldsymbol{h})$.

Because the variables are conditionally independent, Gibbs sampling can be implemented as shown in Algorithm refalgorithmspsRBMspsGibbsspssampling. In this algorithm, sampling $h_j^{(\nu)} \sim \mathbb{P}(h_j \mid \boldsymbol{v}^{(\nu)})$ can be implemented by drawing a sample $u \sim U[0, 1]$ from a uniform distribution and comparing it to the value of the Probability Density Function (PDF) $\mathbb{P}(h_j \mid \boldsymbol{v}^{(\nu)})$. If u is less than or equal to this value, then $h_j = 1$; otherwise, $h_j = 0$. Sampling v_i follows a similar procedure. Alternatively, the inverse of the cumulative distribution function can be used to draw samples (see [14] for further details on sampling methods).

Input: visible dataset $\boldsymbol{v}$, (initialization: optional)
Get initialization or perform random initialization of $\boldsymbol{v}$
while *until burn-in* **do**
  for *j from* 1 *to p* **do**
    $h_i^{(\nu)} \sim \mathbb{P}(h_j \mid \boldsymbol{v}^{(\nu)})$
  for *i from* 1 *to d* **do**
    $v_i^{(\nu+1)} \sim \mathbb{P}(v_i \mid \boldsymbol{h}^{(\nu)})$

Algorithm 1: Gibbs sampling in RBM

3.3.2 Generations and Evaluations by Gibbs Sampling

Gibbs sampling is used to generate both visible and hidden units during the training and evaluation phases of the RBM. Its role in training will be discussed in Sects. 3.4 and 3.5. Once the RBM is trained, any number of p-dimensional hidden variables can be generated as meaningful representations of d-dimensional observations using Gibbs sampling. Conversely, new d-dimensional observations can be generated as representations of p-dimensional hidden variables. This generative capability illustrates that both BMs and RBMs are generative models.

3.4 Training Restricted Boltzmann Machine by Maximum Likelihood Estimation

The weights of the links, namely $\boldsymbol{W}$, $\boldsymbol{b}$, and $\boldsymbol{c}$, must be learned so that they can be used for sampling and generating the hidden and visible units. Consider a dataset of n visible vectors $\{\boldsymbol{v}_i \in \mathbb{R}^d\}_{i=1}^n$. Note that $\boldsymbol{v}_i$ should not be confused with v_i, where the former refers to the ith visible data instance and the latter denotes the ith visible unit. The jth component of $\boldsymbol{v}_i$ is denoted by $\boldsymbol{v}_{i,j}$; that is, $\boldsymbol{v}_i = [\boldsymbol{v}_{i,1}, \ldots, \boldsymbol{v}_{i,d}]^\top$.

The log-likelihood of the visible data is given by:

$$
\begin{aligned}
\ell(\boldsymbol{W}, \boldsymbol{b}, \boldsymbol{c}) &= \sum_{i=1}^{n} \log(\mathbb{P}(\boldsymbol{v}_i)) = \sum_{i=1}^{n} \log\Big(\sum_{\boldsymbol{h}\in\mathbb{R}^p} \mathbb{P}(\boldsymbol{v}_i, \boldsymbol{h})\Big) \\
&\overset{(12)}{=} \sum_{i=1}^{n} \log\Big(\sum_{\boldsymbol{h}\in\mathbb{R}^p} \frac{1}{Z} \exp(-E(\boldsymbol{v}_i, \boldsymbol{h}))\Big) \\
&= \sum_{i=1}^{n} \log\Big(\frac{1}{Z} \sum_{\boldsymbol{h}\in\mathbb{R}^p} \exp(-E(\boldsymbol{v}_i, \boldsymbol{h}))\Big) \\
&= \sum_{i=1}^{n} \Big[\log\Big(\sum_{\boldsymbol{h}\in\mathbb{R}^p} \exp(-E(\boldsymbol{v}_i, \boldsymbol{h}))\Big) - \log Z\Big] \\
&= \sum_{i=1}^{n} \log\Big(\sum_{\boldsymbol{h}\in\mathbb{R}^p} \exp(-E(\boldsymbol{v}_i, \boldsymbol{h}))\Big) - n \log Z \\
&\overset{(13)}{=} \sum_{i=1}^{n} \log\Big(\sum_{\boldsymbol{h}\in\mathbb{R}^p} \exp(-E(\boldsymbol{v}_i, \boldsymbol{h}))\Big) - n \log \sum_{\boldsymbol{v}\in\mathbb{R}^d} \sum_{\boldsymbol{h}\in\mathbb{R}^p} \exp(-E(\boldsymbol{v}, \boldsymbol{h})).
\end{aligned}
\tag{20}
$$

Maximum Likelihood Estimation (MLE) can be used to estimate the parameters $\theta := \{\boldsymbol{W}, \boldsymbol{b}, \boldsymbol{c}\}$. The derivative of the log-likelihood with respect to a parameter θ is

$$
\nabla_\theta \ell(\theta) = \nabla_\theta \sum_{i=1}^{n} \log\Big(\sum_{\boldsymbol{h}\in\mathbb{R}^p} \exp(-E(\boldsymbol{v}_i, \boldsymbol{h}))\Big) - n \nabla_\theta \log \sum_{\boldsymbol{v}\in\mathbb{R}^d} \sum_{\boldsymbol{h}\in\mathbb{R}^p} \exp(-E(\boldsymbol{v}, \boldsymbol{h})). \tag{21}
$$

The first term of this derivative is

$$\begin{aligned}
&\nabla_\theta \sum_{i=1}^{n} \log\Big(\sum_{\boldsymbol{h}\in\mathbb{R}^p} \exp(-E(\boldsymbol{v}_i,\boldsymbol{h}))\Big) \\
&\quad= \sum_{i=1}^{n} \nabla_\theta \log\Big(\sum_{\boldsymbol{h}\in\mathbb{R}^p} \exp(-E(\boldsymbol{v}_i,\boldsymbol{h}))\Big) \\
&\quad= \sum_{i=1}^{n} \frac{\nabla_\theta \sum_{\boldsymbol{h}\in\mathbb{R}^p} \exp(-E(\boldsymbol{v}_i,\boldsymbol{h}))}{\sum_{\boldsymbol{h}\in\mathbb{R}^p} \exp(-E(\boldsymbol{v}_i,\boldsymbol{h}))} \\
&\quad= \sum_{i=1}^{n} \frac{\sum_{\boldsymbol{h}\in\mathbb{R}^p} \exp(-E(\boldsymbol{v}_i,\boldsymbol{h}))\nabla_\theta(-E(\boldsymbol{v}_i,\boldsymbol{h}))}{\sum_{\boldsymbol{h}\in\mathbb{R}^p} \exp(-E(\boldsymbol{v}_i,\boldsymbol{h}))} \\
&\quad\overset{(a)}{=} \sum_{i=1}^{n} \mathbb{E}_{\sim\mathbb{P}(\boldsymbol{h}|\boldsymbol{v}_i)}[\nabla_\theta(-E(\boldsymbol{v}_i,\boldsymbol{h}))],
\end{aligned} \tag{22}$$

where (a) follows from the definition of expectation: $\mathbb{E}_{\sim\mathbb{P}}[\boldsymbol{x}] := \sum_i \mathbb{P}(\boldsymbol{x}_i)\,\boldsymbol{x}_i$. If $\mathbb{P}$ is unnormalized (i.e., it does not sum to one), the expectation is normalized as: $\mathbb{E}_{\sim\mathbb{P}}[\boldsymbol{x}] := (\sum_i \mathbb{P}(\boldsymbol{x}_i)\,\boldsymbol{x}_i)/(\sum_i \mathbb{P}(\boldsymbol{x}_i))$.

The second term of the derivative of log-likelihood is

$$\begin{aligned}
&- n\nabla_\theta \log \sum_{\boldsymbol{v}\in\mathbb{R}^d}\sum_{\boldsymbol{h}\in\mathbb{R}^p} \exp(-E(\boldsymbol{v},\boldsymbol{h})) \\
&\quad= -n\frac{\nabla_\theta \sum_{\boldsymbol{v}\in\mathbb{R}^d}\sum_{\boldsymbol{h}\in\mathbb{R}^p} \exp(-E(\boldsymbol{v},\boldsymbol{h}))}{\sum_{\boldsymbol{v}\in\mathbb{R}^d}\sum_{\boldsymbol{h}\in\mathbb{R}^p} \exp(-E(\boldsymbol{v},\boldsymbol{h}))} \\
&\quad= -n\frac{\sum_{\boldsymbol{v}\in\mathbb{R}^d}\sum_{\boldsymbol{h}\in\mathbb{R}^p} \nabla_\theta\exp(-E(\boldsymbol{v},\boldsymbol{h}))}{\sum_{\boldsymbol{v}\in\mathbb{R}^d}\sum_{\boldsymbol{h}\in\mathbb{R}^p} \exp(-E(\boldsymbol{v},\boldsymbol{h}))} \\
&\quad= -n\frac{\sum_{\boldsymbol{v}\in\mathbb{R}^d}\sum_{\boldsymbol{h}\in\mathbb{R}^p} \exp(-E(\boldsymbol{v},\boldsymbol{h}))\nabla_\theta(-E(\boldsymbol{v},\boldsymbol{h}))}{\sum_{\boldsymbol{v}\in\mathbb{R}^d}\sum_{\boldsymbol{h}\in\mathbb{R}^p} \exp(-E(\boldsymbol{v},\boldsymbol{h}))} \\
&\quad\overset{(a)}{=} -n\,\mathbb{E}_{\sim\mathbb{P}(\boldsymbol{h},\boldsymbol{v})}[\nabla_\theta(-E(\boldsymbol{v},\boldsymbol{h}))],
\end{aligned} \tag{23}$$

where (a) again uses the definition of expectation.

Thus, the gradient of the log-likelihood is

$$\begin{aligned}
\nabla_\theta \ell(\theta) = &\sum_{i=1}^{n} \mathbb{E}_{\sim\mathbb{P}(\boldsymbol{h}|\boldsymbol{v}_i)}[\nabla_\theta(-E(\boldsymbol{v}_i,\boldsymbol{h}))] \\
&- n\,\mathbb{E}_{\sim\mathbb{P}(\boldsymbol{h},\boldsymbol{v})}[\nabla_\theta(-E(\boldsymbol{v},\boldsymbol{h}))].
\end{aligned} \tag{24}$$

Setting this derivative to zero does not yield a closed-form solution, so the parameters must be learned iteratively via gradient ascent.

Now consider each parameter in $\theta = \{\boldsymbol{W}, \boldsymbol{b}, \boldsymbol{c}\}$. The derivatives in Eq. (24) are

$$\begin{aligned}
\nabla_{\boldsymbol{W}}(-E(\boldsymbol{v},\boldsymbol{h})) &\overset{(11)}{=} \frac{\partial}{\partial \boldsymbol{W}}(\boldsymbol{b}^\top\boldsymbol{v} + \boldsymbol{c}^\top\boldsymbol{h} + \boldsymbol{v}^\top\boldsymbol{W}\boldsymbol{h}) = \boldsymbol{v}\boldsymbol{h}^\top, \\
\nabla_{\boldsymbol{b}}(-E(\boldsymbol{v},\boldsymbol{h})) &\overset{(11)}{=} \frac{\partial}{\partial \boldsymbol{b}}(\boldsymbol{b}^\top\boldsymbol{v} + \boldsymbol{c}^\top\boldsymbol{h} + \boldsymbol{v}^\top\boldsymbol{W}\boldsymbol{h}) = \boldsymbol{v}, \\
\nabla_{\boldsymbol{c}}(-E(\boldsymbol{v},\boldsymbol{h})) &\overset{(11)}{=} \frac{\partial}{\partial \boldsymbol{c}}(\boldsymbol{b}^\top\boldsymbol{v} + \boldsymbol{c}^\top\boldsymbol{h} + \boldsymbol{v}^\top\boldsymbol{W}\boldsymbol{h}) = \boldsymbol{h}.
\end{aligned}$$

Thus, Eq. (24) becomes

$$\begin{aligned}
\nabla_{\boldsymbol{W}}\ell(\theta) &= \sum_{i=1}^{n} \mathbb{E}_{\sim\mathbb{P}(\boldsymbol{h}|\boldsymbol{v}_i)}[\boldsymbol{v}\boldsymbol{h}_i^\top] - n\,\mathbb{E}_{\sim\mathbb{P}(\boldsymbol{h},\boldsymbol{v})}[\boldsymbol{v}\boldsymbol{h}^\top] \\
&= \sum_{i=1}^{n} \boldsymbol{v}_i\,\mathbb{E}_{\sim\mathbb{P}(\boldsymbol{h}|\boldsymbol{v}_i)}[\boldsymbol{h}^\top] - n\,\mathbb{E}_{\sim\mathbb{P}(\boldsymbol{h},\boldsymbol{v})}[\boldsymbol{v}\boldsymbol{h}^\top], \\
\nabla_{\boldsymbol{b}}\ell(\theta) &= \sum_{i=1}^{n} \mathbb{E}_{\sim\mathbb{P}(\boldsymbol{h}|\boldsymbol{v}_i)}[\boldsymbol{v}_i] - n\,\mathbb{E}_{\sim\mathbb{P}(\boldsymbol{h},\boldsymbol{v})}[\boldsymbol{v}] \\
&= \sum_{i=1}^{n} \boldsymbol{v}_i - n\,\mathbb{E}_{\sim\mathbb{P}(\boldsymbol{h},\boldsymbol{v})}[\boldsymbol{v}], \\
\nabla_{\boldsymbol{c}}\ell(\theta) &= \sum_{i=1}^{n} \mathbb{E}_{\sim\mathbb{P}(\boldsymbol{h}|\boldsymbol{v}_i)}[\boldsymbol{h}] - n\,\mathbb{E}_{\sim\mathbb{P}(\boldsymbol{h},\boldsymbol{v})}[\boldsymbol{h}].
\end{aligned}$$

We define:

$$\widehat{\boldsymbol{h}}_i := \mathbb{E}_{\sim\mathbb{P}(\boldsymbol{h}|\boldsymbol{v}_i)}[\boldsymbol{h}], \tag{25}$$

and the gradients can be summarized as:

$$\mathbb{R}^{d\times p} \ni \nabla_{\boldsymbol{W}}\ell(\theta) = \sum_{i=1}^{n} \boldsymbol{v}_i\widehat{\boldsymbol{h}}_i^\top - n\,\mathbb{E}_{\sim\mathbb{P}(\boldsymbol{h},\boldsymbol{v})}[\boldsymbol{v}\boldsymbol{h}^\top], \tag{26}$$

$$\mathbb{R}^{d} \ni \nabla_{\boldsymbol{b}}\ell(\theta) = \sum_{i=1}^{n} \boldsymbol{v}_i - n\,\mathbb{E}_{\sim\mathbb{P}(\boldsymbol{h},\boldsymbol{v})}[\boldsymbol{v}], \tag{27}$$

$$\mathbb{R}^{p} \ni \nabla_{\boldsymbol{c}}\ell(\theta) = \sum_{i=1}^{n} \widehat{\boldsymbol{h}}_i - n\,\mathbb{E}_{\sim\mathbb{P}(\boldsymbol{h},\boldsymbol{v})}[\boldsymbol{h}]. \tag{28}$$

Setting these gradients to zero does not yield a closed-form solution, so the parameters are typically optimized iteratively using gradient ascent. Note that the log-likelihood gradient involves two types of expectation: the conditional expectation $\mathbb{E}_{\sim\mathbb{P}(\boldsymbol{h}|\boldsymbol{v}_i)}[\cdot]$ depends on the observed data $\boldsymbol{v}_i$, while the joint expectation $\mathbb{E}_{\sim\mathbb{P}(\boldsymbol{h},\boldsymbol{v})}[\cdot]$ is model-dependent and unrelated to the training data.

3.5 Contrastive Divergence

According to Eq. (22), the conditional expectation used in Eq. (25) includes one summation. Moreover, according to Eq. (23), the joint expectations used in Eqs. (26), (27), and (28) contain two summations. This double summation makes computation of the joint expectation intractable, as it involves summing over all possible values for both hidden and visible units. Therefore, exact computation of MLE is difficult and must be approximated. One common approach to approximate the joint expectations in MLE is *contrastive divergence* [22]. Contrastive divergence improves computational efficiency and reduces the variance of the estimates in RBM training [22, 65].

The core idea behind contrastive divergence is as follows. First, a point $\widetilde{\boldsymbol{v}}$ is generated using Gibbs sampling, starting from the observation $\boldsymbol{v}_i$ (see Sect. 3.3 for Gibbs sampling in RBM). Then, the expectation is approximated using only that sampled point $\widetilde{\boldsymbol{v}}$.

The intuitive rationale for contrastive divergence is as follows. The gradients must be minimized to optimize the MLE objective. In the joint expectations in Eqs. (26), (27), and (28), instead of summing over all possible values of observations, contrastive divergence uses only one sampled observation. If this observation represents an undesired (incorrect) pattern that the RBM should not generate, contrastive divergence acts as a form of *negative sampling* [19]. Rather than attempting to suppress all undesirable outputs at once, negative sampling iteratively teaches the model to avoid generating one such output at a time. Over successive iterations, the model gradually learns to avoid generating incorrect outputs and instead focuses on generating correct ones.

Let $\widetilde{\boldsymbol{h}} = [\widetilde{h}_1, \dots, \widetilde{h}_m]^\top$ denote the corresponding sampled $\boldsymbol{h}$ for $\widetilde{\boldsymbol{v}} = [\widetilde{v}_1, \dots, \widetilde{v}_m]^\top$ in Gibbs sampling. As explained above, contrastive divergence approximates the joint expectation in the derivative of the log-likelihood, Eq. (24), using Monte Carlo approximation [14] evaluated at $\widetilde{\boldsymbol{v}}_i$ and $\widetilde{\boldsymbol{h}}_i$ for the ith training instance. Therefore:

$$\mathbb{E}_{\sim\mathbb{P}(\boldsymbol{h},\boldsymbol{v})}[\nabla_\theta(-E(\boldsymbol{v}, \boldsymbol{h}))] \approx \frac{1}{n}\sum_{i=1}^{n} \nabla_\theta(-E(\boldsymbol{v}_i, \boldsymbol{h}_i))\Big|_{\boldsymbol{v}_i=\widetilde{\boldsymbol{v}}_i, \boldsymbol{h}_i=\widetilde{\boldsymbol{h}}_i}. \tag{29}$$

Empirical studies have shown that a small number of Gibbs sampling iterations are sufficient for contrastive divergence. In fact, [22] uses only a single iteration. This is supported by the literature, as Gibbs sampling is a special case of the Metropolis-Hastings algorithm [14], which is known to be efficient [11].

Input: training data $\{\boldsymbol{x}_i\}_{i=1}^n$
Randomly initialize $\boldsymbol{W}$, $\underline{b}$, $\boldsymbol{c}$
while *not converged* **do**
    Sample a mini-batch $\{\boldsymbol{v}_1, \dots, \boldsymbol{v}_m\}$ from training dataset $\{\boldsymbol{x}_i\}_{i=1}^n$ (n.b. we may set $m = n$)
    // Gibbs sampling for each data point:
    Initialize $\widehat{\boldsymbol{v}}_i^{(0)} \leftarrow \boldsymbol{v}_i$ for all $i \in \{1, \dots, m\}$
    for *i from* 1 *to m* **do**
        Algorithm 1 $\leftarrow \widehat{\boldsymbol{v}}_i^{(0)}$
        $\{h_i\}_{i=1}^p, \{v_i\}_{i=1}^d \leftarrow$ Last iteration of Algorithm 1
        $\widetilde{\boldsymbol{h}}_i \leftarrow [h_1, \dots, h_p]^\top$
        $\widetilde{\boldsymbol{v}}_i \leftarrow [v_1, \dots, v_d]^\top$
        $\widehat{\boldsymbol{h}}_i \leftarrow \mathbb{E}_{\sim\mathbb{P}(\boldsymbol{h}|\boldsymbol{v}_i)}[\boldsymbol{h}]$
    // gradients:
    $\nabla_{\boldsymbol{W}}\ell(\theta) \leftarrow \sum_{i=1}^m \boldsymbol{v}_i \widehat{\boldsymbol{h}}_i^\top - \sum_{i=1}^m \widetilde{\boldsymbol{h}}_i \widetilde{\boldsymbol{v}}_i^\top$
    $\nabla_{\underline{b}}\ell(\theta) \leftarrow \sum_{i=1}^m \boldsymbol{v}_i - \sum_{i=1}^m \widetilde{\boldsymbol{v}}_i$
    $\nabla_{\boldsymbol{c}}\ell(\theta) \leftarrow \sum_{i=1}^m \widehat{\boldsymbol{h}}_i - \sum_{i=1}^m \widetilde{\boldsymbol{h}}_i$
    // gradient ascent for updating solution:
    $\boldsymbol{W} \leftarrow \boldsymbol{W} + \eta\nabla_{\boldsymbol{W}}\ell(\theta)$
    $\underline{b} \leftarrow \underline{b} + \eta\nabla_{\underline{b}}\ell(\theta)$
    $\boldsymbol{c} \leftarrow \boldsymbol{c} + \eta\nabla_{\boldsymbol{c}}\ell(\theta)$
Return $\boldsymbol{W}$, $\underline{b}$, $\boldsymbol{c}$

Algorithm 2: Training RBM using contrastive divergence

By applying the approximation in Eq. (29), Eqs. (26), (27), and (28) become:

$$\nabla_{\boldsymbol{W}} \ell(\theta) = \sum_{i=1}^{n} \boldsymbol{v}_i \widehat{\boldsymbol{h}}_i^\top - \sum_{i=1}^{n} \widetilde{\boldsymbol{v}}_i \widetilde{\boldsymbol{h}}_i^\top, \tag{30}$$

$$\nabla_{\boldsymbol{b}} \ell(\theta) = \sum_{i=1}^{n} \boldsymbol{v}_i - \sum_{i=1}^{n} \widetilde{\boldsymbol{v}}_i, \tag{31}$$

$$\nabla_{\boldsymbol{c}} \ell(\theta) = \sum_{i=1}^{n} \widehat{\boldsymbol{h}}_i - \sum_{i=1}^{n} \widetilde{\boldsymbol{h}}_i. \tag{32}$$

These equations are intuitive: when the observed variable and its reconstruction match, the gradient becomes zero and the training process converges.

Note that some literature restates Eqs. (30), (31), and (32) as follows [19, 22, 63]:

$$\forall i, j: \; \nabla_{w_{ij}} \ell(\theta) = \langle v_i h_j \rangle_{\text{data}} - \langle v_i h_j \rangle_{\text{recon.}}, \tag{33}$$

$$\forall i: \quad \nabla_{b_i} \ell(\theta) = \langle v_i \rangle_{\text{data}} - \langle v_i \rangle_{\text{recon.}}, \tag{34}$$

$$\forall j: \quad \nabla_{c_j} \ell(\theta) = \langle h_j \rangle_{\text{data}} - \langle h_j \rangle_{\text{recon.}}, \tag{35}$$

where $\langle . \rangle_{\text{data}}$ and $\langle . \rangle_{\text{recon.}}$ denote expectations over the data and its reconstruction, respectively.

The training algorithm for the RBM using contrastive divergence is summarized in Algorithm refalgorithmspsRBMspstrainingspsusingspscontrastivespsdivergence. This algorithm uses mini-batch gradient descent with batch size m. If the training dataset is not large, one can set $m = n$ to use full gradient descent. The algorithm proceeds iteratively until convergence. In each iteration, a mini-batch is sampled, and for each ith data point, an observation $\boldsymbol{v}_i \in \mathbb{R}^d$ and corresponding hidden variable $\boldsymbol{h}_i \in \mathbb{R}^p$ are generated. Gibbs sampling is applied for every data point, as shown in Algorithm refalgorithmspsRBMspsGibbsspssampling. After Gibbs sampling, gradients are computed using Eqs. (30), (31), and (32), and the parameters are updated using a gradient ascent step.

3.6 Boltzmann Machine

So far, the RBM has been introduced and explained. A Boltzmann Machine (BM) contains more links compared to an RBM [55]. In this section, we briefly introduce the training of a BM. Its structure is depicted in Fig. 1. As discussed in Sect. 3.1, a BM includes additional links $\boldsymbol{L} = [l_{ij}] \in \mathbb{R}^{d \times d}$ and $\boldsymbol{J} = [j_{ij}] \in \mathbb{R}^{p \times p}$.

The weights $\boldsymbol{W} \in \mathbb{R}^{d \times p}$ and the biases $\boldsymbol{b} \in \mathbb{R}^d$ and $\boldsymbol{c} \in \mathbb{R}^p$ are trained using gradient descent, with gradients given in Eqs. (30), (31), and (32). The additional weights $\boldsymbol{L}$ and $\boldsymbol{J}$ are updated in a similar manner using the following gradients [55]:

$$\nabla_{\boldsymbol{L}} \ell(\theta) = \sum_{i=1}^{n} \boldsymbol{v}_i \boldsymbol{v}_i^\top - \sum_{i=1}^{n} \widetilde{\boldsymbol{v}}_i \widetilde{\boldsymbol{v}}_i^\top, \tag{36}$$

$$\nabla_{\boldsymbol{J}} \ell(\theta) = \sum_{i=1}^{n} \mathbb{E}_{\sim \mathbb{P}(\boldsymbol{h}|\boldsymbol{v}_i)}[\boldsymbol{h}\boldsymbol{h}^\top] - \sum_{i=1}^{n} \widetilde{\boldsymbol{h}}_i \widetilde{\boldsymbol{h}}_i^\top. \tag{37}$$

These equations can also be expressed as:

$$\forall i, j: \nabla_{l_{ij}} \ell(\theta) = \langle v_i v_j \rangle_{\text{data}} - \langle v_i v_j \rangle_{\text{recon.}}, \tag{38}$$

$$\forall i, j: \nabla_{j_{ij}} \ell(\theta) = \langle h_i h_j \rangle_{\text{data}} - \langle h_i h_j \rangle_{\text{recon.}}, \tag{39}$$

where $\langle . \rangle_{\text{data}}$ and $\langle . \rangle_{\text{recon.}}$ denote expectations over the data and its reconstruction, respectively.

4 Distributions of Visible and Hidden Variables

4.1 Modeling with Exponential Family Distributions

According to Proposition 1, the units $\boldsymbol{v} \in \mathbb{R}^d$ and $\boldsymbol{h} \in \mathbb{R}^p$ are conditionally independent, so their distribution factorizes into the product of individual conditional distributions. The distributions for the visible and hidden variables can be selected from the exponential family [65]:

$$\mathbb{P}(\boldsymbol{v}) = \prod_{i=1}^{d} r_i(v_i) \exp\Big(\sum_a \theta_{ia}\, f_{ia}(v_i) - A_i(\{\theta_{ia}\})\Big), \tag{40}$$

$$\mathbb{P}(\boldsymbol{h}) = \prod_{j=1}^{p} s_j(h_j) \exp\Big(\sum_b \lambda_{jb}\, g_{jb}(h_j) - B_j(\{\lambda_{jb}\})\Big), \tag{41}$$

where $\{f_{ia}(v_i), g_{jb}(h_j)\}$ are the sufficient statistics, $\{\theta_i, \lambda_j\}$ are the canonical parameters, $\{A_i, B_j\}$ are the log-normalization functions, and $\{r_i(v_i), s_j(h_j)\}$ are base measures that may include additional scaling constants. The log-normalization terms can often be ignored in practice because they are difficult to compute.

To define the joint distribution of the visible and hidden variables, a quadratic interaction term is introduced to capture the dependencies between the two layers [65]:

$$\mathbb{P}(\boldsymbol{v}, \boldsymbol{h}) \propto \exp\Big(\sum_{i=1}^{d}\sum_a \theta_{ia}\, f_{ia}(v_i) + \sum_{j=1}^{p}\sum_b \lambda_{jb}\, g_{jb}(h_j) + \sum_{i=1}^{d}\sum_{j=1}^{p}\sum_a\sum_b \boldsymbol{W}_{ia}^{jb}\, f_{ia}(v_i)\, g_{jb}(h_j)\Big). \tag{42}$$

As stated in Proposition 1, the visible and hidden units are conditionally independent. Therefore, the conditional distributions can be expressed as products of exponential family distributions [65]:

$$\mathbb{P}(\boldsymbol{v}|\boldsymbol{h}) = \prod_{i=1}^{d} \exp\Big(\sum_a \widehat{\theta}_{ia}\, f_{ia}(v_i) - A_i(\{\widehat{\theta}_{ia}\})\Big), \tag{43}$$

$$\mathbb{P}(\boldsymbol{h}|\boldsymbol{v}) = \prod_{j=1}^{p} \exp\Big(\sum_b \widehat{\lambda}_{jb}\, g_{jb}(h_j) - B_j(\{\widehat{\lambda}_{jb}\})\Big), \tag{44}$$

where:

$$\widehat{\theta}_{ia} := \theta_{ia} + \sum_{j=1}^{p} \sum_{b} \mathbf{W}_{ia}^{jb}\, g_{jb}(h_j), \tag{45}$$

$$\widehat{\lambda}_{jb} := \lambda_{jb} + \sum_{i=1}^{d} \sum_{a} \mathbf{W}_{ia}^{jb}\, f_{ia}(v_i). \tag{46}$$

Thus, one can choose appropriate distributions from the exponential family for modeling the conditional distributions of the visible and hidden units. In the following subsections, different cases are presented where the units may take either discrete or continuous values. In all cases, the distributions are drawn from the exponential family.

4.2 Binary States

The hidden and visible variables can take on a discrete number of values, also called states. Most commonly, inspired by the Hopfield network, BMs and RBMs use binary states. In this case, the hidden and visible units take binary values, i.e., $v_i, h_j \in \{0, 1\}, \forall i, j$. Therefore, we can write:

$$\mathbb{P}(h_j = 1|\boldsymbol{v}) = \frac{\mathbb{P}(h_j = 1, \boldsymbol{v})}{\mathbb{P}(h_j = 0, \boldsymbol{v}) + \mathbb{P}(h_j = 1, \boldsymbol{v})}. \tag{47}$$

In the binary case, the joint probability in Eq. (16) simplifies as follows:

$$\mathbb{P}(h_j = 0, \boldsymbol{v}) = \exp(c_j \times 0 + \boldsymbol{v}^\top \boldsymbol{W}_{:j} \times 0) = \exp(0) = 1,$$
$$\mathbb{P}(h_j = 1, \boldsymbol{v}) = \exp(c_j \times 1 + \boldsymbol{v}^\top \boldsymbol{W}_{:j} \times 1) = \exp(c_j + \boldsymbol{v}^\top \boldsymbol{W}_{:j}).$$

Substituting into Eq. (47), we get:

$$\begin{aligned}\mathbb{P}(h_j = 1|\boldsymbol{v}) &= \frac{\exp(c_j + \boldsymbol{v}^\top \boldsymbol{W}_{:j})}{1 + \exp(c_j + \boldsymbol{v}^\top \boldsymbol{W}_{:j})} \\ &= \frac{1}{1 + \exp\big(-(c_j + \boldsymbol{v}^\top \boldsymbol{W}_{:j})\big)} = \sigma(c_j + \boldsymbol{v}^\top \boldsymbol{W}_{:j}),\end{aligned} \tag{48}$$

where:

$$\sigma(x) := \frac{\exp(x)}{1 + \exp(x)} = \frac{1}{1 + \exp(-x)},$$

is the sigmoid (or logistic) function, and $\boldsymbol{W}_{:j} \in \mathbb{R}^d$ denotes the jth column of matrix $\boldsymbol{W}$.

If the visible units also have binary states, we similarly obtain:

$$\mathbb{P}(v_i = 1|\boldsymbol{h}) = \sigma(b_i + \boldsymbol{W}_{i:}\, \boldsymbol{h}), \tag{49}$$

where $\boldsymbol{W}_{i:} \in \mathbb{R}^p$ denotes the ith row of matrix $\boldsymbol{W}$. Since the state space is binary $\{0, 1\}$, Eqs. (48) and (49) imply:

$$\mathbb{P}(h_j|\boldsymbol{v}) = \sigma(c_j + \boldsymbol{v}^\top \boldsymbol{W}_{:j}) = \sigma(c_j + \sum_{i=1}^{d} v_i\, w_{ij}), \tag{50}$$

$$\mathbb{P}(v_i|\boldsymbol{h}) = \sigma(b_i + \boldsymbol{W}_{i:}\,\boldsymbol{h}) = \sigma(b_i + \sum_{j=1}^{p} w_{ij}\, h_j). \tag{51}$$

According to Proposition 1, the units are conditionally independent, so their distribution is the product of each individual conditional distribution:

$$\mathbb{P}(\boldsymbol{h}|\boldsymbol{v}) = \prod_{j=1}^{p} \mathbb{P}(h_j|\boldsymbol{v}) = \prod_{j=1}^{p} \sigma(c_j + \boldsymbol{v}^\top \boldsymbol{W}_{:j}), \tag{52}$$

$$\mathbb{P}(\boldsymbol{v}|\boldsymbol{h}) = \prod_{i=1}^{d} \mathbb{P}(v_i|\boldsymbol{h}) = \prod_{i=1}^{d} \sigma(b_i + \boldsymbol{W}_{i:}\,\boldsymbol{h}). \tag{53}$$

Therefore, in the Gibbs sampling procedure described in Algorithm refalgorithmspsRBMspsGibbsspssampling, samples are drawn from the distributions given in Eqs. (50) and (51). Since the sigmoid function maps to the interval (0, 1), sampling from it can be implemented using a uniform distribution $u \sim U[0, 1]$, as explained in Sect. 3.3.

Moreover, for binary states, we have $\mathbb{E}_{\sim\mathbb{P}(h_j|\boldsymbol{v})}[h_j] = \sigma(c_j + \boldsymbol{v}^\top \boldsymbol{W}_{:j})$. Therefore, applying the sigmoid function element-wise to $\widehat{\boldsymbol{h}}_i \in \mathbb{R}^p$ gives the following expression for Eq. (25) in training a binary-state RBM:

$$\widehat{\boldsymbol{h}}_i = \mathbb{E}_{\sim\mathbb{P}(\boldsymbol{h}|\boldsymbol{v}_i)}[\boldsymbol{h}] = \sigma(\boldsymbol{c} + \boldsymbol{v}_i^\top \boldsymbol{W}). \tag{54}$$

This equation is also used in Algorithm refalgorithmspsRBMspstrainingspsusingspscontrastivespsdivergence.

4.3 Continuous Values

In some cases, the hidden units are assigned continuous values to serve as continuous representations of the visible units. According to the definition of conditional probability, we have

$$\mathbb{P}(v_i = 1|\boldsymbol{h}) = \frac{\mathbb{P}(v_i = 1, \boldsymbol{h})}{\sum_{v_i \in \mathbb{R}^d} \mathbb{P}(\boldsymbol{h}, v_i)}. \tag{55}$$

From Eq. (17), we obtain:

$$\begin{aligned}\mathbb{P}(\boldsymbol{h}, v_i = 1) &= \exp(b_i \times 1 + 1 \times \boldsymbol{W}_{i:}\,\boldsymbol{h}) \\ &= \exp(b_i + \boldsymbol{W}_{i:}\,\boldsymbol{h}).\end{aligned} \tag{56}$$

Therefore, Eq. (55) becomes

$$\mathbb{P}(v_i = 1|\boldsymbol{h}) = \frac{\exp(b_i + \boldsymbol{W}_{i:}\,\boldsymbol{h})}{\sum_{v_i \in \mathbb{R}^d} \mathbb{P}(\boldsymbol{h}, v_i)}. \tag{57}$$

For continuous-valued visible units, a common approach is to use Gaussian-Bernoulli RBMs, where visible units are modeled as Gaussian distributions with mean given by $b_i + \boldsymbol{W}_{i:}$ and fixed variance. This allows RBMs to handle real-valued data such as pixel intensities. As such, a normal distribution with mean $b_i + \boldsymbol{W}_{i:}$ and a unit variance is used:

$$\mathbb{P}(v_i|\boldsymbol{h}) = \mathcal{N}(b_i + \boldsymbol{W}_{i:}\boldsymbol{h}, 1) = \mathcal{N}(b_i + \sum_{j=1}^{p} w_{ij}h_j, 1). \tag{58}$$

According to Proposition 1, the units are conditionally independent, so the joint conditional distribution factorizes:

$$\mathbb{P}(\boldsymbol{v}|\boldsymbol{h}) = \prod_{i=1}^{d} \mathbb{P}(v_i|\boldsymbol{h}) = \prod_{i=1}^{d} \mathcal{N}(b_i + \boldsymbol{W}_{i:}\boldsymbol{h}, 1). \tag{59}$$

Typically, when the hidden units take continuous values, the visible units are binary [45, 65]. In this case, the conditional distribution $\mathbb{P}(\boldsymbol{h}|\boldsymbol{v})$ is given by Eq. (52). If, instead, the visible units are continuous, their distribution can be expressed similarly:

$$\mathbb{P}(\boldsymbol{h}|\boldsymbol{v}) = \prod_{j=1}^{p} \mathbb{P}(h_j|\boldsymbol{v}) = \prod_{j=1}^{p} \mathcal{N}(c_j + \boldsymbol{v}^\top \boldsymbol{W}_{:j}, 1). \tag{60}$$

These normal distributions can be used to perform sampling in the Gibbs sampling procedure of Algorithm refalgorithmspsRBMspsGibbsspssampling. In this setting, Eq. (25) used in Algorithm refalgorithmspsRBMspstrainingspsusingspscontrastivespsdivergence for RBM training becomes

$$\widehat{\boldsymbol{h}}_i = \mathbb{E}_{\sim\mathbb{P}(\boldsymbol{h}|\boldsymbol{v}_i)}[\boldsymbol{h}] = \mathcal{N}(\boldsymbol{c} + \boldsymbol{v}_i^\top \boldsymbol{W}, \boldsymbol{I}_{p\times p}), \tag{61}$$

where $\boldsymbol{I}$ denotes the identity matrix.

4.4 Discrete Poisson States

In some cases, the units take discrete values with more than two possible states, as discussed in Sect. 4.2. In this scenario, the well-known Poisson distribution can be used to model discrete random variables:

$$\text{Ps}(t, \lambda) = \frac{e^{-\lambda}\lambda^t}{t!}. \tag{62}$$

Assume that each visible unit can take a value $t \in \{0, 1, 2, 3, \dots\}$. If the conditional Poisson distribution is used for the visible units, then [56]:

$$\mathbb{P}(v_i = t|\boldsymbol{h}) = \text{Ps}\left(t, \frac{\exp(b_i + \boldsymbol{W}_{i:}\boldsymbol{h})}{\sum_{k=1}^{d} \exp(b_k + \boldsymbol{W}_{k:}\boldsymbol{h})}\right). \tag{63}$$

Similarly, if the hidden units also take discrete values, their conditional distribution is

$$\mathbb{P}(h_j = t|\boldsymbol{v}) = \text{Ps}\left(t, \frac{\exp(c_j + \boldsymbol{v}^\top \boldsymbol{W}_{:j})}{\sum_{k=1}^{p} \exp(c_k + \boldsymbol{v}^\top \boldsymbol{W}_{:k})}\right). \tag{64}$$

These Poisson distributions can be used for sampling in the Gibbs sampling procedure of Algorithm refalgorithmspsRBMspsGibbsspssampling. In this case, Eq. (25), which is used in Algorithm refalgorithmspsRBMspstrainingspsusingspscontrastivespsdivergence for training the RBM, can be computed using a multivariate Poisson distribution [12].

It is worth noting that RBMs have been applied in *semantic hashing*, where the hidden variables serve as hash-based representations of data [56]. Semantic hashing employs a Poisson distribution for the visible units and a sigmoid function for the conditional hidden variables.

5 Conditional Restricted Boltzmann Machine

When data consist of time-series, a standard RBM does not capture temporal information. In other words, the RBM is suitable for modeling static data. The Conditional RBM (CRBM), introduced in [63], extends the RBM to incorporate temporal information by conditioning on visible variables from previous time steps.

The CRBM adds two sets of directed links to the RBM. The first set consists of autoregressive links from the past $\mathcal{T}_1$ visible units to the visible units at the current time step. The second set includes directed links from the past $\mathcal{T}_2$ visible units to the hidden units at the current time step. Although $\mathcal{T}_1$ and $\mathcal{T}_2$ are generally not equal, it is common to set $\mathcal{T}_1 = \mathcal{T}_2 = \mathcal{T}$ for simplicity [63].

The links from visible units at time $t - \tau$ to visible units at time t are denoted by $\boldsymbol{G}^{(t-\tau)} = [g_{ij}] \in \mathbb{R}^{d\times d}$. The links from visible units at time $t - \tau$ to hidden units at time t are denoted by $\boldsymbol{Q}^{(t-\tau)} = [q_{ij}] \in \mathbb{R}^{d\times p}$. The structure of the CRBM is illustrated in Fig. 2, where each arrow represents a set of links, typically corresponding to a matrix or vector of weights.

The weight updates for $\boldsymbol{W}$ and the biases $\boldsymbol{b}$ and $\boldsymbol{c}$ follow the same rules as given in Eqs. (30), (31), and (32). The newly introduced directed links from previous visible units to current visible and hidden units act as dynamically changing biases. Recall that the biases $\boldsymbol{b}$ and $\boldsymbol{c}$ in a standard RBM are constant values multiplied by one (see Fig. 1). In the CRBM, these bias-like terms are instead multiplied by values from previous time steps (see Fig. 2). The gradients for the added links from previous time steps are given by:

$$\forall i : \mathbb{R}^d \ni \nabla_{\boldsymbol{G}_{i:}^{(t-\tau)}} \ell(\theta) = v_i^{(t-\tau)}\Big(\sum_{k=1}^{n} \boldsymbol{v}_k^{(t)} - \sum_{k=1}^{n} \widetilde{\boldsymbol{v}}_k^{(t)}\Big), \tag{65}$$

$$\forall i : \mathbb{R}^p \ni \nabla_{\boldsymbol{Q}_{i:}^{(t-\tau)}} \ell(\theta) = v_i^{(t-\tau)}\Big(\sum_{k=1}^{n} \widehat{\boldsymbol{h}}_k - \sum_{k=1}^{n} \widetilde{\boldsymbol{h}}_k\Big), \tag{66}$$

where $i \in \{1, \ldots, d\}$, $\tau \in \{1, 2, \ldots, \mathcal{T}\}$, $\boldsymbol{G}_{i:}^{(t-\tau)}$ denotes the ith row of $\boldsymbol{G}^{(t-\tau)}$, and $\boldsymbol{Q}_{i:}^{(t-\tau)}$ denotes the ith row of $\boldsymbol{Q}^{(t-\tau)}$.

These gradients are structurally similar to Eqs. (31) and (32), but they are scaled by the values of visible units at earlier time steps. This scaling arises because, unlike the constant biases in a standard RBM, the CRBM uses previous visible values as input to the dynamic bias terms.

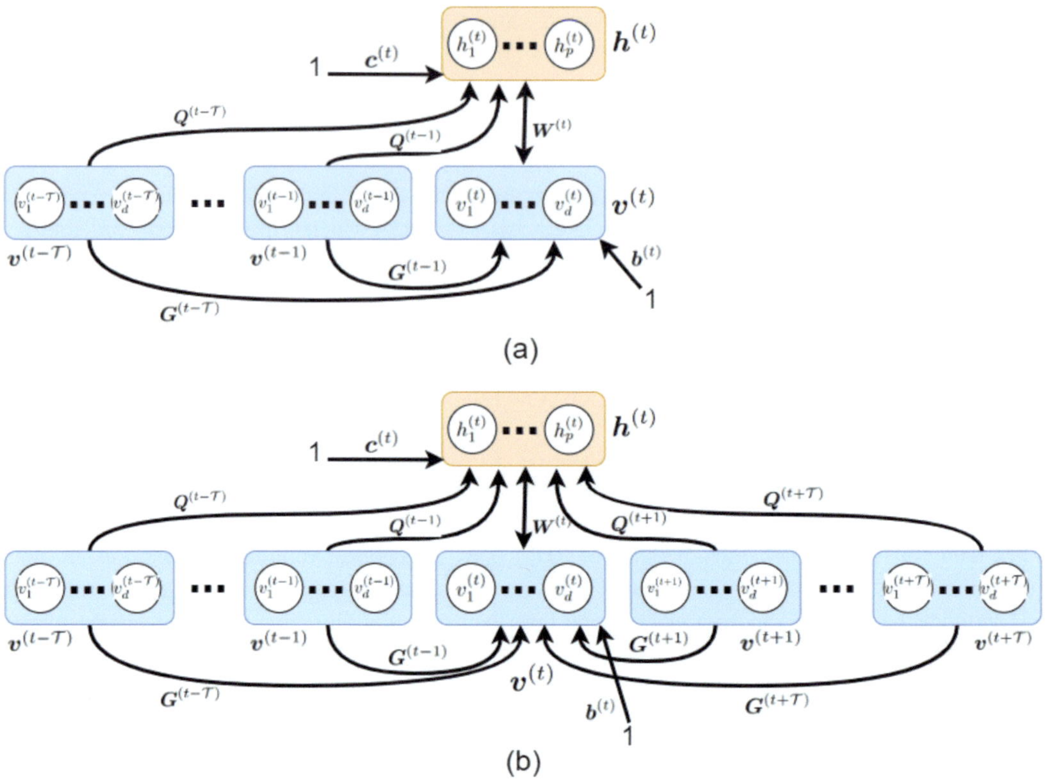

Fig. 2 The structures of **a** a conditional Boltzmann machine and **b** an interpolating conditional restricted Boltzmann machine. Note that each arrow in this figure represents a set of links

These updates can also be expressed as follows [63]:

$$\forall i, j: \nabla_{g_{ij}^{(t-\tau)}} \ell(\theta) = v_i^{(t-\tau)} \Big(\langle v_i^{(t)} \rangle_{\text{data}} - \langle v_i^{(t)} \rangle_{\text{recon.}} \Big), \tag{67}$$

$$\forall i, j: \nabla_{q_{ij}^{(t-\tau)}} \ell(\theta) = v_i^{(t-\tau)} \Big(\langle h_j^{(t)} \rangle_{\text{data}} - \langle h_j^{(t)} \rangle_{\text{recon.}} \Big), \tag{68}$$

where $\langle . \rangle_{\text{data}}$ and $\langle . \rangle_{\text{recon.}}$ represent expectations over the data and its reconstruction, respectively.

The additional parameters introduced in the CRBM can be trained via gradient descent using these gradients. Algorithm refalgorithmspsRBMspstrainingspsusingspscontrastivespsdivergence for RBM training can also be applied to CRBM, provided the updates for the additional links are included.

The Interpolating CRBM (ICRBM) [45] is an extension of the CRBM that includes links from future visible variables. The structure of the ICRBM is shown in Fig. 2. While its formulation and training are similar to the CRBM, the theoretical details of ICRBM are omitted here for brevity.

CRBMs have been applied successfully to various time-series tasks, such as action recognition [63] and acoustic modeling [45].

6 Deep Belief Network

6.1 Stacking RBM Models

A neural network can be trained using RBM-based training [23, 24]. This approach provides an effective initialization of the network weights, which can subsequently be fine-tuned using backpropagation. Prior to the development of ReLU [16] and dropout [58], training deep Multilayer Perceptrons (MLPs) was hindered by the vanishing gradient problem. This occurred because randomly initialized weights were often unsuitable for optimizing deep networks via backpropagation. To address this, a method was proposed for pre-training neural networks by initializing them to a suitable configuration, followed by fine-tuning with backpropagation [23, 24].

A neural network consists of multiple layers. Let ℓ denote the number of layers, with the first layer receiving the input data, and let p_ℓ denote the number of neurons in the ℓth layer. By convention, $p_1 = d$. Every pair of successive layers can be treated as one RBM, as illustrated in Fig. 3. The procedure begins with the first pair of layers, treated as an RBM, where the training dataset $\{\boldsymbol{x}_i \in \mathbb{R}^d\}_{i=1}^n$ is assigned to the visible variables $\{\boldsymbol{v}_i\}_{i=1}^n$. The weights and biases of this RBM are trained using Algorithm refalgorithmspsRBMspstrainingspsusingspscontrastivespsdivergence.

After training, n hidden variables of dimension p_2 are generated using Gibbs sampling, as described in Algorithm refalgorithmspsRBMspsGibbsspssampling. These hidden variables then serve as the visible variables for the second RBM (i.e., the second pair of layers). This RBM is trained using Algorithm refalgorithmspsRBMspstrainingspsusingspscontrastivespsdivergence, followed by sampling of its hidden variables using Gibbs sampling. This process is repeated until all layer pairs in the network have been trained using RBM-based training. This layer-wise training follows a greedy strategy [4], producing a well-initialized set of weights and biases for the entire neural network. These initialized parameters can then be fine-tuned via backpropagation [52].

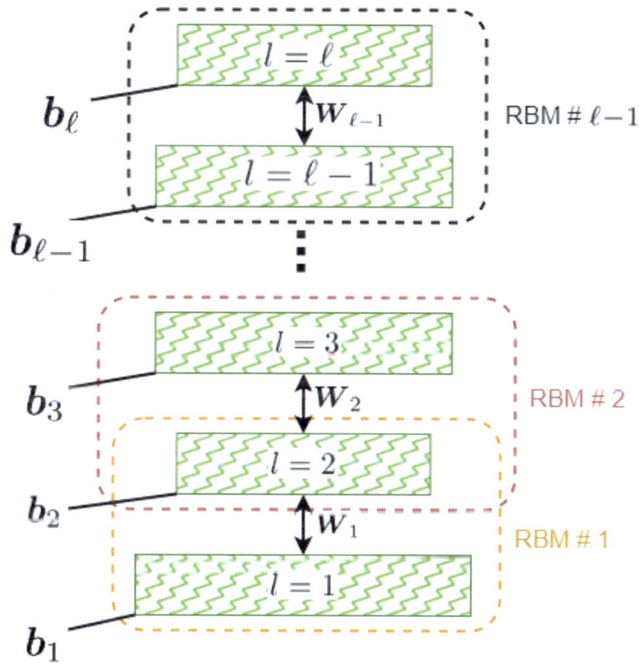

Fig. 3 Pre-training a deep belief network by considering every pair of layers as an RBM

The described training method was first introduced in [23, 24] and was initially applied to dimensionality reduction. By increasing ℓ to any desired number, the network can be made arbitrarily deep. Since each pair of layers is trained individually as an RBM, the depth of the network is no longer constrained by the vanishing gradient problem, as the pretraining provides a favorable initialization. This architecture is known as a Deep Belief Network (DBN) [21, 23], as it forms a stack of RBM models trained via belief propagation. DBNs played an important role in the early revival of deep learning but are now primarily of historical interest, since modern architectures train effectively with backpropagation, improved optimizers, and normalization methods.

The pretraining process for a DBN using RBM-based training is illustrated in Fig. 3, and the algorithm is summarized in Algorithm refalgorithmspstrainingspsdeepspsbeliefspsnetwork. In this algorithm, $\boldsymbol{W}_l \in \mathbb{R}^{p_l \times p_{l+1}}$ denotes the weights connecting layer l to layer $(l+1)$, and $\boldsymbol{b}_l \in \mathbb{R}^{p_l}$ denotes the biases for layer l. As weights exist between successive layers, the complete set of weights is denoted by $\{\boldsymbol{W}_l\}_{l=1}^{\ell-1}$.

It is important to note that DBN pretraining is an unsupervised process, since RBM training is unsupervised. Fine-tuning, however, may be either supervised or unsupervised, depending on the loss function used during backpropagation.

If the DBN is designed as an autoencoder with a low-dimensional central (bottleneck) layer, both the pretraining and fine-tuning stages are unsupervised, with mean squared error used as the loss function. Such a DBN autoencoder can learn compact, low-dimensional representations of the input data and is often used for dimensionality reduction [24]. The architecture of this network is shown in Fig. 4.

Input: training data $\{\boldsymbol{x}_i\}_{i=1}^n$
// pre-training:
for *l from* 1 *to* $\ell - 1$ **do**
  if $l = 1$ **then**
    $\{\boldsymbol{v}_i\}_{i=1}^n \leftarrow \{\boldsymbol{x}_i\}_{i=1}^n$
  else
    // generate n hidden variables of previous RBM:
    $\{\boldsymbol{h}_i\}_{i=1}^n \leftarrow$ Algorithm 1 for $(l-1)$-th RBM $\leftarrow \{\boldsymbol{v}_i\}_{i=1}^n$
    $\{\boldsymbol{v}_i\}_{i=1}^n \leftarrow \{\boldsymbol{h}_i\}_{i=1}^n$
  $\boldsymbol{W}_l, \underline{b}_l, \underline{b}_{l+1} \leftarrow$ Algorithm 2 for l-th RBM $\leftarrow \{\boldsymbol{v}_i\}_{i=1}^n$
// fine-tuning using backpropagation:
Initialize network with weights $\{\boldsymbol{W}_l\}_{l=1}^{\ell-1}$ and biases $\{\underline{b}_l\}_{l=2}^{\ell}$.
$\{\boldsymbol{W}_l\}_{l=1}^{\ell-1}, \{\underline{b}_l\}_{l=1}^{\ell} \leftarrow$ Backpropagate the error of loss fro several epochs.

Algorithm 3: Training a deep belief network

6.2 Other Variants and Improvements of RBM and DBN

Several improvements and extensions of DBNs have been proposed. One notable variant is the Convolutional DBN [34], which applies convolutional operations to handle spatially structured data. Another application involves using DBNs for semantic hashing [56], where learned binary codes provide compact data representations. Greedy training of DBNs via RBM pretraining has also been employed to train the t-SNE network for general degrees of freedom [64].

In addition to the Conditional RBM (CRBM) [63], the Recurrent RBM (RRBM) [62] has been introduced to capture temporal dependencies in sequential data.

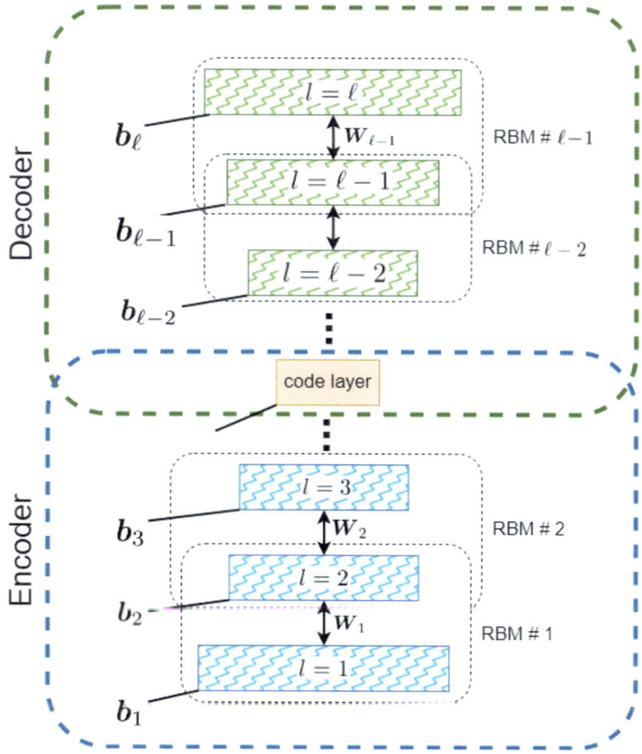

Fig. 4 A DBN autoencoder where the number of neurons in the corresponding layers of encoder and decoder are usually set to be equal. The coder layer is a low-dimensional embedding for representation of data

It is also worth noting that several other energy-based models exist beyond the Boltzmann Machine (BM). One such example is the Helmholtz Machine [10], which employs a generative-recognition pair of networks.

Another important variant is the Deep Boltzmann Machine (DBM), which differs slightly from the DBN. Due to space limitations, DBMs are not covered in detail here; readers interested in DBMs are referred to [55]. Several efficient training methods have been developed for DBMs [17, 25, 42, 48, 53, 54, 57, 59, 60]. DBMs have been successfully applied to various tasks, including document modeling [61] and face modeling [49].

7 Conclusion

7.1 Conclusion of Chapter

This chapter thoroughly explained the theory of energy-based learning models such as Boltzmann Machines (BM), Restricted Boltzmann Machines (RBM), and Deep Belief Networks (DBN). Although BM, RBM, and DBN were influential in the early development of deep learning, today they are mainly of historical or theoretical importance. Modern architectures train effectively with backpropagation, advanced optimizers, normalization methods, and large-scale datasets, making RBMs and DBNs unnecessary in practice. Current energy-based approaches, such as contrastive learning and diffusion models, play a much more central role in modern research and applications.

7.2 Conclusion of the Book Part "Fundamentals of Deep Learning"

Understanding the fundamentals of deep learning requires familiarity with several key concepts. Early neural network models include the McCulloch-Pitts model, the Perceptron, ADALINE, Radial Basis Function (RBF) networks, and self-organizing maps. Backpropagation remains an essential algorithm for training these networks.

Convolutional Neural Networks (CNNs) are particularly suited for image-like data, as they preserve the spatial relationships of nearby pixels without requiring data reshaping. Numerous CNN architectures have been developed, including AlexNet, VGG, Inception, GoogLeNet, U-Net, ResNet, and DenseNet.

A deep understanding of overfitting and generalization is crucial, as a model may perform perfectly on training data but poorly on unseen test data. Regularization techniques help mitigate overfitting, with common methods in deep learning including weight decay, noise injection, early stopping, dropout, and batch normalization.

Theoretical analysis of generalization can be approached through methods such as Probably Approximately Correct (PAC) learning, error bounds, the information bottleneck principle, double descent, grokking, benign overfitting, the implicit bias of stochastic gradient descent, neural network kernel theory, the lottery ticket hypothesis, and scaling laws. While these frameworks provide valuable insights, many do not fully explain why deep neural networks generalize effectively to unseen data.

Historically, Restricted Boltzmann Machines (RBMs) played an important role in the early development of neural networks. Although they are rarely used in modern deep learning, understanding RBMs provides valuable insights into energy-based models and their underlying principles.

8 Exercises

1. (Level: Moderate) Consider a Hopfield network with d binary neurons $x_i \in \{-1, +1\}$ and weights learned by Hebbian learning from a single stored pattern $\boldsymbol{\xi} = (\xi_1, \xi_2, \ldots, \xi_d)$, where each $\xi_i \in \{-1, +1\}$:

$$w_{ij} = \begin{cases} \xi_i \xi_j, & i \neq j, \\ 0, & i = j. \end{cases}$$

   (a) Show that the energy function of this Hopfield network can be written in the Ising model form:

$$E(\boldsymbol{x}) = -\sum_{i<j} w_{ij} x_i x_j,$$

   and explain the equivalence between this Hopfield network and the Ising model.

   (b) If the state vector $\boldsymbol{x}$ matches the stored pattern $\boldsymbol{\xi}$, what is the energy $E(\boldsymbol{\xi})$? Interpret this result in terms of energy minima.

2. (Level: Easy) A Boltzmann machine has three visible units v_1, v_2, v_3 and two hidden units h_1, h_2. The energy of a state $(\boldsymbol{v}, \boldsymbol{h})$ is defined as:

$$E(\boldsymbol{v}, \boldsymbol{h}) = -\sum_i b_i v_i - \sum_j c_j h_j - \sum_{i,j} v_i w_{ij} h_j.$$

Suppose $\boldsymbol{v} = (1, 0, 1)$, $\boldsymbol{h} = (1, 0)$, biases are $\boldsymbol{b} = (0.2, -0.3, 0.1)$, $\boldsymbol{c} = (0.5, -0.1)$, and weights are

$$\boldsymbol{W} = \begin{bmatrix} 0.8 & -0.2 \\ -0.1 & 0.3 \\ 0.5 & 0.6 \end{bmatrix}.$$

Compute the energy $E(\boldsymbol{v}, \boldsymbol{h})$.

3. (Level: Moderate) Consider a Restricted Boltzmann Machine (RBM) with two visible units $v_1, v_2 \in \{0, 1\}$ and 2 hidden units $h_1, h_2 \in \{0, 1\}$. The energy of a configuration (v, h) is defined as:

$$E(\boldsymbol{v}, \boldsymbol{h}) = -\sum_i a_i v_i - \sum_j b_j h_j - \sum_{i,j} v_i w_{ij} h_j.$$

Given:

$$\boldsymbol{a} = (0.4, -0.2), \quad \boldsymbol{b} = (0.1, -0.3), \quad \boldsymbol{W} = \begin{bmatrix} 0.5 & -0.7 \\ 1.0 & 0.8 \end{bmatrix}.$$

where rows correspond to visible units v_1, v_2 and columns correspond to hidden units h_1, h_2. Answer the following questions:

(a) Compute the energy $E(\boldsymbol{v}, \boldsymbol{h})$ for configuration $\boldsymbol{v} = (1, 0)$, $\boldsymbol{h} = (1, 1)$.
(b) Compute the unnormalized probability $\tilde{P}(\boldsymbol{v}, \boldsymbol{h}) = e^{-E(\boldsymbol{v},\boldsymbol{h})}$.
(c) Compute the conditional probability $P(h_1 = 1 \mid \boldsymbol{v} = (1, 0))$.

4. (Level: Difficult) Consider a fully connected Boltzmann machine with three binary units $x_1, x_2, x_3 \in \{0, 1\}$, where all units are visible. The energy of a state $\boldsymbol{x} = (x_1, x_2, x_3)$ is defined as:

$$E(\boldsymbol{x}) = -\sum_i b_i x_i - \sum_{i<j} w_{ij} x_i x_j.$$

Given biases $b = (0.3, -0.2, 0.5)$ and weights:

$$w_{ij} = \begin{bmatrix} \times & 0.4 & -0.6 \\ 0.4 & \times & 0.1 \\ -0.6 & 0.1 & \times \end{bmatrix}.$$

Answer the following questions:

(a) Compute the energy $E(\boldsymbol{x})$ for the state $\boldsymbol{x} = (1, 1, 0)$.
(b) Compute the unnormalized probability $\tilde{P}(\boldsymbol{x}) = e^{-E(\boldsymbol{x})}$.
(c) Compute the partition function $Z = \sum_{\boldsymbol{x} \in \{0,1\}^3} e^{-E(\boldsymbol{x})}$.
(d) Compute the normalized probability $P(\boldsymbol{x} = (1, 1, 0))$.

5. (Level: Moderate) Consider a Deep Belief Network (DBN) composed of two layers of Restricted Boltzmann Machines (RBMs):

- The first RBM has visible units $\boldsymbol{v}$ and hidden units $\boldsymbol{h}^{(1)}$.
- The second RBM takes $\boldsymbol{h}^{(1)}$ as visible units and has hidden units $\boldsymbol{h}^{(2)}$.

Suppose the energy functions are

$$E^{(1)}(\boldsymbol{v}, \boldsymbol{h}^{(1)}) = -\sum_i a_i v_i - \sum_j b_j^{(1)} h_j^{(1)} - \sum_{i,j} v_i w_{ij}^{(1)} h_j^{(1)},$$

and:

$$E^{(2)}(\boldsymbol{h}^{(1)}, \boldsymbol{h}^{(2)}) = -\sum_j b_j^{(1)} h_j^{(1)} - \sum_k b_k^{(2)} h_k^{(2)} - \sum_{j,k} h_j^{(1)} w_{jk}^{(2)} h_k^{(2)}.$$

Given a visible vector $\boldsymbol{v}$, express the joint probability $P(\boldsymbol{v}, \boldsymbol{h}^{(1)}, \boldsymbol{h}^{(2)})$ in terms of the two RBM energy functions.

References

1. David H Ackley, Geoffrey E Hinton, and Terrence J Sejnowski. "A learning algorithm for Boltzmann machines". In: *Cognitive science* 9.1 (1985), pp. 147–169.
2. Diego Alberici, Adriano Barra, Pierluigi Contucci, and Emanuele Mingione. "Annealing and replica-symmetry in deep Boltzmann machines". In: *Journal of Statistical Physics* 180.1 (2020), pp. 665–677.
3. Diego Alberici, Pierluigi Contucci, and Emanuele Mingione. "Deep Boltzmann machines: rigorous results at arbitrary depth". In: *Annales Henri Poincar é*. Springer. 2021, pp. 1–24.
4. Yoshua Bengio, Pascal Lamblin, Dan Popovici, and Hugo Larochelle. "Greedy layer-wise training of deep networks". In: *Advances in neural information processing systems*. 2007, pp. 153–160.
5. Christopher M Bishop. "Pattern recognition". In: *Machine learning* 128.9 (2006).
6. Ludwig Boltzmann. "Studien uber das Gleichgewicht der lebenden Kraft". In: *Wissenschafiliche Abhandlungen* 1 (1868), pp. 49–96.
7. Bernhard E Boser, Isabelle M Guyon, and Vladimir N Vapnik. "A training algorithm for optimal margin classifiers". In: *Proceedings of the fifth annual workshop on Computational learning theory*. 1992, pp. 144–152.
8. Stephen G Brush. "History of the Lenz-Ising model". In: *Reviews of modern physics* 39.4 (1967), p. 883.
9. Sean Carroll. *From eternity to here: the quest for the ultimate theory of time*. Penguin, 2010.
10. Peter Dayan, Geoffrey E Hinton, Radford M Neal, and Richard S Zemel. "The Helmholtz machine". In: *Neural computation* 7.5 (1995), pp. 889–904.
11. Raaz Dwivedi, Yuansi Chen, Martin J Wainwright, and Bin Yu. "Log-concave sampling: Metropolis-Hastings algorithms are fast!" In: *Conference on learning theory*. PMLR. 2018, pp. 793–797.
12. Carol Bates Edwards. *Multivariate and multiple Poisson distributions*. Iowa State University, 1962.
13. Stuart Geman and Donald Geman. "Stochastic relaxation, Gibbs distributions, and the Bayesian restoration of images". In: *IEEE Transactions on pattern analysis and machine intelligence* PAMI-6.6 (1984), pp. 721–741.
14. Benyamin Ghojogh, Hadi Nekoei, Aydin Ghojogh, Fakhri Karray, and Mark Crowley. "Sampling algorithms, from survey sampling to Monte Carlo methods: Tutorial and literature review". In: *arXiv preprint* arXiv:2011.00901 (2020).
15. J Willard Gibbs. *Elementary principles in statistical mechanics*. Courier Corporation, 1902.
16. Xavier Glorot, Antoine Bordes, and Yoshua Bengio. "Deep sparse rectifier neural networks". In: *Proceedings of the fourteenth international conference on artificial intelligence and statistics*. JMLR Workshop and Conference Proceedings. 2011, pp. 315–323.
17. Ian Goodfellow, Mehdi Mirza, Aaron Courville, and Yoshua Bengio. "Multi-prediction deep Boltzmann machines". In: *Advances in Neural Information Processing Systems* 26 (2013), pp. 548–556.
18. Donald Hebb. *The Organization of Behavior*. Wiley & Sons, New York, 1949.
19. Geoffrey E Hinton. "A practical guide to training restricted Boltzmann machines". In: *Neural networks: Tricks of the trade*. Springer, 2012, pp. 599–619.
20. Geoffrey E Hinton. "Boltzmann machine". In: *Scholarpedia* 2.5 (2007), p. 1668.
21. Geoffrey E Hinton. "Deep belief networks". In: *Scholarpedia* 4.5 (2009), p. 5947.
22. Geoffrey E Hinton. "Training products of experts by minimizing contrastive divergence". In: *Neural computation* 14.8 (2002), pp. 1771–1800.
23. Geoffrey E Hinton, Simon Osindero, and Yee-Whye Teh. "A fast learning algorithm for deep belief nets". In: *Neural computation* 18.7 (2006), pp. 1527–1554.
24. Geoffrey E Hinton and Ruslan R Salakhutdinov. "Reducing the dimensionality of data with neural networks". In: *Science* 313.5786 (2006), pp. 504–507.

25. Geoffrey E Hinton and Russ R Salakhutdinov. "A better way to pretrain deep Boltzmann machines". In: *Advances in Neural Information Processing Systems* 25 (2012), pp. 2447–2455.
26. Geoffrey E Hinton and Terrence J Sejnowski. "Optimal perceptual inference". In: *Proceedings of the IEEE conference on Computer Vision and Pattern Recognition*. Vol. 448. IEEE, 1983.
27. John J Hopfield. "Neural networks and physical systems with emergent collective computational abilities". In: *Proceedings of the national academy of sciences* 79.8 (1982), pp. 2554–2558.
28. John J Hopfield. "Neurons with graded response have collective computational properties like those of two-state neurons". In: *Proceedings of the national academy of sciences* 81.10 (1984), pp. 3088–3092.
29. Kerson Huang. *Statistical Mechanics*. John Wiley & Sons, 1987.
30. Ernst Ising. "Beitrag zur theorie des ferromagnetismus". In: *Zeitschrift f ü r Physik* 31.1 (1925), pp. 253–258.
31. Andrej Karpathy and Li Fei-Fei. "Deep visual-semantic alignments for generating image descriptions". In: *Proceedings of the IEEE conference on computer vision and pattern recognition*. 2015, pp. 3128–3137.
32. Scott Kirkpatrick, C Daniel Gelatt, and Mario P Vecchi. "Optimization by simulated annealing". In: *science* 220.4598 (1983), pp. 671–680.
33. Daphne Koller and Nir Friedman. *Probabilistic graphical models: principles and techniques*. MIT press, 2009.
34. Alex Krizhevsky and Geoff Hinton. "Convolutional deep belief networks on CIFAR-10". In: *Unpublished manuscript* 40.7 (2010), pp. 1–9.
35. Dmitry Krotov. "Hierarchical Associative Memory". In: *arXiv preprint* arXiv:2107.06446 (2021).
36. Dmitry Krotov and John Hopfield. "Large associative memory problem in neurobiology and machine learning". In: *International Conference on Learning Representations (ICLR)*. 2021.
37. Dmitry Krotov and John J Hopfield. "Dense associative memory for pattern recognition". In: *Advances in neural information processing systems* 29 (2016), pp. 1172–1180.
38. Yann LeCun, Yoshua Bengio, and Geoffrey Hinton. "Deep learning". In: *nature* 521.7553 (2015), pp. 436–444.
39. Yann LeCun, Sumit Chopra, Raia Hadsell, M Ranzato, and F Huang. "A tutorial on energy-based learning". In: *Predicting structured data* 1 (2006).
40. Wilhelm Lenz. "Beitršge zum verstšndnis der magnetischen eigenschaften in festen kšrpern". In: *Physikalische Z* 21 (1920), pp. 613–615.
41. William A Little. "The existence of persistent states in the brain". In: *Mathematical biosciences* 19.1-2 (1974), pp. 101–120.
42. Jan Melchior, Asja Fischer, and Laurenz Wiskott. "How to center deep Boltzmann machines". In: *The Journal of Machine Learning Research* 17.1 (2016), pp. 3387–3447.
43. Abdel-rahman Mohamed, George Dahl, Geoffrey Hinton, et al. "Deep belief networks for phone recognition". In: *Nips workshop on deep learning for speech recognition and related applications*. Vol. 1. 9. Vancouver, Canada. 2009, p. 39.
44. Abdel-rahman Mohamed, George E Dahl, and Geoffrey Hinton. "Acoustic modeling using deep belief networks". In: *IEEE transactions on audio, speech, and language processing* 20.1 (2011), pp. 14–22.
45. Abdel-rahman Mohamed and Geoffrey Hinton. "Phone recognition using restricted Boltzmann machines". In: *2010 IEEE International Conference on Acoustics, Speech and Signal Processing*. IEEE. 2010, pp. 4354–4357.
46. Mehdi Molkaraie. "Marginal Densities, Factor Graph Duality, and High-Temperature Series Expansions". In: *International Conference on Artificial Intelligence and Statistics*. 2020, pp. 256–265.
47. Mehdi Molkaraie. "The primal versus the dual Ising model". In: *2017 55th Annual Allerton Conference on Communication, Control, and Computing (Allerton)*. IEEE. 2017, pp. 53–60.
48. Grégoire Montavon and Klaus-Robert Müller. "Deep Boltzmann machines and the centering trick". In: *Neural networks: tricks of the trade*. Springer, 2012, pp. 621–637.
49. Chi Nhan Duong, Khoa Luu, Kha Gia Quach, and Tien D Bui. "Beyond principal components: Deep Boltzmann machines for face modeling". In: *Proceedings of the IEEE Conference on Computer Vision and Pattern Recognition*. 2015, pp. 4786–4794.
50. Leandro Aparecido Passos and Joao Paulo Papa. "Temperature-based deep Boltzmann machines". In: *Neural Processing Letters* 48.1 (2018), pp. 95–107.
51. Hubert Ramsauer et al. "Hopfield networks is all you need". In: *arXiv preprint* arXiv:2008.02217 (2020).
52. David E Rumelhart, Geoffrey E Hinton, and Ronald J Williams. "Learning representations by back-propagating errors". In: *Nature* 323.6088 (1986), pp. 533–536.
53. Ruslan Salakhutdinov. "Learning deep Boltzmann machines using adaptive MCMC". In: *Proceedings of the 27th International Conference on Machine Learning*. 2010, pp. 943–950.
54. Ruslan Salakhutdinov and Geoffrey Hinton. "An efficient learning procedure for deep Boltzmann machines". In: *Neural computation* 24.8 (2012), pp. 1967–2006.
55. Ruslan Salakhutdinov and Geoffrey Hinton. "Deep Boltzmann machines". In: *Artificial intelligence and statistics*. PMLR. 2009, pp. 448–455.
56. Ruslan Salakhutdinov and Geoffrey Hinton. "Semantic hashing". In: *International Journal of Approximate Reasoning* 50.7 (2009), pp. 969–978.

57. Ruslan Salakhutdinov and Hugo Larochelle. "Efficient learning of deep Boltzmann machines". In: *Proceedings of the thirteenth international conference on artificial intelligence and statistics*. JMLR Workshop and Conference Proceedings. 2010, pp. 693–700.
58. Nitish Srivastava, Geoffrey Hinton, Alex Krizhevsky, Ilya Sutskever, and Ruslan Salakhutdinov. "Dropout: a simple way to prevent neural networks from overfitting". In: *The journal of machine learning research* 15.1 (2014), pp. 1929–1958.
59. Nitish Srivastava and Ruslan Salakhutdinov. "Multimodal Learning with Deep Boltzmann Machines". In: *Advances in neural information processing systems*. Vol. 1. 2012, p. 2.
60. Nitish Srivastava and Ruslan Salakhutdinov. "Multimodal learning with deep Boltzmann machines". In: *Journal of Machine Learning Research* 15.1 (2014), pp. 2949–2980.
61. Nitish Srivastava, Ruslan R Salakhutdinov, and Geoffrey E Hinton. "Modeling documents with deep Boltzmann machines". In: *arXiv preprint* arXiv:1309.6865 (2013).
62. Ilya Sutskever, Geoffrey E Hinton, and Graham W Taylor. "The recurrent temporal restricted Boltzmann machine". In: *Advances in neural information processing systems*. 2009, pp. 1601–1608.
63. Graham W Taylor, Geoffrey E Hinton, and Sam T Roweis. "Modeling human motion using binary latent variables". In: *Advances in neural information processing systems*. 2007, pp. 1345–1352.
64. Laurens Van Der Maaten. "Learning a parametric embedding by preserving local structure". In: *Artificial Intelligence and Statistics*. 2009, pp. 384–391.
65. Max Welling, Michal Rosen-Zvi, and Geoffrey E Hinton. "Exponential Family Harmoniums with an Application to Information Retrieval." In: *Advances in neural information processing systems*. Vol. 4. 2004, pp. 1481–1488.

Part II
Sequence Models

Recurrent Neural Networks and Long Short-Term Memory Networks

8

Abstract

This chapter presents a comprehensive overview of Recurrent Neural Networks (RNNs) and Long Short-Term Memory (LSTM) networks, and their extensions, which were the dominant sequence modeling approaches before Transformers. Today, they are still used in certain applications (like speech and streaming), and the recurrent principle has re-emerged in modern State Space Models (SSMs) such as S4 and Mamba, which compete with Transformers for long-sequence efficiency. RNNs are dynamic models designed to handle sequential data by incorporating recurrence, where outputs from previous time steps influence future computations. Despite their usefulness, RNNs struggle with long-term dependencies due to the vanishing and exploding gradient problems, which motivated the development of more robust architectures. The chapter begins with the fundamentals of RNNs and the Backpropagation Through Time (BPTT) training algorithm. It then examines key challenges in training RNNs and surveys several mitigation strategies, including leaky units, close-to-identity weight matrices, and echo state networks. To effectively model both short-term and long-term dependencies, LSTM networks were introduced, followed by Gated Recurrent Units (GRUs), which simplify LSTM while maintaining its performance. In contexts where sequences can be processed offline, bidirectional architectures outperform their unidirectional counterparts by leveraging both future and past contexts. This leads to the introduction of bidirectional RNNs and LSTMs, culminating in models like ELMo that demonstrate the power of bidirectional contextual representations. Sections of the chapter detail the core concepts of RNNs, challenges in learning long-term dependencies, architectural innovations such as LSTM and GRU, and finally bidirectional models. This chapter serves as a foundational reference for understanding the evolution and capabilities of recurrent architectures in deep learning.

1 Introduction

Recurrent Neural Networks (RNNs) [51] and Long Short-Term Memory (LSTM) [30, 31] networks and their variants were the primary methods for sequence modeling prior to the advent of Transformers [63] (see Chap. 9). While their usage has declined, they remain relevant in specific applications, such as speech processing and streaming data. Moreover, the recurrent principle has recently resurfaced in modern State Space Models (SSMs), including S4 [25] and Mamba [24], which offer competitive efficiency for modeling long sequences compared to Transformers (see Chap. 10).

B. Ghojogh and A. Ghodsi, *Elements of Deep Learning*,
https://doi.org/10.1007/978-3-032-10738-1_8

Before the era of Transformers in deep learning, regular neural networks could not process sequences, such as sentences (sequences of words) or speech (sequences of phonemes), effectively without recurrence. The Recurrent Neural Network (RNN), proposed in [51], is a dynamical system that incorporates recurrence. In recurrence, the output of a model is fed back as input to the model at the next time step. One of the main training algorithms for RNNs is Backpropagation Through Time (BPTT), developed by several works [44, 50, 64–66], which is similar to the backpropagation algorithm [51] but additionally applies the chain rule through time. RNNs faced problems with vanishing or exploding gradients when modeling long-term dependencies [5, 6]. Several solutions were proposed for this issue, including close-to-identity weight matrices [43], long delays [41], leaky units [37, 62], and echo state networks [34, 36].

Sequence modeling requires capturing both short-term and long-term dependencies. For example, consider the sentence "The police is chasing the thief." In this sentence, the words "police" and "thief" are related by a short-term dependency because they are close to one another in the sequence. Another example is the sentence "I was born in France. My father was working there for many years during my childhood. My family had a great time there while my father was making money in his business there. That is why I know how to speak French." In this second example, the words "France" and "French" are related by a long-term dependency because they are far apart in the sequence. This inspired researchers to propose the Long Short-Term Memory (LSTM) network to handle both short-term and long-term dependencies [30, 31]. Later, the Gated Recurrent Unit (GRU) was proposed [8], which simplified LSTM to reduce its unnecessary complexity.

RNN and LSTM networks are causal models that condition each sequence element on the *previous* elements in the sequence. Later research showed that processing the sequence in both directions can achieve better performance for sequences that can be processed offline; for example, when chunks of the sequence can be saved and processed and the sequence elements do not need to be processed as a stream [21, 22]. Therefore, bidirectional RNNs [2, 57] and bidirectional LSTMs [21, 22] were proposed to process sequences in both directions. The Embeddings from Language Model (ELMo) network [48] is a language model that utilizes bidirectional LSTM.

This chapter introduces RNNs, LSTMs, and their variants. Section 2 explains RNNs. The problems of vanishing or exploding gradients in long-term dependencies, as well as their solutions, are elaborated in Sect. 3. LSTM and its variants are introduced in Sect. 4. Finally, Sect. 5 covers bidirectional RNNs and LSTMs. Note that there exist several tutorials and surveys on RNNs, LSTMs, and their variants. Some of them are [23, 35, 38, 42, 53, 55, 58, 60, 67]. A good survey on the variants of LSTM is [23].

2 Recurrent Neural Network

2.1 Dynamical System

A dynamical system is recursive, and its classical form is as follows:

$$\boldsymbol{h}_t = f_\theta(\boldsymbol{h}_{t-1}), \tag{1}$$

where t denotes the time step, $\boldsymbol{h}_t$ is the state at time t, and $f_\theta(.)$ is a function that remains fixed between the states of all time steps. Figure 1a illustrates such a system. Dynamical systems are widely used in chaos theory [7].

We can also have a dynamical system with an external input signal, where $\boldsymbol{x}_t$ denotes the input signal at time t. This system is modeled as follows:

$$\boldsymbol{h}_t = f_\theta(\boldsymbol{h}_{t-1}, \boldsymbol{x}_t). \tag{2}$$

This system is depicted in Fig. 1b.

2.2 Parameter Sharing

The state $\boldsymbol{h}_t$ can be considered as a summary of the past sequence of inputs and states. If a different function f_θ is defined for each possible sequence length, the model will not generalize. However, if the same parameters are used for any sequence length, the model will have generalization capabilities. Therefore, the parameters are shared across all sequence lengths and between all states. Such a dynamical system with parameter sharing can be implemented as a neural network with weights. This type of network is called a Recurrent Neural Network (RNN), which was proposed in 1986 [51].

An RNN is illustrated in Fig. 1c, where the same weight matrices are used for all time slots. An RNN receives a sequence as input and outputs a sequence as a decision for tasks such as regression or classification. Suppose the input, output, and state at time slot t are denoted by $\boldsymbol{x}_t \in \mathbb{R}^d$, $\boldsymbol{y}_t \in \mathbb{R}^q$, and $\boldsymbol{h}_t \in \mathbb{R}^p$, respectively. Let $\boldsymbol{W} \in \mathbb{R}^{p\times p}$ be the weight matrix between states, $\boldsymbol{U} \in \mathbb{R}^{p\times d}$ be the weight matrix between the inputs and the states, and $\boldsymbol{V} \in \mathbb{R}^{q\times p}$ denote the weight matrix between the states and outputs. The bias weights for the state and the output are denoted by $\boldsymbol{b}_i \in \mathbb{R}^p$ and $\boldsymbol{b}_y \in \mathbb{R}^q$, respectively. As shown in Fig. 1c, we have

$$\mathbb{R}^p \ni \boldsymbol{i}_t = \boldsymbol{W}\boldsymbol{h}_{t-1} + \boldsymbol{U}\boldsymbol{x}_t + \boldsymbol{b}_i, \tag{3}$$

$$[-1, 1]^p \ni \boldsymbol{h}_t = \tanh(\boldsymbol{i}_t) = \tanh(\boldsymbol{W}\boldsymbol{h}_{t-1} + \boldsymbol{U}\boldsymbol{x}_t + \boldsymbol{b}_i), \tag{4}$$

$$\mathbb{R}^q \ni \boldsymbol{y}_t = \boldsymbol{V}\boldsymbol{h}_t + \boldsymbol{b}_y, \tag{5}$$

where $\tanh(.) \in (-1, 1)$ denotes the hyperbolic tangent function, which is used as an element-wise activation function for the states.

If there is an activation function, such as softmax, at the output layer, we denote the output of the activation function by

$$\mathbb{R}^q \ni \widehat{\boldsymbol{y}}_t = \text{softmax}(\boldsymbol{y}_t) = \frac{\exp(y_{t,1})}{\sum_{j=1}^{q} \exp(y_{t,j})}, \tag{6}$$

where $y_{t,j}$ denotes the j-th component of $\boldsymbol{y}_t$.

2.3 Backpropagation Through Time (BPTT)

One of the methods for training RNNs is Backpropagation Through Time (BPTT), which is very similar to the backpropagation algorithm [51] because it is based on gradient descent and the chain rule [15], but it additionally applies the chain rule through time. BPTT was developed by several works [44, 50, 64–66]. Although this algorithm is theoretically sound, it is computationally expensive for long sequences and suffers from vanishing or exploding gradients, which often limit performance in practice.

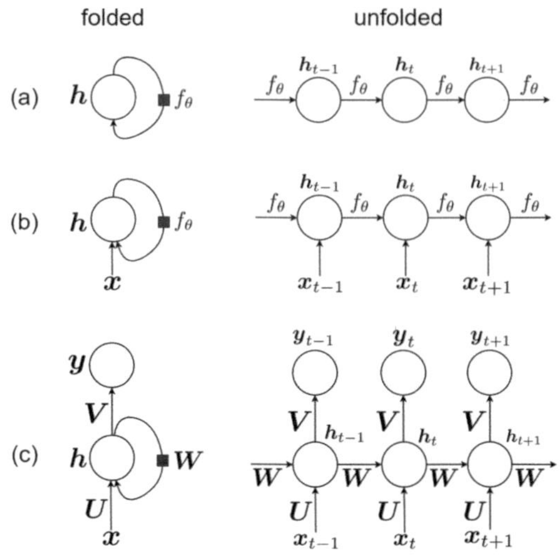

Fig. 1 The folded and unfolded structures of **a** a dynamic system without input, **b** a dynamical system with input, and **c** an RNN. Every square on an edge represents a connection from one time slot before

In BPTT, the loss is considered as the summation of the loss functions at previous time steps up to the current time. As it is impractical to consider all time steps from the start of training (especially after a long training period), only the T previous time steps are considered. In other words, we assume that the RNN has a T-order Markov property [17]. Therefore, the loss function is

$$\mathbb{R} \ni \mathcal{L} = \sum_{t=1}^{T} \mathcal{L}_t, \tag{7}$$

where $\mathcal{L}_1$ is the loss function at the current time slot and $\mathcal{L}_t$ denotes the loss function at the previous $(t-1)$ time slot.

This loss function needs to be optimized using gradient descent and the chain rule. Therefore, we calculate its gradient with respect to the parameters of the RNN. These parameters are $\boldsymbol{y}_t$, $\boldsymbol{h}_t$, $\boldsymbol{V}$, $\boldsymbol{W}$, $\boldsymbol{U}$, $\boldsymbol{b}_i$, and $\boldsymbol{b}_y$, based on Eqs. (3), (4), and (5), and Fig. 1c.

2.3.1 Gradient With Respect to the Output

If there is no activation function at the last layer, the gradient of the loss function of the RNN with respect to the output at time t is

$$\mathbb{R}^q \ni \frac{\partial \mathcal{L}}{\partial \boldsymbol{y}_t} \overset{(a)}{=} \frac{\partial \mathcal{L}}{\partial \mathcal{L}_t} \times \frac{\partial \mathcal{L}_t}{\partial \boldsymbol{y}_t} \overset{(5)}{=} \frac{\partial \mathcal{L}_t}{\partial \boldsymbol{y}_t}, \tag{8}$$

where (a) is due to the chain rule.

The gradient of the loss function at time t with respect to the output at time t, i.e., $\partial \mathcal{L}_t / \partial \boldsymbol{y}_t$, is calculated based on the formula of the loss function. The loss function can be any function used for classification, regression, or other tasks.

If there is an activation function at the last layer (see Eq. (6)), the gradient is

$$\mathbb{R}^q \ni \frac{\partial \mathcal{L}}{\partial \boldsymbol{y}_t} \overset{(a)}{=} \frac{\partial \mathcal{L}}{\partial \mathcal{L}_t} \times \frac{\partial \mathcal{L}_t}{\partial \widehat{\boldsymbol{y}}_t} \times \frac{\partial \widehat{\boldsymbol{y}}_t}{\partial \boldsymbol{y}_t} \overset{(7)}{=} \frac{\partial \mathcal{L}_t}{\partial \widehat{\boldsymbol{y}}_t} \times \frac{\partial \widehat{\boldsymbol{y}}_t}{\partial \boldsymbol{y}_t}, \tag{9}$$

where (a) is due to the chain rule. The derivative $\partial\widehat{\boldsymbol{y}}_t/\partial\boldsymbol{y}_t$ is calculated based on the formula of the activation function. The other derivative, $\partial\mathcal{L}_t/\partial\widehat{\boldsymbol{y}}_t$, is calculated based on the formula of the loss function as a function of the output of the activation function.

2.3.2 Gradient With Respect to the State

The gradient of the loss function of the RNN with respect to the state at time t is

$$\begin{aligned}\mathbb{R}^p \ni \frac{\partial\mathcal{L}}{\partial\boldsymbol{h}_t} &\overset{(a)}{=} \Big(\frac{\partial\mathcal{L}}{\partial\boldsymbol{y}_t} \times \frac{\partial\boldsymbol{y}_t}{\partial\boldsymbol{h}_t}\Big) + \Big(\frac{\partial\mathcal{L}}{\partial\boldsymbol{h}_{t+1}} \times \frac{\partial\boldsymbol{h}_{t+1}}{\partial\boldsymbol{h}_t}\Big) \\ &\overset{(5)}{=} \Big(\frac{\partial\mathcal{L}}{\partial\boldsymbol{y}_t} \times \boldsymbol{V}\Big) + \Big(\frac{\partial\mathcal{L}}{\partial\boldsymbol{h}_{t+1}} \times \frac{\partial\boldsymbol{h}_{t+1}}{\partial\boldsymbol{h}_t}\Big),\end{aligned} \tag{10}$$

where (a) is due to the chain rule, as changing $\boldsymbol{h}_t$ affects both $\boldsymbol{y}_t$ and $\boldsymbol{h}_{t+1}$. We denote $\boldsymbol{\delta}_t := \partial\mathcal{L}/\partial\boldsymbol{h}_t$, so Eq. (10) becomes

$$\boldsymbol{\delta}_t = \Big(\frac{\partial\mathcal{L}}{\partial\boldsymbol{y}_t} \times \boldsymbol{V}\Big) + \Big(\boldsymbol{\delta}_{t+1} \times \frac{\partial\boldsymbol{h}_{t+1}}{\partial\boldsymbol{h}_t}\Big). \tag{11}$$

According to Eqs. (3) and (4), we have

$$\boldsymbol{i}_{t+1} = \boldsymbol{W}\boldsymbol{h}_t + \boldsymbol{U}\boldsymbol{x}_{t+1} + \boldsymbol{b}_i, \tag{12}$$

$$\boldsymbol{h}_{t+1} = \tanh(\boldsymbol{i}_{t+1}) = \tanh(\boldsymbol{W}\boldsymbol{h}_t + \boldsymbol{U}\boldsymbol{x}_{t+1} + \boldsymbol{b}_i). \tag{13}$$

Therefore:

$$\begin{aligned}\mathbb{R}^{p\times p} \ni \frac{\partial\boldsymbol{h}_{t+1}}{\partial\boldsymbol{h}_t} &\overset{(a)}{=} \Big(\frac{\partial\boldsymbol{i}_{t+1}}{\partial\boldsymbol{h}_t}\Big)^\top \times \frac{\partial\boldsymbol{h}_{t+1}}{\partial\boldsymbol{i}_{t+1}} \\ &\overset{(b)}{=} \boldsymbol{W}(1 - \boldsymbol{h}_{t+1}^\top\boldsymbol{h}_{t+1})\boldsymbol{I}_{p\times p},\end{aligned} \tag{14}$$

where $\boldsymbol{I}_{p\times p}$ is the identity matrix of size $(p \times p)$, (a) is due to the chain rule, and (b) holds because $\mathbb{R}^{p\times p} \ni \partial\boldsymbol{i}_{t+1}/\partial\boldsymbol{h}_t = \boldsymbol{W}^\top$ for Eq. (12), and we have

$$\mathbb{R}^{p\times p} \ni \frac{\partial\boldsymbol{h}_{t+1}}{\partial\boldsymbol{i}_{t+1}} = (1 - \boldsymbol{h}_{t+1}^\top\boldsymbol{h}_{t+1})\boldsymbol{I}_{p\times p}, \tag{15}$$

based on Eq. (13) and the formula for the derivative of the hyperbolic tangent function, noting that the state is multidimensional and not a scalar.

For the time slot $t = T$, the derivative $\partial\mathcal{L}/\partial\boldsymbol{h}_T$ is much simpler:

$$\begin{aligned}\mathbb{R}^p \ni \boldsymbol{\delta}_T = \frac{\partial\mathcal{L}}{\partial\boldsymbol{h}_T} &\overset{(a)}{=} \frac{\partial\mathcal{L}}{\partial\boldsymbol{y}_T} \times \frac{\partial\boldsymbol{y}_T}{\partial\boldsymbol{h}_T} \overset{(5)}{=} \frac{\partial\mathcal{L}}{\partial\boldsymbol{y}_T} \times \boldsymbol{V} \\ &\overset{(8)}{=} \frac{\partial\mathcal{L}_T}{\partial\boldsymbol{y}_T} \times \boldsymbol{V},\end{aligned} \tag{16}$$

where (a) is due to the chain rule, and the derivative $\partial\mathcal{L}_T/\partial\boldsymbol{y}_T$ is computed based on the formula of the loss function.

2.3.3 Gradient With Respect to the Weight Matrix Between the States and Outputs

The gradient of the loss function of the RNN with respect to the weight matrix $\boldsymbol{V}$ is

$$\mathbb{R}^{q\times p} \ni \frac{\partial \mathcal{L}}{\partial \boldsymbol{V}} \overset{(a)}{=} \sum_{t=1}^{T} \Big(\frac{\partial \mathcal{L}}{\partial \boldsymbol{y}_t} \times \frac{\partial \boldsymbol{y}_t}{\partial \boldsymbol{V}}\Big) \overset{(b)}{=} \sum_{t=1}^{T} \Big(\frac{\partial \mathcal{L}_t}{\partial \boldsymbol{y}_t} \times \boldsymbol{h}_t^\top\Big), \tag{17}$$

where (a) holds because $\boldsymbol{V}$ exists in all time slots and changing $\boldsymbol{V}$ affects the loss $\mathcal{L}$ in all time slots. Equation (b) follows from Eqs. (8) and (5). The derivative $\partial \mathcal{L}_t / \partial \boldsymbol{y}_t \in \mathbb{R}^q$ is calculated based on the formula of the loss function.

2.3.4 Gradient with Respect to the Weight Matrix Between States

The gradient of the loss function of the RNN with respect to the weight matrix $\boldsymbol{W}$ is

$$\mathbb{R}^{p\times p} \ni \frac{\partial \mathcal{L}}{\partial \boldsymbol{W}} \overset{(a)}{=} \sum_{t=1}^{T} \mathbf{vec}^{-1}_{p\times p}\Big(\Big(\frac{\partial \boldsymbol{h}_t}{\partial \boldsymbol{W}}\Big)^\top \times \frac{\partial \mathcal{L}}{\partial \boldsymbol{h}_t}\Big), \tag{18}$$

where $\mathbf{vec}^{-1}_{p\times p}(.)$ de-vectorizes the vector of length p^2 to a matrix of size $(p \times p)$, and (a) holds because $\boldsymbol{W}$ exists in all time slots and changing $\boldsymbol{W}$ affects the loss $\mathcal{L}$ in all time slots. The derivative $\partial \mathcal{L} / \partial \boldsymbol{h}_t \in \mathbb{R}^p$ in Eq. (18) was computed in Sect. 2.3.2. The derivative $\partial \boldsymbol{h}_t / \partial \boldsymbol{W}$ in Eq. (18) is

$$\mathbb{R}^{p\times p^2} \ni \frac{\partial \boldsymbol{h}_t}{\partial \boldsymbol{W}} = \frac{\partial \boldsymbol{h}_t}{\partial \boldsymbol{i}_t} \times \frac{\partial \boldsymbol{i}_t}{\partial \boldsymbol{W}},$$

due to the chain rule. The first term is

$$\mathbb{R}^{p\times p} \ni \frac{\partial \boldsymbol{h}_t}{\partial \boldsymbol{i}_t} = (1 - \boldsymbol{h}_t^\top \boldsymbol{h}_t)\boldsymbol{I}_{p\times p}, \tag{19}$$

according to Eq. (15). Based on the Magnus–Neudecker convention [15], the second term is calculated as follows:

$$\mathbb{R}^{p\times p^2} \ni \frac{\partial \boldsymbol{i}_t}{\partial \boldsymbol{W}} = \boldsymbol{h}_{t-1}^\top \otimes \boldsymbol{I}_{p\times p},$$

where $\otimes$ denotes the Kronecker product.

2.3.5 Gradient with Respect to the Weight Matrix Between the Inputs and the States

The gradient of the loss function of the RNN with respect to the weight matrix $\boldsymbol{U}$ is

$$\mathbb{R}^{p\times d} \ni \frac{\partial \mathcal{L}}{\partial \boldsymbol{U}} \overset{(a)}{=} \sum_{t=1}^{T} \mathbf{vec}^{-1}_{p\times d}\Big(\Big(\frac{\partial \boldsymbol{h}_t}{\partial \boldsymbol{U}}\Big)^\top \times \frac{\partial \mathcal{L}}{\partial \boldsymbol{h}_t}\Big), \tag{20}$$

where (a) holds because $\boldsymbol{U}$ exists in all time slots and changing $\boldsymbol{U}$ affects the loss $\mathcal{L}$ in all time slots. The derivative $\partial \mathcal{L} / \partial \boldsymbol{h}_t \in \mathbb{R}^p$ in Eq. (18) was computed in Sect. 2.3.2. The derivative $\partial \boldsymbol{h}_t / \partial \boldsymbol{U}$ in Eq. (18) is

$$\mathbb{R}^{p\times(pd)} \ni \frac{\partial \boldsymbol{h}_t}{\partial \boldsymbol{U}} = \frac{\partial \boldsymbol{h}_t}{\partial \boldsymbol{i}_t} \times \frac{\partial \boldsymbol{i}_t}{\partial \boldsymbol{U}},$$

due to the chain rule. The first term was already calculated in Eq. (19). Based on the Magnus–Neudecker convention [15], the second term is calculated as follows:

$$\mathbb{R}^{p\times(pd)} \ni \frac{\partial \boldsymbol{i}_t}{\partial \boldsymbol{U}} = \boldsymbol{x}_t^\top \otimes \boldsymbol{I}_{p\times p},$$

where $\otimes$ denotes the Kronecker product.

2.3.6 Gradient with Respect to the Bias Weight for the State

The gradient of the loss function of the RNN with respect to the bias $\boldsymbol{b}_i$ is

$$\mathbb{R}^p \ni \frac{\partial \mathcal{L}}{\partial \boldsymbol{b}_i} \overset{(a)}{=} \sum_{t=1}^{T} \Big(\big(\frac{\partial \boldsymbol{h}_t}{\partial \boldsymbol{b}_i}\big)^\top \times \frac{\partial \mathcal{L}}{\partial \boldsymbol{h}_t} \Big), \tag{21}$$

where (a) holds because $\boldsymbol{b}_i$ exists in all time slots and changing $\boldsymbol{b}_i$ affects the loss $\mathcal{L}$ in all time slots. The derivative $\partial\mathcal{L}/\partial\boldsymbol{h}_t$ was already calculated in Sect. 2.3.2. The derivative $\partial\boldsymbol{h}_t/\partial\boldsymbol{b}_i$ is calculated as follows:

$$\mathbb{R}^{p\times p} \ni \frac{\partial \boldsymbol{h}_t}{\partial \boldsymbol{b}_i} \overset{(a)}{=} \frac{\partial \boldsymbol{h}_t}{\partial \boldsymbol{i}_t} \times \frac{\partial \boldsymbol{i}_t}{\partial \boldsymbol{b}_i} \overset{(8)}{=} \frac{\partial \boldsymbol{h}_t}{\partial \boldsymbol{i}_t},$$

where (a) is due to the chain rule, and the derivative $\partial\boldsymbol{h}_t/\partial\boldsymbol{i}_t$ was already calculated in Eq. (19).

2.3.7 Gradient with Respect to the Bias Weight for the Output

The gradient of the loss function of the RNN with respect to the bias $\boldsymbol{b}_y$ is

$$\mathbb{R}^q \ni \frac{\partial \mathcal{L}}{\partial \boldsymbol{b}_y} \overset{(a)}{=} \sum_{t=1}^{T} \Big(\frac{\partial \mathcal{L}}{\partial \boldsymbol{y}_t} \times \frac{\partial \boldsymbol{y}_t}{\partial \boldsymbol{b}_y} \Big) \overset{(b)}{=} \sum_{t=1}^{T} \frac{\partial \mathcal{L}_t}{\partial \boldsymbol{y}_t}, \tag{22}$$

where (a) holds because $\boldsymbol{b}_y$ exists in all time slots and changing $\boldsymbol{b}_y$ affects the loss $\mathcal{L}$ in all time slots. Equation (b) follows from Eqs. (8) and (5). The derivative $\partial\mathcal{L}_t/\partial\boldsymbol{y}_t \in \mathbb{R}^q$ is calculated based on the formula of the loss function.

2.3.8 Updates by Gradient Descent

BPTT updates the parameters of the RNN by gradient descent [15] using the calculated gradients:

$$\begin{aligned}
&\boldsymbol{h}_t := \boldsymbol{h}_t - \eta\frac{\partial \mathcal{L}}{\partial \boldsymbol{h}_t}, \quad \forall t \in \{1,\ldots,T\},\\
&\boldsymbol{V} := \boldsymbol{V} - \eta\frac{\partial \mathcal{L}}{\partial \boldsymbol{V}}, \quad \boldsymbol{W} := \boldsymbol{W} - \eta\frac{\partial \mathcal{L}}{\partial \boldsymbol{W}}, \quad \boldsymbol{U} := \boldsymbol{U} - \eta\frac{\partial \mathcal{L}}{\partial \boldsymbol{U}},\\
&\boldsymbol{b}_i := \boldsymbol{b}_i - \eta\frac{\partial \mathcal{L}}{\partial \boldsymbol{b}_i}, \quad \boldsymbol{b}_y := \boldsymbol{b}_y - \eta\frac{\partial \mathcal{L}}{\partial \boldsymbol{b}_y},
\end{aligned}$$

where $\eta > 0$ is the learning rate, and the gradients are calculated by Eqs. (10), (17), (18), (20), (21), and (22).

3 Vanishing or Exploding Gradient in Long-Term Dependencies

In recurrent neural networks, as in deep neural networks, the final output is the composition of a large number of nonlinear transformations. This results in the problem of either vanishing or exploding gradients in recurrent neural networks, especially when capturing long-term dependencies in sequence processing [5, 6]. This problem is explained below. Recall Eq. (2) for a dynamical system:

$$\boldsymbol{h}_t = f_\theta(\boldsymbol{h}_{t-1}, \boldsymbol{x}_t).$$

By induction, the hidden state at time t, i.e., $\boldsymbol{h}_t$, can be written in terms of the previous T time steps. If the subscript t denotes the previous t time steps, we have by induction [28, 29]:

$$\boldsymbol{h}_1 = f_\theta\Big(f_\theta\big(\dots f_\theta(\boldsymbol{h}_T, \boldsymbol{x}_{T+1})\dots, \boldsymbol{x}_2\big), \boldsymbol{x}_1\Big).$$

Then, by the chain rule in derivatives, the derivative of the loss at time T, i.e., $\mathcal{L}_T$, is

$$\begin{aligned}\frac{\partial \mathcal{L}_T}{\partial \theta} &= \sum_{t \leq T} \frac{\partial \mathcal{L}_t}{\partial \boldsymbol{h}_t}\frac{\partial \boldsymbol{h}_t}{\partial \theta} \overset{(a)}{=} \sum_{t \leq T} \frac{\partial \mathcal{L}_t}{\partial \boldsymbol{h}_T}\frac{\partial \boldsymbol{h}_T}{\partial \boldsymbol{h}_t}\frac{\partial \boldsymbol{h}_t}{\partial \theta} \\ &\overset{(2)}{=} \sum_{t \leq T} \frac{\partial \mathcal{L}_t}{\partial \boldsymbol{h}_T}\frac{\partial \boldsymbol{h}_T}{\partial \boldsymbol{h}_t}\frac{\partial f_\theta(\boldsymbol{h}_{t-1}, \boldsymbol{x}_t)}{\partial \theta},\end{aligned} \tag{23}$$

where (a) is due to the chain rule. In this expression, there is the derivative of $\boldsymbol{h}_T$ with respect to $\boldsymbol{h}_t$, which itself can be calculated by the chain rule:

$$\frac{\partial \boldsymbol{h}_T}{\partial \boldsymbol{h}_t} = \frac{\partial \boldsymbol{h}_T}{\partial \boldsymbol{h}_{T-1}} \times \frac{\partial \boldsymbol{h}_{T-1}}{\partial \boldsymbol{h}_{T-2}} \times \dots \times \frac{\partial \boldsymbol{h}_{t+1}}{\partial \boldsymbol{h}_t}. \tag{24}$$

To capture long-term dependencies in the sequence, T should be large. This means that in Eq. (24), and hence in Eq. (23), the number of multiplicative terms becomes huge. On the one hand, if each derivative is slightly smaller than one, the entire derivative in the chain rule becomes very small due to the multiplication of many terms smaller than one. This problem is referred to as the vanishing gradient. On the other hand, if each derivative is slightly larger than one, the entire derivative in the chain rule explodes, resulting in the problem of exploding gradients. Note that the vanishing gradient problem is more common than gradient explosion in recurrent networks.

Various attempts have been made to resolve the problem of vanishing or exploding gradients [4, 28]. Some of these approaches are introduced below.

3.1 Close-to-Identity Weight Matrix

As Eq. (4) shows, the state is multiplied by a weight matrix $\boldsymbol{W}$ at every time step, and if there is a long-term dependency, many of these $\boldsymbol{W}$ matrices are multiplied together. Suppose the eigenvalue decomposition [16] of the matrix $\boldsymbol{W}$ is $\boldsymbol{W} = \boldsymbol{A}\boldsymbol{\Lambda}\boldsymbol{A}^\top$, where $\boldsymbol{A} \in \mathbb{R}^{p\times p}$ and $\boldsymbol{\Lambda} := \textbf{diag}([\lambda_1, \dots, \lambda_p]^\top)$ contain the eigenvectors and eigenvalues of $\boldsymbol{W}$, respectively. Equation (4) can be restated as follows:

$$\boldsymbol{h}_t = \tanh(\boldsymbol{A}\boldsymbol{\Lambda}\boldsymbol{A}^\top \boldsymbol{h}_{t-1} + \boldsymbol{U}\boldsymbol{x}_t + \boldsymbol{b}_i). \tag{25}$$

If a change ε in some element of the state $\boldsymbol{h}_{t-1}$ is aligned with an eigenvector of the weight matrix $\boldsymbol{W}$, then the effect of this change on $\boldsymbol{h}_t$ will be $(\lambda^t\ \varepsilon)$ after t time steps, according to Eq. (25). Two cases may occur:

- If the largest eigenvalue is less than one, i.e., $\lambda < 1$, then the change $(\lambda^t\ \varepsilon)$ is contractive because $\lambda^t \ll 1$ for long-term dependencies. In this case, the vanishing gradient problem occurs, and the network forgets information from the distant past.
- If the largest eigenvalue is greater than one, i.e., $\lambda > 1$, then the change $(\lambda^t\ \varepsilon)$ is diverging because $\lambda^t \gg 1$ for long-term dependencies. In this case, the exploding gradient problem occurs, and remembering information from the distant past dominates short-term memories.

As remembering short-term memories is usually more important than remembering information from the distant past in many tasks, it is recommended to use a weight matrix $\boldsymbol{W}$ whose largest eigenvalue is less than one; this gives the RNN the Markovian property because it forgets information from the distant past after some point. However, if the largest eigenvalue of $\boldsymbol{W}$ is much less than one, i.e., $\lambda \ll 1$, the vanishing gradient problem occurs much sooner than expected. Therefore, it is recommended to use a weight matrix $\boldsymbol{W}$ whose largest eigenvalue is *slightly* less than one, i.e., $\lambda \lesssim 1$. This makes the RNN slightly contractive. One way to mitigate gradient issues is to initialize the weight matrix $\boldsymbol{W}$ carefully, such as using orthogonal or unitary matrices (e.g., to have the largest eigenvalue slightly less than one) [43]. In practice, gradient clipping is also used, and architectures like LSTM, GRU, and modern SSMs (see Chapter 10) largely address these problems.

There exist other ways to determine the weight matrix $\boldsymbol{W}$. For example, the weight matrix can be set to be an orthogonal matrix [1]. Another approach is to copy the previous state directly to the current state. In this approach, Eq. (4) is modified to [32]:

$$\begin{aligned} \boldsymbol{h}_t &= \tanh(\boldsymbol{W}\boldsymbol{h}_{t-1} + \boldsymbol{h}_{t-1} + \boldsymbol{U}\boldsymbol{x}_t + \boldsymbol{b}_i) \\ &= \tanh\big((\boldsymbol{W} + \boldsymbol{I})\boldsymbol{h}_{t-1} + \boldsymbol{U}\boldsymbol{x}_t + \boldsymbol{b}_i\big), \end{aligned} \tag{26}$$

where $\boldsymbol{I}$ is the identity matrix. This prevents the vanishing gradient problem because it brings a copy of the previous state into the current state. This can also be interpreted as strengthening the diagonal of the weight matrix $\boldsymbol{W}$, thereby increasing the largest eigenvalue of $\boldsymbol{W}$ and helping to prevent vanishing gradients.

3.2 Long Delays

As Eq. (4) and Fig. 1c show, in the regular RNN, every state $\boldsymbol{h}_t$ is fed by its previous state $\boldsymbol{h}_{t-1}$ through the weight matrix $\boldsymbol{W}$. As discussed in Sect. 3.1, in the regular RNN, the effect of a change ε in a state results in $(\lambda^t\ \varepsilon)$ after t time steps, where λ is the largest eigenvalue of $\boldsymbol{W}$.

As shown in Fig. 1c, the regular RNN has one-step connections or delays between the states. It is possible to have longer delays between the states in addition to the one-step delays [41]. In other words, it is possible to have higher levels of the Markov property in the network. Let $\boldsymbol{W}_k$ denote the weight matrix for k-step delays between the states. Then, Eq. (4) can be modified to:

$$\boldsymbol{h}_t = \tanh\Big(\sum_k \boldsymbol{W}_k\boldsymbol{h}_{t-k} + \boldsymbol{U}\boldsymbol{x}_t + \boldsymbol{b}_i\Big), \tag{27}$$

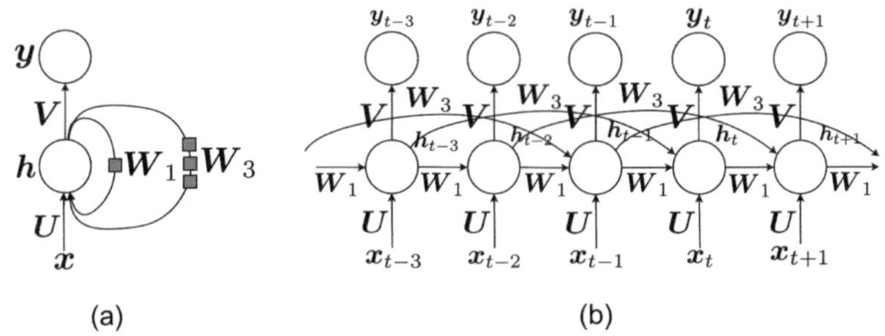

Fig. 2 The **a** folded and **b** unfolded structures of an RNN with long delays. Every square on an edge represents a connection from one time slot before

where the summation is over the k values for the existing delays in the RNN structure. An example of an RNN network with one-step and three-step delays is

$$\boldsymbol{h}_t = \tanh\Big(\boldsymbol{W}_1\boldsymbol{h}_{t-1} + \boldsymbol{W}_3\boldsymbol{h}_{t-3} + \boldsymbol{U}\boldsymbol{x}_t + \boldsymbol{b}_i\Big),$$

which is illustrated in Fig. 2.

Having long delays in RNN is one of the approaches for preventing the vanishing gradient problem [41]. This is justified because every state is influenced not only by the previous state but also by states further back in time. Therefore, in backpropagation through time, there is some skip in the gradient flow from a state to earlier states without the need to propagate through all intermediate states in the chain rule.

3.3 Leaky Units

Another way to resolve the vanishing gradient problem is the use of leaky units [37, 62]. Let $h_{t,j}$ denote the jth element of the state $\boldsymbol{h}_t \in [-1, 1]^p$. In leaky units, Eq. (4) is modified to the following element-wise equation:

$$h_{t,j} = (1 - \frac{1}{\tau_j})\, h_{t-1,j} + \frac{1}{\tau_j}\tanh(\boldsymbol{W}_{j:}\boldsymbol{h}_{t-1} + \boldsymbol{U}_{j:}\boldsymbol{x}_t + b_{i,j}), \tag{28}$$

where $1 \leq \tau_j < \infty$, $\boldsymbol{W}_{j:}$ is the j-th row of $\boldsymbol{W}$, $\boldsymbol{U}_{j:}$ is the j-th row of $\boldsymbol{U}$, and $b_{i,j}$ is the j-th element of $\boldsymbol{b}_i$. When $\tau_i = 1$, Eq. (28) becomes

$$h_{t,j} = \frac{1}{\tau_j}\tanh(\boldsymbol{W}_{j:}\boldsymbol{h}_{t-1} + \boldsymbol{U}_{j:}\boldsymbol{x}_t + b_{i,j}),$$

which reduces to Eq. (4) for the regular RNN. However, when $\tau_i \gg 1$, Eq. (28) becomes

$$h_{t,j} = h_{t-1,j},$$

which means that the previous state is copied to the current state. The larger the τ_i, the easier it is for the gradient to propagate for $h_{t,i}$. Therefore, by tuning τ_i, it is possible to control how much of the past should be directly copied and how much should pass through the weight matrix. This can control the extent of vanishing gradients. Note that leaky units use different τ_i values because there may be a need to keep some of the directions of states (with $\tau_i = 1$) or to forget some of the directions (with $\tau_i \gg 1$). In other words, the model decides about the p directions of states separately.

3.4 Echo State Networks

One approach to handle the vanishing gradient problem in RNNs is the use of echo state networks [34, 36]. These networks consider the recurrent neural network as a black box consisting of hidden units with nonlinear activation functions and connections between them. This black box of recurrent connections is called the *reservoir dynamical system*, which models the internal structure of a computer or brain. The connections in the reservoir system are usually sparse, and the weights of these connections are considered fixed. The output of the reservoir system is connected to an additional linear output layer whose weights are learnable. The echo state network minimizes the mean squared error in the output layer; hence, it performs linear regression in the last layer [36]. This network is shown in Fig. 3. Because the recurrent weights in the reservoir system are not learned and only the weights of the output layer are trained, the echo state network does not face the vanishing gradient problem. A tutorial on this topic is provided in [35].

3.5 Other Methods

There exist other methods to address the vanishing gradient problem in recurrent networks. For example, hierarchical RNNs [27] and deep RNNs [18] have been proposed, which stack several recurrent networks. Another related category of networks is the time-delay neural networks [40, 46], which are used for shift-invariant sequence processing and are especially common in speech recognition [61]. In these networks, every neuron in a layer receives a contextual window of the neurons in the previous layer as well as their delayed outputs from several time slots ago. Moreover, these networks apply backpropagation to several copies of the network shifted across the sequence (time) so that the network becomes time-invariant.

4 Long Short-Term Memory Network

As the examples in Sect. 1 showed, we need short-term relations in some cases and long-term relations in others. An RNN learns the sequence based on one or several previous states, depending on its structure and the level of its Markov property (see Fig. 2). Therefore, we need to decide on the structure of the RNN to handle short-term or long-term dependencies in the sequence effectively. Instead of manually designing the RNN structure or deciding when to clear the state, we can let the neural network learn when to clear the state based on its input sequence. Long Short-Term Memory (LSTM), initially

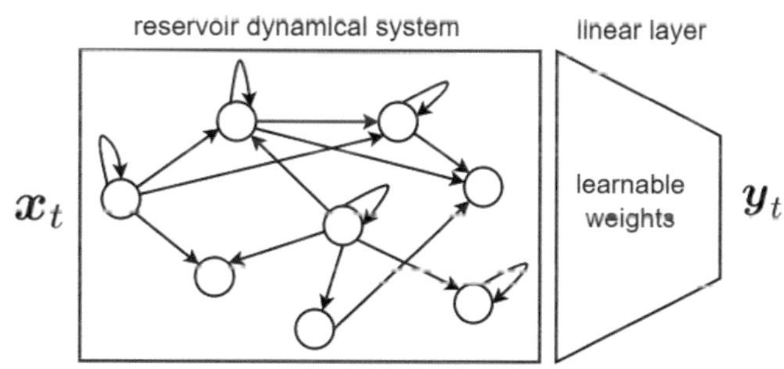

Fig. 3 The echo state network

proposed in [30, 31], is capable of this; it learns from its input sequence when to use short-term dependency (i.e., when to clear the state) and when to use long-term memory (i.e., when not to clear the state).

4.1 LSTM Gates and Cells

LSTM consists of several cells, each corresponding to a time slot (see Fig. 4). Every LSTM cell contains several gates that learn different aspects of the input time series (see Fig. 5). These gates are introduced below.

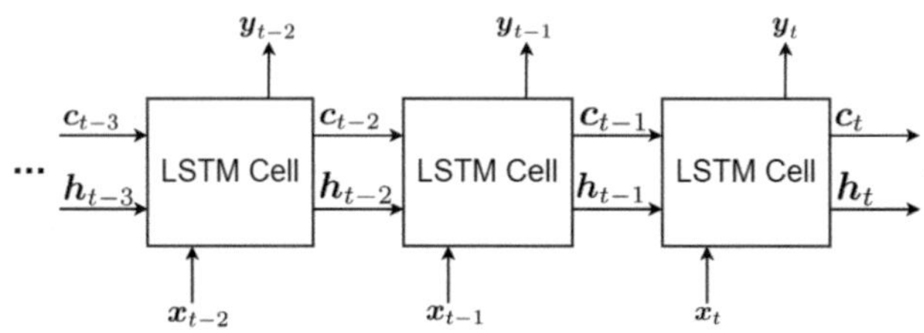

Fig. 4 A sequence of LSTM cells processing the input sequence

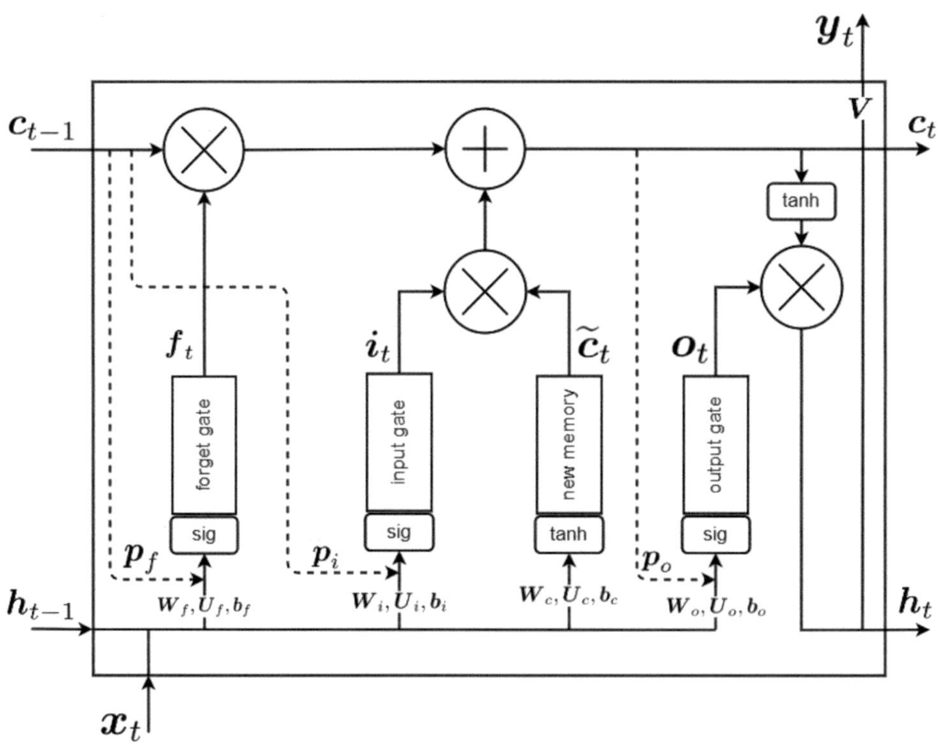

Fig. 5 The conveyor belt in the LSTM cell

4.1.1 The Input Gate

One of the gates in the LSTM cell is the *input gate*, first proposed in [30, 31]. This gate takes the input at the current time slot, $\boldsymbol{x}_t \in \mathbb{R}^d$, and the hidden state of the previous time slot, $\boldsymbol{h}_{t-1} \in [-1, 1]^p$, and outputs the signal $\boldsymbol{i}_t \in [0, 1]^p$:

$$\boldsymbol{i}_t = \text{sig}\big(\boldsymbol{W}_i \boldsymbol{h}_{t-1} + \boldsymbol{U}_i \boldsymbol{x}_t + (\boldsymbol{p}_i \odot \boldsymbol{c}_{t-1}) + \boldsymbol{b}_i\big), \tag{29}$$

where $\boldsymbol{W}_i \in \mathbb{R}^{p \times p}$, $\boldsymbol{U}_i \in \mathbb{R}^{p \times d}$, and the bias $\boldsymbol{b}_i \in \mathbb{R}^p$ are the learnable weights for the input gate, $\odot$ denotes the Hadamard (element-wise) product, $\boldsymbol{c}_{t-1} \in \mathbb{R}^p$ is the final memory from the previous time slot (which will be explained in Sect. 4.1.5), and $\boldsymbol{p}_i \in \mathbb{R}^p$ is the learnable *peephole weight* [13], allowing a possible leak of information from the previous final memory. The function $\text{sig}(.) \in (0, 1)$ is the sigmoid function, which is applied element-wise:

$$\text{sig}(x) = \frac{1}{1 + \exp(-x)}. \tag{30}$$

As Eq. (29) demonstrates, the input gate considers the effects of the input and the previous hidden state. It may also incorporate information from the previous memory through the peephole connection. This gate determines the importance of the input information at the current time slot. The input gate is depicted in Fig. 5.

4.1.2 The Forget Gate

Another gate in the LSTM cell is the *forget gate*, first proposed in [14]. This gate also takes the input at the current time slot, $\boldsymbol{x}_t \in \mathbb{R}^d$, and the hidden state of the previous time slot, $\boldsymbol{h}_{t-1} \in [-1, 1]^p$, and outputs the signal $\boldsymbol{f}_t \in [0, 1]^p$:

$$\boldsymbol{f}_t = \text{sig}\big(\boldsymbol{W}_f \boldsymbol{h}_{t-1} + \boldsymbol{U}_f \boldsymbol{x}_t + (\boldsymbol{p}_f \odot \boldsymbol{c}_{t-1}) + \boldsymbol{b}_f\big), \tag{31}$$

where $\boldsymbol{W}_f \in \mathbb{R}^{p \times p}$, $\boldsymbol{U}_f \in \mathbb{R}^{p \times d}$, and the bias $\boldsymbol{b}_f \in \mathbb{R}^p$ are the learnable weights for the forget gate, and $\boldsymbol{p}_f \in \mathbb{R}^p$ is the learnable *peephole weight* [13], allowing a possible leak of information from the previous final memory.

As Eq. (31) shows, the forget gate considers the effects of the input and the previous hidden state, and possibly incorporates information from the previous memory through the peephole connection. This gate controls the extent to which previous information is forgotten in relation to the incoming new information. The forget gate is illustrated in Fig. 5.

4.1.3 The Output Gate

The next gate in the LSTM cell is the *output gate*, first proposed in [30, 31]. This gate also takes the input at the current time slot, $\boldsymbol{x}_t \in \mathbb{R}^d$, and the hidden state of the previous time slot, $\boldsymbol{h}_{t-1} \in [-1, 1]^p$, and outputs the signal $\boldsymbol{o}_t \in [0, 1]^p$:

$$\boldsymbol{o}_t = \text{sig}\big(\boldsymbol{W}_o \boldsymbol{h}_{t-1} + \boldsymbol{U}_o \boldsymbol{x}_t + (\boldsymbol{p}_o \odot \boldsymbol{c}_t) + \boldsymbol{b}_o\big), \tag{32}$$

where $\boldsymbol{W}_o \in \mathbb{R}^{p \times p}$, $\boldsymbol{U}_o \in \mathbb{R}^{p \times d}$, and the bias $\boldsymbol{b}_o \in \mathbb{R}^p$ are the learnable weights for the output gate, and $\boldsymbol{p}_o \in \mathbb{R}^p$ is the learnable *peephole weight* [13], allowing a possible leak of information from the current final memory.

As shown in Eq. (32), the output gate considers the effects of the input and the previous hidden state, and may also incorporate information from the current memory through the peephole connection. The output gate is shown in Fig. 5.

4.1.4 The New Memory Cell (Block Input)

The LSTM cell includes a gate called the *new memory cell*. This gate takes the input at the current time slot, $\boldsymbol{x}_t \in \mathbb{R}^d$, and the hidden state of the previous time slot, $\boldsymbol{h}_{t-1} \in [-1, 1]^p$, and outputs the signal $\widetilde{\boldsymbol{c}}_t \in [-1, 1]^p$. This gate considers the effects of the input and the previous hidden state to represent the new information from the current input. It is formulated as follows:

$$\widetilde{\boldsymbol{c}}_t = \tanh(\boldsymbol{W}_c \boldsymbol{h}_{t-1} + \boldsymbol{U}_c \boldsymbol{x}_t + \boldsymbol{b}_c), \tag{33}$$

where $\boldsymbol{W}_c \in \mathbb{R}^{p\times p}, \boldsymbol{U}_c \in \mathbb{R}^{p\times d}$, and the bias $\boldsymbol{b}_c$ are the learnable weights for the new memory cell. The new memory cell is also referred to as the *block input* in the literature [23]. The signal $\widetilde{\boldsymbol{c}}_t$ is sometimes denoted by $\boldsymbol{z}_t$ in the literature. The new memory cell is illustrated in Fig. 5.

4.1.5 The Final Memory Calculation

After computing the outputs of the input gate $\boldsymbol{i}_t$, the forget gate $\boldsymbol{f}_t$, and the new memory cell $\widetilde{\boldsymbol{c}}_t$, we calculate the *final memory* $\boldsymbol{c}_t \in \mathbb{R}^p$:

$$\boldsymbol{c}_t = (\boldsymbol{f}_t \odot \boldsymbol{c}_{t-1}) + (\boldsymbol{i}_t \odot \widetilde{\boldsymbol{c}}_t), \tag{34}$$

where $\boldsymbol{c}_{t-1} \in \mathbb{R}^p$ is the final memory from the previous time slot.

As Eq. (34) demonstrates, the final memory considers the effects of the forget gate, the previous memory, the input gate, and the new memory. In the first term, i.e., $\boldsymbol{f}_t \odot \boldsymbol{c}_{t-1}$, the forget gate $\boldsymbol{f}_t \in [0, 1]^p$ controls how much of the previous memory $\boldsymbol{c}_{t-1}$ should be forgotten. The closer $\boldsymbol{f}_t$ is to zero, the more the network forgets the previous memory $\boldsymbol{c}_{t-1}$. In the second term, i.e., $\boldsymbol{i}_t \odot \widetilde{\boldsymbol{c}}_t$, the input gate $\boldsymbol{i}_t \in [0, 1]^p$ and the new memory cell $\widetilde{\boldsymbol{c}}_t \in [-1, 1]^p$ together control how much of the new input information should be used. The closer the input gate $\boldsymbol{i}_t$ is to one and the closer the new memory cell $\widetilde{\boldsymbol{c}}_t$ is to ± 1, the more the input information is used.

In other words, the first and second terms in Eq. (34) determine the trade-off between the use of old and new information in the sequence. The weights of these gates are trained so that they pass or block the input and previous information based on the input sequence and the time step. The final memory calculation is depicted in Fig. 5.

4.1.6 The Hidden State (Block Output)

After computing the output of the output gate $\boldsymbol{o}_t$ and the final memory $\boldsymbol{c}_t$, we calculate the *hidden state* $\boldsymbol{h}_t \in [-1, 1]^p$:

$$\boldsymbol{h}_t = \boldsymbol{o}_t \odot \tanh(\boldsymbol{c}_t). \tag{35}$$

This hidden state is also considered the *block output* of the LSTM cell, as depicted in Fig. 5.

4.1.7 The Output

The output $\boldsymbol{y}_t \in \mathbb{R}^q$ of the LSTM cell is calculated as follows:

$$\boldsymbol{y}_t = \boldsymbol{V}\boldsymbol{h}_t + \boldsymbol{b}_y, \tag{36}$$

where $\boldsymbol{V} \in \mathbb{R}^{q \times p}$ and the bias $\boldsymbol{b}_y \in \mathbb{R}^q$ are the learnable weights for the output. It is possible to use an activation function, such as Eq. (6), after this output signal. Figure 5 shows the output signal in the LSTM cell.

Note that in the literature, the output is sometimes considered to be equal to the hidden state, i.e., $\boldsymbol{y}_t = \boldsymbol{h}_t$, by setting $\boldsymbol{V} = \boldsymbol{I}$ (the identity matrix) and $\boldsymbol{b}_y = \boldsymbol{0}$ (the zero vector).

4.2 History and Variants of LSTM

LSTM has undergone various developments and gradual improvements [23]. Some variants of LSTM do not include peepholes. In this case, Eqs. (29), (31), and (32) are simplified to:

$$\boldsymbol{i}_t = \text{sig}\big(\boldsymbol{W}_i \boldsymbol{h}_{t-1} + \boldsymbol{U}_i \boldsymbol{x}_t + \boldsymbol{b}_i\big), \tag{37}$$

$$\boldsymbol{f}_t = \text{sig}\big(\boldsymbol{W}_f \boldsymbol{h}_{t-1} + \boldsymbol{U}_f \boldsymbol{x}_t + \boldsymbol{b}_f\big), \tag{38}$$

$$\boldsymbol{o}_t = \text{sig}\big(\boldsymbol{W}_o \boldsymbol{h}_{t-1} + \boldsymbol{U}_o \boldsymbol{x}_t + \boldsymbol{b}_o\big), \tag{39}$$

respectively. In the following, we review the history of LSTM network variants.

4.2.1 Original LSTM

LSTM was originally proposed by Hochreiter and Schmidhuber between 1995 and 1997 [30, 31]. This is referred to as the *original LSTM* [31]. The original LSTM included only the input and output gates, introduced in Sects. 4.1.1 and 4.1.3, and did not have a forget gate. It also did not include peepholes; therefore, its gates were defined by Eqs. (37) and (39). The original LSTM trained the network using BPTT (introduced in Sect. 2.3) combined with real-time recurrent learning [50, 65].

4.2.2 Vanilla LSTM

Later, Gers et al. [13, 14] applied some changes to the original LSTM. The forget gate, introduced in Sect. 4.1.2, was proposed for the first time in [14] to allow the network to forget its previous states either completely or partially. The peephole connections, introduced in Sects. 4.1.1, 4.1.2, and 4.1.3, were first proposed in [13]. Peepholes allow a possible leak of information from the previous or current final memory, enabling the memory to control the gates.

These two papers [13, 14] also incorporated *full gate recurrence*, in which all gates receive additional recurrent inputs from all gates at the previous time step. In full gate recurrence, Eqs. (29), (31), and (32) become

$$\begin{aligned}\boldsymbol{i}_t = \text{sig}\big(&\boldsymbol{W}_i \boldsymbol{h}_{t-1} + \boldsymbol{U}_i \boldsymbol{x}_t + (\boldsymbol{p}_i \odot \boldsymbol{c}_{t-1}) + \boldsymbol{b}_i \\ &+ \boldsymbol{R}_{ii}\, \boldsymbol{i}_{t-1} + \boldsymbol{R}_{if}\, \boldsymbol{f}_{t-1} + \boldsymbol{R}_{io}\, \boldsymbol{o}_{t-1}\big),\end{aligned} \tag{40}$$

$$\begin{aligned}\boldsymbol{f}_t = \text{sig}\big(&\boldsymbol{W}_f \boldsymbol{h}_{t-1} + \boldsymbol{U}_f \boldsymbol{x}_t + (\boldsymbol{p}_f \odot \boldsymbol{c}_{t-1}) + \boldsymbol{b}_f \\ &+ \boldsymbol{R}_{fi}\, \boldsymbol{i}_{t-1} + \boldsymbol{R}_{ff}\, \boldsymbol{f}_{t-1} + \boldsymbol{R}_{fo}\, \boldsymbol{o}_{t-1}\big),\end{aligned} \tag{41}$$

$$\begin{aligned}\boldsymbol{o}_t = \text{sig}\big(&\boldsymbol{W}_o \boldsymbol{h}_{t-1} + \boldsymbol{U}_o \boldsymbol{x}_t + (\boldsymbol{p}_o \odot \boldsymbol{c}_t) + \boldsymbol{b}_o \\ &+ \boldsymbol{R}_{oi}\, \boldsymbol{i}_{t-1} + \boldsymbol{R}_{of}\, \boldsymbol{f}_{t-1} + \boldsymbol{R}_{oo}\, \boldsymbol{o}_{t-1}\big),\end{aligned} \tag{42}$$

where $\boldsymbol{R}_{ii}, \boldsymbol{R}_{if}, \boldsymbol{R}_{io}, \boldsymbol{R}_{fi}, \boldsymbol{R}_{ff}, \boldsymbol{R}_{fo}, \boldsymbol{R}_{oi}, \boldsymbol{R}_{of}, \boldsymbol{R}_{oo} \in \mathbb{R}^{p \times p}$ are the learnable recurrent weights. Note that full gate recurrence often disappeared in later papers on LSTM.

Later, Graves and Schmidhuber adapted the original LSTM and proposed the *vanilla LSTM* in 2005 [21], which is one of the most common LSTM variants in the literature. The vanilla LSTM incorporated the structures of the original LSTM [31] and the contributions in [13, 14]. The full BPTT algorithm, introduced in Sect. 2.3, was used for training the vanilla LSTM [21].

4.2.3 Other LSTM Variants

There are other variants of LSTM [23, 38], although the most commonly used LSTM is the vanilla LSTM [21]. BPTT was used for LSTM training in [21]; however, Kalman filtering was used for its training earlier [12]. Another training method for LSTM was evolutionary learning [56]. Context-sensitive evolutionary learning was also used for LSTM training [3].

Later works aimed to improve the performance of LSTM. For example, Sak et al. added a linear layer that projects the output of the LSTM to a smaller number of parameters before the recurrent connections [52]. Doetsch et al. made the slope of the activation functions of the LSTM gates learnable to improve performance [11]. *Dynamic cortex memory* was another LSTM variant that added connections between the gates within each LSTM cell [45]. Finally, in 2014, one of the most significant improvements to LSTM was proposed: the Gated Recurrent Units (GRU) [8]. The philosophy behind GRU was to simplify the LSTM cell, as it may not be necessary to have a highly complex cell to learn sequence information. In other words, GRU questioned whether such flexibility as in LSTM was required for sequence learning. GRU is less flexible than LSTM but is sufficiently powerful for sequence learning.

GRU redesigned the LSTM cell by introducing the reset gate, update gate, and new memory cell; therefore, the number of gates was reduced from four to three. It was empirically shown in [9] that the performance of LSTM improves by using GRU cells. Later, in 2017, GRU was further simplified by merging the reset and update gates into a forget gate [26]. GRUs are simpler and popular in some applications like speech processing, while LSTMs remain common in legacy systems. However, both have been overtaken in most domains by Transformers (see Chap. 9) and, more recently, by SSM-based recurrent architectures (see Chap. 10). Section 4.3 introduces the details of the GRU cell.

4.3 Gated Recurrent Units (GRU)

GRU was first proposed in [8]. As mentioned in Sect. 4.2.3, GRU simplified the LSTM cell to reduce its complexity while retaining its flexibility.

4.3.1 Fully Gated Unit

The main GRU variant is the *fully gated unit* [8], whose gates are introduced below. Its gates are illustrated in Fig. 6.

– The Reset Gate: One of the gates in the GRU cell is the *reset gate*. This gate takes the input at the current time slot, $\boldsymbol{x}_t \in \mathbb{R}^d$, and the hidden state of the previous time slot, $\boldsymbol{h}_{t-1} \in [-1, 1]^p$, and outputs the signal $\boldsymbol{r}_t \in [0, 1]^p$:

$$\boldsymbol{r}_t = \text{sig}(\boldsymbol{W}_r \boldsymbol{h}_{t-1} + \boldsymbol{U}_r \boldsymbol{x}_t + \boldsymbol{b}_r), \tag{43}$$

where $\boldsymbol{W}_r \in \mathbb{R}^{p \times p}$, $\boldsymbol{U}_r \in \mathbb{R}^{p \times d}$, and the bias $\boldsymbol{b}_r \in \mathbb{R}^p$ are the learnable weights for the reset gate. The reset gate considers the effects of the input and the previous hidden state, and it controls the amount of forgetting or resetting the previous information in relation to the incoming new information. Comparing Eqs. (38) and (43) shows that the reset gate in the GRU cell is similar to the forget gate in the LSTM cell. The reset gate is depicted in Fig. 6.

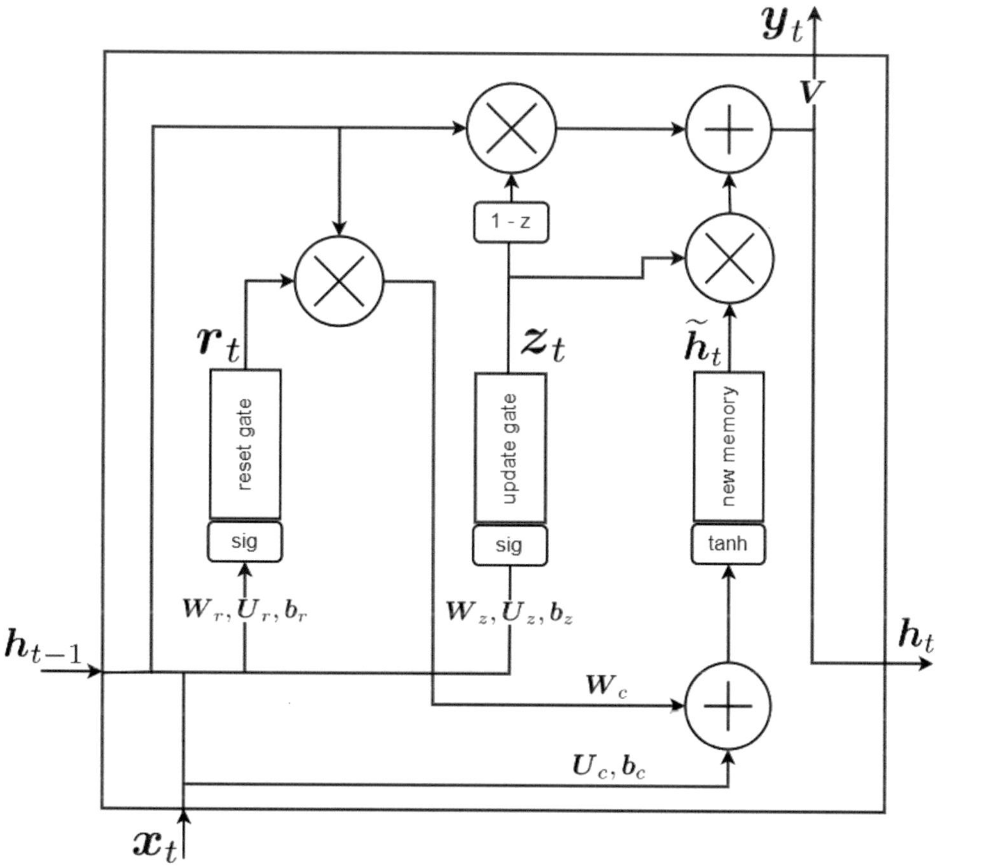

Fig. 6 The conveyor belt in the GRU cell (the fully gated unit)

– **The Update Gate**: Another gate in the GRU cell is the *update gate*. This gate also takes the input at the current time slot, $\boldsymbol{x}_t \in \mathbb{R}^d$, and the hidden state of the previous time slot, $\boldsymbol{h}_{t-1} \in [-1, 1]^p$, and outputs the signal $\boldsymbol{z}_t \in [0, 1]^p$:

$$\boldsymbol{z}_t = \text{sig}(\boldsymbol{W}_z \boldsymbol{h}_{t-1} + \boldsymbol{U}_z \boldsymbol{x}_t + \boldsymbol{b}_z), \tag{44}$$

where $\boldsymbol{W}_z \in \mathbb{R}^{p \times p}$, $\boldsymbol{U}_z \in \mathbb{R}^{p \times d}$, and the bias $\boldsymbol{b}_z \in \mathbb{R}^p$ are the learnable weights for the update gate. The update gate considers the effects of the input and the previous hidden state, and it controls how much the new input data is used to update the cell with incoming sequence information. Comparing Eqs. (37) and (44) shows that the update gate in the GRU cell is similar to the input gate in the LSTM cell. Figure 6 shows the update gate in the GRU cell.

– **The New Memory Cell**: The GRU cell includes a gate called the *new memory cell*. This gate takes the input at the current time slot, $\boldsymbol{x}_t \in \mathbb{R}^d$, and the hidden state of the previous time slot, $\boldsymbol{h}_{t-1} \in [-1, 1]^p$, and outputs the signal $\widetilde{\boldsymbol{h}}_t \in [-1, 1]^p$:

$$\widetilde{\boldsymbol{h}}_t = \tanh\Big(\boldsymbol{W}_c \left(\boldsymbol{r}_t \odot \boldsymbol{h}_{t-1}\right) + \boldsymbol{U}_c \boldsymbol{x}_t + \boldsymbol{b}_c\Big), \tag{45}$$

where $\boldsymbol{W}_c \in \mathbb{R}^{p \times p}$, $\boldsymbol{U}_c \in \mathbb{R}^{p \times d}$, and the bias $\boldsymbol{b}_c$ are the learnable weights for the new memory cell. This gate considers the effects of the input and the previous hidden state to represent the new information from the current input. Comparing Eqs. (33) and (45) shows that the new memory cell in the GRU cell is similar to the new memory cell in the LSTM cell. Note that in the LSTM cell, the hidden state (see Eq. (35)) and the new memory cell (see Eq. (33)) were different; however, in the GRU cell, the hidden state (see Eq. (45)) replaces the new memory signal in the LSTM cell. The new memory cell is shown in Fig. 6.

– **The Final Memory (Hidden State)**: After computing the outputs of the update gate z_t and the new memory cell $\widetilde{\boldsymbol{h}}_t$, we calculate the *final memory* or the *hidden state* $\boldsymbol{h}_t \in \mathbb{R}^p$:

$$\boldsymbol{h}_t = \big((\mathbf{1} - z_t) \odot \boldsymbol{h}_{t-1}\big) + (z_t \odot \widetilde{\boldsymbol{h}}_t), \tag{46}$$

where $\boldsymbol{h}_{t-1} \in \mathbb{R}^p$ is the hidden state from the previous time slot. The final memory block is illustrated in Fig. 6.

As Eq. (46) demonstrates, the final memory considers the effects of the update gate, the previous memory, and the new memory. In the first term, i.e., $(\mathbf{1} - z_t) \odot \boldsymbol{h}_{t-1}$, the update gate $z_t \in [0, 1]^p$ controls how much of the previous state $\boldsymbol{h}_{t-1}$ should be used based on the input data. The closer z_t is to one (resp. zero), the more the network forgets (resp. retains) the previous state $\boldsymbol{h}_{t-1}$. In the second term, i.e., $z_t \odot \widetilde{\boldsymbol{h}}_t$, the update gate $z_t \in [0, 1]^p$ and the new memory cell $\widetilde{\boldsymbol{h}}_t \in [-1, 1]^p$ together control how much of the new input information should be used. In other words, it controls how much the information should be updated with the new input. The closer the update gate z_t is to one and the closer the new memory cell $\widetilde{\boldsymbol{h}}_t$ is to ± 1, the more the input information is used.

Overall, the first and second terms in Eq. (46) determine the trade-off between the use of old and new information in the sequence. The weights of these gates are trained to pass or block the input and previous information based on the input sequence and the time step. Comparing Eqs. (34) and (46) shows that the final memory in the GRU cell has a similar form to that in the LSTM cell; however, they have somewhat different functionalities.

4.3.2 Minimal Gated Unit

The *minimal gated unit* [26] is another variant of GRU that simplifies the gating mechanism by merging the reset and update gates into a single forget gate. This merging is possible because the forget gate can control both the previous and new information in the sequence.

– **The Forget Gate**: The *forget gate* takes the input at the current time slot, $\boldsymbol{x}_t \in \mathbb{R}^d$, and the hidden state of the previous time slot, $\boldsymbol{h}_{t-1} \in [-1, 1]^p$, and outputs the signal $\boldsymbol{f}_t \in [0, 1]^p$:

$$\boldsymbol{f}_t = \text{sig}(\boldsymbol{W}_f \boldsymbol{h}_{t-1} + \boldsymbol{U}_f \boldsymbol{x}_t + \boldsymbol{b}_f), \tag{47}$$

where $\boldsymbol{W}_f \in \mathbb{R}^{p \times p}$, $\boldsymbol{U}_f \in \mathbb{R}^{p \times d}$, and the bias $\boldsymbol{b}_f \in \mathbb{R}^p$ are the learnable weights for the forget gate. The forget gate considers the effects of the input and the previous hidden state, and it controls the extent to which the previous information is forgotten in relation to the incoming new information. Therefore, it controls both the retention or discarding of previous memory and the integration of new information.

– The New Memory Cell and the Final Memory: Because the forget gate replaces both the reset and update gates in the minimal gated unit, Eqs. (45) and (46) are modified to:

$$\widetilde{\boldsymbol{h}}_t = \tanh\Big(\boldsymbol{W}_c\left(\boldsymbol{f}_t \odot \boldsymbol{h}_{t-1}\right) + \boldsymbol{U}_c \boldsymbol{x}_t + \boldsymbol{b}_c\Big), \tag{48}$$

$$\boldsymbol{h}_t = \left((\mathbf{1} - \boldsymbol{f}_t) \odot \boldsymbol{h}_{t-1}\right) + (\boldsymbol{f}_t \odot \widetilde{\boldsymbol{h}}_t), \tag{49}$$

to represent the new memory cell and the final memory in the minimal gated unit, respectively.

5 Bidirectional RNN and LSTM

5.1 Justification of Bidirectional Processing

A bidirectional RNN or LSTM network processes the sequence in both directions: left to right and right to left. At first glance, online causal tasks such as reading text or listening to speech do not have access to future information. Therefore, bidirectional networks may seem to violate causality in these tasks. However, in many such tasks, it is possible to wait for the completion of a part of the sequence, such as a sentence, before making a decision. For example, it is common to wait for the completion of a sentence in speech recognition before recognizing it [21, 22]. In text processing, the entire text is usually available, except in streaming text scenarios. Even in streaming text, it is often possible to wait for a sentence to complete. Therefore, it makes sense to use bidirectional networks for processing sequences, as important related words can sometimes come after a given word rather than before it.

An example of such a case is the sentence "The police is chasing the thief," where the word "thief" is strongly related (opposite) to the word "police." In this sentence, both "thief" and "police" are related, and it is beneficial to process the sentence in both directions.

5.2 Bidirectional RNN

The bidirectional RNN was first proposed in [57] and further explored in [2]. It uses two sets of states, one for each direction in the sequence. Let the states for left-to-right and right-to-left processing be denoted by $\overrightarrow{\boldsymbol{h}}_t$ and $\overleftarrow{\boldsymbol{h}}_t$, respectively. In the bidirectional RNN, Eq. (4) is replaced by two equations [20]:

$$\overrightarrow{\boldsymbol{h}}_t = \tanh(\overrightarrow{\boldsymbol{W}}\,\overrightarrow{\boldsymbol{h}}_{t-1} + \overrightarrow{\boldsymbol{U}}\boldsymbol{x}_t + \overrightarrow{\boldsymbol{b}}_i), \tag{50}$$

$$\overleftarrow{\boldsymbol{h}}_t = \tanh(\overleftarrow{\boldsymbol{W}}\,\overleftarrow{\boldsymbol{h}}_{t+1} + \overleftarrow{\boldsymbol{U}}\boldsymbol{x}_t + \overleftarrow{\boldsymbol{b}}_i), \tag{51}$$

and Eq. (5) is replaced by

$$\boldsymbol{y}_t = \overrightarrow{\boldsymbol{V}}\,\overrightarrow{\boldsymbol{h}}_t + \overleftarrow{\boldsymbol{V}}\,\overleftarrow{\boldsymbol{h}}_t + \boldsymbol{b}_y, \tag{52}$$

where the arrows indicate the parameters for each direction of processing. The unfolded schematic of the bidirectional RNN is illustrated in Fig. 7. As this figure shows, the outputs of both directions are connected to an output layer. In some cases, this output layer may be replaced by an additional multi-layer neural network. All weights of the bidirectional RNN are trained using backpropagation through time, as explained in Sect. 2.3. It is noteworthy that the deep variant of bidirectional RNN was proposed in [20].

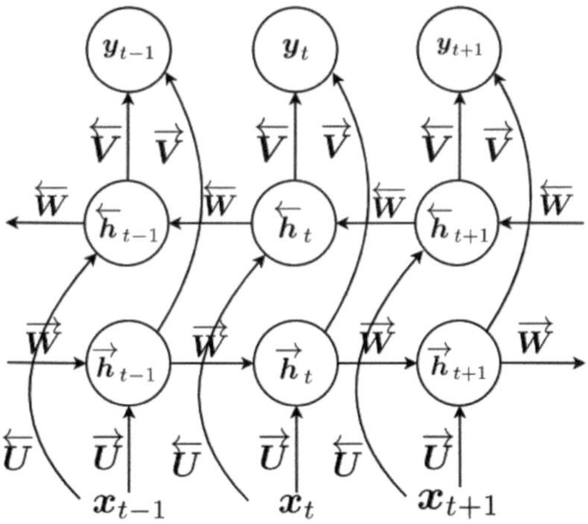

Fig. 7 The bidirectional RNN

5.3 Bidirectional LSTM

Bidirectional LSTM was first proposed in [21, 22]. As its name suggests, the bidirectional LSTM includes two LSTM networks, each processing the sequence in one direction. In other words, there are two LSTM networks that are fed the sequence in opposite orders. This structure is depicted in Fig. 8. Experiments have shown that bidirectional LSTM outperforms unidirectional LSTM [19, 21]. This is expected, as explained in the justification in Sect. 5.1. Because of its advantages, various methods have combined bidirectional LSTM with other approaches. For example, a hybrid of bidirectional LSTM and HMM [17] has shown its merit [19]. Another example is the hybrid of bidirectional LSTM and Conditional Random Fields (CRF) [39] for the sequence tagging task [33]. Other extensions of bidirectional LSTM networks also exist, such as [47]. Today, bidirectional RNNs are mainly used for historical and specialized tasks, while bidirectional Transformers such as BERT [10] dominate sequence modeling (see Chap. 11).

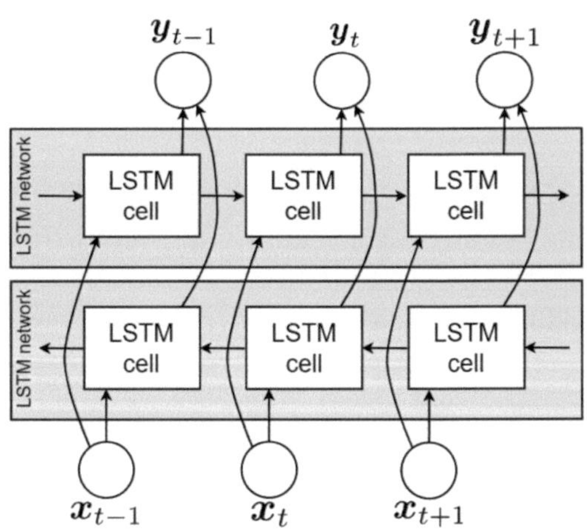

Fig. 8 The bidirectional LSTM

5.4 Embeddings from Language Model (ELMo)

The Embeddings from Language Model (ELMo) network, first proposed in [48], is a language model that makes use of bidirectional LSTM networks. ELMo was an important step toward contextual embeddings and was widely used in various applications, such as medical interview processing, and has been widely used in various applications, such as medical interview processing [54]. However, it has largely been replaced by Transformer-based models such as BERT [10], RoBERTa [59], and GPT [49] (see Chap. 11).

The structure of ELMo is illustrated in Fig. 9. As shown in this figure, ELMo contains L layers of bidirectional LSTM networks, where the output of each bidirectional LSTM is fed to the next bidirectional LSTM. In the bidirectional LSTM networks of ELMo, $\boldsymbol{V} = \boldsymbol{I}$ is set so that the output $\boldsymbol{y}$ becomes equal to the hidden states $\boldsymbol{h}$. At time slot t and layer l, the outputs (or hidden states) of the two directions of LSTM are concatenated to form $\boldsymbol{h}_t^{(l)}$:

$$\boldsymbol{h}_t^{(l)} := [\overrightarrow{\boldsymbol{h}}_t^{(l)\top}, \overleftarrow{\boldsymbol{h}}_t^{(l)\top}]^\top.$$

Then, a linear combination of these hidden states from different layers is considered to be the embedding vector of the ELMo network at time t [48]:

$$\boldsymbol{y}_t^{\text{ELMo}} := \gamma \sum_{l=1}^{L} s_l \, \boldsymbol{h}_t^{(l)}, \tag{53}$$

where γ and $\{s_l\}_{l=1}^{L}$ are scalar hyperparameter weights determined according to the specific task (e.g., question answering, translation, etc.) in natural language processing.

6 Conclusion

This chapter explained the theoretical details of Recurrent Neural Networks (RNNs) [51] and Long Short-Term Memory (LSTM) [30, 31] networks, and their variants. Specifically, RNN, backpropagation through time, vanishing and exploding gradient problems, history of LSTM, Gated Recurrent Unit (GRU), bidirectional RNN and LSTM, and Embeddings from Language Model (ELMo) were introduced.

While classic RNNs, LSTMs, and GRUs are less central today, the recurrent approach itself has returned through modern State Space Models like S4 [25] and Mamba [24] (see Chap. 10), showing that recurrence remains an active area of research alongside Transformers [63] (see Chap. 9).

7 Exercises

1. (Level: Moderate) Consider the following dynamical system:

$$\mathbb{R}^2 \ni \boldsymbol{h}_t - \begin{bmatrix} \boldsymbol{h}_{t-1}^\top \boldsymbol{x}_t \\ \|\boldsymbol{h}_{t-1} \odot \boldsymbol{x}_t\|_2 \end{bmatrix} + 3 \begin{bmatrix} 1 \\ 1 \end{bmatrix},$$
$$\mathbb{R} \ni s_t = \boldsymbol{h}_t^\top \boldsymbol{h}_{t-1} - \|\boldsymbol{h}_t\|_2,$$

where $\odot$ denotes the Hadamard (element-wise) product and $\|.\|_2$ is the ℓ_2 norm.

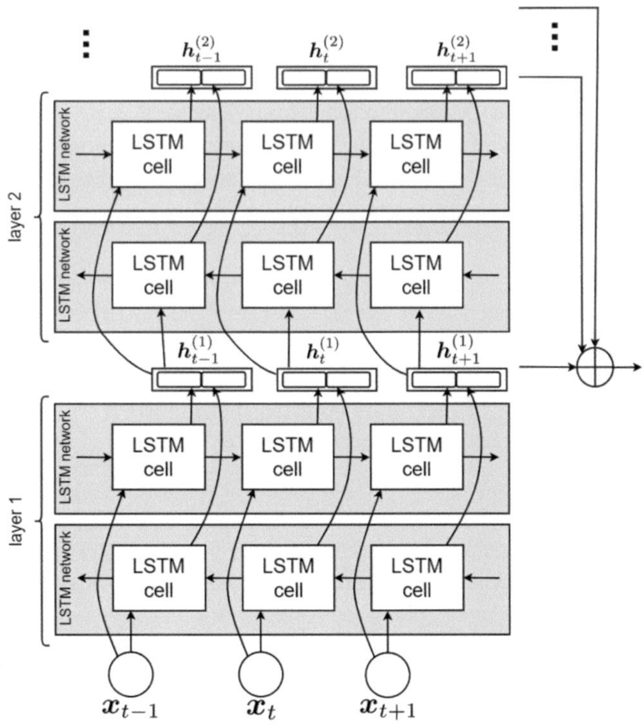

Fig. 9 The ELMo network

Consider the following sequence of data points:

$$x_1 = [1, -1]^\top, \quad x_2 = [0, 1]^\top, \quad x_3 = [-1, 2]^\top,$$

which are fed to the model one by one. With the initial state $\boldsymbol{h}_0 = [1, 1]^\top$, find the states $\boldsymbol{h}_3$ and s_3 after the last fed data point.

2. (Level: Easy) How do Long Short-Term Memory (LSTM) networks and Gated Recurrent Units (GRUs) address the vanishing gradient problem in RNNs? Explain their key mechanisms.
3. (Level: Easy) Answer whether the following statements about Recurrent Neural Networks (RNNs) are true or false.

   (a) RNNs are designed to handle sequential data by maintaining a hidden state that gets updated at each time step.
   (b) RNNs are unable to handle long-term dependencies in data due to the vanishing gradient problem.
   (c) RNNs are generally faster to train than feedforward neural networks because they process sequences in parallel.

4. (Level: Moderate) Let $\boldsymbol{A}$ be a square matrix defining a discrete-time linear system $\boldsymbol{x}_{t+1} = \boldsymbol{A}\boldsymbol{x}_t$. Assume A is diagonalizable. Prove that if the spectral radius of $\boldsymbol{A}$, denoted $\rho(\boldsymbol{A})$, satisfies $\rho(\boldsymbol{A}) < 1$, then the origin $\boldsymbol{x} = 0$ is an asymptotically stable fixed point. That is, show that $\|\boldsymbol{x}_t\| \to 0$ as $t \to \infty$ for any initial condition $\boldsymbol{x}_0$.

5. (Level: Difficult) Consider a state update $h_n = f(h_{n-1}; \theta, c)$, where the transition function is given by

$$f(x; \theta, c) = \tanh(\theta x + c), \quad |\theta| < 1, \quad c \in \mathbb{R}.$$

Assume c is s small number. Find $\lim_{n\to\infty} h_n$, the hidden state of the system in the long run. Hint: Start with the fixed point of the function $f(x; \theta, c)$.

6. (Level: Difficult) Let J be the Jacobian matrix of an Echo State Network (ESN) at a given time step, describing the local linearization of the system's state transition dynamics. That is, given a small change Δs in the state vector, the evolution of this perturbation is given by

$$\Delta s_{t+1} = J \Delta s_t,$$

where s_t represents the perturbation at time t. Prove that if all eigenvalues of J satisfy $|\lambda_i| < 1$ for all i, then the mapping is a contraction.

References

1. Martin Arjovsky, Amar Shah, and Yoshua Bengio. "Unitary evolution recurrent neural networks". In: *International conference on machine learning*. PMLR. 2016, pp. 1120–1128.
2. Pierre Baldi, Søren Brunak, Paolo Frasconi, Giovanni Soda, and Gianluca Pollastri. "Exploiting the past and the future in protein secondary structure prediction". In: *Bioinformatics* 15.11 (1999), pp. 937–946.
3. Justin Bayer, Daan Wierstra, Julian Togelius, and Jürgen Schmidhuber. "Evolving memory cell structures for sequence learning". In: *International conference on artificial neural networks*. Springer. 2009, pp. 755–764.
4. Yoshua Bengio, Nicolas Boulanger-Lewandowski, and Razvan Pascanu. "Advances in optimizing recurrent networks". In: *2013 IEEE international conference on acoustics, speech and signal processing*. IEEE. 2013, pp. 8624–8628.
5. Yoshua Bengio, Paolo Frasconi, and Patrice Simard. "The problem of learning long-term dependencies in recurrent networks". In: *IEEE international conference on neural networks*. IEEE. 1993, pp. 1183–1188.
6. Yoshua Bengio, Patrice Simard, and Paolo Frasconi. "Learning long-term dependencies with gradient descent is difficult". In: *IEEE transactions on neural networks* 5.2 (1994), pp. 157–166.
7. Hendrik Wolter Broer and Floris Takens. *Dynamical systems and chaos*. Vol. 172. Springer, 2011.
8. Kyunghyun Cho, Bart Van Merriënboer, Dzmitry Bahdanau, and Yoshua Bengio. "On the properties of neural machine translation: Encoder-decoder approaches". In: *Eighth Workshop on Syntax, Semantics and Structure in Statistical Translation (SSST-8)*. 2014.
9. Junyoung Chung, Caglar Gulcehre, KyungHyun Cho, and Yoshua Bengio. "Empirical evaluation of gated recurrent neural networks on sequence modeling". In: *arXiv preprint* arXiv:1412.3555 (2014).
10. Jacob Devlin, Ming-Wei Chang, Kenton Lee, and Kristina Toutanova. "BERT: Pre-training of deep bidirectional transformers for language understanding". In: *Proceedings of the 2019 conference of the North American chapter of the association for computational linguistics: human language technologies, volume 1 (long and short papers)*. 2019, pp. 4171–4186.
11. Patrick Doetsch, Michal Kozielski, and Hermann Ney. "Fast and robust training of recurrent neural networks for offline handwriting recognition". In: *2014 14th international conference on frontiers in handwriting recognition*. IEEE. 2014, pp. 279–284.
12. Felix A Gers, Juan Antonio Pérez-Ortiz, Douglas Eck, and Jürgen Schmidhuber. "DEKF-LSTM". In: *10th European Symposium on Artificial Neural Networks (ESANN)*. 2002.
13. Felix A Gers and Jürgen Schmidhuber. "Recurrent nets that time and count". In: *Proceedings of the IEEE-INNS-ENNS International Joint Conference on Neural Networks. IJCNN 2000. Neural Computing: New Challenges and Perspectives for the New Millennium*. Vol. 3. IEEE. 2000, pp. 189–194.
14. Felix A Gers, Jürgen Schmidhuber, and Fred Cummins. "Learning to forget: Continual prediction with LSTM". In: *Neural computation* 12.10 (2000), pp. 2451–2471.
15. Benyamin Ghojogh, Ali Ghodsi, Fakhri Karray, and Mark Crowley. "KKT Conditions, First-Order and Second-Order Optimization, and Distributed Optimization: Tutorial and Survey". In: *arXiv preprint* arXiv:2110.01858 (2021).
16. Benyamin Ghojogh, Fakhri Karray, and Mark Crowley. "Eigenvalue and generalized eigenvalue problems: Tutorial". In: *arXiv preprint* arXiv:1903.11240 (2019).

17. Benyamin Ghojogh, Fakhri Karray, and Mark Crowley. "Hidden Markov Model: Tutorial". In: *Engineering Archive* (2019).
18. Alex Graves. "Generating sequences with recurrent neural networks". In: *arXiv preprint* arXiv:1308.0850 (2013).
19. Alex Graves, Santiago Fernández, and Jürgen Schmidhuber. "Bidirectional LSTM networks for improved phoneme classification and recognition". In: *International conference on artificial neural networks*. Springer. 2005, pp. 799–804.
20. Alex Graves, Abdel-rahman Mohamed, and Geoffrey Hinton. "Speech recognition with deep recurrent neural networks". In: *2013 IEEE international conference on acoustics, speech and signal processing*. IEEE. 2013, pp. 6645–6649.
21. Alex Graves and Jürgen Schmidhuber. "Framewise phoneme classification with bidirectional LSTM and other neural network architectures". In: *Neural networks* 18.5-6 (2005), pp. 602–610.
22. Alex Graves and Jürgen Schmidhuber. "Framewise phoneme classification with bidirectional LSTM networks". In: *Proceedings. 2005 IEEE International Joint Conference on Neural Networks, 2005*. Vol. 4. IEEE. 2005, pp. 2047–2052.
23. Klaus Greff, Rupesh K Srivastava, Jan Koutník, Bas R Steunebrink, and Jürgen Schmidhuber. "LSTM: A search space odyssey". In: *IEEE transactions on neural networks and learning systems* 28.10 (2016), pp. 2222–2232.
24. Albert Gu and Tri Dao. "Mamba: Linear-Time Sequence Modeling with Selective State Spaces". In: *First Conference on Language Modeling*. 2024.
25. Albert Gu, Karan Goel, and Christopher Re. "Efficiently Modeling Long Sequences with Structured State Spaces". In: *International Conference on Learning Representations*. 2022.
26. Joel C Heck and Fathi M Salem. "Simplified minimal gated unit variations for recurrent neural networks". In: *2017 IEEE 60th International Midwest Symposium on Circuits and Systems (MWSCAS)*. IEEE. 2017, pp. 1593–1596.
27. Salah Hihi and Yoshua Bengio. "Hierarchical recurrent neural networks for long-term dependencies". In: *Advances in neural information processing systems* 8 (1995).
28. Sepp Hochreiter. "The vanishing gradient problem during learning recurrent neural nets and problem solutions". In: *International Journal of Uncertainty, Fuzziness and Knowledge-Based Systems* 6.02 (1998), pp. 107–116.
29. Sepp Hochreiter, Yoshua Bengio, Paolo Frasconi, Jürgen Schmidhuber, et al. *Gradient flow in recurrent nets: the difficulty of learning long-term dependencies*. 2001.
30. Sepp Hochreiter and Jürgen Schmidhuber. *Long short-term memory*. Tech. rep. FKI-207-95, Department of Fakultät für Informatik, Technical University of Munich, Munich, Germany, 1995.
31. Sepp Hochreiter and Jürgen Schmidhuber. "Long short-term memory". In: *Neural computation* 9.8 (1997), pp. 1735–1780.
32. Yuhuang Hu, Adrian Huber, Jithendar Anumula, and Shih-Chii Liu. "Overcoming the vanishing gradient problem in plain recurrent networks". In: *arXiv preprint* arXiv:1801.06105 (2018).
33. Zhiheng Huang, Wei Xu, and Kai Yu. "Bidirectional LSTM-CRF models for sequence tagging". In: *arXiv preprint* arXiv:1508.01991 (2015).
34. Herbert Jaeger. "Echo state network". In: *Scholarpedia* 2.9 (2007), p. 2330.
35. Herbert Jaeger. "Tutorial on training recurrent neural networks, covering BPPT, RTRL, EKF and the "echo state network" approach". In: (2002).
36. Herbert Jaeger and Harald Haas. "Harnessing nonlinearity: Predicting chaotic systems and saving energy in wireless communication". In: *Science* 304.5667 (2004), pp. 78–80.
37. Herbert Jaeger, Mantas LukoševiÄ?ius, Dan Popovici, and Udo Siewert. "Optimization and applications of echo state networks with leaky-integrator neurons". In: *Neural networks* 20.3 (2007), pp. 335–352.
38. Rafal Jozefowicz, Wojciech Zaremba, and Ilya Sutskever. "An empirical exploration of recurrent network architectures". In: *International conference on machine learning*. PMLR. 2015, pp. 2342–2350.
39. John Lafferty, Andrew McCallum, and Fernando CN Pereira. "Conditional random fields: Probabilistic models for segmenting and labeling sequence data". In: (2001).
40. Kevin J Lang, Alex H Waibel, and Geoffrey E Hinton. "A time-delay neural network architecture for isolated word recognition". In: *Neural networks* 3.1 (1990), pp. 23–43.
41. Tsungnan Lin, Bill G. Horne, Peter Tino, and C. Lee Giles. "Learning long-term dependencies is not as difficult with NARX recurrent neural networks". In: *Advances in neural information processing systems* (1995).
42. Zachary C Lipton, John Berkowitz, and Charles Elkan. "A critical review of recurrent neural networks for sequence learning". In: *arXiv preprint* arXiv:1506.00019 (2015).
43. Tomas Mikolov, Armand Joulin, Sumit Chopra, Michael Mathieu, and Marc'Aurelio Ranzato. "Learning longer memory in recurrent neural networks". In: *Workshop at the International Conference on Learning Representations* (2015).
44. Michael C Mozer. "A focused backpropagation algorithm for temporal pattern recognition". In: *Backpropagation: Theory, architectures, and applications* 137 (1995).
45. Sebastian Otte, Marcus Liwicki, and Andreas Zell. "Dynamic cortex memory: Enhancing recurrent neural networks for gradient-based sequence learning". In: *International Conference on Artificial Neural Networks*. Springer. 2014, pp. 1–8.

46. Vijayaditya Peddinti, Daniel Povey, and Sanjeev Khudanpur. "A time delay neural network architecture for efficient modeling of long temporal contexts". In: *Sixteenth annual conference of the international speech communication association*. 2015.
47. Matthew E Peters, Waleed Ammar, Chandra Bhagavatula, and Russell Power. "Semi-supervised sequence tagging with bidirectional language models". In: *Association for Computational Linguistics (ACL)*. 2017.
48. Matthew E. Peters et al. "Deep contextualized word representations". In: *North American Chapter of the Association for Computational Linguistics (NAACL)*. 2018.
49. Alec Radford, Karthik Narasimhan, Tim Salimans, and Ilya Sutskever. *Improving language understanding by generative pre-training*. Tech. rep. OpenAI, 2018.
50. AJ Robinson and Frank Fallside. *The utility driven dynamic error propagation network*. Tech. rep. Department of Engineering, University of Cambridge, 1987.
51. David E Rumelhart, Geoffrey E Hinton, and Ronald J Williams. "Learning representations by back-propagating errors". In: *Nature* 323.6088 (1986), pp. 533–536.
52. Hasim Sak, Andrew W Senior, and Françoise Beaufays. "Long short-term memory recurrent neural network architectures for large scale acoustic modeling". In: *INTERSPEECH*. 2014.
53. Hojjat Salehinejad, Sharan Sankar, Joseph Barfett, Errol Colak, and Shahrokh Valaee. "Recent advances in recurrent neural networks". In: *arXiv preprint* arXiv:1801.01078 (2017).
54. Justyna Sarzynska-Wawer et al. "Detecting formal thought disorder by deep contextualized word representations". In: *Psychiatry Research* 304 (2021), p. 114135.
55. Jürgen Schmidhuber. "Deep learning in neural networks: An overview". In: *Neural networks* 61 (2015), pp. 85–117.
56. Jürgen Schmidhuber, Daan Wierstra, Matteo Gagliolo, and Faustino Gomez. "Training recurrent networks by Evolino". In: *Neural computation* 19.3 (2007), pp. 757–779.
57. Mike Schuster and Kuldip K Paliwal. "Bidirectional recurrent neural networks". In: *IEEE transactions on Signal Processing* 45.11 (1997), pp. 2673–2681.
58. Kamilya Smagulova and Alex Pappachen James. "A survey on LSTM memristive neural network architectures and applications". In: *The European Physical Journal Special Topics* 228.10 (2019), pp. 2313–2324.
59. Ieva Staliūnaitė and Ignacio Iacobacci. "Compositional and Lexical Semantics in RoBERTa, BERT and DistilBERT. A Case Study on CoQA". In: *Proceedings of the 2020 Conference on Empirical Methods in Natural Language Processing (EMNLP)*. 2020, pp. 7046–7056.
60. Ralf C Staudemeyer and Eric Rothstein Morris. "Understanding LSTM–a tutorial into long short-term memory recurrent neural networks". In: *arXiv preprint* arXiv:1909.09586 (2019).
61. Masahide Sugiyama, Hidehumi Sawai, and Alexander H Waibel. "Review of tdnn (time delay neural network) architectures for speech recognition". In: *1991 IEEE International Symposium on Circuits and Systems (ISCAS)*. IEEE. 1991, pp. 582–585.
62. Ilya Sutskever and Geoffrey Hinton. "Temporal-kernel recurrent neural networks". In: *Neural Networks* 23.2 (2010), pp. 239–243.
63. Ashish Vaswani et al. "Attention is all you need". In: *Advances in neural information processing systems*. 2017, pp. 5998–6008.
64. Paul J Werbos. "Generalization of backpropagation with application to a recurrent gas market model". In: *Neural networks* 1.4 (1988), pp. 339–356.
65. Ronald J Williams. *Complexity of exact gradient computation algorithms for recurrent neural networks*. Tech. rep. NU-CCS-89-27, Northeastern University, 1989.
66. Ronald J Williams and David Zipser. "Gradient-Based Learning Algorithms for Recurrent Networks and Their Computational Complexity". In: *Backpropagation: Theory, architectures, and applications* 433 (1995), p. 17.
67. Yong Yu, Xiaosheng Si, Changhua Hu, and Jianxun Zhang. "A review of recurrent neural networks: LSTM cells and network architectures". In: *Neural computation* 31.7 (2019), pp. 1235–1270.

9 Attention Mechanism and Transformers

Abstract

This chapter explores the attention mechanism in machine learning, a concept inspired by the way human visual and cognitive systems selectively focus on the most informative parts of a scene or text. Attention enables models to assign varying levels of importance to different parts of the input data, making it a crucial component of modern Natural Language Processing (NLP) systems. The chapter begins by introducing the attention mechanism as a data-weighting strategy and explains its role in enhancing the performance of deep learning models, particularly in sequence modeling. A major breakthrough in this area was the introduction of Transformers—models built from attention modules without any recurrence. This architecture revolutionized NLP by replacing traditional recurrent networks such as Recurrent Neural Networks (RNNs) and Long Short-Term Memory (LSTM) networks, enabling more efficient training and parallelization. The chapter covers the encoder–decoder structure of Transformers and explains how they serve both as sequence-to-sequence autoencoders and as tools for transfer learning. Additionally, the chapter introduces Transformer alternatives, offering a broader perspective on the evolving landscape of attention-based architectures. Key variants of the Transformer model are presented, with a particular focus on the Performer, which improves computational complexity through efficient approximations of attention.

1 Introduction

When looking at a scene or picture, our visual system, similar to a machine learning model [41], focuses on or attends to specific parts of the scene or image that contain more information and are more important, while ignoring the less informative or less important parts. For example, when we look at the Mona Lisa portrait, our visual system attends to Mona Lisa's face and smile, as Fig. 1 illustrates. Moreover, when reading a text, especially when practicing speed reading, one technique is skimming [71], in which our visual system or a model skims the data at high pace and only attends to the more informative words in sentences [75]. Figure 1 shows a sample sentence and highlights the words to which our visual system focuses more during skimming.

The concept of attention can be modeled in machine learning, where attention is implemented as a simple weighting of data. In the attention mechanism explained in this chapter, the more informative

B. Ghojogh and A. Ghodsi, *Elements of Deep Learning*,
https://doi.org/10.1007/978-3-032-10738-1_9

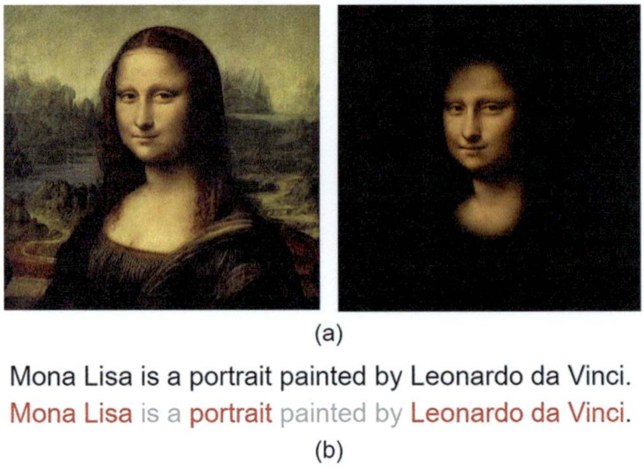

Fig. 1 Attention in the visual system for **a** seeing a picture by attending to more important parts of the scene and **b** reading a sentence by attending to more informative words in the sentence

or important parts of the data are given larger weights to allocate more attention to them. Many state-of-the-art Natural Language Processing (NLP) [30] and deep learning techniques in NLP [58] use attention.

Transformers use an encoder–decoder structure, where the encoder maps inputs into a hidden representation and the decoder generates outputs from this representation. However, unlike classical autoencoders, Transformers are not restricted to reconstructing their inputs; they are primarily designed for sequence-to-sequence tasks such as translation.

Transfer learning is widely used in NLP [69], and Transformers can also be used for transfer learning. Recently, Transformers composed solely of attention modules, excluding recurrence and any recurrent modules, were proposed [65]. This was a significant breakthrough. Prior to the proposal of Transformers using only the attention mechanism, recurrent models such as Long Short-Term Memory (LSTM) [29] and Recurrent Neural Networks (RNN) [36] were mostly used for NLP (see Chap. 8). Moreover, other NLP models such as Word2Vec [22, 45–47] and GloVe [50] were state of the art before the appearance of Transformers. Two well-known NLP models are BERT (Bidirectional Encoder Representations from Transformers) [15], which is built from stacks of Transformer encoders, and GPT (Generative Pre-trained Transformer) [54], which is built from stacks of Transformer decoders. These architectures specialize in different tasks: BERT is optimized for semantic tasks, while GPT is optimized for generative tasks.

This chapter introduces and reviews the attention mechanism and Transformers. Section 2 introduces the sequence-to-sequence model with and without the attention mechanism, as well as self-attention. Section 3 explains the different parts of the encoder and decoder of Transformers. The alternatives and variants of the Transformer, with a focus on Performer [11], are introduced in Sect. 4.

2 Attention Mechanism

2.1 Autoencoder and the Context Vector

Consider an autoencoder with encoder and decoder parts, where the encoder receives an input and converts it into a *context vector*, and the decoder takes this context vector and converts it into an output. The output is related to the input through the context vector in the so-called *hidden space*.

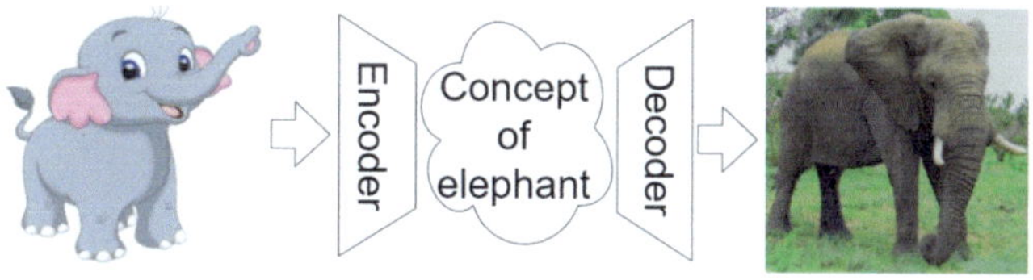

Fig. 2 Using an autoencoder for transformation from one domain to another domain. The images used are taken from the PACS dataset [40]

Figure 2 illustrates an autoencoder with its encoder (left part), hidden space (middle), and decoder (right part). For example, the input can be a sentence or a word in English, and the output is the same sentence or word translated into French. Assume the word "elephant" in English is fed to the encoder and the word "l'éléphant" in French is produced as output. The context vector models the concept of an elephant, which also exists in the human mind when thinking about an elephant. This context is abstract in the mind and can refer to any fat, thin, huge, or small elephant [51].

Another example for a Transformer is transforming a cartoon image of an elephant into a picture of a real elephant (see Fig. 2). As the autoencoder transforms data from one domain to a hidden space and then to another domain, it can be used for domain transformation [68], domain adaptation [6], and domain generalization [16]. Here, every context is modeled as a vector in the hidden space. Let the context vector be denoted by $\boldsymbol{c} \in \mathbb{R}^p$ in the p-dimensional hidden space.

2.2 The Sequence-to-Sequence Model

Consider a sequence of ordered tokens, for example, a sequence of words that form a sentence. The objective is to transform this sequence into another related sequence. For instance, the goal might be to take a sentence in English and translate it into the same sentence in French. This type of model, which transforms a sequence into another related sequence, is called the sequence-to-sequence model [4].

Suppose the number of words in a document or considered sentence is n. Let the ordered input tokens or words of the sequence be denoted by $\{\boldsymbol{x}_1, \boldsymbol{x}_2, \ldots, \boldsymbol{x}_{i-1}, \boldsymbol{x}_i, \ldots, \boldsymbol{x}_n\}$, and the output sequence be denoted by $\{\boldsymbol{y}_1, \boldsymbol{y}_2, \ldots, \boldsymbol{y}_{i-1}, \boldsymbol{y}_i, \ldots, \boldsymbol{y}_n\}$. As Fig. 3a illustrates, there exist latent vectors, denoted by $\{\boldsymbol{l}_i\}_{i=1}^n$, in the decoder part for each word.

In the sequence-to-sequence model, the probability of generating the i-th word conditioned on all the previous words is determined by a function $g(.)$ whose inputs are the immediate previous word $\boldsymbol{y}_{i-1}$, the i-th latent vector $\boldsymbol{l}_i$, and the context vector $\boldsymbol{c}$:

$$\mathbb{P}(\boldsymbol{y}_i | \boldsymbol{y}_1, \ldots, \boldsymbol{y}_{i-1}) = g(\boldsymbol{y}_{i-1}, \boldsymbol{l}_i, \boldsymbol{c}). \tag{1}$$

Figure 3a depicts the sequence-to-sequence model. In this model, each word $\boldsymbol{x}_i$ produces a hidden vector $\boldsymbol{h}_i$ in the encoder part of the autoencoder. The hidden vector of each word, $\boldsymbol{h}_i$, is fed to the next hidden vector, $\boldsymbol{h}_{i+1}$, through a projection matrix $\boldsymbol{W}$. In this model, there is only one context vector $\boldsymbol{c}$ for the entire sequence, which is equal to the last hidden vector of the encoder, i.e., $\boldsymbol{c} = \boldsymbol{h}_n$.

Note that the encoder and decoder in the sequence-to-sequence model can be any sequential model such as a Recurrent Neural Network (RNN) [36] or Long Short-Term Memory (LSTM) network [29].

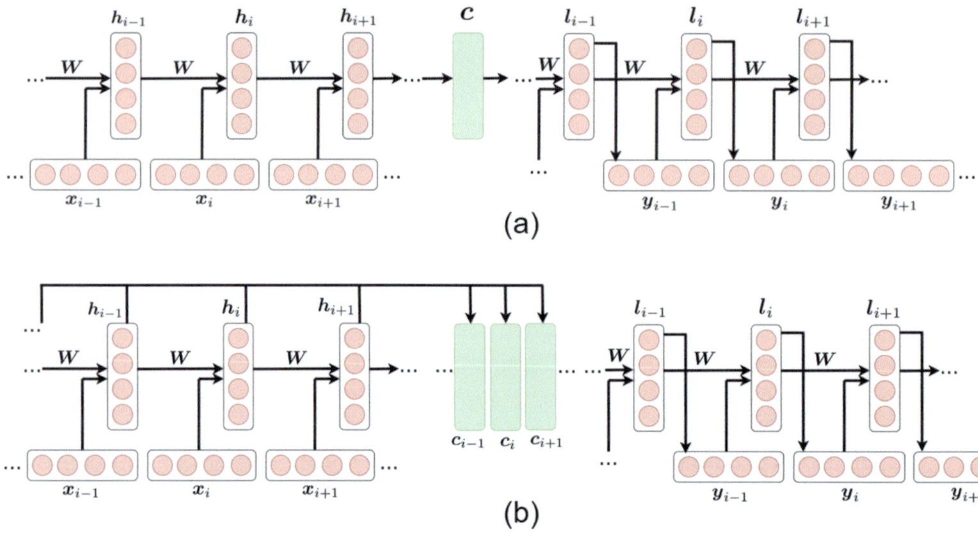

Fig. 3 Sequence-to-sequence model **a** without and **b** with attention

2.3 The Sequence-to-Sequence Model with Attention

The sequence-to-sequence model explained previously can be extended with attention [12, 44]. In the sequence-to-sequence model with attention, the probability of generating the i-th word is determined as [12] follows:

$$\mathbb{P}(\boldsymbol{y}_i | \boldsymbol{y}_1, \ldots, \boldsymbol{y}_{i-1}) = g(\boldsymbol{y}_{i-1}, \boldsymbol{l}_i, \boldsymbol{c}_i). \tag{2}$$

Figure 3b shows the sequence-to-sequence model with attention. In contrast to the sequence-to-sequence model, which has only one context vector for the entire sequence, this model has a context vector for each word. The context vector for each word is a linear combination, or weighted sum, of all the hidden vectors; therefore, the i-th context vector is

$$\boldsymbol{c}_i = \sum_{j=1}^{n} a_{ij} \boldsymbol{h}_j, \tag{3}$$

where $a_{ij} \geq 0$ is the weight of $\boldsymbol{h}_j$ for the i-th context vector.

This weighted sum for a specific word in the sequence determines which words in the sequence have more influence on that word. In other words, it determines which words this specific word "attends" to more. This notion of weighted impact, which identifies the parts with greater influence, is called "attention." It is noteworthy that the original idea of the arithmetic linear combination of vectors for the purpose of word embedding, similar to Eq. (3), appeared in the Word2Vec method [45, 47].

The sequence-to-sequence model with attention considers a notion of similarity between the latent vector $\boldsymbol{l}_{i-1}$ of the decoder and the hidden vector $\boldsymbol{h}_j$ of the encoder [4]:

$$\mathbb{R} \ni s_{ij} := \text{similarity}(\boldsymbol{l}_{i-1}, \boldsymbol{h}_j). \tag{4}$$

The intuition for this similarity score is as follows. The output word $\boldsymbol{y}_i$ depends on the previous latent vector $\boldsymbol{l}_{i-1}$ (see Fig. 3), and the hidden vector $\boldsymbol{h}_j$ depends on the input word $\boldsymbol{x}_j$. Therefore, this similarity score relates to the impact of the input $\boldsymbol{x}_j$ on the output $\boldsymbol{y}_i$. In this way, the score s_{ij} indicates the impact of the j-th word on generating the i-th word in the sequence. This similarity notion can be modeled as a neural network trained by backpropagation [56].

To convert this score into a probability, these scores should sum to one; thus, its softmax form is considered as [12] follows:

$$\mathbb{R} \ni a_{ij} := \frac{e^{s_{ij}}}{\sum_{k=1}^{n} e^{s_{ik}}}. \tag{5}$$

In this way, the score vector $[a_{i1}, a_{i2}, \ldots, a_{in}]^\top$ behaves as a discrete probability distribution (see Sect. 2.8 in Chap. 2 for more information on softmax). Therefore, in Eq. (3), the weights sum to one, and the weights with higher values attend more to their corresponding hidden vectors.

2.4 Self-Attention

2.4.1 The Need for Composite Embedding

Many of the previous methods in NLP, such as Word2Vec [22, 45–47] and GloVe [50], learned a representation for each word individually. However, to understand how words relate to each other, it is useful to have a composite embedding in which compositions of words also have embedding representations [10].

For example, Fig. 4 shows a sentence that highlights the relationships between words. This figure illustrates which previous words in the sentence are remembered more when reading it. This relationship among words indicates that a composite embedding is necessary for effective natural language embedding.

2.4.2 Query-Retrieval Modeling

Consider a database with keys and values, where a query is searched through the keys to retrieve a value [20] (see Fig. 5). This hard definition of query-retrieval can be generalized to a soft query-retrieval, where multiple keys, rather than only one, can correspond to the query. For this, the similarities between

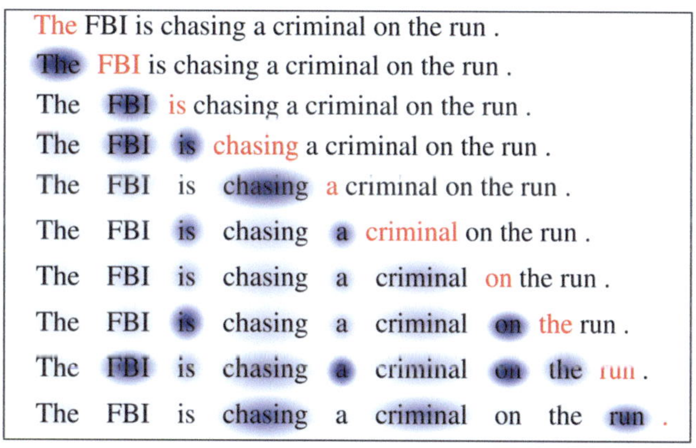

Fig. 4 The relationships between words in a sentence demonstrate the need for composite embedding. The credit for this image goes to [10]

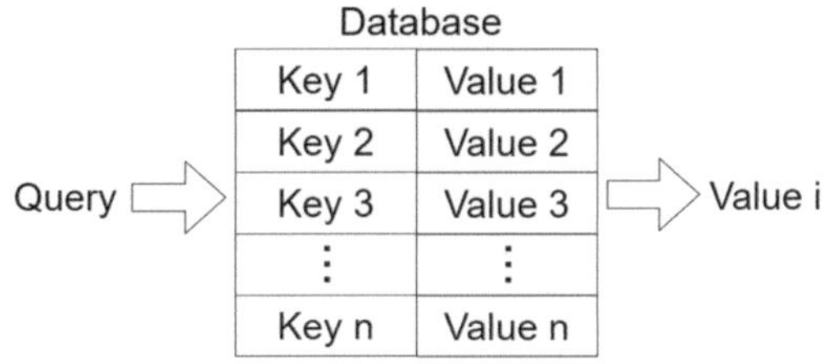

Fig. 5 Query-retrieval in a database

the query and all the keys are calculated to determine which keys are more similar to the query. This soft query-retrieval is formulated as follows:

$$\text{attention}(\boldsymbol{q}, \{\boldsymbol{k}_i\}_{i=1}^n, \{\boldsymbol{v}_i\}_{i=1}^n) := \sum_{i=1}^{n} a_i \boldsymbol{v}_i, \tag{6}$$

where

$$\mathbb{R} \ni s_i := \text{similarity}(\boldsymbol{q}, \boldsymbol{k}_i), \tag{7}$$

$$\mathbb{R} \ni a_i := \text{softmax}(s_i) = \frac{e^{s_i}}{\sum_{k=1}^n e^{s_k}}, \tag{8}$$

and $\boldsymbol{q}$, $\{\boldsymbol{k}_i\}_{i=1}^n$, and $\{\boldsymbol{v}_i\}_{i=1}^n$ denote the query, keys, and values, respectively.

Recall that the context vector of a sequence-to-sequence model with attention, introduced by Eq. (3), was also a linear combination with weights of normalized similarity (see Eq. (5)). The same linear combination is used in Eq. (6), where the weights are the similarities between the query and the keys. An illustration of Eqs. (6), (7), and (8) is shown in Fig. 6.

Note that the similarity s_i can be defined using any similarity measure. Some well-known similarity measures are [65]

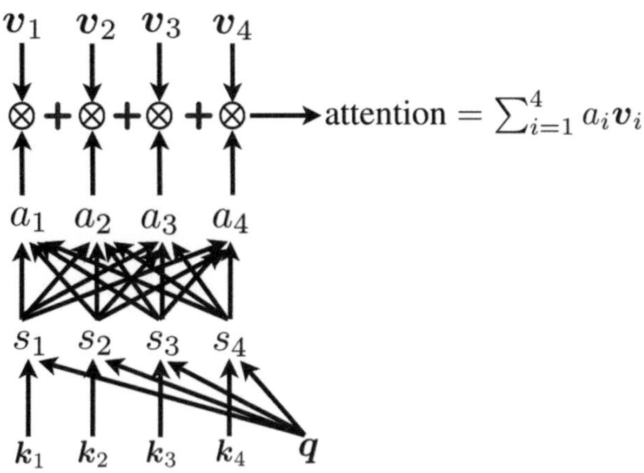

Fig. 6 Illustration of Eqs. (6), (7), and (8) in the attention mechanism. In this example, it is assumed there exist five words in the sequence (four keys and one query)

$$\text{inner product: } s_i = \boldsymbol{q}^\top \boldsymbol{k}_i, \tag{9}$$

$$\text{scaled inner product: } s_i = \frac{\boldsymbol{q}^\top \boldsymbol{k}_i}{\sqrt{p}}, \tag{10}$$

$$\text{general inner product: } s_i = \boldsymbol{q}^\top \boldsymbol{W} \boldsymbol{k}_i, \tag{11}$$

$$\text{additive similarity: } s_i = \boldsymbol{w}_q^\top \boldsymbol{q} + \boldsymbol{w}_k^\top \boldsymbol{k}_i, \tag{12}$$

where $\boldsymbol{W} \in \mathbb{R}^{p \times p}$, $\boldsymbol{w}_q \in \mathbb{R}^p$, and $\boldsymbol{w}_k \in \mathbb{R}^p$ are learnable matrices and vectors. Among these similarity measures, the scaled inner product is used most often.

Equation (6) calculates the attention of a target word (or query) with respect to every input word (or keys), which include the previous and forthcoming words. As Fig. 4 illustrates, when processing a word considered as the query, the other words in the sequence serve as keys. Equation (6) calculates how similar the other words in the sequence are to that word. In other words, it determines how impactful the other previous and forthcoming words are in generating a missing word in the sequence.

An example for Eq. (6) is provided below. Consider the sentence "I am a student." Assume the word "student" is being processed in this sequence. Thus, there is a query corresponding to the word "student." The values correspond to the previous words, which are "I," "am," and "a." Assume the normalized similarities between the query and the values are calculated, yielding weights of 0.7, 0.2, and 0.1 for "I," "am," and "a," respectively, where the weights sum to one. Then, the attention value for the word "student" is $0.7\boldsymbol{v}_{\text{I}} + 0.2\boldsymbol{v}_{\text{am}} + 0.1\boldsymbol{v}_{\text{a}}$.

2.4.3 Attention Formulation

Let the words of a sequence be in a d-dimensional space, i.e., the sequence is $\{\boldsymbol{x}_i \in \mathbb{R}^d\}_{i=1}^n$. This d-dimensional representation of words can be obtained from any word-embedding NLP method such as Word2Vec [22, 45–47] or GloVe [50].

The query, key, and value are projections of the words into p-dimensional, p-dimensional, and r-dimensional subspaces, respectively:

$$\mathbb{R}^p \ni \boldsymbol{q}_i = \boldsymbol{W}_Q^\top \boldsymbol{x}_i, \tag{13}$$

$$\mathbb{R}^p \ni \boldsymbol{k}_i = \boldsymbol{W}_K^\top \boldsymbol{x}_i, \tag{14}$$

$$\mathbb{R}^r \ni \boldsymbol{v}_i = \boldsymbol{W}_V^\top \boldsymbol{x}_i, \tag{15}$$

where $\boldsymbol{W}_Q \in \mathbb{R}^{d \times p}$, $\boldsymbol{W}_K \in \mathbb{R}^{d \times p}$, and $\boldsymbol{W}_V \in \mathbb{R}^{d \times r}$ are the projection matrices into the low-dimensional query, key, and value subspaces, respectively.

Consider the words, queries, keys, and values in matrix form as $\boldsymbol{X} := [\boldsymbol{x}_1, \ldots, \boldsymbol{x}_n] \in \mathbb{R}^{d \times n}$, $\boldsymbol{Q} := [\boldsymbol{q}_1, \ldots, \boldsymbol{q}_n] \in \mathbb{R}^{p \times n}$, $\boldsymbol{K} := [\boldsymbol{k}_1, \ldots, \boldsymbol{k}_n] \in \mathbb{R}^{p \times n}$, and $\boldsymbol{V} := [\boldsymbol{v}_1, \ldots, \boldsymbol{v}_n] \in \mathbb{R}^{r \times n}$, respectively

It is noteworthy that the inner product $\boldsymbol{q}^\top \boldsymbol{k}_i$ appears in similarity measures such as the scaled inner product. As a result, the similarity measures contain $\boldsymbol{q}^\top \boldsymbol{k}_i = \boldsymbol{x}_i^\top \boldsymbol{W}_Q \boldsymbol{W}_K^\top \boldsymbol{x}_i$. This is similar to a kernel matrix, though it is not an exact kernel matrix. If it is assumed that the projection matrices for the query and key are equal, i.e., $\boldsymbol{W}_Q = \boldsymbol{W}_K = \boldsymbol{W}$, then

$$\boldsymbol{q}^\top \boldsymbol{k}_i = \boldsymbol{x}_i^\top \boldsymbol{W}_Q \boldsymbol{W}_K^\top \boldsymbol{x}_i = \boldsymbol{x}_i^\top \boldsymbol{W} \boldsymbol{W}^\top \boldsymbol{x}_i = (\boldsymbol{W}^\top \boldsymbol{x}_i)^\top (\boldsymbol{W}^\top \boldsymbol{x}_i), \tag{16}$$

so its behavior is similar to that of a kernel used for measuring similarity.

Considering Eqs. (7) and (8) along with the above definitions, Eq. (6) can be written in matrix form, for the entire sequence of n words, as follows:

$$\mathbb{R}^{r\times n} \ni \boldsymbol{Z} := \text{attention}(\boldsymbol{Q}, \boldsymbol{K}, \boldsymbol{V}) \\ = \boldsymbol{V}\, \text{softmax}(\frac{1}{\sqrt{p}} \boldsymbol{Q}^\top \boldsymbol{K}), \tag{17}$$

where $\boldsymbol{Z} = [\boldsymbol{z}_1, \ldots, \boldsymbol{z}_n]$ are the attention values for all words, indicating how much each word attends to its previous and forthcoming words. In Eq. (17), the softmax operator applies the softmax function to every row of its input matrix so that each row sums to one.

Note that because the queries, keys, and values are all derived from the same words in the sequence, this attention mechanism is referred to as "self-attention" [10].

2.5 Attention in Other Fields Such as Vision and Speech

Note that the concept of attention can be applied in any field of research, not merely in NLP. Attention has been widely used not only in NLP [12, 44] but also in the field of computer vision [72]. In computer vision, attention means focusing on specific parts of an image that are more important and informative (see Fig. 1). This simulates attention and perception in the human visual system [60], where our brain filters the observed scene to focus on its important parts.

For example, it is possible to generate captions for an input image using an autoencoder with the attention mechanism, as illustrated in Fig. 7a. As shown in this figure, the encoder is a Convolutional Neural Network (CNN) [38] for extracting visual features, and the decoder consists of Long Short-Term Memory (LSTM) [29] or Recurrent Neural Network (RNN) [36] modules for generating the caption text.

Literature has shown that the lower convolutional layers in CNNs capture low-level features and different partitions of input images [39]. Figure 8 shows an example of features extracted by CNN layers trained on facial images, where different facial organs and features have been extracted in the lower layers of the CNN. Therefore, as Fig. 7b shows, one can consider the extracted low-layer features, which correspond to different parts of the image, as the hidden vectors $\{\boldsymbol{h}_j\}_{j=1}^{n'}$ in Eq. (4), where n' is the number of features for the extracted image partitions. The similarity with latent vectors of the decoder (LSTM) is computed using Eq. (4), and the query-retrieval model of the attention mechanism

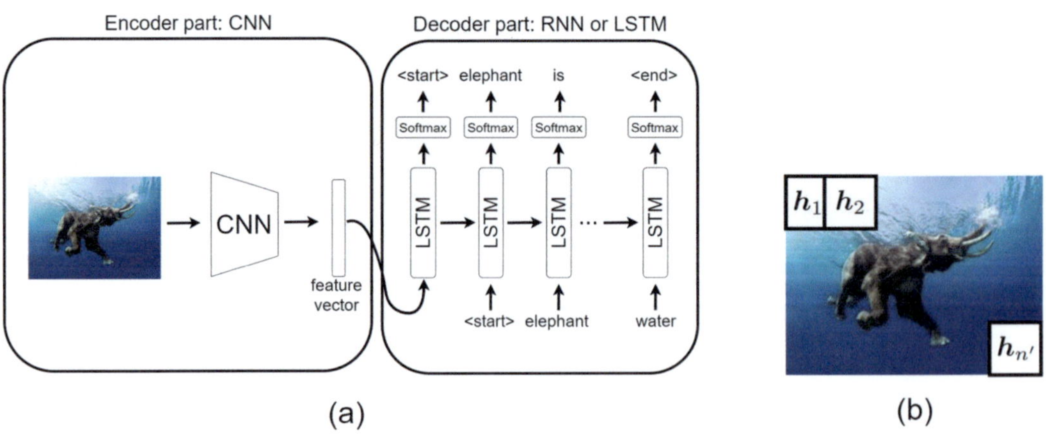

Fig. 7 Using attention in computer vision: **a** Transformer of image to caption with CNN as encoder and RNN or LSTM as decoder. The caption for this image is "Elephant is in water." **b** The convolutional filters extract values from the image for query-retrieval modeling in the attention mechanism

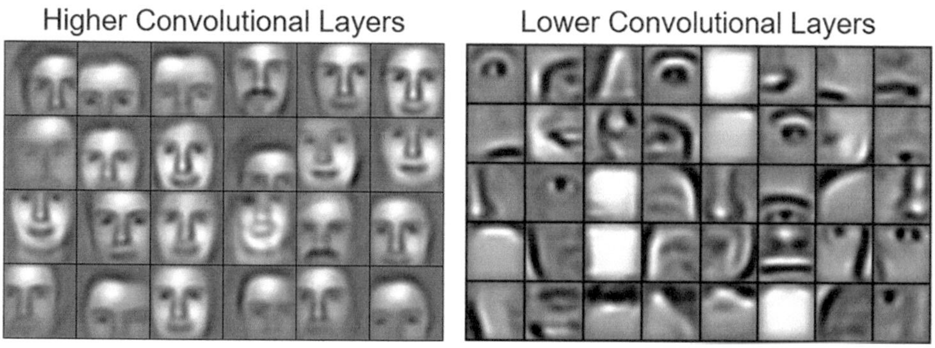

Fig. 8 The low-level and high-level features learned in the low and high convolutional layers, respectively. The credit for this image goes to [39]

introduced earlier is used to learn self-attention on images. Note that, as the partitions of the image are considered hidden variables used for attention, the model attends to the important parts of the input image; see Fig. 1 as an example.

Attention has inspired many applications outside of NLP. For example: "show, attend, and tell" for image captioning [72], "listen, attend, and spell" for speech recognition [8, 9], and "show, attend, and answer" for visual question answering [33]. These patterns highlight how attention enables models to focus on the most relevant parts of input across different modalities.

3 Transformer

3.1 The Concept of Transformation

As explained in Sect. 2.1, an autoencoder takes input data, embeds it into a context vector that simulates a concept in the human mind [51], and generates an output. The input and output are related to each other through this context vector. In other words, the autoencoder transforms the input into a related output. An example of such a transformation is translating a sentence from one language to the same sentence in another language. Another example involving a Transformer is image captioning, where an image is transformed into its caption explaining its content. A purely computer vision example of transformation is converting a daytime input image into the same scene at nighttime.

An autoencoder named *Transformer* was proposed in the literature for the task of transformation [65]. The structure of a Transformer is depicted in Fig. 9. As shown in this figure, a Transformer is an autoencoder consisting of an encoder and a decoder. In the following, the details of the encoder and decoder of a Transformer are introduced.

3.2 Encoder of Transformer

The encoder part of the Transformer, illustrated in Fig. 9, embeds the input sequence of n words $\boldsymbol{X} \in \mathbb{R}^{d \times n}$ into context vectors using the attention mechanism. The different parts of the encoder are explained in the following subsections. The input embedding to the encoder can be any embedding vector of the input tokens. For example, the embedding vectors can be Word2Vec embeddings of the input tokens [22, 45–47]. The dimension d represents the dimensionality of the embedding vectors.

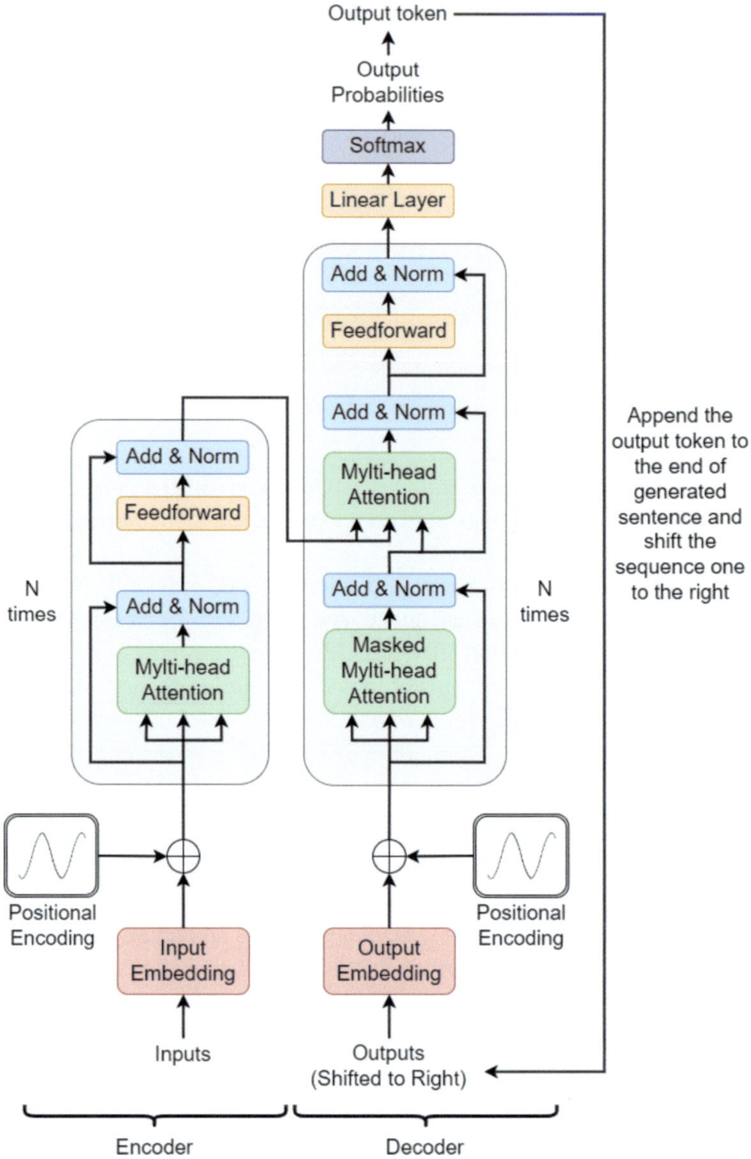

Fig. 9 The encoder and decoder parts of a Transformer

3.2.1 Positional Encoding

It will be explained in Sect. 3.4 that the Transformer introduced here does not include any recurrence, RNN, or LSTM module. As there is no recurrence and no convolution, the model has no inherent sense of order in the sequence. Since the order of words is important for the meaning of sentences[1], a method is required to account for the order of tokens or words in the sequence. For this purpose, a vector encoding the position is added to each input word embedding.

[1] For example, shuffling the order of words in "You are fine" to "Are you fine" changes its meaning.

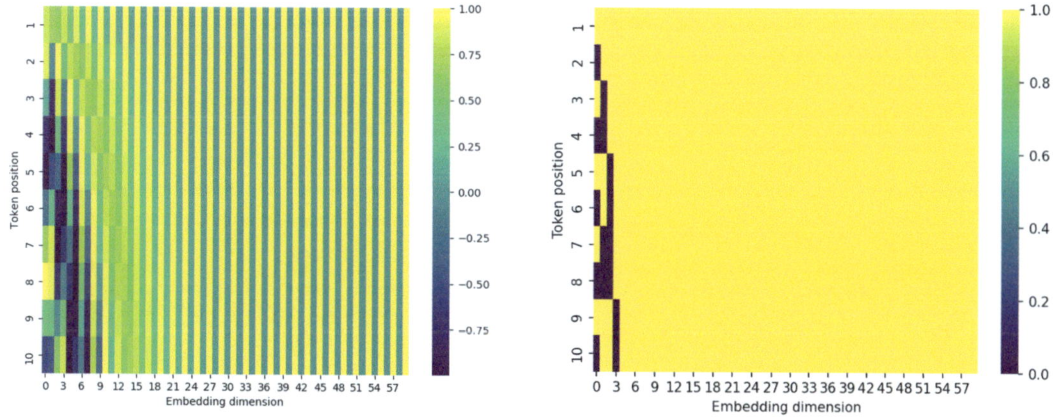

Fig. 10 **a** The vectors of positional encoding. In this example, it is assumed that $n = 10$ (number of word positions), $d = 60$, and $p = 100$. **b** The binary encoding with 60 bits. The bits are flipped in the plot of binary encoding

Consider the embedding of the i-th word in the sequence, denoted by $\boldsymbol{x}_i \in \mathbb{R}^d$. To encode the position of the i-th word in the sequence, the position vector $\boldsymbol{p}_i \in \mathbb{R}^d$ can be defined as follows:

$$\begin{cases} \boldsymbol{p}_i(2j+1) := \cos\left(\frac{i}{10000^{\frac{2j}{p}}}\right), \\ \boldsymbol{p}_i(2j) := \sin\left(\frac{i}{10000^{\frac{2j}{p}}}\right), \end{cases} \tag{18}$$

for all $j \in \{0, 1, \ldots, \lfloor d/2 \rfloor\}$, where $\boldsymbol{p}_i(2j+1)$ and $\boldsymbol{p}_i(2j)$ denote the odd and even elements of $\boldsymbol{p}_i$, respectively.

Figure 10a illustrates the dimensions of the position vectors across different word positions. As shown in this figure, the position vectors for different words are distinct, as expected. Moreover, this figure shows that the differences in position vectors are more concentrated in the initial dimensions of the vectors. This positional encoding is similar to binary encoding (e.g., 0000, 0001, 0010, 0100, 0101, ...) depicted in Fig. 10b.

As shown in Fig. 9, the positional encoding is added to the input embedding to incorporate positional information into the data:

$$\boldsymbol{x}_i \leftarrow \boldsymbol{x}_i + \boldsymbol{p}_i. \tag{19}$$

3.2.2 Multihead Attention with Self-Attention

After positional encoding, the data are fed into a multihead attention module with self-attention. The multihead attention is illustrated in Fig. 11. This module applies the attention mechanism h times.

The reason for these multiple applications of attention is explained as follows. The first attention determines how much each word attends to other words. The second application of attention calculates how much each pair of words attends to other pairs of words. Likewise, the third application of attention determines how much each pair of pairs of words attends to other pairs of pairs of words, and so on.

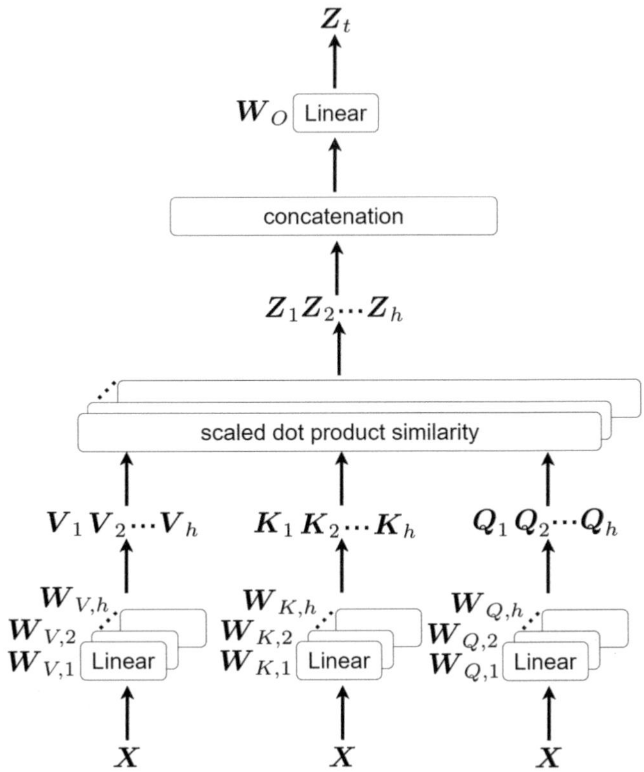

Fig. 11 Multihead attention with h heads

Note that this measure of attention or similarity between hierarchical pairs of words is reminiscent of maximum mean discrepancy [23, 24], which measures similarity between different moments of data distributions.

As shown in Fig. 11, the data, which include positional encoding, are passed through linear layers to obtain the queries, keys, and values. These linear layers model the linear projections introduced in Eqs. (13), (15), and (14), respectively.

There are h of these linear layers to generate h sets of queries, keys, and values as follows:

$$\mathbb{R}^{p \times n} \ni \boldsymbol{Q}_i = \boldsymbol{W}_{Q,i}^\top \boldsymbol{X}, \quad \forall i \in \{1, \ldots, h\}, \tag{20}$$

$$\mathbb{R}^{p \times n} \ni \boldsymbol{K}_i = \boldsymbol{W}_{K,i}^\top \boldsymbol{X}, \quad \forall i \in \{1, \ldots, h\}, \tag{21}$$

$$\mathbb{R}^{r \times n} \ni \boldsymbol{V}_i = \boldsymbol{W}_{V,i}^\top \boldsymbol{X}, \quad \forall i \in \{1, \ldots, h\}. \tag{22}$$

Then, the scaled dot-product similarity, defined in Eq. (10) or (17), is used to generate the h attention values $\{\boldsymbol{Z}_i\}_{i=1}^h$. These h attention values are concatenated to form a new long flattened vector. Finally, a linear layer, which is a linear projection, is used to obtain the total attention value, $\boldsymbol{Z}_t$:

$$\boldsymbol{Z}_t := \boldsymbol{W}_O^\top \operatorname{concat}(\boldsymbol{Z}_1, \boldsymbol{Z}_2, \ldots, \boldsymbol{Z}_h). \tag{23}$$

3.2.3 Layer Normalization

As shown in Fig. 9, the data (containing positional encoding) and the total attention value are added:

$$\boldsymbol{Z}_t' \leftarrow \boldsymbol{Z}_t + \boldsymbol{X}. \tag{24}$$

This addition is inspired by the concept of residual connections introduced by ResNet [28]. After this addition, layer normalization is applied, where for each hidden unit h_i, the following operation is performed:

$$h_i \leftarrow \frac{g}{\sigma}(h_i - \mu), \tag{25}$$

where μ and σ are the empirical mean and standard deviation over H hidden units:

$$\mu := \frac{1}{H}\sum_{i=1}^{H} h_i, \tag{26}$$

$$\sigma := \sqrt{\sum_{i=1}^{H}(h_i - \mu)^2}. \tag{27}$$

This standardization makes the mean zero and the variance one; it is closely related to batch normalization and reduces covariate shift [31].

3.2.4 Feedforward Layer

Henceforth, let $\boldsymbol{Z}_t'$ denote the total attention after both addition and layer normalization. The $\boldsymbol{Z}_t'$ is fed into a feedforward network with nonlinear activation functions, and then, as before, the input to the feedforward network is added to its output:

$$\boldsymbol{Z}_t'' \leftarrow \boldsymbol{R} + \boldsymbol{Z}_t', \tag{28}$$

where $\boldsymbol{R}$ denotes the output of the feedforward network. Again, layer normalization is applied, and the output of the encoder is henceforth denoted by $\boldsymbol{Z}_t''$. This represents the encoding for the entire input sequence or sentence, incorporating information about the attention of words and hierarchical pairs of words to each other.

3.2.5 Stacking

As shown in Fig. 9, the encoder is composed of a stack of N identical layers. This stacking increases the number of learnable parameters, providing sufficient degrees of freedom to learn the entire dictionary of words. Through experiments, a good number of stacks has been found to be $N = 6$ [65].

3.3 Decoder of Transformer

The decoder part of the Transformer is shown in Fig. 9. The different parts of the decoder are introduced in the following subsections. The input to the decoder is the output of the decoder shifted to the right. The output token is appended to the end of the generated sentence, and the generated sequence is shifted to the right. The embedding vectors (e.g., Word2Vec embeddings [22, 45–47]) of the tokens or words in the shifted sequence are fed as input to the decoder.

3.3.1 Masked Multihead Attention with Self-Attention

A component of the decoder is the masked multihead attention module, whose input is the output embeddings $\{\boldsymbol{y}_i\}_{i=1}^n$ shifted one word to the right. Positional encoding is also added to the output embeddings to include information about their positions. Equation (19) is used for this purpose, where $\boldsymbol{x}_i$ is replaced by $\boldsymbol{y}_i$:

$$\boldsymbol{y}_i \leftarrow \boldsymbol{y}_i + \boldsymbol{p}_i. \tag{29}$$

The output embeddings, with positional encodings added, are then fed into the masked multihead attention module. This module is similar to the multihead attention module but masks out the forthcoming words after each word. Therefore, each output word only attends to its previous output words, each pair of output words attends to its previous pairs of output words, each pair of pairs of output words attends to its previous pairs of pairs of output words, and so on. The reason for using the masked version of multihead attention for the output embeddings is that, when generating the output text, there is no access to the next words yet because they have not been generated.

It is noteworthy that this masking imposes a notion of sparsity, which was also introduced by the dropout technique [59], but in a stochastic manner.

Recall Eq. (17), which was used for multihead attention (see Sect. 3.2.2). The masked multihead attention is defined as follows:

$$\begin{aligned}\mathbb{R}^{r\times n} \ni \boldsymbol{Z}_m &:= \text{maskedAttention}(\boldsymbol{Q}, \boldsymbol{K}, \boldsymbol{V})\\ &= \boldsymbol{V}\,\text{softmax}\Big(\frac{1}{\sqrt{p}}(\boldsymbol{Q}^\top \boldsymbol{K} + \boldsymbol{M})\Big),\end{aligned} \tag{30}$$

where the mask matrix $\boldsymbol{M} \in \mathbb{R}^{n\times n}$ is

$$\boldsymbol{M}(i, j) := \begin{cases} 0 & \text{if } j \leq i, \\ -\infty & \text{if } j > i. \end{cases} \tag{31}$$

Since the softmax function includes an exponential operator, the mask has no impact for $j \leq i$ (because it is multiplied by $e^0 = 1$) and masks out positions where $j > i$ (because it is multiplied by $e^{-\infty} = 0$). Note that $j \leq i$ and $j > i$ correspond to previous and next words, respectively, in terms of their positions in the sequence.

As before, the output of the masked multihead attention is normalized and then added to its input.

3.3.2 Multihead Attention with Cross-Attention

As shown in Fig. 9, the output of the masked multihead attention module is fed into a multihead attention module with cross-attention. This module is not self-attention because its values, keys, and queries do not come from the same sequence. Instead, its values and keys are from the output of the encoder, while the queries are from the output of the masked multihead attention module in the decoder. In other words, the values and keys come from the processed input embeddings, and the queries come from the processed output embeddings.

The calculated multihead attention determines how much each output embedding attends to the input embeddings, how much each pair of output embeddings attends to the pairs of input embeddings, how much each pair of pairs of output embeddings attends to the pairs of pairs of input embeddings, and so on. This shows the connection between the input sequence and the generated output sequence.

3.3.3 Feedforward Layer and Softmax Activation

Again, the output of the multihead attention module with cross-attention is normalized and added to its input. Then, it is fed into a feedforward neural network with layer normalization, and its input is added to its output afterward. Note that the masked multihead attention, the multihead attention with cross-attention, and the feedforward network are stacked $N = 6$ times.

The output of the feedforward network then passes through a linear layer by linear projection, and, finally, a softmax activation function is applied. The number of output neurons with the softmax activation function equals the total number of words in the dictionary, which is a large number. The outputs of the decoder sum to one and represent the probability of each word in the dictionary being the next generated word. For sequence generation, the token or word with the largest probability is chosen as the next word.

As explained before, the output token is appended to the sequence, the sequence is shifted to the right, and then the embedding vectors of the tokens are fed as input to the decoder. This explains why GPT-based language models—which are stacks of Transformer decoders (see Chap. 11)—remember the previous sentences they have generated and the previous chats with the user.

3.4 Attention Is All We Need!

3.4.1 No Need for RNN!

As shown in Fig. 9, the output of the decoder is fed into the masked multihead attention module of the decoder with a shift. Note that this does not constitute recurrence, as it can be interpreted through the procedure of teacher-forcing [35]. Therefore, there is no recurrent module like a Recurrent Neural Network (RNN) [36] or Long Short-Term Memory (LSTM) network [29] in the Transformer [65]. It has been shown that a sequence can be learned using the Transformer.

Transformers showed that recurrence is not required for sequence modeling, since self-attention can directly connect all positions in a sequence. This was a major breakthrough in NLP, and most state-of-the-art language models today are built on Transformer-based architectures.

3.4.2 Computational Complexity

Self-attention learns the attention of every word to every other word in a sequence of n words. Additionally, a p-dimensional embedding is learned for each word. Thus, the complexity of operations per layer is $\mathcal{O}(n^2 p)$.

To be more specific, consider Eqs. (20), (21), and (22), where the matrix multiplications have complexities of $\mathcal{O}(ndp)$, $\mathcal{O}(ndp)$, and $\mathcal{O}(ndr)$, respectively, where n is the sequence length, p is the embedding dimensionality of the query and key, and r is the embedding dimensionality of the value. The complexity of the matrix multiplication $\boldsymbol{Q}^\top \boldsymbol{K}$ in Eq. (17) is $\mathcal{O}(n^2 p)$. That equation applies softmax row-wise on the $(n \times n)$ matrix, so its complexity is $\mathcal{O}(n^2)$. Finally, multiplication by V in Eq. (17) has complexity $\mathcal{O}(n^2 r)$. Therefore, the total complexity of self-attention is

$$
\begin{aligned}
&\mathcal{O}(ndp) + \mathcal{O}(ndp) + \mathcal{O}(ndr) + \mathcal{O}(n^2 p) + \mathcal{O}(n^2 r) \\
&= \mathcal{O}\big(\underbrace{nd(2p + r)}_{\text{projection cost}} + \underbrace{n^2(p + r)}_{\text{attention and output cost}}\big) \overset{(a)}{\approx} \mathcal{O}(n^2 p),
\end{aligned}
\tag{32}
$$

Table 1 Comparison of complexities between self-attention and recurrence [65]

	Complexity per layer	Sequential operations	Maximum path length
Self-attention	$\mathcal{O}(n^2 p)$	$\mathcal{O}(1)$	$\mathcal{O}(1)$
Recurrence	$\mathcal{O}(np^2)$	$\mathcal{O}(n)$	$\mathcal{O}(n)$

where (a) is because the input dimension d, query/key dimension p, and value dimension r are all proportional to the model's hidden size, i.e., $d \approx p \approx r$ (e.g., they can be 512).

The sequence length is variable and can grow large. Therefore, both the projection cost and the attention and output cost are reduced to $\mathcal{O}(n^2 p)$. Equation (32) shows that attention, and therefore the Transformer, has quadratic computational complexity with respect to the sequence length n. Likewise, it can be shown that the memory complexity of attention is $\mathcal{O}(n^2 + np)$.

While the complexity per layer in attention is $\mathcal{O}(n^2 p)$, the complexity per layer in recurrence is $\mathcal{O}(np^2)$. Although the complexity per layer in self-attention is worse than recurrence, many of its operations can be performed in parallel because all words in the sequence are processed simultaneously, as also explained below. Therefore, the $\mathcal{O}(n^2 p)$ complexity is not prohibitive, given its parallelizability. In contrast, recurrence cannot be parallelized due to its sequential nature.

Regarding the number of sequential operations, the self-attention mechanism processes all n words simultaneously, so its sequential operations are of order $\mathcal{O}(1)$. As recurrence must process words sequentially, the number of its sequential operations is of order $\mathcal{O}(n)$. Regarding the maximum path length between any two words, self-attention learns attention between every two words; hence, its maximum path length is of order $\mathcal{O}(1)$. However, in recurrence, each word requires a path with a length proportional to a fraction of the sequence (length n in the worst case) to reach the processing of another word, so its maximum path length is $\mathcal{O}(n)$. This shows that attention reduces both sequential operations and maximum path length compared to recurrence.

Table 1 summarizes the computational complexities of operations in the self-attention mechanism and compares them with those in recurrence, such as RNNs.

4 Alternatives and Variants of Transformer

The Transformer model, though powerful, is known for its computational complexity. As discussed in Sect. 3.4.2, the computational complexity of self-attention is $\mathcal{O}(n^2 p)$, where n is the sequence length and p is the embedding dimensionality of the queries and keys. The memory complexity of attention is $\mathcal{O}(n^2 + nd)$. The bottleneck in the complexity of Transformers arises from the n^2 term. This is because n is the sequence length, and having a long sequence, such as a book or a lengthy corpus, significantly increases the computational and memory requirements. This restricts language models from accepting very long manuscripts or prompts as input.

Several alternatives and variants of Transformers have been proposed to improve their computational or memory efficiency. These alternatives and variants can be categorized into the following groups:

- **Linear and quasi-linear Transformers**: Some variants of Transformers attempt to reduce the computational or memory complexity of the attention mechanism to linear or quasi-linear complexity. These methods can be categorized into sparse attention, linear attention, and flash attention methods. The details of these models are introduced in Sect. 4.1.

- **Designing new models as State Space Models (SSMs)**: A new class of models, named State Space Models (SSMs), has been proposed as potential alternatives to Transformers. One well-known SSM model is Mamba [25]. Other important works in the SSM literature include [19, 26, 27]. Chapter 10 describes SSMs in more detail.
  SSM models resemble RNNs in their sequential structure but are formulated with linear state updates, making them easier to train on long sequences (see Chap. 8). Recent works (such as Mamba [25]) suggest that SSMs may offer competitive or complementary alternatives to Transformers, reviving interest in recurrence-inspired architectures [17]. SSMs resolve some problems of RNNs. For example, while RNNs encapsulate all the information of a sequence into a vector, SSMs encapsulate the information into a matrix, providing greater capacity.
  Note that Transformers and SSMs can be converted into each other [14, 73]. As a result, some works initialize the weights of SSMs using pre-trained Transformers [1], thereby transferring the representations learned by Transformers and GPT-based large language models to SSMs.
- **Model compression**: Some methods compress Transformers using neural network compression techniques to reduce memory complexity. These methods fall into three categories. Quantization methods, including [18, 32, 70, 74], quantize the weights of Transformers to consume fewer bits. Other techniques, such as [3, 48, 55, 63], use knowledge distillation, pruning, or weight sharing for compressing Transformers. Refer to Chap. 20 for details on neural network compression techniques including quantization, knowledge distillation, pruning, and weight sharing. Additionally, some methods use KV (Key-Value) cache compression [2, 42, 43, 77], caching the keys and values to reuse them when needed, thus making the Transformer faster.
- **Dynamic models**: Some dynamic models improve Transformer performance by using only a subset of the model. For example, if a Transformer has 20 layers and using all 20 layers is not feasible, the output of the fifth layer may be used, and the remaining layers after the fifth are dropped. This is similar to eigenvalue or singular value decomposition, where if a matrix has rank 20, the best decomposition with rank 5 uses the top 5 eigenvalues. Examples of dynamic models include SortedNet [64], MatFormer [37], SHARCS [57], and FLEXTRON [7].

4.1 Linear and Quasi-Linear Transformers

There are several variants of the Transformer that attempt to improve its computational complexity. These models seek to reduce the complexity from $\mathcal{O}(n^2)$ to linear complexity $\mathcal{O}(n)$ or quasi-linear complexity. This improvement makes the model scalable with respect to sequence length, allowing it to handle long input sequences. Some of these methods also aim to increase inference speed.

These methods can be divided into three sub-categories:

- **Sparse attention**: As discussed earlier, the Transformer uses attention where all tokens (words) attend to all other tokens, resulting in a fully connected graph with tokens as nodes. Sparse attention methods make this attention graph sparse, meaning that each token only attends to a few other tokens that matter most. Examples of sparse attention methods include those proposed in [52, 62, 66, 76].
- **Linear attention**: Linear attention methods typically drop the softmax used in attention (see Eq (17)) to improve computational complexity. Some important methods in this category are as follows:
  - **Reformer** [34]: Introduces a method to reduce memory requirements while maintaining performance. Its key innovation is the use of Locality-Sensitive Hashing (LSH) to reduce the complexity of the attention mechanism. It improves computational complexity to $\mathcal{O}(n \log(n) p)$ and memory complexity to $\mathcal{O}(np)$.

- **Linformer** [67]: Focuses on reducing the time complexity of self-attention, making it more efficient for long sequences. Its key innovation is achieving linear time complexity for self-attention. It improves computational complexity to $\mathcal{O}(np)$ and memory complexity to $\mathcal{O}(np)$.
- **Performer** [11]: Replaces self-attention with a more efficient approximation. Its key innovation is using Fast Attention Via positive Orthogonal Random features (FAVOR+), allowing for faster training and inference. It improves computational complexity to $\mathcal{O}(np)$ and memory complexity to $\mathcal{O}(np)$.
- **Longformer** [5]: Extends the Transformer to handle long sequences more efficiently. Its key innovation is a sparse attention mechanism combining local and global attentions. It improves computational complexity to $\mathcal{O}(np + nw)$ and memory complexity to $\mathcal{O}(np + nw)$, where w is the window size for local attention.
- Other methods in this category include retentive network [61], gated linear attention [73], and TransNormerLLM [53].

Among these, the theory of Performer [11]—as a well-known variant of the Transformer—is explained in Sect. 4.2.

- **Flash attention**: FlashAttention [13] is an optimized implementation of attention that reduces memory usage and speeds up training by computing attention in a more memory-efficient way. It is widely adopted in modern large language models for its practical efficiency.

4.2 Performer

4.2.1 Computational Complexity of Attention

Recall Eq. (17):

$$\mathbb{R}^{r\times n} \ni \boldsymbol{Z} = \underbrace{\boldsymbol{V}}_{r\times n} \text{softmax}\Big(\frac{1}{\sqrt{p}} \underbrace{\boldsymbol{Q}^\top \boldsymbol{K}}_{n\times n}\Big), \tag{33}$$

where $\boldsymbol{Q} \in \mathbb{R}^{p\times n}$, $\boldsymbol{K} \in \mathbb{R}^{p\times n}$, and $\boldsymbol{V} \in \mathbb{R}^{r\times n}$.

The computational complexity of the attention mechanism used in Transformers was discussed in Sect. 3.4.2. The complexity of the matrix multiplication $\boldsymbol{Q}^\top \boldsymbol{K}$ is $\mathcal{O}(n^2 p)$. The dimensionality of $\boldsymbol{Q}^\top \boldsymbol{K}$ is $(n \times n)$, and therefore the softmax on this matrix has complexity $\mathcal{O}(n^2)$. Then, the complexity of multiplying $\boldsymbol{V}$ with the output of the softmax is $\mathcal{O}(n^2 r)$.

Thus, the total complexity is $\mathcal{O}(n^2 p) + \mathcal{O}(n^2) + \mathcal{O}(n^2 r) \approx \mathcal{O}(n^2 p)$, which is quadratic in terms of the sequence length n. The n^2 term creates a bottleneck for handling long sequences with large n.

If there were no softmax, Eq. (17) would become

$$\mathbb{R}^{r\times n} \ni \boldsymbol{Z} = \frac{1}{\sqrt{p}} \underbrace{(\boldsymbol{V}\,\boldsymbol{Q}^\top)}_{r\times p} \underbrace{\boldsymbol{K}}_{p\times n}. \tag{34}$$

In this case, the order of matrix multiplication can be changed to first multiply $\boldsymbol{V}$ and $\boldsymbol{Q}^\top$, and then multiply the result by $\boldsymbol{K}$. If the multiplication is performed in this order, the complexity becomes $\mathcal{O}(np)$. This reduction in complexity is due to avoiding the multiplication of n terms with n terms, as in Eq. (33). This shows that the softmax function in Eq. (33) prevents changing the order of matrix multiplication, which is computationally restrictive.

Performer [11] approximates softmax$(\boldsymbol{Q}^\top \boldsymbol{K})$ to remove the restrictive softmax function:

$$\text{softmax}(\boldsymbol{Q}^\top \boldsymbol{K}) \approx \boldsymbol{Q}'^\top \boldsymbol{K}', \tag{35}$$

enabling the use of Eq. (34) to improve computational complexity. This approximation is performed using Fast Attention Via positive Orthogonal Random features (FAVOR+), which approximates the kernel function of the attention mechanism with random features.

4.2.2 Approximation of Kernel

Most kernels can be approximated by random features [21]. One example of random approximation of kernels is the Nyström approximation [49] (refer to [21] for more information). Another way to approximate kernel functions using random features is explained in the following.

Let $\phi(x) : \boldsymbol{x} \mapsto \phi(\boldsymbol{x})$ be the mapping function to the reproducing kernel Hilbert space. The kernel function is defined as follows:

$$k(\boldsymbol{x}, \boldsymbol{y}) = \phi(\boldsymbol{x})^\top \phi(\boldsymbol{y}). \tag{36}$$

The mapping function can be approximated using s random vectors $\{\boldsymbol{w}_i\}_{i=1}^s$:

$$\phi(\boldsymbol{x}) \approx \frac{h(x)}{\sqrt{s}}[f_1(\boldsymbol{w}_1^\top \boldsymbol{x}), f_1(\boldsymbol{w}_2^\top \boldsymbol{x}), \dots, f_1(\boldsymbol{w}_s^\top \boldsymbol{x}), \dots, f_l(\boldsymbol{w}_1^\top \boldsymbol{x}), \dots, f_l(\boldsymbol{w}_s^\top \boldsymbol{x})]^\top, \tag{37}$$

where $h(x)$ is a deterministic function and $\{f_j\}_{j=1}^l$ are l functions defined based on the kernel function. The dimensionality of $\phi(\boldsymbol{x})$ is $(l \times s)$ because s random vectors are sampled and each is passed through l functions. For many kernel functions, such an approximated $\phi(.)$ can be used where $\phi(\boldsymbol{x})^\top \phi(\boldsymbol{y})$ is approximately equal to the kernel $k(\boldsymbol{x}, \boldsymbol{y})$.

This approximation makes sense because, as shown in Fig. 12, each $\boldsymbol{w}_i^\top \boldsymbol{x}$ is the inner product or similarity of $\boldsymbol{x}$ with the random vector $\boldsymbol{w}_i$. Having many random vectors captures their similarities with $\boldsymbol{x}$, so these features locally represent $\boldsymbol{x}$. The larger the s, the more accurate the approximation becomes.

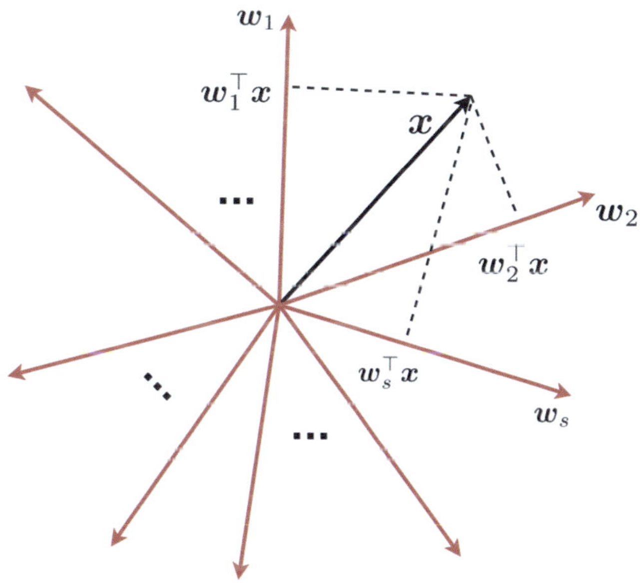

Fig. 12 Local representation of the vector $\boldsymbol{x}$ by its projection onto s random vectors $\{\boldsymbol{w}_i\}_{i=1}^s$

The terms in Eq. (37) are known for approximating certain kernel functions. For example, the Gaussian or RBF kernel can be approximated with $h(x) = 1$, $l = 2$, $f_1(.)$ as the sine function, and $f_2(.)$ as the cosine function:

$$\phi(x) \approx \frac{1}{\sqrt{s}}[\sin(\boldsymbol{w}_1^\top \boldsymbol{x}), \sin(\boldsymbol{w}_2^\top \boldsymbol{x}), \dots, \sin(\boldsymbol{w}_s^\top \boldsymbol{x}), \cos(\boldsymbol{w}_1^\top \boldsymbol{x}), \dots, \cos(\boldsymbol{w}_s^\top \boldsymbol{x})]^\top, \tag{38}$$

where each $\boldsymbol{w}_i$ is randomly sampled from a Gaussian distribution. The Gaussian kernel function can thus be approximated as follows:

$$k(\boldsymbol{x}, \boldsymbol{y}) = \phi(\boldsymbol{x})^\top \phi(\boldsymbol{y}) \approx e^{\frac{-\|x-y\|_2^2}{\gamma}}, \tag{39}$$

where $\phi(.)$ is defined in Eq. (38) and γ is the bandwidth of the kernel.

4.2.3 Approximation of Softmax in Attention

The idea of Performer [11] is to treat the softmax in the attention mechanism as a kernel function. In this way, it can be approximated using Eq. (37). As discussed in Eq. (16), attention can be modeled as a kernel function if it is assumed that $\boldsymbol{W}_Q = \boldsymbol{W}_K$. Performer makes this assumption to enable approximating the softmax in attention using random kernel approximations.

The $\text{softmax}(\boldsymbol{Q}^\top \boldsymbol{K})$ used in the attention mechanism (see Eq. (17)) is in the form:

$$\text{softmax}(\boldsymbol{Q}^\top \boldsymbol{K}) = \frac{e^{[\boldsymbol{Q}^\top \boldsymbol{K}]_i}}{\sum_j e^{[\boldsymbol{Q}^\top \boldsymbol{K}]_j}}, \tag{40}$$

where $[\boldsymbol{Q}^\top \boldsymbol{K}]_i$ is the i-th row of $\boldsymbol{Q}^\top \boldsymbol{K}$. The normalization in Eq. (40) can be performed as follows. Let $\boldsymbol{A} := \boldsymbol{Q}^\top \boldsymbol{K}$. The degree matrix of $\boldsymbol{A}$ is a diagonal matrix whose (i, i)-th element is defined as follows:

$$\boldsymbol{D}_{ii} := \sum_{j=1}^{n} \boldsymbol{A}_{ij}, \tag{41}$$

where normalization in Eq. (40) can then be obtained by $\boldsymbol{A}\boldsymbol{D}^{-1}$, which divides each row of $\boldsymbol{Q}^\top \boldsymbol{K}$ by the sum of that row.

Therefore, $e^{\boldsymbol{Q}^\top \boldsymbol{K}}$ in the form of $e^{\boldsymbol{x}^\top \boldsymbol{y}}$ needs to be approximated, after which normalization can be performed as explained.

Consider the Gaussian kernel function in Eq. (39). Assuming $\gamma = 2$, it can be simplified as follows:

$$\begin{aligned} k(\boldsymbol{x}, \boldsymbol{y}) &= e^{\frac{-\|x-y\|_2^2}{2}} = e^{\frac{-(x-y)^\top (x-y)}{2}} = e^{\frac{-x^\top x}{2}} e^{\frac{-y^\top y}{2}} e^{\boldsymbol{x}^\top \boldsymbol{y}} \\ \implies e^{\boldsymbol{x}^\top \boldsymbol{y}} &= k(\boldsymbol{x}, \boldsymbol{y}) e^{\frac{x^\top x}{2}} e^{\frac{y^\top y}{2}} \overset{(39)}{\approx} \phi(\boldsymbol{x})^\top \phi(\boldsymbol{y}) e^{\frac{x^\top x}{2}} e^{\frac{y^\top y}{2}}, \end{aligned} \tag{42}$$

where $\phi(.)$ is defined in Eq. (38).

Assuming that $e^{\boldsymbol{x}^\top \boldsymbol{y}}$ is a kernel function, it can be expressed as in Eq. (36):

$$e^{\boldsymbol{x}^\top \boldsymbol{y}} = \phi_{\text{SM}}(\boldsymbol{x})^\top \phi_{\text{SM}}(\boldsymbol{y}), \tag{43}$$

where SM stands for SoftMax. Comparing Eqs. (42) and (43) yields

$$e^{\boldsymbol{x}^\top \boldsymbol{y}} = \phi_{\text{SM}}(\boldsymbol{x})^\top \phi_{\text{SM}}(\boldsymbol{y}) = \Big(e^{\frac{\boldsymbol{x}^\top \boldsymbol{x}}{2}} \phi(\boldsymbol{x})\Big)^\top \Big(e^{\frac{\boldsymbol{y}^\top \boldsymbol{y}}{2}} \phi(\boldsymbol{y})\Big). \tag{44}$$

Therefore, $\phi_{\text{SM}}(.)$ can be defined as follows:

$$\begin{aligned}\phi_{\text{SM}}(\boldsymbol{x}) &= e^{\frac{\boldsymbol{x}^\top \boldsymbol{x}}{2}} \phi(\boldsymbol{x}) \\ &\overset{(38)}{\approx} e^{\frac{\boldsymbol{x}^\top \boldsymbol{x}}{2}} \frac{1}{\sqrt{s}} \underbrace{[\sin(\boldsymbol{w}_1^\top \boldsymbol{x}), \sin(\boldsymbol{w}_2^\top \boldsymbol{x}), \dots, \sin(\boldsymbol{w}_s^\top \boldsymbol{x}), \cos(\boldsymbol{w}_1^\top \boldsymbol{x}), \dots, \cos(\boldsymbol{w}_s^\top \boldsymbol{x})]^\top}_{2s \text{ elements}},\end{aligned} \tag{45}$$

where s is a hyperparameter, with larger values resulting in better approximations.

In summary, the softmax function used in the attention mechanism can be approximated using Eqs. (43) and (45). The normalization in the softmax can then be performed as explained using Eq. (41). Thus, the softmax term in attention is approximated as follows:

$$\text{softmax}(\boldsymbol{Q}^\top \boldsymbol{K}) \approx \phi_{\text{SM}}(\boldsymbol{x})^\top \phi_{\text{SM}}(\boldsymbol{y}) = \boldsymbol{Q}'^\top \boldsymbol{K}', \tag{46}$$

where $\boldsymbol{Q}', \boldsymbol{K}' \in \mathbb{R}^{(2s)\times n}$. The embedding dimensionality of the query and key has thus become $p = 2s$.

Using Eq. (46) reduces the computational complexity of attention as discussed in Sect. 4.2.1:

$$\mathbb{R}^{r\times n} \ni \boldsymbol{Z} = \frac{1}{\sqrt{p}} \underbrace{(\boldsymbol{V}\,\boldsymbol{Q}'^\top)}_{r\times 2s}\, \underbrace{\boldsymbol{K}'}_{2s\times n}, \tag{47}$$

whose complexity is $\mathcal{O}(np) = \mathcal{O}(2ns)$, which is linear in terms of the sequence length n. The elimination of the softmax function by this approximation makes it possible to change the order of matrix multiplications, thereby reducing computational complexity.

5 Conclusion

The attention mechanism can be viewed as a form of adaptive weighting. Its introduction marked a major breakthrough in natural language processing, demonstrating that sequential dependencies can be learned without relying on recurrence as in recurrent neural networks. The Transformer architecture builds on attention as its core component, with distinct encoder and decoder structures, as detailed in this chapter. Numerous variants of Transformers, such as the Performer, have since been developed to improve efficiency and scalability. Many state-of-the-art large language models are built upon the Transformer architecture and its attention mechanism. As of 2025, Transformers continue to play a central role across deep learning, natural language processing, and computer vision.

6 Exercises

1. (Level: Easy) Explain how the attention mechanism helps overcome the limitations of traditional sequence models like RNNs and LSTMs in handling long-range dependencies.
2. (Level: Easy) Consider the following sentence: "I am a student."
   Let the embedding vectors of these words be

$$\text{I} : [1, 1, 1]^\top, \quad \text{am} : [0.5, -0.5, 0]^\top, \quad \text{a} : [-1, 1, -1]^\top, \quad \text{student} : [1.5, 0.5, 1]^\top.$$

Take the word "I" as the query and all the words (including the word "I" itself) as the keys and values. Calculate the attention vector of the word "I" to all the words in this sentence. Use the scaled inner product as the similarity measure.

3. (Level: Easy) Consider the following simple example:
   Query vector: $\boldsymbol{Q} = \begin{bmatrix} 1 \\ 0 \end{bmatrix}$,
   Key vector: $\boldsymbol{K} = \begin{bmatrix} 1 \\ 1 \end{bmatrix}$,
   Value vector: $\boldsymbol{V} = \begin{bmatrix} 2 \\ 3 \end{bmatrix}$.
   Perform attention computation using scaled dot-product attention, i.e., compute

$$\text{Attention}(\boldsymbol{Q}, \boldsymbol{K}, \boldsymbol{V}) = \boldsymbol{V} \cdot \text{softmax}\left(\frac{\boldsymbol{Q}^\top \boldsymbol{K}}{\sqrt{p}}\right),$$

   where p is the dimension of the key vector.
4. (Level: Moderate) Consider a machine translation task where the phrase "European Economic Area" is translated into French as "zone économique européenne."
   (a) Describe how the attention mechanism would likely correspond words between the source and target languages in this case.
   (b) How does this correspondence impact the generated translation?
   (c) Given the softmax-based attention mechanism, calculate the attention weight a_{ij} for the i-th target word given the j-th source word using the formula:

$$a_{ij} = \frac{e^{s_{ij}}}{\sum_k e^{s_{ik}}}.$$

   Assume the similarity scores s_{ij} between the hidden state of the i-th target word and the j-th source word are

$$s = \begin{bmatrix} 2.5 & 1.2 & 0.8 \\ 1.0 & 2.2 & 1.8 \\ 0.9 & 1.5 & 2.0 \end{bmatrix},$$

   where rows correspond to target words ("Zone," "économique," "européenne") and columns correspond to source words ("The," "European," "Economic Area").
5. (Level: Easy) The attention score between a query q and key k_i is computed using the dot-product attention formula:

$$s_i = \frac{\boldsymbol{q}^\top \boldsymbol{k}_i}{\sqrt{p}},$$

   where $\boldsymbol{q}$ is the query vector, $\boldsymbol{k}_i$'s are the key vectors, and p is the dimensionality of the key vectors. The attention weights are computed using the softmax function:

$$a_i = \frac{\exp(s_i)}{\sum_j \exp(s_j)}.$$

The final attention output is a weighted sum of the value vectors:

$$c = \sum_i a_i \boldsymbol{v}_i,$$

where $\boldsymbol{v}_i$'s are the value vectors.
Consider a query vector $\boldsymbol{q} = [1, 0.5]^\top$.
The key vectors are

Key index	$\boldsymbol{k_i}$
1	$[0.8,\ 0.6]^\top$
2	$[0.5,\ 0.2]^\top$
3	$[0.1,\ 0.9]^\top$

The value vectors are

Value index	$\boldsymbol{v_i}$
1	$[2,\ 3]^\top$
2	$[1,\ 1]^\top$
3	$[0,\ 4]^\top$

(a) Compute the attention scores s_i's.
(b) Compute the attention weights a_i's.
(c) Compute the attention output c.

6. (Level: Easy) Given the following sequence of words (represented as indices), calculate the positional encoding for each word at a specific dimension. The sequence of words is ["go," "to," "school"]. Let the embedding dimension be $p = 4$. The position indices will be 0, 1, and 2 (corresponding to each word).
7. (Level: Difficult) Consider the following positional encoding for a given position n defined by

$$r_{n,i} = \begin{cases} \sin\left(\frac{n}{L^{i/p}}\right), & \text{if } i \text{ is even,} \\ \cos\left(\frac{n}{L^{(i-1)/p}}\right), & \text{if } i \text{ is odd,} \end{cases}$$

where p is the dimensionality. Show that this positional encoding has the property that, for a fixed offset k, the encoding at position $n + k$ can be represented as a linear combination of the encoding at position n with coefficients that depend only on k and not on n.
8. (Level: Moderate) This exercise is about processing an input through a "multiheaded attention" mechanism in a Transformer. Suppose the architecture contains two attention heads.

(a) Given the following input matrix, $\boldsymbol{X}$, and learned weight matrices, calculate the scaled dot-product attention output, $\boldsymbol{Z}_1$, for the first head:

$$\boldsymbol{X} = \begin{bmatrix} 1 & 1 & 0 \\ 0 & 2 & 1 \\ 2 & 0 & 1 \\ 1 & 2 & 1 \end{bmatrix}, \boldsymbol{W}_Q^1 = \begin{bmatrix} -1 & 1 \\ 1 & 0 \\ 0 & 1 \\ -1 & 0 \end{bmatrix}, \boldsymbol{W}_K^1 = \begin{bmatrix} 0 & 1 \\ 0 & -1 \\ 1 & 0 \\ 1 & 1 \end{bmatrix}, \boldsymbol{W}_V^1 = \begin{bmatrix} 0 & 1 \\ 1 & 1 \\ 1 & -1 \\ 0 & 1 \end{bmatrix}.$$

(b) The output for the second attention head is given as follows:

$$\boldsymbol{Z}_2 = \begin{bmatrix} 1 & 1 & 2 \\ 3 & 1 & 2 \end{bmatrix}.$$

(c) Concatenate $\boldsymbol{Z}_1$ and $\boldsymbol{Z}_2$ and apply the following linear transformation:

$$\boldsymbol{W}_O = \begin{bmatrix} 0 & 1 \\ 1 & -1 \\ 1 & 1 \\ -1 & 0 \end{bmatrix}.$$

References

1. Ido Amos, Jonathan Berant, and Ankit Gupta. "Never Train from Scratch: Fair Comparison of Long-Sequence Models Requires Data-Driven Priors". In: *International Conference on Learning Representations*. 2024.
2. Sotiris Anagnostidis et al. "Dynamic context pruning for efficient and interpretable autoregressive transformers". In: *Advances in Neural Information Processing Systems* 36 (2023), pp. 65202–65223.
3. Saleh Ashkboos, Maximilian L Croci, Marcelo Gennari do Nascimento, Torsten Hoefler, and James Hensman. "Slicegpt: Compress large language models by deleting rows and columns". In: *International Conference on Learning Representations*. 2024.
4. Dzmitry Bahdanau, Kyunghyun Cho, and Yoshua Bengio. "Neural machine translation by jointly learning to align and translate". In: *International Conference on Learning Representations*. 2015.
5. Iz Beltagy, Matthew E Peters, and Arman Cohan. "Longformer: The long-document transformer". In: *arXiv preprint* arXiv:2004.05150 (2020).
6. Shai Ben-David et al. "A theory of learning from different domains". In: *Machine learning* 79.1-2 (2010), pp. 151–175.
7. Ruisi Cai et al. "FLEXTRON: many-in-one flexible large language model". In: *Proceedings of the 41st International Conference on Machine Learning*. 2024, pp. 5298–5311.
8. William Chan, Navdeep Jaitly, Quoc Le, and Oriol Vinyals. "Listen, attend and spell: A neural network for large vocabulary conversational speech recognition". In: *2016 IEEE International Conference on Acoustics, Speech and Signal Processing*. IEEE. 2016, pp. 4960–4964.
9. William Chan, Navdeep Jaitly, Quoc V Le, and Oriol Vinyals. "Listen, attend and spell". In: *arXiv preprint* arXiv:1508.01211 (2015).
10. Jianpeng Cheng, Li Dong, and Mirella Lapata. "Long short-term memory-networks for machine reading". In: *Conference on Empirical Methods in Natural Language Processing*. 2016, pp. 551–561.
11. Krzysztof Marcin Choromanski et al. "Rethinking Attention with Performers". In: *International Conference on Learning Representations*. 2021.
12. Jan Chorowski, Dzmitry Bahdanau, Kyunghyun Cho, and Yoshua Bengio. "End-to-end continuous speech recognition using attention-based recurrent NN: First results". In: *arXiv preprint* arXiv:1412.1602 (2014).
13. Tri Dao, Dan Fu, Stefano Ermon, Atri Rudra, and Christopher Ré. "FlashAttention: Fast and memory-efficient exact attention with io-awareness". In: *Advances in neural information processing systems* 35 (2022), pp. 16344–16359.
14. Tri Dao and Albert Gu. "Transformers are SSMs: generalized models and efficient algorithms through structured state space duality". In: *Proceedings of the 41st International Conference on Machine Learning*. 2024, pp. 10041–10071.
15. Jacob Devlin, Ming-Wei Chang, Kenton Lee, and Kristina Toutanova. "BERT: Pre-training of deep bidirectional transformers for language understanding". In: *Proceedings of the 2019 conference of the North American chapter of the association for computational linguistics: human language technologies, volume 1 (long and short papers)*. 2019, pp. 4171–4186.

16. Qi Dou, Daniel Coelho de Castro, Konstantinos Kamnitsas, and Ben Glocker. "Domain generalization via model-agnostic learning of semantic features". In: *Advances in Neural Information Processing Systems* 32 (2019), pp. 6450–6461.
17. Leo Feng, Frederick Tung, Mohamed Osama Ahmed, Yoshua Bengio, and Hossein Hajimirsadeghi. "Were RNNs all we needed?" In: *arXiv preprint* arXiv:2410.01201 (2024).
18. Elias Frantar, Saleh Ashkboos, Torsten Hoefler, and Dan-Adrian Alistarh. "OPTQ: Accurate post-training quantization for generative pre-trained transformers". In: *11th International Conference on Learning Representations*. 2023.
19. Daniel Y Fu et al. "Hungry Hungry Hippos: Towards Language Modeling with State Space Models". In: *The Eleventh International Conference on Learning Representations*. 2023.
20. Hector Garcia-Molina, Jeffrey D. Ullman, and Jennifer Widom. *Database systems: The complete book*. Prentice Hall, 1999.
21. Benyamin Ghojogh, Mark Crowley, Fakhri Karray, and Ali Ghodsi. "Background on kernels". In: *Elements of Dimensionality Reduction and Manifold Learning* (2023), pp. 43–73.
22. Yoav Goldberg and Omer Levy. "word2vec Explained: deriving Mikolov et al.'s negative-sampling word-embedding method". In: *arXiv preprint* arXiv:1402.3722 (2014).
23. Arthur Gretton, Karsten Borgwardt, Malte Rasch, Bernhard Schölkopf, and Alex J Smola. "A kernel method for the two-sample-problem". In: *Advances in neural information processing systems*. 2007, pp. 513–520.
24. Arthur Gretton, Karsten M Borgwardt, Malte J Rasch, Bernhard Schölkopf, and Alexander Smola. "A kernel two-sample test". In: *The Journal of Machine Learning Research* 13.1 (2012), pp. 723–773.
25. Albert Gu and Tri Dao. "Mamba: Linear-Time Sequence Modeling with Selective State Spaces". In: *First Conference on Language Modeling*. 2024.
26. Albert Gu, Karan Goel, and Christopher Re. "Efficiently Modeling Long Sequences with Structured State Spaces". In: *International Conference on Learning Representations*. 2022.
27. Ankit Gupta, Albert Gu, and Jonathan Berant. "Diagonal state spaces are as effective as structured state spaces". In: *Advances in Neural Information Processing Systems* 35 (2022), pp. 22982–22994.
28. Kaiming He, Xiangyu Zhang, Shaoqing Ren, and Jian Sun. "Deep residual learning for image recognition". In: *Proceedings of the IEEE conference on computer vision and pattern recognition*. 2016, pp. 770–778.
29. Sepp Hochreiter and Jürgen Schmidhuber. "Long short-term memory". In: *Neural computation* 9.8 (1997), pp. 1735–1780.
30. Nitin Indurkhya and Fred J Damerau. *Handbook of natural language processing*. Vol. 2. CRC Press, 2010.
31. Sergey Ioffe and Christian Szegedy. "Batch normalization: Accelerating deep network training by reducing internal covariate shift". In: *arXiv preprint* arXiv:1502.03167 (2015).
32. Benoit Jacob et al. "Quantization and training of neural networks for efficient integer-arithmetic-only inference". In: *Proceedings of the IEEE conference on computer vision and pattern recognition*. 2018, pp. 2704–2713.
33. Vahid Kazemi and Ali Elqursh. "Show, ask, attend, and answer: A strong baseline for visual question answering". In: *arXiv preprint* arXiv:1704.03162 (2017).
34. Nikita Kitaev, Lukasz Kaiser, and Anselm Levskaya. "Reformer: The Efficient Transformer". In: *International Conference on Learning Representations*. 2020.
35. John F Kolen and Stefan C Kremer. *A field guide to dynamical recurrent networks*. John Wiley & Sons, 2001.
36. Stefan Kombrink, Tomáš Mikolov, Martin Karafiát, and Lukáš Burget. "Recurrent neural network based language modeling in meeting recognition". In: *Twelfth annual conference of the international speech communication association*. 2011.
37. Sneha Kudugunta et al. "MatFormer: Nested transformer for elastic inference". In: *arXiv preprint* arXiv:2310.07707 (2023).
38. Yann LeCun, Léon Bottou, Yoshua Bengio, and Patrick Haffner. "Gradient-based learning applied to document recognition". In: *Proceedings of the IEEE* 86.11 (1998), pp. 2278–2324.
39. Honglak Lee, Roger Grosse, Rajesh Ranganath, and Andrew Y Ng. "Convolutional deep belief networks for scalable unsupervised learning of hierarchical representations". In: *International Conference on Machine Learning*. 2009, pp. 609–616.
40. Da Li, Yongxin Yang, Yi-Zhe Song, and Timothy M Hospedales. "Deeper, broader and artier domain generalization". In: *Proceedings of the IEEE international conference on computer vision*. 2017, pp. 5542–5550.
41. Yang Li, Lukasz Kaiser, Samy Bengio, and Si Si. "Area attention". In: *International Conference on Machine Learning*. PMLR. 2019, pp. 3846–3855.
42. Bingli Liao and Danilo Vasconcellos Vargas. "Beyond KV caching: Shared attention for efficient llms". In: *arXiv preprint* arXiv:2407.12866 (2024).
43. Lucas Dax Lingle. "Transformer-VQ: Linear-Time Transformers via Vector Quantization". In: *International Conference on Learning Representations*. 2024.
44. Minh-Thang Luong, Hieu Pham, and Christopher D Manning. "Effective approaches to attention-based neural machine translation". In: *arXiv preprint* arXiv:1508.04025 (2015).

45. Tomas Mikolov, Kai Chen, Greg Corrado, and Jeffrey Dean. "Efficient estimation of word representations in vector space". In: *International Conference on Learning Representations*. 2013.
46. Tomas Mikolov, Kai Chen, Gregory S Corrado, and Jeffrey A Dean. *Computing numeric representations of words in a high-dimensional space*. US Patent 9,037,464. 2015.
47. Tomas Mikolov, Ilya Sutskever, Kai Chen, Greg S Corrado, and Jeff Dean. "Distributed representations of words and phrases and their compositionality". In: *Advances in neural information processing systems*. 2013, pp. 3111–3119.
48. Saurav Muralidharan et al. "Compact language models via pruning and knowledge distillation". In: *Advances in Neural Information Processing Systems* 37 (2024), pp. 41076–41102.
49. Evert J Nyström. "Über die praktische Auflösung von Integralgleichungen mit Anwendungen auf Randwertaufgaben". In: *Acta Mathematica* 54.1 (1930), pp. 185–204.
50. Jeffrey Pennington, Richard Socher, and Christopher D Manning. "GloVe: Global vectors for word representation". In: *Proceedings of the 2014 conference on empirical methods in natural language processing*. 2014, pp. 1532–1543.
51. Leonid I Perlovsky. "Toward physics of the mind: Concepts, emotions, consciousness, and symbols". In: *Physics of Life Reviews* 3.1 (2006), pp. 23–55.
52. Ofir Press, Noah Smith, and Mike Lewis. "Train Short, Test Long: Attention with Linear Biases Enables Input Length Extrapolation". In: *International Conference on Learning Representations*. 2022.
53. Zhen Qin et al. "TransNormerLLM: A Faster and Better Large Language Model with Improved TransNormer". In: *International Conference on Machine Learning*. 2024.
54. Alec Radford, Karthik Narasimhan, Tim Salimans, and Ilya Sutskever. *Improving language understanding by generative pre-training*. Tech. rep. OpenAI, 2018.
55. Hossein Rajabzadeh et al. "EchoAtt: Attend, copy, then adjust for more efficient large language models". In: *arXiv preprint* arXiv:2409.14595 (2024).
56. David E Rumelhart, Geoffrey E Hinton, and Ronald J Williams. "Learning representations by back-propagating errors". In: *Nature* 323.6088 (1986), pp. 533–536.
57. Mohammadreza Salehi, Sachin Mehta, Aditya Kusupati, Ali Farhadi, and Hannaneh Hajishirzi. "SHARCS: Efficient Transformers Through Routing with Dynamic Width Sub-networks". In: *The 2023 Conference on Empirical Methods in Natural Language Processing*.
58. Richard Socher, Yoshua Bengio, and Chris Manning. "Deep learning for NLP". In: *Tutorial at Association of Computational Logistics (ACL)* (2012).
59. Nitish Srivastava, Geoffrey Hinton, Alex Krizhevsky, Ilya Sutskever, and Ruslan Salakhutdinov. "Dropout: a simple way to prevent neural networks from overfitting". In: *The journal of machine learning research* 15.1 (2014), pp. 1929–1958.
60. Christopher Summerfield and Tobias Egner. "Expectation (and attention) in visual cognition". In: *Trends in cognitive sciences* 13.9 (2009), pp. 403–409.
61. Yutao Sun et al. "Retentive Network: A Successor to Transformer for Large Language Models". In: (2024).
62. Yi Tay, Dara Bahri, Liu Yang, Donald Metzler, and Da-Cheng Juan. "Sparse Sinkhorn attention". In: *International conference on machine learning*. PMLR. 2020, pp. 9438–9447.
63. Georgy Tyukin, Gbetondji Jean-Sebastien Dovonon, Jean Kaddour, and Pasquale Minervini. "Attention Is All You Need But You Don't Need All Of It For Inference of Large Language Models". In: *ICML 2024 Workshop on Theoretical Foundations of Foundation Models*.
64. Mojtaba Valipour et al. "SortedNet: A Scalable and Generalized Framework for Training Modular Deep Neural Networks". In: *arXiv preprint* arXiv:2309.00255 (2023).
65. Ashish Vaswani et al. "Attention is all you need". In: *Advances in neural information processing systems*. 2017, pp. 5998–6008.
66. Hanrui Wang, Zhekai Zhang, and Song Han. "SpAtten: Efficient sparse attention architecture with cascade token and head pruning". In: *2021 IEEE International Symposium on High-Performance Computer Architecture (HPCA)*. IEEE. 2021, pp. 97–110.
67. Sinong Wang, Belinda Z Li, Madian Khabsa, Han Fang, and Hao Ma. "Linformer: Self-attention with linear complexity". In: *arXiv preprint* arXiv:2006.04768 (2020).
68. Yong Wang, Longyue Wang, Shuming Shi, Victor OK Li, and Zhaopeng Tu. "Go from the general to the particular: Multi-domain translation with domain transformation networks". In: *Proceedings of the AAAI Conference on Artificial Intelligence*. Vol. 34. 05. 2020, pp. 9233–9241.
69. Thomas Wolf, Victor Sanh, Julien Chaumond, and Clement Delangue. "TransferTransfo: A transfer learning approach for neural network based conversational agents". In: *Proceedings of the AAAI Conference on Artificial Intelligence*. 2019.
70. Guangxuan Xiao et al. "SmoothQuant: Accurate and efficient post-training quantization for large language models". In: *International Conference on Machine Learning*. PMLR. 2023, pp. 38087–38099.
71. Jun Xu. "On the Techniques of English Fast-reading". In: *Theory and Practice in Language Studies*. Vol. 1. 10. Academy Publisher, 2011, pp. 1416–1419.
72. Kelvin Xu et al. "Show, attend and tell: Neural image caption generation with visual attention". In: *International conference on machine learning*. 2015, pp. 2048–2057.

73. Songlin Yang, Bailin Wang, Yikang Shen, Rameswar Panda, and Yoon Kim. "Gated Linear Attention Transformers with Hardware-Efficient Training". In: *International Conference on Machine Learning*. 2024.
74. Zhewei Yao et al. "ZeroQuant: Efficient and affordable post-training quantization for large-scale transformers". In: *Advances in Neural Information Processing Systems* 35 (2022), pp. 27168–27183.
75. Keyi Yu, Yang Liu, Alexander G Schwing, and Jian Peng. "Fast and accurate text classification: Skimming, rereading and early stopping". In: *International Conference on Learning Representations*. 2018.
76. Manzil Zaheer et al. "Big bird: Transformers for longer sequences". In: *Advances in neural information processing systems* 33 (2020), pp. 17283–17297.
77. Zhenyu Zhang et al. "H2O: Heavy-hitter oracle for efficient generative inference of large language models". In: *Advances in Neural Information Processing Systems* 36 (2023), pp. 34661–34710.

State Space Models 10

Abstract

This chapter explores State Space Models (SSMs) as a compelling alternative to traditional sequence modeling methods, particularly when compared to attention-based architectures such as Transformers. SSMs demonstrate strong capabilities in modeling long-range dependencies across various modalities, including time series, audio, and image data. Although early SSMs underperformed in natural language tasks compared to Transformers, recent models such as the Structured State Space Model (S4) have significantly reduced this gap by enabling efficient long-range reasoning. Follow-ups such as Diagonal State Space Models (DSS), H3 (Hungry Hungry Hippos) hybrids, and especially Mamba show that SSMs are now competitive alternatives in language modeling as well as other modalities. Newer architectures such as Mamba dynamically parameterize state matrices based on input, further enhancing adaptability and performance. As the field evolves, SSMs are increasingly recognized for their ability to process extremely long sequences—such as entire books—more effectively than traditional attention-based models. This chapter begins by outlining the advantages of SSMs and introducing their continuous and discrete formulations. It then discusses key architectures, including S4, Mamba, and TTT (Test Time Training) layers, and concludes by formally relating the attention mechanism to SSMs. Prior familiarity with recurrent neural networks and attention-based models is recommended for optimal comprehension of this chapter.

1 Introduction

State Space Models (SSMs) are introduced as powerful alternatives to traditional sequence modeling approaches. These models demonstrate strong performance across various modalities, including time-series analysis, audio generation, and image processing, and they can capture long-range dependencies more efficiently.

SSMs initially struggled to match the performance of Transformers in language modeling tasks, resulting in a performance gap between the two. To address these challenges, recent advances in SSM architectures, such as the Structured State Space Model (S4) [6], have been introduced. S4 succeeded in long-range reasoning tasks and enabled more efficient computation while preserving theoretical strengths. However, its implementation remained complex and computationally demanding.

B. Ghojogh and A. Ghodsi, *Elements of Deep Learning*,
https://doi.org/10.1007/978-3-032-10738-1_10

Consequently, further research led to the development of simplified variants, such as the Diagonal State Space Model (DSS) [7], which achieved comparable performance with a more straightforward formulation.

In parallel, hybrid approaches like the Hungry Hungry Hippos (H3) model [4] integrated SSMs with attention mechanisms to bridge these gaps. For example, H3 aimed to replace almost all attention layers in Transformers with SSMs. More recently, models such as Mamba [5] have pushed the boundaries of SSMs by selectively parameterizing state matrices as functions of the input, allowing for more flexible and adaptive information propagation.

Research in SSMs continues to advance rapidly. Rather than fully replacing Transformers, SSMs are emerging as complementary models that excel in handling very long sequences and streaming data. They can process inputs such as entire books more efficiently than Transformers [13] (see Chap. 9), while hybrid approaches combine the strengths of both families. SSMs will likely play a crucial role in the next generation of sequence modeling frameworks.

This chapter introduces SSMs. The advantages of SSMs are enumerated in Sect. 2. Continuous and discrete SSMs are introduced in Sect. 3. The Structured State Space Model (S4) [6] is explained in Sect. 4. Section 5 describes Mamba [5]. Section 6 introduces the TTT (Test Time Training) layers [10]. Finally, the relation between the attention mechanism and SSM is proved in Sect. 7. It is recommended that the reader studies Chaps. 8 (Recurrent Neural Networks) and 9 (Attention Mechanism and Transformers) before reading this chapter.

2 Advantages of SSMs

There are several advantages of SSMs compared to Transformers and attention-based models:

- Efficient long-range dependency handling: Unlike Transformers, which require $\mathcal{O}(n^2)$ complexity for self-attention, SSMs can process sequences with $\mathcal{O}(n \log n)$ time complexity using efficient matrix-vector multiplications and the Fast Fourier Transform (FFT), or $\mathcal{O}(n)$ in the case of Mamba architectures.
- Effective long-range dependency handling: By leveraging a state update mechanism (parameterized by λ) that preserves and accumulates signals over arbitrarily large time spans, SSMs effectively capture and retain information from distant points in a sequence, enabling robust long-range dependency handling.
- Lower memory requirements: Unlike Transformers, SSMs do not require storing the entire input in memory, resulting in lower memory consumption.
- Parallelization: Modern SSMs such as S4 [6] and Mamba [5] allow efficient GPU parallelization, combining recurrent structure with high throughput. This makes them practical for large-scale training in ways that classic RNNs were not.
- Robustness to irregularly sampled data: Unlike Transformers, which inherently assume regularly spaced inputs, State Space Models naturally handle irregularly sampled data, which is prevalent in real-world applications such as sensor data processing and clinical time series. By explicitly modeling the continuous-time evolution of latent states, SSMs provide robustness against missing or unevenly spaced observations, leading to improved performance in scenarios where input data is incomplete or irregularly sampled.
- Interpretability and diagnostic insight: State Space Models offer a distinct interpretability advantage by representing system behavior explicitly through their state transition dynamics. Unlike black-box models, the learned parameters in SSMs (matrices $\boldsymbol{A}$, $\boldsymbol{B}$, $\boldsymbol{C}$, and $\boldsymbol{D}$) can be analyzed to infer how

inputs influence future states and outputs over time. This interpretability is especially valuable in fields where understanding model behavior is critical, such as financial risk modeling or biological systems analysis.

3 State Space Model

3.1 Review of Transformer and RNN

To better understand SSMs, it is helpful to first examine the limitations of Transformers and Recurrent Neural Networks (RNNs), as well as their relation to SSMs.

In Transformers, during training, a matrix is constructed to compare each token with every preceding token, and the weights in each cell of this matrix indicate their pairwise similarity. The calculation of these weights does not require sequential operations, allowing the relationships between all token pairs to be computed in parallel. However, during inference, when generating the next token, the attention mechanism must be recalculated for the entire sequence, including tokens that have already been generated. As a result, for a sequence of length n, the computational cost is n^2, which becomes a bottleneck for long sequences in Transformers.

RNNs take two inputs at each time step to predict the output $\boldsymbol{y}_t$ and generate the current hidden state ($\boldsymbol{h}_t$). These inputs are the hidden state from the previous time step ($\boldsymbol{h}_{t-1}$) and the input at the current time step ($\boldsymbol{x}_t$). This structure enables inference with linear computational complexity because, unlike Transformers, RNNs do not need to recalculate all previous hidden states. However, a well-known drawback of RNNs is their tendency to forget earlier tokens as the sequence progresses.

A simple RNN updates its hidden state as (see Chap. 8) follows:

$$\boldsymbol{h}_t = \sigma(\boldsymbol{W}\boldsymbol{h}_{t-1} + \boldsymbol{U}\boldsymbol{x}_t), \tag{1}$$

$$\boldsymbol{y}_t = \boldsymbol{V}\boldsymbol{h}_t, \tag{2}$$

where $\boldsymbol{h}_t$ is the hidden state at time step t; $\boldsymbol{x}_t$ is the input at time step t; $\boldsymbol{y}_t$ is the output at time step t; σ is a nonlinear activation function; and $\boldsymbol{W}$, $\boldsymbol{U}$, and $\boldsymbol{V}$ are weight matrices.

3.2 Continuous Space Model

SSMs originate from control theory [9], where they are used for the mathematical representation of systems and for describing their possible states. They define a linear mapping from an input signal $\boldsymbol{x}(t)$ to an output signal $\boldsymbol{y}(t)$ through a latent state representation $\boldsymbol{h}(t)$. There are two main components to an SSM:

1. State equation: Describes how the system's hidden state changes based on the current state and input:

$$\boldsymbol{h}'(t) = \boldsymbol{A}\boldsymbol{h}(t) + \boldsymbol{B}\boldsymbol{x}(t), \tag{3}$$

where $\boldsymbol{x}(t)$ is the input, $\boldsymbol{y}(t)$ is the output, $\boldsymbol{h}(t)$ represents the hidden state, and $\boldsymbol{h}'(t)$ is the derivative of $\boldsymbol{h}(t)$ with respect to time.

2. Output equation: Defines how the system produces output based on the current state and input:

$$\boldsymbol{y}(t) = \boldsymbol{C}\boldsymbol{h}(t) + \boldsymbol{D}\boldsymbol{x}(t). \tag{4}$$

The matrices $\boldsymbol{A}$, $\boldsymbol{B}$, $\boldsymbol{C}$, and $\boldsymbol{D}$ are learnable parameter matrices. Specifically, $\boldsymbol{A}$ is called the state transition matrix, $\boldsymbol{B}$ is the control matrix, and $\boldsymbol{C}$ is the input matrix.

The intuition behind the state equation, i.e., Eq. (3), is that the change in the latent variable is given by the sum of linear transformations of the current latent state and the linear transformation of the input variable. The output equation, i.e., Eq. (4), relates the output to the hidden state and the input. Solving this system of differential equations yields the linear transformations $\boldsymbol{A}, \boldsymbol{B}, \boldsymbol{C}, \boldsymbol{D}$.

3.3 Discrete State Space Model

Equations (3) and (4) use continuous functions and are not applicable to many types of sequential input data, such as sentences, which are discrete. These differential equations can be discretized to accommodate discrete input sequences. This yields

$$\boldsymbol{h}_t = \bar{\boldsymbol{A}}\boldsymbol{h}_{t-1} + \bar{\boldsymbol{B}}\boldsymbol{x}_t, \tag{5}$$

$$\boldsymbol{y}_t = \bar{\boldsymbol{C}}\boldsymbol{h}_t, \tag{6}$$

where $\boldsymbol{h}_t \in \mathbb{R}^s$ is the discrete latent state at step t; $\boldsymbol{x}_t \in \mathbb{R}^d$ is the input at step t; $\boldsymbol{y}_t \in \mathbb{R}^p$ is the output at step t; and the matrices $\bar{\boldsymbol{A}} \in \mathbb{R}^{s\times s}$, $\bar{\boldsymbol{B}} \in \mathbb{R}^{s\times d}$, and $\bar{\boldsymbol{C}} \in \mathbb{R}^{p\times s}$ are discrete versions of matrices $\boldsymbol{A}$, $\boldsymbol{B}$, and $\boldsymbol{C}$, respectively.

Comparing Eq. (4) and Eq. (6), the term $\boldsymbol{D}\boldsymbol{x}_t$ is dropped (or equivalently $\boldsymbol{D} = \boldsymbol{0}$) in Eq. (6) for computational simplicity because $\bar{\boldsymbol{C}}$ can capture the effect of input $\boldsymbol{x}_t$:

$$\boldsymbol{y}_t \overset{(6)}{=} \bar{\boldsymbol{C}}\boldsymbol{h}_t \overset{(5)}{=} \bar{\boldsymbol{C}}(\bar{\boldsymbol{A}}\boldsymbol{h}_{t-1} + \bar{\boldsymbol{B}}\boldsymbol{x}_t) = \bar{\boldsymbol{C}}\bar{\boldsymbol{A}}\boldsymbol{h}_{t-1} + \bar{\boldsymbol{C}}\bar{\boldsymbol{B}}\boldsymbol{x}_t,$$

so $\bar{\boldsymbol{C}}$ also operates on $\boldsymbol{x}_t$.

A question may arise: how does an SSM differ from an RNN, which was introduced in Chap. 8? In fact, the difference lies in the convolutive view of SSMs. Comparing Eqs. (1) and (2) with Eqs. (5) and (6) shows that a basic SSM resembles a linear RNN without nonlinear activations. However, modern SSMs extend far beyond this by using structured parameterizations (S4 [6]) or input-dependent dynamics (Mamba [5]), making them much more expressive than simple linear RNNs.

To understand this, consider a sequence of inputs $\{x_1, \ldots, x_n\}$:

$$\begin{aligned}
\boldsymbol{h}_0 &= \bar{\boldsymbol{B}}\boldsymbol{x}_0, \\
\boldsymbol{h}_1 &= \bar{\boldsymbol{A}}\boldsymbol{h}_0 + \bar{\boldsymbol{B}}\boldsymbol{x}_1 = \bar{\boldsymbol{A}}\bar{\boldsymbol{B}}\boldsymbol{x}_0 + \bar{\boldsymbol{B}}\boldsymbol{x}_1, \\
\boldsymbol{h}_2 &= \bar{\boldsymbol{A}}\boldsymbol{h}_1 + \bar{\boldsymbol{B}}\boldsymbol{x}_2 = \bar{\boldsymbol{A}}^2\bar{\boldsymbol{B}}\boldsymbol{x}_0 + \bar{\boldsymbol{A}}\bar{\boldsymbol{B}}\boldsymbol{x}_1 + \bar{\boldsymbol{B}}\boldsymbol{x}_2, \\
&\vdots
\end{aligned}$$

This shows that SSMs capture the information of previous inputs in the sequence in a convolutive manner.

3.4 Discrete Weight Matrices

It is possible to use the bilinear method [12] to discretize the weight matrices $\boldsymbol{A}$, $\boldsymbol{B}$, and $\boldsymbol{C}$ into $\bar{\boldsymbol{A}}$, $\bar{\boldsymbol{B}}$, and $\bar{\boldsymbol{C}}$ [1, 6]. The trapezoidal rule assumes

$$\boldsymbol{x}_{n+1} - \boldsymbol{x}_n = \frac{\Delta}{2}\big(f(t_{n+1}) + f(t_n)\big), \tag{7}$$

where

$$\Delta := \boldsymbol{x}_{n+1} - \boldsymbol{x}_n. \tag{8}$$

Consider the ordinary differential equation, i.e., Eq. (3):

$$\boldsymbol{h}'(t) = \boldsymbol{A}\boldsymbol{h}(t) + \boldsymbol{B}\boldsymbol{x}(t).$$

By applying the trapezoidal rule, this equation becomes

$$\begin{aligned}
\boldsymbol{h}_{n+1} - \boldsymbol{h}_n &= \frac{\Delta}{2}(\boldsymbol{A}\boldsymbol{h}_{n+1} + \boldsymbol{B}\boldsymbol{x}_{n+1} + \boldsymbol{A}\boldsymbol{h}_n + \boldsymbol{B}\boldsymbol{x}_n),\\
\Longrightarrow \boldsymbol{h}_{n+1} - \frac{\Delta}{2}\boldsymbol{A}\boldsymbol{h}_{n+1} &= \boldsymbol{h}_n + \frac{\Delta}{2}\boldsymbol{A}\boldsymbol{h}_n + \frac{\Delta}{2}\boldsymbol{B}(\boldsymbol{x}_{n+1} + \boldsymbol{x}_n),\\
\Longrightarrow (\boldsymbol{I} - \frac{\Delta}{2}\boldsymbol{A})\boldsymbol{h}_{n+1} &= (\boldsymbol{I} + \frac{\Delta}{2}\boldsymbol{A})\boldsymbol{h}_n + \frac{\Delta}{2}\boldsymbol{B}(\boldsymbol{x}_{n+1} + \boldsymbol{x}_n).
\end{aligned}$$

It is assumed that the control sequence does not change significantly over small Δ, i.e., $\boldsymbol{x}_{n+1} \approx \boldsymbol{x}_n$, leading to

$$(\boldsymbol{I} - \frac{\Delta}{2}\boldsymbol{A})\boldsymbol{h}_{n+1} = (\boldsymbol{I} + \frac{\Delta}{2}\boldsymbol{A})\boldsymbol{h}_n + \Delta\boldsymbol{B}(\boldsymbol{x}_{n+1}).$$

Thus, solving for $\boldsymbol{h}_{n+1}$ gives

$$\boldsymbol{h}_{n+1} = (\boldsymbol{I} - \frac{\Delta}{2}\boldsymbol{A})^{-1}(\boldsymbol{I} + \frac{\Delta}{2}\boldsymbol{A})\boldsymbol{h}_n + (\boldsymbol{I} - \frac{\Delta}{2}\boldsymbol{A})^{-1}\Delta\boldsymbol{B}(\boldsymbol{x}_{n+1}).$$

From this, the discretized state matrices are [6] as follows:

$$\bar{\boldsymbol{A}} = (\boldsymbol{I} - \frac{\Delta}{2}\boldsymbol{A})^{-1}(\boldsymbol{I} + \frac{\Delta}{2}\boldsymbol{A}), \tag{9}$$

$$\bar{\boldsymbol{B}} = (\boldsymbol{I} - \frac{\Delta}{2}\boldsymbol{A})^{-1}\Delta\boldsymbol{B}. \tag{10}$$

The matrix $\bar{\boldsymbol{A}}$ is the discretized state transfer matrix, which is an approximate representation of the continuous state transition matrix $\boldsymbol{A}$. The matrix $\boldsymbol{B}$ is the discrete version of the input matrix. Note that the discrete matrix $\bar{\boldsymbol{C}}$ can be assumed to be equal to the matrix $\boldsymbol{C}$, i.e.:

$$\bar{\boldsymbol{C}} = \boldsymbol{C}. \tag{11}$$

4 Structured State Space (S4)

The Structured State Space (S4) model—proposed in 2021 [6]—was the first model to use SSMs for language modeling. It is specialized for efficient sequence modeling, particularly in scenarios requiring the handling of long-term dependencies. The goal of S4 is to improve computational efficiency, reduce the number of parameters, and enhance learning capability by introducing constraints on the state transition matrices.

To efficiently handle long sequences, the S4 model employs a structured parameterization of its state transition matrices. Specifically, it uses a "Diagonal Plus Low-Rank (DPLR)" parameterization, which combines the simplicity of diagonal matrices with the expressiveness of low-rank updates. This approach allows S4 to learn complex state spaces while maintaining computational efficiency, making it particularly effective for tasks requiring long-range dependencies.

The main innovation of S4 is its ability to track long-range dependencies efficiently. Traditional models struggle with long sequences either because they require excessive memory (like Transformers) or they forget early information (like RNNs). S4 overcomes these limitations by using a mathematical reformulation of state space models, allowing it to process sequences much more efficiently.

4.1 Output Equation in S4

Consider Eqs. (5) and (6) for the state and output equations in an SSM. The S4 model, as an SSM-based model, does not use Eq. (6) for the output equation. Instead, it uses the following formulation to calculate the output:

$$\boldsymbol{y}_k = \sum_{i=0}^{k} \boldsymbol{K}[i]\,\boldsymbol{x}_{k-i}, \tag{12}$$

where $\boldsymbol{K}[i]$ is the learned long-range convolution kernel, defined as follows:

$$\boldsymbol{K}[i] := \bar{\boldsymbol{C}}\bar{\boldsymbol{A}}^i\bar{\boldsymbol{B}}, \tag{13}$$

in which $\bar{\boldsymbol{A}}^i := \bar{\boldsymbol{A}}\bar{\boldsymbol{A}}\ldots\bar{\boldsymbol{A}}$ denotes the i-th power of $\bar{\boldsymbol{A}}$.

The proof of Eqs. (12) and (13) is provided in the following section.

4.2 Proof of Output Equation in S4

Consider Eqs. (5) and (6) as discrete-time state space equations:

$$\begin{aligned}\boldsymbol{h}_{k+1} &= \bar{\boldsymbol{A}}\boldsymbol{h}_k + \bar{\boldsymbol{B}}\boldsymbol{x}_k,\\ \boldsymbol{y}_k &= \bar{\boldsymbol{C}}\boldsymbol{h}_k.\end{aligned}$$

Assume the initial state is zero, i.e., $\boldsymbol{h}_0 = \boldsymbol{0}$. The recurrence can be unrolled to express $\boldsymbol{h}_k$ in terms of the past inputs. Consider the first few unrolled values:

$$\begin{aligned}
\boldsymbol{h}_0 &= \mathbf{0}, \\
\boldsymbol{h}_1 &= \bar{\boldsymbol{B}}\boldsymbol{x}_0, \\
\boldsymbol{h}_2 &= \bar{\boldsymbol{A}}\bar{\boldsymbol{B}}\boldsymbol{x}_0 + \bar{\boldsymbol{B}}\boldsymbol{x}_1, \\
\boldsymbol{h}_3 &= \bar{\boldsymbol{A}}^2\bar{\boldsymbol{B}}\boldsymbol{x}_0 + \bar{\boldsymbol{A}}\bar{\boldsymbol{B}}\boldsymbol{x}_1 + \bar{\boldsymbol{B}}\boldsymbol{x}_2, \\
&\vdots \\
\boldsymbol{h}_k &= \sum_{i=0}^{k-1} \bar{\boldsymbol{A}}^{k-1-i}\bar{\boldsymbol{B}}\boldsymbol{x}_i.
\end{aligned}$$

Now, if $\boldsymbol{h}_k$ is substituted into the output equation, it yields

$$\boldsymbol{y}_k = \bar{\boldsymbol{C}}\boldsymbol{h}_k = \bar{\boldsymbol{C}}\left(\sum_{i=0}^{k-1} \bar{\boldsymbol{A}}^{k-1-i}\bar{\boldsymbol{B}}\boldsymbol{x}_i\right) = \sum_{i=0}^{k-1} \bar{\boldsymbol{C}}\bar{\boldsymbol{A}}^{k-1-i}\bar{\boldsymbol{B}}\boldsymbol{x}_i.$$

The summation can be re-indexed. Let $j = k - 1 - i$, then $i = k - 1 - j$, and the above equation becomes

$$\boldsymbol{y}_k = \sum_{j=0}^{k-1} \bar{\boldsymbol{C}}\bar{\boldsymbol{A}}^j\bar{\boldsymbol{B}}\boldsymbol{x}_{k-1-j}.$$

This is equivalent to a causal 1D convolution:

$$\boldsymbol{y}_k = \sum_{i=0}^{k} \boldsymbol{K}[i]\boldsymbol{x}_{k-i}, \tag{14}$$

where $\boldsymbol{K}[i] := \bar{\boldsymbol{C}}\bar{\boldsymbol{A}}^i\bar{\boldsymbol{B}}$.

This completes the proof of Eqs. (12) and (13).

4.3 State Transition Matrix in S4

The S4 model defines the state transition matrix $\bar{\boldsymbol{A}} \in \mathbb{R}^{s \times s}$ as follows. Note that s is the dimensionality of the hidden state $\boldsymbol{h}_i \in \mathbb{R}^s$.

Suppose $\bar{\boldsymbol{A}}_0$ is a diagonal matrix:

$$\bar{\boldsymbol{A}} = \mathbf{diag}(\boldsymbol{\Lambda}),$$

where $\boldsymbol{\Lambda} = [\Lambda_1, \ldots, \Lambda_s]^\top$ are the eigenvalues of $\bar{\boldsymbol{A}}$. It is straightforward to compute powers:

$$\bar{\boldsymbol{A}}^i = \mathbf{diag}(\boldsymbol{\Lambda}^i),$$

but this lacks expressivity, as it cannot model interactions between state components (it is fully decoupled).

To introduce interactions, S4 adds a low-rank correction [6]:

$$\bar{A} = \mathbf{diag}(\boldsymbol{\Lambda}) + \boldsymbol{P}\boldsymbol{Q}^\top, \tag{15}$$

where $\boldsymbol{P}, \boldsymbol{Q} \in \mathbb{R}^{s \times r}$, with s as the dimensionality of the hidden state and r as the rank of the low-rank correction. The matrices $\boldsymbol{P}$ and $\boldsymbol{Q}$ are low-rank matrices with rank $r \ll s$, thus $\boldsymbol{P}\boldsymbol{Q}^\top$ is a low-rank update. This form is known from matrix theory as a Diagonal Plus Low-Rank (DPLR) matrix.

Using Eq. (15) in Eq. (13) can be computed efficiently using the Woodbury matrix identity and Cauchy kernel tricks to compute convolution kernels. The resolvent (transfer function) of the state space model, in the Laplace transform (with variable z), is

$$\boldsymbol{H}(z) = \bar{\boldsymbol{C}}(z\boldsymbol{I} - \bar{\boldsymbol{A}})^{-1}\bar{\boldsymbol{B}} = \sum_{i=0}^{\infty} \boldsymbol{K}[i]z^{-i}, \tag{16}$$

where $\boldsymbol{I}$ is the identity matrix. With Eq. (15), the term $(z\boldsymbol{I} - \bar{\boldsymbol{A}})^{-1}$ in $\boldsymbol{H}(z)$ can be computed efficiently using the Woodbury matrix identity:

$$\begin{aligned}(z\boldsymbol{I} - \bar{\boldsymbol{A}})^{-1} &\overset{(15)}{=} \left(z\boldsymbol{I} - \text{diag}(\boldsymbol{\Lambda}) - \boldsymbol{P}\boldsymbol{Q}^\top\right)^{-1} \\ &\overset{(a)}{=} (z\boldsymbol{I} - \text{diag}(\boldsymbol{\Lambda}))^{-1} \\ &\quad + (z\boldsymbol{I} - \text{diag}(\boldsymbol{\Lambda}))^{-1}\boldsymbol{P}\left(\boldsymbol{I} - \boldsymbol{Q}^\top(z\boldsymbol{I} - \text{diag}(\boldsymbol{\Lambda}))^{-1}\boldsymbol{P}\right)^{-1}\boldsymbol{Q}^\top(z\boldsymbol{I} - \text{diag}(\boldsymbol{\Lambda}))^{-1},\end{aligned} \tag{17}$$

where (a) follows from the Woodbury matrix identity. The term $(z\boldsymbol{I} - \text{diag}(\boldsymbol{\Lambda}))^{-1}$ is simply the element-wise inversion of a diagonal matrix.

4.4 On Efficiency and Speed of S4

Equations (5) and (6) used in regular RNN-like SSMs have a sequential recurrence; each $\boldsymbol{h}_{t+1}$ depends on $\boldsymbol{h}_t$, so it must be computed step by step, making it slow and not parallelizable across time. However, Eq. (12) is a Linear Time-Invariant (LTI) convolution; if all the inputs $\{\boldsymbol{x}_0, \ldots, \boldsymbol{x}_{n-1}\}$ are known, it is possible to compute all $\{\boldsymbol{y}_0, \ldots, \boldsymbol{y}_{n-1}\}$ in parallel. This reduces the computational complexity from $\mathcal{O}(n^2)$ in the attention mechanism (see Chap. 9) to $\mathcal{O}(n \log n)$ in the S4 model. It also enables vectorization and batching on GPUs, making S4 much faster than Transformers that use attention.

Equations (12) and (13) directly show how the input at lag i affects the output at time k. This provides explicit control over memory (e.g., learning decaying exponentials, oscillations, delays) and theoretical insight into long-range behavior. Moreover, as discussed in Sect. 4.3, S4 computes the convolution kernel $\boldsymbol{K}[i]$ quickly and efficiently using the Woodbury matrix identity and Cauchy kernel tricks.

5 Mamba

Mamba, proposed in 2023 [5], is a technique that builds on the S4 model [6]. It was introduced to increase efficiency for long sequences by leveraging selective attention mechanisms, allowing the model to save memory and computational cost by ignoring irrelevant information.

Mamba [6] builds on the success of structured state space models (S4) by introducing input-dependent state updates, which allow the model to adaptively filter information. This gating-like mechanism makes it competitive with Transformers in language modeling and enables scaling to millions of tokens efficiently. Mamba 2 [3] further improves hardware efficiency, making it widely adopted in 2024–2025.

The mechanism works by turning parameters that usually remain constant (such as the discretization parameter Δ and others controlling the transition between hidden states) into functions of the input. This extra flexibility means that rather than applying a one-size-fits-all transformation to every token, the model tailors its processing dynamically across the sequence. The result is a network that not only maintains the efficiency of linear scaling but also improves accuracy by being content-aware, even for very long sequences.

An important innovation in Mamba is its use of a hardware-aware algorithm to keep the additional computational cost low. Instead of materializing large intermediate states, the algorithm employs optimized memory strategies. In practice, this enables Mamba to run several times faster than comparable Transformer-based models, especially on longer inputs. A further improved variant of Mamba, with additional hardware optimization, is Mamba 2, proposed in 2024 [3].

In summary, Mamba's selective state space models combine the efficiency of classical SSMs with the flexibility of input-dependent gating. This combination allows the architecture to handle extremely long sequences—up to millions of tokens, such as entire books—while still capturing the nuanced patterns essential for tasks ranging from language modeling to audio generation.

Mamba is an SSM architecture like S4 but differs in how it handles input conditioning, parameterization, and efficiency. Mamba is based on a linear SSM with input-dependent dynamics. The main equations in continuous time are

$$\boldsymbol{h}'(t) = \boldsymbol{A}(\boldsymbol{x}(t))\,\boldsymbol{h}(t) + \boldsymbol{B}(\boldsymbol{x}(t))\,\boldsymbol{x}(t), \tag{18}$$

$$\boldsymbol{y}(t) = \boldsymbol{C}\boldsymbol{h}(t), \tag{19}$$

where the matrices $\boldsymbol{A}(\boldsymbol{x}(t))$ and $\boldsymbol{B}(\boldsymbol{x}(t))$ are input-dependent.

In the discrete case, these equations become

$$\boldsymbol{h}_t = \bar{\boldsymbol{A}}_t\boldsymbol{h}_{t-1} + \bar{\boldsymbol{B}}_t\boldsymbol{x}_t, \tag{20}$$

$$\boldsymbol{y}_t = \bar{\boldsymbol{C}}\boldsymbol{h}_t, \tag{21}$$

where $\bar{\boldsymbol{A}}_t$ and $\bar{\boldsymbol{B}}_t$ depend on the input $\boldsymbol{x}_t$:

$$\bar{\boldsymbol{A}}_t := \bar{\boldsymbol{A}}(\boldsymbol{x}_t), \tag{22}$$

$$\bar{\boldsymbol{B}}_t := \bar{\boldsymbol{B}}(\boldsymbol{x}_t). \tag{23}$$

Comparing these with Eqs. (5) and (6) shows that, in contrast to S4, Mamba uses input-dependent state transition matrix $\bar{\boldsymbol{A}}$ and control matrix $\bar{\boldsymbol{B}}$. This allows Mamba to adapt dynamically to each input token.

Mamba's update per timestep is

$$\boldsymbol{h}_t = \mathbf{f}_{\text{gate}}(\boldsymbol{x}_t) \odot \left(\bar{\boldsymbol{A}}_t\boldsymbol{h}_{t-1} + \bar{\boldsymbol{B}}_t\boldsymbol{x}_t\right), \tag{24}$$

where $\mathbf{f}_{\text{gate}}$ is the gating mechanism (e.g., sigmoid activation function) and $\odot$ is the Hadamard (element-wise) product.

6 TTT Layers: Learning-to-Learn Hidden States

Recall that SSMs are based on a linear time-invariant recurrence (see Eqs. (5) and (6)):

$$\begin{aligned}\boldsymbol{h}_t &= \bar{\boldsymbol{A}}\boldsymbol{h}_{t-1} + \bar{\boldsymbol{B}}\boldsymbol{x}_t,\\ \boldsymbol{y}_t &= \bar{\boldsymbol{C}}\boldsymbol{h}_t.\end{aligned}$$

The key idea in TTT (Test Time Training) layers, proposed in 2024 [10], is to replace the equation $\boldsymbol{h}_t = \bar{\boldsymbol{A}}\boldsymbol{h}_{t-1} + \bar{\boldsymbol{B}}\boldsymbol{x}_t$ with a learnable function, such as a feedforward neural network. In TTT layers, the hidden state is a small learnable model (e.g., a feedforward neural network) that adapts over time:

$$\hat{z}_t = f_t(\boldsymbol{x}_t; \theta_t), \tag{25}$$

$$\ell_t = \mathcal{L}(\hat{z}_t, \tilde{\boldsymbol{x}}_t), \tag{26}$$

$$\theta_{t+1} = \theta_t - \eta \nabla_{\theta_t} \ell_t, \tag{27}$$

where f_t is the hidden state function (neural network) with weights θ_t at time t, $\tilde{\boldsymbol{x}}_t$ is a corrupted input at time t used for self-supervised learning (see Chap. 16), $\hat{z}_t$ is the output of the neural network, $\mathcal{L}(.)$ is the loss function, and η is the learning rate. Equation (27) represents backpropagation (see Chap. 3).

As a result, SSMs can be viewed as a special case of TTT layers where f_t is linear, θ_t is fixed, and no gradient update is applied at test time. TTT generalizes the concept of hidden states to small trainable models that adapt online. This idea of adapting hidden states at test time has become a trend, adopted in works such as DeltaNet [16], Gated DeltaNet [14], and Google's Titans [2]. These architectures suggest that combining SSM principles with online adaptation is a promising future direction.

7 Relation of Attention Mechanism and SSM

It is possible to relate the attention mechanism—introduced in Chapter 9—and the SSM introduced in this chapter [15]. Recall the query $\boldsymbol{q}_t$, key $\boldsymbol{k}_t$, and value $\boldsymbol{v}_t$ in the attention mechanism:

$$\mathbb{R}^p \ni \boldsymbol{q}_t = \boldsymbol{W}_Q^\top \boldsymbol{x}_t, \tag{28}$$

$$\mathbb{R}^p \ni \boldsymbol{k}_t = \boldsymbol{W}_K^\top \boldsymbol{x}_t, \tag{29}$$

$$\mathbb{R}^r \ni \boldsymbol{v}_t = \boldsymbol{W}_V^\top \boldsymbol{x}_t, \tag{30}$$

where $\boldsymbol{W}_Q \in \mathbb{R}^{d\times p}$, $\boldsymbol{W}_K \in \mathbb{R}^{d\times p}$, and $\boldsymbol{W}_V \in \mathbb{R}^{d\times r}$ are the projection matrices for projecting $\boldsymbol{x}_t \in \mathbb{R}^d$ into the low-dimensional query, key, and value subspaces, respectively. The attention mechanism is

$$\mathbb{R}^{r\times n} \ni \boldsymbol{Y} = \boldsymbol{V}\,\text{softmax}(\frac{1}{\sqrt{p}}\boldsymbol{Q}^\top \boldsymbol{K}), \tag{31}$$

where $\boldsymbol{Q} := [\boldsymbol{q}_1, \dots, \boldsymbol{q}_n] \in \mathbb{R}^{p\times n}$, $\boldsymbol{K} := [\boldsymbol{k}_1, \dots, \boldsymbol{k}_n] \in \mathbb{R}^{p\times n}$, and $\boldsymbol{V} := [\boldsymbol{v}_1, \dots, \boldsymbol{v}_n] \in \mathbb{R}^{r\times n}$. Noting the formula of the softmax function (see Chap. 2), an element-wise expression of this equation is

$$\mathbb{R}^r \ni \boldsymbol{y}_t = \frac{\sum_{i=1}^t \exp(\boldsymbol{q}_t^\top \boldsymbol{k}_i)\boldsymbol{v}_i}{\sum_{i=1}^t \exp(\boldsymbol{q}_t^\top \boldsymbol{k}_i)}. \tag{32}$$

Some algorithms, such as linear Transformers (see Chap. 9), replace the softmax function with a kernel function:

$$\exp(\boldsymbol{q}_t^\top \boldsymbol{k}) \approx \boldsymbol{k}(\boldsymbol{q}_t, \boldsymbol{k}_i) = \phi(\boldsymbol{q}_t)^\top \phi(\boldsymbol{k}_i), \tag{33}$$

where it is assumed that $r = p$ (see Chap. 9 for the reason). In this case, Eq. (32) becomes

$$\boldsymbol{y}_t = \frac{\sum_{i=1}^t \phi(\boldsymbol{q}_t)^\top \phi(\boldsymbol{k}_i)\boldsymbol{v}_i}{\sum_{i=1}^t \phi(\boldsymbol{q}_t)^\top \phi(\boldsymbol{k}_i)} \overset{(a)}{=} \frac{\phi(\boldsymbol{q}_t)^\top \sum_{i=1}^t \phi(\boldsymbol{k}_i)\boldsymbol{v}_i^\top}{\phi(\boldsymbol{q}_t)^\top \sum_{i=1}^t \phi(\boldsymbol{k}_i)}, \tag{34}$$

where (a) follows because the factorization is mathematically valid given the vector and matrix dimensions, i.e., $\phi(\boldsymbol{k}_i) \in \mathbb{R}^m$, $\boldsymbol{v}_i \in \mathbb{R}^r$, and $\phi(\boldsymbol{k}_i)\boldsymbol{v}_i^\top \in \mathbb{R}^{m \times r}$. Define the following terms:

$$\boldsymbol{h}_t := \sum_{i=1}^t \phi(\boldsymbol{k}_i)\boldsymbol{v}_i^\top, \tag{35}$$

$$z_t := \sum_{i=1}^t \phi(\boldsymbol{k}_i). \tag{36}$$

These equations can be restated as additive models [8]:

$$\boldsymbol{h}_t = \boldsymbol{h}_{t-1} + \phi(\boldsymbol{k}_t)\boldsymbol{v}_t^\top, \tag{37}$$

$$z_t = z_{t-1} + \phi(\boldsymbol{k}_t). \tag{38}$$

Therefore, Eq. (34) can be restated as follows:

$$\boldsymbol{y}_t = \frac{\phi(\boldsymbol{q}_t)^\top \boldsymbol{h}_t}{\phi(\boldsymbol{q}_t)^\top z_t}. \tag{39}$$

Various kernel functions have been used for approximating the softmax function. Recent work has found that a linear kernel (i.e., setting $\phi(.)$ to be the identity function):

$$\phi(\boldsymbol{q}_t) = \boldsymbol{q}_t, \quad \phi(\boldsymbol{k}_t) = \boldsymbol{k}_t,$$

without a normalizer, works well in practice [11]. Without normalization (i.e., dropping the denominator $\phi(\boldsymbol{q}_t)^\top z_t$) and with a linear kernel, Eq. (39) becomes

$$\boldsymbol{y}_t = \boldsymbol{q}_t^\top \boldsymbol{h}_t \tag{40}$$

Moreover, with a linear kernel, Eq. (37) becomes

$$\boldsymbol{h}_t = \boldsymbol{h}_{t-1} + \boldsymbol{k}_t\boldsymbol{v}_t^\top. \tag{41}$$

In summary, combining Eqs. (40) and (41) gives

$$\boldsymbol{h}_t = \boldsymbol{h}_{t-1} + \boldsymbol{k}_t\boldsymbol{v}_t^\top,$$
$$\boldsymbol{y}_t = \boldsymbol{q}_t^\top \boldsymbol{h}_t.$$

These equations correspond to the state update and output equations in SSMs (see Eqs. (5) and (6)). This shows a deep connection between attention and SSMs [15]: when the softmax in attention is approximated with a kernel (such as in linear Transformers), the update equations reduce to a state space form. This duality suggests that SSMs and attention are not opposing paradigms but different views of the same underlying principles [3, 15].

8 Conclusion

Although Transformers remain the backbone of most state-of-the-art systems, State Space Models (SSMs) have re-emerged as a powerful and competitive alternative. From S4 to Mamba and their successors, SSMs offer linear-time scalability, strong performance on long-sequence and streaming data, and hardware efficiency. By 2025, many research groups are exploring hybrids that combine the strengths of attention and SSMs, suggesting that the future of sequence modeling will not be dominated by a single architecture, but by a diverse set of tools where SSMs play a central role alongside Transformers.

9 Exercises

1. (Level: Easy) Consider a discrete-time linear state space model used in sequence modeling:

$$\begin{aligned} \boldsymbol{x}_{t+1} &= \boldsymbol{A}\boldsymbol{x}_t + \boldsymbol{B}u_t, \\ y_t &= \boldsymbol{C}\boldsymbol{x}_t, \end{aligned}$$

where $\boldsymbol{x}_t \in \mathbb{R}^2$ is the hidden state at time t, $u_t \in \mathbb{R}$ is the input at time t, $y_t \in \mathbb{R}$ is the output at time t, and:

$$\boldsymbol{A} = \begin{bmatrix} 0.5 & 0 \\ 0 & 0.25 \end{bmatrix}, \quad \boldsymbol{B} = \begin{bmatrix} 1 \\ 1 \end{bmatrix}, \quad \boldsymbol{C} = \begin{bmatrix} 1 & 1 \end{bmatrix}.$$

The initial state is $\boldsymbol{x}_0 = [0, 0]^\top$, and the input is $u_t = 1$ for all $t \geq 0$. Compute the outputs y_1, y_2, y_3.
2. (Level: Moderate) Let $\bar{\boldsymbol{A}} = \text{diag}(\boldsymbol{\Lambda}) + \boldsymbol{P}\boldsymbol{Q}^\top \in \mathbb{R}^{s\times s}$ be the state transition matrix in the S4 model, where $\boldsymbol{\Lambda} \in \mathbb{C}^s$ is a vector of eigenvalues, and $\boldsymbol{P}, \boldsymbol{Q} \in \mathbb{C}^{s\times r}$ with $r \ll s$.
Using the Woodbury matrix identity, derive the expression for $(z\boldsymbol{I} - \bar{\boldsymbol{A}})^{-1}$, where $z \in \mathbb{C}$ and $\boldsymbol{I}$ is the identity matrix. Then, using this, write the expression for the transfer function:

$$H(z) = \bar{\boldsymbol{C}}(z\boldsymbol{I} - \bar{\boldsymbol{A}})^{-1}\bar{\boldsymbol{B}},$$

in terms of only $\boldsymbol{\Lambda}$, $\boldsymbol{P}$, $\boldsymbol{Q}$, $\bar{\boldsymbol{B}}$, $\bar{\boldsymbol{C}}$, z, where $\bar{\boldsymbol{A}}$, $\bar{\boldsymbol{B}}$, and $\bar{\boldsymbol{C}}$ are the discretized state transition matrix, control matrix, and input matrix, respectively.
3. (Level: Moderate) Given the diagonal state matrix $\boldsymbol{\Lambda} \in \mathbb{C}^s$ (i.e., $\boldsymbol{A} = \text{diag}(\boldsymbol{\Lambda})$) and the control matrix $\boldsymbol{B} \in \mathbb{C}^s$, derive the expression for the convolution kernel $\boldsymbol{K}(t) \in \mathbb{C}$, where

$$\boldsymbol{K}[t] = \boldsymbol{C}\boldsymbol{A}^t\boldsymbol{B},$$

for some output matrix $\boldsymbol{C} \in \mathbb{C}^{1\times s}$, and $t \in \mathbb{N}$ is the discrete time step. Show how this formulation allows computing all $\boldsymbol{K}[t]$ for $t = 0, 1, \ldots, L-1$ in $\mathcal{O}(L \log L)$ time using the Fast Fourier Transform (FFT).

4. (Level: Easy) Consider the Mamba model's discrete-time state update:

$$\boldsymbol{h}_t = f_{\text{gate}}(\boldsymbol{x}_t) \odot (\boldsymbol{A}_t \boldsymbol{h}_{t-1} + \boldsymbol{B}_t \boldsymbol{x}_t),$$

where $\boldsymbol{A}_t = \boldsymbol{A}(\boldsymbol{x}_t)$, $\boldsymbol{B}_t = \boldsymbol{B}(\boldsymbol{x}_t)$, and $f_{\text{gate}}(\boldsymbol{x}_t) = \sigma(\boldsymbol{x}_t)$ is an element-wise sigmoid gating function. Assume $\boldsymbol{x}_t \in \mathbb{R}^2$, $\boldsymbol{h}_t \in \mathbb{R}^2$, $\boldsymbol{A}_t = \textbf{diag}(\boldsymbol{x}_{t,1}, \boldsymbol{x}_{t,2})$, and $\boldsymbol{B}_t = \boldsymbol{I}_2$ (the 2×2 identity matrix). If

$$\boldsymbol{x}_t = \begin{bmatrix} 1 \\ -2 \end{bmatrix}, \quad \boldsymbol{h}_{t-1} = \begin{bmatrix} 2 \\ 1 \end{bmatrix},$$

compute $\boldsymbol{h}_t$.

5. (Level: Moderate) In traditional linear time-invariant (LTI) state space models, the discrete-time update rule is

$$\boldsymbol{h}_t = \boldsymbol{A}\boldsymbol{h}_{t-1} + \boldsymbol{B}\boldsymbol{x}_t,$$

where $\boldsymbol{A} \in \mathbb{R}^{n\times n}$, $\boldsymbol{B} \in \mathbb{R}^{n\times d}$, and both matrices are fixed across time.
Mamba instead uses an input-dependent rule:

$$\boldsymbol{h}_t = f_{\text{gate}}(\boldsymbol{x}_t) \odot (\boldsymbol{A}_t \boldsymbol{h}_{t-1} + \boldsymbol{B}_t \boldsymbol{x}_t), \quad \text{where } \boldsymbol{A}_t = \boldsymbol{A}(\boldsymbol{x}_t),\ \boldsymbol{B}_t = \boldsymbol{B}(\boldsymbol{x}_t).$$

Assume:

- $\boldsymbol{x}_t = \begin{bmatrix} 0 \\ 1 \end{bmatrix}, \boldsymbol{h}_{t-1} = \begin{bmatrix} 1 \\ 2 \end{bmatrix}$.
- LTI case: $\boldsymbol{A} = \begin{bmatrix} 1 & 0 \\ 0 & 1 \end{bmatrix}, \boldsymbol{B} = \begin{bmatrix} 0 & 1 \\ 1 & 0 \end{bmatrix}$.
- Mamba case: $\boldsymbol{A}_t = \text{diag}(\tanh(\boldsymbol{x}_{t,1}), \tanh(\boldsymbol{x}_{t,2}))$, $\boldsymbol{B}_t = \boldsymbol{I}_2$, $f_{\text{gate}}(\boldsymbol{x}_t) = \sigma(\boldsymbol{x}_t)$ element-wise sigmoid function.

Compute and compare the outputs $\boldsymbol{h}_t$ for both the LTI and Mamba update. What does this illustrate about Mamba's input-adaptivity?

6. (Level: Moderate) Consider a TTT (Test Time Training) layer defined by the following update at each time step t:

$$\boldsymbol{\theta}_{t+1} = \boldsymbol{\theta}_t - \eta \nabla_{\boldsymbol{\theta}_t} \ell_t,$$

where

$$\ell_t = \mathcal{L}\big(f_t(\boldsymbol{x}_t; \boldsymbol{\theta}_t), \tilde{\boldsymbol{x}}_t\big),$$

and $f_t(\boldsymbol{x}_t; \boldsymbol{\theta}_t)$ is a neural network with parameters $\boldsymbol{\theta}_t$, input $\boldsymbol{x}_t$, and corrupted input $\tilde{\boldsymbol{x}}_t$.
Suppose at time t, the loss is a simple squared error:

$$\ell_t = \frac{1}{2}\| f_t(\boldsymbol{x}_t; \boldsymbol{\theta}_t) - \boldsymbol{x}_t \|_2^2,$$

and the network output is a linear function $f_t(\boldsymbol{x}_t; \boldsymbol{\theta}_t) = \boldsymbol{\theta}_t^\top \boldsymbol{x}_t$, where $\boldsymbol{\theta}_t, \boldsymbol{x}_t, \tilde{\boldsymbol{x}}_t \in \mathbb{R}^d$.

(a) Write the gradient $\nabla_{\boldsymbol{\theta}_t} \ell_t$ explicitly.
(b) Write the update equation for $\boldsymbol{\theta}_{t+1}$.

7. (Level: Moderate) Suppose an attention mechanism replaces the softmax with a linear kernel, i.e., $\phi(\boldsymbol{q}) = \boldsymbol{q}$ and $\phi(\boldsymbol{k}) = \boldsymbol{k}$, and ignores the normalization denominator. Define the following recursions:

$$\boldsymbol{h}_t = \boldsymbol{h}_{t-1} + \boldsymbol{k}_t \boldsymbol{v}_t^\top, \quad \boldsymbol{y}_t = \boldsymbol{q}_t^\top \boldsymbol{h}_t,$$

where $\boldsymbol{q}_t, \boldsymbol{k}_t \in \mathbb{R}^m$, $\boldsymbol{v}_t \in \mathbb{R}^r$, and $\boldsymbol{h}_t \in \mathbb{R}^{m \times r}$.
(a) Show that $\boldsymbol{y}_t$ can be written as a sum of t outer product terms involving the past key and value pairs weighted by the current query.
(b) Explain how this recursion is equivalent to a first-order discrete-time state space model with additive input updates.

References

1. Malyaban Bal and Abhronil Sengupta. "P-SpikeSSM: Harnessing Probabilistic Spiking State Space Models for Long-Range Dependency Tasks". In: *Workshop on Machine Learning and Compression, NeurIPS 2024*. 2024.
2. Ali Behrouz, Peilin Zhong, and Vahab Mirrokni. "Titans: Learning to memorize at test time". In: *arXiv preprint* arXiv:2501.00663 (2024).
3. Tri Dao and Albert Gu. "Transformers are SSMs: generalized models and efficient algorithms through structured state space duality". In: *Proceedings of the 41st International Conference on Machine Learning*. 2024, pp. 10041–10071.
4. Daniel Y Fu et al. "Hungry Hungry Hippos: Towards Language Modeling with State Space Models". In: *The Eleventh International Conference on Learning Representations*. 2023.
5. Albert Gu and Tri Dao. "Mamba: Linear-Time Sequence Modeling with Selective State Spaces". In: *First Conference on Language Modeling*. 2024.
6. Albert Gu, Karan Goel, and Christopher Re. "Efficiently Modeling Long Sequences with Structured State Spaces". In: *International Conference on Learning Representations*. 2022.
7. Ankit Gupta, Albert Gu, and Jonathan Berant. "Diagonal state spaces are as effective as structured state spaces". In: *Advances in Neural Information Processing Systems* 35 (2022), pp. 22982–22994.
8. Trevor J Hastie and Robert Tibshirani. "Generalized Additive Models". In: *Statistical Science* 1.3 (1986), pp. 297–318.
9. Katsuhiko Ogata. *Modern Control Engineering*. 5th ed. Prentice Hall, 2010.
10. Yu Sun et al. "Learning to (learn at test time): Rnns with expressive hidden states". In: *arXiv preprint* arXiv:2407.04620 (2024).
11. Yutao Sun et al. "Retentive Network: A Successor to Transformer for Large Language Models". In: *International Conference on Learning Representations*. 2024.
12. Arnold Tustin. "A method of analysing the behaviour of linear systems in terms of time series". In: *Journal of the Institution of Electrical Engineers-Part IIA: Automatic Regulators and Servo Mechanisms* 94.1 (1947), pp. 130–142.
13. Ashish Vaswani et al. "Attention is all you need". In: *Advances in neural information processing systems*. 2017, pp. 5998–6008.
14. Songlin Yang, Jan Kautz, and Ali Hatamizadeh. "Gated Delta Networks: Improving Mamba2 with Delta Rule". In: *arXiv preprint* arXiv:2412.06464 (2024).
15. Songlin Yang, Bailin Wang, Yikang Shen, Rameswar Panda, and Yoon Kim. "Gated Linear Attention Transformers with Hardware-Efficient Training". In: *International Conference on Machine Learning*. PMLR. 2024, pp. 56501–56523.
16. Songlin Yang, Bailin Wang, Yu Zhang, Yikang Shen, and Yoon Kim. "Parallelizing Linear Transformers with the Delta Rule over Sequence Length". In: *The Thirty-eighth Annual Conference on Neural Information Processing Systems*. 2024.

11 Large Language Models

Abstract

This chapter provides an in-depth overview of prominent Transformer-based language models, focusing primarily on BERT (Bidirectional Encoder Representations from Transformers) and GPT (Generative Pre-trained Transformer), two foundational architectures that have significantly shaped modern Natural Language Processing (NLP) and Large Language Models (LLMs). BERT, developed by Google, is no longer state of the art by 2025; however, it remains influential for its use of stacked Transformer encoders to produce context-aware token embeddings and for its strong performance on diverse NLP tasks, such as sentiment analysis and question answering. GPT, introduced by OpenAI, is a generative model built on Transformer decoders and serves as the backbone for many state-of-the-art LLMs. The chapter further explores T5, a unified text-to-text model that bridges the design philosophies of BERT and GPT by casting diverse NLP problems into a single framework. Key advancements such as chain-of-thought prompting are also discussed to illustrate emerging techniques in language understanding and generation. Additionally, the chapter surveys notable LLMs and examines research efforts aimed at explaining why LLMs perform so effectively. It then addresses the critical topic of alignment—methods to ensure model outputs align with human values and intentions—detailing strategies such as Supervised Fine-Tuning, Reward Modeling, Reinforcement Learning from Human Feedback (RLHF), and Direct Preference Optimization (DPO). Familiarity with the attention mechanism and Transformers is recommended to fully benefit from this chapter. Note that although large language models are generative models, they are presented in this chapter—following the discussion of the attention mechanism—for improved readability and flow.

1 Introduction

The attention mechanism and Transformers were introduced in Chap. 9. It is recommended to study that chapter before reading this chapter. Although large language models are generative models (covered in Part 3 of the book), they are presented in this chapter—following the discussion of the attention mechanism—for better readability and flow. BERT (Bidirectional Encoder Representations from Transformers) [23] and GPT (Generative Pre-trained Transformer) [78] are both Transformer-based models. They were proposed in 2018 following the introduction of the Transformer in 2017 [98] (see Chap. 9). BERT was proposed by Google, while GPT was proposed by OpenAI. Roughly speaking, BERT is a stack of Transformer encoders, whereas GPT is a stack of Transformer decoders. The main purposes

B. Ghojogh and A. Ghodsi, *Elements of Deep Learning*,
https://doi.org/10.1007/978-3-032-10738-1_11

of BERT and GPT differ. The goal of BERT is to learn context-aware representation embeddings for tokens (words) in sentences, while GPT was proposed to generate text as a generative language model.

These two models serve as backbones for many state-of-the-art Natural Language Processing (NLP) models. BERT has been widely used for various NLP tasks, such as sentiment analysis and question answering. GPT is the backbone of many Large Language Models (LLMs). In fact, most state-of-the-art LLMs are GPT-style models.

This chapter begins with BERT [23] in Sect. 2 and GPT [78] in Sect. 3, as the fundamental models in LLMs. Chain-of-thought prompting is introduced in Sect. 3.3. Then, the T5 model [78] is presented in Sect. 4 as a combination of BERT and GPT, casting NLP tasks as text-to-text problems. Thereafter, important LLMs are listed in Sect. 5 to introduce these models to the reader. Section 6 then discusses lines of research addressing why LLMs work. Subsequently, alignment in LLMs is defined and its strategies are explained in Sect. 7. An approach for aligning LLMs [71] using Supervised Fine-Tuning (SFT), Reward Modeling (RM) [104], and Reinforcement Learning from Human Feedback (RLHF)[1] [18] is explained in Sect. 8. Finally, Direct Preference Optimization (DPO), another method for aligning LLMs, is introduced in Sect. 9.

2 BERT: Bidirectional Encoder Representations from Transformers

BERT [23] is one of the foundational methods for NLP. Although BERT was foundational and remains useful, it is no longer considered state of the art by 2025. Modern models such as RoBERTa [89], DeBERTa [40], and large-scale generative transformers have surpassed BERT in benchmarks. BERT remains one of the useful models for various NLP tasks, including question answering [77], natural language understanding [24], sentiment analysis, and language inference [88]. The goal of BERT is to provide context-aware embedding vectors for the tokens (words) in sequences (sentences).

2.1 Structure of BERT Network

BERT is essentially a stack of Transformer encoders (see Chap. 9), as illustrated in Fig. 1a. In other words, it is built using Transformer encoder blocks.

2.2 Masking and Training in BERT

BERT uses the technique of masked language modeling. BERT typically masks 15% of tokens, but improvements in successors such as RoBERTa [89] included dynamic masking (reshuffling which tokens are masked in each epoch) and removing the Next Sentence Prediction task, leading to stronger performance.

BERT masks some of the words in the input document or corpus and asks the model to predict the missing words. As Fig. 2 depicts, a sentence with a missing word is given to each Transformer encoding block in the stack, and the block is supposed to predict the missing word. BERT's training is self-supervised because it involves masking random words in a sentence and training the model to predict these masked words using the surrounding context. Since it is self-supervised and does not require labels, the vast text data available on the Internet can be used for training the BERT model,

[1] RLHF requires knowledge of reinforcement learning, so it is recommended that the reader studies Chap. 18 before reading Sect. 8.

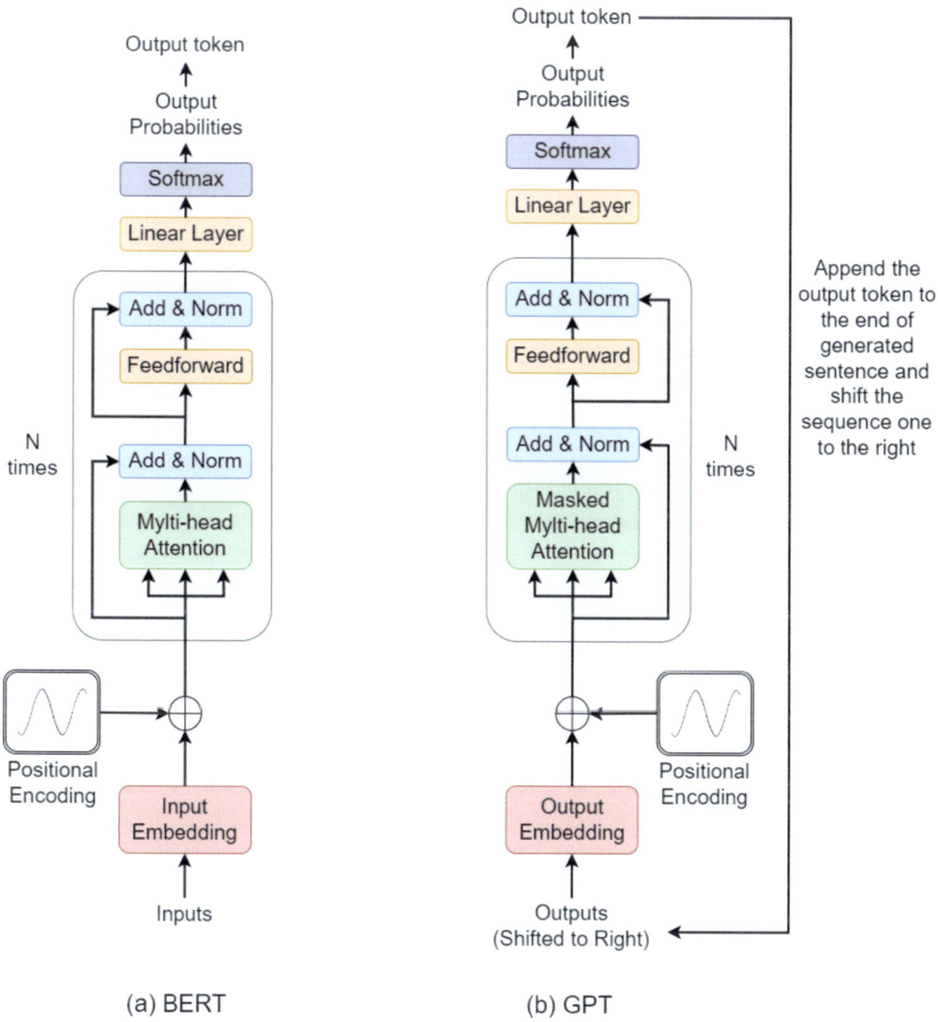

Fig. 1 The architectures of **a** BERT and **b** GPT networks

where words are randomly selected to be masked. In other words, BERT is pre-trained on unlabeled text to capture bidirectional context in language.

Note that BERT learns to predict the missing word based on attention to both its previous and forthcoming words, making it bidirectional. Therefore, BERT jointly conditions on both the left (previous) and right (forthcoming) contexts of every word. Since the missing word is predicted based on the other words in the sentence, BERT embeddings for words are context-aware. Thus, in contrast to Word2Vec [31, 60–62] and GloVe [75], which provide a single embedding per word, each word has different BERT embeddings in various sentences. The BERT embeddings of words differ across sentences based on their context. For example, the word "bank" has different meanings and therefore different embeddings in the sentences "Money is in the bank" and "Some plants grow on the banks of rivers."

As explained, BERT learns embeddings for tokens (words) by masking them during training and predicting the missing token (word). In addition to learning embeddings for individual words and entire

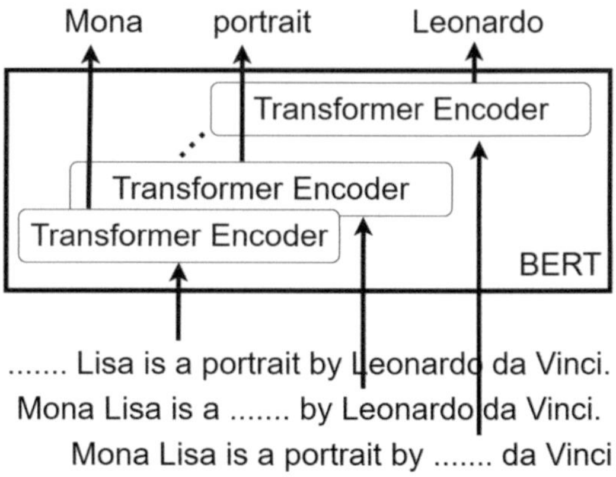

Fig. 2 Feeding sentences with missing words to the BERT model for training

sentences, BERT [23] has also been trained for an additional task. In this task, given two sentences A and B, it predicts whether B is likely to be the sentence that follows A.

2.3 Fine-Tuning BERT

The BERT model is typically not trained from scratch by end users, as its initial training requires significant time and vast amounts of Internet-scale text data. Instead, for various NLP applications such as sentiment analysis, researchers usually apply transfer learning by adding one or more output layers on top of a pre-trained BERT model (i.e., they remove the softmax layer and add new layer(s) with appropriate activation functions on top of the architecture shown in Fig. 1a), and then train the network for their specific task. During this process, one can either freeze the weights of the BERT model and train only the added layer(s) or fine-tune the entire model by backpropagation. Using and fine-tuning BERT generally leads to state-of-the-art performance on a wide range of tasks, such as question answering and natural language inference.

2.4 Sentence Embedding in BERT

It is also noteworthy that, for an input sentence, BERT outputs embeddings for each word as well as a special embedding that represents the entire sentence. Although the [CLS] token embedding may not capture all semantic nuances, it often performs well in practical applications. For example, one can use BERT's sentence embeddings to train a classifier for tasks like spam detection or sentiment analysis.

More specifically, a special [CLS] token is prepended to the input text and passes through the Transformer layers alongside the other tokens. Due to the self-attention mechanism, all tokens, including [CLS], incorporate contextual information from the entire sentence. For sentence-level tasks, the final hidden state of the [CLS] token is used as the sentence representation. During fine-tuning, the model learns to make this representation meaningful and task-specific.

In classification tasks, this [CLS] token representation is typically passed into a classifier to predict the sentence's label.

2.5 Variants of BERT

The original Transformer encoder [98] consists of 6 encoder layers, each with 512 hidden units in the feedforward network and 8 attention heads ($h = 8$). In contrast, BERT [23] comes in two main configurations. BERT Large has 24 encoder layers, 1024 hidden units in the feedforward network, and 16 attention heads ($h = 16$), with a total of 340 million parameters. BERT Base has 12 encoder layers, 768 hidden units, and 12 attention heads ($h = 12$), totaling 110 million parameters.

As the BERT model is large and requires significant memory for storage, it cannot easily be used in embedded systems such as cell phones, which have limited memory and CPU resources. Therefore, many commercial smaller versions of BERT have been proposed with fewer parameters and stacks. These smaller models are usually trained using knowledge distillation from BERT Large or BERT Base (see Chap. 20 for more information on knowledge distillation). Some of these smaller versions of BERT include SmallBERT [95], TinyBERT [45], and DistilBERT [83]. These models are smaller and faster versions of BERT designed for resource-constrained environments, retaining competitive performance with significantly fewer parameters.

RoBERTa [89] is not simply a smaller model, but a re-trained variant of BERT with modifications to the pretraining procedure. It optimizes BERT's training process by using more data, larger batch sizes, dynamic masking, and longer training times, resulting in improved performance on NLP tasks.

There is also a multilingual BERT model [76], which is trained on 104 different languages and is capable of zero-shot adaptation to new language domains (see Chap. 19 for more information on few-shot learning and domain adaptation).

Additionally, there are several domain-specific BERT variants. For example, Clinical BERT [3] and BioBERT [52] have been trained on medical texts for biomedical applications. ProteinBERT [11] has been trained on protein sequences to encode amino acid sequences. SciBERT [8] is trained on over one million published articles. BERTweet [64] is a RoBERTa model trained on 850 million tweets. Finally, FinBERT [43, 56] is a BERT model adapted to the financial domain.

3 GPT: Generative Pre-trained Transformer

GPT [78] is another state-of-the-art method for NLP. The goal of GPT is to predict the next token (word) given all previous tokens (words) in the sequence (sentence). GPT has been used in many chatbots for generating text and conversing with users.

3.1 Structure of GPT Network

GPT is essentially a stack of Transformer decoder blocks (see Chap. 9), as shown in Fig. 1-b. In other words, it is constructed using only the decoder components of the Transformer architecture. However, in GPT, the multihead attention module that performs cross-attention (typically used to attend to encoder outputs) is removed, since GPT does not include an encoder (compare Fig. 1a in this chapter with Fig. 9 in Chap. 9). As a result, the decoder blocks in GPT consist of positional encoding, a masked multihead self-attention module, and a feedforward network—each followed by addition, layer normalization, and activation functions.

As GPT is a generative language model that generates sequences, it only has access to its previous (left) words and not the next (right) words that will be generated. Therefore, GPT is not bidirectional—it processes text in a left-to-right manner and conditions only on the previous (left) tokens, not the future (right) ones. To enforce this, GPT uses masked multihead self-attention, which restricts each token's attention to itself and the tokens before it, preventing access to any future context.

3.2 Sequence Generation by GPT

As explained earlier, BERT is trained using masked language modeling, where random words in a sentence are masked and the model learns to predict them using context from both directions. In contrast, GPT is trained using causal (autoregressive) language modeling, where the goal is to predict the next word (token) in a sequence, given all previous words (tokens).

For example, consider the input sequence: "A storm was approaching." GPT takes this sequence and computes a probability distribution over the entire vocabulary for the next word. Likely candidates such as "quickly," "rapidly," or "from" would receive higher probabilities, while unrelated words like "banana" or "calculator" would receive low probabilities. If the model predicts "quickly," the sequence becomes "A storm was approaching quickly." This extended input is then used to predict the next word, and the process repeats. Since this training only requires raw text (without labels), GPT is trained in a self-supervised manner, allowing it to leverage large volumes of text data from the Internet.

3.3 Chain-of-Thought Prompting and Training

The paper [99], proposed in 2022, introduced Chain-of-Thought (CoT) prompting, a method that improves the reasoning ability of LLMs. Its key idea is as follows. Instead of asking the model to directly give an answer, CoT prompting encourages the model to "think out loud" by showing examples where intermediate reasoning steps are written out. CoT makes the LLM think and reason as a human before answering. An example of the LLM thinking out loud is

- A standard prompt is
  Question: If there are 3 apples and you eat 1, how many are left?
  Answer: 2.
- A CoT prompt is
  Question: If there are 3 apples and you eat 1, how many are left?
  Answer: There are 3 apples. If you eat 1, then $3 - 1 = 2$ apples are left. So, the answer is 2.

CoT prompting changes the way users ask the model to answer by including examples of step-by-step reasoning in the prompt. Instead of asking just the question, the prompt includes examples where both the question and the full reasoning process (not just the answer) are shown. Then the model is given a new question and is expected to follow the same reasoning style. This behavior is expected because LLMs demonstrate few-shot or zero-shot learning, even when such behavior is not explicitly included in their training. For example, consider the following standard and CoT prompts:

- A standard prompt is
  Question: A train travels 100 mi at 50 mph. How long does it take?
  Answer: 2.
- In CoT prompting, the user provides the concatenation of the following question and answer as a prompt to the LLM:
  Question: A train travels 100 mi at 50 mph. How long does it take?
  Answer: The train needs to cover 100 mi. It's going 50 mi per hour. Time = Distance/Speed = 100/50 = 2 h. So the answer is 2.

Then, the user asks the LLM the same or a similar question without an answer concatenated to it: "A car travels 60 mi at 30 mph. How long does it take?" The LLM then answers with intermediate steps of reasoning.

CoT prompting significantly improves performance on complex reasoning tasks such as math word problems and logic puzzles. It works especially well with larger models (e.g., GPT-3 with 100+ billion parameters), showing minimal benefits for smaller models. This demonstrates that LLMs can perform better when prompted in ways that guide them to reason step by step.

Following the success of CoT prompting [99], researchers were encouraged to incorporate CoT in training [42] so that LLMs can reason in a chain of thought before answering. To achieve this, questions and their step-by-step answers were included in the training dataset. Models such as GPT-o3 (OpenAI o3) have been trained with CoT.

3.4 Variants of GPT

3.4.1 GPT-1

The first Generative Pre-trained Transformer (GPT-1) was introduced in 2018 [78]. It uses a 12-layer decoder-only Transformer architecture and contains 117 million parameters. GPT-1 was trained on the BookCorpus dataset, which consists of approximately 7,000 unpublished books. Its main contributions were demonstrating the effectiveness of unsupervised pretraining on large-scale text data and showing that a language model fine-tuned on specific tasks can achieve strong performance with minimal task-specific architectural changes.

3.4.2 GPT-2

The successor to GPT-1 [78] was GPT-2, released in 2019 [79]. GPT-2 extended GPT-1 by significantly increasing the number of Transformer decoder layers, resulting in more learnable parameters and improved capacity for language modeling and inference.

GPT-2 has approximately 1.5 billion parameters and 48 layers in its largest version. It was trained on a much larger and more diverse dataset, called WebText, containing around 8 million web pages and totaling about 40 GB of Internet text. This dataset covers a wide range of general and academic topics, allowing GPT-2 to generate coherent and relevant text across various subjects and styles. The model continued to emphasize the Transformer architecture and self-attention mechanism.

Due to its large size and memory requirements, GPT-2 is not well suited for deployment on embedded systems. To address this, OpenAI released smaller variants—such as GPT-2 Small, Medium, Large, and XL—with fewer layers and parameters. Additionally, DistilGPT-2, a smaller distilled version, was developed for lightweight applications. These versions are publicly available through the Hugging Face Transformers Python library [100].

GPT-2 has been applied in various domains, including dialogue systems [14], patent claim generation [51], and medical text simplification [97]. It has also been combined with BERT for tasks such as question answering [47]. A detailed comparison between GPT and BERT can be found in [28].

3.4.3 GPT-3

GPT-3 was a follow up to the GPT-2 model and was released in 2020 [13]. GPT-3 is a much larger version of GPT, with a significantly greater number of stacks and learnable parameters. For comparison, note that GPT-2, NVIDIA Megatron [87], Microsoft Turing-NLG [59], and GPT-3 [13] have 1.5 billion, 8 billion, 17 billion, and 175 billion learnable parameters, respectively. This large number of parameters allows GPT-3 to be trained on vast amounts of Internet text data covering a wide range of subjects and topics. Thus, GPT-3 has been able to learn almost all topics present in its training data, leading some to discuss whether it can pass Turing's writer's test [27, 29]. Note that GPT-3 has essentially memorized texts across all subjects, but not in a negative way, such as overfitting (see Chap. 5). Rather, this memorization results from the complexity afforded by its huge number of learnable parameters [6], and it avoids overfitting by being trained on sufficiently large Internet-scale data.

GPT-3 consists of up to 96 Transformer decoder layers in its largest version, with 175 billion parameters. Its training dataset comprised 570 GB of multilingual texts. The focus of GPT-3 was on few-shot learning, prompt engineering, and Python support. GPT-3 has had many interesting applications, such as fiction and poetry generation [12]. In this regard, there are discussions on the possible risks associated with the GPT-3 model [57].

3.4.4 GPT-4

GPT-4 [1] was released after GPT-3, but detailed architectural information has not been officially disclosed. It is believed that GPT-4 has significantly more parameters and layers than GPT-3. Claims of GPT-4 having 100 trillion parameters are speculative and not supported by evidence. OpenAI has not disclosed the parameter count, but reliable estimates place it well below that figure. What is known is that GPT-4 was trained on a larger and more diverse dataset than its predecessors.

GPT-4 is a multi-modal model, capable of processing both text and image inputs to generate text outputs. It demonstrates improvements in few-shot learning, reasoning, inference, and both Natural Language Understanding (NLU) and Natural Language Generation (NLG). GPT-3 and GPT-4 are the backbone models behind OpenAI's ChatGPT product [66].

3.4.5 GPT-4o, OpenAI o1, OpenAI o3, and OpenAI o4-Mini

Following the success of CoT (Chain-of-Thought) prompting—introduced in Sect. 3.3—CoT was incorporated into training by including questions and their step-by-step answers. CoT enables the LLM to think and reason as a human before answering. Models such as GPT-4o, OpenAI o1, OpenAI o3, and OpenAI o4-mini are GPT models that were developed using CoT training.

4 T5: Text-to-Text Transfer Transformer

After the introduction of BERT and GPT in 2018, a model called T5 (Text-to-Text Transfer Transformer) was proposed in 2020 [81]. T5 combines both a Transformer encoder and a Transformer decoder, making it a full encoder–decoder architecture—unlike BERT, which uses only the encoder, and GPT, which uses only the decoder.

4.1 NLP Tasks as Text-to-Text Problems

T5 is a unified framework for addressing NLP tasks by converting them into a text-to-text format. It casts any NLP task as a text-to-text problem, generating text2 after text1 as "text1 text2," effectively converting text1 to text2. For example:

- Sentiment analysis in NLP refers to determining the emotional tone or opinion in text (e.g., classifying text as positive or negative). This can be converted into a text-to-text problem:
  - Classifying "I like this product" as positive can be converted to "I like this product positive."
  - Classifying "This product is always having problems" as negative can be converted to "This product is always having problems negative."

- Translation from one language to another can also be converted into a text-to-text problem. For example:
  - Translating the English sentence "I am a student" to French can be converted to "I am a student Je suis étudiant."
  - Translating the French sentence "Je suis étudiant" to English can be converted to "Je suis étudiant I am a student."

4.2 Training of T5

Training of T5 is a combination of supervised and unsupervised (or self-supervised) learning. Supervised training is conducted on downstream tasks provided by the GLUE and SuperGLUE benchmarks, reformatted into text-to-text tasks. Its unsupervised (or self-supervised) training is similar to that of BERT. It employs corrupted tokens, i.e., 15% of tokens are randomly removed and replaced with sentinel tokens.

5 Important Large Language Models

A list of significant LLMs in industry and academia is provided below. These models are language models that generate text, with many being GPT-style next-token predictors. The list is organized roughly in the order of their release, as of mid-2025:

- By OpenAI:
  - GPT-3 (2020) [13]: Introduced with 175 billion parameters. GPT-3 marked a significant advancement in language modeling.
  - GPT-3.5 (2022) [68]: An intermediate version between GPT-3 and GPT-4. GPT-3.5 incorporated refinements and was used in the initial release of ChatGPT in November 2022, often referred to as "ChatGPT" [66] or "GPT-3.5-turbo."
  - GPT-4 (2023) [1]: A multi-modal model capable of processing both text and image inputs to generate text outputs. GPT-4 exhibits human-level performance on various professional and academic benchmarks.
  - GPT-4 Turbo [70] and GPT-4o [67] (2024): Enhanced versions of GPT-4, offering improved performance and efficiency.
  - ChatGPT [66]: A product built on GPT-3.5 or GPT-4, providing conversational AI capabilities. It is not a separate model but utilizes the underlying GPT architecture.
  - GPT-4o (2024), OpenAI o1 (2024), OpenAI o3 (2025), and OpenAI o4-mini (2025): After success of CoT (Chain-of-Thought) prompting—introduced in Sect. 3.3—CoT was used in training by including questions and step-by-step answers to the questions. OpenAI's latest models include GPT-4o, OpenAI o1, o3, and o4-mini. These models extend GPT-style training with CoT reasoning, and GPT-4o in particular added multi-modal capabilities with real-time speech and vision.
- By Meta (Facebook): LLaMA (Large Language Model Meta AI) Series [94]; Meta's series of open-weight models available in various sizes:
  - RAG (Retrieval-Augmented Generation) (2020) [53]: Includes a (1) retriever which searches a large corpus (e.g., Wikipedia, documents, databases) to find relevant information based on a query,

  and (2) a generator which is a language model (e.g., GPT) that uses the retrieved documents as context to generate the final response.
  - LLaMA 1 (2023) [94]: Introduced as an open and efficient foundation language model. It lacked the ability to follow instructions and prompts.
  - LLaMA 2 (2023) [93]: Enhanced version with more training data and fine-tuned chat capabilities. It introduced the ability to follow prompts and was available in 7B, 13B, and 70B parameter sizes.
  - Code LLaMA (2023) [82]: A code-specialized version of LLaMA 2, trained on a large dataset of code, including GitHub repositories. It supports multiple programming languages—including Python, C++, and Java—and is available in 7B, 13B, and 34B parameter sizes.
  - LLaMA 3 (2024) [58]: Released with 8B and 70B parameter sizes, LLaMA 3 demonstrated improvements in performance and scalability, making it competitive with GPT-4 [1].
  - LLaMA 3.1 (2024) [33]: Released in July 2024 with 8B, 70B, and 405B parameter sizes, LLaMA 3.1 further advanced the capabilities of the LLaMA family by significantly extending context length (up to 128k tokens) and improving efficiency across reasoning, coding, and multilingual tasks. It is widely adopted in open-source research.

- By Stanford University; Center for Research on Foundation Models (CRFM):
  - Alpaca (2023) [90]: A fine-tuned version of LLaMA 7B [94], trained on 52K instruction-following samples. Alpaca demonstrated capabilities comparable to OpenAI's text-davinci-003 but was more cost-effective.

- By LAION (Large-scale Artificial Intelligence Open Network):
  - OpenAssistant (2023) [48]: An open-source, open-weight, and open-data chatbot designed to democratize language technologies. It is based on fine-tuning LLaMA and other open LLMs using human feedback and reinforcement learning from human feedback (RLHF) pipelines (see Sect. 7).

- By Technology Innovation Institute (TII):
  - Falcon (2023) [73]: A family of open-source, open-weight, and open-data models trained on high-quality web data. Falcon models are available in 7B, 40B, and 180B parameter sizes. The Falcon 180B model was trained on 3.5 trillion tokens.

- By Anthropic:
  - Claude Series (2023–2024) [5]: Anthropic's GPT-style models with a focus on safety. The series includes Claude 1, Claude 2, and Claude 3, each showing improvements in language understanding and generation.

- By Google DeepMind:
  - T5 (Text-to-Text Transfer Transformer) (2020) [81]: A sequence-to-sequence model, introduced in Sect. 4, that treats all NLP tasks as text-to-text problems.
  - AlphaCode (2022) [54]: A model focused on competitive programming, trained on public GitHub codes.
  - PaLM (Pathways Language Model) Series:
    - PaLM (Pathways Language Model) (2022) [17]: The first large-scale model trained using Google's Pathways system [21], excelling in reasoning, code generation, and few-shot learning.

  - PaLM 2 (2023) [4]: Improved version with enhanced multilingual understanding, reasoning, and coding.

- Gemini Series:

  - Gemini 1.0 (2023) [92]: Introduced native multi-modality, supporting text, code, image, audio, and video.
  - Gemini 2.0 (2024) [22]: Enhanced multi-modal capabilities, including image and audio output.
  - Gemini 2.5 (2025) [32]: Introduced "Deep Think"—a reasoning mode for complex tasks, and emotional dialogue in audio output.

- By Mistral AI:

  - Mistral 7B and Mixtral 8×7B (2023) [63]: Open-weight GPT-style models widely used in open-source applications.

- By Alibaba: Qwen Series:

  - Qwen 1.5 and 2 (2023–2024) [2]: Open-weight GPT-style models developed by Alibaba.
  - Qwen 3 (2025) [102]: A Mixture-of-Experts (MoE) model [16] with 235 billion parameters, designed for efficiency and scalability.

- By DeepSeek: DeepSeek Series:

  - DeepSeek-V2 (2024) [10]: A general-purpose open-weight LLM with 16B parameters, trained from scratch.
  - DeepSeek-V3 (2024) [55]: An improved instruction-following model with better alignment tuning (see Sect. 7) and support for English and Chinese.
  - DeepSeek-Coder (2024) [37]: A code-specialized model fine-tuned on 2 trillion tokens for code generation, completion, and debugging.
  - DeepSeek-MoE (2024) [20]: A Mixture-of-Experts (MoE) variant [16] of DeepSeek.
  - DeepSeekMath (2024) [86]: Specialized in mathematical reasoning, trained on math-related corpora.
  - DeepSeek-R1 (2025) [38]: A reasoning-optimized LLM trained via reinforcement learning (RL) from DeepSeek-V3-Base, achieving performance comparable to OpenAI's o1-1217 on math and code benchmarks.

- By xAI: Grok Series:

  - Grok 1 (2023) [101]: A Mixture-of-Experts (MoE) model [16] with 314B total parameters, of which 80B are active per forward pass. Open sourced in 2024.
  - Grok-1.5 (2024): An improved version of Grok-1 with enhanced reasoning abilities, better mathematical handling, and support for longer contexts.
  - Grok-3 (2025): Combined Transformer-based language modeling with symbolic reasoning modules for advanced logical inference.

Note that apart from the above LLMs, there is also Mamba (2023) [34], which is a State Space Model (SSM) [35, 36] offering a non-attention-based alternative to Transformers (see Chap. 10). It features linear computational complexity for long-context modeling. It is significantly faster and leverages hardware-aware parallelization for high efficiency. Mamba is well suited for long-context tasks, including language modeling, audio processing, and genomics. By 2025, hybrid SSMÃ¢â,¬âŁœTransformer models are an active area of research for efficiency.

6 Why Do Large Language Models Work?

Large language models are becoming increasingly large to improve generalization to unseen data and texts. At the same time, these large models are trained on vast datasets to learn different subjects and domains. There are several lines of research that seek to explain why LLMs work. Classical PAC learning analyses [30] have not explained the generalization of LLMs (see Chap. 6). Current leading explanations focus instead on scaling laws [46], in-context learning, mechanistic interpretability [9, 25], and representational geometry [19, 41]. Some of the works attempting to explain why LLMs work can be categorized into the following groups:

- **Linear representations in LLMs**: Some works, such as [44, 72], discuss that while an LLM is highly nonlinear and learns nonlinear patterns, for the inputs to the model, it finds linear paths in the network from input to output. In this way, it can generate the next words accurately and with high quality.
- **Mechanistic interpretability/circuit analysis**: Some works, such as [9, 25], aim to reverse-engineer LLMs to understand how they compute specific behaviors. This approach is referred to as mechanistic interpretability or circuit analysis. It analyzes real circuits inside models, for example, how models detect induction, perform copying, or detect syntax. There are discussion threads—titled "Transformer Circuits Thread"—about this topic by the Anthropic team in https://transformer-circuits.pub/ (accessed in 2025).
- **Feature and neuron superposition**: Some works explain how concepts are stored within the model's activations via feature and neuron superposition. For example, the paper [26] explains how neural networks store many concepts in overlapping neurons, known as superposition.
- **Causal tracing and attribution**: Some works attempt to trace which parts of the model are responsible for specific outputs. For example, the paper [15] uses causal tracing to find where factual knowledge is stored in the model. The paper [65] dissects how Transformers implement in-context learning via "induction heads."
- **Probing and representational geometry**: Some works use statistical tools to study what information is encoded and where. The paper [19] introduces "probing tasks" to test what linguistic features are encoded. The paper [41] shows that tree-like syntactic structures are linearly recoverable from hidden states.
- **Security and polysemanticity**: Some works examine how models can hide information or become vulnerable. The paper [96] shows that LLMs can have internal beliefs that differ from their outputs, exploring internal monologue. The paper [84] studies how many concepts are packed into a single neuron and how this affects interpretability.

7 Alignment in Large Language Models

7.1 Core Challenge with Large Language Models

LLMs predict the next token (word) based on historical data, lacking an innate ability to follow instructions. At their core, these models predict the next word or token in a sequence based on probabilities derived from pretraining on extensive text corpora. They do not inherently "understand" instructions to follow commands but generate what is statistically likely to come next, given their training.

Moreover, LLMs often exhibit unintended behaviors. For example, they may fabricate facts, generate biased or toxic text, or fail to adhere to user instructions. This occurs because the language modeling objective—predicting the next token or word—is misaligned with the goal of following user instructions helpfully and safely.

7.2 Definition of Alignment

Alignment refers to adjusting the responses of LLMs to match specific user instructions and ethical standards, beyond merely generating probable text. The primary objective is to bridge the gap between these statistical predictions and meaningful adherence to user instructions. This involves ensuring that AI responses are not only contextually appropriate or conversationally relevant but also aligned with the specific intentions, ethical expectations, and task-oriented goals of the user.

The key areas of focus in AI alignment are as follows:

- Learning from human feedback by tailoring AI through human interactions.
- Training to follow instructions by teaching AI-specific task adherence.
- Evaluating AI harms by identifying risks in outputs.
- Modifying behavior to mitigate harms by adjusting AI to prevent negative impacts.

7.3 Alignment Strategies

Multiple strategies have been developed for alignment, i.e., to refine the generation behavior of language models. Some of these approaches are straightforward, while others are more technical. These alignment strategies are introduced below.

7.3.1 Filtering Pretraining Datasets

One straightforward strategy for alignment is removing or excluding certain documents or content from the dataset used for initial training. This prevents the model from learning harmful or undesirable patterns present in those documents. For example, documents containing hate speech or misinformation can be excluded to reduce the chances of the model generating such content. It is similar to the situation where, if you do not want your children to swear, you avoid swearing in front of them.

Filtering pretraining datasets reduces harmful content, but it is insufficient on its own because models can still generalize toxic patterns from benign contexts. Alignment requires multiple strategies, including fine-tuning, preference optimization, and ongoing monitoring.

7.3.2 Fine-Tuning on Value-Targeted Datasets

Another straightforward strategy for alignment is to adjust the model's parameters using a dataset that emphasizes specific values or behaviors. This helps the model better adhere to desired values in its outputs. For example, if you want a model to generate environmentally conscious responses, you can fine-tune it on a dataset full of eco-friendly content. It is similar to the situation where, if you want your children to be moral, you read them moral stories.

7.3.3 Training Strategy by Reinforcement Learning from Human Feedback (RLHF)

InstructGPT is a model introduced in 2022 [71] that applied alignment techniques such as supervised fine-tuning and Reinforcement Learning from Human Feedback (RLHF) [18] to improve how language models follow instructions. These techniques were later further developed and applied in the training of GPT-4 in 2023 [74]. To address the inherent challenges with LLMs discussed earlier, the training strategy of ChatGPT [66] uses the RLHF approach [71], itself an amalgamation of strategies from preceding works. A survey on this approach can be found in [103].

The RLHF approach is one of the most widely used methods for aligning LLMs. It is a three-phased training approach for LLMs [71]:

1. The first step is Supervised Fine-Tuning (SFT), which refines the responses of a pre-trained GPT-style model for specific tasks or guidelines, enhancing its understanding and output relevance.
2. The second step is training a Reward Model (RM) [104], which develops a system that assesses the quality of text generated by the model, guiding it toward human-preferred responses.
3. The third step is Reinforcement Learning (RL) training (see Chap. 18) for Reinforcement Learning from Human Feedback (RLHF) [18], which incorporates user feedback to refine the AI's behavior through direct human input.

The details of these three steps are explained in Sect. 8.

7.3.4 Training Strategy by DPO

An alternative technical method for alignment in LLMs is Direct Preference Optimization (DPO) [80]. It does not require a reward model or reinforcement learning. Instead, it uses a supervised approach to encourage the model to prefer the preferred responses for the prompts. DPO has two steps. The first step is SFT, as in the alignment process using RLHF. The second step involves minimizing a simple supervised loss function. The details of DPO are explained in Sect. 9.

8 Alignment by RLHF

Alignment of LLMs using the RLHF approach involves three steps—supervised fine-tuning, reward modeling, and RLHF—introduced below. These steps are illustrated in Fig. 3.

8.1 Supervised Fine-Tuning

In Supervised Fine-Tuning (SFT), the LLM is fine-tuned using a set of prompts followed by their suitable responses; hence, it is supervised. For this, the authors of the paper [71] hired 40 human labelers who created a dataset consisting of pairs of prompts and their appropriate responses. They used the OpenAI API [69] for labeling. This process is human-in-the-loop to ensure high quality and diversity in the generated data. The goal is to capture a broad spectrum of language patterns and ensure diversity in prompts.

Labelers generated diverse prompts and crafted human-written responses. These prompts were used to train an initial SFT model with instruction-like prompts. This SFT model starts with a pre-trained LLM and is fine-tuned using the collected data.

8.1.1 Types of Prompts

There are three types of prompts—plain, few-shot, and user-based prompts—used for training the SFT model [71]:

- **Plain prompts** generate arbitrary tasks to ensure diversity. An example is "Describe the process of evaporation."

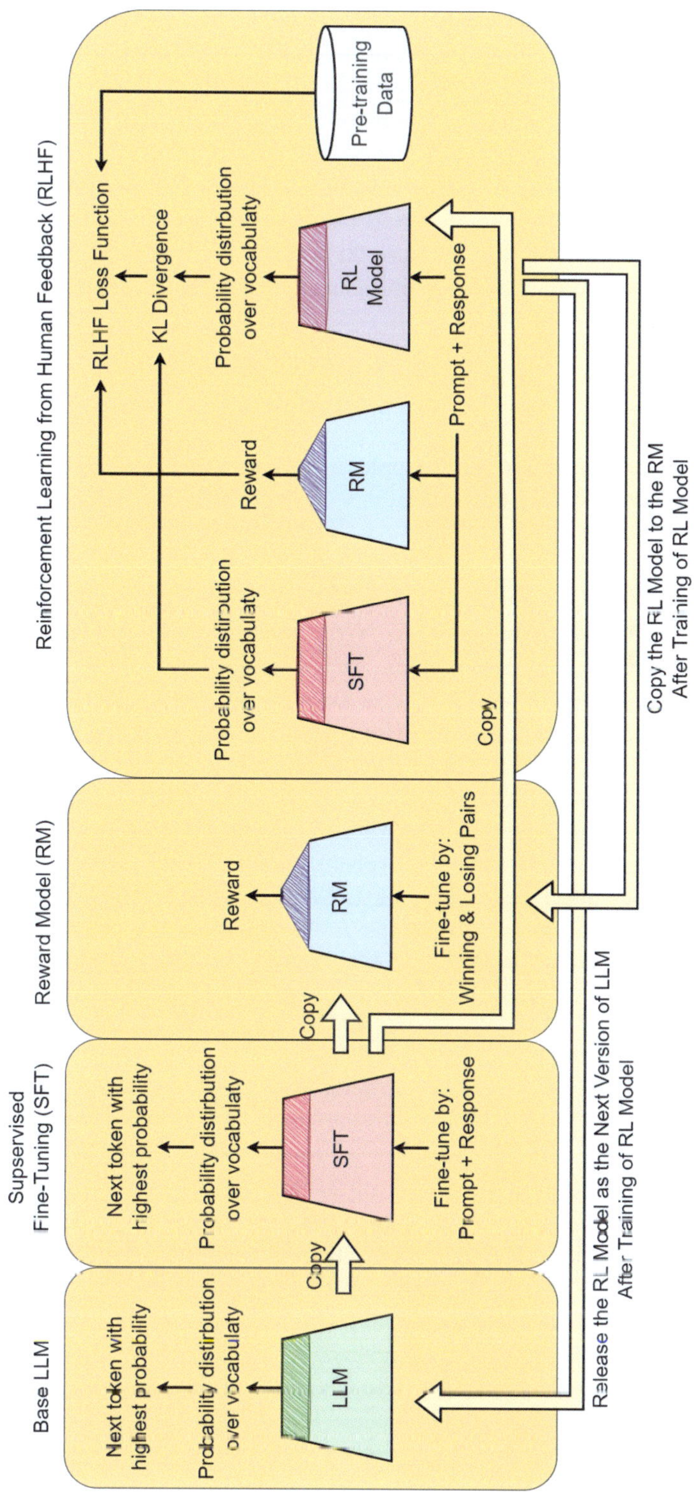

Fig. 3 The three steps of the RLHF approach—supervised fine-tuning, reward model, and RLHF—for alignment in LLMs. This approach has been used in the InstructGPT model

- **Few-shot prompts** provide instructions with multiple query/response pairs. For example, the instruction can be "Give the sentiment for a tweet." The queries can be a few sample tweets, and the responses can be "Positive" or "Negative."
- **User-based prompts** are derived from real-world applications using the OpenAI API [69]. They reflect specific needs or interests of potential users. They are similar to plain prompts, but the difference is that plain prompts are arbitrary prompts created by labelers, while user-based prompts are first collected from the API and then generalized by the labelers. For example, the prompt in the API might be "I want to use the API for a bird identification app." The user-based prompt for this example could be "Describe characteristics of a common bird," which is a more general prompt.

8.1.2 Tasks of Prompts

Prompts are categorized into different tasks. They can belong to one of the following categories [71]:

- **Brainstorming**: For example, a prompt in this category can be "List five ideas for how to regain enthusiasm for my careers."
- **Classification**: An example is "Given the following text, rate, on a scale of 1–10, how sarcastic the person is being."
- **Extract**: An example is "Extract all course titles from this table: | Title | Lecturer | Room |"
- **Generation**: An example is "Write a short story where a brown bear goes to the beach, makes friends with a seal, and then returns home."
- **Chat**: An example is "This is a conversation with an AI assistant. The assistant is helpful, creative, clever, and very friendly. Human: 'Hello, who are you?' AI: 'I am an AI created by OpenAI. How can I help you today?'"
- **Rewrite**: An example is "Translate this sentence to Spanish: <English sentence>"
- **Closed (multiple-choice) question**: An example is "Answer the following question: what shape is the earth? A) a circle, B) a sphere, C) an ellipse, D) a plane"
- **Open question**: An example is "Who built the Statue of Liberty?"
- **Summarization**: An example is "Summarize this for a second-grade student: <text>"
- **Other**: An example is "Look up 'cowboy' on Google and give me the results."

8.1.3 Training of the SFT Model

The prompts and the desired responses written by the labelers are concatenated to prepare the training data for SFT. The SFT model is initialized with the weights of a pre-trained LLM, such as GPT, and it is fine-tuned using this training dataset (see Fig. 3). Each data instance consists of a prompt followed by its response as a sequence of tokens (words). In this way, the model gradually learns to generate the next tokens—which are the words of the response—after the tokens (words) of the prompt.

For example, a prompt in a data instance can be "What's the capital of France?", and the desired response is "Paris." The concatenated input data instance is "What's the capital of France? Paris." After fine-tuning, the model learns to generate concise and accurate responses to a wide range of prompts. Interestingly, the model generalizes well to unseen data. For example, when "What's the capital of France? Paris." is introduced to the model during fine-tuning, it will also learn to answer the prompt "What's the capital of Iran?" with "Tehran" correctly. This is because, in other training corpora, it has seen the words "Iran" and "Tehran," and through this fine-tuning, it learns to respond with Tehran as the capital of Iran.

Note that before fine-tuning, the original LLM could only generate sequences and continue text without interacting with the user. After the SFT model is trained, as a fine-tuned version of the LLM, it can interact with the user as a chatbot, where the user provides prompts and it responds to them. These responses are also expected to align with moral standards because human-written responses have been used for the fine-tuning. However, to ensure that the responses fully align with these standards, the next step—training a reward model—is used.

8.2 Reward Model

The second step in alignment using the RLHF approach [71] is training a Reward Model (RM) [104]. A reward model scores the responses generated by the trained SFT model to assess how well they align with human standards and preferences.

8.2.1 Training Data for the Reward Model

The training dataset for the reward model is prepared as follows. After the SFT model is fully trained, each prompt is passed as input to the model, and the response generated by the SFT model is recorded. This procedure is repeated multiple times (e.g., four to nine times) per prompt. In other words, the same prompt is passed as input multiple times, and the different generated responses are saved for each prompt. For example, the prompt can be "What is ice?" fed four times to the SFT model. Assume it generates four different responses:

A. "It's the solid form of water."
B. "Frozen water that's cold to touch."
C. "Water that has turned solid due to low temperature."
D. "A crystalline substance formed when water freezes."

The responses are then paired with each other. Therefore, if four responses have been recorded, the total number of pairs becomes $C(4, 2) = 6$. In the above example, the pairs can be (A, B), (A, C), (A, D), (B, C), (B, D), and (C, D), where, for example, (A, B) is ("It's the solid form of water.", "Frozen water that's cold to touch."). Then, human labelers are asked to rank the responses within each pair. For example, the human labeler might prefer response A over B. In this way, feedback is obtained from humans. This is why ChatGPT [66] sometimes asks users to select their preferred response; such interactions help collect preferred responses for prompts.

This procedure produces a new dataset. The paper [71] uses 33,000 prompts with four to nine responses per prompt. For four responses per prompt, $33,000 \times C(4, 2) = 198,000$ pairs are prepared, and for nine responses per prompt, $33,000 \times C(9, 2) = 1,188,000$ pairs are gathered. Each of these pairs is also ranked by human labelers, as explained above.

For every pair, the prompt is concatenated once with the winning response and once with the losing response, based on the human labeler rankings. For example, for the pair (A, B) in the example above, the winning and losing pairs are

- Winning pair: "What is ice? It's the solid form of water."
- Losing pair: "What is ice? Frozen water that's cold to touch."

8.2.2 Architecture of the Reward Model

For the neural network of the RM, a copy of the trained SFT model is used. As Fig. 3 shows, the last layer of the SFT model has a softmax activation function (see Sect. 2.8 in Chap. 2), with each neuron representing the probability of a word in the dictionary. This last layer is removed from the copy of the trained SFT model and replaced with a linear layer containing a single output neuron, which outputs a real-valued number (see Fig. 3). Therefore, the model becomes a regression model.

8.2.3 Training of the Reward Model

For training the reward model, the corresponding winning and losing pairs are passed as input to the reward model. The reward model outputs a score for each input pair. The difference between the two scores is calculated and used as the loss function during training via backpropagation (see Chap. 3). This process fine-tunes the weights of the copy of the SFT model and trains the newly added final layer. Assuming that the reward model outputs the rewards r_w and r_l for the winning and losing pairs, respectively, the loss function of the reward model is [71]

$$\underset{\theta}{\text{minimize}} \quad \mathcal{L}(\theta) := -\mathbb{E}\Big[\log\big(\sigma(r_w - r_l)\big)\Big], \tag{1}$$

where θ is the set of weights of the reward model, $\mathbb{E}[.]$ denotes expectation over the mini-batch, and $\sigma(.)$ denotes the sigmoid (logistic) activation function (see Sect. 2.8 in Chap. 2).

Minimizing Eq. (1) trains the reward model to output higher rewards for the winning pairs (i.e., good responses) and lower rewards for the losing pairs (i.e., bad responses). Intuitively, the model aims to maximize the difference between the winning and losing responses. This is because:

$$\begin{aligned}
&\text{If } r_w \gg r_l \implies \sigma(r_w - r_l) \approx 1 \implies -\log(\sigma(r_w - r_l)) \downarrow \\
&\text{If } r_w \ll r_l \implies \sigma(r_w - r_l) \approx 0 \implies -\log(\sigma(r_w - r_l)) \uparrow
\end{aligned}$$

Note that this is not the only way to achieve this goal. One can use contrastive loss [39] or triplet loss [85] for this purpose (see Chap. 16).

After training, during inference, the reward model can provide a scalar reward for any prompt-response combination. This simulates human preference for a given prompt-response pair. For example, in the above case, the trained reward model would assign a high reward to "What is ice? It's the solid form of water." and a low reward to "What is ice? Frozen water that's cold to touch."

8.3 Reinforcement Learning from Human Feedback

So far, there is an LLM that has been fine-tuned in a supervised manner, referred to as the SFT model. The SFT model can generate responses to prompts—an ability not present in the original LLM. There is also a reward model that is trained to assess the quality of the responses generated by the SFT model. The goal now is to fine-tune the SFT model further to produce responses with high reward scores as assessed by the reward model. In this fine-tuning, the reward model remains unchanged, and the SFT model is further fine-tuned. This is where Reinforcement Learning from Human Feedback (RLHF) [7, 18, 50] comes into play.

8.3.1 Policy Gradient in RLHF

It is recommended that the reader studies Chap. 18 on Reinforcement Learning (RL) before reading this section. The aim here is to maximize the reward of the generated response for a prompt using RL. This can be formulated mathematically as an objective maximization problem using the policy gradient algorithm [91] (see Sect. 9 in Chap. 18). The policy should be updated to maximize the total reward.

A copy of the SFT model trained in Sect. 8.1 is considered and denoted as the RL model (see Fig. 3). The policy here is the neural network of the RL model, parameterized by weights w; changing the weights w will change the policy. In this context, the state in reinforcement learning is the set of tokens

$\{x_1, x_2, \ldots, x_{t-1}\}$ and the action is the next generated token x_t. In other words, the model has seen this set of tokens (words) so far and it is supposed to generate the next token (word).

As explained above, the objective is to maximize the total reward:

$$\underset{w}{\text{maximize}} \quad J(w) := \mathbb{E}_{x \sim p_w}[r(x)], \tag{2}$$

where w is the set of weights of the RL model, x is the generated response by the SFT model, the expectation is over the mini-batch, and $r(x)$ is the reward given by the reward model for the response x. Gradient ascent can be used to maximize this objective function:

$$w \leftarrow w + \alpha \nabla_w J(w), \tag{3}$$

where α is the learning rate and $\nabla_w J(w)$ is the gradient of the objective function. As derived in Chap. 18, the gradient expression is

$$\begin{aligned}\nabla_w J(w) &= \mathbb{E}_{x \sim p_w}\big[r(x) \nabla_w \log(p_w(x))\big] \\ &= \frac{1}{|D|} \sum_{D} \sum_{t=0}^{T} \nabla_w r(x) \log\big(p_w(x_t | x_{t-1}, x_{t-2}, \ldots)\big),\end{aligned} \tag{4}$$

where D refers to all the prompt-response pairs, $|D|$ is the number of pairs, and x_t is the token generated at time step t. In the neural network of the RL model, backpropagation minimizes the loss function, so $-J(w)$, with the gradient in Eq. (4), is used as the loss function to be minimized.

The process of fine-tuning the RL model using policy gradient is as follows (see Fig. 3). A copy of the SFT model is designated as the RL model. A prompt is given to the RL model, and it generates a response. The algorithm waits until all tokens of the response are generated by the RL model. This is because the reward model requires the concatenation of the prompt and the complete response to assess it. The generated response is concatenated with the prompt and fed to the reward model. The reward model outputs a reward for the quality of the response. This reward is used in the policy gradient algorithm described above to fine-tune the weights of the RL model. This process is iterative, involving repeatedly generating responses, computing rewards, and updating the model using policy gradient.

Note that during the sequence generation for a prompt by the RL model, the model is unaware of the sequence's reward. It learns the reward only after generation, using it to refine its parameters for better future sequences.

8.3.2 Challenge with Direct Model Updates

Using the initial training method by maximizing Eq. (2) can make the model unpredictable, generating gibberish. These updates can push the model to produce text that does not make sense or is off-topic. This occurs because the RL model is a next-token prediction model, but the reward is only available when the entire set of tokens for a response has been generated. Since the reward model requires a complete response to evaluate it, it is not possible to update the weights of the RL model after generating each token. Thus, updates applied after generating the entire response can drastically change the weights of the RL model, making it unpredictable.

The solution to this problem is to ensure that the updates do not deviate too much from the SFT model trained in Sect. 8.1. To achieve this, the SFT model, which remains unchanged, is used as a reference. The prompt is passed as input to both the SFT and RL models. Each of the SFT and RL models outputs a distribution of probabilities over words using their softmax activation functions. The

Kullback–Leibler (KL) divergence [49] between these two distributions can be used to regularize the objective function in Eq. (2):

$$\underset{w}{\text{maximize}} \quad J(w) := \mathbb{E}_{x \sim p_w}\Big[r(x) - \beta \text{KL}\big(p_w(x) \,\|\, p_{\text{SFT}}(x)\big)\Big], \tag{5}$$

where $p_w(x)$ is the output probability distribution over the vocabulary produced by the RL model for the sequence x, $p_{\text{SFT}}(x)$ is the output probability distribution produced by the SFT model for the sequence x, and $\beta > 0$ is the regularization hyperparameter. This regularization minimizes the KL divergence, ensuring that the output of the RL model does not change drastically compared to the output of the SFT model.

8.3.3 Challenge with Forgetting the Pre-trained Base LLM

After training with the objective function (5), the RL model may generate responses that are not as good as those produced by the pre-trained base LLM before alignment. Due to the extensive fine-tuning steps involved in alignment, the model can forget some of its original training. To resolve this issue, an additional regularization term can be added to the objective function (5) to keep the RL model close to the original pre-trained base LLM before alignment. Thus, the full and final objective function in RLHF becomes

$$\underset{w}{\text{maximize}} \quad J(w) := \mathbb{E}_{x \sim p_w}\Big[r(x) - \beta \text{KL}\big(p_w(x) \,\|\, p_{\text{SFT}}(x)\big)\Big] + \gamma \mathbb{E}_{x \sim \text{pre-train}}\big[\log(p_w(x))\big], \tag{6}$$

where p_w is the RL model with weights w, $x \sim$ pre-train represents sequences observed in the pre-training dataset (used for training the original LLM), and $\gamma > 0$ is the regularization hyperparameter. This regularization ensures that the RL model remains aligned with the original pre-trained base LLM. All three parts of this loss function are depicted in Fig. 3.

The RL model should maximize Eq. (6). In backpropagation, the negative of Eq. (6) is minimized. In conclusion, the model is optimized to (1) maximize the reward obtained using the reward model, (2) minimize divergence from the SFT model, and (3) maximize the likelihood of observing sequences seen during pretraining by the LLM. As shown in Fig. 3, after training of the RL model is complete, it can be released as the next version of the LLM, as it is now further aligned with standards based on human feedback.

8.3.4 Iteration in the Process of Alignment

It is possible to perform the alignment process iteratively to gradually improve the model. After updating the RL model for some steps, a new reward model can be initialized with the updated weights of the RL model (see Fig. 3). Human labelers can then be asked to generate another dataset following the procedure described in Sect. 8.1. Subsequently, another training session can be conducted for the new reward model. Afterward, RLHF can be applied again using the new reward model. This iterative approach gradually makes the LLMs stronger and more aligned with human expectations and standards. This is the approach followed by most GPT-style LLMs introduced in Sect. 5, with slight differences in their experimental setups. This process is illustrated and summarized in Fig. 3.

9 Alignment by DPO

Direct Preference Optimization (DPO) [80] is another method for alignment in LLMs. It is much simpler than the RLHF approach used in InstructGPT. DPO does not require a reward model and avoids reinforcement learning; therefore, it simplifies the fine-tuning process. It has also been empirically

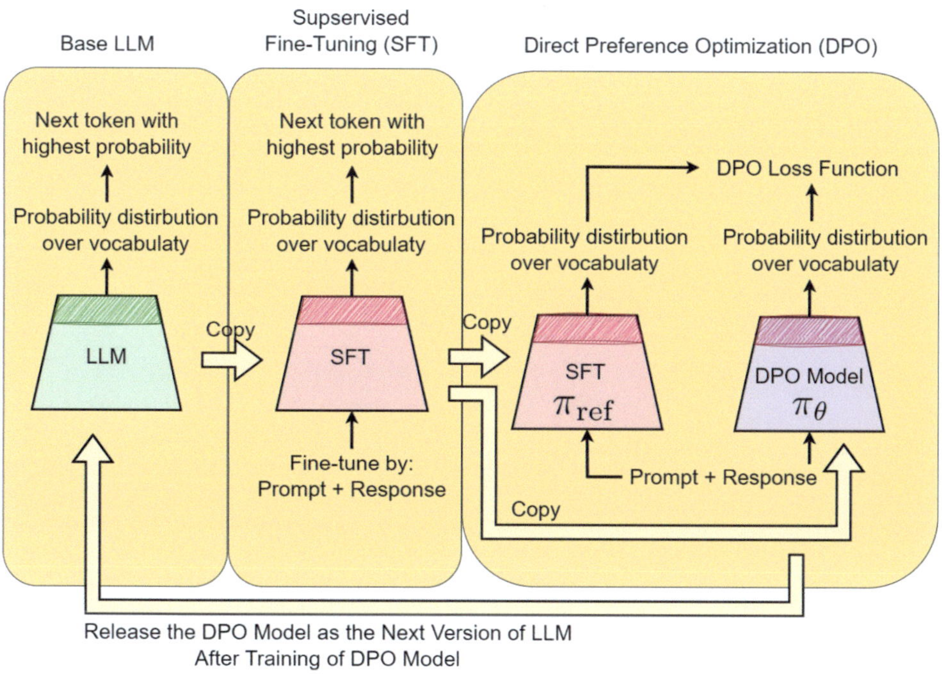

Fig. 4 DPO method for alignment in LLMs

shown that DPO is more stable than RLHF, retaining the model's prior knowledge while aligning with preferences. DPO employs a direct optimization approach by fine-tuning preferences directly using supervised methods.

Alignment using DPO involves a two-stage training approach: Supervised Fine-Tuning (SFT) and Direct Preference Optimization (DPO). The SFT step is the same as described in Sect. 8.1. As explained in that section, SFT refines the responses of a pre-trained LLM for specific tasks or guidelines, enhancing its understanding and output relevance. The DPO step fine-tunes the model to directly optimize preferred behavior using a pairwise comparison loss.

DPO aligns the model's behavior with human preferences without using reinforcement learning. Its training data consist of preferred (winning) and less preferred (losing) responses collected as described in Sect. 8.2.1. It uses pairwise comparisons between the preferred (winning) and less preferred (losing) responses for optimization. DPO avoids the need for a separate reward model; it suggests that the model itself is effectively a reward model.

As depicted in Fig. 4, an SFT model is initialized from an LLM and fine-tuned as explained in Sect. 8.1. Then, a copy of the fine-tuned SFT model is considered as the DPO model. Let the DPO model to be fine-tuned be denoted by π_θ with parameters θ. A frozen SFT model, denoted by π_{ref}, is also used for relative comparison.

Suppose the pairs of winning and losing responses for a prompt $\boldsymbol{x}$ are denoted by (y_w, y_l), with y_w as the preferred (winning) response and y_l as the less preferred (losing) one. These pairs are human-labeled as described in Sect. 8.2.1. For example, the prompt $\boldsymbol{x}$ can be "What is ice?", the preferred response y_w can be "It's solid form of water.", and the less preferred response can be "Water that has turned solid due to low temperature."

Let $\pi(y|\boldsymbol{x})$ denote the output probabilities of model π when its input is the concatenation of the prompt $\boldsymbol{x}$ and the response y. Therefore:

- $\pi_{\text{ref}}(y_w|\boldsymbol{x})$ denotes the output probabilities of the frozen SFT model when its input is the concatenation of the prompt $\boldsymbol{x}$ and the preferred (winning) response y.
- $\pi_{\text{ref}}(y_l|\boldsymbol{x})$ denotes the output probabilities of the frozen SFT model when its input is the concatenation of the prompt $\boldsymbol{x}$ and the less preferred (losing) response y.
- $\pi_\theta(y_w|\boldsymbol{x})$ denotes the output probabilities of the DPO model when its input is the concatenation of the prompt $\boldsymbol{x}$ and the preferred (winning) response y.
- $\pi_\theta(y_l|\boldsymbol{x})$ denotes the output probabilities of the DPO model when its input is the concatenation of the prompt $\boldsymbol{x}$ and the less preferred (losing) response y.

The loss function of the DPO model is

$$\underset{\theta}{\text{minimize}} \;\; \mathcal{L}_{\text{DPO}}(\theta) := -\mathbb{E}_{(\boldsymbol{x},y_w,y_l)}\Big[\log\sigma\Big(\beta\log\big(\frac{\pi_\theta(y_w|\boldsymbol{x})}{\pi_{\text{ref}}(y_w|\boldsymbol{x})}\big) - \beta\log\big(\frac{\pi_\theta(y_l|\boldsymbol{x})}{\pi_{\text{ref}}(y_l|\boldsymbol{x})}\big)\Big)\Big], \tag{7}$$

where $\sigma(.)$ is the sigmoid function (see Sect. 2.8 in Chap. 2) and $\beta > 0$ is the temperature scaling factor.

The loss function (7) can be restated using logarithm properties as follows:

$$\mathcal{L}_{\text{DPO}}(\theta) = -\mathbb{E}_{(\boldsymbol{x},y_w,y_l)}\Big[\log\sigma\Big(\beta\log\Big(\frac{\pi_\theta(y_w|\boldsymbol{x})/\pi_{\text{ref}}(y_w|\boldsymbol{x})}{\pi_\theta(y_l|\boldsymbol{x})/\pi_{\text{ref}}(y_l|\boldsymbol{x})}\Big)\Big)\Big]. \tag{8}$$

Consider three cases:

- If the SFT model π_{ref} is indifferent between the preferred and less preferred responses, i.e., $\pi_{\text{ref}}(y_w|\boldsymbol{x}) \approx \pi_{\text{ref}}(y_l|\boldsymbol{x}) \approx 0.5$, then Eq. (8) simplifies to:

$$\mathcal{L}_{\text{DPO}}(\theta) \approx -\mathbb{E}_{(\boldsymbol{x},y_w,y_l)}\Big[\log\sigma\Big(\beta\log\Big(\frac{\pi_\theta(y_w|\boldsymbol{x})}{\pi_\theta(y_l|\boldsymbol{x})}\Big)\Big)\Big].$$

  Minimizing this loss function maximizes $\pi_\theta(y_w|\boldsymbol{x})$ over $\pi_\theta(y_l|\boldsymbol{x})$, fine-tuning the model to prefer the preferred response over the less preferred response.
- If the SFT model π_{ref} prefers the preferred response over the less preferred response, e.g., $\pi_{\text{ref}}(y_w|\boldsymbol{x}) = 0.8$ and $\pi_{\text{ref}}(y_l|\boldsymbol{x}) = 0.2$, then Eq. (8) simplifies to:

$$\mathcal{L}_{\text{DPO}}(\theta) \approx -\mathbb{E}_{(\boldsymbol{x},y_w,y_l)}\Big[\log\sigma\Big(\beta\log\Big(\frac{0.2}{0.8}\times\frac{\pi_\theta(y_w|\boldsymbol{x})}{\pi_\theta(y_l|\boldsymbol{x})}\Big)\Big)\Big].$$

  The scale $0.2/0.8$ reduces the loss value, so minimizing this loss slightly maximizes $\pi_\theta(y_w|\boldsymbol{x})$ over $\pi_\theta(y_l|\boldsymbol{x})$. In other words, it slightly encourages the model to prefer the preferred response, as it is already preferred.
- If the SFT model π_{ref} does not prefer the preferred response over the less preferred response, e.g., $\pi_{\text{ref}}(y_w|\boldsymbol{x}) = 0.2$ and $\pi_{\text{ref}}(y_l|\boldsymbol{x}) = 0.8$, then Eq. (8) simplifies to:

$$\mathcal{L}_{\text{DPO}}(\theta) \approx -\mathbb{E}_{(\boldsymbol{x},y_w,y_l)}\Big[\log\sigma\Big(\beta\log\Big(\frac{0.8}{0.2}\times\frac{\pi_\theta(y_w|\boldsymbol{x})}{\pi_\theta(y_l|\boldsymbol{x})}\Big)\Big)\Big].$$

  The scale $0.8/0.2$ increases the loss value, so minimizing this loss dramatically maximizes $\pi_\theta(y_w|\boldsymbol{x})$ over $\pi_\theta(y_l|\boldsymbol{x})$. In other words, it strongly encourages the model to prefer the preferred response because it is currently favoring the less preferred response.

In summary, minimizing the loss function (7) or (8) fine-tunes the model to prefer the preferred responses with different degrees of adjustment based on the reference SFT model.

10 Conclusion

10.1 Conclusion of Chapter

BERT and GPT are two foundational models built upon the Transformer architecture. T5 (Text-to-Text Transfer Transformer) extends this idea by combining both a Transformer encoder and a Transformer decoder while casting NLP tasks in a unified text-to-text format. Since then, numerous large language models have emerged, many of which follow the GPT-style decoder-only design. A critical challenge for large language models is alignment, which seeks to ensure that generated text meets quality standards and adheres to human values. Among the different alignment strategies, Reinforcement Learning from Human Feedback (RLHF) and Direct Preference Optimization (DPO) have become two of the most widely adopted approaches. By 2025, large language models are advancing artificial intelligence significantly, both among researchers and in the public sphere.

10.2 Conclusion of the Book Part "Sequence Models"

Sequence models are neural networks designed to process sequential data, such as text or speech. While Recurrent Neural Networks (RNNs) and Long Short-Term Memory (LSTM) networks rely on recurrence to handle sequences, Transformers leverage the attention mechanism, demonstrating that sequences can be effectively processed without recurrence. State Space Models (SSMs), on the other hand, reintroduce recurrence and excel at modeling long-range dependencies and very long sequences.

Large language models (LLMs) are generative models capable of producing coherent text. Foundational models such as BERT and GPT are built on the Transformer architecture. Most modern LLMs follow a GPT-style decoder-only design, although SSM-based language models are also emerging, and hybrid Transformer-SSM models have shown promising results in recent research.

A critical challenge for LLMs is alignment, which ensures that generated text adheres to human values and quality standards. Two widely adopted alignment methods are Reinforcement Learning from Human Feedback (RLHF) and Direct Preference Optimization (DPO). Large language models continue to be a major focus of research and are driving significant advancements in artificial intelligence.

11 Exercises

1. (Level: Easy) Answer the following questions:
   (a) Explain the CLS token in BERT. Why is it important?
   (b) Explain the difference between BERT and GPT.
   (c) How do BERT and GPT perform differently on NLP tasks?
2. (Level: Easy) Answer the following questions:
   (a) How does the size of a Transformer model (e.g., number of parameters) impact its performance and computational cost?
   (b) How does T5 unify different NLP tasks under a single framework?
   (c) How does the training process of BERT differ from that of GPT?

(d) Which of GPT and BERT is bidirectional, and why is that important?
(e) How does fine-tuning impact the performance of large pre-trained Transformer models like BERT, GPT, and T5?

3. (Level: Easy) Answer the following questions:

(a) How do you decide whether a task is more suited for a GPT or a BERT model?
(b) For the following tasks, determine whether a GPT or BERT model would be better suited and explain your choice: Named Entity Recognition (NER), Text Completion, Question Answering, and Machine Translation.

4. (Level: Moderate) The BERT-Large model [23] has a maximum input length of 512 tokens, each of dimensionality $D = 1024$ and taken from a vocabulary of 30,000. It has 24 Transformer layers each with 16 self-attention heads with $D_q = D_k = D_v = 64$, and the MLP position-wise networks have two layers with 4096 hidden nodes. Show that the total number of parameters in the BERT encoder Transformer language model is approximately 340 million.
5. (Level: Easy) Answer the following questions:

(a) What was the key improvement in GPT-2 compared to GPT-1?
(b) How did GPT-3 enhance the capabilities of GPT-2?
(c) In what ways did GPT-4 surpass GPT-3 in terms of performance?

6. (Level: Moderate) In Reinforcement Learning from Human Feedback (RLHF), there is

$$w_{k+1} = w_k + \alpha \nabla_w J(\pi_w),$$

where w_k represents the policy parameters at step k, α is the learning rate, and $\nabla_w J(\pi_w)$ is the policy gradient. However, instead of standard policy gradient updates, a reward model (RM) is used to refine policy updates based on human feedback:

$$\begin{aligned} m_{k+1} &= \beta m_k + (1-\beta)\nabla_w J_{RM}(\pi_w), \\ w_{k+1} &= w_k + \alpha m_{k+1}, \end{aligned}$$

where m_k is the momentum term, which accumulates past gradients, β is the momentum coefficient, and $J_{RM}(\pi_w)$ is the reward-adjusted policy gradient based on human feedback. Given initial policy weight $w_0 = 1.0$, initial momentum: $m_0 = 0.0$, learning rate: $\alpha = 0.1$, momentum coefficient: $\beta = 0.9$, reward-adjusted policy gradient $\nabla_w J_{RM}(\pi_w)$ at each step being 2.0, 1.8, and 1.6, do the following calculations:

(a) Compute the first three iterations (w_1, w_2, w_3) using the momentum-based policy gradient ascent with reward-adjusted updates.
(b) Compare the results with vanilla gradient ascent without momentum. Which method converges faster?
(c) If the learning rate increases to $\alpha = 0.5$, what risks arise, and how does it affect convergence?

7. (Level: Moderate) Reinforcement Learning from Human Feedback (RLHF) is used to align language models with human preferences. The reward model (RM) in RLHF is trained to assign a preference score to different outputs, guiding the model toward more desirable responses. The reward function for a response r given a prompt p is modeled as follows:

$$R(p, r) = f_\theta(p, r),$$

Table 1 Prompt-response pairs and their reward scores $R(p, r)$

Prompt	Response	**R(p, r)**
"What is the capital of France?"	"Paris"	0.98
"What is the capital of France?"	"London"	0.20
"Describe a cat."	"A cat is a small domesticated carnivore."	0.92
"Describe a cat."	"A cat is a fish."	0.10

where f_θ is a learned function parameterized by θ. The policy update in RLHF is performed using policy gradient optimization:

$$\theta' = \theta + \alpha \nabla_\theta J(\theta),$$

where $J(\theta)$ is the expected reward and α is the learning rate.
Consider a language model with a reward model that scores responses on a scale from 0 to 1. A dataset consists of prompt-response pairs with human preference scores as listed in Table 1.
Consider a policy gradient update step with learning rate $\alpha = 0.01$.

(a) Compute the expected reward $J(\theta)$.
(b) Compute the gradient update step for θ.
(c) Explain why RLHF improves the alignment of language models.

References

1. Josh Achiam et al. "GPT-4 technical report". In: *arXiv preprint* arXiv:2303.08774 (2023).
2. Alibaba. *Qwen GitHub*. https://github.com/QwenLM. 2023.
3. Emily Alsentzer et al. "Publicly Available Clinical BERT Embeddings". In: *Proceedings of the 2nd Clinical Natural Language Processing Workshop*. 2019, pp. 72–78.
4. Rohan Anil et al. "PaLM 2 technical report". In: *arXiv preprint* arXiv:2305.10403 (2023).
5. Anthropic. *Introducing the next generation of Claude*. https://www.anthropic.com/news/claude-3-family. 2024.
6. Devansh Arpit et al. "A closer look at memorization in deep networks". In: *International Conference on Machine Learning*. 2017.
7. Yuntao Bai et al. "Training a helpful and harmless assistant with reinforcement learning from human feedback". In: *arXiv preprint* arXiv:2204.05862 (2022).
8. Iz Beltagy, Kyle Lo, and Arman Cohan. "SciBERT: A Pretrained Language Model for Scientific Text". In: *Proceedings of the 2019 Conference on Empirical Methods in Natural Language Processing and the 9th International Joint Conference on Natural Language Processing (EMNLP-IJCNLP)*. 2019, pp. 3615–3620.
9. Adithya Bhaskar, Alexander Wettig, Dan Friedman, and Danqi Chen. "Finding transformer circuits with edge pruning". In: *Advances in Neural Information Processing Systems* 37 (2024), pp. 18506–18534.
10. Xiao Bi et al. "Deepseek LLM: Scaling open-source language models with longtermism". In: *arXiv preprint* arXiv:2401.02954 (2024).
11. Nadav Brandes, Dan Ofer, Yam Peleg, Nadav Rappoport, and Michal Linial. "ProteinBERT: a universal deep-learning model of protein sequence and function". In: *Bioinformatics* 38.8 (2022), pp. 2102–2110.
12. Gwern Branwen. *GPT-3 Creative Fiction*. https://www.gwern.net/GPT-3. 2020.
13. Tom B Brown et al. "Language models are few-shot learners". In: *Advances in neural information processing systems*. 2020.
14. Paweł; Budzianowski and Ivan Vulić. "Hello, It's GPT-2-How Can I Help You? Towards the Use of Pretrained Language Models for Task-Oriented Dialogue Systems". In: *Proceedings of the 3rd Workshop on Neural Generation and Translation*. 2019, pp. 15–22.

15. Collin Burns, Haotian Ye, Dan Klein, and Jacob Steinhardt. "Discovering Latent Knowledge in Language Models Without Supervision". In: *International Conference on Learning Representations*. 2023.
16. Weilin Cai et al. "A survey on mixture of experts in large language models". In: *IEEE Transactions on Knowledge and Data Engineering* (2025).
17. Aakanksha Chowdhery et al. "PaLM: Scaling language modeling with pathways". In: *Journal of Machine Learning Research* 24.240 (2023), pp. 1–113.
18. Paul F Christiano et al. "Deep reinforcement learning from human preferences". In: *Advances in neural information processing systems* 30 (2017).
19. Alexis Conneau, German Kruszewski, Guillaume Lample, Loic Barrault, and Marco Baroni. "What you can cram into a single $&!#* vector: Probing sentence embeddings for linguistic properties". In: *Proceedings of the 56th Annual Meeting of the Association for Computational Linguistics (Volume 1: Long Papers)*. 2018, pp. 2126–2136.
20. Damai Dai et al. "DeepSeekMoE: Towards Ultimate Expert Specialization in Mixture-of-Experts Language Models". In: *Proceedings of the 62nd Annual Meeting of the Association for Computational Linguistics (Volume 1: Long Papers)*. 2024, pp. 1280–1297.
21. Jeff Dean. *Pathways: A next generation AI architecture*. https://blog.google/technology/ai/introducing-pathways-next-generation-ai-architecture/. Accessed: May 2025. 2021.
22. Google DeepMind. *Introducing Gemini 2.0: Our new AI model for the agentic era*. 2024. URL: https://blog.google/technology/google-deepmind/google-gemini-ai-update-december-2024/.
23. Jacob Devlin, Ming-Wei Chang, Kenton Lee, and Kristina Toutanova. "BERT: Pre-training of deep bidirectional transformers for language understanding". In: *Proceedings of the 2019 conference of the North American chapter of the association for computational linguistics: human language technologies, volume 1 (long and short papers)*. 2019, pp. 4171–4186.
24. Li Dong et al. "Unified language model pre-training for natural language understanding and generation". In: *Advances in Neural Information Processing Systems*. 2019, pp. 13063–13075.
25. Nelson Elhage et al. *A mathematical framework for transformer circuits*. Transformer Circuits Thread, https://transformer-circuits.pub/2021/framework/index.html. 2021.
26. Nelson Elhage et al. "Toy models of superposition". In: *arXiv preprint* arXiv:2209.10652 (2022).
27. Katherine Elkins and Jon Chun. "Can GPT-3 pass a writer's Turing Test?" In: *Journal of Cultural Analytics* 2371 (2020), p. 4549.
28. Kawin Ethayarajh. "How Contextual are Contextualized Word Representations? Comparing the Geometry of BERT, ELMo, and GPT-2 Embeddings". In: *Proceedings of the 2019 Conference on Empirical Methods in Natural Language Processing and the 9th International Joint Conference on Natural Language Processing (EMNLP-IJCNLP)*. 2019, pp. 55–65.
29. Luciano Floridi and Massimo Chiriatti. "GPT-3: Its Nature, Scope, Limits, and Consequences". In: *Minds and Machines* (2020), pp. 1–14.
30. Cédric Gerbelot et al. "Applying statistical learning theory to deep learning". In: *Journal of Statistical Mechanics: Theory and Experiment* 2024.10 (2024), p. 104003.
31. Yoav Goldberg and Omer Levy. "word2vec Explained: deriving Mikolov et al.'s negative-sampling word-embedding method". In: *arXiv preprint* arXiv:1402.3722 (2014).
32. Google. *Gemini 2.5: Our most intelligent models are getting even better*. 2025. URL: https://blog.google/technology/google-deepmind/google-gemini-updates-io-2025/.
33. Aaron Grattafiori et al. "The Llama 3 herd of models". In: *arXiv preprint* arXiv:2407.21783 (2024).
34. Albert Gu and Tri Dao. "Mamba: Linear-Time Sequence Modeling with Selective State Spaces". In: *First Conference on Language Modeling*. 2024.
35. Albert Gu, Karan Goel, and Christopher Re. "Efficiently Modeling Long Sequences with Structured State Spaces". In: *International Conference on Learning Representations*. 2022.
36. Albert Gu et al. "Combining recurrent, convolutional, and continuous-time models with linear state space layers". In: *Advances in neural information processing systems* 34 (2021), pp. 572–585.
37. Daya Guo et al. "DeepSeek-Coder: When the Large Language Model Meets Programming–The Rise of Code Intelligence". In: *arXiv preprint* arXiv:2401.14196 (2024).
38. Daya Guo et al. "Deepseek-R1: Incentivizing reasoning capability in llms via reinforcement learning". In: *arXiv preprint* arXiv:2501.12948 (2025).
39. Raia Hadsell, Sumit Chopra, and Yann LeCun. "Dimensionality reduction by learning an invariant mapping". In: *Proceedings of the IEEE conference on computer vision and pattern recognition*. Vol. 2. IEEE. 2006, pp. 1735–1742.
40. Pengcheng He, Xiaodong Liu, Jianfeng Gao, and Weizhu Chen. "DeBERTa: Decoding-enhanced bert with disentangled attention". In: *International Conference on Learning Representations*. 2021.
41. John Hewitt and Christopher D Manning. "A structural probe for finding syntax in word representations". In: *Proceedings of the 2019 Conference of the North American Chapter of the Association for Computational Linguistics: Human Language Technologies, Volume 1 (Long and Short Papers)*. 2019, pp. 4129–4138.
42. Matthew Douglas Hoffman et al. "Training chain-of-thought via latent-variable inference". In: *Advances in Neural Information Processing Systems*. 2023.

43. Allen H Huang, Hui Wang, and Yi Yang. "FinBERT: A large language model for extracting information from financial text". In: *Contemporary Accounting Research* 40.2 (2023), pp. 806–841.
44. Yibo Jiang, Goutham Rajendran, Pradeep Kumar Ravikumar, Bryon Aragam, and Victor Veitch. "On the Origins of Linear Representations in Large Language Models". In: *International Conference on Machine Learning*. PMLR. 2024, pp. 21879–21911.
45. Xiaoqi Jiao et al. "TinyBERT: Distilling BERT for Natural Language Understanding". In: *Findings of the Association for Computational Linguistics: EMNLP 2020*. 2020, pp. 4163–4174.
46. Jared Kaplan et al. "Scaling laws for neural language models". In: *arXiv preprint* arXiv:2001.08361 (2020).
47. Tassilo Klein and Moin Nabi. "Learning to Answer by Learning to Ask: Getting the Best of GPT-2 and BERT Worlds". In: *arXiv preprint* arXiv:1911.02365 (2019).
48. Andreas Köpf et al. "Openassistant conversations-democratizing large language model alignment". In: *Advances in Neural Information Processing Systems* 36 (2023), pp. 47669–47681.
49. Solomon Kullback and Richard A Leibler. "On information and sufficiency". In: *The annals of mathematical statistics* 22.1 (1951), pp. 79–86.
50. Nathan Lambert. "Reinforcement Learning from Human Feedback". In: *arXiv preprint* arXiv:2504.12501 (2025).
51. Jieh-Sheng Lee and Jieh Hsiang. "Patent claim generation by fine-tuning OpenAI GPT-2". In: *World Patent Information* 62 (2020), p. 101983.
52. Jinhyuk Lee et al. "BioBERT: a pre-trained biomedical language representation model for biomedical text mining". In: *Bioinformatics* 36.4 (2020), pp. 1234–1240.
53. Patrick Lewis et al. "Retrieval-augmented generation for knowledge-intensive NLP tasks". In: *Advances in Neural Information Processing Systems*. Vol. 33. 2020, pp. 9459–9474.
54. Yujia Li et al. "Competition-level code generation with alphacode". In: *Science* 378.6624 (2022), pp. 1092–1097.
55. Aixin Liu et al. "Deepseek v3 technical report". In: *arXiv preprint* arXiv:2412.19437 (2024).
56. Zhuang Liu, Degen Huang, Kaiyu Huang, Zhuang Li, and Jun Zhao. "FinBERT: A pre-trained financial language representation model for financial text mining". In: *Proceedings of the twenty-ninth international conference on international joint conferences on artificial intelligence*. 2021, pp. 4513–4519.
57. Kris McGuffie and Alex Newhouse. "The radicalization risks of GPT-3 and advanced neural language models". In: *arXiv preprint* arXiv:2009.06807 (2020).
58. Meta. *Introducing Meta Llama 3: The most capable openly available LLM to date*. https://ai.meta.com/blog/meta-llama-3/. 2024.
59. Microsoft. *Turing-NLG: A 17-billion-parameter language model by Microsoft*. Microsoft Blog. 2020.
60. Tomas Mikolov, Kai Chen, Greg Corrado, and Jeffrey Dean. "Efficient estimation of word representations in vector space". In: *International Conference on Learning Representations*. 2013.
61. Tomas Mikolov, Kai Chen, Gregory S Corrado, and Jeffrey A Dean. *Computing numeric representations of words in a high-dimensional space*. US Patent 9,037,464. 2015.
62. Tomas Mikolov, Ilya Sutskever, Kai Chen, Greg S Corrado, and Jeff Dean. "Distributed representations of words and phrases and their compositionality". In: *Advances in neural information processing systems*. 2013, pp. 3111–3119.
63. Mistral AI. *Mixtral of experts: A high quality Sparse Mixture-of-Experts*. https://mistral.ai/news/mixtral-of-experts. 2023.
64. Dat Quoc Nguyen, Thanh Vu, and Anh-Tuan Nguyen. "BERTweet: A pre-trained language model for English Tweets". In: *Proceedings of the 2020 Conference on Empirical Methods in Natural Language Processing: System Demonstrations*. 2020, pp. 9–14.
65. Catherine Olsson et al. "In-context learning and induction heads". In: *arXiv preprint* arXiv:2209.11895 (2022).
66. OpenAI. *ChatGPT (Mar 14 version) [Large language model]*. https://chat.openai.com. 2023.
67. OpenAI. *GPT-4o System Card*. 2024. URL: https://cdn.openai.com/gpt-4o-system-card.pdf.
68. OpenAI. *Introducing ChatGPT*. 2022. URL: https://openai.com/blog/chatgpt.
69. OpenAI. *OpenAI API*. https://platform.openai.com. Accessed: May 2025. 2020.
70. OpenAI. *OpenAI API Models*. 2023. URL: https://platform.openai.com/docs/models/gpt-4-turbo-and-gpt-4.
71. Long Ouyang et al. "Training language models to follow instructions with human feedback". In: *Advances in neural information processing systems* 35 (2022), pp. 27730–27744.
72. Kiho Park, Yo Joong Choe, and Victor Veitch. "The Linear Representation Hypothesis and the Geometry of Large Language Models". In: *International Conference on Machine Learning*. PMLR. 2024, pp. 39643–39666.
73. Guilherme Penedo et al. "The RefinedWeb dataset for Falcon LLM: outperforming curated corpora with web data, and web data only". In: *arXiv preprint* arXiv:2306.01116 (2023).
74. Baolin Peng, Chunyuan Li, Pengcheng He, Michel Galley, and Jianfeng Gao. "Instruction tuning with GPT-4". In: *arXiv preprint* arXiv:2304.03277 (2023).
75. Jeffrey Pennington, Richard Socher, and Christopher D Manning. "GloVe: Global vectors for word representation". In: *Proceedings of the 2014 conference on empirical methods in natural language processing*. 2014, pp. 1532–1543.
76. Telmo Pires, Eva Schlinger, and Dan Garrette. "How Multilingual is Multilingual BERT?" In: *Proceedings of the 57th Annual Meeting of the Association for Computational Linguistics*. 2019, pp. 4996–5001.

77. Chen Qu et al. "BERT with history answer embedding for conversational question answering". In: *Proceedings of the 42nd International ACM SIGIR Conference on Research and Development in Information Retrieval*. 2019, pp. 1133–1136.
78. Alec Radford, Karthik Narasimhan, Tim Salimans, and Ilya Sutskever. *Improving language understanding by generative pre-training*. Tech. rep. OpenAI, 2018.
79. Alec Radford et al. "Language models are unsupervised multitask learners". In: *OpenAI blog* 1.8 (2019), p. 9.
80. Rafael Rafailov et al. "Direct preference optimization: Your language model is secretly a reward model". In: *Advances in Neural Information Processing Systems* 36 (2023), pp. 53728–53741.
81. Colin Raffel et al. "Exploring the limits of transfer learning with a unified text-to-text transformer". In: *Journal of machine learning research* 21.140 (2020), pp. 1–67.
82. Baptiste Roziere et al. "Code Llama: Open foundation models for code". In: *arXiv preprint* arXiv:2308.12950 (2023).
83. Victor Sanh, Lysandre Debut, Julien Chaumond, and Thomas Wolf. "DistilBERT, a distilled version of BERT: smaller, faster, cheaper and lighter". In: *arXiv preprint* arXiv:1910.01108 (2019).
84. Adam Scherlis, Kshitij Sachan, Adam S Jermyn, Joe Benton, and Buck Shlegeris. "Polysemanticity and capacity in neural networks". In: *arXiv preprint* arXiv:2210.01892 (2022).
85. Florian Schroff, Dmitry Kalenichenko, and James Philbin. "Facenet: A unified embedding for face recognition and clustering". In: *Proceedings of the IEEE conference on computer vision and pattern recognition*. IEEE. 2015, pp. 815–823.
86. Zhihong Shao et al. "Deepseekmath: Pushing the limits of mathematical reasoning in open language models". In: *arXiv preprint* arXiv:2402.03300 (2024).
87. Mohammad Shoeybi et al. "Megatron-LM: Training multi-billion parameter language models using model parallelism". In: *arXiv preprint* arXiv:1909.08053 (2019).
88. Youwei Song, Jiahai Wang, Zhiwei Liang, Zhiyue Liu, and Tao Jiang. "Utilizing BERT intermediate layers for aspect based sentiment analysis and natural language inference". In: *arXiv preprint* arXiv:2002.04815 (2020).
89. Ieva Staliūnaitė and Ignacio Iacobacci. "Compositional and Lexical Semantics in RoBERTa, BERT and DistilBERT: A Case Study on CoQA". In: *Proceedings of the 2020 Conference on Empirical Methods in Natural Language Processing (EMNLP)*. 2020, pp. 7046–7056.
90. Stanford CRFM. *Alpaca GitHub*. https://github.com/tatsu-lab/stanford_alpaca. 2023.
91. Richard S Sutton, David McAllester, Satinder Singh, and Yishay Mansour. "Policy gradient methods for reinforcement learning with function approximation". In: *Advances in neural information processing systems* 12 (1999).
92. Gemini Team et al. "Gemini: a family of highly capable multimodal models". In: *arXiv preprint* arXiv:2312.11805 (2023).
93. Hugo Touvron et al. "Llama 2: Open foundation and fine-tuned chat models". In: *arXiv preprint* arXiv:2307.09288 (2023).
94. Hugo Touvron et al. "Llama: Open and efficient foundation language models". In: *arXiv preprint* arXiv:2302.13971 (2023).
95. Henry Tsai et al. "Small and Practical BERT Models for Sequence Labeling". In: *Proceedings of the 2019 Conference on Empirical Methods in Natural Language Processing and the 9th International Joint Conference on Natural Language Processing (EMNLP-IJCNLP)*. 2019, pp. 3632–3636.
96. Miles Turpin, Julian Michael, Ethan Perez, and Samuel Bowman. "Language models don't always say what they think: Unfaithful explanations in chain-of-thought prompting". In: *Advances in Neural Information Processing Systems* 36 (2023), pp. 74952–74965.
97. Hoang Van, David Kauchak, and Gondy Leroy. "AutoMeTS: The Autocomplete for Medical Text Simplification". In: *Proceedings of the 28th International Conference on Computational Linguistics*. 2020, pp. 1424–1434.
98. Ashish Vaswani et al. "Attention is all you need". In: *Advances in neural information processing systems*. 2017, pp. 5998–6008.
99. Jason Wei et al. "Chain-of-thought prompting elicits reasoning in large language models". In: *Advances in neural information processing systems* 35 (2022), pp. 24824–24837.
100. Thomas Wolf et al. "HuggingFace's Transformers: State-of-the-art Natural Language Processing". In: *arXiv preprint* arXiv:1910.03771 (2019).
101. xAI. *Grok-1*. https://github.com/xai-org/grok-1. 2024.
102. An Yang et al. "Qwen3 technical report". In: *arXiv preprint* arXiv:2505.09388 (2025).
103. Shengyu Zhang et al. "Instruction tuning for large language models: A survey". In: *arXiv preprint* arXiv:2308.10792 (2023).
104. Jialun Zhong et al. "A comprehensive survey of reward models: Taxonomy, applications, challenges, and future". In: *arXiv preprint* arXiv:2504.12328 (2025).

Part III
Generative Models

Variational Models

12

Abstract

Latent variables—unobserved factors that influence observable data—play a critical role in understanding complex phenomena across various fields, including causal inference and machine learning. This chapter explores the concept of latent variables and their significance in modeling hidden structures that affect observed variables. Through illustrative examples, it highlights how latent variables can explain apparent correlations that are not causative, emphasizing their importance in domains such as drug discovery and causal analysis. Focusing on machine learning, this chapter introduces variational inference as a principled framework for estimating distributions over latent variables conditioned on observed data. It covers the theoretical foundations of variational inference, including the Evidence Lower Bound (ELBO) and expectation maximization. This chapter further delves into variational models, with particular emphasis on Variational Autoencoders (VAEs). VAEs combine neural networks with probabilistic inference to learn latent spaces, enabling both data generation and representation learning. They are foundational in modern generative modeling, though newer extensions (such as hierarchical VAEs, adversarial VAEs, and diffusion models) continue to expand the field.

1 Introduction

In life, there are many observable variables that can be seen. However, these observed variables are conditioned on some latent variables, also called latent factors. Latent variables are not easily observed, but they have an impact on the observed variables. Often, phenomena that cannot be easily explained in life are, in fact, influenced by latent variables.

Latent variables have also been used to explain causal effects in causal inference [24]. The following example illustrates the importance of latent variables. It has been observed that most people who sleep with their shoes on wake up with a headache. A naive explanation is that the morning headache is caused by sleeping while wearing shoes. However, as illustrated in Fig. 1a, it has also been observed that the people who sleep wearing shoes were drunk. This explains that the morning headache is due to being drunk. Being drunk is the latent variable that causes the morning headache. Wearing shoes while sleeping merely co-occurs with the morning headache and is not its cause. Similar situations occur in drug discovery, where a drug should be investigated to determine whether it is the actual latent variable treating the disease or if the treatment is an accidental co-occurrence [22].

B. Ghojogh and A. Ghodsi, *Elements of Deep Learning*,
https://doi.org/10.1007/978-3-032-10738-1_12

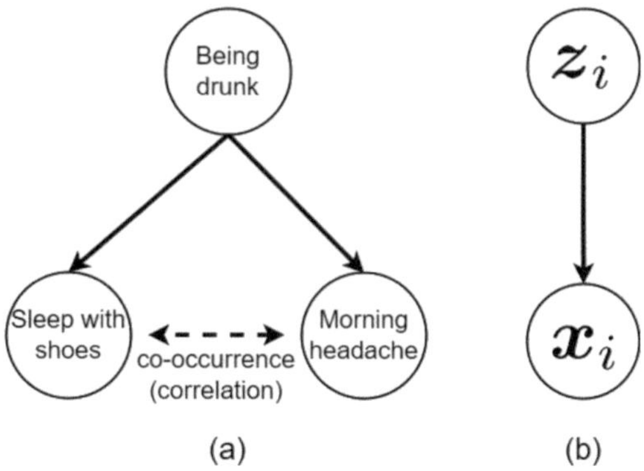

Fig. 1 Latent variables: **a** a real-world example of a latent variable, and **b** latent variable for an observed variable

Observed and latent variables also exist in machine learning and data science. Consider a dataset $\{\boldsymbol{x}_i\}_{i=1}^n$. Every data point $\boldsymbol{x}_i \in \mathbb{R}^d$ is an observed variable that can be seen. However, as shown in Fig. 1b, every data point $\boldsymbol{x}_i$ can be conditioned on some latent variable, also called a latent factor, $z_i \in \mathbb{R}^p$. The observed and latent variables are random variables, each having their own probability distributions. There is also a conditional probability distribution for conditioning the observed variable on the latent variable. In machine learning, variational inference refers to learning these distributions so that one can randomly draw the latent variable from its prior distribution and then randomly draw the observed variable from the conditional distribution. Variational inference was initially developed in factor analysis and probabilistic principal component analysis [7]. It has also been modeled in autoencoder neural networks to form Variational Autoencoders (VAEs) [17]. This chapter introduces variational inference and variational autoencoders as variational models.

This chapter introduces variational inference and variational models in deep learning. Section 2 explains variational inference, the Evidence Lower Bound (ELBO), and expectation maximization. The Variational Autoencoder, which models variational inference in an autoencoder neural network, is introduced in Sect. 3.

2 Variational Inference

Consider a dataset $\{\boldsymbol{x}_i\}_{i=1}^n$. Assume that every data point $\boldsymbol{x}_i \in \mathbb{R}^d$ is conditioned on a latent variable $z_i \in \mathbb{R}^p$. This latent variable has a prior distribution $\mathbb{P}(z_i)$. According to Bayes' rule, we have

$$\mathbb{P}(z_i \mid \boldsymbol{x}_i) = \frac{\mathbb{P}(\boldsymbol{x}_i \mid z_i)\, \mathbb{P}(z_i)}{\mathbb{P}(\boldsymbol{x}_i)}. \tag{1}$$

Suppose the parameter of the conditional distribution of z_i on $\boldsymbol{x}_i$ is denoted by $\boldsymbol{\theta}$; hence, $\mathbb{P}(z_i \mid \boldsymbol{x}_i) = \mathbb{P}(z_i \mid \boldsymbol{x}_i, \boldsymbol{\theta})$. Therefore, it can be written as follows:

$$\mathbb{P}(z_i \mid \boldsymbol{x}_i, \boldsymbol{\theta}) = \frac{\mathbb{P}(\boldsymbol{x}_i \mid z_i, \boldsymbol{\theta})\, \mathbb{P}(z_i \mid \boldsymbol{\theta})}{\mathbb{P}(\boldsymbol{x}_i \mid \boldsymbol{\theta})}. \tag{2}$$

We introduce an approximate distribution $q(z_i)$ to serve as a variational approximation to the true posterior, since the exact posterior $\mathbb{P}(z_i|\boldsymbol{x}_i, \boldsymbol{\theta})$ is often intractable.

2.1 Evidence Lower Bound (ELBO)

Consider the Kullback–Leibler (KL) divergence [19] between the variational distribution $q(z_i)$ and the true posterior $\mathbb{P}(z_i|\boldsymbol{x}_i, \boldsymbol{\theta})$. Minimizing this divergence is equivalent to maximizing the Evidence Lower Bound (ELBO), which will be introduced in this section.

$$\begin{aligned}
\text{KL}\big(q(z_i) \,\|\, \mathbb{P}(z_i \mid \boldsymbol{x}_i, \boldsymbol{\theta})\big) &\overset{(a)}{=} \int q(z_i) \log\Big(\frac{q(z_i)}{\mathbb{P}(z_i \mid \boldsymbol{x}_i, \boldsymbol{\theta})}\Big) dz_i \\
&= \int q(z_i)\big(\log(q(z_i)) - \log(\mathbb{P}(z_i \mid \boldsymbol{x}_i, \boldsymbol{\theta}))\big) dz_i \\
&\overset{(2)}{=} \int q(z_i)\big(\log(q(z_i)) - \log(\mathbb{P}(\boldsymbol{x}_i \mid z_i, \boldsymbol{\theta})) \\
&\qquad - \log(\mathbb{P}(z_i \mid \boldsymbol{\theta})) + \log(\mathbb{P}(\boldsymbol{x}_i \mid \boldsymbol{\theta}))\big)\, dz_i \\
&\overset{(b)}{=} \log(\mathbb{P}(\boldsymbol{x}_i \mid \boldsymbol{\theta})) + \int q(z_i)\big(\log(q(z_i)) \\
&\qquad - \log(\mathbb{P}(\boldsymbol{x}_i \mid z_i, \boldsymbol{\theta})) - \log(\mathbb{P}(z_i \mid \boldsymbol{\theta}))\big)\, dz_i \\
&= \log(\mathbb{P}(\boldsymbol{x}_i \mid \boldsymbol{\theta})) + \int q(z_i) \log\Big(\frac{q(z_i)}{\mathbb{P}(\boldsymbol{x}_i \mid z_i, \boldsymbol{\theta})\mathbb{P}(z_i \mid \boldsymbol{\theta})}\Big)\, dz_i \\
&= \log(\mathbb{P}(\boldsymbol{x}_i \mid \boldsymbol{\theta})) + \int q(z_i) \log\Big(\frac{q(z_i)}{\mathbb{P}(\boldsymbol{x}_i, z_i \mid \boldsymbol{\theta})}\Big)\, dz_i \\
&= \log(\mathbb{P}(\boldsymbol{x}_i \mid \boldsymbol{\theta})) + \text{KL}\big(q(z_i) \,\|\, \mathbb{P}(\boldsymbol{x}_i, z_i \mid \boldsymbol{\theta})\big),
\end{aligned}$$

where (a) is the definition of KL divergence and (b) is because $\log(\mathbb{P}(\boldsymbol{x}_i \mid \boldsymbol{\theta}))$ is independent of z_i and comes out of the integral, and $\int dz_i = 1$. Therefore:

$$\log(\mathbb{P}(\boldsymbol{x}_i \mid \boldsymbol{\theta})) = \text{KL}\big(q(z_i) \,\|\, \mathbb{P}(z_i \mid \boldsymbol{x}_i, \boldsymbol{\theta})\big) - \text{KL}\big(q(z_i) \,\|\, \mathbb{P}(\boldsymbol{x}_i, z_i \mid \boldsymbol{\theta})\big). \tag{3}$$

The *Evidence Lower Bound (ELBO)* is defined as follows:

$$\mathcal{L}(q, \boldsymbol{\theta}) := -\text{KL}\big(q(z_i) \,\|\, \mathbb{P}(\boldsymbol{x}_i, z_i \mid \boldsymbol{\theta})\big). \tag{4}$$

Thus:

$$\log(\mathbb{P}(\boldsymbol{x}_i \mid \boldsymbol{\theta})) = \text{KL}\big(q(z_i) \,\|\, \mathbb{P}(z_i \mid \boldsymbol{x}_i, \boldsymbol{\theta})\big) + \mathcal{L}(q, \boldsymbol{\theta}).$$

Therefore,

$$\mathcal{L}(q, \boldsymbol{\theta}) = \log(\mathbb{P}(\boldsymbol{x}_i \mid \boldsymbol{\theta})) - \underbrace{\text{KL}\big(q(z_i) \,\|\, \mathbb{P}(z_i \mid \boldsymbol{x}_i, \boldsymbol{\theta})\big)}_{\geq 0}. \tag{5}$$

As the second term is negative with its minus, the ELBO is a lower bound on the log-likelihood of the data:

$$\mathcal{L}(q, \boldsymbol{\theta}) \leq \log(\mathbb{P}(\boldsymbol{x}_i \mid \boldsymbol{\theta})). \tag{6}$$

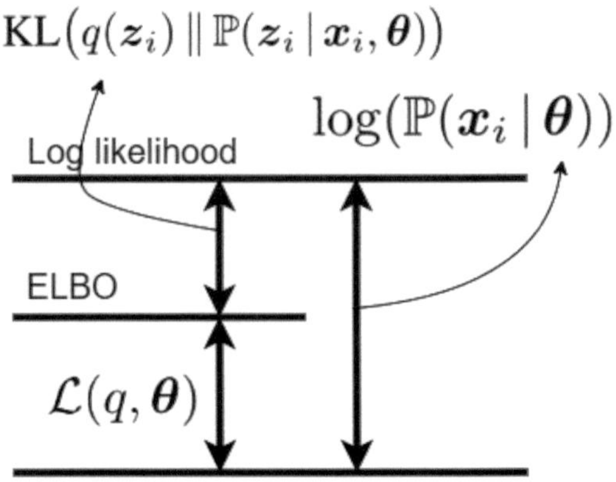

Fig. 2 Depiction of ELBO as the lower bound on log-likelihood. The image is inspired by [2]

The likelihood $\mathbb{P}(\boldsymbol{x}_i \mid \boldsymbol{\theta})$ is also referred to as the *evidence*. Note that this lower bound becomes tight when:

$$
\begin{aligned}
\mathcal{L}(q, \boldsymbol{\theta}) &\approx \log(\mathbb{P}(\boldsymbol{x}_i \mid \boldsymbol{\theta})) \\
&\implies 0 \leq \mathrm{KL}\big(q(z_i) \,\|\, \mathbb{P}(z_i \mid \boldsymbol{x}_i, \boldsymbol{\theta})\big) \overset{\text{set}}{=} 0 \\
&\implies q(z_i) = \mathbb{P}(z_i \mid \boldsymbol{x}_i, \boldsymbol{\theta}).
\end{aligned} \tag{7}
$$

This lower bound is depicted in Fig. 2.

2.2 Expectation Maximization

2.2.1 Background on Expectation Maximization

Sometimes, the data are not fully observable. For example, the data may only be known to be either zero or greater than zero. In such cases, Maximum Likelihood Estimation (MLE) cannot be directly applied because there is no access to complete information and some data are missing. Expectation Maximization (EM) is useful in these scenarios.

The main idea of EM can be summarized in this short friendly conversation [8]:

– *What shall we do? Some data are missing! The log-likelihood is not completely known, so MLE cannot be used.*
– *Hmm, probably we can replace the missing data with something...*
– *Aha! Let us replace it with its mean.*
– *You are right! We can take the mean of the log-likelihood over the possible values of the missing data. Then everything in the log-likelihood will be known, and then...*
– *And then we can do MLE!*

EM consists of two steps: the E-step and the M-step. In the E-step, the expectation of the log-likelihood with respect to the missing data is calculated to obtain a mean estimate. In the M-step, the MLE approach is used, where the log-likelihood is replaced with its expectation. These two steps are iteratively repeated until the estimated parameters converge.

2.2.2 Expectation Maximization in Variational Inference

MLE seeks to maximize the log-likelihood of the data. According to Eq. (6), maximizing the ELBO will also maximize the log-likelihood. Equation (6) holds for any prior distribution q. Therefore, it is desirable to find the best distribution to maximize the lower bound. Consequently, EM for variational inference is performed iteratively as follows:

$$\text{E-step:} \quad q^{(t)} := \arg\max_{q} \quad \mathcal{L}(q, \boldsymbol{\theta}^{(t-1)}), \tag{8}$$

$$\text{M-step:} \quad \boldsymbol{\theta}^{(t)} := \arg\max_{\boldsymbol{\theta}} \quad \mathcal{L}(q^{(t)}, \boldsymbol{\theta}), \tag{9}$$

where t denotes the iteration index.

E-step in EM for Variational Inference: The E-step in expectation maximization is

$$\begin{aligned}
\max_{q} \mathcal{L}(q, \boldsymbol{\theta}^{(t-1)}) &\overset{(5)}{=} \max_{q} \log(\mathbb{P}(\boldsymbol{x}_i \mid \boldsymbol{\theta}^{(t-1)})) \\
&\quad + \max_{q} \big(-\operatorname{KL}\big(q(z_i) \,\|\, \mathbb{P}(z_i \mid \boldsymbol{x}_i, \boldsymbol{\theta}^{(t-1)})\big)\big) \\
&= \max_{q} \log(\mathbb{P}(\boldsymbol{x}_i \mid \boldsymbol{\theta}^{(t-1)})) + \min_{q} \operatorname{KL}\big(q(z_i) \,\|\, \mathbb{P}(z_i \mid \boldsymbol{x}_i, \boldsymbol{\theta}^{(t-1)})\big).
\end{aligned}$$

The second term is always non-negative; therefore, its minimum is zero:

$$\begin{aligned}
&\operatorname{KL}\big(q(z_i) \,\|\, \mathbb{P}(z_i \mid \boldsymbol{x}_i, \boldsymbol{\theta}^{(t-1)})\big) \overset{\text{set}}{=} 0 \\
&\implies q(z_i) = \mathbb{P}(z_i \mid \boldsymbol{x}_i, \boldsymbol{\theta}^{(t-1)}),
\end{aligned}$$

which was already established in Eq. (7). Thus, the E-step assigns

$$q^{(t)}(z_i) \leftarrow \mathbb{P}(z_i \mid \boldsymbol{x}_i, \boldsymbol{\theta}^{(t-1)}). \tag{10}$$

In other words, in Fig. 2, it pushes the middle line toward the upper line by maximizing the ELBO.

M-step in EM for Variational Inference: The M-step in expectation maximization is

$$\begin{aligned}
\max_{\boldsymbol{\theta}} \mathcal{L}(q^{(t)}, \boldsymbol{\theta}) &\overset{(4)}{=} \max_{\boldsymbol{\theta}} \big(-\operatorname{KL}\big(q^{(t)}(z_i) \,\|\, \mathbb{P}(\boldsymbol{x}_i, z_i \mid \boldsymbol{\theta})\big)\big) \\
&\overset{(a)}{=} \max_{\boldsymbol{\theta}} \Big[-\int q^{(t)}(z_i) \log\Big(\frac{q^{(t)}(z_i)}{\mathbb{P}(\boldsymbol{x}_i, z_i \mid \boldsymbol{\theta})}\Big)\, dz_i\Big] \\
&= \max_{\boldsymbol{\theta}} \int q^{(t)}(z_i) \log(\mathbb{P}(\boldsymbol{x}_i, z_i \mid \boldsymbol{\theta}))\, dz_i \\
&\quad - \max_{\boldsymbol{\theta}} \int q^{(t)}(z_i) \log(q^{(t)}(z_i))\, dz_i,
\end{aligned}$$

where (a) is the definition of KL divergence. The second term is constant with respect to $\boldsymbol{\theta}$. Therefore:

$$\max_{\theta} \mathcal{L}(q^{(t)}, \theta) = \max_{\theta} \int q^{(t)}(z_i) \log(\mathbb{P}(x_i, z_i \mid \theta)) \, dz_i$$
$$\overset{(a)}{=} \max_{\theta} \mathbb{E}_{\sim q^{(t)}(z_i)}\big[\log \mathbb{P}(x_i, z_i \mid \theta)\big],$$

where (a) is by the definition of expectation. Thus, the M-step assigns

$$\theta^{(t)} \leftarrow \arg\max_{\theta} \mathbb{E}_{\sim q^{(t)}(z_i)}\big[\log \mathbb{P}(x_i, z_i \mid \theta)\big]. \tag{11}$$

In other words, in Fig. 2, it pushes the upper line higher. The E-step and M-step together can be viewed as a game: the E-step tries to bring the middle line (i.e., the ELBO) closer to the upper line (i.e., the log-likelihood), while the M-step tries to move the upper line higher. This procedure is repeated iteratively, allowing the two steps to help each other improve to higher values.

To summarize, EM in variational inference is

$$q^{(t)}(z_i) \leftarrow \mathbb{P}(z_i \mid x_i, \theta^{(t-1)}), \tag{12}$$
$$\theta^{(t)} \leftarrow \arg\max_{\theta} \mathbb{E}_{\sim q^{(t)}(z_i)}\big[\log \mathbb{P}(x_i, z_i \mid \theta)\big]. \tag{13}$$

It is noteworthy that in variational inference, the parameter θ is sometimes absorbed into the latent variable z_i. According to the chain rule, we have

$$\mathbb{P}(x_i, z_i, \theta) = \mathbb{P}(x_i \mid z_i, \theta)\, \mathbb{P}(z_i \mid \theta)\, \mathbb{P}(\theta).$$

Considering the term $\mathbb{P}(z_i \mid \theta)\, \mathbb{P}(\theta)$ as a single probability term gives

$$\mathbb{P}(x_i, z_i) = \mathbb{P}(x_i \mid z_i)\, \mathbb{P}(z_i),$$

where the parameter θ disappears due to absorption.

3 Variational Autoencoder

The Variational Autoencoder (VAE) [17] applies variational inference, i.e., it maximizes the ELBO, within an autoencoder setup, making it differentiable for backpropagation training [28]. As shown in Fig. 3, a VAE consists of an encoder and a decoder, each of which can have multiple network layers. A latent space is learned between the encoder and decoder, and the latent variable z_i is sampled from this latent space. The encoder and decoder components are explained in the following sections. In a VAE, the input to the encoder is the data point x_i, and the output of the decoder is its reconstruction x_i.

3.1 Parts of Variational Autoencoder

3.1.1 Encoder of Variational Autoencoder

The encoder of a VAE models the distribution $q(z_i) = \mathbb{P}(z_i \mid x_i, \theta_e)$, where the parameters of the distribution, θ_e, are the weights of the encoder layers in the VAE. The input and output of the encoder are $x_i \in \mathbb{R}^d$ and $z_i \in \mathbb{R}^p$, respectively. As shown in Fig. 3, the output neurons of the encoder determine the parameters of the conditional distribution $\mathbb{P}(z_i \mid x_i, \theta_e)$. If this conditional distribution has m parame-

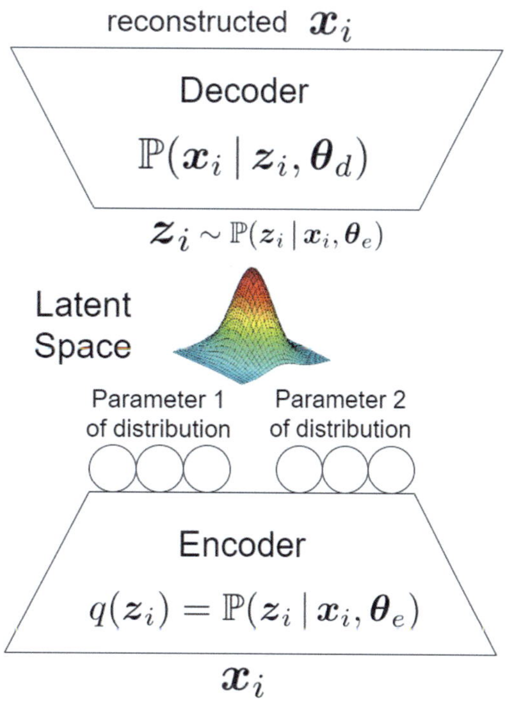

Fig. 3 The structure of a variational autoencoder

ters, there will be m sets of output neurons from the encoder, denoted by $\{\boldsymbol{e}_j\}_{j=1}^m$. The dimensionality of these sets may differ depending on the size of the parameters.

For example, let the latent space be p-dimensional, i.e., $z_i \in \mathbb{R}^p$. If the distribution $\mathbb{P}(z_i \mid \boldsymbol{x}_i, \boldsymbol{\theta}_e)$ is a multivariate Gaussian distribution, there will be two sets of output neurons for the encoder: one set with p neurons representing the mean of this distribution, $\boldsymbol{\mu}_{z|x} = \boldsymbol{e}_1 \in \mathbb{R}^p$, and another set with $(p \times p)$ neurons for the covariance of this distribution, where $\boldsymbol{\Sigma}_{z|x} \in \mathbb{R}^{p \times p}$ is obtained by reshaping $\boldsymbol{e}_2 \in \mathbb{R}^{(p \times p)}$ into a $(p \times p)$ matrix. If the covariance matrix is diagonal, the second set will have p neurons rather than $(p \times p)$, and their values will be the diagonal elements of the covariance matrix. In this case, the covariance is $\boldsymbol{\Sigma}_{z|x} = \mathbf{diag}(\boldsymbol{e}_2) \in \mathbb{R}^{p \times p}$.

Any distribution with any number of parameters can be chosen for $\mathbb{P}(z_i \mid \boldsymbol{x}_i, \boldsymbol{\theta}_e)$, but the multivariate Gaussian with diagonal covariance is often used:

$$q(z_i) = \mathbb{P}(z_i \mid \boldsymbol{x}_i, \boldsymbol{\theta}_e) = \mathcal{N}(z_i \mid \boldsymbol{\mu}_{z|x}, \boldsymbol{\Sigma}_{z|x}). \tag{14}$$

Let the network weights for the output sets of the encoder, $\{\boldsymbol{e}_j\}_{j=1}^m$, be denoted by $\{\boldsymbol{\theta}_{e,j}\}_{j=1}^m$. As the input to the encoder is $\boldsymbol{x}_i$, the j-th output set of the encoder can be written as $\boldsymbol{e}_j(\boldsymbol{x}_i, \boldsymbol{\theta}_{e,j})$. In the case of a multivariate Gaussian distribution for the latent space, the parameters are $\boldsymbol{\mu}_{z|x} = \boldsymbol{e}_1(\boldsymbol{x}_i, \boldsymbol{\theta}_{e,1})$ and $\boldsymbol{\Sigma}_{z|x} = \mathbf{diag}(\boldsymbol{e}_2(\boldsymbol{x}_i, \boldsymbol{\theta}_{e,2}))$.

3.1.2 Sampling the Latent Variable

When the data point $\boldsymbol{x}_i$ is fed as input to the encoder, the parameters of the conditional distribution $q(z_i)$ are obtained; therefore, the distribution of the latent space, which is $q(z_i)$, is determined corresponding to the data point $\boldsymbol{x}_i$. Then, in the latent space, the corresponding latent variable is sampled from this distribution:

$$z_i \sim q(z_i) = \mathbb{P}(z_i \mid \boldsymbol{x}_i, \boldsymbol{\theta}_e). \tag{15}$$

This latent variable is then fed as input to the decoder, which is explained in the following section.

3.1.3 Decoder of Variational Autoencoder

As shown in Fig. 3, the decoder of a VAE models the conditional distribution $\mathbb{P}(\boldsymbol{x}_i \mid z_i, \boldsymbol{\theta}_d)$, where $\boldsymbol{\theta}_d$ represents the weights of the decoder layers in the VAE. The input and output of the decoder are $z_i \in \mathbb{R}^p$ and $\boldsymbol{x}_i \in \mathbb{R}^d$, respectively.

The output neurons of the decoder are designed to either generate the reconstructed data point directly or determine the parameters of the conditional distribution $\mathbb{P}(\boldsymbol{x}_i \mid z_i, \boldsymbol{\theta}_d)$; the former is more common. In the latter case, if this conditional distribution has l parameters, there will be l sets of output neurons from the decoder, denoted by $\{\boldsymbol{d}_j\}_{j=1}^l$. The dimensionality of these sets may differ depending on the size of each parameter, similar to the explanation in Sect. 3.1.1.

Let the network weights for the output sets of the decoder, $\{\boldsymbol{d}_j\}_{j=1}^l$, be denoted by $\{\boldsymbol{\theta}_{d,j}\}_{j=1}^l$. As the input to the decoder is z_i, the j-th output set of the decoder can be written as $\boldsymbol{d}_j(z_i, \boldsymbol{\theta}_{d,j})$.

3.2 Training Variational Autoencoder with Expectation Maximization

While EM provides intuition, in practice, VAEs are trained using backpropagation and the reparameterization trick. The EM description here is illustrative, but actual implementations use stochastic gradient descent (explained in Sect. 3.4).

Recall Eqs. (8) and (9) for EM in variational inference. Inspired by this, the VAE uses EM for training, where the ELBO is a function of the encoder weights $\boldsymbol{\theta}_e$, decoder weights $\boldsymbol{\theta}_d$, and the data point $\boldsymbol{x}_i$:

$$\text{E-step:} \quad \boldsymbol{\theta}_e^{(t)} := \arg\max_q \quad \mathcal{L}(\boldsymbol{\theta}_e, \boldsymbol{\theta}_d^{(t-1)}, \boldsymbol{x}_i), \tag{16}$$

$$\text{M-step:} \quad \boldsymbol{\theta}_d^{(t)} := \arg\max_q \quad \mathcal{L}(\boldsymbol{\theta}_e^{(t)}, \boldsymbol{\theta}_d, \boldsymbol{x}_i). \tag{17}$$

This iterative optimization algorithm can be simplified using alternating optimization [16], where a step of gradient ascent optimization is taken in every iteration. Mini-batch stochastic gradient ascent can be used, where the training data are divided into batches of size b. Therefore, the optimization becomes

$$\text{E-step:} \quad \boldsymbol{\theta}_e^{(t)} := \boldsymbol{\theta}_e^{(t-1)} + \eta_e \frac{\partial \sum_{i=1}^b \mathcal{L}(\boldsymbol{\theta}_e, \boldsymbol{\theta}_d^{(t-1)}, \boldsymbol{x}_i)}{\partial \boldsymbol{\theta}_e}, \tag{18}$$

$$\text{M-step:} \quad \boldsymbol{\theta}_d^{(t)} := \boldsymbol{\theta}_d^{(t-1)} + \eta_d \frac{\partial \sum_{i=1}^b \mathcal{L}(\boldsymbol{\theta}_e^{(t)}, \boldsymbol{\theta}_d, \boldsymbol{x}_i)}{\partial \boldsymbol{\theta}_d}, \tag{19}$$

where η_e and η_d are the learning rates for $\boldsymbol{\theta}_e$ and $\boldsymbol{\theta}_d$, respectively.

The ELBO is simplified as follows:

$$\sum_{i=1}^{b} \mathcal{L}(q, \boldsymbol{\theta}) \overset{(4)}{=} -\sum_{i=1}^{b} \text{KL}\big(q(z_i) \,\|\, \mathbb{P}(\boldsymbol{x}_i, z_i \,|\, \boldsymbol{\theta}_d)\big)$$
$$\overset{(12)}{=} -\sum_{i=1}^{b} \text{KL}\big(\mathbb{P}(z_i \,|\, \boldsymbol{x}_i, \boldsymbol{\theta}_e) \,\|\, \mathbb{P}(\boldsymbol{x}_i, z_i \,|\, \boldsymbol{\theta}_d)\big). \tag{20}$$

Note that the parameter of $\mathbb{P}(\boldsymbol{x}_i, z_i \,|\, \boldsymbol{\theta}_d)$ is $\boldsymbol{\theta}_d$ because z_i is generated after the encoder and before the decoder.

There are different ways to approximate the KL divergence in Eq. (20) [5, 12]. The ELBO can be simplified in at least two different ways, which are explained in the following.

3.2.1 Simplification Type 1

The simplification of the ELBO can be continued as follows:

$$\sum_{i=1}^{b} \mathcal{L}(q, \boldsymbol{\theta}) = -\sum_{i=1}^{b} \text{KL}\big(\mathbb{P}(z_i \,|\, \boldsymbol{x}_i, \boldsymbol{\theta}_e) \,\|\, \mathbb{P}(\boldsymbol{x}_i, z_i \,|\, \boldsymbol{\theta}_d)\big)$$
$$= -\sum_{i=1}^{b} \mathbb{E}_{\sim q^{(t-1)}(z_i)}\Big[\log\Big(\frac{\mathbb{P}(z_i \,|\, \boldsymbol{x}_i, \boldsymbol{\theta}_e)}{\mathbb{P}(\boldsymbol{x}_i, z_i \,|\, \boldsymbol{\theta}_d)}\Big)\Big]$$
$$= -\sum_{i=1}^{b} \mathbb{E}_{\sim \mathbb{P}(z_i \,|\, \boldsymbol{x}_i, \boldsymbol{\theta}_e)}\Big[\log\Big(\frac{\mathbb{P}(z_i \,|\, \boldsymbol{x}_i, \boldsymbol{\theta}_e)}{\mathbb{P}(\boldsymbol{x}_i, z_i \,|\, \boldsymbol{\theta}_d)}\Big)\Big]. \tag{21}$$

This expectation can be approximated using Monte Carlo approximation [9], where ℓ samples $\{z_{i,j}\}_{j=1}^{\ell}$ are drawn for the i-th data point from the conditional distribution:

$$z_{i,j} \sim \mathbb{P}(z_i \,|\, \boldsymbol{x}_i, \boldsymbol{\theta}_e), \quad \forall j \in \{1, \dots, \ell\}. \tag{22}$$

Monte Carlo approximation [9], in general, approximates an expectation as follows:

$$\mathbb{E}_{\sim \mathbb{P}(z_i \,|\, \boldsymbol{x}_i, \boldsymbol{\theta}_e)}\big[f(z_i)\big] \approx \frac{1}{\ell} \sum_{j=1}^{\ell} f(z_{i,j}), \tag{23}$$

where $f(z_i)$ is a function of z_i. Here, the approximation becomes

$$\sum_{i=1}^{b} \mathcal{L}(q, \boldsymbol{\theta}) \approx \sum_{i=1}^{b} \tilde{\mathcal{L}}(q, \boldsymbol{\theta})$$
$$= -\sum_{i=1}^{b} \frac{1}{\ell} \sum_{j=1}^{\ell} \log\Big(\frac{\mathbb{P}(z_{i,j} \,|\, \boldsymbol{x}_i, \boldsymbol{\theta}_e)}{\mathbb{P}(\boldsymbol{x}_i, z_{i,j} \,|\, \boldsymbol{\theta}_d)}\Big)$$
$$= \sum_{i=1}^{b} \frac{1}{\ell} \sum_{j=1}^{\ell} \Big[\log\big(\mathbb{P}(\boldsymbol{x}_i, z_{i,j} \,|\, \boldsymbol{\theta}_d)\big) - \log\big(\mathbb{P}(z_{i,j} \,|\, \boldsymbol{x}_i, \boldsymbol{\theta}_e)\big)\Big]. \tag{24}$$

3.2.2 Simplification Type 2

The ELBO can be simplified using another approach:

$$
\begin{aligned}
\sum_{i=1}^{b} \mathcal{L}(q, \boldsymbol{\theta}) &= -\sum_{i=1}^{b} \text{KL}\big(\mathbb{P}(z_i \,|\, \boldsymbol{x}_i, \boldsymbol{\theta}_e) \,\|\, \mathbb{P}(\boldsymbol{x}_i, z_i \,|\, \boldsymbol{\theta}_d)\big) \\
&= -\sum_{i=1}^{b} \int \mathbb{P}(z_i \,|\, \boldsymbol{x}_i, \boldsymbol{\theta}_e) \log\Big(\frac{\mathbb{P}(z_i \,|\, \boldsymbol{x}_i, \boldsymbol{\theta}_e)}{\mathbb{P}(\boldsymbol{x}_i, z_i \,|\, \boldsymbol{\theta}_d)}\Big)\, dz_i \\
&= -\sum_{i=1}^{b} \int \mathbb{P}(z_i \,|\, \boldsymbol{x}_i, \boldsymbol{\theta}_e) \log\Big(\frac{\mathbb{P}(z_i \,|\, \boldsymbol{x}_i, \boldsymbol{\theta}_e)}{\mathbb{P}(\boldsymbol{x}_i \,|\, z_i, \boldsymbol{\theta}_d)\, \mathbb{P}(z_i)}\Big)\, dz_i \\
&= -\sum_{i=1}^{b} \int \mathbb{P}(z_i \,|\, \boldsymbol{x}_i, \boldsymbol{\theta}_e) \log\Big(\frac{\mathbb{P}(z_i \,|\, \boldsymbol{x}_i, \boldsymbol{\theta}_e)}{\mathbb{P}(z_i)}\Big)\, dz_i \\
&\quad + \sum_{i=1}^{b} \int \mathbb{P}(z_i \,|\, \boldsymbol{x}_i, \boldsymbol{\theta}_e) \log\big(\mathbb{P}(\boldsymbol{x}_i \,|\, z_i, \boldsymbol{\theta}_d)\big)\, dz_i \\
&= -\sum_{i=1}^{b} \text{KL}\big(\mathbb{P}(z_i \,|\, \boldsymbol{x}_i, \boldsymbol{\theta}_e) \,\|\, \mathbb{P}(z_i)\big) \\
&\quad + \sum_{i=1}^{b} \mathbb{E}_{\sim \mathbb{P}(z_i \,|\, \boldsymbol{x}_i, \boldsymbol{\theta}_e)}\Big[\log\big(\mathbb{P}(\boldsymbol{x}_i \,|\, z_i, \boldsymbol{\theta}_d)\big)\Big].
\end{aligned} \tag{25}
$$

The second term in the above equation can be estimated using Monte Carlo approximation [9], where ℓ samples $\{z_{i,j}\}_{j=1}^{\ell}$ are drawn from $\mathbb{P}(z_i \,|\, \boldsymbol{x}_i, \boldsymbol{\theta}_e)$:

$$
\begin{aligned}
\sum_{i=1}^{b} \mathcal{L}(q, \boldsymbol{\theta}) \approx \sum_{i=1}^{b} \widetilde{\mathcal{L}}(q, \boldsymbol{\theta})
&= -\sum_{i=1}^{b} \text{KL}\big(\mathbb{P}(z_i \,|\, \boldsymbol{x}_i, \boldsymbol{\theta}_e) \,\|\, \mathbb{P}(z_i)\big) \\
&\quad + \sum_{i=1}^{b} \frac{1}{\ell} \sum_{j=1}^{\ell} \log\big(\mathbb{P}(\boldsymbol{x}_i \,|\, z_{i,j}, \boldsymbol{\theta}_d)\big).
\end{aligned} \tag{26}
$$

The first term in the above equation can also be converted to an expectation and computed using Monte Carlo approximation [9], where ℓ samples $\{z_{i,j}\}_{j=1}^{\ell}$ are drawn from $\mathbb{P}(z_i \,|\, \boldsymbol{x}_i, \boldsymbol{\theta}_e)$:

$$
\begin{aligned}
\sum_{i=1}^{b} \mathcal{L}(q, \boldsymbol{\theta}) \approx \sum_{i=1}^{b} \widetilde{\mathcal{L}}(q, \boldsymbol{\theta})
&= -\sum_{i=1}^{b} \mathbb{E}_{\sim \mathbb{P}(z_i \,|\, \boldsymbol{x}_i, \boldsymbol{\theta}_e)}\Big[\log\Big(\frac{\mathbb{P}(z_i \,|\, \boldsymbol{x}_i, \boldsymbol{\theta}_e)}{\mathbb{P}(z_i)}\Big)\Big] \\
&\quad + \sum_{i=1}^{b} \frac{1}{\ell} \sum_{j=1}^{\ell} \log\big(\mathbb{P}(\boldsymbol{x}_i \,|\, z_{i,j}, \boldsymbol{\theta}_d)\big)
\end{aligned}
$$

$$\approx -\sum_{i=1}^{b} \frac{1}{\ell} \sum_{j=1}^{\ell} \log \big(\mathbb{P}(z_{i,j} \mid \boldsymbol{x}_i, \boldsymbol{\theta}_e)\big) - \log \big(\mathbb{P}(z_{i,j})\big)$$

$$+ \sum_{i=1}^{b} \frac{1}{\ell} \sum_{j=1}^{\ell} \log \big(\mathbb{P}(\boldsymbol{x}_i \mid z_{i,j}, \boldsymbol{\theta}_d)\big). \tag{27}$$

If specific families of distributions, such as Gaussian distributions, are considered for $\mathbb{P}(z_{i,j} \mid \boldsymbol{x}_i, \boldsymbol{\theta}_e)$ and $\mathbb{P}(z_{i,j})$, the first term in Eq. (26) can be computed analytically. In the following, Eq. (26) is simplified further for Gaussian distributions.

3.2.3 Simplification Type 2 for Special Case of Gaussian Distributions

The KL divergence in the first term of Eq. (26) can be computed analytically for univariate or multivariate Gaussian distributions. For this, the following two lemmas are needed.

Lemma 1 *The KL divergence between two univariate Gaussian distributions* $p_1 \sim \mathcal{N}(\mu_1, \sigma_1^2)$ *and* $p_2 \sim \mathcal{N}(\mu_2, \sigma_2^2)$ *is*

$$KL(p_1 \| p_2) = \log(\frac{\sigma_2}{\sigma_1}) + \frac{\sigma_1^2 + (\mu_1 - \mu_2)^2}{2\sigma_2^2} - \frac{1}{2}. \tag{28}$$

Proof See Appendix A for proof. □

Lemma 2 *The KL divergence between two multivariate Gaussian distributions* $p_1 \sim \mathcal{N}(\boldsymbol{\mu}_1, \boldsymbol{\Sigma}_1)$ *and* $p_2 \sim \mathcal{N}(\boldsymbol{\mu}_2, \boldsymbol{\Sigma}_2)$ *with dimensionality* p *is*

$$KL(p_1 \| p_2) = \frac{1}{2}\Big(\log(\frac{|\boldsymbol{\Sigma}_2|}{|\boldsymbol{\Sigma}_1|}) - p + \boldsymbol{tr}(\boldsymbol{\Sigma}_2^{-1}\boldsymbol{\Sigma}_1) + (\boldsymbol{\mu}_2 - \boldsymbol{\mu}_1)^\top \boldsymbol{\Sigma}_2^{-1}(\boldsymbol{\mu}_2 - \boldsymbol{\mu}_1)\Big). \tag{29}$$

Proof See [5, Section 9] for proof. □

Consider the case with the following distributions:

$$\mathbb{P}(z_i \mid \boldsymbol{x}_i, \boldsymbol{\theta}_e) \sim \mathcal{N}(\boldsymbol{\mu}_{z|x}, \boldsymbol{\Sigma}_{z|x}), \tag{30}$$

$$\mathbb{P}(z_i) \sim \mathcal{N}(\boldsymbol{\mu}_z, \boldsymbol{\Sigma}_z), \tag{31}$$

where $z_i \in \mathbb{R}^p$.

Note that the parameters $\boldsymbol{\mu}_{z|x}$ and $\boldsymbol{\Sigma}_{z|x}$ are trained in the neural network, while the parameters $\mathbb{P}(z_{i,j})$ can be set to $\boldsymbol{\mu}_z = \boldsymbol{0}$ and $\boldsymbol{\Sigma}_z = \boldsymbol{I}$, inspired by factor analysis [7]. According to Lemma 2, the approximation of the ELBO, i.e., Eq. (26), can be simplified to:

$$
\begin{aligned}
\sum_{i=1}^{b} \mathcal{L}(q,\boldsymbol{\theta}) \approx \sum_{i=1}^{b} \widetilde{\mathcal{L}}(q,\boldsymbol{\theta}) \\
= -\sum_{i=1}^{b} \frac{1}{2}\Big(\log(\frac{|\boldsymbol{\Sigma}_z|}{|\boldsymbol{\Sigma}_{z|x}|}) - p + \mathbf{tr}(\boldsymbol{\Sigma}_z^{-1}\boldsymbol{\Sigma}_{z|x}) \\
+ (\boldsymbol{\mu}_z - \boldsymbol{\mu}_{z|x})^\top \boldsymbol{\Sigma}_z^{-1} (\boldsymbol{\mu}_z - \boldsymbol{\mu}_{z|x})\Big) \\
+ \sum_{i=1}^{b} \frac{1}{\ell} \sum_{j=1}^{\ell} \log\big(\mathbb{P}(\boldsymbol{x}_i \,|\, z_{i,j}, \boldsymbol{\theta}_d)\big).
\end{aligned} \tag{32}
$$

3.2.4 Training Variational Autoencoder with Approximations

We can train a VAE using EM, where Monte Carlo approximations are applied to the ELBO. Equations (18) and (19) are replaced by the following equations:

$$
\text{E-step:} \quad \boldsymbol{\theta}_e^{(t)} := \boldsymbol{\theta}_e^{(t-1)} + \eta_e \frac{\partial \sum_{i=1}^{b} \widetilde{\mathcal{L}}(\boldsymbol{\theta}_e, \boldsymbol{\theta}_d^{(t-1)}, \boldsymbol{x}_i)}{\partial \boldsymbol{\theta}_e}, \tag{33}
$$

$$
\text{M-step:} \quad \boldsymbol{\theta}_d^{(t)} := \boldsymbol{\theta}_d^{(t-1)} + \eta_d \frac{\partial \sum_{i=1}^{b} \widetilde{\mathcal{L}}(\boldsymbol{\theta}_e^{(t)}, \boldsymbol{\theta}_d, \boldsymbol{x}_i)}{\partial \boldsymbol{\theta}_d}, \tag{34}
$$

where the approximated ELBO was introduced in the previous sections.

3.2.5 Prior Regularization

Some works regularize the ELBO, Eq. (20), by adding a penalty on the prior distribution $\mathbb{P}(z_i)$. Using this approach, the learned distribution of the latent space, $\mathbb{P}(z_i \,|\, \boldsymbol{x}_i, \boldsymbol{\theta}_e)$, is guided to match a specific prior distribution $\mathbb{P}(z_i)$. Some examples of prior regularization in VAEs include geodesic priors [11] and optimal priors [29].

Note that this regularization can inject domain knowledge into the latent space. It can also be useful for making the latent space more interpretable.

3.3 The Reparameterization Trick

Sampling the ℓ samples for the latent variables, as in Eq. (15), blocks the gradient flow because computing the derivatives through $\mathbb{P}(z_i \,|\, \boldsymbol{x}_i, \boldsymbol{\theta}_e)$ using the chain rule results in a high variance estimate of the gradient. To overcome this problem, the reparameterization technique [17, 27, 30] is used. In this technique, instead of sampling $z_i \sim \mathbb{P}(z_i \,|\, \boldsymbol{x}_i, \boldsymbol{\theta}_e)$, z_i is considered as a random variable that is a deterministic function $g(.)$ of another random variable ϵ_i, as follows:

$$
z_i = g(\epsilon_i, \boldsymbol{x}_i, \boldsymbol{\theta}_e), \tag{35}
$$

where ϵ_i is a stochastic variable sampled from a distribution:

$$
\epsilon_i \sim \mathbb{P}(\epsilon). \tag{36}
$$

Equations (21) and (25) both contain an expectation of a function $f(z_i)$. Using this technique, the expectation is replaced as follows:

$$\mathbb{E}_{\sim \mathbb{P}(z_i \mid \boldsymbol{x}_i, \boldsymbol{\theta}_e)}[f(z_i)] \to \mathbb{E}_{\sim \mathbb{P}(z_i \mid \boldsymbol{x}_i, \boldsymbol{\theta}_e)}[f(g(\boldsymbol{\epsilon}_i, \boldsymbol{x}_i, \boldsymbol{\theta}_e))]. \tag{37}$$

Using the reparameterization technique, the encoder, which implemented $\mathbb{P}(z_i \mid \boldsymbol{x}_i, \boldsymbol{\theta}_e)$, is replaced by $g(\boldsymbol{\epsilon}_i, \boldsymbol{x}_i, \boldsymbol{\theta}_e)$, where in the latent space between encoder and decoder, there is $\boldsymbol{\epsilon}_i \sim \mathbb{P}(\boldsymbol{\epsilon})$ and $z_i = g(\boldsymbol{\epsilon}_i, \boldsymbol{x}_i, \boldsymbol{\theta}_e)$.

A simple example of the reparameterization technique is when z_i and ϵ_i are univariate Gaussian variables:

$$\begin{aligned}
z_i &\sim \mathcal{N}(\mu, \sigma^2), \\
\epsilon_i &\sim \mathcal{N}(0, 1), \\
z_i &= g(\epsilon_i) = \mu + \sigma \epsilon_i.
\end{aligned}$$

For more advanced reparameterization techniques, the reader can refer to [6].

3.4 Training Variational Autoencoder with Backpropagation

In practice, a VAE is trained using backpropagation [27], where the backpropagation algorithm [28] is employed to train the network weights. Recall that in training a VAE with EM, the encoder and decoder are trained separately using the E-step and the M-step of EM, respectively. However, when training a VAE with backpropagation, the entire network is trained together rather than in separate steps. Suppose the complete set of weights of the VAE is denoted by $\boldsymbol{\theta} := \{\boldsymbol{\theta}_e, \boldsymbol{\theta}_d\}$. Backpropagation trains the VAE using mini-batch stochastic gradient descent with the negative ELBO, $\sum_{i=1}^{b} -\widetilde{\mathcal{L}}(\boldsymbol{\theta}, \boldsymbol{x}_i)$, as the loss function:

$$\boldsymbol{\theta}^{(t)} := \boldsymbol{\theta}^{(t-1)} - \eta \frac{\partial \sum_{i=1}^{b} -\widetilde{\mathcal{L}}(\boldsymbol{\theta}, \boldsymbol{x}_i)}{\partial \boldsymbol{\theta}}, \tag{38}$$

where η is the learning rate. Note that minimization is used here, with gradient descent, because neural networks typically minimize the loss function.

3.5 The Test Phase in Variational Autoencoder

In the test phase, the test data point $\boldsymbol{x}_i$ is fed to the encoder to determine the parameters of the conditional distribution of the latent space, i.e., $\mathbb{P}(z_i \mid \boldsymbol{x}_i, \boldsymbol{\theta}_e)$. Then, the latent variable z_i is sampled from this distribution in the latent space, and the corresponding reconstructed data point $\boldsymbol{x}_i$ is generated by the decoder.

Alternatively, one can sample latent variables directly from the prior distribution and decode them to generate new data. This illustrates the generative power of VAEs. In this case, z_i can be sampled from the prior $q(z_i)$, which has converged to $\mathbb{P}(z_i \mid \boldsymbol{x}_i, \boldsymbol{\theta})$ according to Eq. (12). Then, z_i is fed to the decoder to generate the reconstructed data point $\boldsymbol{x}_i$.

3.6 Other Notes and Other Variants of Variational Autoencoder

There exist many improvements to VAEs. A common challenge is that standard VAEs often produce blurry images due to limitations of Gaussian decoders and the ELBO objective. Specifically, this blurriness may be due to several reasons:

- sampling for the Monte Carlo approximations,
- lower bound approximation by ELBO, or
- restrictions on the family of distributions, as simple Gaussian distributions are usually used.

There are other interpretations for the cause of this problem; for example, see [34].

Extensions such as β-VAEs [13], hierarchical VAEs [18], and adversarial VAEs [21] (combination of VAE and GANs [10]) address these issues, while diffusion models [14] (see Chap. 15) and autoregressive transformers [32] (see Chap. 9) now dominate image generation benchmarks.

Variational discriminant analysis [33] has also been proposed for classification and discrimination of classes. Two tutorials on VAEs are [4, 23]. Some other works on VAEs include nearly optimal VAE [1], deep VAE [15], Hamiltonian VAE [3], and Nouveau VAE [31], which is a hierarchical VAE. For image data and image-caption modeling, a fusion of VAE and Convolutional Neural Network (CNN) has also been proposed [25]. The influential factors in VAEs are analyzed in [20].

4 Conclusion

Variational autoencoders remain one of the foundational approaches in probabilistic deep learning, combining neural networks with variational inference to create structured latent spaces. They helped establish the modern field of generative modeling and are still valued for their principled Bayesian foundation, interpretability, and flexibility in applications such as anomaly detection, semi-supervised learning, and scientific modeling. However VAEs are no longer the state of the art for generating high-fidelity images or text. Extensions such as β-VAEs [13], hierarchical VAEs [18], and adversarial VAEs [21] have improved their capabilities, but newer approaches—including GANs [10] (see Chap. 14), normalizing flows [26], autoregressive Transformers [32] (see Chap. 9), and diffusion models [14] (see Chap. 15)—dominate practical generative modeling. Thus, VAEs are best understood today as a powerful conceptual and methodological framework, with enduring importance for understanding latent variable models and probabilistic deep learning, while coexisting with more recent generative methods.

Appendix A: Proof of Lemma 1

$$
\begin{aligned}
\mathrm{KL}(p_1 \| p_2) &= \int p_1(x) \log\big(\frac{p_1(x)}{p_2(x)}\big) dx \\
&= \int p_1(x) \log(p_1(x))\, dx - \int p_1(x) \log(p_2(x))\, dx.
\end{aligned}
$$

According to integration by parts, there is

$$
\int p_1(x) \log(p_1(x))\, dx = -\frac{1}{2}(1 + \log(2\pi\sigma_1^2)).
$$

Moreover:

$$
\begin{aligned}
-\int p_1(x)\log(p_2(x))\,dx &= -\int p_1(x)\log\Big(\frac{1}{\sqrt{2\pi\sigma_2^2}}e^{-\frac{(x-\mu_2)^2}{2\sigma_2^2}}\Big)\,dx \\
&= \frac{1}{2}\log(2\pi\sigma_2^2)\underbrace{\int p_1(x)dx}_{=1} - \int p_1(x)\log\big(e^{-\frac{(x-\mu_2)^2}{2\sigma_2^2}}\big)\,dx \\
&= \frac{1}{2}\log(2\pi\sigma_2^2) + \int p_1(x)\frac{(x-\mu_2)^2}{2\sigma_2^2}\,dx \\
&= \frac{1}{2}\log(2\pi\sigma_2^2) + \frac{1}{2\sigma_2^2}\Big(\int p_1(x)x^2dx \\
&\qquad - \int p_1(x)2x\mu_2dx + \int p_1(x)\mu_2^2dx\Big) \\
&= \frac{1}{2}\log(2\pi\sigma_2^2) + \frac{1}{2\sigma_2^2}\Big(\mathbb{E}_{\sim p_1(x)}[x^2] - 2\mu_2\mathbb{E}_{\sim p_1(x)}[x] + \mu_2^2\Big).
\end{aligned}
$$

It is known that:

$$
\mathbb{V}\text{ar}[x] = \mathbb{E}[x^2] - \mathbb{E}[x]^2 \implies \mathbb{E}[x^2] = \sigma_1^2 + \mu_1^2.
$$

Therefore:

$$
\begin{aligned}
-\int p_1(x)\log(p_2(x))\,dx &= \frac{1}{2}\log(2\pi\sigma_2^2) + \frac{1}{2\sigma_2^2}\Big(\sigma_1^2 + \mu_1^2 - 2\mu_2\mu_1 + \mu_2^2\Big) \\
&= \frac{1}{2}\log(2\pi\sigma_2^2) + \frac{1}{2\sigma_2^2}\big(\sigma_1^2 + (\mu_1 - \mu_2)\big).
\end{aligned}
$$

Therefore, finally:

$$
\begin{aligned}
\text{KL}(p_1\|p_2) &= -\frac{1}{2}(1 + \log(2\pi\sigma_1^2)) + \frac{1}{2}\log(2\pi\sigma_2^2) + \frac{1}{2\sigma_2^2}\big(\sigma_1^2 + (\mu_1 - \mu_2)\big) \\
&= \log(\frac{\sigma_2}{\sigma_1}) + \frac{\sigma_1^2 + (\mu_1 - \mu_2)^2}{2\sigma_2^2} - \frac{1}{2}.
\end{aligned}
$$

5 Exercises

1. (Level: Difficult) Answer the following questions:
   (a) Prove that given two probability distributions, $p(x)$ and $q(x)$, the KL divergence is always non-negative:

   $$
   \text{KL}(p\|q) \geq 0,
   $$

   with equality at 0 if and only if $p(x) = q(x)$.
   (b) Show that the KL divergence is not symmetric; that is, in general:

$$\mathrm{KL}(p\|q) \neq \mathrm{KL}(q\|p).$$

Hint: Give a counterexample of two probability distributions $p(x)$ and $q(x)$ where this inequality holds.

2. (Level: Easy) In a Variational Autoencoder (VAE), the latent space is modeled as a Gaussian distribution. The prior is assumed to be a standard normal distribution $p(z) = \mathcal{N}(0, I)$, while the approximate posterior: $q(z|x) = \mathcal{N}(\mu, \sigma^2)$, where $\mu \in \mathbb{R}^d$ and $\sigma^2 \in \mathbb{R}^d$ are learnable parameters.
   (a) Why is KL divergence used to measure the difference between the approximate posterior $q(z|x)$ and the prior $p(z)$ in VAEs?
   (b) Compute the KL divergence for a one-dimensional latent space with mean $\mu = 0.3$ and variance $\sigma^2 = 1.2$, using the known KL divergence formula:

$$\mathrm{KL}\big(\mathcal{N}(\mu, \sigma^2) \,\|\, \mathcal{N}(0, I)\big) = \frac{1}{2}\left(\sigma^2 + \mu^2 - 1 - \ln \sigma^2\right).$$

3. (Level: Moderate) A probabilistic model has a latent variable z and observed data x. The prior on z is $p(z) = \mathcal{N}(z \mid 0, 1)$ with mean 0 and variance 1. The likelihood of x given z is $p(x \mid z) = \mathcal{N}(x \mid z, 1)$. Since the true posterior $p(z \mid x)$ is intractable, it is approximated using a variational distribution $q(z) = \mathcal{N}(z \mid \mu, \sigma_q^2)$. Given $x = 1.5$, $\mu = 0.2$, $\sigma_q = 0.5$, compute the following:
   (a) Compute the expected log-likelihood $\mathbb{E}_{q(z)}[\log p(x \mid z)]$.
   (b) Compute the KL divergence $\mathrm{KL}(q(z) \,\|\, p(z))$.
   (c) Compute the ELBO.
   (d) Compute the gradients $\frac{\partial ELBO}{\partial \mu}$ and $\frac{\partial ELBO}{\partial \sigma_q}$.
   (e) Update μ and σ_q using gradient descent with learning rate $\eta = 0.1$.
   (f) Compute the variance of the log-likelihood under the variational distribution $q(z)$, i.e., compute $\mathrm{Var}_{q(z)}\left[\log p(x \mid z)\right]$, given $x =$, $\mu = 0.2$, $\sigma_q = 0.5$.

4. (Level: Difficult) Consider a probabilistic model where the latent variable $\boldsymbol{z} \in \mathbb{R}^2$ has a prior $p(\boldsymbol{z}) = \mathcal{N}(\boldsymbol{z} \mid \mathbf{0}, \mathbf{I})$. The likelihood of observed data $\boldsymbol{x} \in \mathbb{R}^2$ given $\boldsymbol{z}$ is $p(\boldsymbol{x} \mid \boldsymbol{z}) = \mathcal{N}(\boldsymbol{x} \mid \boldsymbol{A}\boldsymbol{z}, \mathbf{I})$, where $\boldsymbol{A} \in \mathbb{R}^{2\times 2}$ is a fixed matrix. The posterior is approximated using the variational distribution $q(\boldsymbol{z}) = \mathcal{N}(\boldsymbol{z} \mid \boldsymbol{\mu}, \boldsymbol{\Sigma}_q)$. Given:

$$\boldsymbol{A} = \begin{bmatrix} 1 & 2 \\ 0 & 1 \end{bmatrix}, \quad \boldsymbol{x} = \begin{bmatrix} 1.5 \\ 0.5 \end{bmatrix}, \quad \boldsymbol{\mu} = \begin{bmatrix} 0.1 \\ -0.2 \end{bmatrix}, \quad \boldsymbol{\Sigma}_q = \begin{bmatrix} 0.3 & 0 \\ 0 & 0.4 \end{bmatrix},$$

   compute the following:
   (a) Compute $\mathbb{E}_{q(\boldsymbol{z})}[\log p(\boldsymbol{x} \mid \boldsymbol{z})]$.
   (b) Compute $\mathrm{KL}(q(\boldsymbol{z}) \,\|\, p(\boldsymbol{z}))$.
   (c) Compute the ELBO.
   (d) Compute gradients $\frac{\partial ELBO}{\partial \boldsymbol{\mu}}$ and diagonal elements of $\frac{\partial ELBO}{\partial \boldsymbol{\Sigma}_q}$.
   (e) Update $\boldsymbol{\mu}$ and $\boldsymbol{\Sigma}_q$ using learning rate $\eta = 0.05$.
   (f) Compute $\mathbf{trace}\left(\mathrm{Cov}_{q(\boldsymbol{z})}\left(\log p(\boldsymbol{x} \mid \boldsymbol{z})\right)\right)$.

References

1. Jincheng Bai, Qifan Song, and Guang Cheng. "Nearly Optimal Variational Inference for High Dimensional Regression with Shrinkage Priors". In: *arXiv preprint* arXiv:2010.12887 (2020).
2. Christopher M Bishop. *Pattern recognition and machine learning*. Springer, 2006.
3. Anthony L Caterini, Arnaud Doucet, and Dino Sejdinovic. "Hamiltonian variational auto-encoder". In: *Advances in Neural Information Processing Systems* 31 (2018), pp. 8167–8177.
4. Carl Doersch. "Tutorial on variational autoencoders". In: *arXiv preprint* arXiv:1606.05908 (2016).
5. John Duchi. *Derivations for linear algebra and optimization*. Tech. rep. Berkeley, California, 2007.
6. Mikhail Figurnov, Shakir Mohamed, and Andriy Mnih. "Implicit reparameterization gradients". In: *Advances in Neural Information Processing Systems* 31 (2018), pp. 441–452.
7. Benyamin Ghojogh, Mark Crowley, Fakhri Karray, and Ali Ghodsi. "Factor analysis and probabilistic principal component analysis". In: *Elements of Dimensionality Reduction and Manifold Learning*. Springer, 2023, pp. 355–374.
8. Benyamin Ghojogh, Aydin Ghojogh, Mark Crowley, and Fakhri Karray. "Fitting a mixture distribution to data: tutorial". In: *arXiv preprint* arXiv:1901.06708 (2019).
9. Benyamin Ghojogh, Hadi Nekoei, Aydin Ghojogh, Fakhri Karray, and Mark Crowley. "Sampling Algorithms, from Survey Sampling to Monte Carlo Methods: Tutorial and Literature Review". In: *arXiv preprint* arXiv:2011.00901 (2020).
10. Ian Goodfellow et al. "Generative adversarial nets". In: *Advances in neural information processing systems*. 2014, pp. 2672–2680.
11. Gaëtan Hadjeres, Frank Nielsen, and François Pachet. "GLSR-VAE: Geodesic latent space regularization for variational autoencoder architectures". In: *2017 IEEE Symposium Series on Computational Intelligence*. IEEE. 2017, pp. 1–7.
12. John R Hershey and Peder A Olsen. "Approximating the Kullback Leibler divergence between Gaussian mixture models". In: *2007 IEEE International Conference on Acoustics, Speech and Signal Processing*. Vol. 4. IEEE. 2007, pp. IV–317.
13. Irina Higgins et al. "beta-VAE: Learning basic visual concepts with a constrained variational framework". In: *International conference on learning representations*. 2017.
14. Jonathan Ho, Ajay Jain, and Pieter Abbeel. "Denoising diffusion probabilistic models". In: *Advances in neural information processing systems* 33 (2020), pp. 6840–6851.
15. Xianxu Hou, Linlin Shen, Ke Sun, and Guoping Qiu. "Deep feature consistent variational autoencoder". In: *2017 IEEE Winter Conference on Applications of Computer Vision*. IEEE. 2017, pp. 1133–1141.
16. Prateek Jain and Purushottam Kar. "Non-convex optimization for machine learning". In: *Foundations and Trends ® in Machine Learning* 10.3-4 (2017), pp. 142–336.
17. Diederik P Kingma and Max Welling. "Auto-encoding variational Bayes". In: *International Conference on Learning Representations*. 2014.
18. Durk P Kingma et al. "Improved variational inference with inverse autoregressive flow". In: *Advances in neural information processing systems* 29 (2016).
19. Solomon Kullback and Richard A Leibler. "On information and sufficiency". In: *The annals of mathematical statistics* 22.1 (1951), pp. 79–86.
20. Shiqi Liu et al. "Discovering influential factors in variational autoencoders". In: *Pattern Recognition* 100 (2020), p. 107166.
21. Lars Mescheder, Sebastian Nowozin, and Andreas Geiger. "Adversarial variational Bayes: Unifying variational autoencoders and generative adversarial networks". In: *International Conference on Machine Learning*. 2017.
22. Tom Michoel and Jitao David Zhang. "Causal inference in drug discovery and development". In: *Drug discovery today* 28.10 (2023), p. 103737.
23. Stephen Odaibo. "Tutorial: Deriving the Standard Variational Autoencoder (VAE) Loss Function". In: *arXiv preprint* arXiv:1907.08956 (2019).
24. Jonas Peters, Dominik Janzing, and Bernhard Schölkopf. *Elements of causal inference: foundations and learning algorithms*. The MIT Press, 2017.
25. Yunchen Pu et al. "Variational autoencoder for deep learning of images, labels and captions". In: *Advances in neural information processing systems*. 2016, pp. 2352–2360.
26. Danilo Rezende and Shakir Mohamed. "Variational inference with normalizing flows". In: *International conference on machine learning*. PMLR. 2015, pp. 1530–1538.
27. Danilo Jimenez Rezende, Shakir Mohamed, and Daan Wierstra. "Stochastic backpropagation and approximate inference in deep generative models". In: *International Conference on Machine Learning*. 2014.
28. David E Rumelhart, Geoffrey E Hinton, and Ronald J Williams. "Learning representations by back-propagating errors". In: *Nature* 323.6088 (1986), pp. 533–536.

29. Hiroshi Takahashi, Tomoharu Iwata, Yuki Yamanaka, Masanori Yamada, and Satoshi Yagi. "Variational autoencoder with implicit optimal priors". In: *Proceedings of the AAAI Conference on Artificial Intelligence*. Vol. 33. 2019, pp. 5066–5073.
30. Michalis Titsias and Miguel Lázaro-Gredilla. "Doubly stochastic variational Bayes for non-conjugate inference". In: *International conference on machine learning*. 2014, pp. 1971–1979.
31. Arash Vahdat and Jan Kautz. "NVAE: A deep hierarchical variational autoencoder". In: *Advances in Neural Information Processing Systems* 33 (2020).
32. Ashish Vaswani et al. "Attention is all you need". In: *Advances in neural information processing systems*. 2017, pp. 5998–6008.
33. Weichang Yu, Lamiae Azizi, and John T Ormerod. "Variational nonparametric discriminant analysis". In: *Computational Statistics & Data Analysis* 142 (2020), p. 106817.
34. Shengjia Zhao, Jiaming Song, and Stefano Ermon. "Towards deeper understanding of variational autoencoding models". In: *arXiv preprint* arXiv:1702.08658 (2017).

Generative Moment Matching Networks

13

Abstract

Measuring the difference between probability distributions is a fundamental problem with numerous applications in machine learning and statistics. This chapter focuses on Maximum Mean Discrepancy (MMD), a powerful kernel-based metric for comparing distributions by embedding data into a Reproducing Kernel Hilbert Space (RKHS) and evaluating the distance between their mean embeddings. MMD effectively captures differences across all moments of distributions while relying only on the differences of first moments in RKHS, offering both computational tractability and theoretical guarantees. Generative Moment Matching Networks (GMMNs), proposed in 2015, leverage MMD as a training loss to align the generated data distribution with the training data. This chapter introduces GMMNs as one of the earliest neural generative models trained without likelihood. Developed shortly after GANs, GMMNs offered a non-adversarial, kernel-based approach to distribution matching. While GMMNs are rarely used today, MMD remains widely useful in modern machine learning for tasks such as domain adaptation and statistical testing. This chapter further introduces some extensions of GMMN. Extensions such as Conditional GMMNs enable generation conditioned on specific attributes or classes. Additionally, methods for adversarially optimizing kernel choice to enhance GMMN performance, exemplified by MMD GANs, are presented.

1 Introduction

Calculating the difference between two probability distributions has many applications, such as minimizing the difference between a distribution and a reference distribution. There are several measures for calculating the difference between two distributions, one of which is Maximum Mean Discrepancy (MMD) [12, 13]. Other measures include Kullback–Leibler (KL) divergence [21], Jensen–Shannon (JS) divergence (i.e., symmetric KL divergence) [25], generalized JS divergence [5], Hilbert–Schmidt Independence Criterion (HSIC) [14], quantile–quantile plots [9, 26], and Sinkhorn divergence [6, 22].

MMD [12, 13], also called the kernel two-sample test, is a powerful measure of distributional difference. Its core idea is that if two distributions are equal, all of their moments should be equal. However, because the number of moments is infinite, calculating the difference across all moments is infeasible. It is possible to map data from the input space to the Reproducing Kernel Hilbert Space (RKHS) [7, 8] and measure the difference of their first moments only. Theoretically, the difference of first moments in RKHS is equivalent to the difference of all moments in the input space. Some improved variants of MMD are presented in [15, 32, 36].

B. Ghojogh and A. Ghodsi, *Elements of Deep Learning*,
https://doi.org/10.1007/978-3-032-10738-1_13

The Generative Moment Matching Network (GMMN) [4, 24] uses MMD between the network output and the training dataset as its loss function. This makes the data generated by GMMN have a distribution similar to that of the training data. Conditional GMMN [29] enables generation conditioned on specific class labels or patterns. The MMD loss function used in GMMN employs characteristic kernel functions [8, 19]. It is also possible to find the best characteristic kernel function adversarially for GMMN [16]; MMD GAN [23] is an example of finding the best kernel in GMMN.

GMMNs were proposed around the same time as GANs [11], providing a kernel-based method for distribution matching without adversarial training. While conceptually elegant, they did not achieve the same generative quality or scalability as GANs and were later surpassed by VAEs [20] (see Chap. 12), autoregressive Transformers [35] (see Chap. 9), and diffusion models [18] (see Chap. 15).

This chapter begins by introducing the intuition behind MMD, its definition, its formulation in RKHS, and its use as a metric for distributions in Sect. 2. Then, GMMN and Conditional GMMN are explained in Sects. 3 and 4, respectively. Finally, the problem of finding the best kernel in GMMN is discussed in Sect. 5.

2 Maximum Mean Discrepancy (Kernel Two-Sample Test)

2.1 Intuition of Maximum Mean Discrepancy

Consider the probability distributions P_x and P_y of two random variables X and Y, respectively. Assume there are a few samples drawn from each of these distributions. Let $\{x_i\}_{i=1}^n$ and $\{y_j\}_{j=1}^m$ denote the samples from the distributions P_x and P_y, respectively. We want to compare these distributions to determine whether they are equal, and if not, how different they are.

One may calculate the difference between the means of the samples, i.e., $\mathbb{E}[x] - \mathbb{E}[y]$, as a measure of their difference, where $\mathbb{E}[.]$ denotes the expectation operator. However, two distributions may have equal means but differ significantly in terms of variance (see Fig. 1a). Variance is related to the second moment of the distributions ($\mathbb{E}[x^2]$); in fact, variance is equal to $\mathbb{E}[x^2] - (\mathbb{E}[x])^2$. Therefore, one may first transform the samples as $x \mapsto [x, x^2]^\top$ and then calculate the difference between the means of the transformed samples $[x, x^2]^\top$. In other words, both $\mathbb{E}[x] - \mathbb{E}[y]$ and $\mathbb{E}[x^2] - \mathbb{E}[y^2]$ are considered. However, two distributions may have equal means and variances but still differ in terms of skewness, which is related to the third moment (see Fig. 1b).

As a result, the differences across all moments should be calculated to fully determine the difference between two distributions. This is because

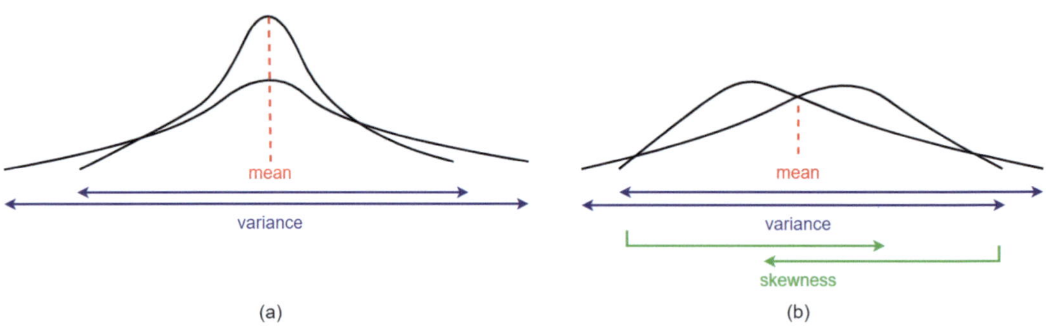

Fig. 1 Distributions with **a** equal mean but different variance, and **b** equal mean and variance but different skewness

Proposition 1 *Two distributions are equal if and only if all of their moments are equal:*

$$P_x = P_y \iff \mathbb{E}[x^i] = \mathbb{E}[y^i], \forall i \in \{1, 2, \dots, \infty\}, \tag{1}$$

where $\mathbb{E}[x^i]$ *and* $\mathbb{E}[y^i]$ *are the i-th moments of* P_x *and* P_y*, respectively.*

However, since the number of moments, i.e., $\mathbb{E}[x]$, $\mathbb{E}[x^2]$, $\mathbb{E}[x^3]$, $\mathbb{E}[x^4]$, …, is infinite, calculating the differences of all moments is not feasible. One approach to make this calculation feasible is to map the data from their input space to the Reproducing Kernel Hilbert Space (RKHS) [7, 8] and calculate the differences of their means (first moments). According to [3, Lemma 9.3.2] and [13, Lemma 1], the means (i.e., first moments) of the mapped data in RKHS are equal if and only if the distributions are equal in the input space.

Proposition 2 ([3, Lemma 9.3.2] and [13, Lemma 1]) *Two distributions are equal if and only if their means (i.e., their first moments) in RKHS are equal:*

$$P_x = P_y \iff \mathbb{E}[f(x)] = \mathbb{E}[f(y)], \forall f \in C(\mathcal{X}), \tag{2}$$

where $\mathcal{X}$ *denotes the input space for x and y, and* $C(\mathcal{X})$ *denotes the space of all bounded continuous functions on* $\mathcal{X}$.

In other words, the difference of means (first moments) in RKHS is equivalent to the difference of all moments in the input space. Therefore, the difference between the means of data mapped to RKHS serves as a measure of distributional difference. This measure is known as Maximum Mean Discrepancy (MMD), also called the kernel two-sample test [12, 13]. MMD measures the difference between two distributions by comparing their moments.

2.2 Formulation of Maximum Mean Discrepancy

2.2.1 Definition of MMD

Definition 1 (*MMD* [12, 13]) Let $\mathcal{F}$ be a class of functions mapping from the input space to scalar outputs. Maximum Mean Discrepancy (MMD) between two distributions P_x and P_y is defined as follows:

$$\text{MMD}(P_x, P_y) := \sup_{f \in \mathcal{F}} \big(\mathbb{E}[f(\boldsymbol{x})] - \mathbb{E}[f(\boldsymbol{y})]\big). \tag{3}$$

Let $\{\boldsymbol{x}_i\}_{i=1}^n$ and $\{\boldsymbol{y}_j\}_{j=1}^m$ denote the samples drawn from the distributions P_x and P_y, respectively. The expectations in Eq. (3) can be approximated using Monte Carlo approximation [10] to obtain the biased empirical MMD [12, 13]:

$$\text{MMD}(P_x, P_y) \approx \sup_{f \in \mathcal{F}} \Big(\frac{1}{n}\sum_{i=1}^{n} f(\boldsymbol{x}_i) - \frac{1}{m}\sum_{j=1}^{m} f(\boldsymbol{y}_j)\Big). \tag{4}$$

2.2.2 MMD in RKHS

The function class $\mathcal{F}$ can be defined as a unit ball in RKHS. Therefore, the function $f \in \mathcal{F}$ can be the mapping function from the input space to RKHS, denoted by $\phi : \boldsymbol{x} \mapsto \phi(\boldsymbol{x})$ [7, 8]. In this setting, the data are mapped to RKHS as follows:

$$\{\boldsymbol{x}_i\}_{i=1}^n \mapsto \{\phi(\boldsymbol{x}_i)\}_{i=1}^n,$$
$$\{\boldsymbol{y}_j\}_{j=1}^m \mapsto \{\phi(\boldsymbol{y}_j)\}_{j=1}^m,$$

and Eq. (4) becomes

$$\text{MMD}(P_x, P_y) \approx \sup_{f \in \mathcal{F}} \Big(\frac{1}{n}\sum_{i=1}^n \phi(\boldsymbol{x}_i) - \frac{1}{m}\sum_{j=1}^m \phi(\boldsymbol{y}_j)\Big), \tag{5}$$

where $(1/n)\sum_{i=1}^n \phi(\boldsymbol{x}_i)$ and $(1/m)\sum_{j=1}^m \phi(\boldsymbol{y}_j)$ are the means of the mapped data in RKHS for X and Y, respectively. Let these means, called mean embeddings in RKHS, be denoted by $\mu_{\phi(\boldsymbol{x})}$ and $\mu_{\phi(\boldsymbol{y})}$, respectively:

$$\mu_{\phi(\boldsymbol{x})} := \frac{1}{n}\sum_{i=1}^n \phi(\boldsymbol{x}_i), \tag{6}$$

$$\mu_{\phi(\boldsymbol{y})} := \frac{1}{m}\sum_{j=1}^m \phi(\boldsymbol{y}_j). \tag{7}$$

Therefore, Eq. (5), squared, becomes

$$\text{MMD}^2(P_x, P_y) \approx \big\|\mu_{\phi(\boldsymbol{x})} - \mu_{\phi(\boldsymbol{y})}\big\|_2^2, \tag{8}$$

where the supremum is absorbed into the mean embeddings according to the properties of RKHS, which are beyond the scope of this chapter. For more details, refer to [13, Lemma 4].

According to Eqs. (6) and (7), Eq. (8) can be expanded as follows:

$$\begin{aligned}
\text{MMD}^2(P_x, P_y) &\approx \Big\|\frac{1}{n}\sum_{i=1}^n \phi(\boldsymbol{x}_i) - \frac{1}{m}\sum_{j=1}^m \phi(\boldsymbol{y}_j)\Big\|_2^2 \\
&= \Big(\frac{1}{n}\sum_{i=1}^n \phi(\boldsymbol{x}_i) - \frac{1}{m}\sum_{j=1}^m \phi(\boldsymbol{y}_j)\Big)^\top \Big(\frac{1}{n}\sum_{i=1}^n \phi(\boldsymbol{x}_i) - \frac{1}{m}\sum_{j=1}^m \phi(\boldsymbol{y}_j)\Big) \\
&= \Big(\frac{1}{n}\sum_{i=1}^n \phi(\boldsymbol{x}_i)^\top - \frac{1}{m}\sum_{j=1}^m \phi(\boldsymbol{y}_j)^\top\Big)\Big(\frac{1}{n}\sum_{i=1}^n \phi(\boldsymbol{x}_i) - \frac{1}{m}\sum_{j=1}^m \phi(\boldsymbol{y}_j)\Big) \\
&\overset{(a)}{=} \frac{1}{n^2}\sum_{i=1}^n\sum_{j=1}^n \phi(\boldsymbol{x}_i)^\top \phi(\boldsymbol{x}_j) + \frac{1}{m^2}\sum_{i=1}^m\sum_{j=1}^m \phi(\boldsymbol{y}_i)^\top \phi(\boldsymbol{y}_j) \\
&\quad - \frac{2}{nm}\sum_{i=1}^n\sum_{j=1}^m \phi(\boldsymbol{x}_i)^\top \phi(\boldsymbol{y}_j),
\end{aligned} \tag{9}$$

where (a) holds because the dummy variable i can be replaced with j, and $\phi(\boldsymbol{y}_j)^\top \phi(\boldsymbol{x}_i) = \phi(\boldsymbol{x}_i)^\top \phi(\boldsymbol{y}_j)$. According to the definition of the kernel function $k(.,.)$ [7, 8]:

$$k(\boldsymbol{x}, \boldsymbol{y}) := \phi(\boldsymbol{x})^\top \phi(\boldsymbol{y}), \tag{10}$$

Equation (9) simplifies to:

$$\mathrm{MMD}^2(P_x, P_y) \approx \frac{1}{n^2}\sum_{i=1}^{n}\sum_{j=1}^{n} k(\boldsymbol{x}_i, \boldsymbol{x}_j) + \frac{1}{m^2}\sum_{i=1}^{m}\sum_{j=1}^{m} k(\boldsymbol{y}_i, \boldsymbol{y}_j) - \frac{2}{nm}\sum_{i=1}^{n}\sum_{j=1}^{m} k(\boldsymbol{x}_i, \boldsymbol{y}_j). \tag{11}$$

Equation (11) represents the biased empirical MMD in RKHS. It is noteworthy that the unbiased empirical MMD in RKHS is given by [13]

$$\mathrm{MMD}^2(P_x, P_y) \approx \frac{1}{n(n-1)}\sum_{i=1}^{n}\sum_{j=1}^{n} k(\boldsymbol{x}_i, \boldsymbol{x}_j) + \frac{1}{m(m-1)}\sum_{i=1}^{m}\sum_{j=1}^{m} k(\boldsymbol{y}_i, \boldsymbol{y}_j) - \frac{2}{nm}\sum_{i=1}^{n}\sum_{j=1}^{m} k(\boldsymbol{x}_i, \boldsymbol{y}_j). \tag{12}$$

2.2.3 MMD in RKHS as a Metric for Distributions

MMD can serve as a distance metric between two distributions under certain conditions. For MMD to qualify as a metric, the RKHS must be a universal RKHS; in other words, the kernel must be a universal kernel function (also called a characteristic kernel function). A universal kernel is defined as follows.

Definition 2 (*Universal kernel* [33, Definition 4], [34, Definition 2]) Let $C(\mathcal{X})$ denote the space of all continuous functions on the input space $\mathcal{X}$. A continuous kernel k on a compact metric space $\mathcal{X}$ is called universal if the RKHS $\mathcal{H}$ with kernel function k is dense in $C(\mathcal{X})$. In other words, for every function $f \in C(\mathcal{X})$ and all $\epsilon > 0$, there exists a function $g \in \mathcal{H}$ such that $\|f - g\|_\infty \leq \epsilon$, where $\|.\|_\infty$ denotes the maximum norm.

According to [33, Corollary 10] and the Stone–Weierstrass theorem [2], a universal kernel has an infinite number of terms in its Taylor series expansion. An example of a universal kernel is the Radial Basis Function (RBF) kernel, also called the Gaussian kernel:

$$k(\boldsymbol{x}, \boldsymbol{y}) := \exp(-\gamma\,\|\boldsymbol{x} - \boldsymbol{y}\|_2^2) = \exp\left(-\frac{\|\boldsymbol{x} - \boldsymbol{y}\|_2^2}{\sigma^2}\right), \tag{13}$$

where $\boldsymbol{x}$ and $\boldsymbol{y}$ are the input vectors to the kernel function, $\gamma := 1/\sigma^2$, and σ^2 is the kernel variance. Another example of a universal kernel is the Laplacian kernel:

$$k(\boldsymbol{x}, \boldsymbol{y}) := \exp(-\gamma\,\|\boldsymbol{x} - \boldsymbol{y}\|_1) = \exp\left(-\frac{\|\boldsymbol{x} - \boldsymbol{y}\|_1}{\sigma^2}\right), \tag{14}$$

where $\|\boldsymbol{x} - \boldsymbol{y}\|_1$ is also called the Manhattan distance. These two kernel functions have an infinite number of terms in their Taylor series expansions due to the exponential operator. In contrast, note that polynomial kernels are not universal.

According to the following proposition, MMD can be used to measure the difference between two distributions when a universal kernel function is used in its calculation. MMD is bounded below but does not have an upper bound. In other words, if two distributions are equal, their MMD is zero. The larger the MMD between two distributions, the more different they are.

Proposition 3 ([12, Theorem 3] and [13, Theorem 5]) *Let $\mathcal{F}$ be a unit ball in a universal RKHS. Then, $MMD(P_x, P_y) = 0$ if and only if P_x and P_y are equal distributions:*

$$MMD(P_x, P_y) = 0 \iff P_x = P_y. \tag{15}$$

Proof If $P_x = P_y$, then it is obvious that $\text{MMD}(P_x, P_y) = 0$. The converse is proved as follows. Assume $\text{MMD}(P_x, P_y) = 0$. According to Eq. (8), we have

$$\begin{aligned} \text{MMD}(P_x, P_y) = 0 &\implies \left\|\mu_{\phi(\boldsymbol{x})} - \mu_{\phi(\boldsymbol{y})}\right\|_2^2 = 0 \\ &\implies \mu_{\phi(\boldsymbol{x})} = \mu_{\phi(\boldsymbol{y})}. \end{aligned} \tag{16}$$

According to Definition 2, due to the universality of the RKHS $\mathcal{H}$, there exists

$$\|f - g\|_\infty \leq \epsilon, \tag{17}$$

for $f \in C(\mathcal{X})$ and $g \in \mathcal{H}$. Consider the terms in the definition of MMD (see Definition 1):

$$\begin{aligned} &|\mathbb{E}[f(\boldsymbol{x})] - \mathbb{E}[f(\boldsymbol{y})]| \\ &= \big|\mathbb{E}[f(\boldsymbol{x})] - \mathbb{E}[g(\boldsymbol{x})] + \mathbb{E}[g(\boldsymbol{x})] - \mathbb{E}[g(\boldsymbol{y})] + \mathbb{E}[g(\boldsymbol{y})] - \mathbb{E}[f(\boldsymbol{y})]\big| \\ &\leq |\mathbb{E}[f(\boldsymbol{x})] - \mathbb{E}[g(\boldsymbol{x})]| + |\mathbb{E}[g(\boldsymbol{x})] - \mathbb{E}[g(\boldsymbol{y})]| + |\mathbb{E}[g(\boldsymbol{y})] - \mathbb{E}[f(\boldsymbol{y})]|. \end{aligned} \tag{18}$$

Consider the first and third terms in Eq. (18):

$$|\mathbb{E}[f(\boldsymbol{x})] - \mathbb{E}[g(\boldsymbol{x})]| \leq \mathbb{E}\big[|f(\boldsymbol{x}) - g(\boldsymbol{x})|\big] \overset{(17)}{\leq} \epsilon,$$

$$|\mathbb{E}[g(\boldsymbol{y})] - \mathbb{E}[f(\boldsymbol{y})]| \leq \mathbb{E}\big[|g(\boldsymbol{y}) - f(\boldsymbol{y})|\big] = \mathbb{E}\big[|f(\boldsymbol{y}) - g(\boldsymbol{y})|\big] \overset{(17)}{\leq} \epsilon.$$

Consider the second term in Eq. (18):

$$\mathbb{E}[g(\boldsymbol{x})] - \mathbb{E}[g(\boldsymbol{y})] \overset{(a)}{=} \langle g, \mu_{\phi(\boldsymbol{x})} - \mu_{\phi(\boldsymbol{y})}\rangle_{\mathcal{H}} \overset{(16)}{=} 0.$$

Therefore, Eq. (18) simplifies to

$$|\mathbb{E}[f(\boldsymbol{x})] - \mathbb{E}[f(\boldsymbol{y})]| \leq \epsilon + 0 + \epsilon = 2\epsilon,$$

for all $f \in C(\mathcal{X})$ and $\epsilon > 0$. According to Eq. (2) in Proposition 2, this implies that $P_x = P_y$. □

Remark 1 (*Equivalence of MMD and HSIC* [31]) After the development of MMD [12, 13] and HSIC [14], it was found that they are equivalent.

3 Generative Moment Matching Network

Recall the Generative Adversarial Network (GAN) [11] (see Chap. 14), where a generative neural network takes random noise as input and generates data, while a discriminator attempts to classify the data as real or fake (i.e., generated by the generator); see Fig. 2a. The generator and discriminator are trained simultaneously in a minimax optimization framework.

Generative Moment Matching Network (GMMN) [4, 24] replaces the discriminator with an MMD loss function between the generated data and the real data (see Fig. 2b). GMMN is a generative neural network that takes a random noise vector z as input and outputs generated data. The random noise can follow any distribution, such as a uniform distribution where each input neuron receives a random value drawn from a univariate uniform distribution. Therefore, the dimensions of the input random noise vector $z = [z_1, \ldots, z_p]^\top$ are i.i.d. uniform random numbers, where p is the dimensionality of the noise vector, i.e., the number of neurons in the input layer of GMMN. In other words, $z \in [-1, 1]^p$ [24] or $z \in [0, 1]^p$ [29] has the following joint probability distribution:

$$\mathbb{P}(z) = \prod_{j=1}^{p} U(z_j), \tag{19}$$

where $U(.)$ denotes the continuous uniform distribution in the range $[-1, 1]$ or $[0, 1]$.

Assume the training dataset is $\{x_i \in \mathbb{R}^d\}_{i=1}^n$, where n is the number of training instances. Suppose GMMN generates, or outputs, m data instances $\{\widehat{x}_j \in \mathbb{R}^d\}_{j=1}^m$. GMMN uses Eq. (11) as the loss function for training the neural network:

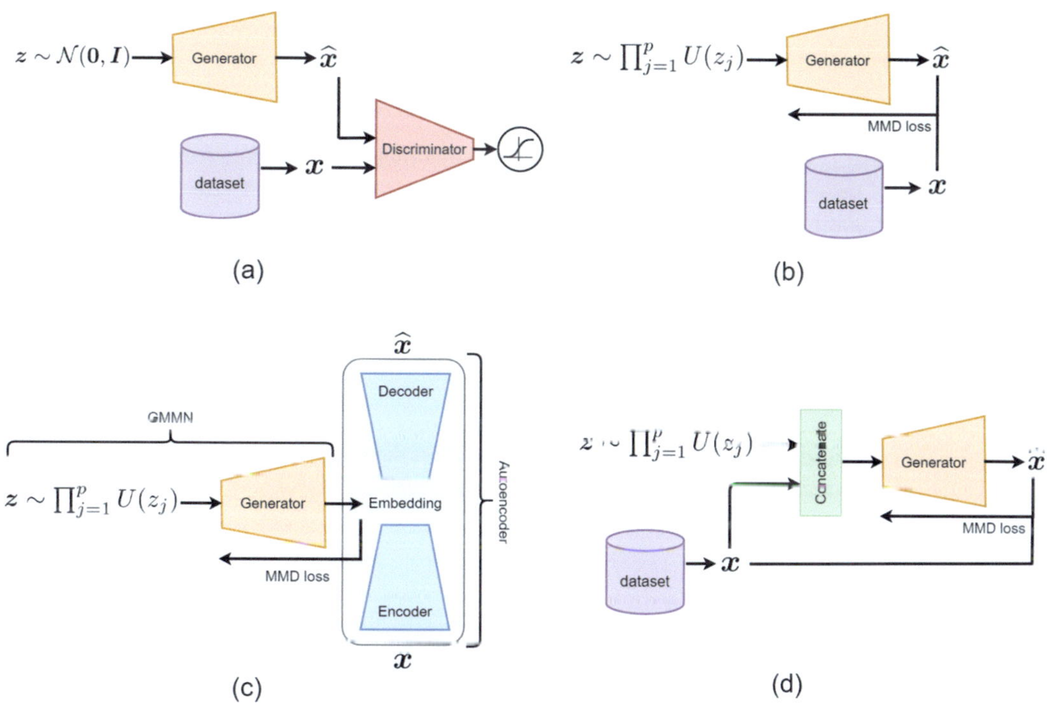

Fig. 2 The structures of **a** GAN, **b** GMMN, **c** GMMN in the latent space of an autoencoder, and **d** conditional GMMN

$$\mathcal{L} := \text{MMD}^2(P_x, P_{\widehat{x}})$$
$$\approx \frac{1}{n^2} \sum_{i=1}^{n} \sum_{j=1}^{n} k(\boldsymbol{x}_i, \boldsymbol{x}_j) + \frac{1}{m^2} \sum_{i=1}^{m} \sum_{j=1}^{m} k(\widehat{\boldsymbol{x}}_i, \widehat{\boldsymbol{x}}_j) - \frac{2}{nm} \sum_{i=1}^{n} \sum_{j=1}^{m} k(\boldsymbol{x}_i, \widehat{\boldsymbol{x}}_j). \tag{20}$$

Note that this loss can be minimized in batches so that in each batch, m data instances can be generated. Minimizing this loss function pushes $\text{MMD}(P_x, P_{\widehat{x}})$ toward zero. Therefore, according to Proposition 3, minimizing this loss function changes the distribution of the generated data, $P_{\widehat{x}}$, to become similar to the distribution of the training data, P_x. By making the distributions of the generated data and the training dataset similar, the network learns to generate data resembling the training data.

There is another variant of GMMN, depicted in Fig. 2c, which learns to generate data that has the distribution of the latent space (also called the code space) of a reconstruction autoencoder. In this variant, the output of the GMMN is the latent middle layer of a reconstruction autoencoder. First, the autoencoder is trained either by backpropagation [30] (see Chap. 3), greedy layer-wise training [1], or belief propagation [17]. Then, the GMMN is trained using Eq. (11) as its loss function, aiming for the distribution of its output to become similar to the distribution of the autoencoder's latent embeddings.

There are several considerations for GMMN. Firstly, according to Proposition 3, the kernel function used in the loss function of GMMN, i.e., Eq. (20), must be a universal kernel. Therefore, for example, RBF (Eq. (13)) or Laplacian kernels (Eq. (14)) can be used, but polynomial kernels cannot be used for training GMMN. Secondly, it has been empirically observed [24] that using $\sqrt{\text{MMD}^2(P_x, P_{\widehat{x}})}$ as the loss function of GMMN tends to generate data of better quality. Thirdly, it is recommended to use ReLU activation functions [28] in the hidden layers of GMMN, while the last layer can use a sigmoid activation function (generating values in the range [0, 1]) or any other activation required for the range of generated values.

4 Conditional Generative Moment Matching Network

GMMN generates data from the domains of the training dataset regardless of any class or label. Conditional GMMN [29] incorporates labels or side information to generate data from specific classes or sub-domains of the training data. For example, conditional GMMN can generate data from specific classes or data similar to the pattern of a given data instance.

Conditional GMMN receives side information in addition to the noise as input to the neural network. The p-dimensional random noise $\boldsymbol{z}$ has a distribution as described in Eq. (19). To generate data similar to the pattern of a data instance $\boldsymbol{x} \in \mathbb{R}^d$, it is possible to concatenate the vectors $\boldsymbol{x}$ and $\boldsymbol{z}$ [29]:

$$\mathbb{R}^{d+p} \ni \widetilde{\boldsymbol{x}} = [\boldsymbol{x}, \boldsymbol{z}]^\top, \tag{21}$$

and input it to the conditional GMMN network. See Fig. 2d for the structure of the conditional GMMN network. If p is set equal to d, another option is to add these vectors together [29]:

$$\mathbb{R}^d \ni \widetilde{\boldsymbol{x}} = \boldsymbol{x} + \boldsymbol{z}, \tag{22}$$

and use it as input to the neural network. Assume the generated output of the conditional GMMN neural network is denoted by $\widehat{\boldsymbol{x}} \in \mathbb{R}^d$. Then, minimizing the loss function, as in Eq. (20), makes the distribution of the generated output similar to the distribution of the training dataset. Again, ReLU activation functions [28] are used in the hidden layers of conditional GMMN.

In the test phase, the user can choose a data instance $\boldsymbol{x}$ to be used as input in Eqs. (21) or (22). The neural network will then generate a data instance similar to the pattern of $\boldsymbol{x}$. For example, if $\boldsymbol{x}$ is a picture of a house, the network will generate a picture of a new house that may not exist in the training dataset, although it is an interpolation of all house pictures in the training data.

Note that it is also possible to generate data from a specific class label, inspired by conditional generative adversarial network [27]. Let $\boldsymbol{\ell} \in \{0, 1\}^c$ denote the one-hot encoding of the class label, where c is the number of classes. Then, the label and the noise can be concatenated:

$$\mathbb{R}^{c+p} \ni \widetilde{\boldsymbol{x}} = [\boldsymbol{\ell}, z]^\top, \tag{23}$$

and fed as input to the neural network. In this setting, the MMD loss function, i.e., Eq. (20), should be minimized for each class label. For every class label, MMD should be minimized between the training dataset of that class and the generated data for that class label. In the test phase, the user selects a class label, which is fed as input in Eq. (23), and the network generates data from that class label. For example, the user may choose to generate data from the class of houses, and the network generates a new house picture.

5 Finding the Best Kernel in Generative Moment Matching Network

GMMN uses a fixed RBF kernel for the calculation of the MMD loss:

$$\underset{\theta}{\text{minimize}} \quad \text{MMD}^2\big(P_x, P_{\widetilde{x}}(\theta)\big), \tag{24}$$

where θ denotes the weights of the GMMN neural network (see Eq. (20) for the detailed loss function of GMMN). However, one might ask why the kernel function should be fixed; it is possible to find the best universal kernel (also called characteristic kernel function) [16].

MMD GAN [23] unfixes the kernel function and attempts to adversarially learn the best characteristic kernel function for the network. It combines the ideas of GMMN [4, 24] and adversarial learning in Generative Adversarial Networks (GAN) [11] (see Chap. 14). MMD GAN finds a characteristic kernel function that maximizes the loss function adversarially, so that the GMMN network minimizes the MMD loss in its worst-case scenario for better learning [23]:

$$\underset{\theta}{\text{minimize}} \quad \underset{k \in \mathcal{K}}{\text{maximize}} \quad \text{MMD}_k^2\big(P_x, P_{\widetilde{x}}(\theta)\big), \tag{25}$$

where MMD_k denotes MMD calculated using the kernel function k, and $\mathcal{K}$ is a set of characteristic kernel functions, e.g., RBF and Laplacian kernels with different bandwidths.

It is difficult to minimize the loss function (25) by evaluating all possible universal kernel functions in $\mathcal{K}$. According to [13, 16], if f is an injective function and k is a characteristic kernel function, then the function composition $\widetilde{k} = k \circ f$ is also characteristic. Consider a second neural network, denoted by f_ϕ, which models f with weights ϕ. Therefore, the loss function (25) can be reformulated as [23] follows:

$$\underset{\theta}{\text{minimize}} \quad \underset{\phi}{\text{maximize}} \quad \text{MMD}_{k \circ f_\phi}^2\big(P_x, P_{\widetilde{x}}(\theta)\big). \tag{26}$$

An example of $\widetilde{k} = k \circ f$ is combining the RBF kernel with an injective function f_ϕ:

$$\widetilde{k}(\boldsymbol{x}, \boldsymbol{y}) := \exp(-\gamma \, \|f_\phi(\boldsymbol{x}) - f_\phi(\boldsymbol{y})\|_2^2), \tag{27}$$

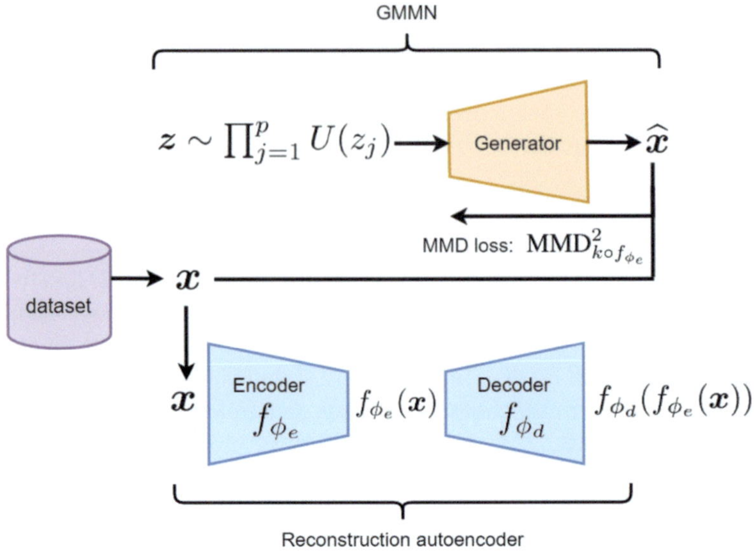

Fig. 3 The structure of MMD GAN

for vectors $\boldsymbol{x}$ and $\boldsymbol{y}$.

Because the function f should be injective, there must exist its inverse f^{-1} such that $f^{-1}(f(\boldsymbol{x})) = \boldsymbol{x}, \forall \boldsymbol{x}$. This can be modeled by a reconstruction autoencoder where f and f^{-1} correspond to the encoder and decoder, respectively. Let the encoder and decoder of the reconstruction autoencoder be denoted by $f_{\phi_e} = f$ and $f_{\phi_d} \approx f^{-1}$, with weights ϕ_e and ϕ_d, respectively. The loss function (26) can thus be reformulated as [23] follows:

$$\underset{\theta}{\text{minimize}} \underset{\phi:=\{\phi_e,\phi_d\}}{\text{maximize}} \Big(\text{MMD}^2_{k \circ f_{\phi_e}}\big(P_x, P_{\widehat{x}}(\theta)\big) - \lambda \mathbb{E}\Big[\|\boldsymbol{x} - f_{\phi_d}(f_{\phi_e}(\boldsymbol{x}))\|_2^2\Big]\Big), \tag{28}$$

where $\lambda > 0$ is the regularization parameter, and the inner maximization adversarially finds the worst-case characteristic kernel function for the outer minimization. The inner maximization also minimizes the reconstruction error in the reconstruction autoencoder.

In summary, MMD GAN employs two neural networks: one for the GMMN and the other for the reconstruction autoencoder. These two neural networks are trained simultaneously, as in GAN, using the loss function (28). Figure 3 depicts the structure of MMD GAN.

6 Conclusion

While Generative Moment Matching Networks introduced an elegant kernel-based approach to training generative models, they did not achieve the scalability or sample quality of other methods. By 2025, GMMNs are mainly of historical and conceptual interest. However, the Maximum Mean Discrepancy (MMD) criterion itself remains influential and continues to be applied in domain adaptation, distribution alignment, and statistical testing. Thus, MMD is a lasting contribution, even if GMMNs are no longer central to generative modeling.

7 Exercises

1. (Level: Moderate) Consider two samples:

$$X = \{1, 2\}, \quad Y = \{2, 3\}.$$

Using the Gaussian (RBF) kernel with bandwidth $\sigma = 1$:

$$k(a, b) = \exp\left(-\frac{(a-b)^2}{2\sigma^2}\right),$$

calculate the biased estimate of the squared Maximum Mean Discrepancy (MMD^2) between X and Y.

2. (Level: Difficult) Let P and Q be distributions over $\mathbb{R}$ such that:

$$P = \mathcal{N}(\mu_P = 0, \sigma_P^2 = 1), \quad Q = \mathcal{N}(\mu_Q = 1, \sigma_Q^2 = 2).$$

Consider the Gaussian kernel:

$$k(x, y) = \exp\left(-\frac{(x-y)^2}{2\gamma^2}\right),$$

with bandwidth $\gamma = 1$. Compute the squared Maximum Mean Discrepancy (MMD) between P and Q, defined as follows:

$$\text{MMD}^2(P, Q) = \mathbb{E}_{x,x' \sim P}[k(x, x')] + \mathbb{E}_{y,y' \sim Q}[k(y, y')] - 2\mathbb{E}_{x \sim P, y \sim Q}[k(x, y)].$$

Hint: There is a closed-form expression for expectation of Gaussian kernels with Gaussian inputs:

$$\mathbb{E}_{x,x' \sim \mathcal{N}(\mu,\sigma^2)}[k(x, x')] = \frac{1}{\sqrt{1 + \frac{2\sigma^2}{\gamma^2}}}.$$

3. (Level: Moderate) Consider a Generative Moment Matching Network (GMMN) that aims to generate samples from a target distribution P_{data} using a neural network generator $G(z)$ where $z \sim P_z$ is noise input. Suppose you have two empirical samples:

$$X = \{1, 2\}, \quad Y = \{1.5, 2.5\}.$$

Using the linear kernel:

$$k(a, b) = a \cdot b,$$

compute the squared Maximum Mean Discrepancy (MMD^2) between X and Y, which is used as the loss to train the GMMN.

4. (Level: Difficult) Given the following two conditional inputs and corresponding real and generated samples:

- Conditioning input 1: $\boldsymbol{x}^{(1)} = [1, 0]$.

- Real samples: $\boldsymbol{y}_1 = [1.0, 0.0]$, $\boldsymbol{y}_2 = [1.5, 0.5]$.
- Generated samples: $\widehat{\boldsymbol{y}}_1 = [0.7, 0.0]$, $\widehat{\boldsymbol{y}}_2 = [1.1, 0.3]$.

- Conditioning input 2: $\boldsymbol{x}^{(2)} = [0, 1]$.
  - Real samples: $\boldsymbol{y}_1 = [0.0, 1.0]$, $\boldsymbol{y}_2 = [-0.5, 1.5]$.
  - Generated samples: $\widehat{\boldsymbol{y}}_1 = [-0.3, 0.8]$, $\widehat{\boldsymbol{y}}_2 = [-0.6, 1.0]$.

Compute the squared Conditional Maximum Mean Discrepancy (CMMD^2) with the unbiased empirical estimate and the linear kernel:

$$k(\boldsymbol{a}, \boldsymbol{b}) = \boldsymbol{a}^\top \boldsymbol{b}.$$

Hint: Compute the MMD^2 loss for each condition and then calculate the average loss across the conditions.

References

1. Yoshua Bengio, Pascal Lamblin, Dan Popovici, and Hugo Larochelle. "Greedy layer-wise training of deep networks". In: *Advances in neural information processing systems* 19 (2006).
2. Louis De Branges. "The Stone-Weierstrass theorem". In: *Proceedings of the American Mathematical Society* 10.5 (1959), pp. 822–824.
3. Richard M Dudley. *Real analysis and probability*. Cambridge University Press, Cambridge, UK, 2002.
4. Gintare Karolina Dziugaite, Daniel M Roy, and Zoubin Ghahramani. "Training generative neural networks via maximum mean discrepancy optimization". In: *Proceedings of the Thirty-First Conference on Uncertainty in Artificial Intelligence*. 2015, pp. 258–267.
5. Nielsen Frank. "On a generalization of the Jensen-Shannon divergence and the JS-symmetrization of distances relying on abstract means". In: *arXiv preprint* arXiv:1904.04017 (2019).
6. Aude Genevay, Gabriel Peyré, and Marco Cuturi. "Learning generative models with Sinkhorn divergences". In: *International Conference on Artificial Intelligence and Statistics*. PMLR. 2018, pp. 1608–1617.
7. Benyamin Ghojogh, Mark Crowley, Fakhri Karray, and Ali Ghodsi. "Background on Kernels". In: *Elements of Dimensionality Reduction and Manifold Learning* (2023), pp. 43–73.
8. Benyamin Ghojogh, Ali Ghodsi, Fakhri Karray, and Mark Crowley. "Reproducing Kernel Hilbert Space, Mercer's Theorem, Eigenfunctions, Nyström Method, and Use of Kernels in Machine Learning: Tutorial and Survey". In: *arXiv preprint* arXiv:2106.08443 (2021).
9. Benyamin Ghojogh, Fakhri Karray, and Mark Crowley. "Quantile–Quantile Embedding for distribution transformation and manifold embedding with ability to choose the embedding distribution". In: *Machine Learning with Applications* 6 (2021), p. 100088.
10. Benyamin Ghojogh, Hadi Nekoei, Aydin Ghojogh, Fakhri Karray, and Mark Crowley. "Sampling algorithms, from survey sampling to Monte Carlo methods: Tutorial and literature review". In: *arXiv preprint* arXiv:2011.00901 (2020).
11. Ian Goodfellow et al. "Generative adversarial networks". In: *Advances in neural information processing systems*. Vol. 27. 2014.
12. Arthur Gretton, Karsten Borgwardt, Malte Rasch, Bernhard Schölkopf, and Alex Smola. "A kernel method for the two-sample-problem". In: *Advances in neural information processing systems* 19 (2006).
13. Arthur Gretton, Karsten M Borgwardt, Malte J Rasch, Bernhard Schölkopf, and Alexander Smola. "A kernel two-sample test". In: *The Journal of Machine Learning Research* 13.1 (2012), pp. 723–773.
14. Arthur Gretton, Olivier Bousquet, Alex Smola, and Bernhard Schölkopf. "Measuring statistical dependence with Hilbert-Schmidt norms". In: *International conference on algorithmic learning theory*. Springer. 2005, pp. 63–77.
15. Arthur Gretton, Kenji Fukumizu, Zaid Harchaoui, and Bharath K Sriperumbudur. "A fast, consistent kernel two-sample test". In: *Advances in neural information processing systems* 22 (2009).
16. Arthur Gretton et al. "Optimal kernel choice for large-scale two-sample tests". In: *Advances in neural information processing systems* 25 (2012).

17. Geoffrey E Hinton and Ruslan R Salakhutdinov. "Reducing the dimensionality of data with neural networks". In: *Science* 313.5786 (2006), pp. 504–507.
18. Jonathan Ho, Ajay Jain, and Pieter Abbeel. "Denoising diffusion probabilistic models". In: *Advances in neural information processing systems* 33 (2020), pp. 6840–6851.
19. Thomas Hofmann, Bernhard Schölkopf, and Alexander J Smola. "Kernel methods in machine learning". In: *The annals of statistics* (2008), pp. 1171–1220.
20. Diederik P Kingma and Max Welling. "Auto-encoding variational Bayes". In: *International Conference on Learning Representations*. 2014.
21. Solomon Kullback and Richard A Leibler. "On information and sufficiency". In: *The annals of mathematical statistics* 22.1 (1951), pp. 79–86.
22. Hugo Lavenant, Jonas Luckhardt, Gilles Mordant, Bernhard Schmitzer, and Luca Tamanini. "The Riemannian geometry of Sinkhorn divergences". In: *arXiv preprint* arXiv:2405.04987 (2024).
23. Chun-Liang Li, Wei-Cheng Chang, Yu Cheng, Yiming Yang, and Barnabás Póczos. "MMD GAN: Towards deeper understanding of moment matching network". In: *Advances in neural information processing systems* 30 (2017).
24. Yujia Li, Kevin Swersky, and Rich Zemel. "Generative moment matching networks". In: *International conference on machine learning*. PMLR. 2015, pp. 1718–1727.
25. Jianhua Lin. "Divergence measures based on the Shannon entropy". In: *IEEE Transactions on Information theory* 37.1 (1991), pp. 145–151.
26. Adam Loy, Lendie Follett, and Heike Hofmann. "Variations of Q–Q plots: The power of our eyes!" In: *The American Statistician* 70.2 (2016), pp. 202–214.
27. Mehdi Mirza. "Conditional generative adversarial nets". In: *arXiv preprint* arXiv:1411.1784 (2014).
28. Vinod Nair and Geoffrey E Hinton. "Rectified linear units improve restricted boltzmann machines". In: *Proceedings of the 27th international conference on machine learning (ICML-10)*. 2010, pp. 807–814.
29. Yong Ren, Jun Zhu, Jialian Li, and Yucen Luo. "Conditional generative moment-matching networks". In: *Advances in Neural Information Processing Systems* 29 (2016).
30. David E Rumelhart, Geoffrey E Hinton, and Ronald J Williams. "Learning representations by back-propagating errors". In: *nature* 323.6088 (1986), pp. 533–536.
31. Dino Sejdinovic, Bharath Sriperumbudur, Arthur Gretton, and Kenji Fukumizu. "Equivalence of distance-based and RKHS-based statistics in hypothesis testing". In: *The Annals of Statistics* (2013), pp. 2263–2291.
32. Shubhanshu Shekhar, Ilmun Kim, and Aaditya Ramdas. "A permutation-free kernel two-sample test". In: *Advances in Neural Information Processing Systems* 35 (2022), pp. 18168–18180.
33. Ingo Steinwart. "On the influence of the kernel on the consistency of support vector machines". In: *Journal of machine learning research* 2.Nov (2001), pp. 67–93.
34. Ingo Steinwart. "Support vector machines are universally consistent". In: *Journal of Complexity* 18.3 (2002), pp. 768–791.
35. Ashish Vaswani et al. "Attention is all you need". In: *Advances in neural information processing systems*. 2017, pp. 5998–6008.
36. George Wynne and Andrew B Duncan. "A kernel two-sample test for functional data". In: *Journal of Machine Learning Research* 23.73 (2022), pp. 1–51.

Generative Adversarial Networks 14

Abstract

Generative Adversarial Networks (GANs) are a class of generative models where two networks are trained together: a generator that produces synthetic samples from noise, and a discriminator that attempts to distinguish real samples from synthetic ones. Through this adversarial game, the generator gradually improves at producing realistic outputs. This adversarial process enables the generator to enhance the quality of generated samples without relying on human judgment, making GANs highly effective for tasks such as image synthesis, data augmentation, and beyond. This chapter provides a comprehensive introduction to GANs and their underlying principles. Key challenges, such as mode collapse, are discussed alongside theoretical perspectives including maximum likelihood estimation within GANs. The chapter also surveys important GAN variants, sampling and interpolation techniques, and highlights diverse applications. Finally, it explores extensions of adversarial learning within autoencoder architectures. The content aims to equip readers with a foundational understanding as well as insights into recent developments and practical applications of adversarial generative modeling.

1 Introduction

Suppose there is a generative model that takes random noise as input and generates a data point. The generated data point is desired to be of good quality; therefore, its quality must be assessed in some way. One approach is to observe the generated sample and evaluate its quality visually. In this case, the judge is a human. However, it is not possible to calculate the derivative of a human's judgment for optimization. The Generative Adversarial Network (GAN), proposed in [36], employs the same idea but enables the calculation of derivatives for optimization by using a classifier as the judge rather than a human. Therefore, there is a generator that produces a sample and a binary classifier (or discriminator) that classifies the generated sample as either real or generated.

In principle, one could use a pre-trained classifier to evaluate generated samples. However, the key idea of GANs is to train the discriminator simultaneously with the generator, creating a dynamic competition that drives both models to improve. This is the core concept of adversarial learning, where the classifier, also called the discriminator, and the generator compete with each other. Consequently, they gradually strengthen one another through this competition [37].

B. Ghojogh and A. Ghodsi, *Elements of Deep Learning*,
https://doi.org/10.1007/978-3-032-10738-1_14

The term "adversarial" has two distinct meanings in machine learning: (1) adversarial examples and attacks, where inputs are perturbed to fool a model (e.g., [48, 58, 59, 71, 84]), and (2) adversarial learning for generation, where a generator and discriminator are trained in competition (e.g., [36]). This chapter focuses on the second meaning.

A good tutorial on GAN is provided in [33], although it does not cover the most recent methods in adversarial learning. An honorary introduction to GAN by several of its main contributors is presented in [37]. Other existing surveys on GAN include [15, 32, 47, 101, 126].

This chapter discusses adversarial learning, GAN, and its variants. Section 2 introduces GAN. The mode collapse problem in GAN is explained in Sect. 3. Maximum likelihood estimation in GAN is covered in Sect. 4. A few variants of GAN are introduced in Sect. 5. Section 6 explains sampling and interpolation in GAN. Some applications of GAN are enumerated in Sect. 7. Finally, Sect. 8 elaborates on autoencoders based on adversarial learning.

2 Generative Adversarial Network (GAN)

2.1 Adversarial Learning: The Adversarial Game

The original GAN, often referred to as "vanilla GAN," was introduced by Goodfellow et al. in 2014 [36]. It remains the foundation upon which most subsequent GAN variants are built.

Consider a d-dimensional dataset with n data points, i.e., $\{\boldsymbol{x}_i \in \mathbb{R}^d\}_{i=1}^n$. In GAN, there is a generator G that takes a p-dimensional random noise $\boldsymbol{z} \in \mathbb{R}^p$ as input and outputs a d-dimensional generated point $\boldsymbol{x} \in \mathbb{R}^d$. Therefore, it defines the mapping $G : \boldsymbol{z} \to \boldsymbol{x}$ where:

$$G(\boldsymbol{z}) = \boldsymbol{x}. \tag{1}$$

The random noise can be seen as a latent factor on which the generated data point is conditioned. The probabilistic graphical model of the generator represents a variable $\boldsymbol{x}$ conditioned on a latent variable $\boldsymbol{z}$ (see Fig. 1 for its visualization).

Let the distribution of random noise be denoted by $\boldsymbol{z} \sim p_z(\boldsymbol{z})$. The generated $\widehat{\boldsymbol{x}}$ is intended to be very similar to some original (or real) data point $\boldsymbol{x}$ in the dataset. A module is required to evaluate the quality of the generated point to determine how similar it is to the real point. This module could be a human, but it is not possible to compute the derivative of a human's judgment for optimization. A suitable candidate for this evaluation is a classifier, also called the discriminator. The discriminator (also known as the critic), denoted by $D : \boldsymbol{x} \to [0, 1]$, is a binary classifier that classifies a data point as real or generated:

$$D(\boldsymbol{x}) := \begin{cases} 1 \text{ if } \boldsymbol{x} \text{ is real,} \\ 0 \text{ if } \boldsymbol{x} \text{ is generated (fake).} \end{cases} \tag{2}$$

Fig. 1 The probabilistic graphical model of a variable $\boldsymbol{x}$ conditioned on a latent variable $\boldsymbol{z}$

The perfect discriminator outputs one for real points and zero for generated points. The discriminator's output lies in the range [0, 1], where outputs for real data are closer to one and outputs for fake data are closer to zero. If the generated point is of high quality and closely resembles a real data point, the classifier may make a mistake and output a value close to one for it. Therefore, if the classifier makes a mistake for the generated point, the generator has succeeded in producing a realistic data point.

The discriminator can be a pre-trained classifier, but the problem can be made more sophisticated, as explained below. It is possible to train the discriminator simultaneously with the generator. This approach enables both the discriminator D and the generator G to become stronger gradually as they compete against each other. On the one hand, the generator attempts to produce realistic points to fool the discriminator, making it difficult for the discriminator to distinguish the generated point from a real one. On the other hand, the discriminator seeks to differentiate the fake (i.e., generated) points from real ones. When one becomes stronger during training, the other also strives to improve to remain competitive. Therefore, there exists an adversarial game between the generator and the discriminator. This game is zero-sum because whatever one of them loses, the other gains.

2.2 Optimization and Loss Function

The probability distributions of the dataset and noise are denoted by $p_{\text{data}}(\boldsymbol{x})$ and $p_z(\boldsymbol{z})$, respectively. The structure of GAN is depicted in Fig. 2. As the figure shows, the discriminator is trained using real points from the dataset as well as generated points from the generator. The discriminator and generator are trained simultaneously. The optimization loss function for both the discriminator and generator is

$$\min_G \max_D \; V(D, G) := \mathbb{E}_{\boldsymbol{x}\sim p_{\text{data}}(\boldsymbol{x})}\Big[\log\big(D(\boldsymbol{x})\big)\Big] + \mathbb{E}_{\boldsymbol{z}\sim p_z(\boldsymbol{z})}\Big[\log\Big(1 - D\big(G(\boldsymbol{z})\big)\Big)\Big], \tag{3}$$

where $\mathbb{E}[.]$ denotes the expectation operator and the loss function $V(D, G)$ is also called the value function of the game. In practice, the Monte Carlo approximation [30] of expectation can be used, where the expectations are replaced with averages over the mini-batch. This loss function is in the form of a cross-entropy loss.

The first term in Eq. (3) is the expectation over the real data. This term is only used for the discriminator while it is constant for the generator. According to Eq. (2), $D(\boldsymbol{x})$ outputs one (the higher label) for the real data; therefore, the discriminator maximizes this term because it assigns the higher label to real data.

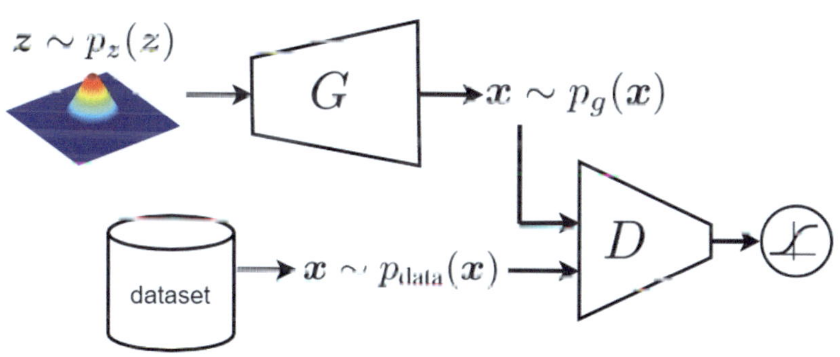

Fig. 2 The structure of GAN

The second term in Eq. (3) is the expectation over noise. The noise z is input to the generator to produce $G(z)$. The output of the generator, which is the generated point, is then fed as input to the discriminator (see Fig. 2) to compute $D\big(G(z)\big)$. The discriminator aims to minimize $D\big(G(z)\big)$ because, according to Eq. (2), the smaller label is assigned to the generated data. In other words, the discriminator seeks to maximize $1 - D\big(G(z)\big)$. As the logarithm is a monotonic function, it can be said that the discriminator wants to maximize $\mathbb{E}_{z \sim p_z(z)}[\log(1 - D(G(z)))]$, which is the second term in Eq. (3). In contrast, the generator minimizes $\mathbb{E}_{z \sim p_z(z)}[\log(1 - D(G(z)))]$, as it seeks to fool the discriminator into labeling the generated data as real.

Equation (3) defines a minimax optimization problem [19], which can be solved using alternating optimization [27]. This approach optimizes over D and G iteratively until convergence (i.e., Nash equilibrium). The original GAN [36] employs stochastic gradient descent [27] for updates of each variable in the alternating optimization. If the loss function in Eq. (3) is denoted by $V(D, G)$, the alternating optimization is performed as:

$$D^{(k+1)} := D^{(k)} + \eta^{(k)} \frac{\partial}{\partial D}\Big(V(D, G^{(k)})\Big), \tag{4}$$

$$G^{(k+1)} := G^{(k)} - \eta^{(k)} \frac{\partial}{\partial G}\Big(V(D^{(k+1)}, G)\Big), \tag{5}$$

where k is the iteration index and $\eta^{(k)}$ is the learning rate at iteration k. Throughout this chapter, derivatives with respect to D and G refer to derivatives with respect to the parameters (weights) of the D and G networks, respectively. Equations (4) and (5) represent one step of gradient ascent and gradient descent, respectively. Note that these gradients are averages over the mini-batch, which includes both real and generated data. The paper [36] suggests performing Eq. (4) several times before Eq. (5); however, its experiments perform Eq. (4) only once before Eq. (5). Another approach to solving the GAN optimization problem is simultaneous optimization [80], in which Eqs. (4) and (5) are performed simultaneously rather than sequentially.

Remark 1 (*Minimax versus maximin in GAN* [33, Sect. 5]) As shown in Eq. (3), the optimization of GAN is a minimax problem:

$$\min_{G} \max_{D} \quad V(D, G). \tag{6}$$

By changing the order of optimization, GAN can also be viewed as a maximin problem [33]:

$$\max_{D} \min_{G} \quad V(D, G). \tag{7}$$

In fact, under certain conditions, Eqs. (6) and (7) are equivalent [19]. According to the minimax theorem [105], a minimax equilibrium exists in pure or mixed strategies for every two-player zero-sum game and is identical to the maximin equilibrium. As GAN is a two-player zero-sum game, minimax and maximin formulations are equivalent in this context.

2.3 Network Structure of GAN

In practice, the discriminator and generator are implemented as two (deep) neural networks. The structure of GAN is depicted in Fig. 2. The first layer of the discriminator network is d-dimensional, and its last layer is one-dimensional with a scalar output. The last layer of discriminator employs the sigmoid activation function to output a probability as described in Eq. (2). The closer the output of D is to one, the more probable it is that the input is real.

The original GAN architecture used simple multi-layer perceptrons with maxout activations [35] in intermediate layers and a sigmoid output in the discriminator. Later variants such as DCGAN [104]—which will be introduced later in the chapter—replaced this with convolutional architectures and more effective activations, leading to major improvements.

The generator network has a p-dimensional input layer for noise and a d-dimensional output layer for generating data. Within the generator, a combination of ReLU [89] and sigmoid activation functions is used. The space of noise input to the generator is called the latent space or latent factor. Each of Eqs. (4) and (5) is implemented using backpropagation within the neural networks.

2.4 Optimal Solution of GAN

Theorem 1 ([36, Proposition 1]) *For a fixed generator G, the optimal discriminator is*

$$D^*(\boldsymbol{x}) = \frac{p_{data}(\boldsymbol{x})}{p_{data}(\boldsymbol{x}) + p_g(\boldsymbol{x})}, \tag{8}$$

where $p_{data}(\boldsymbol{x})$ is the probability distribution of the real dataset evaluated at point $\boldsymbol{x}$, and $p_g(\boldsymbol{x})$ is the probability distribution of the generator output evaluated at point $\boldsymbol{x}$.

Proof According to the definition of expectation, the loss function in Eq. (3) can be stated as:

$$V(D, G) = \int_{\boldsymbol{x}} p_{\text{data}}(\boldsymbol{x}) \log(D(\boldsymbol{x}))d\boldsymbol{x} + \int_{z} p_z(z) \log(1 - D(G(z)))dz.$$

According to Eq. (1), there is

$$G(z) = \boldsymbol{x} \implies z = G^{-1}(\boldsymbol{x}) \implies dz = (G^{-1})'(\boldsymbol{x})d\boldsymbol{x},$$

where $(G^{-1})'(\boldsymbol{x})$ is the derivative of $(G^{-1})(\boldsymbol{x})$ with respect to $\boldsymbol{x}$. Therefore:

$$V(D, G) = \int_{\boldsymbol{x}} p_{\text{data}}(\boldsymbol{x}) \log(D(\boldsymbol{x}))d\boldsymbol{x} + \int_{z} p_z(G^{-1}(\boldsymbol{x})) \log(1 - D(\boldsymbol{x}))(G^{-1})'(\boldsymbol{x})d\boldsymbol{x}.$$

The relation of distributions of input and output of generator is

$$p_g(\boldsymbol{x}) = p_z(z) \times G^{-1}(\boldsymbol{x}) = p_z(G^{-1}(\boldsymbol{x}))\, G^{-1}(\boldsymbol{x}), \tag{9}$$

where $G^{-1}(\boldsymbol{x})$ is the Jacobian of distribution at point $\boldsymbol{x}$. Hence:

$$\begin{aligned} V(D, G) &= \int_{\boldsymbol{x}} p_{\text{data}}(\boldsymbol{x}) \log(D(\boldsymbol{x}))d\boldsymbol{x} + \int_{z} p_g(\boldsymbol{x}) \log(1 - D(\boldsymbol{x}))d\boldsymbol{x} \\ &= \int_{\boldsymbol{x}} \Big(p_{\text{data}}(\boldsymbol{x}) \log(D(\boldsymbol{x})) + p_g(\boldsymbol{x}) \log(1 - D(\boldsymbol{x}))\Big)d\boldsymbol{x}. \end{aligned} \tag{10}$$

For optimization in Eq. (3), taking derivative with respect to $D(\boldsymbol{x})$ gives

$$\begin{aligned}
\frac{\partial V(D,G)}{\partial D(\boldsymbol{x})} &\overset{(a)}{=} \frac{\partial}{\partial D(\boldsymbol{x})}\Big(p_{\text{data}}(\boldsymbol{x})\log(D(\boldsymbol{x})) + p_g(\boldsymbol{x})\log(1-D(\boldsymbol{x}))\Big) \\
&= \frac{p_{\text{data}}(\boldsymbol{x})}{D(\boldsymbol{x})} - \frac{p_g(\boldsymbol{x})}{1-D(\boldsymbol{x})} \\
&= \frac{p_{\text{data}}(\boldsymbol{x})(1-D(\boldsymbol{x})) - p_g(\boldsymbol{x})D(\boldsymbol{x})}{D(\boldsymbol{x})(1-D(\boldsymbol{x}))} \overset{\text{set}}{=} 0 \\
\implies & p_{\text{data}}(\boldsymbol{x}) - p_{\text{data}}(\boldsymbol{x})D(\boldsymbol{x}) - p_g(\boldsymbol{x})D(\boldsymbol{x}) = 0 \\
\implies & D(\boldsymbol{x}) = \frac{p_{\text{data}}(\boldsymbol{x})}{p_{\text{data}}(\boldsymbol{x}) + p_g(\boldsymbol{x})},
\end{aligned}$$

where (a) is because taking derivative with respect to $D(\boldsymbol{x})$ considers a specific $\boldsymbol{x}$ and hence it removes the integral (summation). □

Theorem 2 ([36, Theorem 1]) *The optimal solution of GAN is achieved when the distribution of the generated data becomes equal to the distribution of the real data:*

$$p_{g^*}(\boldsymbol{x}) = p_{data}(\boldsymbol{x}). \tag{11}$$

Proof Putting the optimum $D^*(\boldsymbol{x})$, i.e., Eq. (8), in Eq. (10) gives

$$\begin{aligned}
V(D^*, G) &= \int_{\boldsymbol{x}} \Big(p_{\text{data}}(\boldsymbol{x})\log(D^*(\boldsymbol{x})) + p_g(\boldsymbol{x})\log(1-D^*(\boldsymbol{x}))\Big)d\boldsymbol{x} \\
&\overset{(8)}{=} \int_{\boldsymbol{x}} \Big[p_{\text{data}}(\boldsymbol{x})\log\Big(\frac{p_{\text{data}}(\boldsymbol{x})}{p_{\text{data}}(\boldsymbol{x}) + p_g(\boldsymbol{x})}\Big) + p_g(\boldsymbol{x})\log\Big(\frac{p_g(\boldsymbol{x})}{p_{\text{data}}(\boldsymbol{x}) + p_g(\boldsymbol{x})}\Big)\Big]d\boldsymbol{x} \\
&= \int_{\boldsymbol{x}} \Big[p_{\text{data}}(\boldsymbol{x})\log\Big(\frac{p_{\text{data}}(\boldsymbol{x})}{2\times\frac{p_{\text{data}}(\boldsymbol{x})+p_g(\boldsymbol{x})}{2}}\Big) + p_g(\boldsymbol{x})\log\Big(\frac{p_g(\boldsymbol{x})}{2\times\frac{p_{\text{data}}(\boldsymbol{x})+p_g(\boldsymbol{x})}{2}}\Big)\Big]d\boldsymbol{x} \\
&= \int_{\boldsymbol{x}} \Big[p_{\text{data}}(\boldsymbol{x})\log\Big(\frac{p_{\text{data}}(\boldsymbol{x})}{\frac{p_{\text{data}}(\boldsymbol{x})+p_g(\boldsymbol{x})}{2}}\Big) \\
&\qquad + p_g(\boldsymbol{x})\log\Big(\frac{p_g(\boldsymbol{x})}{\frac{p_{\text{data}}(\boldsymbol{x})+p_g(\boldsymbol{x})}{2}}\Big)\Big]d\boldsymbol{x} + \log(\frac{1}{2}) + \log(\frac{1}{2}) \\
&= \int_{\boldsymbol{x}} \Big[p_{\text{data}}(\boldsymbol{x})\log\Big(\frac{p_{\text{data}}(\boldsymbol{x})}{\frac{p_{\text{data}}(\boldsymbol{x})+p_g(\boldsymbol{x})}{2}}\Big) + p_g(\boldsymbol{x})\log\Big(\frac{p_g(\boldsymbol{x})}{\frac{p_{\text{data}}(\boldsymbol{x})+p_g(\boldsymbol{x})}{2}}\Big)\Big]d\boldsymbol{x} - \log(4) \\
&\overset{(a)}{=} \text{KL}\Big(p_{\text{data}}(\boldsymbol{x})\Big\|\frac{p_{\text{data}}(\boldsymbol{x}) + p_g(\boldsymbol{x})}{2}\Big) + \text{KL}\Big(p_g(\boldsymbol{x})\Big\|\frac{p_{\text{data}}(\boldsymbol{x}) + p_g(\boldsymbol{x})}{2}\Big) - \log(4),
\end{aligned} \tag{12}$$

where (a) is because of the definition of KL divergence. The Jensen-Shannon Divergence (JSD) is defined as [94]:

$$\text{JSD}(P\|Q) := \frac{1}{2}\text{KL}(P\|\frac{1}{2}(P+Q)) + \frac{1}{2}\text{KL}(Q\|\frac{1}{2}(P+Q)), \tag{13}$$

where P and Q denote the probability densities. In contrast to KL divergence, the JSD is symmetric. The obtained $V(D^*, G)$ can be restated as:

$$V(D^*, G) = 2\,\mathrm{JSD}\big(p_{\mathrm{data}}(\boldsymbol{x}) \,\|\, p_g(\boldsymbol{x})\big) - \log(4), \tag{14}$$

According to Eq. (3), the generator minimizes $V(D^*, G)$. As the JSD is non-negative, the above loss function is minimized if we have

$$\mathrm{JSD}\big(p_{\mathrm{data}}(\boldsymbol{x}) \,\|\, p_{g^*}(\boldsymbol{x})\big) = 0 \implies p_{\mathrm{data}}(\boldsymbol{x}) = p_{g^*}(\boldsymbol{x}).$$

□

Corollary 1 ([36, Theorem 1]) *From Eqs. (11) and (14), it can be concluded that the optimal loss function in GAN is*

$$V(D^*, G^*) = -\log(4). \tag{15}$$

It is noteworthy that Eq. (13) can be generalized in GAN to [49]:

$$\mathrm{JSD}_\pi(P\|Q) := \pi\,\mathrm{KL}\big(P\|\pi P + (1-\pi)Q\big) + (1-\pi)\,\mathrm{KL}\big(Q\|\pi P + (1-\pi)Q\big), \tag{16}$$

where $\pi \in (0, 1)$. A special case is Eq. (13) with $\pi = 0.5$.

Corollary 2 *From Eqs. (8) and (11), it can be concluded that at convergence (i.e., Nash equilibrium), the discriminator cannot distinguish between generated and real data:*

$$\begin{aligned} D^*(\boldsymbol{x}) &= 0.5, \quad \forall \boldsymbol{x} \sim p_{data}(\boldsymbol{x}), \\ D^*(\boldsymbol{x}) &= 0.5, \quad \boldsymbol{x} = G^*(z), \forall z \sim p_z(z). \end{aligned} \tag{17}$$

Lemma 1 (Label smoothing in GAN [111, Sect. 3.4]) *It has been shown that replacing labels* 0 *and* 1 *with smoother values* 0.1 *and* 0.9*, respectively [119], can improve neural networks against adversarial attacks [43]. If the labels of the discriminator D are smoothed for real and generated data to be α and β, respectively, the optimal discriminator becomes [111]:*

$$D^*(\boldsymbol{x}) = \frac{\alpha p_{data}(\boldsymbol{x}) + \beta p_g(\boldsymbol{x})}{p_{data}(\boldsymbol{x}) + p_g(\boldsymbol{x})}, \tag{18}$$

which generalizes Eq. (8). The presence of $p_g(\boldsymbol{x})$ can cause a problem because, for an $\boldsymbol{x}$ with small $p_{data}(\boldsymbol{x})$ and large $p_g(\boldsymbol{x})$, the point does not sufficiently influence the generator to move it closer to the real data. Hence, it is recommended to set $\beta = 0$ to achieve one-sided label smoothing. In this case, the optimal discriminator is

$$D^*(\boldsymbol{x}) = \frac{\alpha p_{data}(\boldsymbol{x})}{p_{data}(\boldsymbol{x}) + p_g(\boldsymbol{x})}. \tag{19}$$

Proof (sketch) Using α and β in the proof of Theorem 1 results in Eq. (18). □

2.5 Convergence and Equilibrium Analysis of GAN

Theorem 3 ([36, Proposition 2]) *If the discriminator and generator have sufficient capacity and, at every iteration of the alternating optimization, the following conditions hold:*

- *the discriminator is allowed to reach its optimum value as given in Eq. (8), and*
- *$p_g(\boldsymbol{x})$ is updated to minimize $V(D^*, G)$, as stated in Eq. (14),*

then $p_g(\boldsymbol{x})$ converges to $p_{data}(\boldsymbol{x})$, as stated in Eq. (11).

Proof The KL divergences in Eq. (12) are convex functions with respect to $p_g(\boldsymbol{x})$. Thus, with sufficiently small updates of $p_g(\boldsymbol{x})$, it converges to $p_{\text{data}}(\boldsymbol{x})$. Note that Eq. (12), used here, holds if Eq. (8) holds, i.e., if the discriminator is allowed to reach its optimum value. □

The GAN loss, i.e., Eq. (3), can be restated as [87]:

$$\min_G \max_D \; V(D, G) := \mathbb{E}_{\boldsymbol{x} \sim p_{\text{data}}(\boldsymbol{x})}\Big[f\big(D(\boldsymbol{x})\big)\Big] + \mathbb{E}_{\boldsymbol{z} \sim p_z(\boldsymbol{z})}\Big[f\Big(-D\big(G(\boldsymbol{z})\big)\Big)\Big], \tag{20}$$

where f is the negative logistic function, i.e., $f(x) := -\log(1 + \exp(-x))$. In fact, the function $f(.)$ can be any concave function. This formulation differs slightly from the original GAN in that, here, the discriminator D outputs a real-valued scalar (without any activation function), while the discriminator in Eq. (3) outputs values in the range $(0, 1)$ after a sigmoid activation function. If D outputs 0.5 and 0, it means it is completely confused in Eqs. (3) and (20), respectively. Equation (20) is a concave-concave loss function in most of the domain of the discriminator [87, Proposition 3.1].

Theorem 4 ([87, Theorem 3.1]) *After satisfying several reasonable assumptions (see [87] for details), a GAN with the loss function of Eq. (20) is locally exponentially stable.*

Lemma 2 (Nash equilibrium in GAN [22]) *Nash equilibrium is the state of the game where no player can improve its gain by choosing a different strategy. At the Nash equilibrium of GAN, we have*

$$V(D, G^*) \leq V(D^*, G^*) \leq V(D^*, G), \tag{21}$$

which is evident because the generator and discriminator are minimizing and maximizing $V(G, D)$, respectively, as shown in Eq. (3).

Empirical experiments have shown that GAN may not reach its Nash equilibrium in practice [22]. Regularization can help the convergence of GAN to the Nash equilibrium [78]. It has been shown in [78] that an effective regularization for GAN is noise injection (see Chap. 5), where independent Gaussian noise is added to the training data points.

Definition 1 (*Proximal equilibrium* [22]) The proximal operator [27] can be used in the loss function of GAN:

$$\min_G \max_D \Big(V_{\text{prox}}(D, G) := \max_{\widetilde{D}} (V(\widetilde{D}, G) - \lambda \|\widetilde{D} - D\|_2^2) \Big),$$

where $V(D, G)$ is defined in Eq. (3) and $\lambda > 0$ is the regularization parameter. The equilibrium of the game using this loss function is called the proximal equilibrium.

Theorem 5 (Convergence of GAN based on the Jacobian [80]) *Let the updated solution of GAN optimization at every iteration be obtained by some operator $F(D, G)$, such as a step of gradient descent. The convergence of GAN can be analyzed based on the Jacobian of $F(D, G)$ with respect to D and G. If the absolute values of some eigenvalues of the Jacobian are larger than one, GAN will not converge to the Nash equilibrium. If all eigenvalues have absolute values less than one, GAN will converge to the Nash equilibrium with a linear rate $\mathcal{O}(|\lambda_{\max}|^k)$, where $\lambda_{\max}$ is the eigenvalue with the largest absolute value and k is the iteration index. If all eigenvalues have unit absolute value, GAN may or may not converge.*

Readers can refer to [23] for duality in GAN, which is not explained here for brevity. Moreover, some papers have specifically combined GAN with game theory. Interested readers can refer to [3, 99, 120, 123].

2.6 Conditional GAN

As explained previously, the original GAN randomly draws noise z from a prior distribution $p_z(z)$ and feeds it to the generator, which outputs a point $\boldsymbol{x}$ from the noise z. Assume that the dataset used to train the GAN has c classes. The original GAN generates points from any class without control over generating a point from a specific class. However, note that after the GAN is trained, the latent space for z becomes meaningful in the sense that different regions of the latent space correspond to the generation of specific points from certain classes. Nevertheless, the user cannot explicitly choose which class to generate points from.

Conditional GAN [83], also called the conditional adversarial network, allows the user to choose the class for generating points. For the dataset $\{\boldsymbol{x}_i \in \mathbb{R}^d\}_{i=1}^n$, let the one-hot encoded class labels be $\{\boldsymbol{y}_i \in \mathbb{R}^c\}_{i=1}^n$. Conditional GAN uses the following loss function instead of Eq. (3):

$$\min_G \max_D \; V_C(D, G) := \mathbb{E}_{\boldsymbol{x} \sim p_{\text{data}}(\boldsymbol{x})}\Big[\log\big(D(\boldsymbol{x}|\boldsymbol{y})\big)\Big] + \mathbb{E}_{z \sim p_z(z)}\Big[\log\Big(1 - D\big(G(z|\boldsymbol{y})\big)\Big)\Big], \tag{22}$$

where both the discriminator and generator are conditioned on the labels. In practice, to implement the loss function (22), the one-hot encoded label $\boldsymbol{y}$ is concatenated with the point $\boldsymbol{x}$ for the input to the discriminator. Additionally, the one-hot encoded label $\boldsymbol{y}$ is concatenated with the noise z for the input to the generator. Consequently, the input layers of the discriminator and generator are enlarged to accept these concatenated inputs. During the test phase, the user selects the desired class label, and the generator produces a new point from that class.

2.7 Deep Convolutional GAN (DCGAN)

Deep Convolutional GAN (DCGAN), proposed in [104], extended GANs by using deeper architectures and generating higher-resolution images than the original GAN. It also demonstrated that training GANs is inherently challenging.

2.7.1 Network Structure

DCGAN employs an all-convolutional network [117], which replaces pooling functions with strided convolutions. This allows the network to learn its own spatial downsampling. The same architecture is used for both the generator and discriminator. Additionally, DCGAN eliminates fully connected layers in the input layer of the generator and the output layer of the discriminator, using only convolutional layers. This design choice is inspired by [85].

Moreover, batch normalization [51] is applied to all layers except the last layer of the generator and the first layer of the discriminator. This is because batch normalization enforces zero mean and unit variance for inputs to each neuron, while the mean and variance of the data should be learned by the first layer of the discriminator and reproduced by the last layer of the generator. Batch normalization mitigates the problem of mode collapse (introduced later in Sect. 3) at the cost of causing some fluctuations and instability [104].

Remark 2 (*Virtual batch normalization* [111, Sect. 3.5]) Batch normalization has a limitation: it makes the output for each input $\boldsymbol{x}$ dependent on other inputs in the mini-batch. To avoid this dependency, Virtual Batch Normalization (VBN) fixes mini-batches initially before training begins; these are called reference batches. Each reference batch is normalized using only its own statistics (mean and covariance). VBN has been found effective in training generators [111].

In DCGAN, the last layer of the generator uses the hyperbolic tangent activation function, while its other layers use the ReLU activation function [89]. Similar to GAN, the penultimate layer of the discriminator is flattened and connected to a single neuron with a sigmoid activation function. However, unlike GAN, which uses the maxout activation function [35] for discriminator layers (see Sect. 2.3), DCGAN uses the leaky ReLU activation function for the discriminator.

2.7.2 Vector Arithmetic in the Latent Space

DCGAN showed that when trained on domain-specific datasets (such as bedrooms or faces), GANs can generate realistic samples from that domain. It also revealed that the latent space encodes semantic directions, enabling vector arithmetic on features (e.g., adding or removing a smile). Vector arithmetic in the latent space was previously demonstrated for Word2Vec embeddings [82]. DCGAN extended this concept to image generation.

An example of vector arithmetic using DCGAN is shown in Fig. 3. In this example, latent variables corresponding to “man with glasses,” “man without glasses,” and “woman without glasses” are used. Using a single latent vector for each concept does not work well; instead, averaging three latent vectors

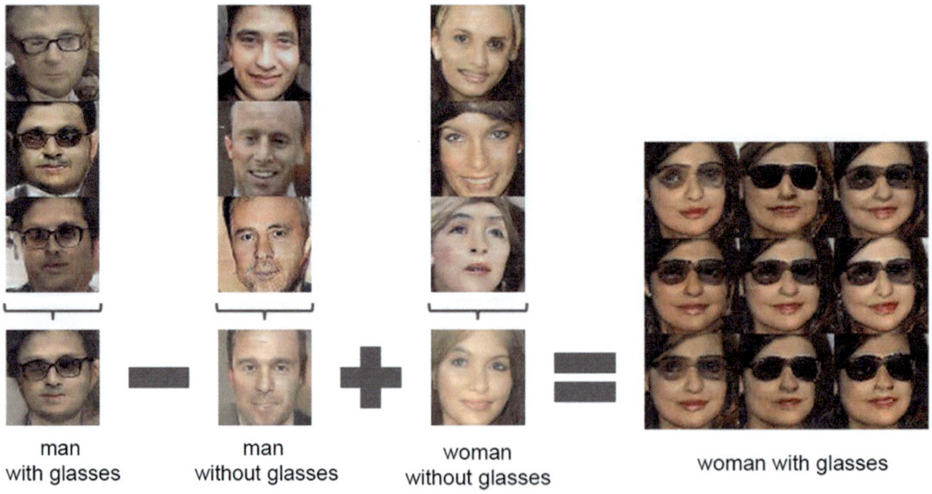

Fig. 3 Vector arithmetic in the latent space on images by DCGAN. Credit for the image: [104]

for each concept yields better results in practice. As shown in Fig. 3, vector arithmetic works because "man with glasses" minus "man without glasses" plus "woman without glasses" results in "woman with glasses."

3 Mode Collapse Problem in GAN

3.1 Mode Collapse Problem

It is expected that a GAN learns a meaningful latent space for z such that each specific value of z maps to a specific generated data point $\boldsymbol{x}$. Moreover, nearby z values in the latent space should map to similar but slightly different generated samples. The mode collapse problem [81], also known as the Helvetica scenario [33], is a common issue in GAN models. It occurs when the generator fails to learn a perfectly meaningful latent space as described above and instead maps several different z values to the same generated data point. Mode collapse typically arises when the training data distribution, $p_{\text{data}}(\boldsymbol{x})$, has multiple modes.

An example of mode collapse is illustrated in Fig. 4, which shows the training steps of a GAN model when the training data is a mixture of Gaussians [81]. During different training steps, GAN learns to map all z values to one of the modes of the mixture. When the discriminator learns to reject the generation of a particular mode, the generator switches to mapping all z values to another mode. However, it never learns to generate all modes of the mixture simultaneously. Ideally, GAN should map different regions of the latent space to different modes such that the entire latent space covers all modes.

Another explanation of mode collapse is as follows [130, Fig. 1]. Assume $p_{\text{data}}(\boldsymbol{x})$ is multi-modal while the latent space $p_z(z)$ has only one mode. Consider two points $\boldsymbol{x}_1$ and $\boldsymbol{x}_2$ from two different data modes, with corresponding latent noises z_1 and z_2, respectively. According to the mean value theorem, there exists a latent noise with an absolute gradient value of $\|\boldsymbol{x}_2 - \boldsymbol{x}_1\|/\|z_2 - z_1\|$, where $\|.\|$ denotes a norm. As this gradient is Lipschitz continuous, if the two modes are very far apart resulting in a large $\|\boldsymbol{x}_2 - \boldsymbol{x}_1\|$, a problem arises. In this case, latent noises between z_1 and z_2 generate data points between $\boldsymbol{x}_1$ and $\boldsymbol{x}_2$, which do not belong to any mode of the data and thus are invalid.

Various methods have been proposed to resolve the mode collapse problem in GANs and adversarial learning. Some approaches design the latent space as a mixture distribution to emulate the multi-modal nature of the training data, while others employ different strategies. The following subsections introduce methods that address the mode collapse problem.

3.2 Minibatch GAN

One way to resolve the mode collapse problem is through minibatch discrimination [111, Sect. 3.2]. In this method, the discriminator evaluates multiple data points in combination rather than individually, which helps prevent mode collapse in the generator. Suppose $\boldsymbol{f}(\boldsymbol{x}_i) \in \mathbb{R}^a$ is the feature vector from one

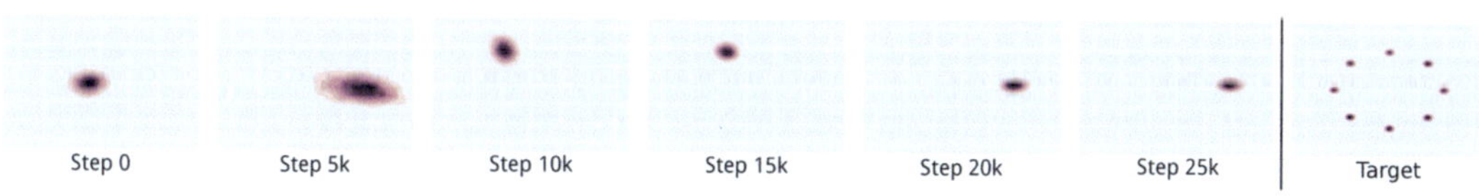

Fig. 4 An example of mode collapse in GAN. Image credit: [81]

of the intermediate layers, with a neurons, in the discriminator for the data point $\boldsymbol{x}_i$. The data point $\boldsymbol{x}_i$ can be either real or generated (fake). The feature vector $\boldsymbol{f}(\boldsymbol{x}_i)$ is multiplied by a tensor $\boldsymbol{T} \in \mathbb{R}^{a\times b\times c}$ to obtain $\boldsymbol{M}_i \in \mathbb{R}^{b\times c}$ as follows: $\boldsymbol{M}_i := (\boldsymbol{T}^\top \boldsymbol{f}(\boldsymbol{x}_i))^\top$. If there are $|\mathcal{B}|$ points in the mini-batch $\mathcal{B}$, then $\{\boldsymbol{M}_i\}_{i=1}^{|\mathcal{B}|}$ can be computed. Let $(\boldsymbol{M}_i)_{l:}$ denote the l-th row of $\boldsymbol{M}_i$. The following are defined [111]:

$$\begin{aligned}
&\mathbb{R} \ni c_l(\boldsymbol{x}_i, \boldsymbol{x}_j) := \exp(-\|(\boldsymbol{M}_i)_{l:} - (\boldsymbol{M}_j)_{l:}\|_1), \quad \forall l \in \{1, \dots, b\}, \\
&\mathbb{R} \ni o(\boldsymbol{x}_i)_l := \sum_{j=1}^{|\mathcal{B}|} c_l(\boldsymbol{x}_i, \boldsymbol{x}_j), \quad \forall i \in \{1, \dots, |\mathcal{B}|\}, \\
&\mathbb{R}^b \ni \boldsymbol{o}(\boldsymbol{x}_i) := [o(\boldsymbol{x}_i)_1, \dots, o(\boldsymbol{x}_i)_b]^\top, \quad \forall i, \\
&\mathbb{R}^{|\mathcal{B}|\times b} \ni \boldsymbol{o}(\boldsymbol{X}) := [\boldsymbol{o}(\boldsymbol{x}_1), \dots, \boldsymbol{o}(\boldsymbol{x}_{|\mathcal{B}|})]^\top.
\end{aligned} \tag{23}$$

For every point $\boldsymbol{x}_i$ within the mini-batch, $\boldsymbol{o}(\boldsymbol{x}_i)$ is concatenated with $\boldsymbol{f}(\boldsymbol{x}_i)$ before being fed to the next layer, rather than feeding only $\boldsymbol{f}(\boldsymbol{x}_i)$. In other words, the additional features $\boldsymbol{o}(\boldsymbol{x}_i)$ provide side information that improves the training of the discriminator, which in turn strengthens the generator during adversarial training. This procedure is applied to the discriminator for both mini-batches of real and generated data.

3.3 Unrolled GAN

Unrolled GAN [81] uses Δ levels of unrolling of the discriminator when updating the generator. The alternating optimization in unrolled GAN is

$$D^{(k+1)} := D^{(k)} + \eta \frac{\partial}{\partial D}\Big(V(D, G^{(k)})\Big), \tag{24}$$

$$\begin{cases}
D_0 := D^{(k+1)}, \\
D_{(\delta+1)} := D_{(\delta)} + \eta^{(\delta)} \frac{\partial}{\partial D}\Big(V(D, G^{(k)})\Big), \quad \forall \delta \in \{0, \dots, \Delta - 1\}, \\
V_\Delta(D^{(k+1)}, G) := V(D_{(\Delta)}, G),
\end{cases} \tag{25}$$

$$G^{(k+1)} := G^{(k)} - \eta \frac{\partial}{\partial G}\Big(V_\Delta(D^{(k+1)}, G)\Big), \tag{26}$$

where Eq. (25) unrolls the parameters of the discriminator Δ times before using it to update the generator. The loss function $V(D_{(\Delta)}, G)$ is called the surrogate loss. As mentioned in Sect. 2.2, the original GAN can update the discriminator several times before updating the generator [36]. However, Eq. (25) is different from repeatedly updating the discriminator, as done in GAN, because it does not actually update the discriminator but instead uses the unrolled discriminator for updating the generator in Eq. (26).

The gradient for the generator in Eq. (26) can be simplified as:

$$\frac{\partial}{\partial G}\Big(V_\Delta(D^{(k+1)}, G)\Big) \overset{(25)}{=} \frac{\partial}{\partial G}\Big(V_\Delta(D_{(\Delta)}, G)\Big) + \frac{\partial}{\partial D_{(\Delta)}}\Big(V_\Delta(D_{(\Delta)}, G)\Big)\frac{\partial D_{(\Delta)}}{\partial G}.$$

The gradient for the generator in the original GAN includes only the first term. The second term in the above equation captures the effect of changes in the discriminator with respect to changes in the generator. This additional term reduces the mode collapse problem that exists in GANs.

3.4 Bourgain GAN (BourGAN)

Bourgain GAN (BourGAN) [130] learns a mixture distribution [29] in the latent space to resolve the mode collapse problem and to enable the generation of multi-modal data. If the dataset size n is large, it samples m points from the dataset $\{\boldsymbol{x}_i\}_{i=1}^n$ where $m \ll n$. Let $\{\widetilde{\boldsymbol{x}}_i\}_{i=1}^m$ denote the sampled data. The idea of BourGAN is based on the Bourgain embedding [5], which enables embedding a dataset of size m into a $\mathcal{O}(\log^2(m))$-dimensional ℓ_2 norm embedding space with high probability. An improved Bourgain embedding is described as follows.

Theorem 6 ([130, Corollary 2]) *For a dataset $\{\widetilde{\boldsymbol{x}}_i\}_{i=1}^m$ in a space with norm $\|.\|$, there exists a mapping f from the data space to a $\mathcal{O}(\log(m))$-dimensional embedding space that preserves local distances:*

$$\|\widetilde{\boldsymbol{x}}_i - \widetilde{\boldsymbol{x}}_j\| \leq \|f(\widetilde{\boldsymbol{x}}_i) - f(\widetilde{\boldsymbol{x}}_j)\| \leq \alpha\, \|\widetilde{\boldsymbol{x}}_i - \widetilde{\boldsymbol{x}}_j\|,$$

where $\alpha \leq \mathcal{O}(\log(m))$. This embedding is achieved by applying Bourgain embedding [5] followed by random projection [26, 53]. The combined embedding procedure is detailed in [130, Appendix A].

This embedding requires computing pairwise distances. Sampling m points from the n data points makes this embedding computationally feasible. Such sampling does not affect the characteristics of pairwise distances in the data if m is sufficiently large [130, Theorem 4]. All sampled points are embedded to obtain $\{f(\widetilde{\boldsymbol{x}}_i)\}_{i=1}^m$. The distance characteristics of the data are preserved by this embedding [130, Theorem 5].

As the next step, BourGAN samples uniformly from the embedding points, i.e., $\boldsymbol{\mu} \sim \{f(\widetilde{\boldsymbol{x}}_i)\}_{i=1}^m$. Then, it samples the latent noise from a multivariate Gaussian distribution with mean $\boldsymbol{\mu}$, i.e., $z \sim \mathcal{N}(\boldsymbol{\mu}, 0.01\boldsymbol{I})$. This procedure models a multi-modal mixture of Gaussians in the latent space of GAN. Note that this does not imply a mixture of exactly m modes, because if several $f(\widetilde{\boldsymbol{x}}_i)$ are close to each other, they can be considered as representing a single mode. The multi-modality of the latent space is thus preserved to match the multi-modality of the data [130, Eq. (6)].

The loss function of BourGAN regularizes the generator to preserve the distances between generated points in accordance with the distances between their corresponding latent noises. In this way, the multi-modality of the latent space is reflected in the distribution of generated data $p_g(\boldsymbol{x})$:

$$\begin{aligned}
\max_{D} \quad & V_{\text{BourGAN}}(D) := V(D, G), \\
\min_{G} \quad & V_{\text{BourGAN}}(G) := V(D, G) + \lambda\, \mathbb{E}_{z_i, z_j \sim p_z(z)}\Big[\big(\log(\|G(z_i) - G(z_j)\|) \\
& \qquad\qquad - \log(\|z_i - z_j\|)\big)^2\Big],
\end{aligned} \tag{27}$$

where $V(D, G)$ is defined in Eq. (3), $\lambda > 0$ is the regularization parameter, and $\|.\|$ is a norm such as the ℓ_2 norm.

3.5 Mixture GAN (MGAN)

Mixture GAN (MGAN) [45] addresses the mode collapse issue by assuming that the latent space distribution, from which noise is sampled, is a mixture distribution [29] rather than a single distribution. It also increases the divergence among the latent distributions in the mixture so that each covers a different mode in the data distribution. The structure of MGAN is shown in Fig. 5. It comprises

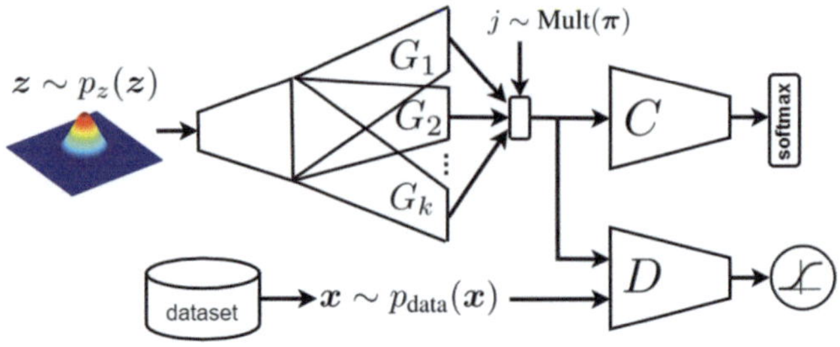

Fig. 5 The structure of MGAN

k generators $\{G_j\}_{j=1}^k$, a discriminator D, and a classifier C. In terms of incorporating a classifier, it is similar to Triple GAN [64] (see Sect. 5.9). Each generator G_j corresponds to the j-th latent distribution in the mixture. The key difference between MGAN and BourGAN (see Sect. 3.4) is that MGAN assigns a separate generator to each mode, while BourGAN uses a single generator with a mixture latent distribution.

Let π_j denote the mixing probability for the j-th component in the mixture. The distributions of generated data from the mixture latent distribution and from the j-th component are denoted by $p_g(\boldsymbol{x})$ and $p_{g_j}(\boldsymbol{x})$, respectively. The discriminator aims to determine whether a data point is real, $\boldsymbol{x} \sim p_{\text{data}}(\boldsymbol{x})$, or generated by one of the mixture components, i.e., $\boldsymbol{x} = G_j(\boldsymbol{z})$. At every iteration of the alternating optimization, the index j of the selected generator G_j for input to the discriminator D is sampled from a multinomial distribution Mult($\boldsymbol{\pi}$), where $\boldsymbol{\pi} := [\pi_1, \dots, \pi_k]^\top$. The classifier C identifies which of the k generators produced the generated data. The loss function of MGAN, formulated as a multi-player minimax game, is

$$
\begin{aligned}
\min_{\{G_j\}_{j=1}^k, C} \max_{D} \quad & V(D, C, G_1, \dots, G_k) := \\
& \mathbb{E}_{\boldsymbol{x} \sim p_{\text{data}}(\boldsymbol{x})}\Big[\log\big(D(\boldsymbol{x})\big)\Big] + \mathbb{E}_{\boldsymbol{x} \sim p_g(\boldsymbol{x})}\Big[\log\big(1 - D(\boldsymbol{x})\big)\Big] \\
& - \lambda \sum_{j=1}^{k} \pi_j \, \mathbb{E}_{\boldsymbol{x} \sim p_{g_j}(\boldsymbol{x})}\Big[\log\big(C_j(\boldsymbol{x})\big)\Big],
\end{aligned}
\tag{28}
$$

where $\lambda > 0$ is the regularization parameter and $C_j(\boldsymbol{x})$ is the probability that $\boldsymbol{x}$ was generated by G_j. The last layer of the classifier C contains k neurons with a softmax activation function, and C_j is the output of the j-th neuron after activation. The last term in the loss function maximizes the entropy of the classification, encouraging the generators and classifier to separate the generated data by mode. This ensures that different generators cover different modes of the data, thus resolving the mode collapse problem.

Theorem 7 ([45, Theorem 2]) *At convergence (i.e., Nash equilibrium) of MGAN, the following hold:*

$$
\begin{aligned}
& p_{g^*}(\boldsymbol{x}) = p_{data}(\boldsymbol{x}), \\
& D^*(\boldsymbol{x}) = \frac{p_{data}(\boldsymbol{x})}{p_{data}(\boldsymbol{x}) + p_g(\boldsymbol{x})}, \\
& C_j^*(\boldsymbol{x}) = \frac{\pi_j \, p_{g_j^*}(\boldsymbol{x})}{\sum_{l=1}^{k} \pi_l \, p_{g_l^*}(\boldsymbol{x})}, \\
& G^* := \arg\min_{G} \big(2\,JSD(p_{data} \| p_g) - \lambda\, JSD_\pi(p_{G_1}, \dots, p_{G_k})\big),
\end{aligned}
\tag{29}
$$

where:

$$JSD_{\pi}(p_{G_1}, \ldots, p_{G_k}) := \sum_{j=1}^{k} \pi_j \mathbb{E}_{\boldsymbol{x} \sim p_{g_j}}[\log(\frac{\pi_j\, p_{g_j}(\boldsymbol{x})}{\sum_{l=1}^{k} \pi_l\, p_{g_l}(\boldsymbol{x})})] - \sum_{j=1}^{k} \pi_j \log(\pi_j). \tag{30}$$

This theorem implies that the JSD between the mixture distribution p_g and the data distribution p_{data} is minimized, while the JSD among the components of the mixture is maximized so that the components capture distinct modes in the data.

The following theorem shows that if the data distribution is itself a mixture distribution, the optimal generation distribution at Nash equilibrium reproduces that mixture exactly.

Theorem 8 ([45, Theorem 3]) *If the data distribution is a mixture $p_{data} = \sum_{j=1}^{k} \pi_j\, q_j(\boldsymbol{x})$, where the components $q_j(\boldsymbol{x})$ are well-separated, the optimal generation distribution in MGAN is*

$$\begin{aligned} p_{g_j^*}(\boldsymbol{x}) &= q_j(\boldsymbol{x}), \quad \forall j = 1, \ldots, k, \\ p_{g^*}(\boldsymbol{x}) &= p_{data} = \sum_{j=1}^{k} \pi_j\, q_j(\boldsymbol{x}). \end{aligned} \tag{31}$$

3.6 Dual Discriminator GAN (D2GAN)

It has been empirically observed [33, 49, 121] that the JSD used in GAN (see Eq. (14)) behaves similarly to the reverse KL divergence, $\text{KL}(p_g(\boldsymbol{x}) \| p_{\text{data}}(\boldsymbol{x}))$. This reverse KL divergence tends to cause mode collapse because it captures a single mode very well but fails to cover multiple modes effectively. In contrast, the KL divergence $\text{KL}(p_{\text{data}}(\boldsymbol{x}) \| p_g(\boldsymbol{x}))$ can cover multiple modes and avoids mode collapse; however, it may include undesirable samples [92]. Dual Discriminator GAN (D2GAN) [92] combines the benefits of both KL and reverse KL divergences in its formulation, thus avoiding mode collapse while preventing the generation of undesirable samples.

In D2GAN, there are two discriminators D_1 and D_2, and one generator G. The discriminators do not share weights. Unlike the original GAN, the outputs of the discriminators are non-negative and unbounded rather than being in the range [0, 1]. The discriminator D_1 assigns high scores to real data and low scores to generated (fake) data. Conversely, discriminator D_2 assigns high scores to generated data and low scores to real data. The generator G aims to fool both discriminators. D2GAN formulates a three-player game with the following loss function:

$$\begin{aligned} \min_{G} \max_{D_1, D_2} \quad & V(D_1, D_2, G) := \\ & \alpha \mathbb{E}_{\boldsymbol{x} \sim p_{\text{data}}(\boldsymbol{x})}\Big[\log\big(D_1(\boldsymbol{x})\big)\Big] + \mathbb{E}_{z \sim p_z(z)}\Big[-D_1\big(G(z)\big)\Big] \\ & + \mathbb{E}_{\boldsymbol{x} \sim p_{\text{data}}(\boldsymbol{x})}\Big[-D_2(\boldsymbol{x})\Big] + \beta \mathbb{E}_{z \sim p_z(z)}\Big[\log\big(D_2\big(G(z)\big)\big)\Big], \end{aligned} \tag{32}$$

where $\alpha, \beta \in (0, 1]$ are hyperparameters. Alternating optimization [27] among D_1, D_2, and G is used to solve this optimization problem.

Theorem 9 ([92, Proposition 1 and Theorem 2]) *At convergence (i.e., Nash equilibrium) of D2GAN, the following hold:*

$$D_1^*(\boldsymbol{x}) = \frac{\alpha p_{data}(\boldsymbol{x})}{p_g(\boldsymbol{x})}, \quad D_2^*(\boldsymbol{x}) = \frac{\beta p_g(\boldsymbol{x})}{p_{data}(\boldsymbol{x})}. \tag{33}$$

The loss function at the optimal discriminators is

$$\begin{aligned} V(D_1^*, D_2^*, G) &:= \alpha(\log(\alpha) - 1) + \beta(\log(\beta) - 1) \\ &\quad + \alpha KL(p_{data}(\boldsymbol{x}) \| p_g(\boldsymbol{x})) + \beta KL(p_g(\boldsymbol{x}) \| p_{data}(\boldsymbol{x})). \\ \implies V(D_1^*, D_2^*, G^*) &:= \alpha(\log(\alpha) - 1) + \beta(\log(\beta) - 1). \end{aligned} \tag{34}$$

Therefore,

$$p_{g^*}(\boldsymbol{x}) = p_{data}(\boldsymbol{x}), \quad D_1^*(\boldsymbol{x}) = \alpha, \quad D_2^*(\boldsymbol{x}) = \beta. \tag{35}$$

According to Eq. (34), the parameters α and β control the weights of the KL divergence $\mathrm{KL}(p_{\text{data}}(\boldsymbol{x}) \| p_g(\boldsymbol{x}))$ and the reverse KL divergence $\mathrm{KL}(p_g(\boldsymbol{x}) \| p_{\text{data}}(\boldsymbol{x}))$, respectively. Increasing α promotes generation that covers multiple modes, thus mitigating mode collapse, but it may include undesirable samples. Increasing β encourages generation focusing on a single mode, potentially missing other modes. Therefore, it is important to balance the parameters α and β appropriately.

3.7 Wasserstein GAN (WGAN)

Wasserstein GAN (WGAN) was proposed in [2], developed from [1]. The Wasserstein-1, or Earth-Mover, distance between two distributions $p_{\text{data}}(\boldsymbol{x})$ and $p_g(\boldsymbol{x})$ is defined as:

$$W(p_{\text{data}}(\boldsymbol{x}), p_g(\boldsymbol{x})) := \inf_{\gamma \in \Pi(p_{\text{data}}(\boldsymbol{x}), p_g(\boldsymbol{x}))} \mathbb{E}_{(\boldsymbol{x}_i, \boldsymbol{x}_j) \sim \gamma}[\|\boldsymbol{x}_i - \boldsymbol{x}_j\|], \tag{36}$$

where $\Pi(p_{\text{data}}(\boldsymbol{x}), p_g(\boldsymbol{x}))$ is the set of all joint distributions with marginals $p_{\text{data}}(\boldsymbol{x})$ and $p_g(\boldsymbol{x})$. By the Kantorovich-Rubinstein duality [125], the Wasserstein-1 distance is equivalent to:

$$W(p_{\text{data}}(\boldsymbol{x}), p_g(\boldsymbol{x})) = \sup_{\|D\|_L \leq 1} \big(\mathbb{E}_{\boldsymbol{x} \sim p_{\text{data}}(\boldsymbol{x})}[D(\boldsymbol{x})] - \mathbb{E}_{\boldsymbol{x} \sim p_g(\boldsymbol{x})}[D(\boldsymbol{x})]\big), \tag{37}$$

where $\|D\|_L \leq 1$ denotes that D is a 1-Lipschitz function with $D : \mathcal{X} \to \mathbb{R}$. The gradient of the Wasserstein-1 distance with respect to the generator parameters is [2, Theorem 3]:

$$\frac{\partial W(p_{\text{data}}(\boldsymbol{x}), p_g(\boldsymbol{x}))}{\partial G} = -\mathbb{E}_{z \sim p_z(z)}\Big[\frac{\partial D(G(z))}{\partial G}\Big]. \tag{38}$$

The function $D(.)$ plays the role of the discriminator in WGAN. In alternating optimization, the loss in Eq. (37) is maximized for the discriminator and minimized for the generator using the gradient in Eq. (38). Thus, the loss function of WGAN is

$$\min_G \max_{\|D\|_L \leq 1} \mathbb{E}_{\boldsymbol{x} \sim p_{\text{data}}(\boldsymbol{x})}[D(\boldsymbol{x})] - \mathbb{E}_{\boldsymbol{x} \sim p_g(\boldsymbol{x})}[D(\boldsymbol{x})], \tag{39}$$

where $\boldsymbol{x} \sim p_g(\boldsymbol{x})$ is generated from the generator, i.e., $\boldsymbol{x} = G(z)$ with z as the latent noise. The weights of the discriminator are clipped to $[-0.01, 0.01]$. The constraint $\|D\|_L \leq 1$ can also be enforced by regularization using a gradient penalty [41]:

$$\begin{aligned}\min_G \max_{\|D\|_L \leq 1} \; & \mathbb{E}_{\boldsymbol{x}\sim p_{\text{data}}(\boldsymbol{x})}[D(\boldsymbol{x})] - \mathbb{E}_{\boldsymbol{x}\sim p_g(\boldsymbol{x})}[D(\boldsymbol{x})] \\ & - \lambda(\|\nabla_{\widehat{\boldsymbol{x}}} D(\widehat{\boldsymbol{x}})\|_2 - 1)^2,\end{aligned} \tag{40}$$

where $\lambda > 0$ is the regularization parameter and $\widehat{\boldsymbol{x}}$ is uniformly sampled as an interpolation between real and generated data:

$$\widehat{\boldsymbol{x}} := \rho \boldsymbol{x} + (1 - \rho) G(z), \tag{41}$$

with $\rho \sim U(0, 1)$.

Experiments showed that WGAN improves training stability and reduces mode collapse compared to vanilla GANs, though it does not eliminate the problem entirely [2]. Its use of the Wasserstein distance provides smoother gradients for the generator.

4 Maximum Likelihood Estimation in GAN

The following introduces methods that relate GAN and Maximum Likelihood Estimation (MLE).

4.1 Comparison of MLE and GAN

GAN is related to Noise-Contrastive Estimation (NCE) [42] and MLE in the sense that they all optimize a distinguishability game value function [34]:

$$V(p_c, p_g) := \mathbb{E}_{\boldsymbol{x}\sim p_{\text{data}}}[\log(p_c(y = 1|\boldsymbol{x}))] + \mathbb{E}_{\boldsymbol{x}\sim p_g}[\log(p_c(y = 0|\boldsymbol{x}))], \tag{42}$$

where p_g and p_{data} denote the distributions of generated and real data, respectively, and $p_c(y|\boldsymbol{x})$ is the output probability of a classifier that judges whether a data point $\boldsymbol{x}$ is generated or real. Let $\boldsymbol{\theta}$ denote the parameters of the $p_g(\boldsymbol{x})$ distribution. If $f(\boldsymbol{x}) := \log(p_c(y = 0|\boldsymbol{x}))$, GAN performs the following optimization for its generator [34]:

$$\min_{\boldsymbol{\theta}} \mathbb{E}_{\boldsymbol{x}\sim p_g}[f(\boldsymbol{x})], \tag{43}$$

which implies

$$\begin{aligned}\frac{\partial}{\partial \boldsymbol{\theta}} \mathbb{E}_{\boldsymbol{x}\sim p_g}[f(\boldsymbol{x})] &= \frac{\partial}{\partial \boldsymbol{\theta}} \int f(\boldsymbol{x}) p_g(\boldsymbol{x}) d\boldsymbol{x} \\ &= \int f(\boldsymbol{x}) \frac{\partial}{\partial \boldsymbol{\theta}} p_g(\boldsymbol{x}) d\boldsymbol{x} \\ &= \int f(\boldsymbol{x}) p_g(\boldsymbol{x}) \frac{1}{p_g(\boldsymbol{x})} \frac{\partial}{\partial \boldsymbol{\theta}} p_g(\boldsymbol{x}) d\boldsymbol{x} \\ &= \int f(\boldsymbol{x}) p_g(\boldsymbol{x}) \frac{\partial}{\partial \boldsymbol{\theta}} \log(p_g(\boldsymbol{x})) d\boldsymbol{x}.\end{aligned} \tag{44}$$

On the other hand, MLE has the following optimization:

$$\max_{\boldsymbol{\theta}} \mathbb{E}_{\boldsymbol{x}\sim p_g}[p_{\text{data}}(\boldsymbol{x})],$$
$$\implies \frac{\partial}{\partial \boldsymbol{\theta}} \mathbb{E}_{\boldsymbol{x}\sim p_g}[p_{\text{data}}(\boldsymbol{x})] = \int p_{\text{data}}(\boldsymbol{x}) \frac{\partial}{\partial \boldsymbol{\theta}} \log(p_g(\boldsymbol{x}))d\boldsymbol{x}. \tag{45}$$

Equations (44) and (45) correspond to minimization in GAN and maximization in MLE, respectively. By comparing them, it is possible to achieve MLE in GAN if we set:

$$f(\boldsymbol{x}) = -\frac{p_{\text{data}}(\boldsymbol{x})}{p_g(\boldsymbol{x})}. \tag{46}$$

The discriminator of GAN is modeled as a classifier predicting whether $\boldsymbol{x}$ is real or generated (fake); therefore:

$$D(\boldsymbol{x}) = p_c(y = 1|\boldsymbol{x}) = \sigma(D'(\boldsymbol{x})), \tag{47}$$

where $\sigma(.)$ is the sigmoid activation function and $D'(\boldsymbol{x})$ denotes the discriminator network excluding its final sigmoid layer.

Theorem 10 ([33, 34]) *The loss function for the generator of GAN can be expressed in any of the following forms:*

$$\min_G \mathbb{E}_{z\sim p_z(z)}\Big[\log\Big(1 - D\big(G(z)\big)\Big)\Big], \tag{48}$$
$$\min_G -\mathbb{E}_{z\sim p_z(z)}\Big[\log\Big(D\big(G(z)\big)\Big)\Big], \tag{49}$$
$$\min_G -\mathbb{E}_{z\sim p_z(z)}\Big[\sigma^{-1}\Big(D\big(G(z)\big)\Big)\Big], \tag{50}$$

where $\sigma(.)$ is the sigmoid function.

Proof Equation (48) is the generator part of the loss function (3). The generator seeks to fool the discriminator, so it aims for $D(G(z))$ to be close to one (see Eq. (2)). Thus, instead of minimizing $\log(1 - D(G(z)))$ as in Eq. (48), it is possible to maximize $\log(D(G(z)))$, or equivalently minimize its negation, as in Eq. (49) [33]. The proof of Eq. (50) is as follows [34]. Assuming the discriminator is optimal for a given generator, then according to Eq. (8):

$$p_c(y = 1|\boldsymbol{x}) \overset{(8)}{=} \frac{p_{\text{data}}(\boldsymbol{x})}{p_{\text{data}}(\boldsymbol{x}) + p_g(\boldsymbol{x})} = \frac{1}{1 + \frac{p_g(\boldsymbol{x})}{p_{\text{data}}(\boldsymbol{x})}}$$
$$\overset{(47)}{=} \sigma(D'(\boldsymbol{x})) = \frac{1}{1 + \exp(-D'(\boldsymbol{x}))}$$
$$\implies \frac{p_g(\boldsymbol{x})}{p_{\text{data}}(\boldsymbol{x})} = \exp(-D'(\boldsymbol{x}))$$
$$\overset{(46)}{\implies} f(\boldsymbol{x}) = -\exp(D'(\boldsymbol{x})) \overset{(47)}{=} -\exp\big(\sigma^{-1}(D(\boldsymbol{x}))\big).$$

Therefore, for the generated data $\boldsymbol{x} = G(z)$ from latent noise $z \sim p_z(z)$, Eq. (43) becomes Eq. (50).

□

4.2 f-GAN

f-GAN [95] uses f-divergence in the formulation of GAN. The f-divergence between two distributions $p(\boldsymbol{x})$ and $q(\boldsymbol{x})$ is computed as [68]:

$$D_f(P\|Q) := \int q(\boldsymbol{x}) f\big(\frac{p(\boldsymbol{x})}{q(\boldsymbol{x})}\big) d\boldsymbol{x}, \tag{51}$$

where the convex generator function $f : \mathbb{R}_+ \to \mathbb{R}$ satisfies $f(1) = 0$. Two special cases of f-GAN are KL-divergence and JS-divergence. The data space is denoted by $\mathcal{X}$.

Lemma 3 ([93, Lemma 1]) *A lower bound on the f-divergence is*

$$D_f(P\|Q) \geq \sup_{T \in \mathcal{T}} (\mathbb{E}_{\boldsymbol{x}\sim p(\boldsymbol{x})}[T(\boldsymbol{x})] - \mathbb{E}_{\boldsymbol{x}\sim q(\boldsymbol{x})}[f^*(T(\boldsymbol{x}))]), \tag{52}$$

where f^* is the convex conjugate of f and $\mathcal{T} : \mathcal{X} \to \mathbb{R}$ is an arbitrary class of functions.

Proof The convex conjugate of function f is defined as [27]:

$$\begin{aligned} f^*(t) &:= \sup_{u \in \mathrm{dom}(f)} (ut - f(u)) \\ \implies f(u) &:= \sup_{t \in \mathrm{dom}(f^*)} (tu - f^*(t)). \end{aligned} \tag{53}$$

Therefore:

$$\begin{aligned} D_f(P\|Q) &\overset{(53)}{=} \int q(\boldsymbol{x}) \sup_{t \in \mathrm{dom}(f^*)} \big(t\frac{p(\boldsymbol{x})}{q(\boldsymbol{x})} - f^*(t)\big) d\boldsymbol{x} \\ &\overset{(a)}{\geq} \sup_{T \in \mathcal{T}} \Big(\int p(\boldsymbol{x}) T(\boldsymbol{x}) d\boldsymbol{x} - \int q(\boldsymbol{x}) f^*(T(\boldsymbol{x})) d\boldsymbol{x}\Big) \\ &= \sup_{T \in \mathcal{T}} (\mathbb{E}_{\boldsymbol{x}\sim p(\boldsymbol{x})}[T(\boldsymbol{x})] - \mathbb{E}_{\boldsymbol{x}\sim q(\boldsymbol{x})}[f^*(T(\boldsymbol{x}))]), \end{aligned}$$

where (a) follows because the sum of maxima is greater than or equal to the maximum of sums. □

Variational Divergence Minimization (VDM) [95] optimizes the f-divergence by optimizing the bound in Eq. (52). In this sense, it is similar to variational inference (see Chap. 12). Suppose $p(\boldsymbol{x}) = p_{\text{data}}(\boldsymbol{x})$ and $q(\boldsymbol{x}) = p_g(\boldsymbol{x})$ in Eq. (52), where $p_{\text{data}}(\boldsymbol{x})$ and $p_g(\boldsymbol{x})$ are the distributions of real and generated data, respectively. Let $T(\boldsymbol{x}) = o_f(V(\boldsymbol{x}))$, where $V : \mathcal{X} \to \mathbb{R}$ is the network mapping from input to its output neuron (before activation), and $o_f : \mathbb{R} \to \mathrm{dom}(f^*)$ is the output activation function. VDM can be used to optimize various f-divergences. Its loss function is

$$\min_G \max_V \; \mathbb{E}_{\boldsymbol{x}\sim p_{\text{data}}(\boldsymbol{x})}\Big[o_f(V(\boldsymbol{x}))\Big] + \mathbb{E}_{\boldsymbol{x}\sim p_g(\boldsymbol{x})}\Big[-f^*\big(o_f(V(\boldsymbol{x}))\big)\Big]. \tag{54}$$

Readers can refer to [95, Table 2] for a complete list of expressions for $o_f(v)$ and f^* in different special cases of f-divergence.

A special case of VDM is f-GAN, in which $o_f(v) = -\log(1+\exp(-v))$, $f^*(t) = -\log(1-\exp(t))$, and the discriminator is the sigmoid function of $V(\boldsymbol{x})$, i.e., $D(\boldsymbol{x}) = 1/(1+\exp(-V(\boldsymbol{x})))$. Therefore, in f-GAN:

$$o_f(V(\boldsymbol{x})) = -\log(1+\exp(-V(\boldsymbol{x}))) = \log(D(\boldsymbol{x})),$$
$$f^*\big(o_f(V(\boldsymbol{x}))\big) = -\log(1-\exp(\log(D(\boldsymbol{x})))) = -\log(1-D(\boldsymbol{x})).$$

Substituting these into Eq. (54) yields the loss of f-GAN:

$$\min_G \max_D \; \mathbb{E}_{\boldsymbol{x}\sim p_{\text{data}}(\boldsymbol{x})}\Big[\log\big(D(\boldsymbol{x})\big)\Big] + \mathbb{E}_{\boldsymbol{x}\sim p_g(\boldsymbol{x})}\Big[\log\big(1-D(\boldsymbol{x})\big)\Big]. \tag{55}$$

4.3 Adversarial Variational Bayes (AVB)

Adversarial Variational Bayes (AVB) [79] combines the ideas of variational and adversarial training. Variational inference (see Chap. 12) maximizes the evidence lower bound defined as:

$$\max_\theta \max_\phi \; \mathbb{E}_{p(\boldsymbol{x})}\mathbb{E}_{q_\phi(z|\boldsymbol{x})}\big[\log(p(z)) - \log(q_\phi(z|\boldsymbol{x})) + \log(p_\theta(\boldsymbol{x}|z))\big], \tag{56}$$

where θ and ϕ are parameters corresponding to $p_\theta(\boldsymbol{x}|z)$ and $q_\phi(z|\boldsymbol{x})$, respectively. On the other hand, adversarial learning employs a discriminator during training. AVB integrates a discriminator $D(\boldsymbol{x}, z)$ with a single output neuron and sigmoid activation function into variational inference. The loss function for the discriminator is

$$\max_D \; \mathbb{E}_{p(\boldsymbol{x})}\mathbb{E}_{q_\phi(z|\boldsymbol{x})}\big[\log(D(\boldsymbol{x}, z))\big] + \mathbb{E}_{p(\boldsymbol{x})}\mathbb{E}_{p_z(z)}\big[\log(1-D(\boldsymbol{x}, z))\big], \tag{57}$$

whose solution is $D^*(\boldsymbol{x}, z) = -\log(p(z)) + \log(q_\phi(z|\boldsymbol{x}))$ [79, Proposition 1]. Therefore, Eq. (56) becomes

$$\max_\theta \max_\phi \; \mathbb{E}_{p(\boldsymbol{x})}\mathbb{E}_{q_\phi(z|\boldsymbol{x})}\big[-D^*(\boldsymbol{x}, z) + \log(p_\theta(\boldsymbol{x}|z))\big], \tag{58}$$

which is optimized using backpropagation, after applying the reparameterization trick (see Chap. 12).

4.4 Bayesian GAN (BGAN)

Bayesian GAN (BGAN) [110] models GAN using Bayesian analysis. Let G and D denote the parameters of the generator and discriminator, respectively, $\boldsymbol{x}$ be the real data, z be the noise sample, and b be the mini-batch size. Consider the following distributions:

$$p(G|z, D) \propto \Big(\prod_{i=1}^{b} D(G(z_i))\Big)\, p(G),$$
$$p(D|z, \boldsymbol{x}) \propto \Big(\prod_{i=1}^{b} D(\boldsymbol{x}_i)\Big)\Big(\prod_{i=1}^{b}(1-D(G(z_i)))\Big)\, p(D).$$

These distributions can be marginalized as:

$$p(G|D) = \int p(G, z|D)\, dz = \int p(G|z, D)\, p(z|D)\, dz$$
$$\overset{(a)}{=} \int p(G|z, D)\, p_z(z)\, dz \overset{(b)}{\approx} \frac{1}{b} \sum_{i=1}^{b} p(G|z_i), \tag{59}$$

where $z_i \sim p_z(z)$, (a) follows because the noise z is independent of the discriminator D, and (b) uses the Monte Carlo approximation [30]. Similarly, we have

$$p(D|G) = \frac{1}{b} \sum_{i=1}^{b} p(D|z_i, \boldsymbol{x}_i). \tag{60}$$

Sampling from the distributions in Eqs. (59) and (60) will converge to the joint distribution of the generator and discriminator, based on Gibbs sampling [30]. Stochastic Gradient Hamiltonian Monte Carlo (SGHMC) is a technique used to train neural networks using these posteriors. In BGAN, the discriminator and generator are trained alternately using SGHMC and the posteriors in Eqs. (59) and (60).

Note that another GAN model incorporating variational inference and Bayesian analysis is the Variational Bayesian GAN [12].

5 Other Variants of GAN

GANs have many extensions such as InfoGAN [11], LSGAN [76, 77], and EBGAN [134]. Later, StyleGAN [55] and BigGAN [6] became especially influential: StyleGAN set new standards for photorealistic face generation, while BigGAN demonstrated how scaling GANs leads to high-quality samples across diverse categories. In the following, some of the variants of GAN are introduced.

5.1 Feature Matching in GAN

During training, the layers of the discriminator D learn discriminative features that differentiate between real and generated data. Therefore, for better training of the generator G and to enable it to fool the discriminator, the features from an intermediate layer of the discriminator can be utilized [111, Sect. 3.1]. The generator is trained to match the expected values of these intermediate features for both real and generated data inputs. Thus, the optimization objective for the generator is

$$\min_{G} \; \|\mathbb{E}_{\boldsymbol{x}\sim p_{\text{data}}(\boldsymbol{x})}[\boldsymbol{f}(\boldsymbol{x})] - \mathbb{E}_{z\sim p_z(z)}\big[\boldsymbol{f}\big(G(z)\big)\big]\|_2^2, \tag{61}$$

where $\boldsymbol{f}(\boldsymbol{x})$ and $\boldsymbol{f}(G(z))$ denote the features from an intermediate layer of the discriminator for inputs $\boldsymbol{x}$ (real data) and $G(z)$ (generated data), respectively. The discriminator is trained as in the original GAN, i.e., by maximizing Eq. (3).

5.2 InfoGAN

Information Maximizing GAN (InfoGAN), proposed in [11], is an information-theoretic extension of GAN. It maximizes the mutual information between latent variables and generated data. InfoGAN introduces two sets of latent variables, $\boldsymbol{z}$ and $\boldsymbol{c}$. The generator takes these two latent variables as input and outputs $G(\boldsymbol{z}, \boldsymbol{c})$. The optimization problem in InfoGAN is formulated as:

$$\min_{G} \max_{D} \quad V_I(D, G) := V(D, G) - \lambda I(\boldsymbol{c}; G(\boldsymbol{z}, \boldsymbol{c})), \tag{62}$$

where $V(D, G)$ is defined in Eq. (3), $\lambda > 0$ is the regularization parameter, and $I(.; .)$ is the mutual information defined as $I(\boldsymbol{c}; G(\boldsymbol{z}, \boldsymbol{c})) := H(\boldsymbol{c}) - H(\boldsymbol{c}|G(\boldsymbol{z}, \boldsymbol{c}))$, with $H(.)$ denoting entropy. Note that the regularization term depends only on G and not on D. The generator aims to maximize the mutual information $I(\boldsymbol{c}; G(\boldsymbol{z}, \boldsymbol{c}))$.

Computing this mutual information directly is difficult in practice. It can be simplified by introducing an auxiliary distribution $Q(\boldsymbol{c}|\boldsymbol{x})$:

$$\begin{aligned}
I(\boldsymbol{c}; G(\boldsymbol{z}, \boldsymbol{c})) &:= H(\boldsymbol{c}) - H(\boldsymbol{c}|G(\boldsymbol{z}, \boldsymbol{c})) \\
&\overset{(a)}{=} H(\boldsymbol{c}) - \big(-\mathbb{E}_{\boldsymbol{x}\sim G(\boldsymbol{z},\boldsymbol{c})}[\log P(\boldsymbol{c}|\boldsymbol{x})]\big) \\
&= H(\boldsymbol{c}) + \mathbb{E}_{\boldsymbol{x}\sim G(\boldsymbol{z},\boldsymbol{c})}\big[\mathbb{E}_{\boldsymbol{c}'\sim P(\boldsymbol{c}|\boldsymbol{x})}[\log P(\boldsymbol{c}'|\boldsymbol{x})]\big] \\
&\overset{(b)}{=} H(\boldsymbol{c}) + \mathbb{E}_{\boldsymbol{x}\sim G(\boldsymbol{z},\boldsymbol{c})}\big[\mathrm{KL}(P(.|\boldsymbol{x})\|Q(.|\boldsymbol{x})) + \mathbb{E}_{\boldsymbol{c}'\sim P(\boldsymbol{c}|\boldsymbol{x})}[\log Q(\boldsymbol{c}'|\boldsymbol{x})]\big] \\
&\overset{(c)}{\geq} H(\boldsymbol{c}) + \mathbb{E}_{\boldsymbol{x}\sim G(\boldsymbol{z},\boldsymbol{c})}\big[\mathbb{E}_{\boldsymbol{c}'\sim P(\boldsymbol{c}|\boldsymbol{x})}[\log Q(\boldsymbol{c}'|\boldsymbol{x})]\big] \\
&\overset{(d)}{=} H(\boldsymbol{c}) + \mathbb{E}_{\boldsymbol{c}\sim P(\boldsymbol{c}),\, \boldsymbol{x}\sim G(\boldsymbol{z},\boldsymbol{c})}[\log Q(\boldsymbol{c}'|\boldsymbol{x})] \\
&\overset{(e)}{=} L_I(G, Q),
\end{aligned}$$

where (a) follows from the definition of entropy, (b) uses the definition of KL divergence, (c) holds because KL divergence is non-negative, (d) applies [11, Lemma 5.1], and (e) defines $L_I(G, Q)$ as the final expression. Therefore, $L_I(G, Q)$ is a lower bound for $I(\boldsymbol{c}; G(\boldsymbol{z}, \boldsymbol{c}))$. Using this lower bound in Eq. (62) yields:

$$\min_{G,Q} \max_{D} \quad V_I(D, G) := V(D, G) - \lambda L_I(G, Q), \tag{63}$$

where:

$$L_I(G, Q) := H(\boldsymbol{c}) + \mathbb{E}_{\boldsymbol{c}\sim P(\boldsymbol{c}),\, \boldsymbol{x}\sim G(\boldsymbol{z},\boldsymbol{c})}[\log Q(\boldsymbol{c}'|\boldsymbol{x})],$$

which can be estimated using Monte Carlo approximation [30].

5.3 Generative Recurrent Adversarial Network (GRAN)

Generative Recurrent Adversarial Network (GRAN) [50] is inspired by the Deep Recurrent Attentive Writer (DRAW) [38]. DRAW uses variational inference to draw images on a canvas incrementally over time. GRAN adopts a similar approach but employs adversarial learning, thus combining GANs with recurrent networks.

In GRAN, the generator G incorporates a recurrent feedback loop that takes as input a sequence of noise samples $\{z_t\}_{t=1}^T$. The recurrent loop of the generator produces a sequence of canvas drawings, i.e., $\{\Delta C_1, \Delta C_2, \ldots, \Delta C_T\}$. At each time step $t \in \{1, \ldots, T\}$, the recurrent loop functions like an autoencoder with encoder $f(.)$ and decoder $g(.)$. The coding layer between the encoder and decoder receives the concatenation of the latent coding $\boldsymbol{h}_{z,t}$ and the canvas coding $\boldsymbol{h}_{c,t}$. This concatenated coding is fed to the decoder f to produce the canvas drawing ΔC_t.

At each time step $t \in \{1, \ldots, T\}$, the following computations are performed:

$$\begin{aligned} z_t &\sim p_z(z), \\ \boldsymbol{h}_{c,t} &:= g(\Delta C_{t-1}), \\ \boldsymbol{h}_{z,t} &:= \tanh(\boldsymbol{W} z_t + \boldsymbol{b}), \\ \Delta C_t &:= f([\boldsymbol{h}_{z,t}^\top, \boldsymbol{h}_{c,t}^\top]^\top), \end{aligned} \tag{64}$$

where $\boldsymbol{W}$ and $\boldsymbol{b}$ are the weights and bias for the latent variable z_t. DCGAN [104] (see Sect. 2.7) is used for the encoder f and decoder g at each recurrent loop to generate the canvas drawings $\{\Delta C_t\}_{t=1}^T$.

The final canvas drawing is obtained by summing the drawings across all time steps. A logistic function $\sigma(.)$ is applied to scale the drawing to the range (0, 1) for pixel visualization:

$$C = \sigma\Big(\sum_{t=1}^{T} \Delta C_t\Big).$$

5.4 Least Squares GAN (LSGAN)

The GAN loss function has a limitation: in the discriminator, the gradient vanishes for generated data points that fall on the correct side of the decision boundary but are still different from real data. Least Squares GAN (LSGAN) [76, 77] addresses this issue by using a least squares cost in the adversarial loss function.

For the discriminator D in LSGAN, two scalar labels a and b are used for generated (fake) and real data points, respectively. For the generator G, a scalar label c is used, which the generator wants the discriminator to predict for its generated data. The loss functions in LSGAN are

$$\begin{aligned} \min_D \quad V_{\text{LSGAN}}(D) &:= \frac{1}{2}\mathbb{E}_{\boldsymbol{x}\sim p_{\text{data}}(\boldsymbol{x})}\big[(D(\boldsymbol{x}) - b)^2\big] + \frac{1}{2}\mathbb{E}_{z\sim p_z(z)}\big[(D(G(z)) - a)^2\big], \\ \min_G \quad V_{\text{LSGAN}}(G) &:= \frac{1}{2}\mathbb{E}_{z\sim p_z(z)}\big[(D(G(z)) - c)^2\big]. \end{aligned} \tag{65}$$

Lemma 4 ([77, Proposition 1]) *For a fixed generator G, the optimal discriminator in LSGAN is*

$$D^*(\boldsymbol{x}) = \frac{b p_{data}(\boldsymbol{x}) + a p_g(\boldsymbol{x})}{p_{data}(\boldsymbol{x}) + p_g(\boldsymbol{x})}, \tag{66}$$

where $p_{data}(\boldsymbol{x})$ is the probability distribution of the real dataset evaluated at point $\boldsymbol{x}$, and $p_g(\boldsymbol{x})$ is the probability distribution of the generator output evaluated at point $\boldsymbol{x}$.

Proof

$$
\begin{aligned}
V_{\text{LSGAN}}(D) &\overset{(65)}{=} \frac{1}{2}\mathbb{E}_{\boldsymbol{x}\sim p_{\text{data}}(\boldsymbol{x})}\big[(D(\boldsymbol{x})-b)^2\big] + \frac{1}{2}\mathbb{E}_{z\sim p_z(z)}\big[(D(G(z))-a)^2\big] \\
&= \int_{\boldsymbol{x}} \frac{1}{2} p_{\text{data}}(\boldsymbol{x})(D(\boldsymbol{x})-b)^2 d\boldsymbol{x} + \int_{\boldsymbol{x}} \frac{1}{2} p_z(z)(D(G(z))-a)^2 d\boldsymbol{x} \\
&\overset{(9)}{=} \int_{\boldsymbol{x}} \frac{1}{2}\Big(p_{\text{data}}(\boldsymbol{x})(D(\boldsymbol{x})-b)^2 + p_g(\boldsymbol{x})(D(\boldsymbol{x})-a)^2\Big) d\boldsymbol{x}.
\end{aligned}
$$

For optimization in Eq. (65), taking derivative with respect to $D(\boldsymbol{x})$ gives

$$
\begin{aligned}
\frac{\partial V_{\text{LSGAN}}(D)}{\partial D(\boldsymbol{x})} &\overset{(a)}{=} \frac{\partial}{\partial D(\boldsymbol{x})}\Big(\frac{1}{2}\big(p_{\text{data}}(\boldsymbol{x})(D(\boldsymbol{x})-b)^2 + p_g(\boldsymbol{x})(D(\boldsymbol{x})-a)^2\big)\Big) \\
&= p_{\text{data}}(\boldsymbol{x})(D(\boldsymbol{x})-b) + p_g(\boldsymbol{x})(D(\boldsymbol{x})-a) \overset{\text{set}}{=} 0 \\
\implies D(\boldsymbol{x}) &= \frac{b p_{\text{data}}(\boldsymbol{x}) + a p_g(\boldsymbol{x})}{p_{\text{data}}(\boldsymbol{x}) + p_g(\boldsymbol{x})},
\end{aligned}
$$

where (a) is because taking derivative with respect to $D(\boldsymbol{x})$ considers a specific $\boldsymbol{x}$ and hence it removes the integral (summation). □

Theorem 11 ([77, Theorem 1]) *The optimization of LSGAN is equivalent to minimizing the Pearson* χ^2 *divergence between* $p_{data}(\boldsymbol{x}) + p_g(\boldsymbol{x})$ *and* $2p_g(\boldsymbol{x})$*, if the following conditions hold:*

$$
b - c = 1, \quad b - a = 2. \tag{67}
$$

Proof

$$
\begin{aligned}
2V_{\text{LSGAN}}(G) &\overset{(65)}{=} \mathbb{E}_{\boldsymbol{x}\sim p_z(z)}\big[(D^*(G(z))-c)^2\big] \\
&\overset{(a)}{=} \mathbb{E}_{\boldsymbol{x}\sim p_{\text{data}}(\boldsymbol{x})}\big[(D^*(\boldsymbol{x})-c)^2\big] + \mathbb{E}_{z\sim p_z(z)}\big[(D^*(G(z))-c)^2\big] \\
&\overset{(9)}{=} \mathbb{E}_{\boldsymbol{x}\sim p_{\text{data}}(\boldsymbol{x})}\big[(D^*(\boldsymbol{x})-c)^2\big] + \mathbb{E}_{\boldsymbol{x}\sim p_g(\boldsymbol{x})}\big[(D^*(\boldsymbol{x})-c)^2\big] \\
&\overset{(66)}{=} \mathbb{E}_{\boldsymbol{x}\sim p_{\text{data}}(\boldsymbol{x})}\Big[\Big(\frac{b p_{\text{data}}(\boldsymbol{x}) + a p_g(\boldsymbol{x})}{p_{\text{data}}(\boldsymbol{x}) + p_g(\boldsymbol{x})} - c\Big)^2\Big] \\
&\quad + \mathbb{E}_{\boldsymbol{x}\sim p_g(\boldsymbol{x})}\Big[\Big(\frac{b p_{\text{data}}(\boldsymbol{x}) + a p_g(\boldsymbol{x})}{p_{\text{data}}(\boldsymbol{x}) + p_g(\boldsymbol{x})} - c\Big)^2\Big] \\
&\overset{(b)}{=} \int_{\boldsymbol{x}} p_{\text{data}}(\boldsymbol{x})\Big(\frac{(b-c)p_{\text{data}}(\boldsymbol{x}) + (a-c)p_g(\boldsymbol{x})}{p_{\text{data}}(\boldsymbol{x}) + p_g(\boldsymbol{x})}\Big)^2 d\boldsymbol{x} \\
&\quad + \int_{\boldsymbol{x}} p_g(\boldsymbol{x})\Big(\frac{(b-c)p_{\text{data}}(\boldsymbol{x}) + (a-c)p_g(\boldsymbol{x})}{p_{\text{data}}(\boldsymbol{x}) + p_g(\boldsymbol{x})}\Big)^2 d\boldsymbol{x} \\
&\overset{(c)}{=} \int_{\boldsymbol{x}} \frac{\big((b-c)p_{\text{data}}(\boldsymbol{x}) + (a-c)p_g(\boldsymbol{x})\big)^2}{p_{\text{data}}(\boldsymbol{x}) + p_g(\boldsymbol{x})} d\boldsymbol{x}
\end{aligned}
$$

$$
\begin{aligned}
&= \int_{x} \frac{\big((b-c)(p_{\text{data}}(\boldsymbol{x}) + p_g(\boldsymbol{x})) - (b-a)p_g(\boldsymbol{x})\big)^2}{p_{\text{data}}(\boldsymbol{x}) + p_g(\boldsymbol{x})} d\boldsymbol{x} \\
&\overset{(67)}{=} \int_{x} \frac{\big(2p_g(\boldsymbol{x}) - (p_{\text{data}}(\boldsymbol{x}) + p_g(\boldsymbol{x}))\big)^2}{p_{\text{data}}(\boldsymbol{x}) + p_g(\boldsymbol{x})} d\boldsymbol{x} \\
&\overset{(d)}{=} \chi^2(p_{\text{data}}(\boldsymbol{x}) + p_g(\boldsymbol{x}) \,\|\, 2p_g(\boldsymbol{x})),
\end{aligned}
$$

where (a) is because $\mathbb{E}_{\boldsymbol{x}\sim p_{\text{data}}(\boldsymbol{x})}\big[(D^*(\boldsymbol{x}) - c)^2\big]$ is constant with respect to G, (b) is because of the definition of expectation, (c) is because of simplification of terms, and (d) is because of definition of Pearson χ^2 divergence. □

As seen, the labels a, b, and c in LSGAN should satisfy Eq. (67). One option to satisfy these conditions is

$$a = -1, \quad b = 1, \quad c = 0, \tag{68}$$

which means that the real and fake labels for the discriminator are $+1$ and -1, respectively, while the generator attempts to fool the discriminator by targeting the label 0. In other words, the generator does not aim to completely mimic the real label but instead chooses a moderate value.

Another possible option for the labels is

$$a = 0, \quad b = c = 1, \tag{69}$$

which does not satisfy Eq. (67) but makes the generator fully fool the discriminator by targeting the real data label directly.

Experiments have shown that both of these options perform equally well in practice [76].

5.5 Energy-Based GAN (EBGAN)

In energy-based learning [61], a function is learned that maps data points to energy values such that incorrectly labeled data points are assigned higher energy values. In unsupervised energy-based learning, higher energy is assigned to data points that are far from the data manifold or data cloud. Energy-based GAN (EBGAN) [134] incorporates energy-based learning into adversarial training. The loss functions in EBGAN are

$$
\begin{aligned}
&\min_{D} \quad V_{\text{EBGAN}}(D) := D(\boldsymbol{x}) + [m - D(G(\boldsymbol{z}))]_+, \\
&\min_{G} \quad V_{\text{EBGAN}}(G) := D(G(\boldsymbol{z})),
\end{aligned} \tag{70}
$$

where $[.]_+ := \max(., 0)$ denotes the standard hinge loss, and $m > 0$ is the margin parameter. The discriminator minimizes the energy for real data, $D(\boldsymbol{x})$, while ensuring that the energy of generated data, $D(G(\boldsymbol{z}))$, is at least m, thus not being fooled by the generator. The generator, on the other hand, minimizes $D(G(\boldsymbol{z}))$ to fool the discriminator.

Theorem 12 ([134, Theorem 1]) *Let:*

$$Q(D, G) := \int_{x,z} V_{EBGAN}(D)\, p_{data}(\boldsymbol{x})\, p_z(z)\, d\boldsymbol{x}\, dz. \tag{71}$$

Optimization of EBGAN results in $p_g(\boldsymbol{x}) = p_{data}(\boldsymbol{x})$ *and* $Q(D^*, G^*) = m$ *after convergence (i.e., Nash equilibrium).*

Proof

$$\begin{aligned}
Q(D, G^*) &\overset{(70)}{=} \int_{x,z} \big(D(\boldsymbol{x}) + [m - D(G^*(z))]_+\big)\, p_{\text{data}}(\boldsymbol{x})\, p_z(z)\, d\boldsymbol{x}\, dz \\
&= \int_x D(\boldsymbol{x})\, p_{\text{data}}(\boldsymbol{x})\, d\boldsymbol{x} \underbrace{\int_z p_z(z)\, dz}_{=1} \\
&\quad + \underbrace{\int_x p_{\text{data}}(\boldsymbol{x})\, d\boldsymbol{x}}_{=1} \int_z [m - D(G^*(z))]_+\, p_z(z)\, dz \\
&\overset{(9)}{=} \int_x \big(p_{\text{data}}(\boldsymbol{x})\, D(\boldsymbol{x}) + p_{g^*}(\boldsymbol{x})\, [m - D(\boldsymbol{x})]_+\big)\, d\boldsymbol{x}.
\end{aligned} \tag{72}$$

The function inside the integral is in the form $at + b[m - t]_+$ whose minimum occurs if $a < b$. Therefore, the minimum of $Q(D, G^*)$ is

$$\begin{aligned}
Q(D^*, G^*) &= m \int_x \mathbb{I}(p_{\text{data}}(\boldsymbol{x}) < p_{g^*}(\boldsymbol{x}))\, p_{\text{data}}(\boldsymbol{x}) d\boldsymbol{x} \\
&\quad + m \int_x \mathbb{I}(p_{\text{data}}(\boldsymbol{x}) \geq p_g(\boldsymbol{x}))\, p_{g^*}(\boldsymbol{x}) d\boldsymbol{x} \\
&= m \int_x \Big(\mathbb{I}(p_{\text{data}}(\boldsymbol{x}) < p_{g^*}(\boldsymbol{x}))\, p_{\text{data}}(\boldsymbol{x}) \\
&\quad + \big(1 - \mathbb{I}(p_{\text{data}}(\boldsymbol{x}) < p_{g^*}(\boldsymbol{x}))\big)\, p_{g^*}(\boldsymbol{x})\Big) d\boldsymbol{x} \\
&= m \underbrace{\int_x p_{g^*}(\boldsymbol{x}) d\boldsymbol{x}}_{=1} + m \int_x \mathbb{I}(p_{\text{data}}(\boldsymbol{x}) < p_{g^*}(\boldsymbol{x}))\, \big(p_{\text{data}}(\boldsymbol{x}) - p_{g^*}(\boldsymbol{x})\big) d\boldsymbol{x}.
\end{aligned} \tag{73}$$

As the probability of generated data $p_{g^*}(\boldsymbol{x})$ is upper-bounded by the probability of data $p_{\text{data}}(\boldsymbol{x})$, the second term is non-positive. Thus, $Q(D, G) \leq m$. On the other hand, as $p_{g^*}(\boldsymbol{x}) \leq p_{\text{data}}(\boldsymbol{x})$, there is

$$\int_x p_{g^*}(\boldsymbol{x}) D^*(\boldsymbol{x}) d\boldsymbol{x} \leq \int_x p_{\text{data}}(\boldsymbol{x}) D^*(\boldsymbol{x}) d\boldsymbol{x}.$$

Using this in Eq. (72) gives

$$\begin{aligned} Q(D^*, G^*) &\geq \int_{\boldsymbol{x}} \big(p_{g^*}(\boldsymbol{x}) D^*(\boldsymbol{x}) + p_{g^*}(\boldsymbol{x})\, [m - D^*(\boldsymbol{x})]_+\big)\, d\boldsymbol{x} \\ &\overset{(a)}{=} \int_{\boldsymbol{x}} p_{g^*}(\boldsymbol{x}) D^*(\boldsymbol{x}) d\boldsymbol{x} + \int_{\boldsymbol{x}} p_{g^*}(\boldsymbol{x})\, (m - D^*(\boldsymbol{x}))\, d\boldsymbol{x} \\ &= m \underbrace{\int_{\boldsymbol{x}} p_{g^*}(\boldsymbol{x})\, d\boldsymbol{x}}_{=1} = m, \end{aligned}$$

where (a) is because $D^*(\boldsymbol{x}) \leq m$ almost everywhere at the Nash equilibrium (since discriminator is trained at the convergence to not violate the margin). It was shown that $m \leq Q(D^*, G^*) \leq m$ so $Q(D^*, G^*) = m$. From $Q(D^*, G^*) = m$ and Eq. (73), it is concluded that $\int_{\boldsymbol{x}} \mathbb{I}(p_{\text{data}}(\boldsymbol{x}) < p_{g^*}(\boldsymbol{x}))d\boldsymbol{x} = 0$. As there is $p_{g^*}(\boldsymbol{x}) \leq p_{\text{data}}(\boldsymbol{x})$, this only holds when $p_{\text{data}}(\boldsymbol{x}) = p_{g^*}(\boldsymbol{x})$. □

5.6 Semi-Supervised GAN

In the following, semi-supervised methods in the GAN literature are introduced.

5.6.1 Categorical GAN (CatGAN)

– Unsupervised CatGAN: Categorical GAN (CatGAN) [116] extends GAN by having the discrim inator classify data into c categories instead of performing binary classification. Thus, the last layer of the discriminator has c neurons with softmax activation functions. Let $D_k(\boldsymbol{x})$ denote the k-th logit, i.e., the softmax output. The conditional probabilities for the categories are modeled as:

$$p(y = k \mid \boldsymbol{x}) = \frac{\exp(D_k(\boldsymbol{x}))}{\sum_{k=1}^{c} \exp(D_k(\boldsymbol{x}))}, \quad \forall k = \{1, \ldots, c\}. \tag{74}$$

Note that the dataset is unlabeled (unsupervised), and the categories are determined by the model through the discriminator logits. The discriminator seeks to be confident in classifying real data into the c categories; thus, it minimizes the entropy H of the conditional probabilities for real data:

$$\begin{aligned} \mathbb{E}_{\boldsymbol{x} \sim p_{\text{data}}(\boldsymbol{x})}[H(p(y = k \mid \boldsymbol{x}))] &\overset{(a)}{\approx} \frac{1}{n} \sum_{i=1}^{n} H(p(y = k \mid \boldsymbol{x}_i)) \\ &\overset{(b)}{=} \frac{1}{n} \sum_{i=1}^{n} \Big(- \sum_{k=1}^{c} p(y = k|\boldsymbol{x}_i) \log \big(p(y = k|\boldsymbol{x}_i)\big)\Big), \end{aligned} \tag{75}$$

where n is the number of real data points, (a) uses the Monte Carlo approximation [30], and (b) follows from the definition of entropy.

Then, n noise samples $\boldsymbol{z} \sim p_z(\boldsymbol{z})$ are drawn and fed to the generator to produce generated data points $G(\boldsymbol{z})$. The discriminator aims to be uncertain about the classification of generated (fake) data into the c categories, hence, it maximizes their entropy:

$$\mathbb{E}_{\boldsymbol{z} \sim p_z(\boldsymbol{z})}[H(p(y = k \mid G(\boldsymbol{z})))] \approx \frac{1}{n} \sum_{i=1}^{n} H(p(y = k \mid G(\boldsymbol{z}_i))). \tag{76}$$

Conversely, the generator aims to minimize this entropy to fool the discriminator. A uniform prior $p(y)$ is assumed for the categories so that both the discriminator and generator use all categories equally. To achieve this, they maximize the entropy of the marginal category distributions:

$$H_{\text{data}}(p(y)) = H\Big(\frac{1}{n}\sum_{i=1}^{n} p(y\,|\,\boldsymbol{x}_i)\Big), \tag{77}$$

$$H_g(p(y)) = H\Big(\frac{1}{n}\sum_{i=1}^{n} p(y\,|\,G(\boldsymbol{z}_i))\Big). \tag{78}$$

Overall, based on the above explanations, the loss functions in CatGAN are

$$\begin{aligned} \max_D \quad V(D) := H_{\text{data}}(p(y)) - \mathbb{E}_{\boldsymbol{x}\sim p_{\text{data}}(\boldsymbol{x})}[H(p(y=k\,|\,\boldsymbol{x}))] \\ + \mathbb{E}_{\boldsymbol{z}\sim p_z(\boldsymbol{z})}[H(p(y=k\,|\,G(\boldsymbol{z})))], \end{aligned} \tag{79}$$

$$\min_G \quad V(D) := -H_g(p(y)) + \mathbb{E}_{\boldsymbol{z}\sim p_z(\boldsymbol{z})}[H(p(y=k\,|\,G(\boldsymbol{z})))]. \tag{80}$$

– Semi-supervised CatGAN: The above loss function for CatGAN is used in the unsupervised setting. CatGAN can be extended to semi-supervised scenarios [116]. Suppose there are n_ℓ labeled data points in addition to the n unlabeled data points. The number of categories c is set equal to the number of classes in the labeled data. The labeled dataset is denoted by $\mathcal{X}_L := \{(\boldsymbol{x}_i^\ell, \boldsymbol{y}_i^\ell)\}_{i=1}^{n_\ell}$, where $\boldsymbol{y}_i^\ell \in \mathbb{R}^c$ is the one-hot encoded label for the i-th labeled data point. The discriminator should maximize the cross-entropy for the labeled data to accurately classify them, in addition to performing category discrimination. This cross-entropy is defined as:

$$\text{CE}(\boldsymbol{y}, p(y|\boldsymbol{x})) := -\sum_{k=1}^{c} y_i \log(p(y = y_i|\boldsymbol{x})), \tag{81}$$

where $p(y = y_i|\boldsymbol{x})$ is the discriminator logit for the labeled data input. This cross-entropy term is incorporated into the discriminator loss as a regularization:

$$\begin{aligned} \max_D \quad V(D) := H_{\text{data}}(p(y)) - \mathbb{E}_{\boldsymbol{x}\sim p_{\text{data}}(\boldsymbol{x})}[H(p(y=k\,|\,\boldsymbol{x}))] \\ + \mathbb{E}_{\boldsymbol{z}\sim p_z(\boldsymbol{z})}[H(p(y=k\,|\,G(\boldsymbol{z})))] + \lambda \text{CE}(\boldsymbol{y}, p(y|\boldsymbol{x})), \end{aligned} \tag{82}$$

where $\lambda > 0$ is the regularization parameter.

5.6.2 Generated Data as a New Class

Generated data can be treated as belonging to an additional class $(c + 1)$. This idea was proposed independently in two papers: [111, Sect. 5] and [96]. Here, the method in [111, Sect. 5] is described.

In this approach, the discriminator D classifies which class a data point $\boldsymbol{x}$ belongs to, unlike the original GAN discriminator, which has a single neuron with a sigmoid activation function in its last layer. Here, the discriminator's last layer has $(c + 1)$ neurons with a softmax activation function, where the j-th neuron outputs the probability of $\boldsymbol{x}$ belonging to the jth class. The optimization objective for the discriminator minimizes the sum of two cross-entropy costs:

$$\min_{D} \big(V_{\text{supervised}}(D) + V_{\text{unsupervised}}(D)\big), \tag{83}$$

where:

$$\begin{aligned} V_{\text{supervised}}(D) &:= -\mathbb{E}_{\boldsymbol{x},y\sim p_{\text{data}}(\boldsymbol{x},y)}\big[\log(p_d(y|\boldsymbol{x}))\big], \quad \forall y < c+1, \\ V_{\text{unsupervised}}(D) &:= -\mathbb{E}_{\boldsymbol{x}\sim p_{\text{data}}(\boldsymbol{x})}\big[\log(1 - p_d(y|\boldsymbol{x}))\big] \\ &\quad + \mathbb{E}_{\boldsymbol{x}\sim p_g(\boldsymbol{x})}\big[\log(p_d(y|\boldsymbol{x}))\big], \quad \text{for } y = c+1, \end{aligned} \tag{84}$$

where $p_d(y|\boldsymbol{x})$ is the softmax output at the discriminator's last layer.

With this cost, the discriminator learns to classify generated (fake) data points as a new class, while the generator attempts to fool the discriminator to prevent correct classification into this new class. The generator's cost remains the same as in Eq. (3).

A general function can be subtracted from every class label. Therefore, the output corresponding to the generated data labels can be subtracted from all labels to set the fake data label to zero, $\ell_{c+1} = 0$. Thus, the softmax output for generated data becomes $\exp(\ell_{c+1} = 0) = 1$. Consequently, according to Eq. (8) and the fact that probabilities are obtained from softmax outputs (logits), we have

$$\begin{aligned} D(\boldsymbol{x}) &\overset{(8)}{=} \frac{\sum_{j=1}^{c} \exp(\ell_j(\boldsymbol{x}))}{\sum_{j=1}^{c} \exp(\ell_j(\boldsymbol{x})) + \exp(\ell_{c+1}(\boldsymbol{x}))} \\ &= \frac{\sum_{j=1}^{c} \exp(\ell_j(\boldsymbol{x}))}{\sum_{j=1}^{c} \exp(\ell_j(\boldsymbol{x})) + 1}. \end{aligned} \tag{85}$$

5.7 MMD GAN

MMD GAN [65] combines the ideas of moment matching networks [67] (see Chap. 13) and GAN [36] by incorporating adversarial learning into Maximum Mean Discrepancy (MMD). MMD [39] measures the divergence between two distributions using distances in the Reproducing Kernel Hilbert Space (RKHS) to compare their moments [28] (see Chap. 13). The MMD between two distributions $p(\boldsymbol{x})$ and $q(\boldsymbol{x})$ is

$$\begin{aligned} M_k(p, q) := &\,\mathbb{E}_{\boldsymbol{x}_i,\boldsymbol{x}_j\sim p(\boldsymbol{x})}[k(\boldsymbol{x}_i, \boldsymbol{x}_j)] + \mathbb{E}_{\boldsymbol{x}_i,\boldsymbol{x}_j\sim q(\boldsymbol{x})}[k(\boldsymbol{x}_i, \boldsymbol{x}_j)] \\ &- 2\mathbb{E}_{\boldsymbol{x}_i\sim p(\boldsymbol{x}),\boldsymbol{x}_j\sim q(\boldsymbol{x})}[k(\boldsymbol{x}_i, \boldsymbol{x}_j)], \end{aligned}$$

where $k(.,.)$ is a kernel function such as the Gaussian kernel. If $p(\boldsymbol{x}) = p_{\text{data}}(\boldsymbol{x})$ and $q(\boldsymbol{x}) = p_g(\boldsymbol{x})$ represent the distributions of real and generated data, respectively, minimizing this MMD ensures that the generated data distribution approximates the real data distribution.

The optimal kernel, yielding the largest MMD in the worst-case scenario, can be found from a set of valid kernel functions $\mathcal{K}$:

$$\min_{G} \max_{k\in\mathcal{K}} \; M_k(p_{\text{data}}(\boldsymbol{x}), p_g(\boldsymbol{x})).$$

However, this optimization is computationally challenging. In MMD GAN, instead of using a fixed kernel such as the Gaussian kernel, the kernel function is learned via adversarial training. Specifically, a function $D(.)$ is learned to define the kernel as:

$$k_D(\boldsymbol{x}_i, \boldsymbol{x}_j) = \exp(-\|D(\boldsymbol{x}_i) - D(\boldsymbol{x}_j)\|^2).$$

An autoencoder is used for $D(.)$, with $D_e(.)$ and $D_d(.)$ as its encoder and decoder, respectively. This autoencoder plays the role of the discriminator in adversarial learning, and the generator is denoted by $G(.)$. The autoencoder is trained to reconstruct both real data, $\boldsymbol{x} \in \mathcal{X}$, and generated data, $\boldsymbol{x} \in G(\boldsymbol{z})$. The loss function of MMD GAN is

$$\min_{G} \max_{D} \; M_{k_D}(p_{\text{data}}(\boldsymbol{x}), p_g(\boldsymbol{x})) - \lambda \mathbb{E}_{\boldsymbol{x} \in \mathcal{X} \cup G(z)} \big[\|\boldsymbol{x} - D_d(D_e(\boldsymbol{x}))\|_2^2 \big], \tag{86}$$

where $\lambda > 0$ is the regularization parameter. Both terms depend on the autoencoder D, while the first term also depends on the generator G. Further theoretical analysis of MMD GAN can be found in [86].

5.8 Additive GANs

In the following, additive GAN models, which use a hierarchical or additive approach, are introduced.

5.8.1 Laplacian GAN (LapGAN)

Laplacian GAN (LapGAN) [17] was one of the first extensions of GAN, enabling generation of higher-resolution images compared to GAN and conditional GAN. Inspired by the Laplacian pyramid for images [8], LapGAN employs a Laplacian pyramid structure.

The training structure of LapGAN is illustrated in Fig. 6a. Let the pyramid have k levels. It begins with the image itself at level zero, i.e., $I = I_0$. The image is downsampled to I_1 by a factor of two, halving its rows and columns. Then, I_1 is upsampled back to l_0 by a factor of two, where l_0 is the low-pass (low-resolution) version of I_0. A conditional GAN [83] (see Sect. 2.6), denoted by G_0, receives the noise z_0 as its input noise and the low-pass l_0 as its conditional input. The generator produces $\widetilde{h}_0$. Define $h_0 := I_0 - l_0$. The inputs to the discriminator D_0 are h_0, $\widetilde{h}_0$, and l_0, and its last layer is a neuron with sigmoid activation function. The discriminator determines whether the image at this level is real

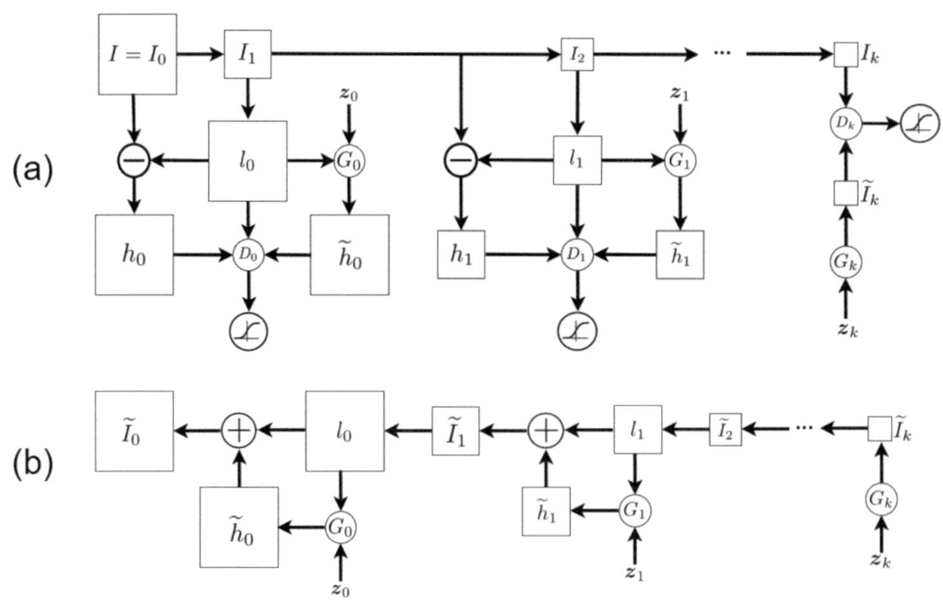

Fig. 6 The structure of LapGAN for **a** training and **b** testing, i.e., sampling

or fake (generated). This procedure is repeated for all levels up to level $(k-1)$, with each level using a conditional GAN. At the last level k, a GAN (non-conditional) is used, which takes noise z_k as input and generates $\widetilde{I}_k$. The discriminator D_k receives $\widetilde{I}_k$ and the downsampled I_k as inputs and judges the generated image at that level.

The test or sampling phase of LapGAN is shown in Fig. 6b. Similar to the training phase, all levels except the last level k use conditional GANs, while the last level uses a GAN. At the j-th level, the noise z_j and the low-pass image l_j are input to the generator G_j as its noise and conditional input, respectively. The generator outputs $\widetilde{h}_j$. The generated image at the j-th level is obtained as $\widetilde{I}_j := \widetilde{h}_j + l_j$. The generated image at level zero, $\widetilde{I}_0$, is the final output image produced by LapGAN.

5.8.2 Progressive GAN

Progressive GAN [54] begins with shallow networks for both the generator G and the discriminator D, and progressively adds new layers to these networks during training. Initially, a small convolutional layer with low spatial resolution is used in both G and D, generating low-resolution images. As training progresses, convolutional layers with higher spatial resolutions are gradually added to G and D, enabling the generation of higher-resolution images. The training of the GAN and the addition of layers occur simultaneously.

5.9 Triple GAN

Triple GAN [64] incorporates a discriminator D, a classifier C, and a generator G. In terms of including a classifier, it is similar to MGAN [45] (see Sect. 3.5). The generator models the conditional distribution of data given the label, $p_g(\boldsymbol{x}|y)$, while the classifier models the opposite conditional distribution, i.e., $p_c(y|\boldsymbol{x})$. The discriminator determines whether the data-label pair $(\boldsymbol{x}, y)$ is real or generated (fake). The classifier predicts the class label y for either real or generated data $\boldsymbol{x}$.

Let $p_{\text{data}}(\boldsymbol{x}, y)$ denote the joint distribution of real data and labels. The joint distributions for data-labels in the generator and classifier are $p_g(\boldsymbol{x}, y) = p_g(\boldsymbol{x}|y)p(y)$ and $p_c(\boldsymbol{x}, y) = p_c(\boldsymbol{x}|y)p(y)$, respectively, where $p(y)$ is the marginal distribution of labels. The generator takes noise $z \sim p_z(z)$ and label y as input and generates a data point $\boldsymbol{x} = G(z, y)$, resulting in $(G(z, y), y) \sim p_g(\boldsymbol{x}, y)$.

Triple GAN optimizes the loss function for a three-player game:

$$\begin{aligned}\min_{G,C}\max_{D}\ V(D, C, G) := \;& \mathbb{E}_{(\boldsymbol{x},y)\sim p_{\text{data}}(\boldsymbol{x},y)}\big[\log(D(\boldsymbol{x}, y))\big] \\ &+ \alpha\mathbb{E}_{(\boldsymbol{x},y)\sim p_c(\boldsymbol{x},y)}\big[\log(1 - D(\boldsymbol{x}, y))\big] \\ &+ (1-\alpha)\mathbb{E}_{z\sim p_z(z), y\sim p(y)}\Big[\log\Big(1 - D\big(G(z, y), y\big)\Big)\Big] \\ &+ \mathbb{E}_{(\boldsymbol{x},y)\sim p_{\text{data}}(\boldsymbol{x},y)}\big[-\log(p_c(y|\boldsymbol{x}))\big],\end{aligned} \tag{87}$$

where $\alpha \in (0, 1)$ is the regularization parameter (with $\alpha = 0.5$ recommended). The last term in the loss, where $p_c(y|\boldsymbol{x})$ is the predicted label by the classifier, models the KL-divergence between $p_c(\boldsymbol{x}, y)$ and $p_{\text{data}}(\boldsymbol{x}, y)$. The discriminator, classifier, and generator are trained using alternating optimization [27], allowing them to strengthen gradually.

Theorem 13 ([64, Lemma 3.1 and Theorem 3.3]) *The optimal discriminator in Triple GAN is*

$$D^*(\boldsymbol{x}, y) = \frac{p(\boldsymbol{x}, y)}{p(\boldsymbol{x}, y) + (1-\alpha)p_g(\boldsymbol{x}, y) + \alpha p_c(\boldsymbol{x}, y)}. \tag{88}$$

After convergence (i.e., Nash equilibrium) of Triple GAN, the following holds:

$$p_{g^*}(\boldsymbol{x}, y) = p_{c^*}(\boldsymbol{x}, y) = p(\boldsymbol{x}, y) \stackrel{(88)}{\Longrightarrow} D^*(\boldsymbol{x}, y) = 0.5. \tag{89}$$

5.10 Latent Adversarial Generator (LAG)

Latent Adversarial Generator (LAG) [4] can generate high-resolution images using a corresponding low-resolution image as an input cue. In terms of receiving a cue, it is related to the conditional GAN (see Sect. 2.6).

Let z, $\boldsymbol{x}$, and $\widetilde{\boldsymbol{x}}$ denote the noise sample, the real data point, and the cue low-resolution data point, respectively. The generator G takes $\widetilde{\boldsymbol{x}}$ and z as inputs and generates the high-resolution image $G(\widetilde{\boldsymbol{x}}, z)$. The discriminator D consists of two parts. First, a projection operator Π projects data onto a low-dimensional space called the perceptual latent space. The operator Π is a nonlinear neural network that receives both the high- and low-dimensional data points as input. Then, additional layers, denoted by the mapping $F(.)$, map the projected data from the perceptual latent space to a scalar output after applying a sigmoid activation function. Thus, the discriminator is defined as $D(\boldsymbol{x}) = F(\Pi(\boldsymbol{x}, \widetilde{\boldsymbol{x}}))$.

To ensure that $\Pi(G(\widetilde{\boldsymbol{x}}, z = \boldsymbol{0}), \widetilde{\boldsymbol{x}})$ is similar to $\Pi(\boldsymbol{x}, \widetilde{\boldsymbol{x}})$, a regularization term is used. LAG employs WGAN (see Sect. 3.7) with a regularized loss function. Equation (40) is regularized as:

$$\begin{aligned} \min_{G} \max_{\|D\|_L \leq 1} \quad & D(\boldsymbol{x}, \widetilde{\boldsymbol{x}}) - D(G(\widetilde{\boldsymbol{x}}, z), \widetilde{\boldsymbol{x}}) - \lambda_1 (\|\nabla_{\widehat{\boldsymbol{x}}} D(\widehat{\boldsymbol{x}})\|_2 - 1)^2 \\ & + \lambda_2 \|\Pi(G(\widetilde{\boldsymbol{x}}, z = \boldsymbol{0}), \widetilde{\boldsymbol{x}}) - \Pi(\boldsymbol{x}, \widetilde{\boldsymbol{x}})\|_F^2, \end{aligned} \tag{90}$$

where $\lambda_1, \lambda_2 > 0$ are regularization parameters, $\|.\|_F$ denotes the Frobenius norm, and $\widehat{\boldsymbol{x}}$ is defined in Eq. (41).

5.11 Ensembles of GAN Models

In the following, several GAN models that utilize an ensemble of generators and/or discriminators are introduced. Some of these models were already discussed, such as MGAN (see Sect. 3.5) and D2GAN (see Sect. 3.6). Here, other ensemble GAN methods are introduced.

5.11.1 Generative Multi-adversarial Network (GMAN)

Generative Multi-Adversarial Network (GMAN) [21] accelerates the training of GANs by employing multiple discriminators. Assume there are n_d discriminators. The loss function of GMAN is formulated as:

$$\max_{D_i} V(D_i, G), \quad \forall i \in \{1, \ldots, n_d\}, \tag{91}$$

$$\min_{G} F\big(V(D_1, G), \ldots, V(D_{n_d}, G)\big), \tag{92}$$

where each $V(D_i, G)$ is defined in Eq. (3), and the function $F(.)$ is an aggregation function such as $F(.) = \max(.)$ or $F(.) = \text{mean}(.)$. If $F(.)$ is the maximum function, the generator is trained using the most challenging discriminator at each iteration of the alternating optimization. If $F(.)$ is the mean function, the generator is trained based on the average effect of all discriminators.

5.11.2 AdaGAN: Boosting GANs

Boosting refers to using weak models additively, where each subsequent model places more weight on data points that were not correctly classified or regressed by previous models [25]. One of the most well-known boosting methods for classification and regression is AdaBoost [24]. AdaGAN [122] applies the concept of boosting to GAN models for data generation.

Let n be the number of data points. AdaGAN starts with the first GAN where the weights of all data points are initialized to $1/n$. Let the generator of the j-th GAN be denoted by G_j. Only one discriminator D is used as the classifier, whose scalar output after the sigmoid activation function is $D(\boldsymbol{x})$. For the j-th GAN model, the discriminator D discriminates between the true data and the generated data $G_{j-1}(z)$, where z is the latent noise.

The weights of data points are updated as:

$$w_{i,j} := \frac{1}{n\beta_j}\big[\lambda - (1-\beta_j)h(D(\boldsymbol{x}_i))\big]_+, \tag{93}$$

where $w_{i,j}$ is the weight of $\boldsymbol{x}_i$ for the j-th GAN, $[.]_+ := \max(., 0)$, $\beta_j := 1/j$ (or a fixed number in the range $[0, 1]$), $h(D(\boldsymbol{x})) := (1 - D(\boldsymbol{x}))/D(\boldsymbol{x})$, and λ is obtained by iteratively updating:

$$\lambda := \frac{\beta_j}{\sum_{i=1}^{k}(1/n)}\Big(1 + \frac{1-\beta_j}{\beta_j}\sum_{i=1}^{k}(1/n)h\big(D(\boldsymbol{x}_i)\big)\Big),$$

where k is the iteration index of the iterative updating procedure.

The generator of the jth weak GAN, G'_t, is trained using the weighted data points from Eq. (93). Finally, the jth GAN is computed as a linear combination of G'_t and the previous GAN:

$$G_j := (1-\beta_j)G_{j-1} + \beta_j G'_t.$$

The proofs for these expressions can be found in [122].

5.11.3 Boosted Generative Model (BGM)

Another similar method for boosting GAN models is the Boosted Generative Model (BGM) [40]. Its idea is briefly introduced here. BGM starts by assigning equal weights of $1/n$ to all data points and trains the first generative model G_1. For the j-th GAN, it uses the lower bound of the f-divergence in Eq. (52) to estimate the next generative model based on the previous model. Its formulation is inspired by AdaBoost [24].

5.12 Coupled GAN (CoGAN)

Coupled GAN (CoGAN) [69] is a generative model for multiple domains, enabling generation of several data points from different domains that are related. For example, one domain could be images and another domain could be text, allowing generation of an image and its corresponding caption. Another example is generating two related images from different domains, such as facial and nature images. If tuples of corresponding data points are available, CoGAN can learn to generate related images across domains; otherwise, it generates not necessarily related data points from each domain.

Assume there are two domains. In this case, CoGAN employs two coupled GAN structures, as illustrated in Fig. 7. Let G_1/D_1 and G_2/D_2 denote the generators/discriminators for the first and second

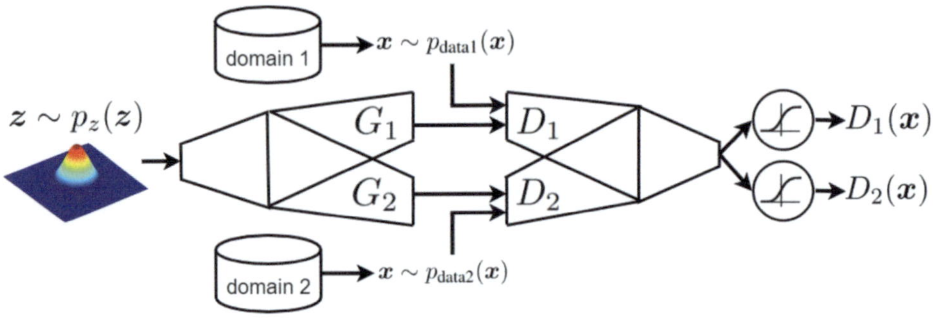

Fig. 7 The structure of CoGAN

domains, respectively. In a generator, the first and last layers extract high-level and low-level features, respectively [69]. Conversely, in a discriminator, the first and last layers extract low-level and high-level features, respectively [57]. To share high-level semantic features while capturing domain-specific details, the first layers of the generators and the last layers of the discriminators are shared, as shown in Fig. 7.

Let the datasets of the first and second domains be denoted by $p_{\text{data1}}(\boldsymbol{x})$ and $p_{\text{data2}}(\boldsymbol{x})$, respectively. The loss function of CoGAN is

$$\begin{aligned} \min_{G_1,G_2} \max_{D_1,D_2} \quad V(D_1, D_2, G_1, G_2) := &\mathbb{E}_{\boldsymbol{x}\sim p_{\text{data1}}(\boldsymbol{x})}\Big[\log\big(D_1(\boldsymbol{x})\big)\Big] \\ &+ \mathbb{E}_{\boldsymbol{z}\sim p_z(\boldsymbol{z})}\Big[\log\Big(1 - D_1\big(G_1(\boldsymbol{z})\big)\Big)\Big] \\ &+ \mathbb{E}_{\boldsymbol{x}\sim p_{\text{data2}}(\boldsymbol{x})}\Big[\log\big(D_2(\boldsymbol{x})\big)\Big] \\ &+ \mathbb{E}_{\boldsymbol{z}\sim p_z(\boldsymbol{z})}\Big[\log\Big(1 - D_2\big(G_2(\boldsymbol{z})\big)\Big)\Big], \end{aligned} \tag{94}$$

subject to the constraint that some layers of the generators and some layers of the discriminators are shared, as shown in Fig. 7.

Note that, although [69] focused on coupling two GAN structures, CoGAN can be easily extended to any number of structures and thus any number of domains.

5.13 Inverse GAN Models

In GANs, data generation can be inverted to obtain a latent noise sample $\boldsymbol{z}$ from a given data point $\boldsymbol{x}$. This latent noise corresponds to $\boldsymbol{x}$ such that feeding it to the generator reproduces $\boldsymbol{x}$. Some existing methods for inverse generation in GANs are adversarial autoencoders, BiGAN, ALI, and inverse techniques. The adversarial autoencoder will be introduced later in Sect. 8.1, while the other methods are explained in the following.

5.13.1 Bidirectional GAN (BiGAN)

In GAN, the generator receives a latent noise $\boldsymbol{z}$ and generates a data point $\boldsymbol{x}$. However, the inverse process–outputting a latent variable from a data point $\boldsymbol{x}$–does not exist in GAN. Bidirectional GAN (BiGAN) [18] extends GAN by including this inverse mapping. Its structure is depicted in Fig. 8.

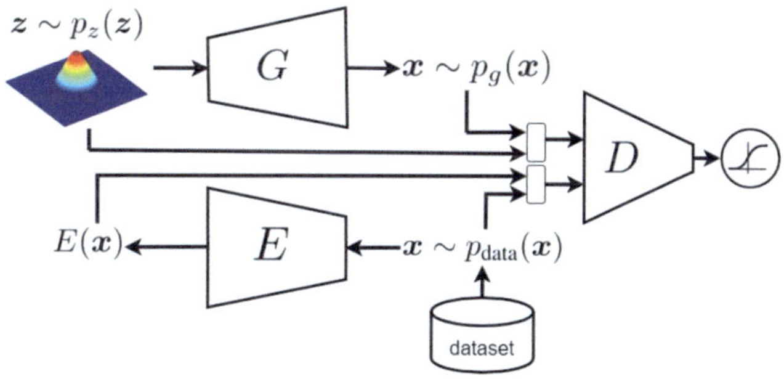

Fig. 8 The structure of BiGAN

In BiGAN, the generator G takes noise z as input and generates $G(z)$, while the encoder E, serving as the inverse of G, takes $\boldsymbol{x}$ as input and outputs $E(\boldsymbol{x})$. Recall that the discriminator in GAN receives either real data $\boldsymbol{x}$ or generated data $G(z)$ as input (see Fig. 2). In contrast, the discriminator in BiGAN receives all four inputs: $G(z)$, z, $E(\boldsymbol{x})$, and $\boldsymbol{x}$, and judges whether the input pair is real or generated (fake). It assigns label one to each pair $(\boldsymbol{x}, E(\boldsymbol{x}))$ and label zero to each pair $(z, G(z))$.

The loss function of BiGAN is

$$\begin{aligned} \min_{G,E} \max_{D} \quad V(D, G, E) &:= \mathbb{E}_{\boldsymbol{x} \sim p_{\text{data}}(\boldsymbol{x})} \Big[\mathbb{E}_{z \sim p_E(.|\boldsymbol{x})} \big[\log(D(\boldsymbol{x}, z)) \big] \Big] \\ &\quad + \mathbb{E}_{z \sim p_z(z)} \Big[\mathbb{E}_{\boldsymbol{x} \sim p_G(.|z)} \big[\log(1 - D(\boldsymbol{x}, z)) \big] \Big] \\ &= \mathbb{E}_{\boldsymbol{x} \sim p_{\text{data}}(\boldsymbol{x})} \Big[\log\Big(D\big(\boldsymbol{x}, E(\boldsymbol{x})\big)\Big) \Big] \\ &\quad + \mathbb{E}_{z \sim p_z(z)} \Big[\log\Big(1 - D\big(G(z), z\big)\Big) \Big]. \end{aligned} \tag{95}$$

Alternating optimization [27] is used by alternately optimizing D, G, and E.

Theorem 14 ([18, Theorem 2]) *At convergence (i.e., Nash equilibrium) of BiGAN, the optimal encoder E and generator G are inverses of each other:*

$$E^* = (G^*)^{-1}, \quad G^*(E^*(\boldsymbol{x})) = \boldsymbol{x}, \quad E^*(G^*(z)) = z. \tag{96}$$

5.13.2 Adversarially Learned Inference (ALI)

Adversarially Learned Inference (ALI) [20] is a method for incorporating inference into GANs to enable inversion. The generator G in ALI is structured as an autoencoder, where its encoder $G_x(z)$ and decoder $G_z(\boldsymbol{x})$ are referred to as the generator network and the inference network, respectively.

The generator network $G_x(z)$ maps a latent noise sample z to a generated data point $\widetilde{\boldsymbol{x}} := G_x(z)$. The inference network $G_z(\boldsymbol{x})$ maps a data point $\boldsymbol{x}$ to its corresponding latent noise sample $\widetilde{z} := G_z(\boldsymbol{x})$. The discriminator $D(\boldsymbol{x}, z)$ attempts to distinguish between the pairs $(\widetilde{\boldsymbol{x}}, z)$ from the generator network and $(\boldsymbol{x}, \widetilde{z})$ from the inference network.

The loss function of ALI is

$$\min_{G} \max_{D} \mathbb{E}_{\boldsymbol{x} \sim p_{\text{data}}(\boldsymbol{x})} \big[\log(D(\boldsymbol{x}, G_z(\boldsymbol{x}))) \big] + \mathbb{E}_{z \sim p_z(z)} \big[\log(1 - D(G_x(z), z)) \big]. \tag{97}$$

5.13.3 The Inversion Technique

Another approach for incorporating inversion in GANs is the inversion technique [14]. In this method, after training a GAN model, a noise sample z is found that generates a desired data point by solving:

$$\max_{z} \mathbb{E}\big[\log(G(z))\big] + \lambda \log(p_z(z)), \tag{98}$$

where $p_z(z)$ is the prior distribution of the latent space (e.g., $\mathcal{N}(\mathbf{0}, \boldsymbol{I})$) and $\lambda > 0$ is a regularization parameter. This optimization problem can be solved using gradient descent.

5.14 Self-Attention GAN (SAGAN)

The attention mechanism [124] assigns weights to data features such that the model attends more to important features by giving them larger weights (see Chap. 9). These weights are computed by measuring the similarity between features using inner products. In self-attention, the similarities of features within each data point are calculated, i.e., each feature is compared to other features of the same data point. These inner products are implemented within the convolutional layers of the network.

Self-Attention GAN (SAGAN) [131] incorporates the self-attention mechanism into both the generator and discriminator networks.

For the mathematical details of the attention mechanism and SAGAN, refer to Chap. 9 and [131], respectively.

5.15 Few-Shot GAN Models

In the following, GAN models that learn from a small number of training data points are introduced.

5.15.1 Transfer Learning in GAN

Consider a GAN (G_s, D_s) that has been trained on data from a source domain. Few-shot GAN [98] enables transfer learning, where a GAN trained on the source domain can also generate images from a target domain. This method uses an adapted generator $G_{s\to t}$ aimed at generating data points from the target domain. As the target domain has few data points in few-shot learning, it is prone to overfitting (see Chap. 5). Thus, pairwise similarities before and after adaptation should be preserved.

To achieve this, a mini-batch of $(b+1)$ noise samples $\{z_i\}_{i=1}^{b+1}$ is drawn from the latent space. These noise samples are fed to the generators G_s and $G_{s\to t}$. At the ℓ-th layer, the following are calculated:

$$\begin{aligned}
y_{s,i}^{\ell} &:= \text{softmax}(\text{sim}(G_s^{\ell}(z_i), G_s^{\ell}(z_j))), \\
y_{s\to t,i}^{\ell} &:= \text{softmax}(\text{sim}(G_{s\to t}^{\ell}(z_i), G_{s\to t}^{\ell}(z_j))),
\end{aligned}$$

for all $i \neq j, i, j \in \{1, \dots, b+1\}$, where sim(.) denotes cosine similarity. To ensure that the adapted generator has similar distributions across layers, the following loss is defined:

$$V(G_{s\to t}, G_s) := \mathbb{E}_{z_i \sim p_z(z)}\Big[\sum_{\ell}\sum_{i} \text{KL}(y_{s\to t,i}^{\ell} \| y_{s,i}^{\ell})\Big],$$

where KL(.) denotes KL-divergence.

Then, k random noise samples are drawn as anchor points Z_{anchor}, a subset of the latent space Z. Two discriminators are used: D_{image} for judging the whole image and D_{patch} for judging an image patch. Define:

$$V(D_{\text{image}}, D_{\text{patch}}, G_{s\to t}) := \mathbb{E}_{\boldsymbol{x}\sim\mathcal{D}_t}\big[\mathbb{E}_{z\sim Z_{\text{anchor}}}[D_{\text{image}}(G_{s\to t}(z)) - D_{\text{image}}(\boldsymbol{x}_{\text{image}})] + \mathbb{E}_{z\sim p_z(z)}[D_{\text{patch}}(G_{s\to t}(z)) - D_{\text{patch}}(\boldsymbol{x}_{\text{patch}})]\big],$$

where $\mathcal{D}_t$ denotes the target domain.

The overall loss function is

$$\min_{G_{s\to t}} \max_{D_{\text{image}}, D_{\text{patch}}} V(D_{\text{image}}, D_{\text{patch}}, G_{s\to t}) + \lambda V(G_{s\to t}, G_s), \tag{99}$$

where $\lambda > 0$ is a regularization parameter. In this loss, the first term gives flexibility to patch structures within the image, while the second term enforces transfer learning regularization.

5.15.2 GAN with Single Image (SinGAN)

GAN with Single Image (SinGAN) [112] learns to generate images by being trained on only one image. It generates images that are all related in texture to the training image. SinGAN learns the distribution of patches within the image at different scales using multi-scale adversarial learning. In terms of employing multiple scales in a Laplacian pyramid, it is similar to LapGAN [17] (see Sect. 5.8.1).

Assume there are $(k + 1)$ levels $\{0, \dots, k\}$ in the Laplacian pyramid, where level 0 is the original image and subsequent levels are downsampled versions. At each j-th level, there is a GAN (G_j, D_j). Training proceeds from level k to level 0. If z_j is the latent noise at level j, the generations are

$$\boldsymbol{x}_k = G_k(z_k),$$
$$\boldsymbol{x}_j = G_j(z_j, \boldsymbol{x}'_{j+1}), \quad \forall j < k,$$

where $\boldsymbol{x}'_{j+1}$ is the upsampled version of $\boldsymbol{x}_{j+1}$. The GANs are trained sequentially, keeping previously trained GANs fixed while training the next. The loss function is regularized with a reconstruction error to improve image generation quality.

5.16 Training Triplet Network with GAN

A Siamese network [7] is composed of multiple networks sharing their weights. When the number of networks is three, the Siamese network is called a triplet network. Adversarial learning can be used to train a triplet network [137].

Consider triplets $(\boldsymbol{x}_a, \boldsymbol{x}_p, \boldsymbol{x}_n)$ where $\boldsymbol{x}_a$ is the anchor point, $\boldsymbol{x}_p$ is the positive point belonging to the same class as the anchor, and $\boldsymbol{x}_n$ is the negative point belonging to a different class from the anchor. For this setup, the loss function can be defined as:

$$\min_{\theta} \; -\log\left(\frac{\exp(\|\boldsymbol{x}_a - \boldsymbol{x}_p\|_2^2)}{\exp(\|\boldsymbol{x}_a - \boldsymbol{x}_p\|_2^2) + \exp(\|\boldsymbol{x}_a - \boldsymbol{x}_n\|_2^2)}\right) - V(D, G), \tag{100}$$

where θ denotes the network weights, the first term is the Neighborhood Component Analysis (NCA) loss [31], and the second term is the adversarial loss function. The paper [137] uses Eq. (85) for the discriminator D.

6 Sampling and Interpolation in GAN

After training a GAN, new data points can be generated by sampling noise from the latent space and feeding it to the generator. However, two issues may arise when sampling from the latent space [128]. First, it is important to avoid sampling from locations in the latent space that are highly unlikely. Secondly, since the latent space is usually high dimensional, there may exist dead-zone locations that were not trained during training [75].

In the following, several techniques for sampling and interpolation in the latent space are introduced. Note that these techniques can also be applied to other generative models such as variational autoencoders [56] (see Chap. 12).

6.1 Interpolation in the Latent Space

To demonstrate that a GAN model has not merely memorized the training data and that its latent space is meaningful, different locations in the latent space are traversed to observe the generated data points. This process is called interpolation in the latent space.

A limitation of linear interpolation, which uses a fixed step size, is that it may traverse highly unlikely priors, resulting in unrealistic generated data points. To address this, spherical linear interpolation [128], called *slerp*, can be used to traverse a path on a p-dimensional hypersphere within the p-dimensional latent space. If interpolating between locations z_1 and z_2 in the latent space, the interpolated locations are computed as [113]:

$$\text{slerp}(z_1, z_2, \mu) := \frac{\sin((1-\mu)\theta)}{\sin(\theta)} z_1 + \frac{\sin(\mu\theta)}{\sin(\theta)} z_2, \tag{101}$$

where μ is swept in the range $[0, 1]$ and $\theta := \cos^{-1}(z_1^\top z_2)$.

Generated data points can then be obtained from these sampled noises by interpolation in the latent space. If interpolation is performed along two perpendicular axes in the latent space, the generations can be arranged into a two-dimensional grid. An example of linear interpolation is shown in Fig. 9a. Interpolation reveals how the latent space covers the generation of various data points and highlights shared features among them.

6.2 Manifold Interpolated Neighbor Embedding

Instead of reporting generated data points directly, one can find the nearest neighbor of each generated point within the training dataset. Displaying the nearest neighbors of generated points in a two-dimensional grid is known as *Manifold Interpolated Neighbor Embedding* (MINE) [128]. An example grid for MINE is shown in Fig. 9b.

6.3 Analogy and J-Diagram

Vector arithmetic can be performed in the latent space (see Sect. 2.7.2). These operations reveal analogy relationships between vectors. Let $\boldsymbol{a}$, $\boldsymbol{b}$, $\boldsymbol{c}$, and $\boldsymbol{d}$ be latent vectors corresponding to four generated data points. It is desired to find $\boldsymbol{d}$ such that the following analogy holds:

$$\boldsymbol{a} : \boldsymbol{b} :: \boldsymbol{c} : \boldsymbol{d} \implies (\boldsymbol{b} - \boldsymbol{a}) = (\boldsymbol{d} - \boldsymbol{c}). \tag{102}$$

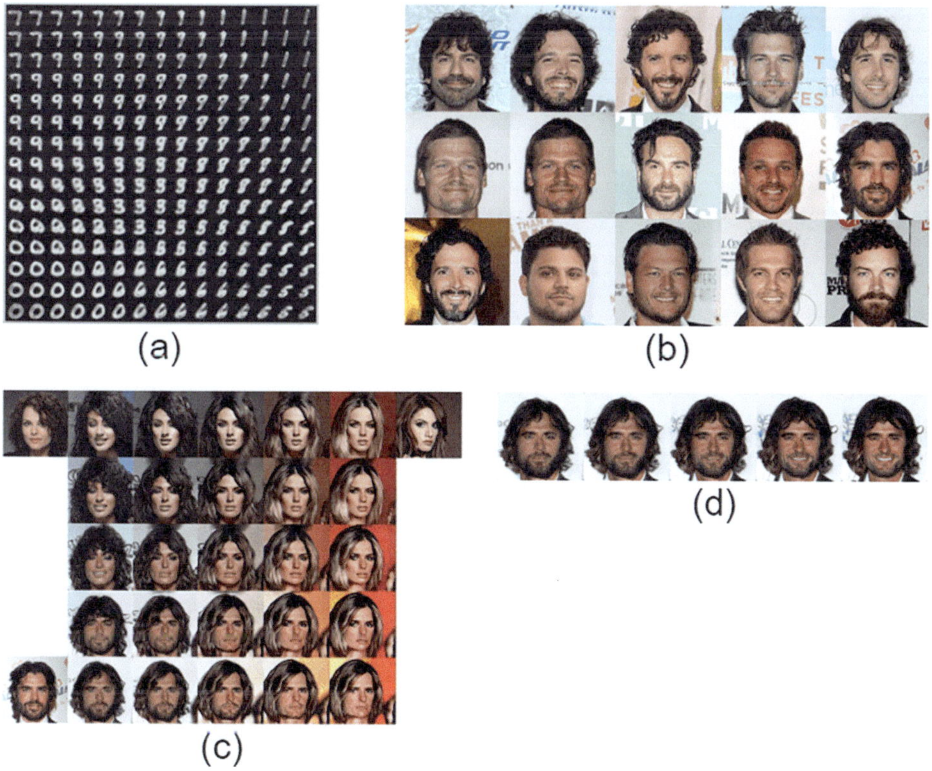

Fig. 9 **a** Interpolation in the latent space of a VAE trained on MNIST data (image from https://blog.keras.io/building-autoencoders-in-keras.html), **b** MINE for a VAE trained on the CelebA dataset [70], **c** J-diagram by interpolation in the latent space of a GAN trained on the CelebA dataset, and **d** traversal along the smile vector for a GAN trained on the CelebA dataset (Images for (**b**), (**c**), and (**d**) are from [128])

In natural language processing, a well-known analogy is "man : king :: woman : queen" [82].

The *J-diagram* [128] is a diagram arranged in a J-shape where the top-left, top-right, bottom-left, and bottom-right corners represent the generated images for vectors $\boldsymbol{a}$, $\boldsymbol{b}$, $\boldsymbol{c}$, and $\boldsymbol{d}$, respectively. The images within the diagram are obtained by linear or slerp interpolation between these vectors. This diagram illustrates how an image is transformed into another by changing its features. An example J-diagram from a GAN trained on the CelebA dataset [70] is shown in Fig. 9c. In this example, the vertical axis modifies gender, while the horizontal axis alters hair color, hair type, and facial pose.

6.4 Attribute Vector

Attribute vectors can be derived for an embedding space as follows [128]. For instance, a smile vector [60] can be computed by subtracting the latent vector for a neutral face from that of the same person smiling. The resulting vector represents the latent encoding for smiling. Other attribute vectors can be derived similarly. An attribute vector can be used to modify an image to include the corresponding attribute. For example, adding the smile vector $z_s \in \mathbb{R}^p$ to a neutral face latent vector $z_n \in \mathbb{R}^p$ results in a new latent vector that generates a smiling version of that face. Let $\eta \in \mathbb{R}$ be the weight for the attribute. Then, $z_n + \eta z_s$ yields a latent vector encoding varying degrees of smiling. A negative η can reduce or neutralize the smile. An example traversal along the smile vector is shown in Fig. 9d.

6.5 Evaluation of Generated Images

The *Inception score* [111, Sect. 4] assesses the quality of images generated by GAN models. For this, the generated images $\boldsymbol{x}$ are fed to the Inception network [119], producing predicted labels $p(y|\boldsymbol{x})$. On one hand, this conditional label distribution is desired to have low entropy, indicating confident classification. On the other hand, the generator should produce diverse images, implying that the marginal distribution $p(y) = \int p(y|\boldsymbol{x})dz$ (for $\boldsymbol{x} = G(z)$) should be large. The Inception score combines these criteria as:

$$\text{Inception score} = \exp\Big(\mathbb{E}_{\boldsymbol{x}}\Big[\text{KL}\big(p(y|\boldsymbol{x})\|p(y)\big)\Big]\Big). \tag{103}$$

A higher Inception score indicates higher quality generated images. It has been observed that this score closely aligns with human evaluation of generated images [111].

Note that another method for quantitative analysis of GAN results [129] is based on annealed importance sampling [90].

7 Applications of GAN

As seen earlier, GAN can be used for data generation across various data types, such as images. In the following, additional applications of GAN are introduced.

7.1 Image-to-Image Translation by GAN

There exist several GAN-based methods for image-to-image translation, where an image is generated corresponding to an input image. The correspondence can vary depending on the specific application. These methods are introduced below.

7.1.1 PatchGAN

PatchGAN [52] employs a conditional GAN [83] (see Sect. 2.6) with a regularized loss function. It uses the ℓ_1 norm between the data and generated data as a regularizer, since ℓ_1 encourages less blurring compared to ℓ_2. The loss function is

$$\min_{G} \max_{D} \; V'_C(D, G) + \lambda\, \mathbb{E}_{\boldsymbol{x},z,\boldsymbol{y}}\big[\|\boldsymbol{x} - G(z, \boldsymbol{y})\|_1\big], \tag{104}$$

where $\lambda > 0$ is the regularization parameter and $V'_C(D, G)$ is a slightly modified version of Eq. (22):

$$V'_C(D, G) := \mathbb{E}_{\boldsymbol{x},\boldsymbol{y}}\Big[\log\big(D(\boldsymbol{x}, \boldsymbol{y})\big)\Big] + \mathbb{E}_{z,\boldsymbol{y}}\Big[\log\Big(1 - D\big(G(z, \boldsymbol{y}), \boldsymbol{y}\big)\Big)\Big], \tag{105}$$

where $\boldsymbol{x}$ is the data, $\boldsymbol{y}$ is the label of the data, and $z \sim p_z(z)$ is the noise. The generator G takes the noise z and label $\boldsymbol{y}$ as input and generates data $G(z, \boldsymbol{y})$. The discriminator takes the data point $\boldsymbol{x}$ and its label $\boldsymbol{y}$ as input to judge whether $\boldsymbol{x}$ is real or generated.

For the generator G, PatchGAN uses skip connections between each layer ℓ and layer $(L - \ell)$, where L is the total number of layers. This design is inspired by the U-Net architecture [109]. Moreover, the ℓ_1 norm in Eq. (104) addresses the low-frequency features of the generated image [52], while the discriminator focuses on the high-frequency features. To achieve this, the discriminator D classifies

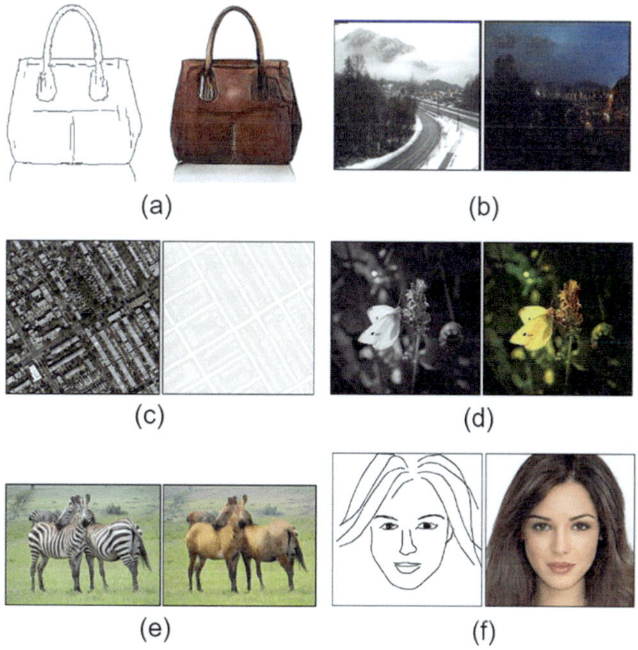

Fig. 10 Image-to-image translation: **a** coloring a sketch, **b** changing daylight to night darkness in an image, **c** transforming an aerial image to a map, **d** coloring a black-and-white image, **e** transforming zebra to horse and vice versa, and **f** generating a facial image from a facial sketch. Transformations in (**a**), (**b**), (**c**), and (**d**) are by PatchGAN (credit: [52]). Transformations in (**e**) and (**f**) are by CycleGAN (credit: [136]) and DeepFaceDrawing (credit: [10]), respectively

images in a patch-wise manner rather than as a whole. Each patch is classified as real or fake, and the results are averaged to produce the final decision for the entire image. This patch-wise classification models the image as a Markov random field, assuming that each patch is independent of others.

PatchGAN has been used for image-to-image translation $I_1 \mapsto I_2$, translating image I_1 to image I_2. For this task, $\boldsymbol{x} = I_2$, $\boldsymbol{y} = I_1$, and $z \sim p_z(z)$ in Eqs. (104) and (105). In other words, I_1 is used as the label in the conditional GAN, while I_2 is the data point. The generator takes I_1 and noise as inputs to generate $\widetilde{I}_2$. The discriminator takes I_1 and I_2 as input and judges whether I_2 is a real translation of I_1 or a generated one. Through adversarial training, both the generator and discriminator improve. Training PatchGAN requires a dataset with pairs of (I_1, I_2) images. Some results of PatchGAN are shown in Fig. 10a–d.

7.1.2 CycleGAN

CycleGAN (Cycle-Consistent Generative Adversarial Networks) [136] is a method for image-to-image translation without requiring paired training images, in contrast to PatchGAN, which needs image pairs. Let the two image domains be X and Y. In CycleGAN, there are two generators: $G : X \to Y$ and $F : Y \to X$.

There are also two discriminators: D_X, which judges between images in X and $F(Y)$, and D_Y, which judges between images in Y and $G(X)$. Thus, there are two GAN loss functions:

$$V(D_Y, G, X, Y) := \mathbb{E}_{\boldsymbol{y}\sim p_{\text{data}}(\boldsymbol{y})}\Big[\log\big(D_Y(\boldsymbol{y})\big)\Big] + \mathbb{E}_{\boldsymbol{x}\sim p_{\text{data}}(\boldsymbol{x})}\Big[\log\Big(1 - D_Y\big(G(\boldsymbol{x})\big)\Big)\Big],$$
$$V(D_X, F, X, Y) := \mathbb{E}_{\boldsymbol{x}\sim p_{\text{data}}(\boldsymbol{x})}\Big[\log\big(D_X(\boldsymbol{x})\big)\Big] + \mathbb{E}_{\boldsymbol{y}\sim p_{\text{data}}(\boldsymbol{y})}\Big[\log\Big(1 - D_X\big(F(\boldsymbol{y})\big)\Big)\Big].$$

Additionally, the cycle consistency loss is defined to enforce $F(G(\boldsymbol{x})) \approx \boldsymbol{x}$ and $G(F(\boldsymbol{y})) \approx \boldsymbol{y}$:

$$V_{\text{cyc}}(G, F) := \mathbb{E}_{\boldsymbol{x}\sim p_{\text{data}}(\boldsymbol{x})}\Big[\|F(G(\boldsymbol{x})) - \boldsymbol{x}\|_1\Big] + \mathbb{E}_{\boldsymbol{y}\sim p_{\text{data}}(\boldsymbol{y})}\Big[\|G(F(\boldsymbol{y})) - \boldsymbol{y}\|_1\Big].$$

The overall loss function of CycleGAN is

$$\min_{G,F} \max_{D_X,D_Y} \quad V(D_Y, G, X, Y) + V(D_X, F, X, Y) + \lambda V_{\text{cyc}}(G, F), \tag{106}$$

where $\lambda > 0$ is the regularization parameter. An example result of CycleGAN is shown in Fig. 10e.

7.1.3 Deep Face Drawing

DeepFaceDrawing [10] generates high-quality facial images from input sketches of faces. For its training data, automatic sketches are created using the Canny edge detection algorithm [9]. DeepFaceDrawing consists of three modules. The first module is the component embedding module, which takes different facial patches as input and learns embedding vectors for them. These vectors are then fed to the feature mapping module, which transforms the embedding vectors into 2D facial feature patches. Finally, these feature patches are input to the image synthesis module, which is a conditional GAN (see Sect. 2.6) that generates facial images from the feature patches. An example result of DeepFaceDrawing is shown in Fig. 10f.

7.1.4 Simulated GAN (SimGAN)

Simulated GAN (SimGAN) [114] is an unsupervised method for transforming simulated images into real-world images while preserving annotation information such as image landmarks and pose. This transformation is performed by a refiner network $R(.)$.

Let $\boldsymbol{y}_j$ denote the training unlabeled real-world images, $\boldsymbol{x}_i$ denote the training simulated images, and $\widetilde{\boldsymbol{x}}_i = R(\boldsymbol{x}_i)$ denote the refined (transformed) simulated images. SimGAN trains a discriminator D by minimizing the following loss:

$$\min_{D} \; -\sum_{i} \log(D(\widetilde{\boldsymbol{x}}_i)) - \sum_{j} \log(1 - D(\boldsymbol{y}_j)), \tag{107}$$

so that D outputs labels close to one for real-world images and close to zero for refined simulated images.

After training the discriminator, it is used in the loss function of the refiner. The refiner, acting as the generator in GAN, tries to fool the discriminator. Therefore, the loss function for the refiner is

$$\min_{R} \; -\sum_{i} \log(1 - D(R(\boldsymbol{x}_i))) + \lambda \, \|\psi(R(\boldsymbol{x}_i)) - \psi(\boldsymbol{x}_i)\|_1, \tag{108}$$

where $\lambda > 0$ is the regularization parameter, $\psi(.)$ is a mapping from the pixel space to a feature space, and the second term minimizes the reconstruction error in the feature space to preserve annotation-relevant features.

7.1.5 Interactive GAN (iGAN)

Interactive GAN (iGAN) [135] enables users to edit images interactively while ensuring that the edited images remain realistic. iGAN first projects an input image onto the image manifold, which is the manifold of latent noise in GAN. This projection is performed by finding the closest latent noise z^* that can generate the image:

$$z^* := \arg\min_{z} \|G(z) - \boldsymbol{x}\|_2^2. \tag{109}$$

In this sense, this projection is similar to the approach used in inverse GAN models (see Sect. 5.13).

After projection, the latent vector z^* can be edited using different brushing and editing tools. The geometric and color changes are then applied to obtain the edited image while ensuring it remains on the learned image manifold, thus maintaining its realism.

7.2 Text-to-Image Generation

There exist several methods for text-to-image generation, where an image is generated from a descriptive caption. Some of these methods include [91, 106–108, 132, 133]. Here, Stacked GAN (StackGAN) [132] is introduced for text-to-image generation.

StackGAN first generates text embeddings using a pre-trained autoencoder. Let the text and its embedding be denoted by $\boldsymbol{t}$ and ϕ_t, respectively. StackGAN consists of two stages of adversarial learning. The first stage generates a low-resolution image capturing the general shapes and colors. The loss function for the first stage is

$$\min_D \mathbb{E}_{(\boldsymbol{x},\boldsymbol{t})\sim p_{\text{data}}(\boldsymbol{x},\boldsymbol{t})}[\log(D(\boldsymbol{x}, \phi_t))] + \mathbb{E}_{\boldsymbol{t}\sim p_{\text{data}}(\boldsymbol{t}), z\sim p_z(z)}[\log(1 - D(G(z), \phi_t))], \tag{110}$$

$$\min_G \mathbb{E}_{\boldsymbol{t}\sim p_{\text{data}}(\boldsymbol{t}), z\sim p_z(z)}[\log(1 - D(G(z), \phi_t))] + \lambda\, \text{KL}(q_z(z)\| p_z(z)), \tag{111}$$

where $q_z(z)$ is the distribution of the latent code from the encoder of an autoencoder, and $p_z(z)$ is the prior on the latent noise.

The second stage takes the low-resolution generated image from the first stage, denoted by $\boldsymbol{s}$, along with the text embedding, and generates a high-resolution image. The adversarial loss for the second stage is the same as Eqs. (110) and (111), but uses $G(\boldsymbol{s})$ instead of $G(z)$ since the low-resolution image is fed to its generator.

Some results of StackGAN are shown in Fig. 11. An improved version of StackGAN is StackGAN++ [133].

Fig. 11 Text-to-image translation by StackGAN. Images are from [132]

7.3 Mixing Image Characteristics

7.3.1 FineGAN

FineGAN [115] is an unsupervised GAN model that disentangles the features of generated images into background, shape, and color/texture. For this purpose, it uses three separate latent noise samples: the background code $\boldsymbol{b}$, the parent code $\boldsymbol{p}$, and the child code $\boldsymbol{c}$, which are responsible for background, shape, and color/texture, respectively.

It is assumed that there are n_b, n_p, and n_c unknown categories (classes) for background, shape, and color/texture, respectively, which are learned by FineGAN. The priors for the latent codes follow a categorical distribution where the probability of each class is $1/n_b$, $1/n_p$, and $1/n_c$, respectively. Since each object shape may have multiple textures across images, it is assumed that $n_p < n_c$.

FineGAN generates an image hierarchically, starting with the background generation. For training data, a pre-trained detector is used to identify background patches. A continuous latent code z_b is also used to control background details within each category. The generator G_b takes both $\boldsymbol{b}$ and z_b as input, and D_b is the discriminator for judging the generated background. Additionally, a discriminator D'_b is used as a binary classifier to distinguish foreground from background patches. This discriminator is pre-trained using cross-entropy on background and foreground patches.

The loss for the background stage is

$$\begin{aligned}\min_{G_b}\max_{D_b}\ &\mathbb{E}_{\boldsymbol{x}}[\log(D_b(\boldsymbol{x}))] + \mathbb{E}_{\boldsymbol{b},z_b}[\log(1 - D_b(G_b(\boldsymbol{b}, z_b)))] \\ &+ \lambda\,\mathbb{E}_{\boldsymbol{b},z_b}[\log(1 - D'_b(G_b(\boldsymbol{b}, z_b)))],\end{aligned} \tag{112}$$

where $\lambda > 0$ is a regularization parameter.

In the parent stage, two generators $G_{p,m}$ and $G_{p,f}$ generate the mask and initial texture of the object, respectively. A network G_p takes the categorical $\boldsymbol{p}$ and continuous z_p as input and outputs z'_p, which is the input code for $G_{p,m}$ and $G_{p,f}$. Here, z_p controls the initial texture. The outputs of $G_{p,m}$ and $G_{p,f}$ are combined to produce the object shape with an initial texture, denoted by $\boldsymbol{x}_p$, which is then composited onto the background. If D_p is the discriminator for this stage, the loss maximizes the mutual information between $\boldsymbol{p}$ and $\boldsymbol{x}_p$:

$$\max_{D_p, G_{p,m}, G_{p,f}}\ \mathbb{E}_{\boldsymbol{p},z_p}[\log(D_p(\boldsymbol{p}|\boldsymbol{x}_p))]. \tag{113}$$

In the child stage, two generators $G_{c,m}$ and $G_{c,f}$ generate the mask and color/texture of the object, respectively. A network G_c takes $\boldsymbol{c}$ and z'_p as input and outputs z'_c, which is the input code for $G_{c,m}$ and $G_{c,f}$. Their outputs are combined to obtain the object's shape with color/texture, denoted by $\boldsymbol{x}_c$, which is then composited onto $\boldsymbol{x}_p$ to produce the final generated image $\boldsymbol{x}_f$.

The loss for the child stage is

$$\begin{aligned}\min_{G_c}\max_{D_c}\ &\mathbb{E}_{\boldsymbol{x}}[\log(D_c(\boldsymbol{x}))] + \mathbb{E}_{\boldsymbol{c},\boldsymbol{p},z_p}[\log(1 - D_c(\boldsymbol{x}_f))] \\ &+ \max_{D_c, G_{c,m}, G_{c,f}}\ \mathbb{E}_{\boldsymbol{c},z_c}[\log(1 - D'_c(\boldsymbol{c}|\boldsymbol{x}_c))],\end{aligned} \tag{114}$$

where the first two terms are for adversarial learning and the last term maximizes the mutual information.

Some results of FineGAN are illustrated in Fig. 12a.

Fig. 12 Mixing image characteristics using GAN: **a** Generating an image with background, shape, and color characteristics by FineGAN, **b** generating an image by borrowing its characteristics from three images using MixNMatch, and **c** generating an image by borrowing its characteristics from different domains using improved MixNMatch. Images are from [66, 97, 115], respectively

7.3.2 MixNMatch

MixNMatch [66] builds upon FineGAN introduced in Sect. 7.3.1. It allows users to choose the background, shape, and color/texture from three different images and generates a new image with the selected characteristics. For this purpose, an encoder network $E(\boldsymbol{x})$ is used, which receives three images corresponding to their background, shape, and color/texture characteristics and outputs the three latent codes $\boldsymbol{b}$, $\boldsymbol{p}$, and $\boldsymbol{c}$. These codes are then fed into FineGAN.

MixNMatch utilizes the idea of inverse GAN models (see Sect. 5.13) to provide inputs for the encoder and FineGAN networks. The input/output pair of the encoder is $(\boldsymbol{x} \sim p_{\text{data}}(\boldsymbol{x}), \widetilde{\boldsymbol{y}} = E(\boldsymbol{x}))$, where $\widetilde{\boldsymbol{y}}$ comprises the codes $\boldsymbol{b}$, $\boldsymbol{p}$, and $\boldsymbol{c}$. The output/input pair of FineGAN is $(\widetilde{\boldsymbol{x}} = G(\boldsymbol{y}), \boldsymbol{y} \sim p_{\text{code}}(\boldsymbol{y}))$, where $G(.)$ denotes FineGAN and $p_{\text{code}}(\boldsymbol{y})$ is the prior distribution of the latent codes $\boldsymbol{b}$, $\boldsymbol{p}$, and $\boldsymbol{c}$.

MixNMatch employs a discriminator D that takes an image-code pair and judges whether it is produced by the encoder or by FineGAN. The loss function of MixNMatch is

$$\begin{aligned}\min_{G,E}\max_{D}\ & \mathbb{E}_{\boldsymbol{x}\sim p_{\text{data}}(\boldsymbol{x})}\big[\mathbb{E}_{\widetilde{\boldsymbol{y}}=E(\boldsymbol{x})}[\log(D(\boldsymbol{x},\widetilde{\boldsymbol{y}}))]\big] \\ & + \mathbb{E}_{\boldsymbol{y}\sim p_{\text{code}}(\boldsymbol{y})}\big[\mathbb{E}_{\widetilde{\boldsymbol{x}}=G(\boldsymbol{y})}[\log(1 - D(\widetilde{\boldsymbol{x}},\boldsymbol{y}))]\big].\end{aligned} \tag{115}$$

MixNMatch allows users to select image characteristics from a specific domain, and the generated image will belong to that domain. An example result of MixNMatch is shown in Fig. 12b.

Recently, an improved version of MixNMatch [97] has been proposed, which can take characteristics from multiple domains and generate a new image combining those characteristics. An example result of this version is also shown in Fig. 12c.

7.4 Other Applications

There are several other applications for GANs. One application is image inpainting, where GAN is used to fill in missing parts of an image based on the available pixels [103].

A medical application of GANs is generating histopathology images, which can provide insights into cancer diagnosis from pathology whole slide images [62].

GANs have also been applied in various other domains, including natural language processing [63, 127], speech processing [102, 118], network embedding [16], logic [88], and sketch retrieval [13].

8 Autoencoders Based on Adversarial Learning

Previously, variational Bayes was used in an autoencoder setting to develop the variational autoencoder [56] (see Chap. 12). Likewise, adversarial learning can be integrated into an autoencoder framework [73]. Several adversarial-based autoencoders exist, which are introduced in the following.

8.1 Adversarial Autoencoder (AAE)

8.1.1 Unsupervised AAE

Adversarial Autoencoder (AAE) was proposed in [75]. In contrast to the variational autoencoder [56], which uses KL divergence and the evidence lower bound (see Chap. 12), AAE employs adversarial learning to impose a specific distribution on the latent variable in its coding layer.

The structure of AAE is depicted in Fig. 13. Each of the blocks B_1, B_2, and B_3 in this figure contains several network layers with nonlinear activation functions. AAE comprises an encoder (block B_1) and a decoder (block B_2). The input to the encoder is a real data point $\boldsymbol{x} \in \mathbb{R}^d$, and the output of the decoder is the reconstructed data point $\widehat{\boldsymbol{x}} \in \mathbb{R}^d$. One of the low-dimensional intermediate layers is the latent (or code) layer, denoted by $z \in \mathbb{R}^p$, where $p \ll d$. The encoder and decoder model the conditional distributions $p(z|\boldsymbol{x})$ and $p(\boldsymbol{x}|z)$, respectively. Let the distribution of the latent variable in the autoencoder be denoted by $q(z)$, representing the posterior distribution of the latent variable.

Blocks B_1 and B_3 correspond to the generator G and discriminator D of the adversarial network, respectively. There is also a prior distribution, denoted by $p(z)$, on the latent variable, which is chosen by the user (e.g., a p-dimensional normal distribution $\mathcal{N}(\boldsymbol{0}, \boldsymbol{I})$). The encoder of the autoencoder (block B_1) serves as the generator G, generating the latent variable from the posterior distribution:

$$G(\boldsymbol{x}) = z \sim q(z). \tag{116}$$

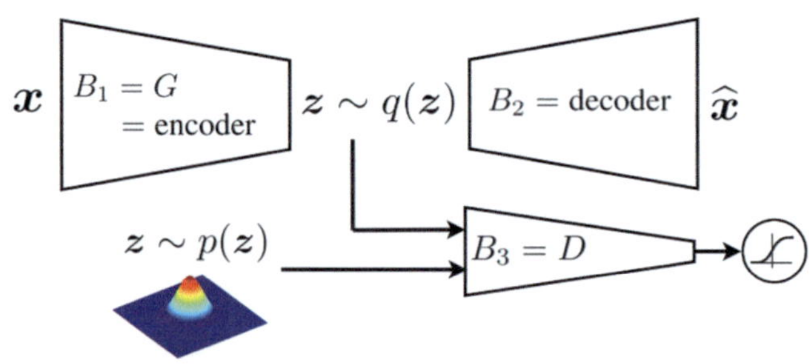

Fig. 13 The structure of unsupervised AAE

The discriminator D (block B_3) has a single output neuron with a sigmoid activation function. It classifies the latent variable z as either a real latent variable from the prior distribution $p(z)$ or a generated latent variable from the encoder:

$$D(z) := \begin{cases} 1 \text{ if } z \text{ is real, i.e., } z \sim p(z), \\ 0 \text{ if } z \text{ is generated, i.e., } z \sim q(z). \end{cases} \tag{117}$$

As described, block B_1 is shared between the autoencoder and the adversarial network. This adversarial learning process makes both the autoencoder and the adversarial network stronger gradually, as the autoencoder tries to generate latent variables similar to those from the prior distribution to fool the discriminator, while the discriminator becomes stronger to distinguish between them.

AAE uses alternating optimization [27], where reconstruction and regularization phases are performed iteratively. In the reconstruction phase, the mean squared error between the data $\boldsymbol{x}$ and the reconstructed data $\widehat{\boldsymbol{x}}$ is minimized. In the regularization phase, the discriminator and generator are updated using the GAN approach. Each update uses stochastic gradient descent [27] with backpropagation.

Overall, the two phases are performed as follows:

$$B_1', B_2^{(k+1)} := \arg\min_{B_1, B_2} \|\widehat{\boldsymbol{x}} - \boldsymbol{x}\|_2^2, \tag{118}$$

$$\begin{cases} B_3^{(k+1)} := B_3^{(k)} - \eta^{(k)} \frac{\partial}{\partial B_3}\Big(V(B_3, B_1')\Big), \\ B_1^{(k+1)} := B_1' - \eta^{(k)} \frac{\partial}{\partial B_1}\Big(V(B_3^{(k+1)}, B_1)\Big), \end{cases} \tag{119}$$

where $B_1 = G$ and $B_3 = D$ (see Fig. 13). Equation (118) is the reconstruction phase, and Eq. (119) is the regularization phase.

8.1.2 Sampling the Latent Variable

There are several approaches for sampling the latent variable z from the coding layer of an autoencoder with posterior $q(z)$. These approaches are explained below [75]:

- **Deterministic approach**: The latent variable is taken directly as the output of the encoder, i.e., $z_i = B_1(\boldsymbol{x}_i)$. The stochasticity in $q(z)$ arises from the distribution of the dataset, $p_{\text{data}}(\boldsymbol{x})$.
- **Gaussian posterior**: This approach is similar to that used in variational autoencoders [56] (see Chap. 12). The encoder outputs the mean $\boldsymbol{\mu}$ and covariance $\boldsymbol{\Sigma}$, and the latent variable is sampled from the Gaussian distribution, i.e., $z_i \sim \mathcal{N}(\boldsymbol{\mu}(\boldsymbol{x}_i), \boldsymbol{\Sigma}(\boldsymbol{x}_i))$. The stochasticity in $q(z)$ comes from both $p_{\text{data}}(\boldsymbol{x})$ and the Gaussian distribution output by the encoder.
- **Universal approximator posterior**: The data point $\boldsymbol{x}$ is concatenated with noise η, sampled from a fixed distribution such as Gaussian, and the concatenated vector is fed as input to the encoder. Therefore, the latent variable is $z_i = B_1(\boldsymbol{x}_i, \eta_i)$, where $\eta_i \sim \mathcal{N}(\boldsymbol{0}, \boldsymbol{I})$. The stochasticity in $q(z)$ arises from both $p_{\text{data}}(\boldsymbol{x})$ and the noise η.

8.1.3 Supervised AAE

There are two variants of supervised AAE [75], where class labels are utilized. These two structures are illustrated in Fig. 14. Let c denote the number of classes.

In the first variant, the one-hot encoded label $\boldsymbol{y} \in \mathbb{R}^c$ is fed to the discriminator (block B_3) by concatenating it with the latent variable z. In this way, the discriminator learns both the label of the

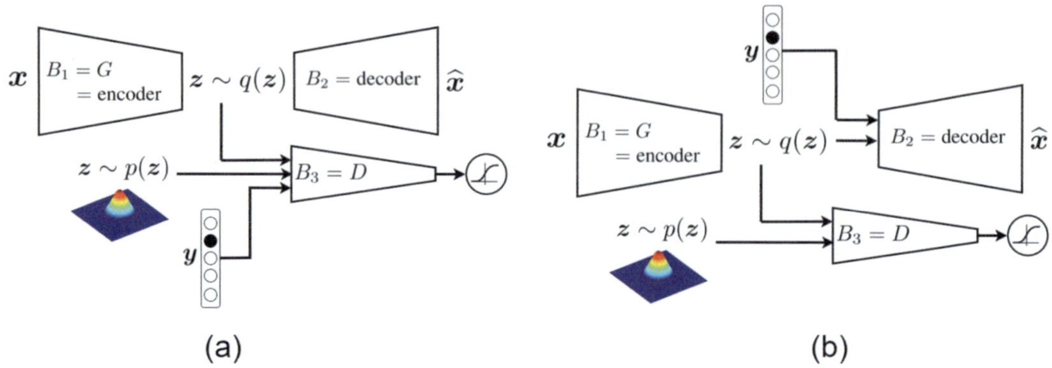

Fig. 14 Two structures for supervised AAE

point $\boldsymbol{x}$ and the discrimination between real and generated latent variables. This setup encourages the generator (i.e., the encoder) to generate latent variables corresponding to the label of the data point to fool the discriminator.

In the second variant, the one-hot encoded label $\boldsymbol{y}$ is fed to the decoder (block B_2) by concatenating it with the latent variable $\boldsymbol{z}$. This allows the decoder to reconstruct the data point using its label. Consequently, the encoder, which also acts as the generator, is encouraged to generate latent variables $\boldsymbol{z}$ that are consistent with the label of the data point. This, in turn, strengthens the discriminator in its competition with the generator during adversarial learning.

Note that these two variants can also be combined, where the one-hot encoded label is fed to both the discriminator and the decoder.

8.1.4 Semi-Supervised AAE

Consider a partially labeled dataset, where the labeled portion has c classes. AAE can be used for semi-supervised learning with such partially labeled data. The structure for semi-supervised AAE is depicted in Fig. 15.

This structure includes an autoencoder (blocks B_1 and B_2), adversarial learning for generating the latent variable (blocks B_1 and B_3), and adversarial learning for generating class labels (blocks B_1 and B_4). The encoder generates both the label $\boldsymbol{y} \in \mathbb{R}^c$ and the latent variable $\boldsymbol{z} \in \mathbb{R}^p$. The last layer of the encoder for labels has a softmax activation function, outputting a c-dimensional vector whose entries sum to one, behaving as probabilities. The last layer of the encoder for latent variables has a linear activation function.

Semi-supervised AAE consists of three phases: reconstruction, regularization, and semi-supervised classification. In the reconstruction phase, the reconstruction error is minimized. The regularization phase trains the discriminator and generator for generating the latent variable $\boldsymbol{z}$. The semi-supervised classification phase generates the one-hot encoded class label $\boldsymbol{y}$ for the point $\boldsymbol{x}$. If $\boldsymbol{x}$ has a label, its label is used for training B_1 and B_4. However, if $\boldsymbol{x}$ does not have a label, a label $\boldsymbol{y} \in \mathbb{R}^c$ is randomly sampled from a categorical distribution, i.e., $\boldsymbol{y} \sim \text{Cat}(\boldsymbol{y})$. This categorical distribution yields a one-hot encoded vector where the prior probability of each class is estimated by the proportion of the class's population to the total number of labeled points.

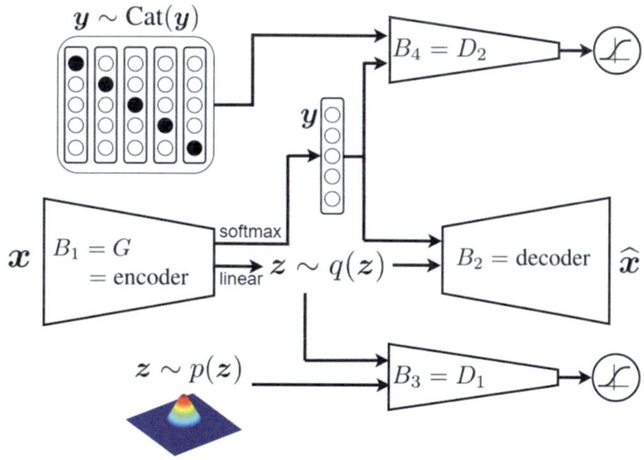

Fig. 15 The structure of semi-supervised AAE

An iteration of the alternating optimization for semi-supervised learning is

$$B_1', B_2^{(k+1)} := \arg\min_{B_1, B_2} \|\widehat{\boldsymbol{x}} - \boldsymbol{x}\|_2^2, \tag{120}$$

$$\begin{cases} B_3^{(k+1)} := B_3^{(k)} - \eta^{(k)} \frac{\partial}{\partial B_3}\Big(V_z(B_3, B_1')\Big), \\ B_1'' := B_1' - \eta^{(k)} \frac{\partial}{\partial B_1}\Big(V_z(B_3^{(k+1)}, B_1)\Big), \end{cases} \tag{121}$$

$$\begin{cases} B_4^{(k+1)} := B_4^{(k)} - \eta^{(k)} \frac{\partial}{\partial B_4}\Big(V_y(B_4, B_1'')\Big), \\ B_1^{(k+1)} := B_1'' - \eta^{(k)} \frac{\partial}{\partial B_1}\Big(V_y(B_4^{(k+1)}, B_1)\Big), \end{cases} \tag{122}$$

where $V_z(D, G)$ and $V_y(D, G)$ are the loss functions defined in Eq. (3), with the generated variables being the latent variable z and the one-hot encoded label $\boldsymbol{y}$, respectively.

8.1.5 Unsupervised Clustering with AAE

The structure shown in Fig. 15 can be used for clustering by assuming there are c clusters instead of classes. In this setting, there is no partially labeled portion of the dataset; all data points are unlabeled, and the cluster indices are sampled randomly from a categorical distribution. The cluster labels and the latent code are both trained using the three phases described in Sect. 8.1.4.

8.1.6 Dimensionality Reduction with AAE

AAE can also be used for dimensionality reduction and representation learning. The structure of AAE for this purpose is depicted in Fig. 16. The encoder generates both the label $\boldsymbol{y} \in \mathbb{R}^c$ and the latent variable $z \in \mathbb{R}^p$, where $p \ll d$.

Everything is similar to what was explained previously, but a network layer $\boldsymbol{W} \in \mathbb{R}^{c \times p}$ is added after the generated label from the encoder. The low-dimensional representation $\widetilde{\boldsymbol{x}} \in \mathbb{R}^p$ is obtained as:

$$\mathbb{R}^p \ni \widetilde{\boldsymbol{x}} = \boldsymbol{W}^\top \boldsymbol{y} + z, \tag{123}$$

where z is the latent variable generated by the encoder. The three phases described in Sect. 8.1.4 are used to train the AAE for dimensionality reduction.

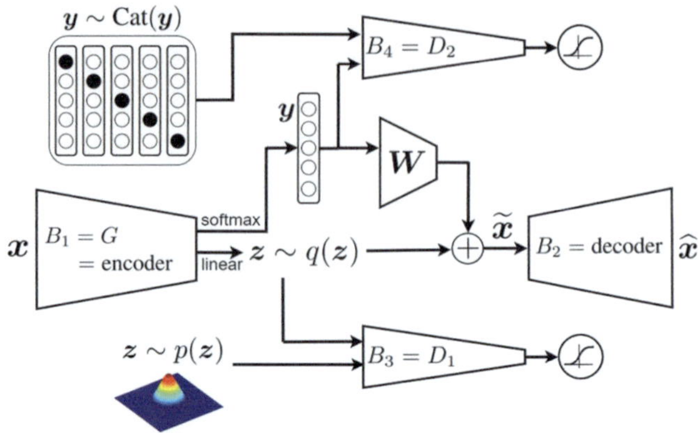

Fig. 16 The structure of AAE for dimensionality reduction

8.2 PixelGAN Autoencoder

In variational inference (see Chap. 12), the Evidence Lower Bound (ELBO) can be restated as [46]:

$$\mathbb{E}_{\boldsymbol{x}\sim p_{\text{data}}(\boldsymbol{x})}[\log(p(\boldsymbol{x}))] > -\mathbb{E}_{\boldsymbol{x}\sim p_{\text{data}}(\boldsymbol{x})}[\mathbb{E}_{q(z|\boldsymbol{x})}[-\log(p(\boldsymbol{x}|z))]] \\ - \text{KL}(q(z)\|p(z)) - \mathbb{I}(z;\boldsymbol{x}), \tag{124}$$

where $\mathbb{I}(.;.)$ denotes mutual information.

The first and second terms in this lower bound correspond to the reconstruction error and the marginal KL divergence on the latent space, respectively. The PixelGAN autoencoder [74] utilizes this lower bound but ignores the third term, which is the mutual information, as optimizing that term would make z independent of $\boldsymbol{x}$. The reconstruction error is minimized during a reconstruction phase of training, while the KL divergence term is addressed using adversarial learning.

The structure of PixelGAN is shown in Fig. 17. Block B_1 is the encoder, which receives the data point $\boldsymbol{x}$ added with some noise $\boldsymbol{n}$ as input and outputs the latent code $z \sim q(z|\boldsymbol{x})$. Block B_2 is the decoder, which is implemented as a PixelCNN network [100], from which PixelGAN derives its name. This decoder outputs the reconstructed data $\widehat{\boldsymbol{x}}$. The generated latent code z is used as adaptive biases in the layers of the PixelCNN.

Blocks B_1 and B_3 serve as the generator and discriminator in adversarial learning, respectively, aiming to align the distribution of the generated latent code z with a prior distribution $p(z)$.

In summary, blocks B_1 and B_2 are used during the reconstruction phase, while blocks B_1 and B_3 are involved in the adversarial learning phase.

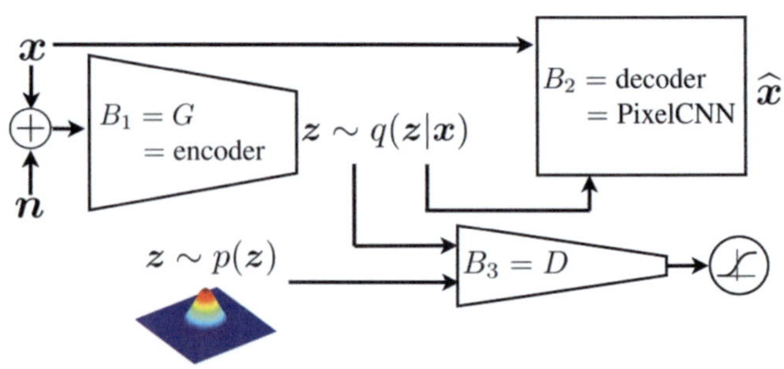

Fig. 17 The structure of PixelGAN

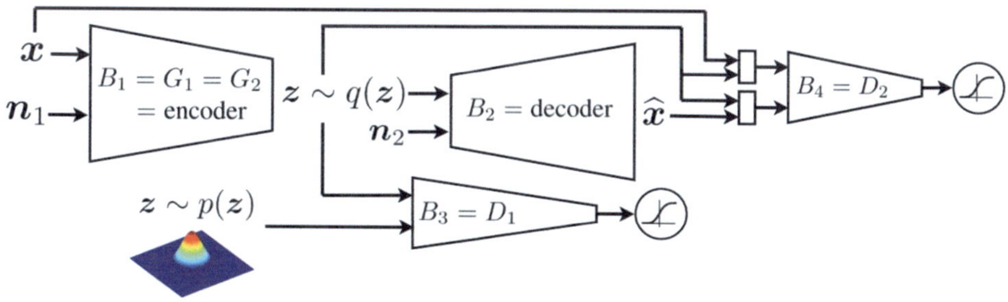

Fig. 18 The structure of IAE

8.3 Implicit Autoencoder (IAE)

In variational inference (see Chap. 12), the Evidence Lower Bound (ELBO) can be restated as [72]:

$$\mathbb{E}_{x \sim p_{\text{data}}(x)}[\log(p(x))] \geq - \text{KL}(q(x, z) \| q(\widehat{x}, z)) \\ - \text{KL}(q(z) \| p(z)) - H_{\text{data}}(x), \tag{125}$$

where $H_{\text{data}}(\boldsymbol{x})$ is the entropy of the data, $\widehat{\boldsymbol{x}}$ is the reconstructed data, and z is some latent factor. The proof is straightforward and can be found in [72, Appendix A]. The first and second terms correspond to the reconstruction and regularization terms, respectively.

The Implicit Autoencoder (IAE) [72] implements the distributions in Eq. (125) implicitly using networks. The structure of IAE is shown in Fig. 18. Block B_1 is the encoder, which takes data $\boldsymbol{x}$ and noise $\boldsymbol{n}_1$ as input and outputs the latent code $z \sim q(z)$. Block B_2 takes the generated latent code z along with some noise $\boldsymbol{n}_2$ and outputs the reconstructed data $\widehat{\boldsymbol{x}}$.

Blocks B_1 and B_3 serve as the generator G_1 and discriminator D_1 for the first adversarial learning phase, which aligns the distribution of the latent code z with a prior distribution $p(z)$. Blocks B_1 and B_4 act as the generator G_2 and discriminator D_2 for the second adversarial learning phase, ensuring that the distribution of the reconstructed data $\widehat{\boldsymbol{x}}$ matches that of the data $\boldsymbol{x}$. The inputs to B_4 are the pairs $(\boldsymbol{x}, z)$ and $(\widehat{\boldsymbol{x}}, z)$, modeling the distributions $q(\boldsymbol{x}, z)$ and $q(\widehat{\boldsymbol{x}}, z)$ in Eq. (125). In summary, three phases of training are performed: the reconstruction phase and two adversarial learning phases.

9 Conclusion

GANs were a groundbreaking innovation in generative modeling, introducing adversarial training and inspiring a huge research wave. Variants such as DCGAN [104], WGAN [2], and StyleGAN [55] pushed generative quality to new heights, with StyleGAN in particular setting benchmarks in photorealistic image synthesis. GANs remain important both conceptually and in applications such as image generation, video synthesis, and data augmentation. However, by 2025, GANs are no longer the dominant generative approach. Diffusion models [44] (see Chap. 15) and autoregressive Transformers [124] (see Chap. 9) now achieve higher sample quality and broader applicability, while GANs are best understood as a milestone technology that reshaped generative modeling and continues to influence new research directions.

10 Exercises

1. (Level: Easy) In a standard Generative Adversarial Network (GAN), the generator G tries to generate realistic samples, while the discriminator D attempts to distinguish between real and generated samples. The objective function for a GAN is

$$\min_G \max_D V(D, G) = \mathbb{E}_{x \sim p_{data}(x)}[\log D(x)] + \mathbb{E}_{z \sim p(z)}[\log(1 - D(G(z)))],$$

where $p_{data}(x)$ is the true data distribution and $p(z)$ is the prior distribution used as input for the generator.

(a) Explain the intuition behind this objective function. Why is the minimax game important for training a GAN?

(b) Suppose the discriminator $D(x)$ outputs the following probabilities for real and generated samples:

Sample type	**D(x)**
Real sample	0.85
Fake sample	0.20

Compute the GAN loss for both D and G using the objective function above.

2. (Level: Easy) Consider a simple GAN with the following generator and discriminator functions:

$$G(z) = 2z + 1,$$
$$D(x) = \frac{e^x}{1 + e^x}.$$

Assume the noise variable z follows a uniform distribution $U(0, 1)$.

(a) Compute the expected value of a generated sample $G(z)$.

(b) Compute the expected value of the discriminator output $D(G(z))$.

(c) Explain why the minimax game leads to equilibrium.

3. (Level: Moderate) Consider a GAN setting where the real data distribution is $\mathcal{N}(0, 1)$, and the generator produces outputs from $\mathcal{N}(\mu_g, \sigma_g^2)$. Compute the following:

(a) Compute the Jensen-Shannon (JS) divergence between the real data distribution p_{data} and the generator distribution p_g.

(b) Explain the significance of this JS divergence value in the context of GAN training.

(c) Let $\mu_g = 1, \sigma_g = 1$. Compute the JS divergence numerically.

(d) How would training be affected if $\text{JS}(p_{\text{data}} \| p_g) = \log 2$? What does this imply about the generator and discriminator behavior?

(e) Propose a method or loss function to mitigate the issues caused when p_g and p_{data} have little or no overlap.

4. (Level: Difficult) Show that the optimal discriminator predicts fake and real inputs each with probability 0.5.

5. (Level: Moderate) Consider a generative adversarial network. Suppose the real data distribution is $p_{data}(x) = \mathcal{N}(0, 1)$ and the generator produces samples from $p_g(x) = \mathcal{N}(\mu, 1)$.

   (a) Compute the optimal discriminator, $D^*(x)$.
   (b) What happens when $\mu \to 0$?
   (c) In the above example, let $\mu = 0.5$. Sketch or plot the decision boundaries of this optimal discriminator. In other words, you should plot both the real data distribution and the generator distribution, and show which values of x are predicted as real and which ones are predicted as generated.
   (d) Consider instead that $p_{data}(x) = 0.5\mathcal{N}(2, 1) + 0.5\mathcal{N}(-2, 1)$ and that the generator samples from $\mathcal{N}(1, 1)$. Sketch or plot the decision boundaries of the optimal discriminator.

References

1. Martin Arjovsky and Léon Bottou. "Towards principled methods for training generative adversarial networks". In: *International Conference on Machine Learning*. 2017.
2. Martin Arjovsky, Soumith Chintala, and Léon Bottou. "Wasserstein generative adversarial networks". In: *International conference on machine learning*. 2017, pp. 214–223.
3. Sanjeev Arora, Rong Ge, Yingyu Liang, Tengyu Ma, and Yi Zhang. "Generalization and equilibrium in generative adversarial nets (GANs)". In: *International Conference on Machine Learning*. 2017, pp. 224–232.
4. David Berthelot, Peyman Milanfar, and Ian Goodfellow. "Creating high resolution images with a latent adversarial generator". In: *arXiv preprint* arXiv:2003.02365 (2020).
5. Jean Bourgain. "On Lipschitz embedding of finite metric spaces in Hilbert space". In: *Israel Journal of Mathematics* 52.1-2 (1985), pp. 46–52.
6. Andrew Brock, Jeff Donahue, and Karen Simonyan. "Large scale GAN training for high fidelity natural image synthesis". In: *International Conference on Learning Representations*. 2019.
7. Jane Bromley et al. "Signature verification using a "Siamese" time delay neural network". In: *International Journal of Pattern Recognition and Artificial Intelligence* 7.04 (1993), pp. 669–688.
8. Peter J Burt and Edward H Adelson. "The Laplacian pyramid as a compact image code". In: *IEEE Transactions on Communications* 31.4 (1983), pp. 532–540.
9. John Canny. "A computational approach to edge detection". In: *IEEE Transactions on pattern analysis and machine intelligence* 6 (1986), pp. 679–698.
10. Shu-Yu Chen, Wanchao Su, Lin Gao, Shihong Xia, and Hongbo Fu. "DeepFaceDrawing: Deep generation of face images from sketches". In: *ACM Transactions on Graphics (TOG)* 39.4 (2020), pp. 72–1.
11. Xi Chen et al. "InfoGAN: Interpretable representation learning by information maximizing generative adversarial nets". In: *Proceedings of the 30th International Conference on Neural Information Processing Systems*. 2016, pp. 2180–2188.
12. Jen-Tzung Chien and Chun-Lin Kuo. "Variational Bayesian GAN". In: *2019 27th European Signal Processing Conference (EUSIPCO)*. IEEE. 2019, pp. 1–5.
13. Antonia Creswell and Anil Anthony Bharath. "Adversarial training for sketch retrieval". In: *European Conference on Computer Vision*. Springer. 2016, pp. 798–809.
14. Antonia Creswell and Anil Anthony Bharath. "Inverting the generator of a generative adversarial network". In: *IEEE transactions on neural networks and learning systems* 30.7 (2018), pp. 1967–1974.
15. Antonia Creswell et al. "Generative adversarial networks: An overview". In: *IEEE Signal Processing Magazine* 35.1 (2018), pp. 53–65.
16. Quanyu Dai, Qiang Li, Jian Tang, and Dan Wang. "Adversarial network embedding". In: *Proceedings of the AAAI Conference on Artificial Intelligence*. Vol. 32. 1. 2018
17. Emily Denton, Soumith Chintala, Arthur Szlam, and Rob Fergus. "Deep generative image models using a laplacian pyramid of adversarial networks". In: *arXiv preprint* arXiv:1506.05751 (2015).
18. Jeff Donahue, Philipp Krähenbühl, and Trevor Darrell. "Adversarial feature learning". In: *International Conference on Learning Representations*. 2017.
19. Ding-Zhu Du and Panos M Pardalos. *Minimax and applications*. Vol. 4. Springer Science & Business Media, 2013.
20. Vincent Dumoulin et al. "Adversarially learned inference". In: *International Conference on Learning Representations*. 2017.

21. Ishan Durugkar, Ian Gemp, and Sridhar Mahadevan. "Generative multi-adversarial networks". In: *International Conference on Learning Representations*. 2017.
22. Farzan Farnia and Asuman Ozdaglar. "Do GANs always have Nash equilibria?" In: *International Conference on Machine Learning*. 2020, pp. 3029–3039.
23. Farzan Farnia and David Tse. "A convex duality framework for GANs". In: *Advances in neural information processing systems*. Vol. 31. 2018.
24. Yoav Freund and Robert E Schapire. "A decision-theoretic generalization of on-line learning and an application to boosting". In: *Journal of computer and system sciences* 55.1 (1997), pp. 119–139.
25. Benyamin Ghojogh and Mark Crowley. "The theory behind overfitting, cross validation, regularization, bagging, and boosting: tutorial". In: *arXiv preprint* arXiv:1905.12787 (2019).
26. Benyamin Ghojogh, Ali Ghodsi, Fakhri Karray, and Mark Crowley. "Johnson-Lindenstrauss Lemma, Linear and Nonlinear Random Projections, Random Fourier Features, and Random Kitchen Sinks: Tutorial and Survey". In: *arXiv preprint* arXiv:2108.04172 (2021).
27. Benyamin Ghojogh, Ali Ghodsi, Fakhri Karray, and Mark Crowley. "KKT Conditions, First-Order and Second-Order Optimization, and Distributed Optimization: Tutorial and Survey". In: *arXiv preprint* arXiv:2110.01858 (2021).
28. Benyamin Ghojogh, Ali Ghodsi, Fakhri Karray, and Mark Crowley. "Reproducing Kernel Hilbert Space, Mercer's Theorem, Eigenfunctions, Nyström Method, and Use of Kernels in Machine Learning: Tutorial and Survey". In: *arXiv preprint* arXiv:2106.08443 (2021).
29. Benyamin Ghojogh, Aydin Ghojogh, Mark Crowley, and Fakhri Karray. "Fitting a mixture distribution to data: tutorial". In: *arXiv preprint* arXiv:1901.06708 (2019).
30. Benyamin Ghojogh, Hadi Nekoei, Aydin Ghojogh, Fakhri Karray, and Mark Crowley. "Sampling algorithms, from survey sampling to Monte Carlo methods: Tutorial and literature review". In: *arXiv preprint* arXiv:2011.00901 (2020).
31. Jacob Goldberger, Geoffrey E Hinton, Sam Roweis, and Russ R Salakhutdinov. "Neighbourhood components analysis". In: *Advances in neural information processing systems* 17 (2004).
32. Liang Gonog and Yimin Zhou. "A review: Generative adversarial networks". In: *2019 14th IEEE Conference on Industrial Electronics and Applications (ICIEA)*. IEEE. 2019, pp. 505–510.
33. Ian Goodfellow. "NIPS 2016 tutorial: Generative adversarial networks". In: *Advances in neural information processing systems, Tutorial rack*. 2016.
34. Ian Goodfellow. "On distinguishability criteria for estimating generative models". In: *International Conference on Learning Representations, Workshop track*. 2015.
35. Ian Goodfellow, David Warde-Farley, Mehdi Mirza, Aaron Courville, and Yoshua Bengio. "Maxout networks". In: *International conference on machine learning*. 2013, pp. 1319–1327.
36. Ian Goodfellow et al. "Generative adversarial nets". In: *Advances in neural information processing systems*. Vol. 27. 2014.
37. Ian Goodfellow et al. "Generative adversarial networks". In: *Communications of the ACM* 63.11 (2020), pp. 139–144.
38. Karol Gregor, Ivo Danihelka, Alex Graves, Danilo Rezende, and Daan Wierstra. "DRAW: A recurrent neural network for image generation". In: *International Conference on Machine Learning*. 2015, pp. 1462–1471.
39. Arthur Gretton, Karsten Borgwardt, Malte Rasch, Bernhard Schölkopf, and Alex Smola. "A kernel method for the two-sample-problem". In: *Advances in neural information processing systems* 19 (2006), pp. 513–520.
40. Aditya Grover and Stefano Ermon. "Boosted generative models". In: *Proceedings of the AAAI Conference on Artificial Intelligence*. Vol. 32. 1. 2018.
41. Ishaan Gulrajani, Faruk Ahmed, Martin Arjovsky, Vincent Dumoulin, and Aaron Courville. "Improved training of Wasserstein GANs". In: *Advances in neural information processing systems*. 2017.
42. Michael Gutmann and Aapo Hyvärinen. "Noise-contrastive estimation: A new estimation principle for unnormalized statistical models". In: *Proceedings of the thirteenth international conference on artificial intelligence and statistics*. JMLR Workshop and Conference Proceedings. 2010, pp. 297–304.
43. Tamir Hazan, George Papandreou, and Daniel Tarlow. "Adversarial Perturbations of Deep Neural Networks". In: (2017).
44. Jonathan Ho, Ajay Jain, and Pieter Abbeel. "Denoising diffusion probabilistic models". In: *Advances in neural information processing systems* 33 (2020), pp. 6840–6851.
45. Quan Hoang, Tu Dinh Nguyen, Trung Le, and Dinh Phung. "MGAN: Training generative adversarial nets with multiple generators". In: *International Conference on Learning Representations*. 2018.
46. Matthew D Hoffman and Matthew J Johnson. "ELBO surgery: yet another way to carve up the variational evidence lower bound". In: *Workshop in Advances in Approximate Bayesian Inference, NIPS*. 2016.
47. Yongjun Hong, Uiwon Hwang, Jaeyoon Yoo, and Sungroh Yoon. "How generative adversarial networks and their variants work: An overview". In: *ACM Computing Surveys (CSUR)* 52.1 (2019), pp. 1–43.
48. Ling Huang, Anthony D Joseph, Blaine Nelson, Benjamin IP Rubinstein, and J Doug Tygar. "Adversarial machine learning". In: *Proceedings of the 4th ACM workshop on Security and artificial intelligence*. 2011, pp. 43–58.

49. Ferenc Huszár. "How (not) to train your generative model: Scheduled sampling, likelihood, adversary?" In: *arXiv preprint* arXiv:1511.05101 (2015).
50. Daniel Jiwoong Im, Chris Dongjoo Kim, Hui Jiang, and Roland Memisevic. "Generating images with recurrent adversarial networks". In: *arXiv preprint* arXiv:1602.05110 (2016).
51. Sergey Ioffe and Christian Szegedy. "Batch normalization: Accelerating deep network training by reducing internal covariate shift". In: *International conference on machine learning*. 2015, pp. 448–456.
52. Phillip Isola, Jun-Yan Zhu, Tinghui Zhou, and Alexei A Efros. "Image-to-image translation with conditional adversarial networks". In: *Proceedings of the IEEE conference on computer vision and pattern recognition*. 2017, pp. 1125–1134.
53. William B Johnson and Joram Lindenstrauss. "Extensions of Lipschitz mappings into a Hilbert space". In: *Contemporary mathematics* 26 (1984).
54. Tero Karras, Timo Aila, Samuli Laine, and Jaakko Lehtinen. "Progressive growing of GANs for improved quality, stability, and variation". In: *International Conference on Learning Representations*. 2018.
55. Tero Karras, Samuli Laine, and Timo Aila. "A style-based generator architecture for generative adversarial networks". In: *Proceedings of the IEEE/CVF conference on computer vision and pattern recognition*. 2019, pp. 4401–4410.
56. Diederik P Kingma and Max Welling. "Auto-encoding variational Bayes". In: *International Conference on Learning Representations*. 2014.
57. Alex Krizhevsky, Ilya Sutskever, and Geoffrey E Hinton. "Imagenet classification with deep convolutional neural networks". In: *Advances in neural information processing systems* 25 (2012), pp. 1097–1105.
58. Alexey Kurakin, Ian Goodfellow, Samy Bengio, et al. "Adversarial examples in the physical world". In: *International Conference on Learning Representations, Workshop Track*. 2017.
59. Alexey Kurakin, Ian Goodfellow, and Samy Bengio. "Adversarial machine learning at scale". In: *International Conference on Learning Representations*. 2017.
60. Anders Boesen Lindbo Larsen, Søren Kaae Sønderby, Hugo Larochelle, and Ole Winther. "Autoencoding beyond pixels using a learned similarity metric". In: *International conference on machine learning*. 2016, pp. 1558–1566.
61. Yann LeCun, Sumit Chopra, Raia Hadsell, M Ranzato, and F Huang. "A tutorial on energy-based learning". In: *Predicting Structured Data* 1 (2006).
62. Adrian B Levine et al. "Synthesis of diagnostic quality cancer pathology images by generative adversarial networks". In: *The Journal of pathology* 252.2 (2020), pp. 178–188.
63. Changliang Li, Yixin Su, and Wenju Liu. "Text-to-text generative adversarial networks". In: *2018 International Joint Conference on Neural Networks (IJCNN)*. IEEE. 2018, pp. 1–7.
64. Chongxuan Li, Kun Xu, Jun Zhu, and Bo Zhang. "Triple generative adversarial nets". In: *Advances in neural information processing systems*. 2017.
65. Chun-Liang Li, Wei-Cheng Chang, Yu Cheng, Yiming Yang, and Barnabás Póczos. "MMD GAN: Towards deeper understanding of moment matching network". In: *Advances in neural information processing systems*. 2017.
66. Yuheng Li, Krishna Kumar Singh, Utkarsh Ojha, and Yong Jae Lee. "MixNMatch: Multifactor disentanglement and encoding for conditional image generation". In: *Proceedings of the IEEE/CVF Conference on Computer Vision and Pattern Recognition*. 2020, pp. 8039–8048.
67. Yujia Li, Kevin Swersky, and Rich Zemel. "Generative moment matching networks". In: *International Conference on Machine Learning*. PMLR. 2015, pp. 1718–1727.
68. Friedrich Liese and Igor Vajda. "On divergences and informations in statistics and information theory". In: *IEEE Transactions on Information Theory* 52.10 (2006), pp. 4394–4412.
69. Ming-Yu Liu and Oncel Tuzel. "Coupled generative adversarial networks". In: *Advances in neural information processing systems* 29 (2016), pp. 469–477.
70. Ziwei Liu, Ping Luo, Xiaogang Wang, and Xiaoou Tang. "Deep learning face attributes in the wild". In: *Proceedings of the IEEE international conference on computer vision*. 2015, pp. 3730–3738.
71. Aleksander Madry, Aleksandar Makelov, Ludwig Schmidt, Dimitris Tsipras, and Adrian Vladu. "Towards deep learning models resistant to adversarial attacks". In: *International Conference on Learning Representations*. 2018.
72. Alireza Makhzani. "Implicit autoencoders". In: *arXiv preprint* arXiv:1805.09804 (2018).
73. Alireza Makhzani. "Unsupervised representation learning with autoencoders". PhD thesis. University of Toronto, 2018.
74. Alireza Makhzani and Brendan Frey. "PixelGAN autoencoders". In: *Advances in neural information processing systems*. 2017.
75. Alireza Makhzani, Jonathon Shlens, Navdeep Jaitly, Ian Goodfellow, and Brendan Frey. "Adversarial autoencoders". In: *arXiv preprint* arXiv:1511.05644 (2015).
76. Xudong Mao et al. "Least squares generative adversarial networks". In: *Proceedings of the IEEE international conference on computer vision*. 2017, pp. 2794–2802.
77. Xudong Mao et al. "On the Effectiveness of Least Squares Generative Adversarial Networks". In: *IEEE Transactions on Pattern Analysis & Machine Intelligence* 41.12 (2019), pp. 2947–2960.

78. Lars Mescheder, Andreas Geiger, and Sebastian Nowozin. "Which training methods for GANs do actually converge?" In: *International conference on machine learning*. PMLR. 2018, pp. 3481–3490.
79. Lars Mescheder, Sebastian Nowozin, and Andreas Geiger. "Adversarial variational Bayes: Unifying variational autoencoders and generative adversarial networks". In: *International Conference on Machine Learning*. 2017, pp. 2391–2400.
80. Lars Mescheder, Sebastian Nowozin, and Andreas Geiger. "The numerics of GANs". In: *Advances in neural information processing systems*. 2017.
81. Luke Metz, Ben Poole, David Pfau, and Jascha Sohl-Dickstein. "Unrolled generative adversarial networks". In: *International Conference on Learning Representations*. 2017.
82. Tomas Mikolov, Ilya Sutskever, Kai Chen, Greg S Corrado, and Jeff Dean. "Distributed representations of words and phrases and their compositionality". In: *Advances in neural information processing systems*. 2013, pp. 3111–3119.
83. Mehdi Mirza and Simon Osindero. "Conditional generative adversarial nets". In: *arXiv preprint* arXiv:1411.1784 (2014).
84. Seyed-Mohsen Moosavi-Dezfooli, Alhussein Fawzi, and Pascal Frossard. "DeepFool: a simple and accurate method to fool deep neural networks". In: *Proceedings of the IEEE conference on computer vision and pattern recognition*. 2016, pp. 2574–2582.
85. Alexander Mordvintsev, Christopher Olah, and Mike Tyka. *Inceptionism: Going deeper into neural networks*. Google AI Blog. 2015.
86. Youssef Mroueh and Truyen Nguyen. "On the convergence of gradient descent in GANs: MMD GAN as a gradient flow". In: *International Conference on Artificial Intelligence and Statistics*. 2021, pp. 1720–1728.
87. Vaishnavh Nagarajan and J Zico Kolter. "Gradient descent GAN optimization is locally stable". In: *Advances in neural information processing systems*. 2017.
88. Vineel Nagisetty, Laura Graves, Joseph Scott, and Vijay Ganesh. "xAI-GAN: Enhancing Generative Adversarial Networks via Explainable AI Systems". In: (2021).
89. Vinod Nair and Geoffrey E Hinton. "Rectified linear units improve restricted Boltzmann machines". In: *International Conference on Machine Learning*. 2010.
90. Radford M Neal. "Annealed importance sampling". In: *Statistics and computing* 11.2 (2001), pp. 125–139.
91. Anh Nguyen, Jeff Clune, Yoshua Bengio, Alexey Dosovitskiy, and Jason Yosinski. "Plug & play generative networks: Conditional iterative generation of images in latent space". In: *Proceedings of the IEEE Conference on Computer Vision and Pattern Recognition*. 2017, pp. 4467–4477.
92. Tu Dinh Nguyen, Trung Le, Hung Vu, and Dinh Phung. "Dual discriminator generative adversarial nets". In: *Advances in neural information processing systems* (2017).
93. XuanLong Nguyen, Martin J Wainwright, and Michael I Jordan. "Estimating divergence functionals and the likelihood ratio by convex risk minimization". In: *IEEE Transactions on Information Theory* 56.11 (2010), pp. 5847–5861.
94. Frank Nielsen. "A family of statistical symmetric divergences based on Jensen's inequality". In: *arXiv preprint* arXiv:1009.4004 (2010).
95. Sebastian Nowozin, Botond Cseke, and Ryota Tomioka. "f-GAN: Training generative neural samplers using variational divergence minimization". In: *Proceedings of the 30th International Conference on Neural Information Processing Systems*. 2016, pp. 271–279.
96. Augustus Odena. "Semi-supervised learning with generative adversarial networks". In: *International conference on machine learning, Data Efficient Machine Learning workshop*. 2016.
97. Utkarsh Ojha, Krishna Kumar Singh, and Yong Jae Lee. "Generating Furry Cars: Disentangling Object Shape & Appearance across Multiple Domains". In: *International Conference on Learning Representations*. 2021.
98. Utkarsh Ojha et al. "Few-shot Image Generation via Cross-domain Correspondence". In: *Proceedings of the IEEE/CVF Conference on Computer Vision and Pattern Recognition*. 2021, pp. 10743–10752.
99. Frans A Oliehoek et al. "GANGs: Generative adversarial network games". In: *arXiv preprint* arXiv:1712.00679 (2017).
100. Aaron van den Oord et al. "Conditional image generation with PixelCNN decoders". In: *Advances in neural information processing systems*. 2016, pp. 4790–4798.
101. Zhaoqing Pan et al. "Recent progress on generative adversarial networks (GANs): A survey". In: *IEEE Access* 7 (2019), pp. 36322–36333.
102. Santiago Pascual, Antonio Bonafonte, and Joan Serra. "SEGAN: Speech enhancement generative adversarial network". In: *Conference of the International Speech Communication Association (INTERSPEECH)*. 2017.
103. Deepak Pathak, Philipp Krahenbuhl, Jeff Donahue, Trevor Darrell, and Alexei A Efros. "Context encoders: Feature learning by inpainting". In: *Proceedings of the IEEE conference on computer vision and pattern recognition*. 2016, pp. 2536–2544.
104. Alec Radford, Luke Metz, and Soumith Chintala. "Unsupervised representation learning with deep convolutional generative adversarial networks". In: *International Conference on Learning Representations*. 2016.
105. Eric Rasmusen. *Games and information: An introduction to game theory*. 4th ed. Wiley-Blackwell, 2007.

106. Scott Reed et al. "Generating interpretable images with controllable structure". In: *International Conference on Learning Representations, Workshop track*. 2017.
107. Scott Reed et al. "Generative adversarial text to image synthesis". In: *International Conference on Machine Learning*. 2016, pp. 1060–1069.
108. Scott Reed et al. "Learning what and where to draw". In: *Advances in neural information processing systems* 29 (2016), pp. 217–225.
109. Olaf Ronneberger, Philipp Fischer, and Thomas Brox. "U-net: Convolutional networks for biomedical image segmentation". In: *International Conference on Medical image computing and computer-assisted intervention*. Springer. 2015, pp. 234–241.
110. Yunus Saatci and Andrew Wilson. "Bayesian GAN". In: *Advances in neural information processing systems*. 2017, pp. 3624–3633.
111. Tim Salimans et al. "Improved techniques for training GANs". In: *Advances in neural information processing systems* 29 (2016), pp. 2234–2242.
112. Tamar Rott Shaham, Tali Dekel, and Tomer Michaeli. "SinGAN: Learning a generative model from a single natural image". In: *Proceedings of the IEEE/CVF International Conference on Computer Vision*. 2019, pp. 4570–4580.
113. Ken Shoemake. "Animating rotation with quaternion curves". In: *Proceedings of the 12th annual conference on Computer graphics and interactive techniques*. 1985, pp. 245–254.
114. Ashish Shrivastava et al. "Learning from simulated and unsupervised images through adversarial training". In: *Proceedings of the IEEE conference on computer vision and pattern recognition*. 2017, pp. 2107–2116.
115. Krishna Kumar Singh, Utkarsh Ojha, and Yong Jae Lee. "FineGAN: Unsupervised hierarchical disentanglement for fine-grained object generation and discovery". In: *Proceedings of the IEEE/CVF Conference on Computer Vision and Pattern Recognition*. 2019, pp. 6490–6499.
116. Jost Tobias Springenberg. "Unsupervised and semi-supervised learning with categorical generative adversarial networks". In: *International Conference on Learning Representations*. 2016.
117. Jost Tobias Springenberg, Alexey Dosovitskiy, Thomas Brox, and Martin Riedmiller. "Striving for simplicity: The all convolutional net". In: *International Conference on Learning Representations, Workshop Track*. 2015.
118. Anuroop Sriram, Heewoo Jun, Yashesh Gaur, and Sanjeev Satheesh. "Robust speech recognition using generative adversarial networks". In: *2018 IEEE international conference on acoustics, speech and signal processing (ICASSP)*. IEEE. 2018, pp. 5639–5643.
119. Christian Szegedy, Vincent Vanhoucke, Sergey Ioffe, Jon Shlens, and Zbigniew Wojna. "Rethinking the inception architecture for computer vision". In: *Proceedings of the IEEE conference on computer vision and pattern recognition*. 2016, pp. 2818–2826.
120. Hamidou Tembine. "Deep learning meets game theory: Bregman-based algorithms for interactive deep generative adversarial networks". In: *IEEE transactions on cybernetics* 50.3 (2019), pp. 1132–1145.
121. Lucas Theis, Aäron van den Oord, and Matthias Bethge. "A note on the evaluation of generative models". In: *International Conference on Learning Representations*. 2016.
122. Ilya Tolstikhin, Sylvain Gelly, Olivier Bousquet, Carl-Johann Simon-Gabriel, and Bernhard Schölkopf. "AdaGAN: Boosting generative models". In: *arXiv preprint* arXiv:1701.02386 (2017).
123. Thomas Unterthiner et al. "Coulomb GANs: Provably optimal Nash equilibria via potential fields". In: *International Conference on Learning Representations*. 2018.
124. Ashish Vaswani et al. "Attention is all you need". In: *Advances in neural information processing systems*. 2017, pp. 5998–6008.
125. Cédric Villani. *Optimal transport: old and new*. Vol. 338. Springer, 2009.
126. Kunfeng Wang et al. "Generative adversarial networks: introduction and outlook". In: *IEEE/CAA Journal of Automatica Sinica* 4.4 (2017), pp. 588–598.
127. William Yang Wang, Sameer Singh, and Jiwei Li. "Deep adversarial learning for NLP". In: *Proceedings of the 2019 Conference of the North American Chapter of the Association for Computational Linguistics: Tutorials*. 2019, pp. 1–5.
128. Tom White. "Sampling generative networks". In: *Advances in neural information processing systems*. 2016.
129. Yuhuai Wu, Yuri Burda, Ruslan Salakhutdinov, and Roger Grosse. "On the quantitative analysis of decoder-based generative models". In: (2017).
130. Chang Xiao, Peilin Zhong, and Changxi Zheng. "BourGAN: Generative networks with metric embeddings". In: *Advances in neural information processing systems*. 2018.
131. Han Zhang, Ian Goodfellow, Dimitris Metaxas, and Augustus Odena. "Self-attention generative adversarial networks". In: *International conference on machine learning*. 2019, pp. 7354–7363.
132. Han Zhang et al. "StackGAN: Text to photo-realistic image synthesis with stacked generative adversarial networks". In: *Proceedings of the IEEE international conference on computer vision*. 2017, pp. 5907–5915.
133. Han Zhang et al. "StackGAN++: Realistic image synthesis with stacked generative adversarial networks". In: *IEEE transactions on pattern analysis and machine intelligence* 41.8 (2018), pp. 1947–1962.
134. Junbo Zhao, Michael Mathieu, and Yann LeCun. "Energy-based generative adversarial network". In: *International Conference on Learning Representations*. 2017.

135. Jun-Yan Zhu, Philipp Krähenbühl, Eli Shechtman, and Alexei A Efros. "Generative visual manipulation on the natural image manifold". In: *European conference on computer vision*. Springer. 2016, pp. 597–613.
136. Jun-Yan Zhu, Taesung Park, Phillip Isola, and Alexei A Efros. "Unpaired image-to-image translation using cycle-consistent adversarial networks". In: *Proceedings of the IEEE international conference on computer vision*. 2017, pp. 2223–2232.
137. Maciej Zieba and Lei Wang. "Training triplet networks with GAN". In: *International Conference on Learning Representations, Workshop track*. 2017.

Diffusion Models

15

Abstract

Diffusion models, inspired by non-equilibrium thermodynamics, provide a powerful generative framework that progressively adds noise to data through a fixed forward Markov process and then learns to reverse this noising process to generate realistic samples from pure noise. Diffusion models can be conceptualized as an autoencoder: the forward noising process acts like a fixed encoder, and the learned denoising network serves as a decoder that reconstructs data from noise. While this analogy is useful, diffusion models are more generally understood as iterative denoising or score-matching processes. This chapter presents a comprehensive introduction to diffusion models, beginning with the foundational Denoising Diffusion Probabilistic Model (DDPM) and its various extensions. It further explores conditional diffusion models that enable controlled generation, as well as continuous formulations employing stochastic differential equations for noise scheduling. Through this coverage, the chapter aims to equip readers with both theoretical understanding and practical insights into the rapidly evolving field of diffusion-based generative modeling.

1 Introduction

Diffusion models, introduced in their modern form in 2020 [10], have rapidly become the leading family of generative models. By 2025, they dominate applications in image, video, and audio generation, powering systems such as DALL·E (2021) [22], Stable Diffusion (2022) [23], and Imagen (2022) [24].

Diffusion models are inspired by the principles of non-equilibrium thermodynamics. They employ a Markov chain (Markovian process) to gradually introduce noise into data and then learn to reverse this process to reconstruct the desired data from the noise.

As shown in Fig. 1, a diffusion model consists of two processes in opposite directions. In the fixed forward diffusion process, the model starts with the data and progressively adds noise until it becomes complete noise. Then, in the generative reverse denoising process, it reverses this forward process and generates data from the noise by gradually denoising it.

As depicted in Fig. 2, the forward and reverse processes can be modeled by the encoder and decoder of an autoencoder, respectively. The forward noising process resembles a fixed encoder, while the learned denoising network acts as a decoder that reconstructs data from noise.[1] Since the forward

[1] While this analogy is useful, diffusion models are more accurately characterized as iterative denoising or score-matching processes.

B. Ghojogh and A. Ghodsi, *Elements of Deep Learning*,
https://doi.org/10.1007/978-3-032-10738-1_15

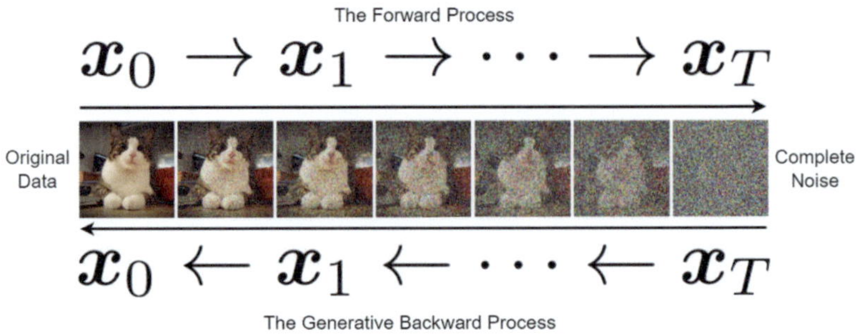

Fig. 1 The forward and backward processes of the diffusion model. *Image credit* Arash Vahdat

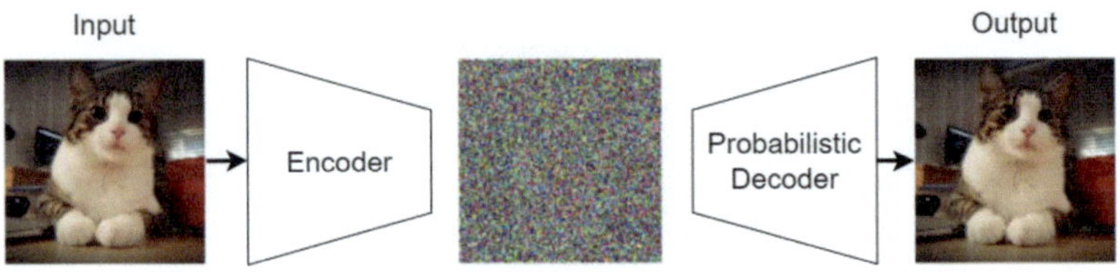

Fig. 2 The diffusion model as an autoencoder

process only needs to add noise to the data, it does not involve any learning and is thus fixed. Therefore, only the decoder needs to be learned in diffusion models. This is one of the differences between diffusion models and Variational Autoencoders (VAEs) [13] (see Chap. 12), in which both the encoder and decoder are learned.

This chapter introduces diffusion models. The Denoising Diffusion Probabilistic Model (DDPM) is explained in Sect. 2. Section 3 covers modifications to the original DDPM. The conditional diffusion model is introduced in Sect. 4. Finally, Sect. 5 discusses the continuous noise schedule using stochastic differential equations.

2 Denoising Diffusion Probabilistic Model (DDPM)

The first and original diffusion model is the Denoising Diffusion Probabilistic Model (DDPM), proposed in [10]. This section explains DDPM in detail.

The process of adding noise to data can be viewed as a sequence or chain $\{\boldsymbol{x}_0, \boldsymbol{x}_1, \ldots, \boldsymbol{x}_T\}$, where $\boldsymbol{x}_0 \in \mathbb{R}^d$ and $\boldsymbol{x}_T \in \mathbb{R}^d$ represent the original data and the data completely corrupted with noise, respectively. Figure 3 depicts such a chain. In this notation, $\boldsymbol{x}_{t+1}$ is $\boldsymbol{x}_t$ plus some noise. As the noise is random,

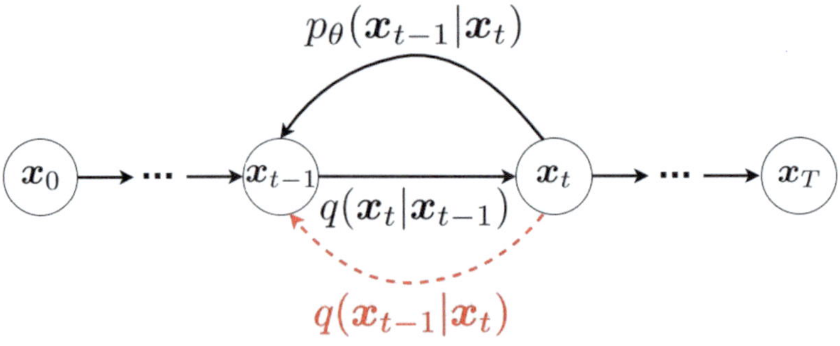

Fig. 3 The conditional distributions in the forward and backward processes of the diffusion model

a conditional distribution $q(\boldsymbol{x}_{t+1}|\boldsymbol{x}_t)$ can model the probability of obtaining $\boldsymbol{x}_{t+1}$ from $\boldsymbol{x}_t$. The goal of diffusion models is to learn the conditional distribution $q(\boldsymbol{x}_t|\boldsymbol{x}_{t+1})$, which models how to generate data from noise. This is because when $q(\boldsymbol{x}_t|\boldsymbol{x}_{t+1})$ is learned, t can be swept from T to 0 to generate a non-noisy data instance from complete noise.

The distribution $q(\boldsymbol{x}_t|\boldsymbol{x}_{t+1})$ is intractable and must be approximated using a neural network. Let $p_\theta(\boldsymbol{x}_t|\boldsymbol{x}_{t+1})$ denote the neural network with weights θ that approximates this conditional distribution. It is noteworthy that $q(\boldsymbol{x}_t|\boldsymbol{x}_{t+1}, \boldsymbol{x}_0)$ is obviously tractable because it conditions on the original data $\boldsymbol{x}_0$. However, modeling $q(\boldsymbol{x}_t|\boldsymbol{x}_{t+1}, \boldsymbol{x}_0)$ is not useful because the goal of diffusion models is to find $\boldsymbol{x}_0$, and this goal cannot be used in the model itself. In fact, this represents a chicken-and-egg problem. Although this distribution is not practical, it will appear later in the derivations of this chapter.

2.1 The Forward Process

The forward process gradually adds noise to the data:

$$\boldsymbol{x}_0 \to \boldsymbol{x}_1 \to \cdots \to \boldsymbol{x}_T. \tag{1}$$

It is defined by the following Gaussian conditional distribution:

$$q(\boldsymbol{x}_t|\boldsymbol{x}_{t-1}) = \mathcal{N}(\boldsymbol{x}_t; \sqrt{1-\beta_t}\boldsymbol{x}_{t-1}, \beta_t \boldsymbol{I}), \tag{2}$$

where $\sqrt{1-\beta_t}\boldsymbol{x}_{t-1}$ is the mean and $\beta_t \boldsymbol{I}$ is the covariance. Here, $\boldsymbol{I}$ denotes the identity matrix, and $\beta_t \in (0, 1)$ is the noise level parameter in a predetermined noise schedule:

$$\beta_1 < \beta_2 < \cdots < \beta_T. \tag{3}$$

The β_t parameter increases gradually in the forward process, causing the mean $\sqrt{1-\beta_t}\boldsymbol{x}_{t-1}$ to deviate further from the previously noisy data $\boldsymbol{x}_{t-1}$, while the variance $\beta_t \boldsymbol{I}$ increases. The closer t gets to T, the noisier the data $\boldsymbol{x}_t$ becomes.

The forward process is considered a Markov chain with the first Markovian property. In a Markov chain, each state depends only on its previous state and not on the entire previous sequence [7]:

$$q(\boldsymbol{x}_t|\boldsymbol{x}_{t-1}, \boldsymbol{x}_{t-2}, \ldots, \boldsymbol{x}_0) = q(\boldsymbol{x}_t|\boldsymbol{x}_{t-1}). \tag{4}$$

Therefore, the joint distribution of the forward process is

$$q(\boldsymbol{x}_{1:T}|\boldsymbol{x}_0) = \prod_{t=1}^{T} q(\boldsymbol{x}_t|\boldsymbol{x}_{t-1}), \tag{5}$$

where $\boldsymbol{x}_{1:T}$ represents all latent states from $\boldsymbol{x}_1$ to $\boldsymbol{x}_T$, and $\boldsymbol{x}_0$ is the initial state, i.e., the original data. The probabilistic graphical model [15] of the diffusion model is illustrated in Fig. 4, where the states $\boldsymbol{x}_{1:T}$ are the latent (or hidden) variables for the observed data $\boldsymbol{x}_0$. This figure also compares the diffusion model with the Variational Autoencoder (VAE), where a variable z is the latent variable for the observed data $\boldsymbol{x}$.

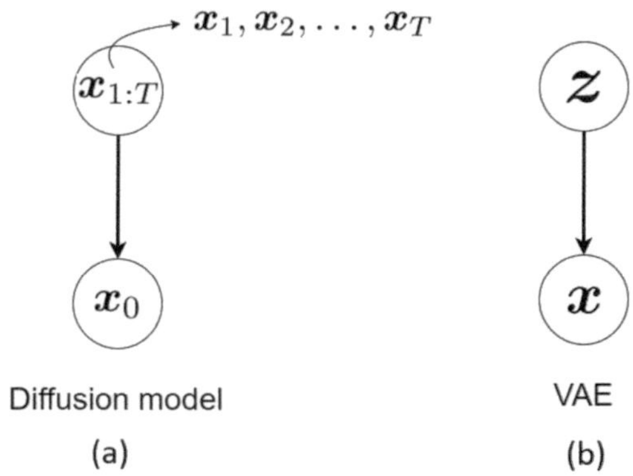

Fig. 4 The probabilistic graphical models of **a** the diffusion model and **b** Variational Autoencoder (VAE)

2.2 The Generative Backward Process

The diffusion process, or the generative backward process, is the reverse process that gradually generates data from noise:

$$\boldsymbol{x}_T \rightarrow \boldsymbol{x}_{T-1} \rightarrow \cdots \rightarrow \boldsymbol{x}_0. \tag{6}$$

Since the forward process uses Gaussian distributions, as shown in Eq. (2), it makes sense to use Gaussian distributions for the backward process as well. A generator machine learning model, with parameters or weights θ, is used to model the backward process:

$$p_\theta(\boldsymbol{x}_{t-1}|\boldsymbol{x}_t) = \mathcal{N}\big(\boldsymbol{x}_{t-1}; \boldsymbol{\mu}_\theta(\boldsymbol{x}_t; t), \boldsymbol{\Sigma}_\theta(\boldsymbol{x}_t; t)\big), \tag{7}$$

where $\boldsymbol{\mu}_\theta(\boldsymbol{x}_t; t) \in \mathbb{R}^d$ and $\boldsymbol{\Sigma}_\theta(\boldsymbol{x}_t; t) \in \mathbb{R}^{d\times d}$ are the mean and covariance of the distribution at iteration t, respectively. For simplicity, the covariance is usually set to be isotropic and diagonal:

$$\boldsymbol{\Sigma}_\theta(\boldsymbol{x}_t; t) = \sigma_t^2 \boldsymbol{I}. \tag{8}$$

The goal of the diffusion model is to learn the parameters of this Gaussian distribution, namely, the mean and variance.

The generative process is also considered a Markov chain with the first Markovian property [7]:

$$q(\boldsymbol{x}_{t-1}|\boldsymbol{x}_t, \boldsymbol{x}_{t+1}, \ldots, \boldsymbol{x}_T) = q(\boldsymbol{x}_{t-1}|\boldsymbol{x}_t). \tag{9}$$

Therefore, the joint distribution of the generative process is

$$p_\theta(\boldsymbol{x}_{0:T}) = p(\boldsymbol{x}_T) \prod_{t=1}^{T} p_\theta(\boldsymbol{x}_{t-1}|\boldsymbol{x}_t), \tag{10}$$

where $\boldsymbol{x}_{0:T}$ represents all states from $\boldsymbol{x}_T$ to $\boldsymbol{x}_0$, and $\boldsymbol{x}_T$ is the state corresponding to complete noise. Usually, $p(\boldsymbol{x}_T)$ is defined as standard Gaussian noise:

$$p(\boldsymbol{x}_T) = \mathcal{N}(\boldsymbol{0}, \boldsymbol{I}). \tag{11}$$

2.3 Variational Lower Bound of the Likelihood

As discussed earlier, Variational Autoencoders (VAEs) consider a latent variable z as the latent factor for the observed data $\boldsymbol{x}$. Similarly, diffusion models consider the latent variables $\{\boldsymbol{x}_1, \boldsymbol{x}_2, \dots, \boldsymbol{x}_T\}$ for the observed data $\boldsymbol{x}_0$ (see Fig. 4). The likelihood of data in VAEs and diffusion models is given by

$$p_\theta(\boldsymbol{x}) = \int p_\theta(\boldsymbol{x}, z)\, dz, \tag{12}$$

$$p_\theta(\boldsymbol{x}_0) = \int p_\theta(\boldsymbol{x}_0, \boldsymbol{x}_{1:T})\, d\boldsymbol{x}_{1:T}, \tag{13}$$

respectively. Recall that in variational inference, maximizing the likelihood is not tractable, so a lower bound of the likelihood, called the Evidence Lower Bound (ELBO), is maximized instead (see Chap. 12):

$$\mathbb{E}_{q(z|\boldsymbol{x})}\big[\log p_\theta(\boldsymbol{x}|z)\big] - \mathrm{KL}\big(q(z|\boldsymbol{x}) \| p_\theta(z)\big) \leq p_\theta(\boldsymbol{x}), \tag{14}$$

where $\mathbb{E}[.]$ denotes expectation and $\mathrm{KL}(.\|.)$ denotes the Kullback–Leibler (KL) divergence [16]. Refer to Chap. 12 for the derivation of the ELBO in Eq. (14).

Similarly, diffusion models maximize the variational lower bound of the likelihood of the data:

$$\mathcal{L} := \mathbb{E}_{q(\boldsymbol{x}_{1:T}|\boldsymbol{x})}\big[\log p_\theta(\boldsymbol{x}_0|\boldsymbol{x}_{1:T})\big] - \mathrm{KL}\big(q(\boldsymbol{x}_{1:T}|\boldsymbol{x}_0) \| p_\theta(\boldsymbol{x}_{1:T})\big) \leq p_\theta(\boldsymbol{x}_0). \tag{15}$$

The variational lower bound of the likelihood in diffusion models can be simplified as follows:

$$\begin{aligned}
\mathcal{L} &= \mathbb{E}_{q(\boldsymbol{x}_{1:T}|\boldsymbol{x}_0)}\big[\log p_\theta(\boldsymbol{x}_0|\boldsymbol{x}_{1:T})\big] - \mathrm{KL}\big(q(\boldsymbol{x}_{1:T}|\boldsymbol{x}_0) \| p_\theta(\boldsymbol{x}_{1:T})\big) \\
&\overset{(a)}{=} \mathbb{E}_{q(\boldsymbol{x}_{1:T}|\boldsymbol{x}_0)}\big[\log p_\theta(\boldsymbol{x}_0|\boldsymbol{x}_{1:T})\big] - \int q(\boldsymbol{x}_{1:T}|\boldsymbol{x}_0) \log \frac{q(\boldsymbol{x}_{1:T}|\boldsymbol{x}_0)}{p_\theta(\boldsymbol{x}_{1:T})} d\boldsymbol{x}_{1:T} \\
&\overset{(b)}{=} \mathbb{E}_{q(\boldsymbol{x}_{1:T}|\boldsymbol{x}_0)}\big[\log p_\theta(\boldsymbol{x}_0|\boldsymbol{x}_{1:T})\big] - \mathbb{E}_{q(\boldsymbol{x}_{1:T}|\boldsymbol{x}_0)}\Big[\log \frac{q(\boldsymbol{x}_{1:T}|\boldsymbol{x}_0)}{p_\theta(\boldsymbol{x}_{1:T})}\Big] \\
&\overset{(c)}{=} \mathbb{E}_{q(\boldsymbol{x}_{1:T}|\boldsymbol{x}_0)}\Big[\log p_\theta(\boldsymbol{x}_0|\boldsymbol{x}_{1:T}) - \log \frac{q(\boldsymbol{x}_{1:T}|\boldsymbol{x}_0)}{p_\theta(\boldsymbol{x}_{1:T})}\Big] \\
&\overset{(d)}{=} \mathbb{E}_{q(\boldsymbol{x}_{1:T}|\boldsymbol{x}_0)}\Big[\log p_\theta(\boldsymbol{x}_0|\boldsymbol{x}_{1:T}) + \log \frac{p_\theta(\boldsymbol{x}_{1:T})}{q(\boldsymbol{x}_{1:T}|\boldsymbol{x}_0)}\Big] \\
&\overset{(e)}{=} \mathbb{E}_{q(\boldsymbol{x}_{1:T}|\boldsymbol{x}_0)}\Big[\log \frac{p_\theta(\boldsymbol{x}_0|\boldsymbol{x}_{1:T})\, p_\theta(\boldsymbol{x}_{1:T})}{q(\boldsymbol{x}_{1:T}|\boldsymbol{x}_0)}\Big] \\
&\overset{(f)}{=} \mathbb{E}_{q(\boldsymbol{x}_{1:T}|\boldsymbol{x}_0)}\Big[\log \frac{p_\theta(\boldsymbol{x}_{0:T})}{q(\boldsymbol{x}_{1:T}|\boldsymbol{x}_0)}\Big],
\end{aligned} \tag{16}$$

where (a) follows from the definition of the KL divergence, (b) is due to the definition of the expectation in the second term of the expression, (c) holds because expectation is a linear operator, (d) and (e) follow from the properties of logarithms, and (f) is because $p_\theta(\boldsymbol{x}_0|\boldsymbol{x}_{1:T})\, p_\theta(\boldsymbol{x}_{1:T}) = p_\theta(\boldsymbol{x}_0, \boldsymbol{x}_{1:T}) = p_\theta(\boldsymbol{x}_{0:T})$.

Equation (16) can be simplified further:

$$
\begin{aligned}
\mathcal{L} &= \mathbb{E}_{q(\boldsymbol{x}_{1:T}|\boldsymbol{x}_0)}\Big[\log \frac{p_\theta(\boldsymbol{x}_{0:T})}{q(\boldsymbol{x}_{1:T}|\boldsymbol{x}_0)}\Big] \\
&\overset{(10)}{=} \mathbb{E}_{q(\boldsymbol{x}_{1:T}|\boldsymbol{x}_0)}\Bigg[\log \frac{p(\boldsymbol{x}_T)\prod_{t=1}^{T} p_\theta(\boldsymbol{x}_{t-1}|\boldsymbol{x}_t)}{q(\boldsymbol{x}_{1:T}|\boldsymbol{x}_0)}\Bigg] \\
&\overset{(a)}{=} \mathbb{E}_{q(\boldsymbol{x}_{1:T}|\boldsymbol{x}_0)}\big[\log p(\boldsymbol{x}_T)\big] + \sum_{t=1}^{T} \mathbb{E}_{q(\boldsymbol{x}_{1:T}|\boldsymbol{x}_0)}\big[\log p_\theta(\boldsymbol{x}_{t-1}|\boldsymbol{x}_t)\big] \\
&\quad - \mathbb{E}_{q(\boldsymbol{x}_{1:T}|\boldsymbol{x}_0)}\big[\log q(\boldsymbol{x}_{1:T}|\boldsymbol{x}_0)\big] \\
&\overset{(5)}{=} \mathbb{E}_{q(\boldsymbol{x}_{1:T}|\boldsymbol{x}_0)}\big[\log p(\boldsymbol{x}_T)\big] + \sum_{t=1}^{T} \mathbb{E}_{q(\boldsymbol{x}_{1:T}|\boldsymbol{x}_0)}\Big[\log p_\theta(\boldsymbol{x}_{t-1}|\boldsymbol{x}_t)\Big] \\
&\quad - \mathbb{E}_{q(\boldsymbol{x}_{1:T}|\boldsymbol{x}_0)}\Big[\log \prod_{t=1}^{T} q(\boldsymbol{x}_t|\boldsymbol{x}_{t-1})\Big] \\
&\overset{(b)}{=} \mathbb{E}_{q(\boldsymbol{x}_{1:T}|\boldsymbol{x}_0)}\big[\log p(\boldsymbol{x}_T)\big] + \sum_{t=1}^{T} \mathbb{E}_{q(\boldsymbol{x}_{1:T}|\boldsymbol{x}_0)}\big[\log p_\theta(\boldsymbol{x}_{t-1}|\boldsymbol{x}_t)\big] \\
&\quad - \sum_{t=1}^{T} \mathbb{E}_{q(\boldsymbol{x}_{1:T}|\boldsymbol{x}_0)}\big[\log q(\boldsymbol{x}_t|\boldsymbol{x}_{t-1})\big] \\
&\overset{(c)}{=} \mathbb{E}_{q(\boldsymbol{x}_{1:T}|\boldsymbol{x}_0)}\big[\log p(\boldsymbol{x}_T)\big] + \sum_{t=1}^{T} \mathbb{E}_{q(\boldsymbol{x}_{1:T}|\boldsymbol{x}_0)}\Big[\log \frac{p_\theta(\boldsymbol{x}_{t-1}|\boldsymbol{x}_t)}{q(\boldsymbol{x}_t|\boldsymbol{x}_{t-1})}\Big] \\
&\overset{(d)}{=} \mathbb{E}_{q(\boldsymbol{x}_{1:T}|\boldsymbol{x}_0)}\Big[\log p(\boldsymbol{x}_T) + \sum_{t=1}^{T} \log \frac{p_\theta(\boldsymbol{x}_{t-1}|\boldsymbol{x}_t)}{q(\boldsymbol{x}_t|\boldsymbol{x}_{t-1})}\Big],
\end{aligned}
\tag{17}
$$

where (a) follows from the properties of logarithms and the linearity of expectation, (b) is because the logarithm of a multiplication becomes the summation of logarithms, (c) is due to the properties of logarithms, and (d) holds because expectation is a linear operator.

According to the definition of the KL divergence, Eq. (17), i.e., the variational lower bound, can be stated as follows:

$$\mathcal{L} = \mathbb{E}_{q(\boldsymbol{x}_{1:T}|\boldsymbol{x}_0)}\big[\log p(\boldsymbol{x}_T)\big] - \mathrm{KL}\big(q(\boldsymbol{x}_t|\boldsymbol{x}_{t-1}) \| p_\theta(\boldsymbol{x}_{t-1}|\boldsymbol{x}_t)\big).$$

This is the variational lower bound that needs to be maximized; therefore, the KL divergence is minimized. Here, the KL divergence is between the forward process $q(\boldsymbol{x}_t|\boldsymbol{x}_{t-1})$ and the generative process $p_\theta(\boldsymbol{x}_{t-1}|\boldsymbol{x}_t)$. Minimizing this KL divergence is intuitive because the distribution of the generative process should mimic the distribution of the forward process.

Since the sequence of states in the diffusion process forms a Markov chain, the following holds:

$$q(\boldsymbol{x}_t|\boldsymbol{x}_{t-1}) = q(\boldsymbol{x}_t|\boldsymbol{x}_{t-1}, \boldsymbol{x}_0). \tag{18}$$

According to Bayes' rule, we have

$$q(\boldsymbol{x}_t|\boldsymbol{x}_{t-1}, \boldsymbol{x}_0) = \frac{q(\boldsymbol{x}_{t-1}|\boldsymbol{x}_t, \boldsymbol{x}_0)\, q(\boldsymbol{x}_t|\boldsymbol{x}_0)}{q(\boldsymbol{x}_{t-1}|\boldsymbol{x}_0)}, \tag{19}$$

where $q(\boldsymbol{x}_{t-1}|\boldsymbol{x}_t, \boldsymbol{x}_0)$ is tractable, in contrast to $q(\boldsymbol{x}_{t-1}|\boldsymbol{x}_t)$, because it conditions on the original data. In fact, Eqs. (18) and (19) are used to make $q(\boldsymbol{x}_{t-1}|\boldsymbol{x}_t)$ tractable. The tractable distribution $q(\boldsymbol{x}_{t-1}|\boldsymbol{x}_t, \boldsymbol{x}_0)$ is called the *ground-truth denoising distribution* because it has access to the original data $\boldsymbol{x}_0$.

Plugging Eqs. (18) and (19) in Eq. (17) results in:

$$
\begin{aligned}
\mathcal{L} &= \mathbb{E}_{q(\boldsymbol{x}_{1:T}|\boldsymbol{x}_0)}\Big[\log p(\boldsymbol{x}_T) + \sum_{t=1}^{T} \log \frac{p_\theta(\boldsymbol{x}_{t-1}|\boldsymbol{x}_t)\, q(\boldsymbol{x}_{t-1}|\boldsymbol{x}_0)}{q(\boldsymbol{x}_{t-1}|\boldsymbol{x}_t, \boldsymbol{x}_0)\, q(\boldsymbol{x}_t|\boldsymbol{x}_0)}\Big] \\
&\overset{(a)}{=} \mathbb{E}_{q(\boldsymbol{x}_{1:T}|\boldsymbol{x}_0)}\Big[\log p(\boldsymbol{x}_T) + \sum_{t=2}^{T} \log \frac{p_\theta(\boldsymbol{x}_{t-1}|\boldsymbol{x}_t)}{q(\boldsymbol{x}_{t-1}|\boldsymbol{x}_t, \boldsymbol{x}_0)} \\
&\quad + \sum_{t=2}^{T} \log \frac{q(\boldsymbol{x}_{t-1}|\boldsymbol{x}_0)}{q(\boldsymbol{x}_t|\boldsymbol{x}_0)} + \log \frac{p_\theta(\boldsymbol{x}_0|\boldsymbol{x}_1)\, q(\boldsymbol{x}_0|\boldsymbol{x}_0)}{q(\boldsymbol{x}_0|\boldsymbol{x}_1, \boldsymbol{x}_0)\, q(\boldsymbol{x}_1|\boldsymbol{x}_0)}\Big] \\
&\overset{(b)}{=} \mathbb{E}_{q(\boldsymbol{x}_{1:T}|\boldsymbol{x}_0)}\Big[\log p(\boldsymbol{x}_T) + \sum_{t=2}^{T} \log \frac{p_\theta(\boldsymbol{x}_{t-1}|\boldsymbol{x}_t)}{q(\boldsymbol{x}_{t-1}|\boldsymbol{x}_t, \boldsymbol{x}_0)} \\
&\quad + \sum_{t=2}^{T} \log \frac{q(\boldsymbol{x}_{t-1}|\boldsymbol{x}_0)}{q(\boldsymbol{x}_t|\boldsymbol{x}_0)} + \log \frac{p_\theta(\boldsymbol{x}_0|\boldsymbol{x}_1)}{q(\boldsymbol{x}_1|\boldsymbol{x}_0)}\Big],
\end{aligned} \tag{20}
$$

where (a) breaks down the logarithm and separates out the $t = 1$ term from the summation, and (b) holds because $q(\boldsymbol{x}_0|\boldsymbol{x}_0) = q(\boldsymbol{x}_0|\boldsymbol{x}_1, \boldsymbol{x}_0) = 1$. The third term in the expectation is a telescoping summation and can be simplified as follows:

$$
\begin{aligned}
\sum_{t=2}^{T} \log \frac{q(\boldsymbol{x}_{t-1}|\boldsymbol{x}_0)}{q(\boldsymbol{x}_t|\boldsymbol{x}_0)} &= \log q(\boldsymbol{x}_{T-1}|\boldsymbol{x}_0) + \dots + \log q(\boldsymbol{x}_2|\boldsymbol{x}_0) + \log q(\boldsymbol{x}_1|\boldsymbol{x}_0) \\
&\quad - \log q(\boldsymbol{x}_T|\boldsymbol{x}_0) - \log q(\boldsymbol{x}_{T-1}|\boldsymbol{x}_0) - \dots - \log q(\boldsymbol{x}_2|\boldsymbol{x}_0) \\
&= -\log q(\boldsymbol{x}_T|\boldsymbol{x}_0) + \log q(\boldsymbol{x}_1|\boldsymbol{x}_0) = \log \frac{q(\boldsymbol{x}_1|\boldsymbol{x}_0)}{q(\boldsymbol{x}_T|\boldsymbol{x}_0)}.
\end{aligned}
$$

Therefore, the variational lower bound in Eq. (20) becomes

$$
\begin{aligned}
\mathcal{L} &= \mathbb{E}_{q(\boldsymbol{x}_{1:T}|\boldsymbol{x}_0)}\Big[\log p(\boldsymbol{x}_T) + \sum_{t=2}^{T} \log \frac{p_\theta(\boldsymbol{x}_{t-1}|\boldsymbol{x}_t)}{q(\boldsymbol{x}_{t-1}|\boldsymbol{x}_t, \boldsymbol{x}_0)} \\
&\quad + \log \frac{q(\boldsymbol{x}_1|\boldsymbol{x}_0)}{q(\boldsymbol{x}_T|\boldsymbol{x}_0)} + \log \frac{p_\theta(\boldsymbol{x}_0|\boldsymbol{x}_1)}{q(\boldsymbol{x}_1|\boldsymbol{x}_0)}\Big] \\
&\overset{(a)}{=} \mathbb{E}_{q(\boldsymbol{x}_{1:T}|\boldsymbol{x}_0)}\Big[\log \frac{p(\boldsymbol{x}_T)}{q(\boldsymbol{x}_T|\boldsymbol{x}_0)} + \sum_{t=2}^{T} \log \frac{p_\theta(\boldsymbol{x}_{t-1}|\boldsymbol{x}_t)}{q(\boldsymbol{x}_{t-1}|\boldsymbol{x}_t, \boldsymbol{x}_0)} + \log p_\theta(\boldsymbol{x}_0|\boldsymbol{x}_1)\Big] \\
&\overset{(b)}{=} \mathbb{E}_{q(\boldsymbol{x}_{1:T}|\boldsymbol{x}_0)}\Big[\log \frac{p(\boldsymbol{x}_T)}{q(\boldsymbol{x}_T|\boldsymbol{x}_0)}\Big] + \sum_{t=2}^{T} \mathbb{E}_{q(\boldsymbol{x}_{1:T}|\boldsymbol{x}_0)}\Big[\log \frac{p_\theta(\boldsymbol{x}_{t-1}|\boldsymbol{x}_t)}{q(\boldsymbol{x}_{t-1}|\boldsymbol{x}_t, \boldsymbol{x}_0)}\Big] \\
&\quad + \mathbb{E}_{q(\boldsymbol{x}_{1:T}|\boldsymbol{x}_0)}\big[\log p_\theta(\boldsymbol{x}_0|\boldsymbol{x}_1)\big]
\end{aligned}
$$

$$\overset{(c)}{=} -\mathrm{KL}\big(q(\boldsymbol{x}_T|\boldsymbol{x}_0)\|p(\boldsymbol{x}_T)\big) - \sum_{t=2}^{T} \mathrm{KL}\big(q(\boldsymbol{x}_{t-1}|\boldsymbol{x}_t, \boldsymbol{x}_0)\|p_\theta(\boldsymbol{x}_{t-1}|\boldsymbol{x}_t)\big)$$
$$+ \mathbb{E}_{q(\boldsymbol{x}_{1:T}|\boldsymbol{x}_0)}\big[\log p_\theta(\boldsymbol{x}_0|\boldsymbol{x}_1)\big],$$

where (a) follows from the properties of logarithms, (b) holds because expectation is a linear operator, and (c) follows from the definition of the KL divergence.

The variational lower bound of the likelihood should be maximized with respect to the parameters θ. The first KL divergence is constant with respect to θ, so it can be dropped. Maximizing the lower bound is equivalent to minimizing its negative:

$$\underset{\theta}{\text{minimize}} \quad \sum_{t=2}^{T} \mathrm{KL}\big(q(\boldsymbol{x}_{t-1}|\boldsymbol{x}_t, \boldsymbol{x}_0)\|p_\theta(\boldsymbol{x}_{t-1}|\boldsymbol{x}_t)\big) - \mathbb{E}_{q(\boldsymbol{x}_{1:T}|\boldsymbol{x}_0)}\big[\log p_\theta(\boldsymbol{x}_0|\boldsymbol{x}_1)\big]. \tag{21}$$

Therefore, this can serve as the loss function for a neural network with weights θ. This loss function is intuitive because it trains the backward distribution $p_\theta(\boldsymbol{x}_{t-1}|\boldsymbol{x}_t)$ to become similar to the ground-truth denoising distribution $q(\boldsymbol{x}_{t-1}|\boldsymbol{x}_t, \boldsymbol{x}_0)$ by minimizing their KL divergence. That is why the focus of the loss function is on the KL divergence, and the second term in Eq. (21) can be ignored. Remember this loss function, as we will revisit it later in the chapter.

2.4 Parameters of the Distributions

2.4.1 Sampling in Forward Process at Arbitrary Iterations

Recall the Gaussian distributions of the forward process in Eq. (2). That conditional distribution was for sampling $\boldsymbol{x}_t$ given $\boldsymbol{x}_{t-1}$. However, it is also possible to sample $\boldsymbol{x}_t$ at any arbitrary iteration t in closed form. Consider the following definitions:

$$\alpha_t := 1 - \beta_t, \tag{22}$$

$$\bar{\alpha}_t = \prod_{i=1}^{t} \alpha_i. \tag{23}$$

Therefore, Eq. (2) can be restated as follows:

$$q(\boldsymbol{x}_t|\boldsymbol{x}_{t-1}) = \mathcal{N}\big(\boldsymbol{x}_t; \sqrt{\alpha_t}\boldsymbol{x}_{t-1}, (1-\alpha_t)\boldsymbol{I}\big). \tag{24}$$

Here, the reparameterization technique [13] can be used. In this technique, to sample from a non-standard Gaussian distribution $z \sim \mathcal{N}(\boldsymbol{\mu}, \sigma^2\boldsymbol{I})$, one can sample from a standard distribution instead, i.e., $\boldsymbol{\epsilon} \sim \mathcal{N}(\boldsymbol{0}, \boldsymbol{I})$, and obtain the sample z as $z = \boldsymbol{\mu} + \sigma\boldsymbol{\epsilon}$. Here, using the reparameterization technique, $\boldsymbol{\epsilon} \sim \mathcal{N}(\boldsymbol{0}, \boldsymbol{I})$ is sampled, and then $\{\boldsymbol{x}_t, \boldsymbol{x}_{t-1}, \ldots, \boldsymbol{x}_1\}$ are each obtained as follows:

$$\boldsymbol{x}_t = \sqrt{\alpha_t}\boldsymbol{x}_{t-1} + \sqrt{1-\alpha_t}\boldsymbol{\epsilon}, \tag{25}$$

$$\boldsymbol{x}_{t-1} = \sqrt{\alpha_{t-1}}\boldsymbol{x}_{t-2} + \sqrt{1-\alpha_{t-1}}\boldsymbol{\epsilon}, \tag{26}$$

$$\boldsymbol{x}_{t-2} = \sqrt{\alpha_{t-2}}\boldsymbol{x}_{t-3} + \sqrt{1-\alpha_{t-2}}\boldsymbol{\epsilon}, \tag{27}$$

$$\vdots$$

Plugging $\boldsymbol{x}_{t-1}$ from Eq. (26) into $\boldsymbol{x}_t$ in Eq. (25) gives

$$\begin{aligned} \boldsymbol{x}_t &= \sqrt{\alpha_t}\big(\sqrt{\alpha_{t-1}}\boldsymbol{x}_{t-2} + \sqrt{1-\alpha_{t-1}}\boldsymbol{\epsilon}\big) + \sqrt{1-\alpha_t}\boldsymbol{\epsilon} \\ &= \sqrt{\alpha_t\alpha_{t-1}}\boldsymbol{x}_{t-2} + \sqrt{\alpha_t(1-\alpha_{t-1})}\boldsymbol{\epsilon} + \sqrt{1-\alpha_t}\boldsymbol{\epsilon}. \end{aligned} \tag{28}$$

Note that the $\boldsymbol{\epsilon}$ in the second and third terms are two different samples from the standard Gaussian distribution. Therefore, the second and third terms are two independent random Gaussian distributions:

$$\begin{aligned} &\sqrt{\alpha_t(1-\alpha_{t-1})}\boldsymbol{\epsilon} \sim \mathcal{N}\big(\mathbf{0}, \underbrace{\alpha_t(1-\alpha_{t-1})}_{=\alpha_t-\alpha_t\alpha_{t-1}}\big), \\ &\sqrt{1-\alpha_t}\boldsymbol{\epsilon} \sim \mathcal{N}(\mathbf{0}, 1-\alpha_t), \end{aligned}$$

so their summation becomes a Gaussian distribution whose variance is the sum of the two variances:

$$\sqrt{\alpha_t(1-\alpha_{t-1})}\boldsymbol{\epsilon} + \sqrt{1-\alpha_t}\boldsymbol{\epsilon} \sim \mathcal{N}(\mathbf{0}, 1-\alpha_t\alpha_{t-1}).$$

As a result, Eq. (28) can be written as follows:

$$\boldsymbol{x}_t = \sqrt{\alpha_t\alpha_{t-1}}\boldsymbol{x}_{t-2} + \sqrt{1-\alpha_t\alpha_{t-1}}\,\boldsymbol{\epsilon}.$$

Plugging $\boldsymbol{x}_{t-2}$ from Eq. (27) into this expression gives

$$\begin{aligned} \boldsymbol{x}_t &= \sqrt{\alpha_t\alpha_{t-1}}\Big(\sqrt{\alpha_{t-2}}\boldsymbol{x}_{t-3} + \sqrt{1-\alpha_{t-2}}\boldsymbol{\epsilon}\Big) + \sqrt{1-\alpha_t\alpha_{t-1}}\boldsymbol{\epsilon} \\ &= \sqrt{\alpha_t\alpha_{t-1}\alpha_{t-2}}\boldsymbol{x}_{t-3} + \sqrt{\alpha_t\alpha_{t-1} - \alpha_t\alpha_{t-1}\alpha_{t-2}}\boldsymbol{\epsilon} + \sqrt{1-\alpha_t\alpha_{t-1}}\boldsymbol{\epsilon}. \end{aligned} \tag{29}$$

Again, the second and third terms are two independent random Gaussian distributions:

$$\begin{aligned} &\sqrt{\alpha_t\alpha_{t-1} - \alpha_t\alpha_{t-1}\alpha_{t-2}}\boldsymbol{\epsilon} \sim \mathcal{N}(\mathbf{0}, \alpha_t\alpha_{t-1} - \alpha_t\alpha_{t-1}\alpha_{t-2}), \\ &\sqrt{1-\alpha_t\alpha_{t-1}}\boldsymbol{\epsilon} \sim \mathcal{N}(\mathbf{0}, 1-\alpha_t\alpha_{t-1}), \end{aligned}$$

so their summation becomes a Gaussian distribution whose variance is the sum of the two variances:

$$\sqrt{\alpha_t\alpha_{t-1} - \alpha_t\alpha_{t-1}\alpha_{t-2}}\boldsymbol{\epsilon} + \sqrt{1-\alpha_t\alpha_{t-1}}\boldsymbol{\epsilon} \sim \mathcal{N}\big(\mathbf{0}, 1-\alpha_t\alpha_{t-1}\alpha_{t-2}\big).$$

As a result, Eq. (29) can be written as follows:

$$\boldsymbol{x}_t = \sqrt{\alpha_t\alpha_{t-1}\alpha_{t-2}}\,\boldsymbol{x}_{t-3} + \sqrt{1-\alpha_t\alpha_{t-1}\alpha_{t-2}}\,\boldsymbol{\epsilon}.$$

By induction, continuing this recursion until $\boldsymbol{x}_0$ gives $\boldsymbol{x}_t$ in terms of the original data $\boldsymbol{x}_0$ and the noise $\boldsymbol{\epsilon}$:

$$\begin{aligned} \boldsymbol{x}_t &= \sqrt{\prod_{i=1}^{t}\alpha_i}\,\boldsymbol{x}_0 + \sqrt{1-\prod_{i=1}^{t}\alpha_i}\,\boldsymbol{\epsilon} \\ &\overset{(23)}{=} \sqrt{\bar{\alpha}_t}\boldsymbol{x}_0 + \sqrt{1-\bar{\alpha}_t}\boldsymbol{\epsilon}, \end{aligned} \tag{30}$$

which expresses $\boldsymbol{x}_t$ in terms of the original data $\boldsymbol{x}_0$ and noise $\boldsymbol{\epsilon}$. Since $\boldsymbol{x}_0$ is not random and $\boldsymbol{\epsilon}$ has a standard Gaussian distribution, the distribution of $\boldsymbol{x}_t$ in this equation has mean $\sqrt{\bar{\alpha}_t}\boldsymbol{x}_0$ and covariance $(1-\bar{\alpha}_t)\boldsymbol{I}$. Therefore:

$$q(\boldsymbol{x}_t|\boldsymbol{x}_0) = \mathcal{N}\big(\boldsymbol{x}_t; \sqrt{\bar{\alpha}_t}\boldsymbol{x}_0, (1-\bar{\alpha}_t)\boldsymbol{I}\big). \tag{31}$$

Using this conditional distribution, the noisy data $\boldsymbol{x}_t$ at any iteration t can be sampled based on the original data $\boldsymbol{x}_0$.

2.4.2 The Ground-Truth Denoising Distribution

A neural network is required to learn the parameters, i.e., the mean and covariance, of the backward process in Eq. (7). Similar to Eq. (7), the ground-truth denoising distribution can also be represented as a Gaussian distribution:

$$q(\boldsymbol{x}_{t-1}|\boldsymbol{x}_t, \boldsymbol{x}_0) = \mathcal{N}\big(\boldsymbol{x}_{t-1}; \widetilde{\boldsymbol{\mu}}(\boldsymbol{x}_t, \boldsymbol{x}_0), \widetilde{\boldsymbol{\Sigma}}(\boldsymbol{x}_t, \boldsymbol{x}_0)\big), \tag{32}$$

where $\widetilde{\boldsymbol{\mu}}(\boldsymbol{x}_t, \boldsymbol{x}_0) \in \mathbb{R}^d$ and $\widetilde{\boldsymbol{\Sigma}}(\boldsymbol{x}_t, \boldsymbol{x}_0) \in \mathbb{R}^{d\times d}$ are the mean and covariance of the distribution, respectively. For simplicity, similar to Eq. (8), the covariance is usually set to be isotropic and diagonal, i.e., $\widetilde{\boldsymbol{\Sigma}}(\boldsymbol{x}_t, \boldsymbol{x}_0) = \widetilde{\sigma}_t^2\boldsymbol{I}$.

By Bayes' rule, we have

$$q(\boldsymbol{x}_{t-1}|\boldsymbol{x}_t, \boldsymbol{x}_0) = \frac{q(\boldsymbol{x}_t|\boldsymbol{x}_{t-1}, \boldsymbol{x}_0)\, q(\boldsymbol{x}_{t-1}|\boldsymbol{x}_0)}{q(\boldsymbol{x}_t|\boldsymbol{x}_0)}. \tag{33}$$

According to Eqs. (18), (22), (24), and (31), the probabilities in this Bayes' rule are

$$\begin{aligned}
q(\boldsymbol{x}_t|\boldsymbol{x}_{t-1}, \boldsymbol{x}_0) &\overset{(18)}{=} q(\boldsymbol{x}_t|\boldsymbol{x}_{t-1}) \overset{(22),(24)}{=} \mathcal{N}(\boldsymbol{x}_t; \sqrt{\alpha_t}\boldsymbol{x}_{t-1}, \beta_t\boldsymbol{I}),\\
&\propto \exp\Big(-\frac{1}{2}\Big(\frac{(\boldsymbol{x}_t-\sqrt{\alpha_t}\boldsymbol{x}_{t-1})^2}{\beta_t}\Big)\Big),\\
q(\boldsymbol{x}_{t-1}|\boldsymbol{x}_0) &\overset{(31)}{=} \mathcal{N}\big(\boldsymbol{x}_{t-1}; \sqrt{\bar{\alpha}_{t-1}}\boldsymbol{x}_0, (1-\bar{\alpha}_{t-1})\boldsymbol{I}\big),\\
&\propto \exp\Big(-\frac{1}{2}\Big(\frac{(\boldsymbol{x}_{t-1}-\sqrt{\bar{\alpha}_{t-1}}\boldsymbol{x}_0)^2}{1-\bar{\alpha}_{t-1}}\Big)\Big),\\
q(\boldsymbol{x}_t|\boldsymbol{x}_0) &\overset{(31)}{=} \mathcal{N}\big(\boldsymbol{x}_t; \sqrt{\bar{\alpha}_t}\boldsymbol{x}_0, (1-\bar{\alpha}_t)\boldsymbol{I}\big)\\
&\propto \exp\Big(-\frac{1}{2}\Big(\frac{(\boldsymbol{x}_t-\sqrt{\bar{\alpha}_t}\boldsymbol{x}_0)^2}{1-\bar{\alpha}_t}\Big)\Big).
\end{aligned}$$

Therefore:

$$\begin{aligned}
&q(\boldsymbol{x}_{t-1}|\boldsymbol{x}_t, \boldsymbol{x}_0)\\
&\overset{(33)}{\propto} \exp\Big(-\frac{1}{2}\Big(\frac{(\boldsymbol{x}_t-\sqrt{\alpha_t}\boldsymbol{x}_{t-1})^2}{\beta_t} + \frac{(\boldsymbol{x}_{t-1}-\sqrt{\bar{\alpha}_{t-1}}\boldsymbol{x}_0)^2}{1-\bar{\alpha}_{t-1}}\\
&\qquad - \frac{(\boldsymbol{x}_t-\sqrt{\bar{\alpha}_t}\boldsymbol{x}_0)^2}{1-\bar{\alpha}_t}\Big)\Big)
\end{aligned}$$

$$
\begin{aligned}
&\overset{(a)}{=} \exp\Bigg(-\frac{1}{2}\Bigg(\frac{\boldsymbol{x}_t^2 - 2\sqrt{\alpha_t}\boldsymbol{x}_t\boldsymbol{x}_{t-1} + \alpha_t\boldsymbol{x}_{t-1}^2}{\beta_t} \\
&\qquad + \frac{\boldsymbol{x}_{t-1}^2 - 2\sqrt{\bar{\alpha}_{t-1}}\boldsymbol{x}_0\boldsymbol{x}_{t-1} + \bar{\alpha}_{t-1}\boldsymbol{x}_0^2}{1-\bar{\alpha}_{t-1}} - \frac{(\boldsymbol{x}_t - \sqrt{\bar{\alpha}_t}\boldsymbol{x}_0)^2}{1-\bar{\alpha}_t}\Bigg)\Bigg) \\
&\overset{(b)}{=} \exp\Bigg(-\frac{1}{2}\Bigg(\Big(\frac{\alpha_t}{\beta_t} + \frac{1}{1-\bar{\alpha}_{t-1}}\Big)\boldsymbol{x}_{t-1}^2 \\
&\qquad - 2\Big(\frac{\sqrt{\alpha_t}}{\beta_t}\boldsymbol{x}_t + \frac{\sqrt{\bar{\alpha}_{t-1}}}{1-\bar{\alpha}_{t-1}}\boldsymbol{x}_0\Big)\boldsymbol{x}_{t-1} + C(\boldsymbol{x}_t, \boldsymbol{x}_0)\Bigg)\Bigg) \\
&\overset{(c)}{\propto} \exp\Bigg(-\frac{1}{2}\frac{\Bigg(g\boldsymbol{x}_{t-1} - \frac{\frac{\sqrt{\alpha_t}}{\beta_t}\boldsymbol{x}_t + \frac{\sqrt{\bar{\alpha}_{t-1}}}{1-\bar{\alpha}_{t-1}}\boldsymbol{x}_0}{\frac{\alpha_t}{\beta_t} + \frac{1}{1-\bar{\alpha}_{t-1}}}\Bigg)^2}{1/(\frac{\alpha_t}{\beta_t} + \frac{1}{1-\bar{\alpha}_{t-1}})}\Bigg),
\end{aligned}
\tag{34}
$$

where (a) and (c) follow from the binomial theorem, (b) is due to factoring out the terms $\boldsymbol{x}_{t-1}^2$ and $\boldsymbol{x}_{t-1}$, and $C(\boldsymbol{x}_t, \boldsymbol{x}_0)$ is a function that does not include $\boldsymbol{x}_{t-1}$ and thus can be dropped and ignored for optimization.

2.4.3 Mean of the Ground-Truth Denoising Distribution

According to Eq. (34), the mean of the ground-truth denoising distribution, in Eq. (32), is as follows:

$$
\begin{aligned}
\widetilde{\boldsymbol{\mu}}(\boldsymbol{x}_t, \boldsymbol{x}_0) &= \frac{\frac{\sqrt{\alpha_t}}{\beta_t}\boldsymbol{x}_t + \frac{\sqrt{\bar{\alpha}_{t-1}}}{1-\bar{\alpha}_{t-1}}\boldsymbol{x}_0}{\frac{\alpha_t}{\beta_t} + \frac{1}{1-\bar{\alpha}_{t-1}}} = \frac{\frac{\sqrt{\alpha_t}}{\beta_t}\boldsymbol{x}_t + \frac{\sqrt{\bar{\alpha}_{t-1}}}{1-\bar{\alpha}_{t-1}}\boldsymbol{x}_0}{\frac{\alpha_t - \alpha_t\bar{\alpha}_{t-1} + \beta_t}{\beta_t(1-\bar{\alpha}_{t-1})}} \\
&\overset{(a)}{=} \frac{\frac{\sqrt{\alpha_t}}{\beta_t}\boldsymbol{x}_t + \frac{\sqrt{\bar{\alpha}_{t-1}}}{1-\bar{\alpha}_{t-1}}\boldsymbol{x}_0}{\frac{1-\bar{\alpha}_t}{\beta_t(1-\bar{\alpha}_{t-1})}} \\
&= \Big(\frac{\sqrt{\alpha_t}}{\beta_t}\boldsymbol{x}_t + \frac{\sqrt{\bar{\alpha}_{t-1}}}{1-\bar{\alpha}_{t-1}}\boldsymbol{x}_0\Big)\frac{1-\bar{\alpha}_{t-1}}{1-\bar{\alpha}_t}\beta_t \\
&= \frac{\sqrt{\alpha_t}(1-\bar{\alpha}_{t-1})}{1-\bar{\alpha}_t}\boldsymbol{x}_t + \frac{\sqrt{\bar{\alpha}_{t-1}}\beta_t}{1-\bar{\alpha}_t}\boldsymbol{x}_0,
\end{aligned}
\tag{35}
$$

where (a) is because $\alpha_t + \beta_t = 1$ and $\alpha_t\bar{\alpha}_{t-1} = \bar{\alpha}_t$ by Eqs. (22) and (23). According to Eq. (30), we have

$$
\boldsymbol{x}_t = \sqrt{\bar{\alpha}_t}\boldsymbol{x}_0 + \sqrt{1-\bar{\alpha}_t}\boldsymbol{\epsilon} \implies \boldsymbol{x}_0 = \frac{1}{\sqrt{\bar{\alpha}_t}}(\boldsymbol{x}_t - \sqrt{1-\bar{\alpha}_t}\boldsymbol{\epsilon}).
\tag{36}
$$

Substituting Eq. (36) in Eq. (35) gives

$$
\begin{aligned}
\widetilde{\boldsymbol{\mu}}(\boldsymbol{x}_t, \boldsymbol{x}_0) &= \frac{\sqrt{\alpha_t}(1-\bar{\alpha}_{t-1})}{1-\bar{\alpha}_t}\boldsymbol{x}_t + \frac{\sqrt{\bar{\alpha}_{t-1}}\beta_t}{1-\bar{\alpha}_t}\frac{1}{\sqrt{\bar{\alpha}_t}}(\boldsymbol{x}_t - \sqrt{1-\bar{\alpha}_t}\boldsymbol{\epsilon}) \\
&\overset{(a)}{=} \frac{1}{\sqrt{\alpha_t}}\Big(\boldsymbol{x}_t - \frac{1-\alpha_t}{\sqrt{1-\bar{\alpha}_t}}\boldsymbol{\epsilon}\Big) \overset{(22)}{=} \frac{1}{\sqrt{\alpha_t}}\Big(\boldsymbol{x}_t - \frac{\beta_t}{\sqrt{1-\bar{\alpha}_t}}\boldsymbol{\epsilon}\Big),
\end{aligned}
\tag{37}
$$

where (a) results from simplifying the expression while noting Eqs. (22) and (23). Equation (37) represents the mean of the tractable backward distribution $q(\boldsymbol{x}_{t-1}|\boldsymbol{x}_t, \boldsymbol{x}_0)$ as defined in Eq. (32).

Let $\boldsymbol{\mu}_\theta(\boldsymbol{x}_t, t)$ denote the mean of the distribution of the backward process, i.e., $p_\theta(\boldsymbol{x}_{t-1}|\boldsymbol{x}_t)$. This mean has a similar form to $\widetilde{\boldsymbol{\mu}}(\boldsymbol{x}_t, \boldsymbol{x}_0)$; therefore:

$$\boldsymbol{\mu}_\theta(\boldsymbol{x}_t, t) = \frac{1}{\sqrt{\alpha_t}}\Big(\boldsymbol{x}_t - \frac{\beta_t}{\sqrt{1-\bar{\alpha}_t}}\boldsymbol{\epsilon}_\theta(\boldsymbol{x}_t, t)\Big), \tag{38}$$

where $\boldsymbol{\epsilon}_\theta(\boldsymbol{x}_t, t)$ is a function approximator (i.e., a neural network) designed to predict $\boldsymbol{\epsilon}$ from $\boldsymbol{x}_t$. In summary, the backward process $p_\theta(\boldsymbol{x}_{t-1}|\boldsymbol{x}_t)$ has a Gaussian distribution as in Eq. (7), with its mean and covariance given by Eqs. (38) and (8), respectively.

2.5 The Loss Function of the Diffusion Model

Recall the KL divergence in the loss function of Eq. (21):

$$\text{KL}\big(q(\boldsymbol{x}_{t-1}|\boldsymbol{x}_t, \boldsymbol{x}_0)\|p_\theta(\boldsymbol{x}_{t-1}|\boldsymbol{x}_t)\big),$$

where $q(\boldsymbol{x}) = \mathcal{N}(\boldsymbol{x}; \widetilde{\boldsymbol{\mu}}, \widetilde{\boldsymbol{\Sigma}})$ and $p(\boldsymbol{x}) = \mathcal{N}(\boldsymbol{x}; \boldsymbol{\mu}_\theta, \boldsymbol{\Sigma}_\theta)$. There is a closed-form expression for the KL divergence between two Gaussian distributions:

$$\text{KL}(q\|p) = \frac{1}{2}\Big(\textbf{tr}(\boldsymbol{\Sigma}_\theta^{-1}\widetilde{\boldsymbol{\Sigma}}) + (\boldsymbol{\mu}_\theta - \widetilde{\boldsymbol{\mu}})^\top\boldsymbol{\Sigma}_\theta^{-1}(\boldsymbol{\mu}_\theta - \widetilde{\boldsymbol{\mu}}) - d + \ln\big(\frac{\textbf{det}(\boldsymbol{\Sigma}_\theta)}{\textbf{det}(\widetilde{\boldsymbol{\Sigma}})}\big)\Big), \tag{39}$$

where d is the dimensionality of $\boldsymbol{x}$, and $\textbf{tr}(.)$ and $\textbf{det}(.)$ denote the trace and determinant of a matrix, respectively. For simplicity, assume the covariances of the two distributions are diagonal and equal, i.e., $\boldsymbol{\Sigma}_\theta = \boldsymbol{\Sigma}_q = \lambda\boldsymbol{I}$ (see Eq. (8)). Then, Eq. (39) simplifies to:

$$\text{KL}(q\|p) = \frac{1}{2}(\boldsymbol{\mu}_\theta - \widetilde{\boldsymbol{\mu}})^\top\boldsymbol{\Sigma}_\theta^{-1}(\boldsymbol{\mu}_\theta - \widetilde{\boldsymbol{\mu}}),$$

which is the squared Mahalanobis distance between the means of the distributions. This result makes sense because the two distributions p and q are normal distributions with equal covariances, which are related to the second moment. Therefore, to make them identical, their first moments, i.e., their means, must also become identical.

As a result, the loss function can be written as the squared Mahalanobis distance between the means of the distributions:

$$\begin{aligned}
\ell_t &:= \mathbb{E}_{\boldsymbol{x}_0,\boldsymbol{\epsilon}}\Big[\frac{1}{2\|\boldsymbol{\Sigma}_\theta(\boldsymbol{x}_t, t)\|_2^2}\|\widetilde{\boldsymbol{\mu}}(\boldsymbol{x}_t, \boldsymbol{x}_0) - \boldsymbol{\mu}_\theta(\boldsymbol{x}_t, t)\|_2^2\Big] \\
&\overset{(a)}{=} \mathbb{E}_{\boldsymbol{x}_0,\boldsymbol{\epsilon}}\Big[\frac{1}{2\|\boldsymbol{\Sigma}_\theta(\boldsymbol{x}_t, t)\|_2^2}\Big\|\frac{1}{\sqrt{\alpha_t}}\Big(\boldsymbol{x}_t - \frac{1-\alpha_t}{\sqrt{1-\bar{\alpha}_t}}\boldsymbol{\epsilon}\Big) \\
&\qquad - \frac{1}{\sqrt{\alpha_t}}\Big(\boldsymbol{x}_t - \frac{1-\alpha_t}{\sqrt{1-\bar{\alpha}_t}}\boldsymbol{\epsilon}_\theta\Big)\Big\|_2^2\Big] \\
&= \mathbb{E}_{\boldsymbol{x}_0,\boldsymbol{\epsilon}}\Big[\frac{(1-\alpha_t)^2}{2\alpha_t(1-\bar{\alpha}_t)\|\boldsymbol{\Sigma}_\theta(\boldsymbol{x}_t, t)\|_2^2}\|\boldsymbol{\epsilon} - \boldsymbol{\epsilon}_\theta(\boldsymbol{x}_t, t)\|_2^2\Big] \\
&\overset{(30)}{=} \mathbb{E}_{\boldsymbol{x}_0,\boldsymbol{\epsilon}}\Big[\frac{(1-\alpha_t)^2}{2\alpha_t(1-\bar{\alpha}_t)\|\boldsymbol{\Sigma}_\theta(\boldsymbol{x}_t, t)\|_2^2} \times \big\|\boldsymbol{\epsilon} - \boldsymbol{\epsilon}_\theta(\sqrt{\bar{\alpha}_t}\boldsymbol{x}_0 + \sqrt{1-\bar{\alpha}_t}\boldsymbol{\epsilon}, t)\big\|_2^2\Big],
\end{aligned} \tag{40}$$

while *not converged* **do**
  $x_0 \sim q_0(x_0)$
  $t \sim \text{Uniform}\{1, \ldots, T\}$
  $\epsilon \sim \mathcal{N}(\mathbf{0}, \boldsymbol{I})$
  Do backpropagation by the gradient of the loss function (41)

Algorithm 1: Training algorithm in DDPM

where (a) follows from Eqs. (37) and (38), while noting Eq. (22).

Ho et al. [10] discovered empirically that the diffusion model produces better quality images when using a simplified loss function that omits the weighting term:

$$\ell = \mathbb{E}_{t\sim[1,T],x_0,\epsilon}\Big[\big\|\epsilon - \epsilon_\theta(\sqrt{\bar{\alpha}_t}x_0 + \sqrt{1-\bar{\alpha}_t}\epsilon, t)\big\|_2^2\Big], \tag{41}$$

where $t \sim [1, T]$ means sampling t uniformly from the set $\{1, 2, \ldots, T\}$. Recall that $\bar{\alpha}_t$ is calculated by Eq. (22) based on the selected noise schedule β_t in Eq. (3). Note that the expectation can be approximated by an average according to the Monte Carlo approximation [8].

2.6 Training Phase of DDPM

The training algorithm of DDPM is presented in Algorithm 1. In each iteration, an image x_0 is sampled from the training dataset. Then, the time step t is sampled uniformly from $\{1, \ldots, T\}$. For the sampled time step t, the noise ϵ is sampled from the standard Gaussian distribution. Finally, a backpropagation step is performed using the gradient of the loss function in Eq. (41). This process trains the neural network to learn the noise. In other words, it learns how much noise should be present in the backward process to reverse the forward process.

This procedure is repeated until the weights θ of the neural network converge. One of the advantages of DDPM compared to Generative Adversarial Networks (GANs) [9] is that it is simple to train and stable, whereas training GANs is challenging and often unstable.

2.7 Inference (Sampling) Phase of DDPM

In the inference phase, the network has been trained, meaning that the distribution p_θ has been learned. This distribution can be used to start with complete noise and generate (or sample) data from it through the backward process.

The inference phase begins with complete noise, i.e., $x_T \sim \mathcal{N}(\mathbf{0}, \boldsymbol{I})$. Let $\sigma_t \boldsymbol{I}$ denote the covariance of x_t. By using the reparameterization technique [13] and according to Eq. (38), while noting Eq. (22), we have

$$x_{t-1} = \frac{1}{\sqrt{\alpha_t}}\Big(x_t - \frac{1-\alpha_t}{\sqrt{1-\bar{\alpha}_t}}\epsilon_\theta(x_t, t)\Big) + \sigma_t\epsilon, \tag{42}$$

where $\epsilon \sim \mathcal{N}(\mathbf{0}, \boldsymbol{I})$, and α_t and $\bar{\alpha}_t$ are calculated by Eqs. (22) and (23), based on the selected noise schedule β_t in Eq. (3). Note that $\epsilon_\theta(x_t, t)$ is the output of the trained network. Iterating over t from T to 1 yields the generated data x_0. The inference phase of DDPM is summarized in Algorithm 2.

$\boldsymbol{x}_T \sim \mathcal{N}(\mathbf{0}, \boldsymbol{I})$
for t *from* T *to* 1 **do**
  $\epsilon \sim \mathcal{N}(\mathbf{0}, \boldsymbol{I})$
  $\boldsymbol{x}_{t-1} = \frac{1}{\sqrt{\alpha_t}}\Big(\boldsymbol{x}_t - \frac{1-\alpha_t}{\sqrt{1-\bar{\alpha}_t}}\epsilon_\theta(\boldsymbol{x}_t, t)\Big) + \sigma_t \epsilon$
Return $\boldsymbol{x}_0$

Algorithm 2: Inference algorithm for generating data in DDPM

3 Modifications to the Original DDPM

3.1 Non-fixed Covariance Matrix

Recall Eq. (8), in which $\boldsymbol{\Sigma}_\theta$ is assumed to be fixed and diagonal, i.e., $\boldsymbol{\Sigma}_\theta(\boldsymbol{x}_t; t) = \sigma_t^2 \boldsymbol{I}$, in the original DDPM [10]. We define:

$$\widetilde{\beta}_t := \frac{1 - \bar{\alpha}_{t-1}}{1 - \bar{\alpha}_t}\beta_t. \tag{43}$$

Nichol et al. [20] found empirically that the covariance $\boldsymbol{\Sigma}_\theta(\boldsymbol{x}_t; t)$ used in Eq. (40) yields better results when learned as a linear combination of β_t and $\widetilde{\beta}_t$ in the log domain:

$$\boldsymbol{\Sigma}_\theta(\boldsymbol{x}_t; t) = \mathbf{diag}\Big(\exp\big(\boldsymbol{v} \log \beta_t + (\mathbf{1} - \boldsymbol{v}) \log \widetilde{\beta}_t\big)\Big), \tag{44}$$

where $\mathbf{diag}(.)$ creates a diagonal matrix with its input as the diagonal, exp is the exponential operator, $\mathbf{1} \in \mathbb{R}^d$ is the vector of ones, and $\boldsymbol{v} \in \mathbb{R}^d$ is a vector outputted by the DDPM network and learned through backpropagation.

The loss function of the network is modified from Eq. (41) to:

$$\ell + \lambda \sum_{t=0}^{T} \ell_t, \tag{45}$$

where ℓ_t and ℓ are defined in Eqs. (40) and (41), respectively, and $\lambda = 0.001$ is the regularization hyperparameter. Note that $\boldsymbol{\Sigma}_\theta(\boldsymbol{x}_t; t)$, as defined in Eq. (44), is used in ℓ_t, so its $\boldsymbol{v}$ is learned by backpropagation.

3.2 Reducing the Gradient Noise

One might wonder why a regularized loss function, as in Eq. (45), is required when ℓ is a simplification of the summation of the ℓ_t terms, implying that the information in the first term is contained in the second term of Eq. (45). Using the concept of gradient noise scale [19], Nichol et al. [20] empirically observed that the gradient noise of ℓ_t is much higher than that of ℓ.

They found that this gradient noise is caused by the uniform sampling of t in Algorithm 1. By replacing uniform sampling with importance sampling [8], which records a history of previous values, they were able to reduce this gradient noise. When using importance sampling, they could simply use $\sum_{t=0}^{T} \ell_t$ as the loss function.

3.3 Improving the Noise Schedule

Recall the noise schedule stated in Eq. (3). While the linear noise schedule used in the original DDPM [10] works well on high-resolution images, Nichol et al. [20] observed that it does not perform well on smaller 32×32 or 64×64 images. This is because the linear noise schedule makes $\boldsymbol{x}_T$ excessively noisy at the end of the forward process.

To address this, they proposed a cosine noise schedule, which also works well on smaller images [20]:

$$\bar{\alpha}_t = \frac{f(t)}{f(0)}, \tag{46}$$

$$f(t) := \cos\Big(\frac{t/T + s}{1+s} \times \frac{\pi}{2}\Big)^2, \tag{47}$$

where $s = 0.008$ is the offset. Note that according to Eqs. (22) and (23), we have

$$\beta_t = 1 - \alpha_t = 1 - \frac{\bar{\alpha}_t}{\bar{\alpha}_{t-1}}. \tag{48}$$

Using Eq. (46) in Eq. (48) yields β_t at time t in the schedule.

3.4 Non-standard Gaussian Noise Distribution

The original DDPM [10] uses a standard Gaussian distribution for the noise ϵ. However, the corresponding data distribution may be more complex than the standard Gaussian, making this noise distribution not necessarily optimal for the process.

PriorGrad [17] addresses this concern by modifying DDPM to use a data-dependent prior distribution. Suppose the mean $\boldsymbol{\mu}$ and the covariance $\boldsymbol{\Sigma}$ are obtained from some data-dependent prior; for example, they can be the mean and variance of the training data. In PriorGrad, the noise is sampled from a Gaussian distribution with covariance $\boldsymbol{\Sigma}$, i.e., $\boldsymbol{\epsilon} \sim \mathcal{N}(\boldsymbol{0}, \boldsymbol{\Sigma})$. This changes Eq. (30) to:

$$\boldsymbol{x}_t = \sqrt{\bar{\alpha}_t}(\boldsymbol{x}_0 - \boldsymbol{\mu}) + \sqrt{1 - \bar{\alpha}_t}\boldsymbol{\epsilon}, \tag{49}$$

where the mean is subtracted from the data $\boldsymbol{x}_0$.

This algorithm also changes the loss function from Eq. (41) to:

$$\ell = \mathbb{E}_{t\sim[1,T],\boldsymbol{x}_0,\boldsymbol{\epsilon}}\Big[\big\|\boldsymbol{\epsilon} - \boldsymbol{\epsilon}_\theta(\boldsymbol{x}_t, t)\big\|^2_{\boldsymbol{\Sigma}^{-1}}\Big], \tag{50}$$

where $\boldsymbol{x}_t$ is obtained from Eq. (49) and $\|\boldsymbol{x}\|^2_{\boldsymbol{\Sigma}^{-1}} := \boldsymbol{x}^\top \boldsymbol{\Sigma}^{-1}\boldsymbol{x}$. In the inference phase, it simply adds back the subtracted mean $\boldsymbol{\mu}$ to the generated data $\boldsymbol{x}_0$.

The main ideas of PriorGrad [17] are two-fold: (1) subtracting the mean of the data during training and adding it back during sampling and (2) using the inverse of the covariance matrix in the loss function. Therefore, it enables the use of a non standard Gaussian distribution in the diffusion model.

$x_T \sim \mathcal{N}(\mathbf{0}, \boldsymbol{I})$
for *t from T to* 1 **do**
  $\epsilon \sim \mathcal{N}(\mathbf{0}, \boldsymbol{I})$
  $x_{t-1} = \frac{1}{\sqrt{\alpha_t}}\Big(x_t - \frac{1-\alpha_t}{\sqrt{1-\bar{\alpha}_t}}\epsilon_\theta(x_t, t)\Big) + \sigma_t \epsilon$
  $y_{t-1} = \sqrt{\bar{\alpha}_{t-1}}y + \sqrt{1-\bar{\alpha}_{t-1}}\epsilon$
  $x_{t-1} \leftarrow x_{t-1} - \phi_n(x_{t-1}) + \phi_n(y_{t-1})$
Return x_0

Algorithm 3: Inference algorithm for generating data in ILVR (conditional diffusion model)

3.5 Some Other Improvements

Diffusion models achieve state-of-the-art quality but require many iterative denoising steps, making inference slow compared to GANs. In practice, accelerated variants such as Denoising Diffusion Implicit Model (DDIM) [26], consistency models [27], and distillation [25] reduce sampling time significantly.

Another improvement is the Discrete Denoising Diffusion Probabilistic Model (D3PM) [2], which extends DDPM from generating continuous-state data to discrete-state data. In other words, D3PM operates on categorical (discrete) data rather than the conventional float data used in DDPM.

4 Conditional Diffusion Model

DDPM [10] is unconditional, meaning that it generates data from the entire data distribution during the inference (sampling) phase. Iterative Latent Variable Refinement (ILVR) [4] is a conditional diffusion model that can generate data from a specific category or class chosen by the user. In this algorithm, a reference data instance (e.g., a reference image) is provided, and the model generates a data instance similar to, or within the category of, the reference.

The training phase of ILVR is the same as that of DDPM. In fact, ILVR modifies the inference (sampling) phase of DDPM to act as a conditional generator. The inference algorithm of ILVR is shown in Algorithm 3.

Let $\phi_n(.)$ be a low-pass filter that downsamples data by a factor of n and then upsamples it again by the same factor. This preserves only the low frequencies of the data and removes its high frequencies. For example, in images, it results in a blurred image. Obviously, if $x \in \mathbb{R}^d$, then $\phi_n(x) \in \mathbb{R}^d$.

Suppose the unconditional sample $x_{t-1} \in \mathbb{R}^d$ has been calculated by Eq. (42). Let the reference data instance be denoted by $y \in \mathbb{R}^d$. Similar to Eq. (30), y_{t-1} can be obtained from y and the noise ϵ as follows:

$$y_{t-1} = \sqrt{\bar{\alpha}_{t-1}}y + \sqrt{1-\bar{\alpha}_{t-1}}\epsilon. \tag{51}$$

The unconditional sample x_{t-1} can then be modified to resemble the noisy reference y_{t-1} at time step $(t-1)$. To achieve this, the low-frequency component of x_{t-1}, i.e., $\phi_n(x_{t-1}) \in \mathbb{R}^d$, is subtracted, and the low-frequency component of y_{t-1} is added back. This replaces the low frequency of x_{t-1} with that of the noisy reference at time step $(t-1)$. Performing this iteratively from $t = T$ to $t = 1$ produces a sample similar to the reference.

It is noteworthy that n is a positive integer selected by the user. The smaller the n, the more similar the sample will be to the reference. This is because a smaller n results in a larger bandwidth for the low-pass filter, replacing more frequencies of the sample with those of the reference data.

The explained ILVR is one approach for conditional diffusion, but in practice the most widely used methods are classifier guidance [5], classifier-free guidance [11], and cross-attention conditioning [23]. These methods enable text-to-image systems like Stable Diffusion [23] and Imagen [24], where natural language controls the generation process.

5 Continuous Noise Schedule by Stochastic Differential Equation

Diffusion models can be reformulated as stochastic differential equations, which unify discrete-time DDPMs with continuous-time score-based models. This perspective has broadened their applications to scientific modeling and physics-inspired simulations.

The noise schedule in DDPM [10] is defined in the discrete domain, meaning there are T discrete steps from the original data to complete noise and back. However, it is possible to define the noise schedule of diffusion models in a continuous domain, meaning that $t \in [0, T]$ can be continuous rather than discrete $t \in \{1, 2, \dots, T\}$. The continuous noise schedule converts the Markov chain of the diffusion model into a Stochastic Differential Equation (SDE) [28].

Consider Eq. (25):

$$x_t = \sqrt{\alpha_t} x_{t-1} + \sqrt{1 - \alpha_t} \epsilon \overset{(22)}{=} \sqrt{1 - \beta_t} x_{t-1} + \sqrt{\beta_t} \epsilon,$$

in DDPM, where $t \in \{1, 2, \dots, T\}$. If $T \to \infty$, the number of steps in the noise schedule becomes extremely large, making the flow of time effectively continuous. As $T \to \infty$, Eq. (25) converges to the following SDE [28]:

$$dx = -\frac{1}{2}\beta(t)\, x\, dt + \sqrt{\beta(t)}\, dw, \tag{52}$$

where $w \in \mathbb{R}^d$ is the standard Wiener process (Brownian motion), $x \in \mathbb{R}^d$ is the data variable, and dx and dw are the differentials of x and w, respectively. As a result, DDPM [10] is a special case of diffusion models with SDEs [28].

In a more general case, the forward process in the diffusion model with SDE is given by [28]

$$dx = f(x, t)\, dt + g(t)\, dw, \tag{53}$$

where $w \in \mathbb{R}^d$ is the standard Wiener process (Brownian motion), $f(., t) : \mathbb{R}^d \to \mathbb{R}^d$ is a function called the drift coefficient of $x(t)$, and $g(.) : \mathbb{R} \to \mathbb{R}$ is a function called the diffusion coefficient of $x(t)$. The backward process of the diffusion model with SDE is the following reverse-time SDE [28]:

$$dx = \Big(f(x, t) - g(t)^2 \nabla_x \log p_t(x)\Big) dt + g(t)\, d\bar{w}, \tag{54}$$

where $\bar{w}$ is the standard Wiener process when time flows backward from T to 0, and dt is the infinitesimal negative time. The operator ∇_x denotes the derivative with respect to x, $p_t(x)$ is the probability distribution of x at time t, and $\nabla_x \log p_t(x)$ is the time-dependent gradient field, often called the score. As long as the score is estimated, this reverse-time SDE can be solved to generate samples from noise in the backward process of the diffusion model.

According to [28], the loss function of the original DDPM, for maximizing the ELBO, can be expressed as a weighted sum of denoising score matching [30]:

$$\ell = \sum_{t=1}^{T}(1 - \alpha_t)\,\mathbb{E}_{p_{\text{data}}(\boldsymbol{x})}\mathbb{E}_{q(\boldsymbol{x}_t|\boldsymbol{x}_0)}\Big[\big\|\boldsymbol{s}_\theta(\boldsymbol{x}_t, t) - \nabla_{\boldsymbol{x}_t}\log\big(q(\boldsymbol{x}_t|\boldsymbol{x}_0)\big)\big\|_2^2\Big], \tag{55}$$

where $\boldsymbol{s}_\theta(\boldsymbol{x}_t, t)$ is the output of the neural network with weights θ. Likewise, the loss function in the diffusion model with SDE is [28]

$$\ell = \mathbb{E}_t\Big[\lambda(t)\,\mathbb{E}_{\boldsymbol{x}(0)}\mathbb{E}_{\boldsymbol{x}(t)|\boldsymbol{x}(0)}\Big[\Big\|\boldsymbol{s}_\theta(\boldsymbol{x}_t, t) - \nabla_{\boldsymbol{x}(t)}\log\Big(q\big(\boldsymbol{x}(t)|\boldsymbol{x}(0)\big)\Big)\Big\|_2^2\Big]\Big], \tag{56}$$

where $\lambda : [0, T] \to \mathbb{R}_{>0}$ is a positive weighting function, t is sampled from the uniform distribution over $[0, T]$, $\boldsymbol{x}(0)$ is a sampled data instance from the training dataset, and $\boldsymbol{x}(t) \sim q\big(\boldsymbol{x}(t)|\boldsymbol{x}(0)\big)$. This loss function is minimized by backpropagation until the network output $\boldsymbol{s}_\theta(\boldsymbol{x}_t, t)$ becomes a good estimator of the score $\nabla_{\boldsymbol{x}(t)}\log\big(q\big(\boldsymbol{x}(t)|\boldsymbol{x}(0)\big)\big)$.

After training, the neural network becomes a function approximator for the score. Therefore, the reverse-time SDE in Eq. (54) can be solved using numerical SDE solvers to generate data in the backward process of the diffusion model. Various numerical methods exist for solving SDEs, including Euler–Maruyama and stochastic Runge–Kutta methods [14]. Any SDE solver can be used to solve the reverse-time SDE of the diffusion model to generate data.

It is noteworthy that continuous noise schedules using SDEs have also been extended to discrete-state (categorical) data [3].

6 Conclusion

6.1 Conclusion of Chapter

Diffusion models have emerged as the most influential generative modeling approach of the 2020s. They combine stability, sample quality, and flexible conditioning, enabling breakthroughs in text-to-image, audio, and video synthesis. Their main limitations are slow inference and high computational cost, which ongoing research addresses through acceleration methods and hybrid architectures. By 2025, diffusion models stand alongside autoregressive Transformers [29] (see Chap. 9) as the leading paradigm in generative AI.

6.2 Conclusion of the Book Part "Generative Models"

The evolution of generative models reflects the broader story of deep learning progress. Variational autoencoders [13] (see Chap. 12) introduced a probabilistic latent space framework with clear theoretical grounding, though their outputs were often blurry. Generative moment matching networks [6, 18] (see Chap. 13) explored distribution matching via kernels but did not scale well and they remain mainly of historical interest. Generative Adversarial Networks (GANs) [9] (see Chap. 14) revolutionized the field with adversarial training, producing sharp and realistic images, and variants like DCGAN [21], WGAN [1], and StyleGAN [12] set milestones in generative quality. However, by the mid-2020s, diffusion models [10] have become the dominant paradigm (see this chapter), powering systems like DALL·E [22], Stable Diffusion [23], and Imagen [24], and showing unmatched stability and fidelity across modalities. Each of these families contributed unique ideas—VAEs with probabilistic inference, GMMNs with kernel methods, GANs with adversarial learning, and diffusion models with iterative denoising—that together shaped today's generative AI. By 2025, diffusion models and autoregressive Transformers [29] (see Chap. 9) stand at the forefront, while earlier models are best appreciated for their conceptual contributions and specialized applications.

7 Exercises

1. (Level: Difficult) A diffusion process transforms a data point $\boldsymbol{x}_0$ into a noisy version $\boldsymbol{x}_t$ using $\boldsymbol{x}_t = \sqrt{\bar{\alpha}_t} x_0 + \sqrt{1 - \bar{\alpha}_t}\boldsymbol{\epsilon}$, where $\boldsymbol{\epsilon} \sim \mathcal{N}(\boldsymbol{0}, \boldsymbol{I})$ is Gaussian noise and $\bar{\alpha}_t$ is a variance decay factor.

   (a) Show that the posterior mean of $\boldsymbol{x}_0$ given $\boldsymbol{x}_t$ is

   $$\mathbb{E}[\boldsymbol{x}_0|\boldsymbol{x}_t] = \frac{\boldsymbol{x}_t - \sqrt{1 - \bar{\alpha}_t}\boldsymbol{\epsilon}}{\sqrt{\bar{\alpha}_t}}.$$

   (b) Derive the conditional probability distribution $q(\boldsymbol{x}_0|\boldsymbol{x}_t)$ and show that it follows a Gaussian distribution.

   (c) What happens to $q(\boldsymbol{x}_0|\boldsymbol{x}_t)$ as $t \to \infty$? Interpret the result in the context of diffusion models.

2. (Level: Moderate) In Denoising Diffusion Probabilistic Models (DDPMs), the forward process gradually adds Gaussian noise to a data point $\boldsymbol{x}_0$, leading to a noisy sample $\boldsymbol{x}_t$ at time step t. This process is defined as $\boldsymbol{x}_t = \sqrt{\bar{\alpha}_t}\boldsymbol{x}_0 + \sqrt{1 - \bar{\alpha}_t}\boldsymbol{\epsilon}$, where $\boldsymbol{\epsilon} \sim \mathcal{N}(\boldsymbol{0}, \boldsymbol{I})$ and $\bar{\alpha}_t = \prod_{i=1}^{t} \alpha_i$ represents the accumulated noise decay factor.

   (a) Compute the variance of $\boldsymbol{x}_t$ given $\boldsymbol{x}_0$, i.e., $\text{Var}(\boldsymbol{x}_t|\boldsymbol{x}_0)$.

   (b) Derive the entropy of $q(\boldsymbol{x}_t|\boldsymbol{x}_0)$.

   Hint: The entropy of a multivariate Gaussian distribution $\mathcal{N}(\boldsymbol{\mu}, \boldsymbol{\Sigma})$ in d dimensions is given by

   $$H(\mathcal{N}(\boldsymbol{\mu}, \boldsymbol{\Sigma})) = \frac{1}{2}\log((2\pi e)^d \mathbf{det}(\Sigma)),$$

   where $\mathbf{det}(.)$ denotes the determinant of matrix and e is Napier's number.

3. (Level: Moderate) Diffusion models learn to generate high-quality data samples by iteratively denoising random noise. The forward process in a Denoising Diffusion Probabilistic Model (DDPM) follows a Markovian structure:

   $$q(\boldsymbol{x}_t|\boldsymbol{x}_{t-1}) = \mathcal{N}(\boldsymbol{x}_t; \sqrt{\alpha_t}\boldsymbol{x}_{t-1}, (1 - \alpha_t)\boldsymbol{I}),$$

   where $\boldsymbol{x}_t$ is the latent variable at step t, α_t controls the variance schedule, and $\mathcal{N}$ represents a Gaussian distribution. The reverse process learns to recover the data by estimating the posterior:

   $$p_\theta(\boldsymbol{x}_{t-1}|\boldsymbol{x}_t) = \mathcal{N}(\boldsymbol{x}_{t-1}; \boldsymbol{\mu}_\theta(\boldsymbol{x}_t, t), \boldsymbol{\Sigma}_\theta(\boldsymbol{x}_t, t)),$$

   where $\boldsymbol{\mu}_\theta$ and $\boldsymbol{\Sigma}_\theta$ are learned parameters. Assume a linear variance schedule with $\alpha_t = 0.9^t$. The initial data point is $\boldsymbol{x}_0 = 1.0$. The noise model follows Gaussian noise.

   (a) Compute $\boldsymbol{x}_1$ using the forward process.

   (b) Compute $\boldsymbol{x}_2$ using the forward process iteratively.

(c) Explain the significance of the variance schedule α_t in training diffusion models.

4. (Level: Difficult) Denoising Diffusion Probabilistic Models (DDPMs) define their training objective as a variational evidence lower bound (ELBO), similar to variational autoencoders but involving a series of latent Markov variables $\boldsymbol{x}_1, \boldsymbol{x}_2, ..., \boldsymbol{x}_T$. Answer the following questions:

(a) Given the forward process:

$$q(\boldsymbol{x}_{1:T} \mid \boldsymbol{x}_0) = \prod_{t=1}^{T} q(\boldsymbol{x}_t \mid \boldsymbol{x}_{t-1}),$$
$$q(\boldsymbol{x}_t \mid \boldsymbol{x}_{t-1}) = \mathcal{N}(\boldsymbol{x}_t; \sqrt{1-\beta_t}\boldsymbol{x}_{t-1}, \beta_t \boldsymbol{I}),$$

derive the joint distribution $q(\boldsymbol{x}_0, \boldsymbol{x}_{1:T})$ and express the general ELBO for DDPMs.

(b) Assume a fixed β_t schedule. Show how the ELBO simplifies to the commonly used ℓ_2-norm loss:

$$\mathcal{L} = \mathbb{E}_{t,\boldsymbol{x}_0,\epsilon}\left[\|\epsilon - \epsilon_\theta(\boldsymbol{x}_t, t)\|_2^2\right].$$

References

1. Martin Arjovsky, Soumith Chintala, and Léon Bottou. "Wasserstein generative adversarial networks". In: *International conference on machine learning*. 2017, pp. 214–223.
2. Jacob Austin, Daniel D Johnson, Jonathan Ho, Daniel Tarlow, and Rianne van den Berg. "Structured Denoising Diffusion Models in Discrete State-Spaces". In: *Advances in Neural Information Processing Systems*. 2021.
3. Andrew Campbell et al. "A Continuous Time Framework for Discrete Denoising Models". In: *Advances in Neural Information Processing Systems*. 2022.
4. Jooyoung Choi, Sungwon Kim, Yonghyun Jeong, Youngjune Gwon, and Sungroh Yoon. "ILVR: Conditioning Method for Denoising Diffusion Probabilistic Models". In: *Proceedings of the IEEE/CVF International Conference on Computer Vision*. 2021, pp. 14367–14376.
5. Prafulla Dhariwal and Alexander Nichol. "Diffusion models beat gans on image synthesis". In: *Advances in neural information processing systems* 34 (2021), pp. 8780–8794.
6. Gintare Karolina Dziugaite, Daniel M Roy, and Zoubin Ghahramani. "Training generative neural networks via maximum mean discrepancy optimization". In: *Proceedings of the Thirty-First Conference on Uncertainty in Artificial Intelligence*. 2015, pp. 258–267.
7. Benyamin Ghojogh, Fakhri Karray, and Mark Crowley. "Hidden Markov Model: Tutorial". In: *Engineering Archive* (2019).
8. Benyamin Ghojogh, Hadi Nekoei, Aydin Ghojogh, Fakhri Karray, and Mark Crowley. "Sampling algorithms, from survey sampling to Monte Carlo methods: Tutorial and literature review". In: *arXiv preprint* arXiv:2011.00901 (2020).
9. Ian Goodfellow et al. "Generative adversarial nets". In: *Advances in neural information processing systems* 27 (2014).
10. Jonathan Ho, Ajay Jain, and Pieter Abbeel. "Denoising diffusion probabilistic models". In: *Advances in neural information processing systems* 33 (2020), pp. 6840–6851.
11. Jonathan Ho and Tim Salimans. "Classifier-Free Diffusion Guidance". In: *NeurIPS 2021 Workshop on Deep Generative Models and Downstream Applications*. 2021.
12. Tero Karras, Samuli Laine, and Timo Aila. "A style-based generator architecture for generative adversarial networks". In: *Proceedings of the IEEE/CVF conference on computer vision and pattern recognition*. 2019, pp. 4401–4410.
13. Diederik P Kingma and Max Welling. "Auto-encoding variational Bayes". In: *International Conference on Learning Representations*. 2014.
14. Peter E Kloeden and Eckhard Platen. *Stochastic differential equations*. Springer, 1992.
15. Daphne Koller and Nir Friedman. *Probabilistic graphical models: principles and techniques*. MIT press, 2009.

16. Solomon Kullback and Richard A Leibler. "On information and sufficiency". In: *The annals of mathematical statistics* 22.1 (1951), pp. 79–86.
17. Sang-gil Lee et al. "PriorGrad: Improving Conditional Denoising Diffusion Models with Data-Dependent Adaptive Prior". In: *International Conference on Learning Representations*. 2022.
18. Yujia Li, Kevin Swersky, and Rich Zemel. "Generative moment matching networks". In: *International conference on machine learning*. PMLR. 2015, pp. 1718–1727.
19. Sam McCandlish, Jared Kaplan, Dario Amodei, and OpenAI Dota Team. "An empirical model of large-batch training". In: *arXiv preprint* arXiv:1812.06162 (2018).
20. Alexander Quinn Nichol and Prafulla Dhariwal. "Improved denoising diffusion probabilistic models". In: *International Conference on Machine Learning*. PMLR. 2021, pp. 8162–8171.
21. Alec Radford, Luke Metz, and Soumith Chintala. "Unsupervised representation learning with deep convolutional generative adversarial networks". In: *International Conference on Learning Representations*. 2016.
22. Aditya Ramesh et al. "Zero-shot text-to-image generation". In: *International conference on machine learning*. PMLR. 2021, pp. 8821–8831.
23. Robin Rombach, Andreas Blattmann, Dominik Lorenz, Patrick Esser, and Björn Ommer. "High-resolution image synthesis with latent diffusion models". In: *Proceedings of the IEEE/CVF conference on computer vision and pattern recognition*. 2022, pp. 10684–10695.
24. Chitwan Saharia et al. "Photorealistic text-to-image diffusion models with deep language understanding". In: *Advances in neural information processing systems* 35 (2022), pp. 36479–36494.
25. Tim Salimans and Jonathan Ho. "Progressive Distillation for Fast Sampling of Diffusion Models". In: *International Conference on Learning Representations*. 2022.
26. Jiaming Song, Chenlin Meng, and Stefano Ermon. "Denoising Diffusion Implicit Models". In: *International Conference on Learning Representations*. 2021.
27. Yang Song, Prafulla Dhariwal, Mark Chen, and Ilya Sutskever. "Consistency Models". In: *International Conference on Machine Learning*. PMLR. 2023, pp. 32211–32252.
28. Yang Song et al. "Score-Based Generative Modeling through Stochastic Differential Equations". In: *International Conference on Learning Representations*. 2021.
29. Ashish Vaswani et al. "Attention is all you need". In: *Advances in neural information processing systems*. 2017, pp. 5998–6008.
30. Pascal Vincent. "A connection between score matching and denoising autoencoders". In: *Neural computation* 23.7 (2011), pp. 1661–1674.

Part IV
Emerging Topics in Deep Learning

Deep Metric Learning and Self-supervised Learning

16

Abstract

Deep metric learning is a class of deep learning techniques focused on learning an embedding space where similar data points are mapped close together while dissimilar points are pushed apart. Utilizing neural networks, these methods transform high-dimensional input data into a lower-dimensional space optimized for a chosen distance metric. This learned embedding facilitates tasks such as classification and retrieval by emphasizing meaningful relationships between data points. This chapter provides a thorough overview of deep metric learning approaches, beginning with foundational models such as reconstruction and denoising autoencoders. It then explores methods that leverage classification loss functions to guide metric learning, followed by a detailed discussion on Siamese neural networks designed for pairwise similarity learning. The chapter concludes with an explanation of self-supervised learning and its key techniques, highlighting recent advances in metric learning without explicit labels.

1 Introduction

Deep metric learning is a family of methods within deep learning. These methods learn an embedding space using a neural network. Consider a dataset of n data points (instances), each being d-dimensional, i.e., $\{\boldsymbol{x}_i \in \mathbb{R}^d\}_{i=1}^n$. The goal of deep metric learning is to learn a p-dimensional embedding space, via a neural network, that discriminates between classes by bringing similar points closer together and pushing dissimilar points further apart. The network embeds data in the embedding space (or subspace) of a distance metric used in the neural network's loss function. Any distance metric $d(.,.) : \mathbb{R}^p \times \mathbb{R}^p \to \mathbb{R}$ can be used in this embedding space. In the network's loss function, the distance function $d(.,.)$ operates within the embedding space. For example, one option for the distance function is the squared ℓ_2 norm or squared Euclidean distance:

$$d\big(\mathbf{f}(\boldsymbol{x}_i^1), \mathbf{f}(\boldsymbol{x}_i^2)\big) := \|\mathbf{f}(\boldsymbol{x}_i^1) - \mathbf{f}(\boldsymbol{x}_i^2)\|_2^2, \tag{1}$$

where $\mathbf{f}(\boldsymbol{x}_i) \in \mathbb{R}^p$ denotes the output of the network for input $\boldsymbol{x}_i$ as its p-dimensional embedding. The network is trained using mini-batch methods such as mini-batch stochastic gradient descent (see Chap. 3). The mini-batch size is denoted by b, and the weights of the network are denoted by the learnable parameter θ.

B. Ghojogh and A. Ghodsi, *Elements of Deep Learning*,
https://doi.org/10.1007/978-3-032-10738-1_16

This chapter introduces deep metric learning methods.[1] First, reconstruction autoencoders and denoising autoencoders are introduced in Sect. 2. Then, deep metric learning using classification loss functions is explained in Sect. 3. Section 4 describes Siamese neural networks. Finally, self-supervised learning is elaborated in Sect. 5.

2 Reconstruction Autoencoders

Reconstruction and denoising autoencoders are not metric learning in the strict sense, but they represent early attempts at learning compact latent representations. These models influenced the later development of metric learning by showing how networks can structure embedding spaces.

2.1 Types of Autoencoders

An autoencoder is a model consisting of an encoder $E(.)$ and a decoder $D(.)$. There are several types of autoencoders. All types learn a code layer, or latent layer, positioned between the encoder and decoder. Inferential autoencoders learn a stochastic latent space in this code layer. Variational autoencoder [29] (see Chap. 12) and adversarial autoencoder [32] (see Chap. 14) are two important types of inferential autoencoders. Another type is the reconstruction autoencoder, which consists of an encoder that transforms data into a code and a decoder that transforms the code back into the data. Therefore, the decoder reconstructs the input data fed to the encoder. The code serves as a representation of the data. Each of the encoder and decoder can consist of multiple layers of neural networks with activation functions.

2.2 Reconstruction Loss

The input data point to the encoder is denoted by $\boldsymbol{x} \in \mathbb{R}^d$, where d is the dimensionality of the data. The reconstructed data point, which is the output of the decoder, is denoted by $\widehat{\boldsymbol{x}} \in \mathbb{R}^d$. The representation code, which is the output of the encoder and the input to the decoder, is denoted by $\mathbf{f}(\boldsymbol{x}) := E(\boldsymbol{x}) \in \mathbb{R}^p$. The reconstruction is performed as follows:

$$\widehat{\boldsymbol{x}} = D(E(\boldsymbol{x})) = D(\mathbf{f}(\boldsymbol{x})). \tag{2}$$

If the dimensionality of the code is greater than that of the input data, i.e., $p > d$, the autoencoder is called an over-complete autoencoder. Otherwise, if $p < d$, it is an under-complete autoencoder [19] (see Fig. 1). The loss function of a reconstruction autoencoder aims to make the reconstructed data as close as possible to the input data:

$$\underset{\theta}{\text{minimize}} \sum_{i=1}^{b} \Big(d\big(\boldsymbol{x}_i, \widehat{\boldsymbol{x}}_i\big) + \lambda\Omega(\theta) \Big), \tag{3}$$

where $\lambda \geq 0$ is the regularization parameter and $\Omega(\theta)$ is a penalty or regularization term on the weights (see Chap. 5). Here, the distance function $d(.,.)$ is defined on $\mathbb{R}^d \times \mathbb{R}^d$. Note that the penalty term can

[1] Note that deep metric learning is different from spectral and probabilistic metric learning. For spectral and probabilistic metric learning, refer to [14, 15], respectively.

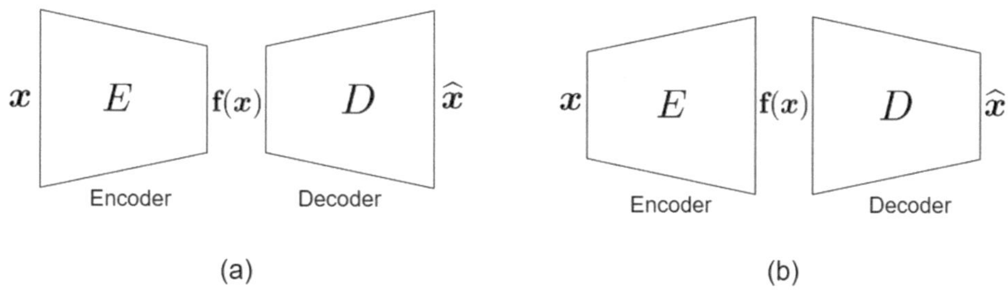

Fig. 1 The structures of **a** an under-complete autoencoder and **b** an over-complete autoencoder. Each block may have multiple layers of neural networks

also be a regularization on the code $\mathbf{f}(\boldsymbol{x}_i)$. If the distance metric used is the squared Euclidean distance, this loss is referred to as the regularized Mean Squared Error (MSE) loss.

2.3 Denoising Autoencoder

A problem with the over-complete autoencoder is that during training it merely copies each feature of the input data to one of the neurons in the code layer and then copies it back to the corresponding feature in the output layer (see Fig. 2a). This occurs because the number of neurons in the code layer is greater than the number of neurons in the input and output layers. In other words, the network simply memorizes the input data, leading to overfitting. This copying is achieved by setting some of the weights equal to one (or scaled versions of one, depending on the activation functions) and the remaining weights equal to zero.

To avoid this problem in over-complete autoencoders, noise can be added to the input data, and the network is trained to reconstruct the data without noise. This forces the over-complete autoencoder to learn meaningful representations rather than simply copying data to the code layer. Such an autoencoder can be used for denoising, as it reconstructs clean data from noisy inputs (see Fig. 2b). This network is called the Denoising Autoencoder (DAE) [19].

2.4 Metric Learning by Reconstruction Autoencoder

The under-complete reconstruction autoencoder can be used for metric learning and dimensionality reduction, especially when $p \ll d$. The loss function for learning a low-dimensional representation

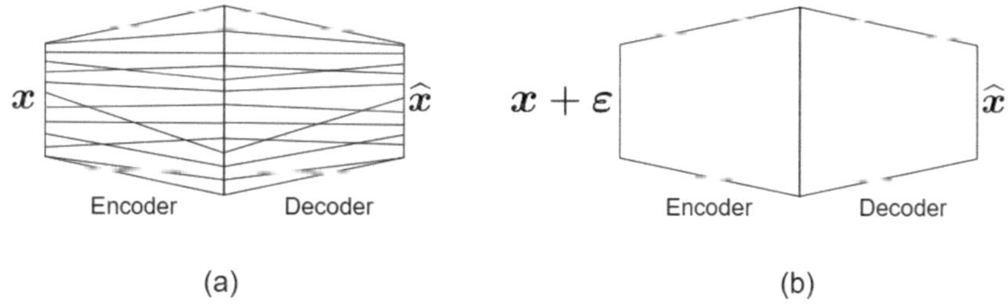

Fig. 2 Use of over-complete autoencoder for denoising: **a** If data $\boldsymbol{x}$ are not corrupted with noise, the autoencoder learns to copy the values of the input to the code layer and then back to the output layer, but **b** if the data $\boldsymbol{x}$ are corrupted with noise ε, the autoencoder can behave as a denoising autoencoder

code and reconstructing data using the autoencoder is given by Eq. (3). The code layer between the encoder and decoder serves as the embedding space of the metric.

Note that if the activation functions of all layers are linear, the under-complete autoencoder reduces to principal component analysis [13]. Let $\boldsymbol{U}_l$ denote the weight matrix of the lth layer of the network, ℓ_e be the number of layers in the encoder, and ℓ_d be the number of layers in the decoder. With linear activation functions, the encoder and decoder are

$$\begin{aligned}
&\text{encoder:} \quad \mathbb{R}^p \ni \mathbf{f}(\boldsymbol{x}_i) = \underbrace{\boldsymbol{U}_{\ell_e}^\top \boldsymbol{U}_{\ell_e-1}^\top \dots \boldsymbol{U}_1^\top}_{\boldsymbol{U}_e^\top} \boldsymbol{x}_i, \\
&\text{decoder:} \quad \mathbb{R}^d \ni \widehat{\boldsymbol{x}}_i = \underbrace{\boldsymbol{U}_1 \dots \boldsymbol{U}_{\ell_d-1} \boldsymbol{U}_{\ell_d}}_{\boldsymbol{U}_d} \mathbf{f}(\boldsymbol{x}_i),
\end{aligned} \tag{4}$$

where the ℓ concatenated linear projections can be replaced by linear projections with projection matrices $\boldsymbol{U}_e$ and $\boldsymbol{U}_d$ in the encoder and decoder, respectively. For learning complex data patterns, nonlinear activation functions can be used between layers of the encoder and decoder to achieve nonlinear metric learning and dimensionality reduction.

It is noteworthy that a nonlinear neural network can be viewed as an ensemble or concatenation of dimensionality reduction (or feature extraction) and kernel methods. The justification for this claim is as follows. Let the dimensionality for a layer of the network be $\boldsymbol{U} \in \mathbb{R}^{d_1 \times d_2}$, connecting d_1 neurons to d_2 neurons. Two cases can occur:

- If $d_1 \geq d_2$, this layer acts as dimensionality reduction or feature extraction because it reduces the dimensionality of its input data. If this layer has a nonlinear activation function, the dimensionality reduction is nonlinear; otherwise, it is linear.
- If $d_1 < d_2$, this layer acts as a kernel method, mapping its input data to a high-dimensional feature space in some Reproducing Kernel Hilbert Space (RKHS). This kernelization can facilitate nonlinear separation of classes that are not linearly separable [11]. An example of kernelization in machine learning is the kernel support vector machine [44].

Therefore, a neural network can be considered a complex feature extraction method, functioning as a concatenation of dimensionality reduction and kernel methods, where each layer learns its own features from the data.

3 Deep Metric Learning by Classification Loss Functions

Various loss functions exist for supervised metric learning using neural networks and classification. Supervised classification loss functions, such as cross-entropy, can train the network to separate classes in the embedding space [41]. Refer to Sect. 2.9, in Chap. 2 for a detailed discussion on classification loss functions.

A network can be designed such that its last layer performs classification of data points (see Fig. 3). The features of the penultimate layer can be used for feature embedding. This penultimate layer is referred to as the embedding layer, while the final layer after the embedding features is called the classification layer.

Let the ith point in the mini-batch be denoted by $\boldsymbol{x}_i \in \mathbb{R}^d$, and its label be denoted by $y_i \in \mathbb{R}$. Suppose the network has one output neuron, and its output for input $\boldsymbol{x}_i$ is denoted by $\mathbf{f}_o(\boldsymbol{x}_i) \in \mathbb{R}$. This output represents the estimated class label by the network. The output of the penultimate layer is denoted by $\mathbf{f}(\boldsymbol{x}_i) \in \mathbb{R}^p$, where p is the number of neurons in the embedding layer, which is equivalent

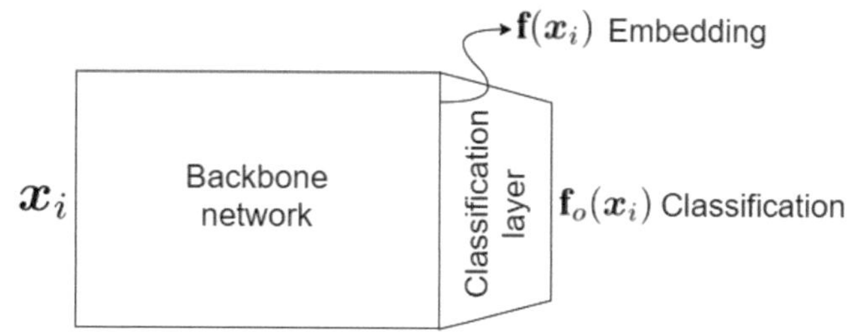

Fig. 3 The embedding and classification layers in a neural network with a supervised loss function

to the dimensionality of the embedding space. The final layer of the network, connecting the p neurons to the output neuron, is a fully connected layer.

The network up to the penultimate layer can be any feedforward or Convolutional Neural Network (CNN), depending on the type of data. If the network is convolutional, it should be flattened at the penultimate (embedding) layer. The network learns to classify the classes using supervised loss functions, resulting in features from the penultimate layer that are discriminative and suitable for embedding. This network is illustrated in Fig. 3.

4 Deep Metric Learning by Siamese Networks

4.1 Siamese and Triplet Networks

One of the important deep metric learning methods is the Siamese network, which is widely used for feature extraction. The Siamese network, originally proposed in [2], consists of several equivalent sub-networks that share their weights. The number of sub-networks in a Siamese network can vary, but it is usually two or three. A Siamese network with three sub-networks is also called a triplet network [27].

The weights of the sub-networks in a Siamese network are trained such that intra-class variance is decreased while inter-class variance is increased. In other words, similar points are pushed closer together, while dissimilar points are pulled apart. The core idea reflects the classic principle of reducing intra-class variation and increasing inter-class separation (e.g., in Fisher discriminant analysis [8, 12]), which has been adapted into modern neural loss functions. Siamese networks have been used in various applications such as computer vision [37] and natural language processing [49].

4.2 Pairs and Triplets of Data Points

Depending on the number of sub-networks in the Siamese network, different loss functions are used for training. The loss functions of Siamese networks typically require pairs or triplets of data points. Siamese networks do not process data points individually; instead, pairs or triplets must be created from the dataset for training.

To create pairs or triplets, every data point is considered an anchor point, denoted by $\boldsymbol{x}_i^a$. Then, one of the similar points to the anchor is chosen as the positive (or neighbor) point, denoted by $\boldsymbol{x}_i^p$. One of the dissimilar points relative to the anchor is selected as the negative (or distant) point, denoted by $\boldsymbol{x}_i^n$. If class labels are available, they can be used to find the positive point as one of the points in the same class as the anchor, and to find the negative point as one of the points in a different class from the anchor's class. Another approach is to augment the anchor point using augmentation methods to generate positive points for the anchor [4, 28].

For Siamese networks with two sub-networks, pairs of anchor-positive points $\{(\boldsymbol{x}_i^a, \boldsymbol{x}_i^p)\}_{i=1}^{n_t}$ and anchor-negative points $\{(\boldsymbol{x}_i^a, \boldsymbol{x}_i^n)\}_{i=1}^{n_t}$ are created, where n_t is the number of pairs. For Siamese networks with three sub-networks, triplets of anchor-positive-negative points $\{(\boldsymbol{x}_i^a, \boldsymbol{x}_i^p, \boldsymbol{x}_i^n)\}_{i=1}^{n_t}$ are prepared, where n_t is the number of triplets. If every point in the dataset is considered an anchor, the number of pairs or triplets equals the number of data points, i.e., $n_t = n$.

Various loss functions for Siamese networks use these pairs or triplets to push the positive point toward the anchor and pull the negative point away from it. Performing this iteratively for all pairs or triplets reduces intra-class variance and increases inter-class variance, leading to better discrimination of classes or clusters. Some of the loss functions for training Siamese networks are introduced in the following sections.

4.3 Implementation of Siamese Networks

A Siamese network with two and three sub-networks is depicted in Fig. 4. The output of a Siamese network for input $\boldsymbol{x} \in \mathbb{R}^d$ is denoted by $\mathbf{f}(\boldsymbol{x}) \in \mathbb{R}^p$, where p is the dimensionality of the embedding (or the number of neurons in the last layer of the network), which is usually much less than the dimensionality of the data, i.e., $p \ll d$. Note that the sub-networks of a Siamese network can be any fully connected or Convolutional Neural Network (CNN), depending on the type of data. The network structure used for the sub-networks is usually referred to as the backbone network.

The weights of the sub-networks are shared, meaning that the values of their weights are identical. Implementation of a Siamese network can be done in two ways:

1. Several sub-networks can be implemented in memory. During the training phase, each data point in the pairs or triplets is fed to one of the sub-networks, and the outputs of the sub-networks are $\mathbf{f}(\boldsymbol{x}_i^a)$, $\mathbf{f}(\boldsymbol{x}_i^p)$, and $\mathbf{f}(\boldsymbol{x}_i^n)$. These outputs are used in the loss function, and the weights of only one of the sub-networks are updated via backpropagation (see Chap. 3). The updated weights are then copied to the other sub-networks. This process is repeated for all mini-batches and epochs until convergence. During the test phase, the test point $\boldsymbol{x}$ is fed to only one of the sub-networks, and the output $\mathbf{f}(\boldsymbol{x})$ is considered its embedding.

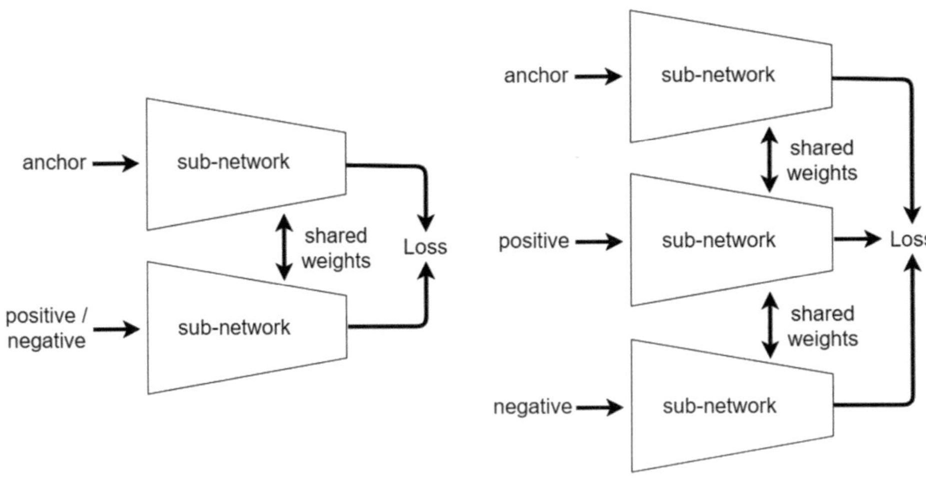

Fig. 4 The structure of Siamese network with **a** two and **b** three sub-networks

2. It is also possible to implement only one sub-network in memory. During the training phase, the data points in the pairs or triplets are fed to the sub-network one by one, and the outputs of the sub-network are $\mathbf{f}(\boldsymbol{x}_i^a)$, $\mathbf{f}(\boldsymbol{x}_i^p)$, and $\mathbf{f}(\boldsymbol{x}_i^n)$. These outputs are used in the loss function, and the weights of the sub-network are updated via backpropagation (see Chap. 3). This process is repeated for all mini-batches and epochs until convergence. During the test phase, the test point $\boldsymbol{x}$ is fed to the sub-network, and the output $\mathbf{f}(\boldsymbol{x})$ is taken as its embedding.

The advantage of the first approach is that all the sub-networks are readily available, eliminating the need to feed the points of pairs or triplets one by one. Its disadvantage is higher memory usage. Since the number of points in the pairs or triplets is small (i.e., only two or three), the second approach is generally recommended as it is more memory-efficient.

4.4 Important Loss Functions for Siamese Networks

There are various loss functions for training Siamese networks. Two of the most important loss functions are contrastive loss [22], proposed by Facebook, and triplet loss [37], proposed by Google. As shown in Fig. 5a, these losses push the embedding of the positive point toward the embedding of the anchor point while pulling the embedding of the negative point away from the embedding of the anchor point.

As mentioned earlier, both of these loss functions derive their core idea from Fisher Discriminant Analysis [8, 12], originally proposed by Fisher about a century ago. These loss functions are introduced in the following sections.

4.5 Contrastive Loss

One loss function for Siamese networks is the contrastive loss, which uses anchor-positive and anchor-negative pairs of points. Suppose that in each mini-batch there are b pairs of points $\{(\boldsymbol{x}_i^1, \boldsymbol{x}_i^2)\}_{i=1}^b$, some of which are anchor-positive pairs and some are anchor-negative pairs. The points in an anchor-positive pair are similar, i.e., $(\boldsymbol{x}_i^1, \boldsymbol{x}_i^2) \in \mathcal{S}$, and the points in an anchor-negative pair are dissimilar, i.e., $(\boldsymbol{x}_i^1, \boldsymbol{x}_i^2) \in \mathcal{D}$, where $\mathcal{S}$ and $\mathcal{D}$ denote the sets of similar and dissimilar pairs, respectively.

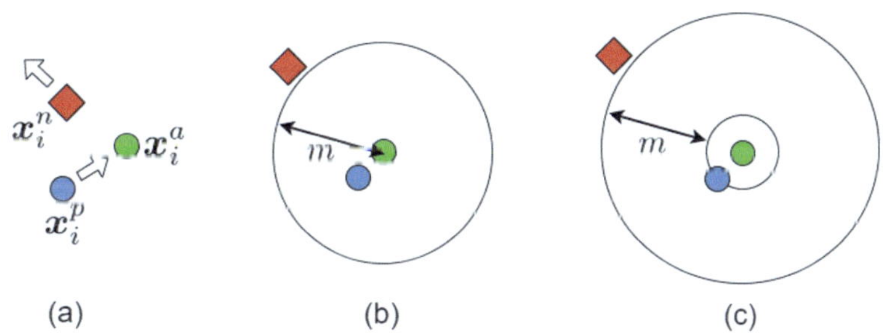

Fig. 5 Visualization of what contrastive and triplet losses do: **a** a triplet of anchor (green circle), positive (blue circle), and negative (red diamond) points, **b** the effect of contrastive loss creating a margin between the anchor and negative point, and **c** the effect of triplet loss creating a margin between the positive and negative points

4.5.1 Contrastive Loss

Suppose:

$$y_i := \begin{cases} 0 \text{ if } (\boldsymbol{x}_i^1, \boldsymbol{x}_i^2) \in \mathcal{S}, \\ 1 \text{ if } (\boldsymbol{x}_i^1, \boldsymbol{x}_i^2) \in \mathcal{D}, \end{cases} \quad \forall i \in \{1, \dots, n_t\}. \tag{5}$$

The main contrastive loss, proposed in 2006 [22], is

$$\underset{\theta}{\text{minimize}} \quad \sum_{i=1}^{b} \Big((1 - y_i) d\big(\mathbf{f}(\boldsymbol{x}_i^1), \mathbf{f}(\boldsymbol{x}_i^2)\big) + y_i \big[- d\big(\mathbf{f}(\boldsymbol{x}_i^1), \mathbf{f}(\boldsymbol{x}_i^2)\big) + m \big]_+ \Big), \tag{6}$$

where $m > 0$ is the margin and $[.]_+ := \max(., 0)$ is the standard hinge loss. The first term of the loss minimizes the embedding distances of similar points, and the second term maximizes the embedding distances of dissimilar points. As shown in Fig. 5b, it aims to make the distances of similar points as small as possible and the distances of dissimilar points at least greater than a margin m (because the term inside the hinge loss should approach zero).

4.5.2 Generalized Contrastive Loss

The y_i defined in Eq. (5) is used in the contrastive loss in Eq. (6). This variable is binary, representing a hard measure of similarity and dissimilarity. Instead of this hard measure, it is possible to use a soft measure of similarity and dissimilarity, denoted by ψ_i, indicating how similar $\boldsymbol{x}_i^1$ and $\boldsymbol{x}_i^2$ are. This measure ranges between zero (completely similar) and one (completely dissimilar). It can either be provided by the dataset as a predefined measure or computed using a similarity function such as the cosine function:

$$[0, 1] \ni \psi_i := \frac{1}{2}\big(-\cos(\boldsymbol{x}_i^1, \boldsymbol{x}_i^2) + 1 \big). \tag{7}$$

In this case, the pairs $\{(\boldsymbol{x}_i^1, \boldsymbol{x}_i^2)\}_{i=1}^b$ need not be completely similar or dissimilar but can be any two random points from the dataset with varying levels of similarity or dissimilarity. The generalized contrastive loss extends the contrastive loss using this soft measure of similarity [30]:

$$\underset{\theta}{\text{minimize}} \quad \sum_{i=1}^{b} \Big((1 - \psi_i) d\big(\mathbf{f}(\boldsymbol{x}_i^1), \mathbf{f}(\boldsymbol{x}_i^2)\big) + \psi_i \big[- d\big(\mathbf{f}(\boldsymbol{x}_i^1), \mathbf{f}(\boldsymbol{x}_i^2)\big) + m \big]_+ \Big). \tag{8}$$

4.6 Triplet Loss

One of the loss functions for Siamese networks with three sub-networks is the triplet loss, proposed in 2015 [37], which uses triplets in mini-batches, denoted by $\{(\boldsymbol{x}_i^a, \boldsymbol{x}_i^p, \boldsymbol{x}_i^n)\}_{i=1}^b$. It is defined as follows:

$$\underset{\theta}{\text{minimize}} \sum_{i=1}^{b} \Big[d\big(\mathbf{f}(\boldsymbol{x}_i^a), \mathbf{f}(\boldsymbol{x}_i^p)\big) - d\big(\mathbf{f}(\boldsymbol{x}_i^a), \mathbf{f}(\boldsymbol{x}_i^n)\big) + m \Big]_+, \tag{9}$$

where $m > 0$ is the margin and $[.]_+ := \max(., 0)$ is the standard hinge loss. As shown in Fig. 5c, due to the hinge loss, this loss function ensures that the distances to dissimilar points are greater than the distances to similar points by at least a margin m. In other words, there will be a distance of at least margin m between the positive and negative points. This loss function aims to eventually achieve

$$d\big(\mathbf{f}(\boldsymbol{x}_i^a), \mathbf{f}(\boldsymbol{x}_i^p)\big) + m \leq d\big(\mathbf{f}(\boldsymbol{x}_i^a), \mathbf{f}(\boldsymbol{x}_i^n)\big), \tag{10}$$

for all triplets.

Note that the triplet loss is closely related to the cost function used in spectral large margin metric learning [46, 47]. It is also noteworthy that using the triplet loss as a regularization term alongside cross-entropy loss has been shown to increase the robustness of networks to certain adversarial attacks [33].

4.7 Triplet Mining

In every mini-batch containing data points from c classes, triplets of data points can be selected and used in different ways. For example, all similar and dissimilar points can be used for every anchor point as positive and negative points, respectively. Another approach is to use only some of the similar and dissimilar points within the mini-batch. These approaches for selecting and using triplets are called triplet mining [40].

In the following, some of the most important triplet mining methods are reviewed. Triplet mining methods are used with the triplet loss, i.e., Eq. (9). Suppose b is the mini-batch size, $c(\boldsymbol{x}_i)$ is the class index of $\boldsymbol{x}_i$, $\mathcal{X}_j$ denotes the points of the jth class in the mini-batch, and $\mathcal{X}$ denotes all data points in the mini-batch.

- **Batch-all**: Batch-all triplet mining [7] considers every point in the mini-batch as an anchor point. All points in the mini-batch that are in the same class as the anchor point are used as positive points. All points in the mini-batch that are in a different class from the anchor point are used as negative points:

$$\underset{\theta}{\text{minimize}} \sum_{i=1}^{b} \sum_{\boldsymbol{x}_j \in \mathcal{X}_{c(\boldsymbol{x}_i)}} \sum_{\boldsymbol{x}_k \in \mathcal{X} \setminus \mathcal{X}_{c(\boldsymbol{x}_i)}} \Big[d\big(\mathbf{f}(\boldsymbol{x}_i), \mathbf{f}(\boldsymbol{x}_j)\big) - d\big(\mathbf{f}(\boldsymbol{x}_i), \mathbf{f}(\boldsymbol{x}_k)\big) + m \Big]_+ . \tag{11}$$

  Batch-all mining utilizes all data points in the mini-batch to maximize the available information.
- **Batch-hard**: Batch-hard triplet mining [25] considers every point in the mini-batch as an anchor point. The hardest positive, which is the farthest point from the anchor point within the same class, is used as the positive point. The hardest negative, which is the closest point to the anchor point from another class, is used as the negative point:

$$\underset{\theta}{\text{minimize}} \sum_{i=1}^{b} \Big[\max_{\boldsymbol{x}_j \in \mathcal{X}_{c(\boldsymbol{x}_i)}} d\big(\mathbf{f}(\boldsymbol{x}_i), \mathbf{f}(\boldsymbol{x}_j)\big) - \min_{\boldsymbol{x}_k \in \mathcal{X} \setminus \mathcal{X}_{c(\boldsymbol{x}_i)}} d\big(\mathbf{f}(\boldsymbol{x}_i), \mathbf{f}(\boldsymbol{x}_k)\big) + m \Big]_+ . \tag{12}$$

  Batch-hard mining selects the hardest points to ensure the network learns the most challenging cases. By learning these hardest cases, the network is expected to generalize well to easier cases. This approach is also supported by opposition-based learning [42]. Batch-hard mining has been used in many applications such as person re-identification [45].

– **Batch-semi-hard**: Batch-semi-hard triplet mining [37] considers every point in the mini-batch as an anchor point. All points in the mini-batch that are in the same class as the anchor point are used as positive points. The hardest negative, which is the closest point to the anchor point from another class but still farther than the positive point, is used as the negative point:

$$
\begin{aligned}
\underset{\theta}{\text{minimize}} \sum_{i=1}^{b} \sum_{\boldsymbol{x}_j \in \mathcal{X}_{c(\boldsymbol{x}_i)}} \Big[& d\big(\mathbf{f}(\boldsymbol{x}_i), \mathbf{f}(\boldsymbol{x}_j)\big) \\
& - \min_{\boldsymbol{x}_k \in \mathcal{X} \setminus \mathcal{X}_{c(\boldsymbol{x}_i)}} \big\{ d\big(\mathbf{f}(\boldsymbol{x}_i), \mathbf{f}(\boldsymbol{x}_k)\big) \mid d\big(\mathbf{f}(\boldsymbol{x}_i), \mathbf{f}(\boldsymbol{x}_k)\big) > d\big(\mathbf{f}(\boldsymbol{x}_i), \mathbf{f}(\boldsymbol{x}_j)\big) \big\} + m \Big]_+ .
\end{aligned} \tag{13}
$$

4.8 Triplet Sampling

Rather than using only the extreme (hardest or easiest) positive and negative points [40], it is possible to sample positive and negative points from those available in the mini-batch or from certain distributions. There are several approaches for sampling positive and negative points [10]:

- Sampling based on extreme distances between points.
- Random sampling from classes.
- Sampling by distribution but from existing points.
- Stochastic sampling from class distributions.

The first, second, and third approaches sample the positive and negative points from the set of points in the mini-batch. This type of sampling is called survey sampling [18]. The third and fourth approaches sample points stochastically from distributions.

An example of the third approach, i.e., sampling by distribution but from existing points, is distance weighted sampling [48]. An example of the fourth approach is sampling triplets using the Bayesian updating theorem [39].

5 Deep Metric Learning by Self-supervised Learning

If deep metric learning uses class labels, it is a supervised algorithm. However, labeling is a difficult task that is both time-consuming and costly. Due to the lack of labels in many applications, it is often preferable to let the algorithm generate the labels itself in unsupervised cases. Such algorithms, which create labels from unlabeled data, are referred to as Self-supervised Learning (SSL). SSL is a family of algorithms that includes various loss functions and methods. It can be used for feature extraction and for learning a rich embedding space, making it useful for deep metric learning.

5.1 The Core Idea of Self-supervised Learning

The goal of SSL is to learn an embedding space where similar instances are close together and dissimilar instances are far apart, but without using labels. As mentioned earlier, this idea was initially proposed by Ronald A. Fisher [8, 12] a century ago. Moreover, the contrastive [22] and triplet [37] losses used in Siamese networks—introduced in Sect. 4—also have this goal. Therefore, SSL can be used with these loss functions when data are unlabeled.

5.2 Augmentation for Creating Positive Samples

Most SSL algorithms use augmentation to create labels. The idea is simple: these methods generate augmentations of a data instance and consider these augmentations as having the same label. They do this for different data instances to create multiple sets (e.g., pairs or triplets) where the data instances in each set share the same label. For example, they may take a picture of a dog and rotate it, then consider the original and rotated images to have the same label. As another example, they may create two augmentations of a picture of a house, one by cropping and another by blurring. These different augmentations should belong to the same label because they are all images of the same house.

The data instances obtained by augmentations of the same instance are called positive (or corresponding) samples. Instances that are not positive are considered negative (non-corresponding) samples. Note that SSL can even outperform supervised learning [20], as data augmentation can teach the network to generalize better to unseen data.

Consider a d-dimensional dataset $\{\boldsymbol{x}_\ell \in \mathbb{R}^d\}_{\ell=1}^n$, where n is the total number of training data instances. In SSL, each mini-batch with size b contains two sets of instances, $\{\boldsymbol{x}_i\}_{i=1}^b$ and $\{\boldsymbol{x}'_i\}_{i=1}^b$, where each pair $(\boldsymbol{x}_i, \boldsymbol{x}'_i)$ consists of different augmentations of the same original sample. Formally:

$$\boldsymbol{x}_i := \mathcal{A}(\boldsymbol{x}_\ell), \quad \boldsymbol{x}'_i := \mathcal{A}(\boldsymbol{x}_\ell), \quad i \in \{1, \dots, b\}, \tag{14}$$

with ℓ randomly chosen from $\{1, \dots, n\}$ and $\mathcal{A}(\cdot)$ denoting an augmentation such as rotation, cropping, scaling, or translation. As a result, each pair $(\boldsymbol{x}_i, \boldsymbol{x}'_i)$ shares the same underlying pattern, producing positive (corresponding) embeddings (z_i, z'_i) without using class labels.

5.3 Negative Samples

As mentioned in Sect. 5.1, the loss functions of different SSL algorithms often share the same goal as deep metric learning. They usually push the embeddings of same-label instances toward each other. In addition to bringing same-label instances closer, some SSL algorithms also pull apart the embeddings of instances that do not have the same label.

However, it should be noted that this approach carries the risk of pulling apart embeddings of two sets that accidentally belong to the same class. This is because, although it is certain that each set contains same-label instances, there is no guarantee that two different sets have different labels; both may be augmentations of samples from the same class (e.g., augmentations of two different dog pictures). This issue cannot be detected because the data are unlabeled.

Nevertheless, this risk is reduced in cases with a large number of classes, as the chances of sampling multiple instances from the same class decrease. When the number of classes c is much larger than the mini-batch size b, the probability of sampling at least k instances from the same class in a mini-batch is [38]

$$\mathbb{P} = \sum_{i=k}^{b} \binom{b}{i} \left(\frac{1}{c}\right)^i \left(1 - \frac{1}{c}\right)^{b-i}, \tag{15}$$

which becomes negligible for datasets with large c. Therefore, negative pairs $(\boldsymbol{x}_i, \boldsymbol{x}_j)$ are likely to originate from different classes, leading to distinct embeddings (z_i, z_j). This assumption underlies many contrastive SSL methods [31, 35, 38] and few-shot learning approaches [28], although it may break down when the number of classes is small.

5.4 Architecture of Self-supervised Learning Networks

A typical SSL pipeline includes an encoder followed by a projector (also called a projection head). Both the encoder and projector can have multiple layers of neural networks. The encoder f_θ maps inputs to a q-dimensional latent space, and the projector h_ϕ projects them to a higher-dimensional or lower-dimensional space $\mathbb{R}^p$:

$$\text{encoder: } f_\theta : \mathbb{R}^d \to \mathbb{R}^q, \quad \boldsymbol{y} := f_\theta(\boldsymbol{x}), \tag{16}$$

$$\text{projector: } h_\phi : \mathbb{R}^q \to \mathbb{R}^p, \quad \boldsymbol{z} := h_\phi(\boldsymbol{y}), \tag{17}$$

where θ and ϕ are learnable parameters.

Some SSL networks use an expander as the projector on top of the encoder, while others use a reducer after the encoder. If the projector is an expander, it projects the encodings $\boldsymbol{y} \in \mathbb{R}^q$ into higher-dimensional embeddings $\boldsymbol{z} \in \mathbb{R}^p$, where $p \gg q$. If it is a reducer, then $p < q$.

The rationale behind using an expander is that it behaves like a kernel method, mapping the encodings into a higher-dimensional embedding space, similar to a Reproducing Kernel Hilbert Space. This transformation simplifies the nonlinear patterns of the encodings into more linear embeddings [11]. Conversely, the rationale for using a reducer is that it acts as a dimensionality reduction technique for feature extraction (see the discussion in Sect. 2.4).

The embeddings $\boldsymbol{z}$—either normalized or unnormalized—are used in the loss functions for training SSL networks. Many SSL methods normalize the embeddings $\boldsymbol{z}$ before using them in the loss functions. Some perform feature-wise normalization, while others perform batch-wise (or sample-wise) normalization. After training the SSL method, during the test phase, the expander is discarded and the trained encodings $\boldsymbol{y}$ are used as the final embeddings of the data instances.

5.5 Evaluating the Trained Self-supervised Learning Networks

Evaluation of SSL networks is commonly done with linear probing (training a simple classifier on frozen embeddings), but fine-tuning on downstream tasks such as classification, detection, or segmentation is also standard. Together these provide a fuller measure of the learned representation quality.

For linear probing, the expander is usually dropped and a linear layer with softmax or logistic activation function is added on top of the encoder (see Sect. 2.8 in Chap. 2). This added final layer acts as a linear classifier, also called logistic regression (see Sect. 3.2 in Chap. 2).

During evaluation, the target class labels are used to train the weights of the linear classifier, which is fed by the encodings $\boldsymbol{y}$. After training the linear classifier, a held-out test dataset is fed to the encoder, and the resulting encodings $\boldsymbol{y}$ are passed through the trained linear classifier. The classification accuracy (usually the top-k accuracy) is reported; higher accuracy indicates that the SSL algorithm has learned better embeddings or encodings.

In some cases, the encodings $\boldsymbol{y}$ are also visualized using UMAP [17, 34] or t-SNE [16, 43] to illustrate the separation of classes in the learned encoder space. The dataset commonly used to train and test SSL methods is the ImageNet dataset [6] because it is large and contains diverse classes. The labels of the dataset are not used for training the SSL algorithm; they are only used for evaluation.

SSL papers also evaluate their methods using downstream tasks such as domain adaptation. In these experiments, the SSL model is trained on one dataset, and the expander network is discarded after training. A linear classifier is then trained on top of the learned representations without fine-tuning to assess generalization to other unseen datasets. Better generalization indicates better domain adaptation between datasets, demonstrating the quality of the embeddings or encodings learned by the SSL algorithm.

5.6 Important Self-supervised Learning Methods

There are various SSL methods in the literature [21]. SSL methods include SimCLR [4], BYOL [20], SimSiam [5], Barlow Twins [50], and VICReg [1], alongside other influential approaches such as Masked Autoencoders (MAE) [24] and multi-modal frameworks like Contrastive Language-Image Pretraining (CLIP) [36], which have become central in practice by 2025. Some of the SSL methods are introduced in the following.

5.6.1 SimCLR

SimCLR (Simple framework for Contrastive Learning), proposed in 2020 [4], uses contrastive learning, i.e., it maximizes agreement between two augmented views of the same image. Its backbone encoder, with weights θ, can be any neural network, such as ResNet [23] (see Chap. 4). Its projector, with weights ϕ, is a reducer implemented as a multi-layer perceptron, usually with two layers (see Chap. 2). The neural network architecture of SimCLR is illustrated in Fig. 6a.

Consider a mini-batch of size b containing the corresponding positive samples $\{(\boldsymbol{x}_i, \boldsymbol{x}'_i)\}_{i=1}^b$. The embeddings of the mini-batch are $\{(z_i, z'_i)\}_{i=1}^b$. SimCLR uses the InfoNCE loss function [35]:

$$\underset{\theta,\phi}{\text{minimize}} \quad \mathcal{L}(z, z') := -\sum_{i=1}^{b} \log \frac{\exp\left(\cos(z_i, z'_i)/\tau\right)}{\sum\limits_{j=1, j\neq i}^{b} \exp\left(\cos(z_i, z'_k)/\tau\right)}, \tag{18}$$

where $\tau > 0$ is a temperature parameter, and cos(.) is the similarity measure computed by the cosine function:

$$\cos(\boldsymbol{u}, \boldsymbol{v}) := \frac{\boldsymbol{u}^\top \boldsymbol{v}}{\|\boldsymbol{u}\|_2 \|\boldsymbol{v}\|_2}. \tag{19}$$

Note that the feature-wise normalized embeddings z are used in the loss function of SimCLR. This contrastive loss encourages corresponding (similar) samples to be close in the embedding space and non-corresponding (dissimilar) samples to be far apart.

5.6.2 BYOL

BYOL (Bootstrap Your Own Latent), proposed in 2020 [20], is an SSL method that does not use negative samples, making it an improvement over SimCLR [4]. It uses only positive samples, i.e., two augmentations of the same data instance. BYOL avoids collapse—without needing contrastive loss or negative samples—by using asymmetry, a momentum encoder, and stop-gradient, as explained below.

As depicted in Fig. 6b, BYOL consists of two branches, each being a Siamese neural network (see Sect. 4): an online network and a target network. The online network contains the encoder, projector, and predictor heads:

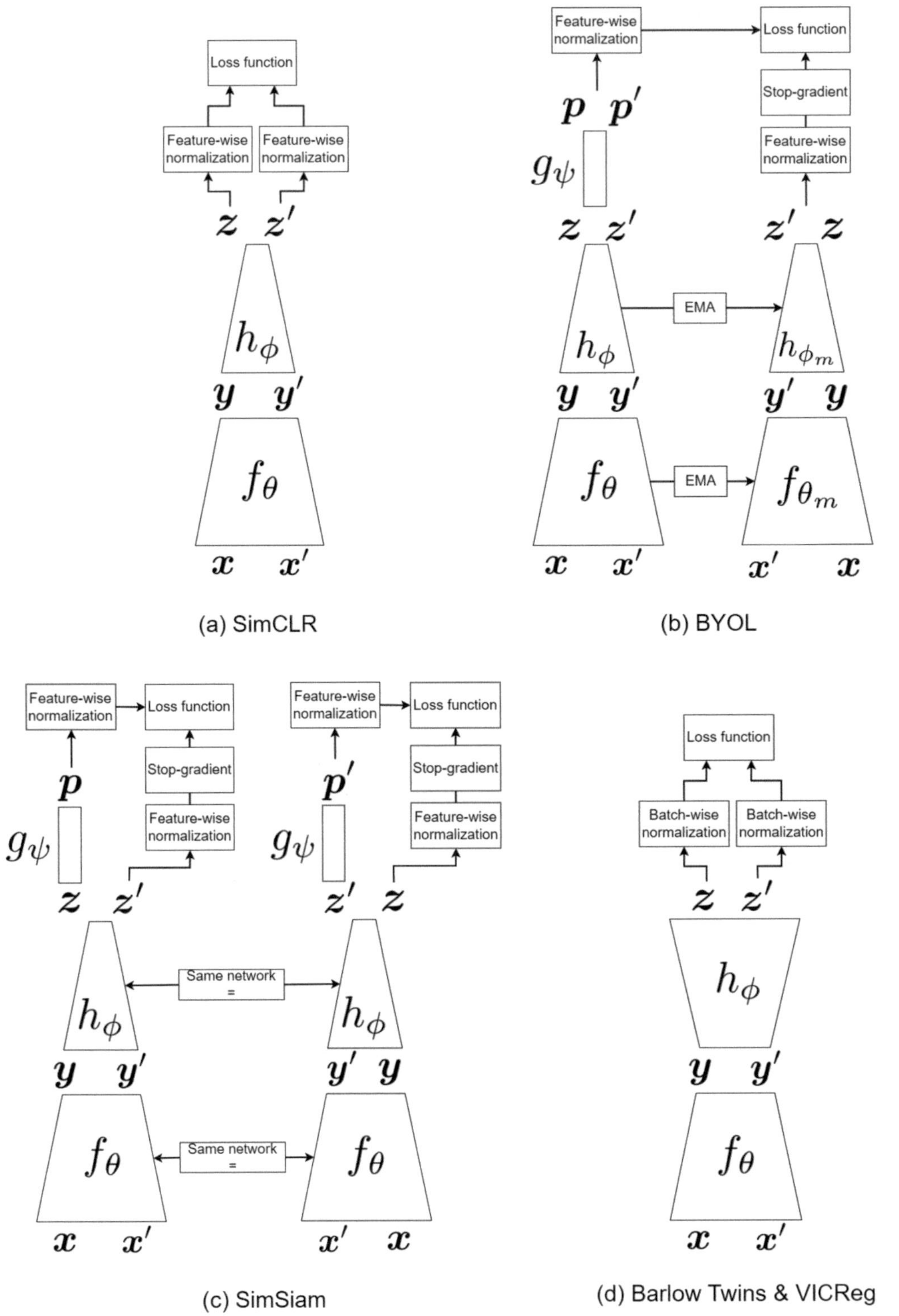

Fig. 6 The neural network architectures of **a** SimCLR, **b** BYOL, **c** SimSiam, and **d** Barlow Twins and VICReg methods

$$\text{encoder: } f_\theta : \mathbb{R}^d \to \mathbb{R}^q, \quad \boldsymbol{y} := f_\theta(\boldsymbol{x}), \tag{20}$$
$$\text{projector: } h_\phi : \mathbb{R}^q \to \mathbb{R}^p, \quad \boldsymbol{z} := h_\phi(\boldsymbol{y}), \tag{21}$$
$$\text{predictor: } g_\psi : \mathbb{R}^p \to \mathbb{R}^p, \quad \boldsymbol{p} := g_\psi(\boldsymbol{z}), \tag{22}$$

where θ, ϕ, and ψ are the weights of the encoder, projector, and predictor in the online network, respectively. Each of these sub-networks may have multiple layers.

The target network does not have a predictor; it contains only the encoder and projector:

$$\text{encoder: } f_{\theta_m} : \mathbb{R}^d \to \mathbb{R}^q, \quad \boldsymbol{y} := f_{\theta_m}(\boldsymbol{x}), \tag{23}$$
$$\text{projector: } h_{\phi_m} : \mathbb{R}^q \to \mathbb{R}^p, \quad \boldsymbol{z} := h_{\phi_m}(\boldsymbol{y}), \tag{24}$$

where θ_m and ϕ_m are the weights of the encoder and projector in the target network, respectively.[2] The architectures of the encoder and projector are the same in both the online and target networks. Note that the input and output of the predictor head have the same dimensionality because the BYOL loss function compares the prediction $\boldsymbol{p}$ from the online network with the embedding $\boldsymbol{z}$ from the target network.

The mini-batch of size b contains the corresponding positive samples $\{(\boldsymbol{x}_i, \boldsymbol{x}'_i)\}_{i=1}^b$. Each $\boldsymbol{x}_i$ is fed to the online network to obtain $\boldsymbol{p}_i$, and each $\boldsymbol{x}'_i$ is fed to the online network to obtain $\boldsymbol{z}'_i$. Then, $\boldsymbol{x}_i$ is fed to the target network to get $\boldsymbol{z}_i$, and each $\boldsymbol{x}'_i$ is fed to the online network to get $\boldsymbol{p}'_i$. These vectors $\boldsymbol{p}_i$, $\boldsymbol{z}'_i$, $\boldsymbol{z}_i$, and $\boldsymbol{p}'_i$ undergo feature-wise normalization before being used in the loss function. The BYOL loss function is

$$\underset{\theta,\phi,\psi}{\text{minimize}} \quad \mathcal{L}(\boldsymbol{p}, \boldsymbol{p}', \boldsymbol{z}, \boldsymbol{z}') := \sum_{i=1}^{b} \Big(\mathcal{D}\big(\boldsymbol{p}_i, \text{stopgrad}(\boldsymbol{z}'_i)\big) + \mathcal{D}\big(\boldsymbol{p}'_i, \text{stopgrad}(\boldsymbol{z}_i)\big) \Big), \tag{25}$$

where stopgrad(.) is the stop-gradient operation, and $\mathcal{D}(.,.)$ is the cosine similarity loss, defined as follows:

$$\mathcal{D}(\boldsymbol{u}, \boldsymbol{v}) := \left\| \frac{\boldsymbol{u}}{\|\boldsymbol{u}\|_2} - \frac{\boldsymbol{v}}{\|\boldsymbol{v}\|_2} \right\|_2^2 = 2 - 2\cos(\boldsymbol{u}, \boldsymbol{v}), \tag{26}$$

where $\cos(.,.)$ is defined in Eq. (19).

BYOL uses stop-gradient on the target network to prevent its weights from being updated by backpropagation; in other words, its weights are frozen during backpropagation. Thus, minimizing Eq. (25) by gradient descent only updates the weights of the online network, i.e., θ, ϕ, and ψ. The weights of the target network, i.e., θ_m and ϕ_m, are updated using momentum (see Sect. 2.4 in Chap. 3):

$$\theta_m \leftarrow \tau\theta_m + (1-\tau)\theta, \tag{27}$$
$$\phi_m \leftarrow \tau\phi_m + (1-\tau)\phi, \tag{28}$$

where $\tau \in (0, 1)$ is the momentum coefficient, typically close to one (e.g., 0.99). This update by momentum is also called Exponential Moving Average (EMA), and the target network, which uses EMA, is referred to as the momentum encoder. The EMA updates the weights of the target network slowly, allowing them to retain previous values while incorporating slight updates from the online network.

[2] The subscript m stands for momentum here.

According to Eq. (25), the online network follows the target network, while the target network is updated slowly. This behavior is similar to knowledge distillation [3, 26] (see Sect. 3 in Chap. 20), where a student network follows a teacher network, with the teacher network possibly being updated slowly during student training. The stop-gradient in the target network ensures it serves as a stable reference for the online network to match.

The use of momentum results in different update speeds between the online and target networks. Additionally, the use of stop-gradient causes the weight updates of these networks to differ. As a result, there is an asymmetry between the online and target networks, which prevents trivial solutions and collapse of the embeddings.

5.6.3 SimSiam

SimSiam (Simple Siamese), proposed in 2021 [5], is an SSL method that, like BYOL [20], does not use negative samples. Unlike BYOL, it also does not require a momentum encoder (or target network), making it an improvement over BYOL [20].

SimSiam has only one encoder and reducer projector, which are used twice through two passes (see Fig. 6c). The encoder, with weights θ, accepts $\boldsymbol{x}$ and $\boldsymbol{x}'$ as inputs and then passes the encodings $\boldsymbol{y}$ and $\boldsymbol{y}'$ to the projector, with weights ϕ. Inspired by BYOL, in one of the passes, there is a prediction head $g_\psi(.)$, with weights ψ, on top of the projector to produce predictions $\boldsymbol{p}$. These predictions undergo feature-wise normalization.

In the other pass, the embeddings $\boldsymbol{z}$ are normalized feature-wise. Again inspired by BYOL, there is a stop-gradient in the second pass, preventing the weights in that branch from being updated by backpropagation. As explained in Sect. 5.6.2, this prevents collapse of the embeddings.

Consider a positive pair $(\boldsymbol{x}_1, \boldsymbol{x}_2)$. These two data instances are passed through each branch to obtain $\{\boldsymbol{p}_1, \boldsymbol{z}_2\}$ and $\{\boldsymbol{p}_2, \boldsymbol{z}_1\}$. The loss function is

$$\underset{\theta,\phi,\psi}{\text{minimize}} \quad \mathcal{L}(\boldsymbol{p}, \boldsymbol{p}', \boldsymbol{z}, \boldsymbol{z}') := -\frac{1}{2}\sum_{i=1}^{b}\Big(\cos\big(\boldsymbol{p}_i, \text{stopgrad}(\boldsymbol{z}_i')\big) + \cos\big(\boldsymbol{p}_i', \text{stopgrad}(\boldsymbol{z}_i)\big)\Big), \tag{29}$$

where stopgrad(.) is the stop-gradient operation, and the cosine function is defined in Eq. (19). This loss maximizes the similarity between the positive embeddings and predictions without requiring negative samples.

There is no need for a momentum encoder in SimSiam because asymmetry alone, combined with the predictor, is sufficient to prevent collapse. Comparing SimSiam and BYOL shows that SimSiam is simpler than BYOL (introduced in Sect. 5.6.2) because it does not require two network branches; this simplicity may be one reason behind its naming. Another inspiration for its naming may be its conceptual relation to SimCLR [4].

5.6.4 Barlow Twins

Barlow Twins, proposed in 2021 [50], matches embeddings from two augmented views while minimizing redundancy by decorrelating the dimensions of the embeddings. It has an encoder followed by an expander projector, as shown in Fig. 6d. The embeddings $\boldsymbol{z}$ undergo batch-wise normalization.

Let the normalized positive pairs in the mini-batch of size b be denoted by $\{\boldsymbol{z}_i, \boldsymbol{z}_i'\}_{i=1}^{b}$, where $\boldsymbol{z}_i \in \mathbb{R}^p$ and $\boldsymbol{z}_i' \in \mathbb{R}^p$ correspond to each $i \in \{1, \dots, b\}$. Suppose $z_{i,j} \in \mathbb{R}$ and $z_{i,j}' \in \mathbb{R}$ denote the jth elements of $\boldsymbol{z}_i$ and $\boldsymbol{z}_i'$, respectively. The (j, k)th element of the Pearson correlation coefficient (cross-correlation) matrix of the embeddings, $\boldsymbol{C} \in \mathbb{R}^{p\times p}$, denoted by c_{jk}, is

$$\mathbb{R} \ni c_{jk} := \frac{\sum_{i=1}^{b} z_{i,j} z'_{i,k}}{\sqrt{\sum_{i=1}^{b} (z_{i,j})^2} \sqrt{\sum_{i=1}^{b} (z'_{i,k})^2}}. \tag{30}$$

The loss function is

$$\underset{\theta,\phi}{\text{minimize}} \quad \mathcal{L}(z, z') := \sum_{j=1}^{p} (1 - c_{jj})^2 + \lambda \sum_{j=1}^{p} \sum_{k=1, k \neq j}^{p} c_{jk}^2, \tag{31}$$

where $\lambda > 0$ is the regularization parameter. This loss function maximizes the correlation of corresponding embedding features of positive samples to one while minimizing the correlation between different features, thus reducing redundancy and decorrelating dimensions.

5.6.5 VICReg

VICReg (Variance-Invariance-Covariance Regularization), proposed in 2022 [1], balances three components—invariance, variance, and covariance—in its loss function. Similar to Barlow Twins [50], it has an encoder followed by an expander projector (see Fig. 6d).

Consider a mini-batch (with size b) of p-dimensional embeddings $\mathbf{Z} = [z_1, \dots, z_b]^\top \in \mathbb{R}^{b \times p}$ and $\mathbf{Z}' = [z'_1, \dots, z'_b]^\top \in \mathbb{R}^{b \times p}$, where z_i and z'_i are embeddings of corresponding positive samples. The invariance term in the loss function is

$$\mathcal{L}_{\text{inv}}(z, z') := \frac{1}{b} \sum_{i=1}^{b} \|z_i - z'_i\|_2^2, \tag{32}$$

where z_i and z'_i are embeddings of the ith positive samples in the mini-batch.

Minimizing the invariance term pushes the positive embeddings toward each other. However, this minimization alone would collapse all embeddings to zero vectors; in other words, the network would undergo undesired shortcut learning [9] by making all vectors zero to minimize the loss function. Therefore, two regularization terms on the variance and covariance of embedding features are added to the loss function.

Let $z^j \in \mathbb{R}^b$ and $z'^j \in \mathbb{R}^b$ denote the jth columns (features) of $\mathbf{Z}$ and $\mathbf{Z}'$, respectively. The variance regularization terms in the loss function are

$$\mathcal{L}_{\text{var}}(z) := \frac{1}{p} \sum_{j=1}^{p} \max(0, \gamma - \sqrt{\mathbf{Var}(z^j) + \epsilon}), \tag{33}$$

$$\mathcal{L}_{\text{var}}(z') := \frac{1}{p} \sum_{j=1}^{p} \max(0, \gamma - \sqrt{\mathbf{Var}(z'^j) + \epsilon}), \tag{34}$$

where $\gamma > 0$ is a constant target value for the standard deviation, $\epsilon > 0$ is a small scalar preventing numerical instabilities, and $\mathbf{Var}(.)$ calculates the variance of the elements of its input vector. Minimizing the variance regularization terms maximizes the variance of embedding features up to the value γ, making the learned embedding features variant and uncorrelated, thus preventing collapse.

The covariance matrices of embeddings z and z' are

$$\mathbb{R}^{p\times p} \ni \boldsymbol{C}(\mathbf{Z}) = \frac{1}{b-1}\sum_{i=1}^{b}(z_i - \bar{z})(z_i - \bar{z})^\top, \tag{35}$$

$$\mathbb{R}^{p\times p} \ni \boldsymbol{C}(\mathbf{Z}') = \frac{1}{b-1}\sum_{i=1}^{b}(z_i' - \bar{z}')(z_i' - \bar{z}')^\top, \tag{36}$$

where $\bar{z}$ and $\bar{z}'$ are the mean embeddings in the mini-batch:

$$\bar{z} := \frac{1}{b}\sum_{i=1}^{b} z_i, \quad \bar{z}' := \frac{1}{b}\sum_{i=1}^{b} z_i'. \tag{37}$$

Suppose that $c_{jk}(\mathbf{Z})$ and $c_{jk}(\mathbf{Z}')$ denote the (j, k)th elements of $\boldsymbol{C}(\mathbf{Z})$ and $\boldsymbol{C}(\mathbf{Z}')$, respectively. The covariance regularization terms are

$$\mathcal{L}_{\text{cov}}(z) := \frac{1}{p}\sum_{j=1}^{p}\sum_{k=1,k\neq j}^{p} c_{jk}^2(\mathbf{Z}), \tag{38}$$

$$\mathcal{L}_{\text{cov}}(z') := \frac{1}{p}\sum_{j=1}^{p}\sum_{k=1,k\neq j}^{p} c_{jk}^2(\mathbf{Z}'). \tag{39}$$

According to the expression for the Pearson correlation coefficient:

$$\mathbf{Cor}(X, Y) = \frac{\mathbf{Cov}(X, Y)}{\sqrt{\mathbf{Var}(X)}\sqrt{\mathbf{Var}(Y)}}, \tag{40}$$

correlation $\mathbf{Cor}(X, Y)$ is directly proportional to covariance $\mathbf{Cov}(X, Y)$. Therefore, minimizing the covariance regularization terms reduces the correlation of the embedding features, decorrelating the learned embeddings and preventing collapse.

The overall loss function in VICReg is

$$\begin{aligned}\underset{\theta,\phi}{\text{minimize}} \quad \mathcal{L}(z, z') := \lambda_{\text{inv}}\mathcal{L}_{\text{inv}}(z, z') + \lambda_{\text{var}}\big(\mathcal{L}_{\text{var}}(z) + \mathcal{L}_{\text{var}}(z')\big) \\ + \lambda_{\text{cov}}\big(\mathcal{L}_{\text{cov}}(z) + \mathcal{L}_{\text{cov}}(z')\big),\end{aligned} \tag{41}$$

where $\lambda_{\text{inv}} > 0$, $\lambda_{\text{var}} > 0$, and $\lambda_{\text{cov}} > 0$ are regularization parameters. Minimizing this loss function pushes the embeddings of positive samples toward each other while preventing possible collapse by increasing the variance of embedding features and decreasing their correlation.

6 Conclusion

Deep metric learning and self-supervised learning form a bridge between supervised training and fully unsupervised representation learning. From autoencoders to contrastive methods and modern SSL frameworks, these approaches have reshaped how embeddings are learned without labels. Now, SSL has become the default paradigm for pretraining across vision, language, and multi-modal domains, with metric learning losses providing the foundation for robust, generalizable representations. Taken together, autoencoders, metric learning, and self-supervised methods show how representation learning has matured into a central pillar of deep learning, setting the stage for specialized domains such as graph neural networks.

7 Exercises

1. (Level: Easy) Consider a simple autoencoder trained to reconstruct three-dimensional binary vectors. The autoencoder consists of:

   - Encoder: Compresses input $\boldsymbol{x} = [x_1, x_2, x_3]^\top \in \mathbb{R}^3$ to latent space $z \in \mathbb{R}^2$.
   - Decoder: Reconstructs $\widehat{\boldsymbol{x}} = [\widehat{x}_1, \widehat{x}_2, \widehat{x}_3]^\top \in \mathbb{R}^3$ from z.

   Given the input:

$$\boldsymbol{x} = \begin{bmatrix} 1 \\ 0 \\ 1 \end{bmatrix}, \quad \widehat{\boldsymbol{x}} = \begin{bmatrix} 0.9 \\ 0.2 \\ 0.7 \end{bmatrix}.$$

   Assume the reconstruction loss is the binary cross-entropy loss.

$$\mathcal{L}_{\text{BCE}}(\boldsymbol{x}, \widehat{\boldsymbol{x}}) = -\sum_{i=1}^{3} \left[x_i \log(\hat{x}_i) + (1 - x_i) \log(1 - \hat{x}_i) \right].$$

   Compute the binary cross-entropy loss.
2. (Level: Moderate) Suppose you are training a Siamese neural network to embed data points into a 2D space. You have the following three embeddings for anchor, positive, and negative points:

$$f(x_a) = [1, 2]^\top, \quad f(x_p) = [2, 3]^\top, \quad f(x_n) = [5, 6]^\top.$$

   Assume the margin $\alpha = 1$.

   (a) Compute the triplet loss:

$$\mathcal{L}_{\text{triplet}} = \max\left(\|f(x_a) - f(x_p)\|_2^2 - \|f(x_a) - f(x_n)\|_2^2 + \alpha,\ 0 \right).$$

   (b) Compute the contrastive loss. Use the following formula for contrastive loss:

$$\mathcal{L}_{\text{contrastive}} = (1 - y) \cdot \frac{1}{2} \left(\max(0, m - D)^2 \right) + y \cdot \frac{1}{2} D^2,$$

   where D is the Euclidean distance between the two embeddings, $y = 1$ for a positive (similar) pair and $y = 0$ for a negative (dissimilar) pair, and $m = 1$ is the contrastive margin.
3. (Level: Moderate) In a SimCLR-style self-supervised learning setup, a batch of $b = 4$ images is augmented twice, creating $2b = 8$ encoded vectors $z_1, z_2, \ldots, z_8 \in \mathbb{R}^2$. Each z_i is already ℓ_2-normalized (i.e., $\|z_i\|_2 = 1$), so the cosine similarity between two vectors simplifies to the dot product:

$$\text{sim}(z_i, z_j) = z_i^\top z_j.$$

Assume z_1 and z_2 form a positive pair, and all others z_3 through z_8 are negatives with respect to z_1. Let the temperature be $\tau = 0.5$, and you are given the following normalized vectors:

$$z_1 = \begin{bmatrix} 0.6 \\ 0.8 \end{bmatrix} \quad z_2 = \begin{bmatrix} 0.9 \\ 0.4 \end{bmatrix}$$
$$z_3 = \begin{bmatrix} 0.2 \\ 0.98 \end{bmatrix} \quad z_4 = \begin{bmatrix} -0.6 \\ 0.8 \end{bmatrix}$$
$$z_5 = \begin{bmatrix} 1.0 \\ 0.0 \end{bmatrix} \quad z_6 = \begin{bmatrix} -1.0 \\ 0.0 \end{bmatrix}$$
$$z_7 = \begin{bmatrix} 0.707 \\ 0.707 \end{bmatrix} \quad z_8 = \begin{bmatrix} 0.0 \\ 1.0 \end{bmatrix}.$$

Compute the following loss function for anchor z_1 with positive z_2, defined as follows:

$$\ell(z_1, z_2) = -\log\left(\frac{\exp(\text{sim}(z_1, z_2)/\tau)}{\sum_{k=1,k\neq 1}^{8} \exp(\text{sim}(z_1, z_k)/\tau)}\right).$$

References

1. Adrien Bardes, Jean Ponce, and Yann LeCun. “VICReg: Variance-Invariance-Covariance Regularization for Self-supervised Learning”. In: *International Conference on Learning Representations*. 2022.
2. Jane Bromley et al. “Signature verification using a “Siamese” time delay neural network”. In: *International Journal of Pattern Recognition and Artificial Intelligence* 7.04 (1993), pp. 669–688.
3. Cristian BuciluČŽ, Rich Caruana, and Alexandru Niculescu-Mizil. “Model compression”. In: *Proceedings of the 12th ACM SIGKDD international conference on Knowledge discovery and data mining*. 2006, pp. 535–541.
4. Ting Chen, Simon Kornblith, Mohammad Norouzi, and Geoffrey Hinton. “A simple framework for contrastive learning of visual representations”. In: *International conference on machine learning*. 2020, pp. 1597–1607.
5. Xinlei Chen and Kaiming He. “Exploring simple siamese representation learning”. In: *Proceedings of the IEEE/CVF conference on computer vision and pattern recognition*. 2021, pp. 15750–15758.
6. Jia Deng et al. “ImageNet: A large-scale hierarchical image database”. In: *2009 IEEE Conference on Computer Vision and Pattern Recognition*. IEEE. 2009, pp. 248–255.
7. Shengyong Ding, Liang Lin, Guangrun Wang, and Hongyang Chao. “Deep feature learning with relative distance comparison for person re-identification”. In: *Pattern Recognition* 48.10 (2015), pp. 2993–3003.
8. Ronald A Fisher. “The use of multiple measurements in taxonomic problems”. In: *Annals of eugenics* 7.2 (1936), pp. 179–188.
9. Robert Geirhos et al. “Shortcut learning in deep neural networks”. In: *Nature Machine Intelligence* 2.11 (2020), pp. 665–673.
10. Benyamin Ghojogh. “Data Reduction Algorithms in Machine Learning and Data Science”. PhD thesis. University of Waterloo, 2021.
11. Benyamin Ghojogh, Mark Crowley, Fakhri Karray, and Ali Ghodsi. “Background on kernels”. In: *Elements of Dimensionality Reduction and Manifold Learning*. Springer, 2023, pp. 43–73.
12. Benyamin Ghojogh, Mark Crowley, Fakhri Karray, and Ali Ghodsi. “Fisher Discriminant Analysis”. In: *Elements of Dimensionality Reduction and Manifold Learning*. Springer, 2023, pp. 155–184.
13. Benyamin Ghojogh, Mark Crowley, Fakhri Karray, and Ali Ghodsi. “Principal component analysis”. In: *Elements of Dimensionality Reduction and Manifold Learning*. Springer, 2023, pp. 123–154.
14. Benyamin Ghojogh, Mark Crowley, Fakhri Karray, and Ali Ghodsi. “Probabilistic Metric Learning”. In: *Elements of Dimensionality Reduction and Manifold Learning*. Springer, 2023, pp. 375–394.
15. Benyamin Ghojogh, Mark Crowley, Fakhri Karray, and Ali Ghodsi. “Spectral Metric Learning”. In: *Elements of Dimensionality Reduction and Manifold Learning*. Springer, 2022, pp. 313–351.

16. Benyamin Ghojogh, Mark Crowley, Fakhri Karray, and Ali Ghodsi. "Stochastic neighbour embedding". In: *Elements of Dimensionality Reduction and Manifold Learning*. Springer, 2023, pp. 455–477.
17. Benyamin Ghojogh, Mark Crowley, Fakhri Karray, and Ali Ghodsi. "Uniform manifold approximation and projection (UMAP)". In: *Elements of Dimensionality Reduction and Manifold Learning*. Springer, 2023, pp. 479–497.
18. Benyamin Ghojogh, Hadi Nekoei, Aydin Ghojogh, Fakhri Karray, and Mark Crowley. "Sampling algorithms, from survey sampling to Monte Carlo methods: Tutorial and literature review". In: *arXiv preprint* arXiv:2011.00901 (2020).
19. Ian Goodfellow, Yoshua Bengio, and Aaron Courville. *Deep learning*. MIT press, 2016.
20. Jean-Bastien Grill et al. "Bootstrap your own latent-a new approach to self-supervised learning". In: *Advances in neural information processing systems* 33 (2020), pp. 21271–21284.
21. Jie Gui et al. "A survey on self-supervised learning: Algorithms, applications, and future trends". In: *IEEE Transactions on Pattern Analysis and Machine Intelligence* (2024).
22. Raia Hadsell, Sumit Chopra, and Yann LeCun. "Dimensionality reduction by learning an invariant mapping". In: *2006 IEEE Computer Society Conference on Computer Vision and Pattern Recognition (CVPR'06)*. Vol. 2. IEEE. 2006, pp. 1735–1742.
23. Kaiming He, Xiangyu Zhang, Shaoqing Ren, and Jian Sun. "Deep residual learning for image recognition". In: *Proceedings of the IEEE conference on computer vision and pattern recognition*. 2016, pp. 770–778.
24. Kaiming He et al. "Masked autoencoders are scalable vision learners". In: *Proceedings of the IEEE/CVF conference on computer vision and pattern recognition*. 2022, pp. 16000–16009.
25. Alexander Hermans, Lucas Beyer, and Bastian Leibe. "In defense of the triplet loss for person re-identification". In: *arXiv preprint* arXiv:1703.07737 (2017).
26. Geoffrey Hinton, Oriol Vinyals, and Jeff Dean. "Distilling the Knowledge in a Neural Network". In: *NIPS 2014 Deep Learning Workshop*. 2015.
27. Elad Hoffer and Nir Ailon. "Deep metric learning using triplet network". In: *International workshop on similarity-based pattern recognition*. Springer. 2015, pp. 84–92.
28. Siavash Khodadadeh, Ladislau Bölöni, and Mubarak Shah. "Unsupervised meta-learning for few-shot image classification". In: *Advances in neural information processing systems*. Vol. 32. 2019.
29. Diederik P Kingma and Max Welling. "Auto-encoding variational Bayes". In: *International Conference on Learning Representations*. 2014.
30. María Leyva-Vallina, Nicola Strisciuglio, and Nicolai Petkov. "Generalized Contrastive Optimization of Siamese Networks for Place Recognition". In: *arXiv preprint* arXiv:2103.06638 (2021).
31. Mengjuan Li, Wenming Cao, Lei Zhang, Man Li, and Mingming Yang. "Enhancing Image Clustering with CLIP". In: *2024 International Conference on Cyber-Physical Social Intelligence (ICCSI)*. IEEE. 2024, pp. 1–6.
32. Alireza Makhzani, Jonathon Shlens, Navdeep Jaitly, Ian Goodfellow, and Brendan Frey. "Adversarial autoencoders". In: *arXiv preprint* arXiv:1511.05644 (2015).
33. Chengzhi Mao, Ziyuan Zhong, Junfeng Yang, Carl Vondrick, and Baishakhi Ray. "Metric learning for adversarial robustness". In: *Advances in neural information processing systems* (2019).
34. Leland McInnes, John Healy, and James Melville. "UMAP: Uniform manifold approximation and projection for dimension reduction". In: *arXiv preprint* arXiv:1802.03426 (2018).
35. Aaron van den Oord, Yazhe Li, and Oriol Vinyals. "Representation learning with contrastive predictive coding". In: *arXiv preprint* arXiv:1807.03748 (2018).
36. Alec Radford et al. "Learning transferable visual models from natural language supervision". In: *International conference on machine learning*. PmLR. 2021, pp. 8748–8763.
37. Florian Schroff, Dmitry Kalenichenko, and James Philbin. "FaceNet: A unified embedding for face recognition and clustering". In: *Proceedings of the IEEE conference on computer vision and pattern recognition*. 2015, pp. 815–823.
38. M Hadi Sepanj, Benyamin Ghojogh, and Paul Fieguth. "Self-Supervised Learning Using Nonlinear Dependence". In: *arXiv preprint* arXiv:2501.18875 (2025).
39. Milad Sikaroudi, Benyamin Ghojogh, Fakhri Karray, Mark Crowley, and Hamid R Tizhoosh. "Batch-incremental triplet sampling for training triplet networks using Bayesian updating theorem". In: *2020 25th International Conference on Pattern Recognition (ICPR)*. IEEE. 2021, pp. 7080–7086.
40. Milad Sikaroudi et al. "Offline versus online triplet mining based on extreme distances of histopathology patches". In: *International Symposium on Visual Computing*. Springer. 2020, pp. 333–345.
41. Milad Sikaroudi et al. "Supervision and source domain impact on representation learning: A histopathology case study". In: *2020 42nd Annual International Conference of the IEEE Engineering in Medicine & Biology Society (EMBC)*. IEEE. 2020, pp. 1400–1403.
42. Hamid R Tizhoosh. "Opposition-based learning: a new scheme for machine intelligence". In: *International conference on computational intelligence for modelling, control and automation and international conference on intelligent agents, web technologies and internet commerce (CIMCA-IAWTIC'06)*. Vol. 1. IEEE. 2005, pp. 695–701.
43. Laurens Van der Maaten and Geoffrey Hinton. "Visualizing data using t-SNE". In: *Journal of machine learning research* 9.11 (2008).

44. Vladimir Vapnik. *The nature of statistical learning theory*. Springer science & business media, 1995.
45. Xiao Wang, Ziliang Chen, Rui Yang, Bin Luo, and Jin Tang. "Improved Hard Example Mining by Discovering Attribute-based Hard Person Identity". In: *arXiv preprint* arXiv:1905.02102 (2019).
46. Kilian Q Weinberger, John Blitzer, and Lawrence K Saul. "Distance metric learning for large margin nearest neighbor classification". In: *Advances in neural information processing systems*. 2006, pp. 1473–1480.
47. Kilian Q Weinberger and Lawrence K Saul. "Distance metric learning for large margin nearest neighbor classification". In: *Journal of machine learning research* 10.2 (2009).
48. Chao-Yuan Wu, R Manmatha, Alexander J Smola, and Philipp Krahenbuhl. "Sampling matters in deep embedding learning". In: *Proceedings of the IEEE International Conference on Computer Vision*. 2017, pp. 2840–2848.
49. Liu Yang, Mingyang Zhang, Cheng Li, Michael Bendersky, and Marc Najork. "Beyond 512 tokens: Siamese multi-depth transformer-based hierarchical encoder for long-form document matching". In: *Proceedings of the 29th ACM International Conference on Information & Knowledge Management*. 2020, pp. 1725–1734.
50. Jure Zbontar, Li Jing, Ishan Misra, Yann LeCun, and Stéphane Deny. "Barlow twins: Self-supervised learning via redundancy reduction". In: *International conference on machine learning*. PMLR. 2021, pp. 12310–12320.

Graph Neural Networks

17

Abstract

Graphs provide a versatile representation for a wide range of real-world data, including social networks, biological molecules, text, and images. This chapter explores Graph Neural Networks (GNNs) as powerful tools for processing graph-structured data. It covers key tasks in graph analysis, such as graph-level property prediction, node classification, and edge prediction, highlighting their relevance across diverse applications. Beginning with foundational concepts like the graph Fourier transform, the chapter introduces spectral methods, including ChebNet and Graph Convolutional Networks (GCNs). It then discusses advanced architectures such as Graph Attention Networks (GATs) and general GNN frameworks. The chapter concludes with an examination of graph autoencoders for unsupervised learning on graphs, providing a comprehensive overview of modern techniques in graph-based deep learning.

1 Introduction

Many real-world datasets are in the form of graphs. Examples of graph data include social networks, protein interaction networks, the Internet (World Wide Web), and molecules. Image data can also be considered as graphs, where each pixel is a node connected to its neighbors [1, 2]. This shows that even familiar data like images can be seen as graphs, helping readers connect with the idea of Graph Neural Networks (GNNs). In practice, though, the most impactful applications of GNNs are in molecules, proteins, social networks, recommendation systems, and knowledge graphs.

For example, text data can be represented as graphs [3], where each token (word) is a node connected by an edge to its subsequent token (word). Another example is the use of graphs in biology, where proteins and antibodies are modeled as graphs with amino acids as the nodes. For instance, GearNet [4] and PECAN [5] are graph-based representations of proteins in neural networks.

There are different tasks in graph processing:

- **Graph-level task**: It predicts properties of the entire graph. For example, it can predict whether an antibody protein binds to an antigen protein [6].
- **Node-level task**: It predicts the identity or role of each node in the graph. In this task, each node has some features and there is a label for every node. For instance, if the nodes correspond to people, the label could indicate whether a person lives in a specific city.

B. Ghojogh and A. Ghodsi, *Elements of Deep Learning*,
https://doi.org/10.1007/978-3-032-10738-1_17

- **Edge-level task**: It predicts the identity or role of each edge in the graph. For example, in recommender systems for movie suggestions, some nodes represent users while others represent movies [7]. An edge between a user and a movie exists if the user has rated that movie, with the label of the edge being the rating score. It is also possible to predict the label (score) of non-existing edges between a user and a movie.

GNNs, Graph Convolutional Networks (GCNs), and their extensions are widely used for graph tasks, but they face challenges such as oversmoothing (node features becoming indistinguishable across layers) [8] and oversquashing (long-range information being compressed) [9]. Techniques like residual connections [10], skip layers [11], and sampling-based methods such as GraphSAGE [12] and PinSAGE [13] are used in practice to address these limitations.

In addition, recent advances include Graph Transformers (e.g., Graphormer) [14] and equivariant GNNs for 3D molecular modeling [15, 16]. These architectures extend GNNs to large-scale graphs and geometric tasks that are central in 2025 applications.

This chapter explains Graph Neural Networks (GNNs) for working with graph data. Section 2 introduces the graph Fourier transform. ChebNet is discussed in Sect. 3. Graph Convolutional Networks (GCNs) are explained in Sect. 4. Section 5 covers more general frameworks of graph convolutional networks. Section 6 introduces Graph Attention Networks (GATs). Finally, graph autoencoders are elaborated in Sect. 7.

2 Graph Fourier Transform

As mentioned, images are special cases of graphs. The graph of an image is called a Euclidean graph or grid graph (see Fig. 1a). In a Convolutional Neural Network (CNN) [17], convolution is performed between a filter kernel and the Euclidean graph of the image. However, what about graphs with arbitrary structures or irregular shapes (see Fig. 1b)? The question is how to define the convolution of a filter kernel with an arbitrary graph. This section addresses that question.

2.1 Laplacian of Graph

Consider a graph $\mathcal{G}(\mathcal{V}, \mathcal{E})$ with nodes (vertices) $\mathcal{V}$ and edges $\mathcal{E}$. Let the number of nodes be n. The adjacency matrix $\boldsymbol{A} \in \mathbb{R}^{n \times n}$ is a matrix whose (i, j)-th element is one if node i is connected to node j, and zero otherwise. The degree matrix of $\boldsymbol{A}$ is a diagonal matrix whose (i, i)-th element is the summation of the ith row of $\boldsymbol{A}$, i.e.:

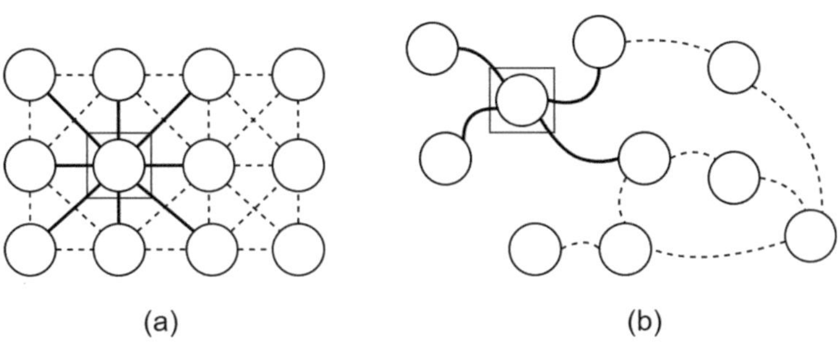

Fig. 1 (a) Euclidean graph (grid graph) such as pixels of an image, (b) arbitrary graph with an irregular shape

$$D(i,i) := \sum_{j=1}^{n} A(i,j), \tag{1}$$

where $A(i, j)$ denotes the (i, j)th element of A. The Laplacian matrix of the graph $\mathcal{G}$ is defined as [18] follows:

$$\mathbb{R}^{n\times n} \ni L := D - A. \tag{2}$$

It is noteworthy that there exist other variants of the Laplacian matrix, such as [19, 20]:

$$L \leftarrow D^{-\alpha} A D^{-\alpha}, \tag{3}$$

where $\alpha \geq 0$ is a parameter. A common value for this parameter is $\alpha = 0.5$:

$$L = D^{-1/2} A D^{-1/2}. \tag{4}$$

This matrix is also referred to as the normalized Laplacian matrix. In this context, the normalized Laplacian is used.

2.2 Graph Fourier Transform

Consider the eigenvalue decomposition of the normalized Laplacian matrix [21]:

$$L = U \Lambda U^\top, \tag{5}$$

where $U = [u_1, \ldots, u_n] \in \mathbb{R}^{n\times n}$ and $\Lambda = \mathbf{diag}([\lambda_1, \ldots, \lambda_n]^\top) \in \mathbb{R}^{n\times n}$ contain the eigenvectors and eigenvalues of the normalized Laplacian matrix, respectively. The eigenvectors of the (normalized) Laplacian, i.e., $u_1, \ldots, u_n$, are called the Fourier functions. The Fourier transform involves projecting a signal x onto the Fourier functions, resulting in the coefficients of the Fourier series [22].

The Graph Fourier Transform projects the input graph signal onto a space whose orthonormal bases are the eigenvectors of the graph's normalized Laplacian. For now, assume that every node of the graph has a scalar feature value. Let $\mathbb{R}^n \ni x = [x_1, \ldots, x_n]^\top$ be the vector of features for all nodes in the graph, where $x_i \in \mathbb{R}$ is the feature value of the ith node. The graph Fourier transform of x is its projection onto the column space of the matrix U [23]:

$$f(x) = \widehat{x} = U^\top x \tag{6}$$

The inverse graph Fourier transform reconstructs the signal from its projection:

$$f^{-1}(\widehat{x}) = U f(x) = U U^\top x. \tag{7}$$

2.3 Graph Convolution

The graph convolution of the input signal $\boldsymbol{x}$ with the filter $\boldsymbol{g} \in \mathbb{R}^n$ is defined as follows:

$$\begin{aligned} \boldsymbol{x} * \boldsymbol{g} &:= f^{-1}\big(f(\boldsymbol{x})f(\boldsymbol{g})\big) \overset{(6)}{=} f^{-1}\big(\boldsymbol{U}^\top \boldsymbol{x} \boldsymbol{U}^\top \boldsymbol{g}\big) \\ &\overset{(7)}{=} \boldsymbol{U}\big(\boldsymbol{U}^\top \boldsymbol{x} \boldsymbol{U}^\top \boldsymbol{g}\big). \end{aligned} \tag{8}$$

We define

$$\mathbb{R}^{n\times n} \ni \boldsymbol{G} := \mathbf{diag}(\boldsymbol{U}^\top \boldsymbol{g}) = \begin{bmatrix} \boldsymbol{u}_1^\top \boldsymbol{g} & 0 & \cdots & 0 \\ 0 & \boldsymbol{u}_2^\top \boldsymbol{g} & \cdots & 0 \\ 0 & 0 & \ddots & 0 \\ 0 & 0 & \cdots & \boldsymbol{u}_n^\top \boldsymbol{g} \end{bmatrix}. \tag{9}$$

Hence, the graph convolution in Eq. (8) can be written as follows:

$$\mathbb{R}^n \ni \boldsymbol{x} * \boldsymbol{g} = \boldsymbol{U}\boldsymbol{G}\boldsymbol{U}^\top \boldsymbol{x}. \tag{10}$$

If each node has a feature vector rather than a scalar feature value, the features form a matrix $\boldsymbol{X} \in \mathbb{R}^{n\times d}$, where each row is the d-dimensional feature vector of a node. In this case, the graph convolution becomes

$$\mathbb{R}^{n\times d} \ni \boldsymbol{X} * \boldsymbol{g} = \boldsymbol{U}\boldsymbol{G}\boldsymbol{U}^\top \boldsymbol{X}. \tag{11}$$

3 ChebNet

Convolutional graph neural networks have been built upon two main approaches:

- Spectral methods, which adopt a graph signal processing perspective, and
- Spatial methods, which define graph convolution through information propagation.

Graph Convolutional Networks (GCNs) [24] bridge the gap between spectral and spatial approaches. Recall Eq. (11). If the input of the ℓth layer is denoted by $\boldsymbol{H}^{(\ell-1)}$ and the output of the ℓth layer is $\boldsymbol{H}^{(\ell)}$, then Eq. (11) becomes

$$\boldsymbol{H}^{(\ell)} = \sigma(\boldsymbol{U}\boldsymbol{G}\boldsymbol{U}^\top \boldsymbol{H}^{(\ell-1)}), \tag{12}$$

where the activation function $\sigma(.)$ is applied to the result of the graph convolution. The first layer accepts the data features as input:

$$\boldsymbol{H}^{(0)} = \boldsymbol{X}. \tag{13}$$

A major limitation of Eq. (12) is that $\boldsymbol{U}$ in this equation is the matrix of eigenvectors of the Laplacian of its input graph. The computational complexity of the eigenvalue decomposition of the $n \times n$ Laplacian matrix is $\mathcal{O}(n^3)$ [25]. Chebyshev Network (ChebNet) [26] addresses this computational bottleneck by approximating the filter $\boldsymbol{g}$ using Chebyshev polynomials of the diagonal matrix of eigenvalues, i.e., $\boldsymbol{\Lambda}$.

The Chebyshev polynomials are defined as [27] follows:

$$\begin{aligned} T_0(x) &= 1, \\ T_1(x) &= x, \\ T_i(x) &= 2xT_{i-1}(x) - T_{i-2}(x). \end{aligned} \tag{14}$$

The domain of input x for Chebyshev polynomials is $[-1, 1]$. For example, Chebyshev polynomials are widely used in cosine expressions:

$$\cos(i\alpha) = T_i\big(\cos(\alpha)\big).$$

ChebNet approximates the filter $\boldsymbol{G}$ by a linear combination of Chebyshev polynomials of the eigenvalues $\boldsymbol{\Lambda}$:

$$\boldsymbol{G} = \sum_{i=0}^{k} \theta_i T_i(\boldsymbol{\Lambda}),$$

where k is the order of the Chebyshev polynomials. However, there is a problem with the domain of the Chebyshev polynomials in this equation. The eigenvalues, i.e., the diagonal elements of $\boldsymbol{\Lambda}$, lie between zero and the largest eigenvalue $\lambda_{\max}$. Therefore, the eigenvalues need to be normalized as follows:

$$\mathbb{R}^{n\times n} \ni \widetilde{\boldsymbol{\Lambda}} := \frac{2}{\lambda_{\max}}\boldsymbol{\Lambda} - \boldsymbol{I}_n, \tag{15}$$

where $\boldsymbol{I}_n$ is the $(n \times n)$ identity matrix. The values in the normalized eigenvalue matrix are in the range $[-1, 1]$, as required by the domain of Chebyshev polynomials. Hence, the approximation of the filter $\boldsymbol{g}$ is

$$\boldsymbol{G} = \sum_{i=0}^{k} \theta_i T_i(\widetilde{\boldsymbol{\Lambda}}). \tag{16}$$

Recall Eq. (10):

$$\begin{aligned} \boldsymbol{x} * \boldsymbol{g} = \boldsymbol{U}\boldsymbol{G}\boldsymbol{U}^\top \boldsymbol{x} &\overset{(16)}{=} \boldsymbol{U}\Big(\sum_{i=0}^{k} \theta_i T_i(\widetilde{\boldsymbol{\Lambda}})\Big)\boldsymbol{U}^\top \boldsymbol{x} \\ &= \sum_{i=0}^{k} \theta_i \boldsymbol{U} T_i(\widetilde{\boldsymbol{\Lambda}})\boldsymbol{U}^\top \boldsymbol{x}. \end{aligned} \tag{17}$$

The matrix $\boldsymbol{U}$ is orthogonal, i.e., its columns are orthonormal, because it is the matrix of eigenvectors. For an orthonormal transformation, the following holds:

$$\boldsymbol{U} T_i(\widetilde{\boldsymbol{\Lambda}})\boldsymbol{U}^\top = T_i(\boldsymbol{U}\widetilde{\boldsymbol{\Lambda}}\boldsymbol{U}^\top). \tag{18}$$

Similar to Eq. (15), we define

$$\widetilde{\boldsymbol{L}} := \frac{2}{\lambda_{\max}} \boldsymbol{L} - \boldsymbol{I}_n, \tag{19}$$

where $\lambda_{\max}$ is the largest eigenvalue of the normalized Laplacian $\boldsymbol{L}$. Then, according to Eq. (15) and similar to Eq. (5), the eigenvalue decomposition of $\widetilde{\boldsymbol{L}}$ becomes

$$\widetilde{\boldsymbol{L}} = \boldsymbol{U} \widetilde{\boldsymbol{\Lambda}} \boldsymbol{U}^\top. \tag{20}$$

Combining Eqs. (18) and (20) gives

$$\boldsymbol{U} T_i(\widetilde{\boldsymbol{\Lambda}}) \boldsymbol{U}^\top = T_i(\widetilde{\boldsymbol{L}}). \tag{21}$$

Substituting Eq. (21) into Eq. (17) yields

$$\boldsymbol{x} * \boldsymbol{g} = \sum_{i=0}^{k} \theta_i T_i(\widetilde{\boldsymbol{L}}) \boldsymbol{x}. \tag{22}$$

Comparing Eqs. (12) and (22) shows that ChebNet overcomes the limitation of eigenvalue decomposition of the Laplacian, as it uses the approximation of Chebyshev polynomials without requiring eigenvalue decomposition.

4 Graph Convolutional Network

Graph Convolutional Network (GCN) [24] is the first-order approximation of ChebNet. In Eq. (22), GCN approximates the Chebyshev polynomials to first order ($k = 1$):

$$T_i(\widetilde{\boldsymbol{L}}) \approx T_0(\widetilde{\boldsymbol{L}}) + T_1(\widetilde{\boldsymbol{L}}). \tag{23}$$

In other words:

$$\begin{aligned} \boldsymbol{x} * \boldsymbol{g} &\approx \sum_{i=0}^{1} \theta_i T_i(\widetilde{\boldsymbol{L}}) \boldsymbol{x} = \theta_0 T_0(\widetilde{\boldsymbol{L}}) \boldsymbol{x} + \theta_1 T_1(\widetilde{\boldsymbol{L}}) \boldsymbol{x} \\ &\overset{(14)}{=} \theta_0 \boldsymbol{x} + \theta_1 \widetilde{\boldsymbol{L}} \boldsymbol{x}. \end{aligned}$$

A larger number of learnable parameters may result in overfitting (see Chap. 5). To reduce the number of parameters and avoid overfitting, it is assumed that $\theta_0 = \theta_1 = \theta$, so:

$$\begin{aligned} \boldsymbol{x} * \boldsymbol{g} &= \theta \boldsymbol{x} + \theta \widetilde{\boldsymbol{L}} \boldsymbol{x} = \theta(\boldsymbol{I} + \widetilde{\boldsymbol{L}}) \boldsymbol{x} \\ &\overset{(19)}{=} \theta \Big(\boldsymbol{I} + \frac{2}{\lambda_{\max}} \boldsymbol{L} - \boldsymbol{I} \Big) \boldsymbol{x} = \theta \frac{2}{\lambda_{\max}} \boldsymbol{L} \boldsymbol{x}. \end{aligned}$$

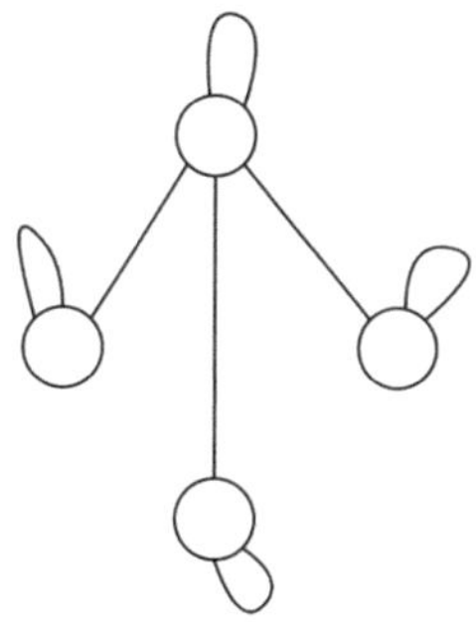

Fig. 2 Self-loops of nodes in a graph for GCN

It is possible to absorb the constant $2/\lambda_{\max}$ into the learnable parameters and simplify the graph convolution as follows:

$$\boldsymbol{x} * \boldsymbol{g} = \theta \boldsymbol{L}\boldsymbol{x} \overset{(4)}{=} \theta \boldsymbol{D}^{-1/2}\boldsymbol{A}\boldsymbol{D}^{-1/2}\boldsymbol{x}. \tag{24}$$

It has been empirically observed that this results in some instability during the training of GCN [24]. Therefore, an additional assumption is introduced to include self-loops for the nodes, meaning that every node has an edge connecting it to itself (see Fig. 2).

Mathematically, this means that the main diagonal of the adjacency matrix should become one by adding the identity matrix to it. Therefore, we define

$$\begin{aligned}
\widetilde{\boldsymbol{A}} &:= \boldsymbol{A} + \boldsymbol{I}, \\
\widetilde{\boldsymbol{D}}(i, j) &:= \sum_{j=1}^{n} \widetilde{\boldsymbol{A}}(i, j), \\
\bar{\boldsymbol{L}} &:= \widetilde{\boldsymbol{D}}^{-1/2}\widetilde{\boldsymbol{A}}\widetilde{\boldsymbol{D}}^{-1/2}.
\end{aligned} \tag{25}$$

As a result, Eq. (24) is replaced by

$$\boldsymbol{x} * \boldsymbol{g} = \theta \widetilde{\boldsymbol{D}}^{-1/2}\widetilde{\boldsymbol{A}}\widetilde{\boldsymbol{D}}^{-1/2}\boldsymbol{x} = \theta \bar{\boldsymbol{L}}\boldsymbol{x}.$$

In matrix form, if each row of $\boldsymbol{X} \in \mathbb{R}^{n \times d}$ is the d-dimensional feature vector of a node, this equation becomes $\boldsymbol{x} * \boldsymbol{g} = \bar{\boldsymbol{L}}\boldsymbol{X}\boldsymbol{\theta}$ where $\boldsymbol{\theta} \in \mathbb{R}^{d}$. If there is a need to have f feature maps after the convolution, then this equation can become $\boldsymbol{x} * \boldsymbol{g} = \bar{\boldsymbol{L}}\boldsymbol{X}\boldsymbol{\Theta}$ where $\boldsymbol{\Theta} \in \mathbb{R}^{d \times f}$. As a result, if the input of the ℓth layer is denoted by $\boldsymbol{H}^{(\ell-1)}$ and the output of the ℓth layer is $\boldsymbol{H}^{(\ell)}$, then Eq. (11) becomes

$$\boldsymbol{H}^{(\ell)} = \sigma(\bar{\boldsymbol{L}}\boldsymbol{H}^{(\ell-1)}\boldsymbol{\Theta}), \tag{26}$$

where the activation function $\sigma(.)$ is applied to the result of the graph convolution. The first layer accepts the data features as input, as stated in Eq. (13).

Equation (26) represents the graph convolution performed in every layer of GCN, where $\boldsymbol{\Theta}$ is the matrix of learnable weights in the layer. Comparing Eqs. (12) and (26)

$$\begin{aligned}
\boldsymbol{H}^{(\ell)} &= \sigma(\boldsymbol{U}\boldsymbol{G}\boldsymbol{U}^{\top}\boldsymbol{H}^{(\ell-1)}), \\
\boldsymbol{H}^{(\ell)} &= \sigma(\bar{\boldsymbol{L}}\boldsymbol{H}^{(\ell-1)}\boldsymbol{\Theta})
\end{aligned}$$

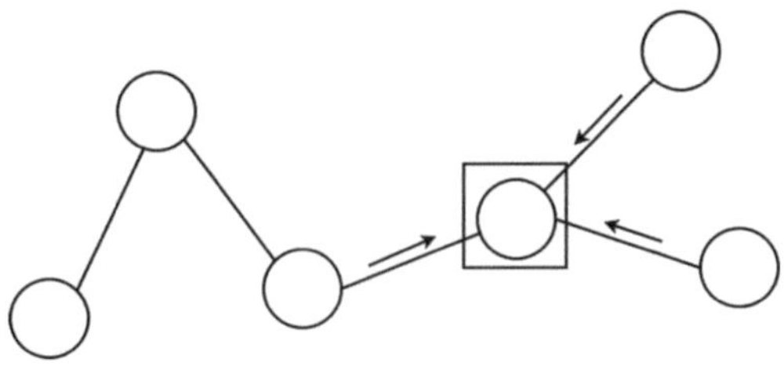

Fig. 3 A node being impacted by its connected nodes before linear transformation in a graph neural network

shows that GCN resolves the limitation of eigenvalue decomposition of the Laplacian by using the approximation of Chebyshev polynomials without performing eigenvalue decomposition.

In a fully connected layer of a feedforward neural network, the operation of the layer is

$$\boldsymbol{H}^{(\ell)} = \sigma(\boldsymbol{H}^{(\ell-1)}\boldsymbol{\Theta}). \tag{27}$$

However, according to Eqs. (25) and (26), the operation of convolution in a layer of GCN is

$$\boldsymbol{H}^{(\ell)} = \sigma(\widetilde{\boldsymbol{D}}^{-1/2}\widetilde{\boldsymbol{A}}\widetilde{\boldsymbol{D}}^{-1/2}\boldsymbol{H}^{(\ell-1)}\boldsymbol{\Theta}). \tag{28}$$

Comparing Eqs. (27) and (28) highlights the relation between GCNs and feedforward neural networks. In a fully connected layer of a feedforward network, all the features of the previous layer $\boldsymbol{H}^{(\ell-1)}$ are passed to the next layer through a linear transformation by the weight matrix $\boldsymbol{\Theta}$, followed by a nonlinear activation function. In a graph neural network, however, the adjacency matrix first determines which nodes (or features) are connected, and then the linear transformation by the weight matrix $\boldsymbol{\Theta}$ is performed, followed by a nonlinear activation function. In other words, the adjacency matrix determines which nodes influence the features of each node (see Fig. 3).

5 More General Frameworks of Graph Convolutional Network

5.1 The General Framework

The update rule of each layer, i.e., Eq. (28), can be restated as follows:

$$\boldsymbol{h}_i^{(\ell)} = \sigma\Big(\sum_{j\in\mathcal{N}_i} \boldsymbol{\Theta}\boldsymbol{h}_j^{(\ell-1)}\Big), \tag{29}$$

for the ith neuron in the ℓth layer, where $\mathcal{N}_i$ denotes the neighbors of node i in the input of the layer. This update rule is called sum pooling because it sums over the neighbors. However, there is an issue with sum pooling: summing the contents of neighboring nodes increases the scale of the output features gradually over multiple layers. To resolve this, it is possible to normalize the input to the activation function by $\widetilde{\boldsymbol{D}}^{-1}$:

$$\boldsymbol{H}^{(\ell)} = \sigma\Big(\widetilde{\boldsymbol{D}}^{-1}\widetilde{\boldsymbol{A}}\boldsymbol{H}^{(\ell-1)}\boldsymbol{\Theta}\Big), \tag{30}$$

where $\widetilde{\boldsymbol{D}}$ is defined in Eq. (25). Equation (30) can also be expressed for each node i as follows:

$$\boldsymbol{h}_i^{(\ell)} = \sigma\Big(\sum_{j\in\mathcal{N}_i} \frac{1}{|\mathcal{N}_i|}\boldsymbol{\Theta}\boldsymbol{h}_j^{(\ell-1)}\Big), \tag{31}$$

where $|\mathcal{N}_i|$ denotes the number of neighbors of node i. This is because the degree matrix counts the number of neighbors for each node. The update rule in Eq. (30) or (31) is called mean pooling.

Alternatively, symmetric normalization can be used in mean pooling:

$$\boldsymbol{H}^{(\ell)} = \sigma\Big(\widetilde{\boldsymbol{D}}^{-1/2}\widetilde{\boldsymbol{A}}\widetilde{\boldsymbol{D}}^{-1/2}\boldsymbol{H}^{(\ell-1)}\boldsymbol{\Theta}\Big). \tag{32}$$

Equation (32) can be stated for each node i as follows:

$$\boldsymbol{h}_i^{(\ell)} = \sigma\Big(\sum_{j\in\mathcal{N}_i} \frac{1}{\sqrt{|\mathcal{N}_i||\mathcal{N}_j|}}\boldsymbol{\Theta}\boldsymbol{h}_j^{(\ell-1)}\Big), \tag{33}$$

which is called mean pooling with symmetric normalization. Comparing Eqs. (28) and (32) shows that the original GCN uses mean pooling with symmetric normalization. Equation (33) indicates that for each node i, if node j is its neighbor, its impact on node i should be greater if node j has few neighbors ($|\mathcal{N}_j|$ is small). Conversely, its impact should be less if node j has many neighbors ($|\mathcal{N}_j|$ is large), because node i would be just one of many neighbors of node j. Note that this effect is not considered in Eq. (31).

5.2 Machine Learning Tasks in Graph Neural Networks

There exist different machine learning tasks in graph neural networks:

- Node classification/regression: After multiple layers of convolution, the $\boldsymbol{h}_i$'s from the last layer are used in the loss function for node classification or regression.
- Graph classification/regression: After multiple layers of convolution, all the $\boldsymbol{h}_i$'s from the last layer are aggregated and used in the loss function for graph classification or regression.
- Link classification/regression: After multiple layers of convolution, the $\boldsymbol{h}_i$'s and the edges $\boldsymbol{e}_{ij}$ from the last layer are used in the loss function for link classification or regression.

The use of features in these tasks is depicted in Fig. 4.

6 Graph Attention Network

6.1 Formulation of Graph Attention Network

As observed in Eqs. (29), (31), and (33), the linear combination in pooling can include weights. Graph Attention Network (GAT) [28] adopts attention mechanisms to learn the relative weights between two connected nodes. In the pooling operation, the attention weights are incorporated:

$$\boldsymbol{h}_i^{(\ell)} = \sigma\Big(\sum_{j\in\mathcal{N}_i} \alpha_{ij}\boldsymbol{h}_j^{(\ell-1)}\Big), \tag{34}$$

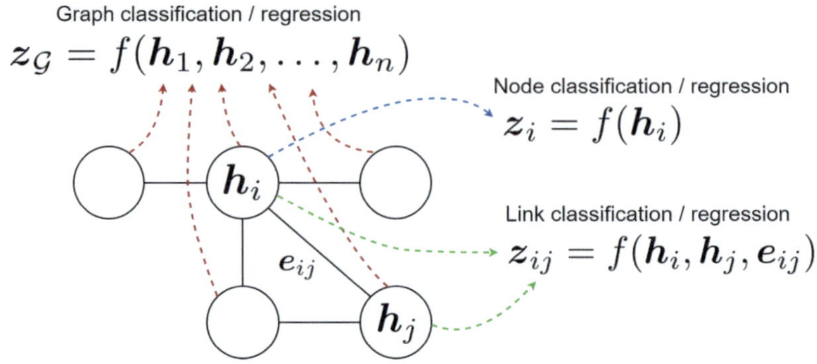

Fig. 4 Use of features for machine learning tasks in graph neural networks

where the attention weight α_{ij} measures the influence of node j on node i (see Chap. 9):

$$\alpha_{ij} = \text{attention}(\boldsymbol{h}_i^{(\ell-1)}, \boldsymbol{h}_j^{(\ell-1)}). \tag{35}$$

The attention weight can be computed by an attention function $a(.)$ between $\boldsymbol{h}_i^{(\ell-1)}$ and $\boldsymbol{h}_j^{(\ell-1)}$:

$$a_{ij} = a(\boldsymbol{h}_i^{(\ell-1)}, \boldsymbol{h}_j^{(\ell-1)}). \tag{36}$$

This attention function may also consider the edge between nodes i and j:

$$a_{ij} = a(\boldsymbol{h}_i^{(\ell-1)}, \boldsymbol{h}_j^{(\ell-1)}, e_{ij}). \tag{37}$$

This attention function $a(.)$ can be a Transformer autoencoder [29] (see Chap. 9). However, GAT models the attention function $a(.)$ as a single-layer feedforward neural network, which calculates the attention between nodes. Finally, the attention values for each node are normalized in a softmax form to obtain the attention weights:

$$\alpha_{ij} = \frac{e^{a_{ij}}}{\sum_{k \in \mathcal{N}_i} e^{a_{ik}}}, \tag{38}$$

where the summation in the denominator is over the neighbors of node i. Figure 5 shows the attention of nodes in a graph.

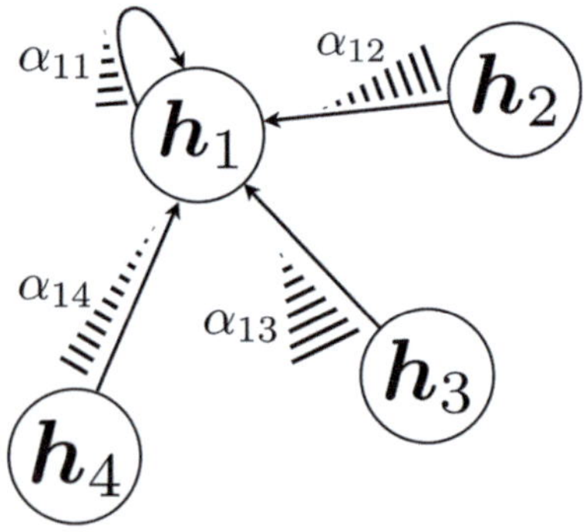

Fig. 5 Illustration of attention between nodes in a graph

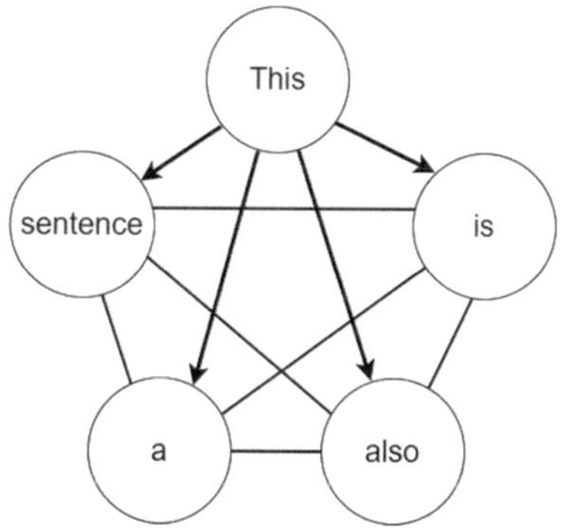

Fig. 6 The graph for the sentence "This is also a sentence." The arrows show the attention of the word "This" to other words in the sentence

It is noteworthy that Transformers [29] (see Chap. 9) are special cases of graph neural networks. In fact, every sentence or sequence can be considered as a graph where GAT calculates the attention between the tokens in the sequence. For example, the graph for the sentence "This is also a sentence" is illustrated in Fig. 6.

6.2 Comparison of Graph Attention Network and Transformer

In the following, GAT and Transformer (see Chap. 9) are compared. Firstly, in GAT, the attention is defined as $a_{ij} = a(\boldsymbol{h}_i^{(\ell-1)}, \boldsymbol{h}_j^{(\ell-1)})$, where $\boldsymbol{h}_i^{(\ell-1)}$ and $\boldsymbol{h}_j^{(\ell-1)}$ are passed through a single-layer network with a weight matrix $\boldsymbol{W}$. Therefore, the attention is calculated between $\boldsymbol{W}^\top \boldsymbol{h}_i^{(\ell-1)}$ and $\boldsymbol{W}^\top \boldsymbol{h}_j^{(\ell-1)}$ after they are fed to the single-layer network. In contrast, in Transformer, the attention is $a_{ij} = a(\boldsymbol{q}_i, \boldsymbol{k}_j)$, where the query $\boldsymbol{q}_i$ and key $\boldsymbol{k}_j$ are different linear transformations of the same tokens, i.e., $\boldsymbol{q}_i = \boldsymbol{W}_Q^\top \boldsymbol{x}$ and $\boldsymbol{k}_i = \boldsymbol{W}_K^\top \boldsymbol{x}$ (see Chap. 9). Therefore, the difference between GAT and Transformer is that GAT uses shared learnable weights for the query and key, whereas Transformer uses different learnable weights for them.

Secondly, another difference between GAT and Transformer is that GAT uses a single-layer feed-forward neural network as the attention function $a(.)$. However, in Transformer, this function is defined as (see Chap. 9) follows:

$$a(\boldsymbol{q}_i, \boldsymbol{k}_j) = \frac{1}{\sqrt{p}} \boldsymbol{q}_i^\top \boldsymbol{k}_j,$$

where p is the dimensionality of the query and the key.

Thirdly, the final difference between GAT and Transformer is in the softmax formulation. GAT sums over the neighbors in the denominator of the softmax (see Eq. (38)), whereas Transformer sums over all tokens in the sequence:

$$\alpha_{ij} = \frac{e^{a_{ij}}}{\sum_{k=1}^{n} e^{a_{ik}}}.$$

7 Graph Autoencoder

Consider an autoencoder, shown in Fig. 7, where the encoder has two layers. This autoencoder accepts a graph as its input; hence, it is called a Graph Autoencoder (GAE) [30].

According to Eq. (26), the first layer of the encoder is

$$\boldsymbol{H}^{(1)} = \sigma(\bar{\boldsymbol{L}}\boldsymbol{X}\boldsymbol{\Theta}_1), \tag{39}$$

where $\boldsymbol{\Theta}_1$ is the learnable weight matrix of the first layer, $\boldsymbol{X} \in \mathbb{R}^{n \times d}$ contains the feature vectors of nodes stacked row-wise, $\bar{\boldsymbol{L}}$ is defined in Eq. (25) based on the adjacency matrix of the graph, $\sigma(.)$ is usually the ReLU activation function [31], and $\boldsymbol{H}^{(1)}$ is the output of the first layer.

Again, according to Eq. (26), the second layer of the encoder is

$$\boldsymbol{H}^{(2)} = \bar{\boldsymbol{L}}\boldsymbol{H}^{(1)}\boldsymbol{\Theta}_2, \tag{40}$$

where $\boldsymbol{\Theta}_2$ is the learnable weight matrix of the second layer, $\boldsymbol{H}^{(2)}$ is the output of the second layer, and the second layer is assumed not to have an activation function.

Substituting Eq. (39) into Eq. (40) yields the following function, denoted by $\mathrm{GCN}(\boldsymbol{X}, \boldsymbol{A}; \boldsymbol{\Theta}_1, \boldsymbol{\Theta}_2)$:

$$\mathrm{GCN}(\boldsymbol{X}, \boldsymbol{A}; \boldsymbol{\Theta}_1, \boldsymbol{\Theta}_2) := \bar{\boldsymbol{L}}\,\sigma(\bar{\boldsymbol{L}}\boldsymbol{X}\boldsymbol{\Theta}_1)\boldsymbol{\Theta}_2. \tag{41}$$

There are two types of GAE, namely, graph reconstruction autoencoder and graph variational autoencoder [30]. These autoencoders are introduced in the following subsections.

7.1 Graph Reconstruction Autoencoder

In the graph reconstruction autoencoder, also called the non-probabilistic GAE, the encoder is given by Eq. (41) with two layers. The p-dimensional latent embeddings of nodes, denoted by $\boldsymbol{Z} \in \mathbb{R}^{n \times p}$, are obtained as follows:

$$\boldsymbol{Z} = \mathrm{GCN}(\boldsymbol{X}, \boldsymbol{A}; \boldsymbol{\Theta}_1, \boldsymbol{\Theta}_2) := \bar{\boldsymbol{L}}\,\sigma(\bar{\boldsymbol{L}}\boldsymbol{X}\boldsymbol{\Theta}_1)\boldsymbol{\Theta}_2.$$

The decoder of the graph reconstruction autoencoder does not contain any layers; instead, it models the similarity between the embedding vectors of the nodes (see Fig. 7). It uses the sigmoid function of $z_i^\top z_j$ to indicate the similarity score (inner product) between the latent variables z_i and z_j. In other

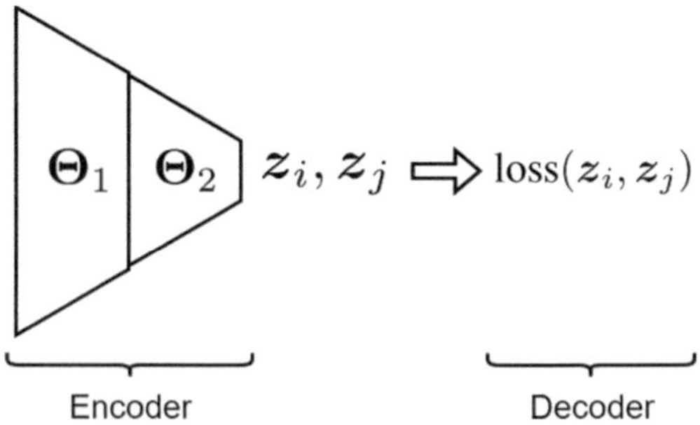

Fig. 7 Graph reconstruction autoencoder

words, it reconstructs the adjacency matrix using the latent embeddings of the nodes rather than the original node features:

$$\widehat{\boldsymbol{A}} = \text{sigmoid}(\boldsymbol{Z}\boldsymbol{Z}^\top), \text{ or} \tag{42}$$

$$\widehat{\boldsymbol{A}}(i, j) = \frac{1}{1 + e^{-z_i^\top z_j}}. \tag{43}$$

The graph reconstruction autoencoder is depicted in Fig. 7.

The loss function is the mean squared error between the adjacency matrix and the reconstructed adjacency matrix:

$$\underset{\boldsymbol{\theta}}{\text{minimize}} \quad \|\widehat{\boldsymbol{A}} - \boldsymbol{A}\|_F^2, \tag{44}$$

where $\|.\|_F$ denotes the Frobenius norm and $\boldsymbol{\theta} := \{\boldsymbol{\Theta}_1, \boldsymbol{\Theta}_2\}$ represents the learnable parameters. This loss function is minimized using backpropagation [32].

7.2 Graph Variational Autoencoder

The graph variational autoencoder uses two GCN modules to estimate the mean and variance in the latent space through the encoder:

$$\text{GCN}_\mu(\boldsymbol{X}, \boldsymbol{A}; \boldsymbol{\Theta}_1, \boldsymbol{\Theta}_2) := \bar{\boldsymbol{L}}\,\sigma(\bar{\boldsymbol{L}}\boldsymbol{X}\boldsymbol{\Theta}_1)\boldsymbol{\Theta}_2, \tag{45}$$

$$\text{GCN}_\sigma(\boldsymbol{X}, \boldsymbol{A}; \boldsymbol{\Theta}_1, \boldsymbol{\Theta}_3) := \bar{\boldsymbol{L}}\,\sigma(\bar{\boldsymbol{L}}\boldsymbol{X}\boldsymbol{\Theta}_1)\boldsymbol{\Theta}_3, \tag{46}$$

where the first layer is shared between them, as shown in Fig. 8.

Since the latent variables of the nodes are independent, the encoder of the graph variational autoencoder models the following conditional distribution:

$$q(\boldsymbol{Z}|\boldsymbol{X}, \boldsymbol{A}) = \prod_{i=1}^{n} q(z_i|\boldsymbol{X}, \boldsymbol{A}), \tag{47}$$

Fig. 8 Graph variational autoencoder

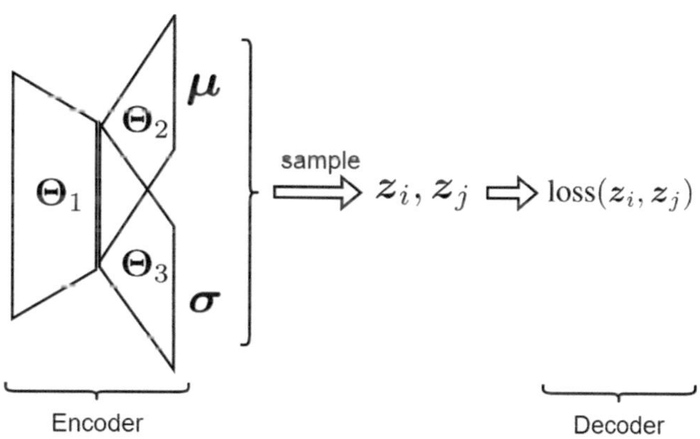

where $\boldsymbol{Z} \in \mathbb{R}^{n \times p}$ contains the p-dimensional latent variables and $z_i \in \mathbb{R}^p$ is the latent variable of the i-th node, whose conditional distribution is a Gaussian:

$$q(z_i|\boldsymbol{X}, \boldsymbol{A}) = \mathcal{N}(z_i \mid \boldsymbol{\mu}_i, \mathbf{diag}(\sigma_i^2)), \tag{48}$$

where $\mathbf{diag}(.)$ creates a diagonal matrix with its input as the diagonal elements. The latent variables $\{z_i\}_{i=1}^n$ are sampled from the multivariate joint distribution in Eq. (47).

The decoder models the following conditional distribution:

$$q(\boldsymbol{A}|\boldsymbol{Z}) = \prod_{i=1}^{n}\prod_{j=1}^{n} p(\boldsymbol{A}(i, j)|z_i, z_j), \tag{49}$$

where $p(\boldsymbol{A}(i, j)|z_i, z_j)$ is the sigmoid function of $z_i^\top z_j$, representing the probability of similarity (inner product) between the latent variables z_i and z_j:

$$p(\boldsymbol{A}(i, j) = 1|z_i, z_j) = \frac{1}{1 + e^{-z_i^\top z_j}}. \tag{50}$$

Thus, the decoder does not contain any layers but models the similarity between the sampled latent variables in the latent space (see Fig. 8).

The graph variational autoencoder maximizes the Evidence Lower Bound (ELBO) in variational inference (see Chap. 12):

$$\underset{\boldsymbol{\theta}}{\text{maximize}} \quad \mathbb{E}_{q(\boldsymbol{Z}|\boldsymbol{X},\boldsymbol{A})}\big[\log(p(\boldsymbol{A} \mid \boldsymbol{Z}))\big] - \text{KL}\big(q(\boldsymbol{Z}|\boldsymbol{X}, \boldsymbol{A}) \| \, p(\boldsymbol{Z})\big), \tag{51}$$

where KL(.‖.) denotes the Kullback–Leibler (KL) divergence [33], $p(\boldsymbol{Z})$ is the desired prior distribution such as a Gaussian, and $\boldsymbol{\theta} := \{\boldsymbol{\Theta}_1, \boldsymbol{\Theta}_2, \boldsymbol{\Theta}_3\}$ are the learnable parameters. The graph variational autoencoder is trained using backpropagation [32]. Since backpropagation minimizes the loss function, the loss is defined as the negative ELBO:

$$\underset{\boldsymbol{\theta}}{\text{minimize}} \quad -\mathbb{E}_{q(\boldsymbol{Z}|\boldsymbol{X},\boldsymbol{A})}\big[\log(p(\boldsymbol{A} \mid \boldsymbol{Z}))\big] + \text{KL}\big(q(\boldsymbol{Z}|\boldsymbol{X}, \boldsymbol{A}) \| \, p(\boldsymbol{Z})\big). \tag{52}$$

Minimizing this loss function trains the model to generate the adjacency matrix $\boldsymbol{A}$ given the sampled latent variables $\boldsymbol{Z}$, while ensuring that the conditional distribution of the latent variables given the graph and its adjacency matrix becomes similar to the desired prior distribution of the latent space.

8 Conclusion

Graph neural networks unify deep learning with relational data. From spectral methods like ChebNet to spatial approaches like GCN and GAT, they provide flexible tools for graph-structured prediction. At the same time, they face challenges of oversmoothing, scalability, and long-range dependency modeling. By 2025, GNNs are widely deployed in chemistry, biology, social networks, and recommendation systems, while Graph Transformers and equivariant architectures extend them to large-scale and geometric domains. GNNs remain a cornerstone of deep learning for structured data, bridging rigorous theory with impactful applications.

In practice, libraries like PyTorch Geometric[1] [34] and Deep Graph Library (DGL)[2] [35] provide efficient implementations of GCNs, GATs, and autoencoders, making it possible to train and deploy GNNs on large graphs in real-world applications.

9 Exercises

1. (Level: Moderate) Graph Neural Networks (GNNs) generalize neural networks to graph-structured data. The Graph Convolutional Network (GCN) aggregates information from neighboring nodes using:

$$\boldsymbol{H}^{(l+1)} = \sigma\left(\tilde{\boldsymbol{D}}^{-1/2}\tilde{\boldsymbol{A}}\tilde{\boldsymbol{D}}^{-1/2}\boldsymbol{H}^{(l)}\boldsymbol{W}^{(l)}\right),$$

where $\boldsymbol{H}^{(l)}$ is the node feature matrix at layer l, $\tilde{\boldsymbol{A}} = \boldsymbol{A} + \boldsymbol{I}$ is the adjacency matrix with self-loops, $\tilde{\boldsymbol{D}}$ is the degree matrix of $\tilde{\boldsymbol{A}}$, $\boldsymbol{W}^{(l)}$ is the weight matrix, and $\sigma(\cdot)$ is the activation function. Consider a simple graph with the following adjacency matrix:

$$\boldsymbol{A} = \begin{bmatrix} 0 & 1 & 0 \\ 1 & 0 & 1 \\ 0 & 1 & 0 \end{bmatrix}.$$

The initial node feature matrix is

$$\boldsymbol{H}^{(0)} = \begin{bmatrix} 1 & 2 \\ 3 & 4 \\ 5 & 6 \end{bmatrix}.$$

Assume the weight matrix is

$$\boldsymbol{W} = \begin{bmatrix} 0.1 & 0.2 \\ 0.3 & 0.4 \end{bmatrix}.$$

Use ReLU activation $\sigma(x) = \max(0, x)$.

(a) Compute the normalized adjacency matrix $\tilde{\boldsymbol{D}}^{-1/2}\tilde{\boldsymbol{A}}\tilde{\boldsymbol{D}}^{-1/2}$.
(b) Compute the first-layer output $\boldsymbol{H}^{(1)}$ after applying the GCN update rule.
(c) Explain how GCNs leverage graph structure to improve learning.

2. (Level: Moderate) You are given a simple undirected graph with the following adjacency matrix:

$$\boldsymbol{A} = \begin{bmatrix} 0 & 1 & 1 & 0 \\ 1 & 0 & 1 & 0 \\ 1 & 1 & 0 & 1 \\ 0 & 0 & 1 & 0 \end{bmatrix}.$$

[1] See https://pytorch-geometric.readthedocs.io/en/latest/get_started/introduction.html, accessed February 2024.
[2] See https://www.dgl.ai/, accessed February 2024.

Each node has an initial feature vector:

$$X = \begin{bmatrix} 1 \\ 0 \\ 1 \\ 0 \end{bmatrix}.$$

Let $\boldsymbol{L} = \boldsymbol{D} - \boldsymbol{A}$ be the unnormalized graph Laplacian. Let $g_\theta(\boldsymbol{L}) = \theta_0 \boldsymbol{I} + \theta_1 \boldsymbol{L}$ represent a simple graph convolution filter using a first-order polynomial. Compute the following:

(a) Compute the graph Laplacian L.
(b) Compute $g_\theta(\boldsymbol{L})\boldsymbol{X}$ using $\theta_0 = 1.0$ and $\theta_1 = -0.5$.
(c) Interpret the result of the convolution.
(d) Briefly explain how this operation resembles spatial graph convolution and how it is generalized in ChebNet.

3. (Level: Moderate) Suppose the input feature vector $\boldsymbol{x} \in \mathbb{R}^4$ and the Laplacian is

$$x = \begin{bmatrix} 1 \\ 2 \\ 3 \\ 4 \end{bmatrix}, \quad L = \begin{bmatrix} 1 & -1 & 0 & 0 \\ -1 & 2 & -1 & 0 \\ 0 & -1 & 2 & -1 \\ 0 & 0 & -1 & 1 \end{bmatrix}.$$

Consider the coefficients $\theta_0 = 1, \theta_1 = 0.5, \theta_2 = -0.2$. Compute each polynomial terms $T_k(\boldsymbol{L})\,\boldsymbol{x}$ for $k \in \{0, 1, 2\}$ and the output of this ChebNet filter.

4. (Level: Moderate) Graph Attention Networks (GAT) extend Graph Convolutional Networks (GCNs) by introducing self-attention to compute the importance of neighboring nodes. The attention weight between node i and node j is given by

$$\alpha_{ij} = \frac{\exp(a(\boldsymbol{h}_i, \boldsymbol{h}_j))}{\sum_{k \in \mathcal{N}(i)} \exp(a(\boldsymbol{h}_i, \boldsymbol{h}_k))},$$

where $\boldsymbol{h}_i$ and $\boldsymbol{h}_j$ are the node feature vectors, $a(\boldsymbol{h}_i, \boldsymbol{h}_j)$ is a learnable attention function, and $\mathcal{N}(i)$ represents the neighborhood of node i. The updated node representation is

$$\boldsymbol{h}_i' = \sigma\left(\sum_{j \in \mathcal{N}(i)} \alpha_{ij} \boldsymbol{W} \boldsymbol{h}_j \right),$$

where $\boldsymbol{W}$ is a learnable weight matrix.

Consider a graph with three nodes and the following adjacency matrix:

$$\boldsymbol{A} = \begin{bmatrix} 1 & 1 & 0 \\ 1 & 1 & 1 \\ 0 & 1 & 1 \end{bmatrix}.$$

The initial node feature matrix is

$$\boldsymbol{H} = \begin{bmatrix} 1 & 2 \\ 3 & 4 \\ 5 & 6 \end{bmatrix}.$$

Assume weight matrix:

$$\boldsymbol{W} = \begin{bmatrix} 0.5 & 0.2 \\ 0.1 & 0.7 \end{bmatrix}.$$

The attention function is computed as $a(\boldsymbol{h}_i, \boldsymbol{h}_j) = \boldsymbol{h}_i^T \boldsymbol{h}_j$.

(a) Compute the attention scores $a(\boldsymbol{h}_i, \boldsymbol{h}_j)$ for all pairs of connected nodes.
(b) Compute the attention weights α_{ij}.
(c) Compute the updated node representations $\boldsymbol{h}'_i$.

References

1. Gene Cheung, Enrico Magli, Yuichi Tanaka, and Michael K Ng. "Graph spectral image processing". In: *Proceedings of the IEEE* 106.5 (2018), pp. 907–930.
2. Erik B Sudderth and William T Freeman. "Signal and image processing with belief propagation [DSP applications]". In: *IEEE Signal Processing Magazine* 25.2 (2008), pp. 114–141.
3. Rik Koncel-Kedziorski, Dhanush Bekal, Yi Luan, Mirella Lapata, and Hannaneh Hajishirzi. "Text generation from knowledge graphs with graph transformers". In: *North American Chapter of the Association for Computational Linguistics (NAACL)*. 2019.
4. Zuobai Zhang et al. "Protein representation learning by geometric structure pretraining". In: *International Conference on Learning Representations (ICLR)*. 2023.
5. Srivamshi Pittala and Chris Bailey-Kellogg. "Learning context-aware structural representations to predict antigen and antibody binding interfaces". In: *Bioinformatics* 36.13 (2020), pp. 3996–4003.
6. Yoochan Myung, Douglas EV Pires, and David B Ascher. "CSM-AB: graph-based antibody–antigen binding affinity prediction and docking scoring function". In: *Bioinformatics* 38.4 (2022), pp. 1141–1143.
7. Shoujin Wang et al. "Graph learning based recommender systems: A review". In: *International Joint Conference on Artificial Intelligence (IJCAI)*. 2021.
8. Qimai Li, Zhichao Han, and Xiao-Ming Wu. "Deeper insights into graph convolutional networks for semi-supervised learning". In: *Proceedings of the AAAI conference on artificial intelligence*. Vol. 32. 1. 2018.
9. Uri Alon and Eran Yahav. "On the Bottleneck of Graph Neural Networks and its Practical Implications". In: *International Conference on Learning Representations*. 2021.
10. Guohao Li, Chenxin Xiong, Guocheng Qian, Ali Thabet, and Bernard Ghanem. "DeeperGCN: All You Need to Train Deeper GCNs". In: *International Conference on Learning Representations*. 2022.
11. Keyulu Xu et al. "Representation learning on graphs with jumping knowledge networks". In: *International conference on machine learning*. PMLR. 2018, pp. 5453–5462.
12. Will Hamilton, Zhitao Ying, and Jure Leskovec. "Inductive representation learning on large graphs". In: *Advances in neural information processing systems* 30 (2017).
13. Rex Ying et al. "Graph convolutional neural networks for web-scale recommender systems". In: *Proceedings of the 24th ACM SIGKDD international conference on knowledge discovery & data mining*. 2018, pp. 974–983.
14. Chengxuan Ying et al. "Do transformers really perform badly for graph representation?" In: *Advances in neural information processing systems* 34 (2021), pp. 28877–28888.
15. Victor Garcia Satorras, Emiel Hoogeboom, and Max Welling. "E(n) equivariant graph neural networks". In: *International conference on machine learning*. PMLR. 2021, pp. 9323–9332.
16. Kristof Schütt et al. "SchNet: A continuous-filter convolutional neural network for modeling quantum interactions". In: *Advances in neural information processing systems* 30 (2017).
17. Yann LeCun, Léon Bottou, Yoshua Bengio, and Patrick Haffner. "Gradient-based learning applied to document recognition". In: *Proceedings of the IEEE* 86.11 (1998), pp. 2278–2324.

18. Benyamin Ghojogh, Mark Crowley, Fakhri Karray, and Ali Ghodsi. "Laplacian-Based Dimensionality Reduction". In: *Elements of Dimensionality Reduction and Manifold Learning*. Springer, 2023, pp. 249–284.
19. Andrew Ng, Michael Jordan, and Yair Weiss. "On spectral clustering: Analysis and an algorithm". In: *Advances in neural information processing systems* 14 (2001), pp. 849–856.
20. Yair Weiss. "Segmentation using eigenvectors: a unifying view". In: *Proceedings of the seventh IEEE international conference on computer vision*. Vol. 2. IEEE. 1999, pp. 975–982.
21. Benyamin Ghojogh, Fakhri Karray, and Mark Crowley. "Eigenvalue and generalized eigenvalue problems: Tutorial". In: *arXiv preprint* arXiv:1903.11240 (2019).
22. Roald M Trigub and Eduard S Belinsky. *Fourier analysis and approximation of functions*. Springer Science & Business Media, 2012.
23. Benyamin Ghojogh, Mark Crowley, Fakhri Karray, and Ali Ghodsi. "Background on Linear Algebra". In: *Elements of Dimensionality Reduction and Manifold Learning* (2023), pp. 17–41.
24. Thomas N Kipf and Max Welling. "Semi-supervised classification with graph convolutional networks". In: *International Conference on Learning Representations*. 2017.
25. Gene H Golub and Charles F Van Loan. *Matrix computations*. JHU press, 2013.
26. Michaël Defferrard, Xavier Bresson, and Pierre Vandergheynst. "Convolutional neural networks on graphs with fast localized spectral filtering". In: *Advances in neural information processing systems*. Vol. 29. 2016, pp. 3844–3852.
27. John C Mason and David C Handscomb. *Chebyshev polynomials*. Chapman and Hall/CRC, 2002.
28. Petar Veličković et al. "Graph attention networks". In: *International Conference on Learning Representations*. 2017.
29. Ashish Vaswani et al. "Attention is all you need". In: *Advances in neural information processing systems*. Vol. 30. 2017.
30. Thomas N Kipf and Max Welling. "Variational graph auto-encoders". In: *arXiv preprint* arXiv:1611.07308 (2016).
31. Vinod Nair and Geoffrey E Hinton. "Rectified linear units improve restricted boltzmann machines". In: *Proceedings of the 27th international conference on machine learning (ICML-10)*. 2010, pp. 807–814.
32. David E Rumelhart, Geoffrey E Hinton, and Ronald J Williams. "Learning representations by back-propagating errors". In: *Nature* 323.6088 (1986), pp. 533–536.
33. Solomon Kullback and Richard A Leibler. "On information and sufficiency". In: *The annals of mathematical statistics* 22.1 (1951), pp. 79–86.
34. Matthias Fey and Jan Eric Lenssen. "Fast graph representation learning with PyTorch Geometric". In: *International Conference on Learning Representations (ICLR), RLGM Workshop*. 2019.
35. Minjie Wang et al. "Deep graph library: A graph-centric, highly-performant package for graph neural networks". In: *arXiv preprint* arXiv:1909.01315 (2019).

Deep Reinforcement Learning 18

Abstract

Reinforcement Learning (RL) is a branch of machine learning that focuses on training agents to make optimal decisions through interactions with an environment to maximize cumulative rewards. This chapter provides a comprehensive introduction to RL and deep reinforcement learning, covering fundamental concepts such as the elements of RL, Markov decision processes, and the Bellman equation. It explores key solution techniques, including value iteration, policy iteration, and temporal difference methods. The chapter also discusses essential algorithms such as Q-learning, Deep Q Network (DQN), policy gradient methods, and the REINFORCE algorithm. Additionally, it highlights multi-agent reinforcement learning and practical applications, exemplified by the AlphaGo system, to illustrate the power and versatility of deep RL in complex decision-making tasks.

1 Introduction

Reinforcement Learning (RL) [31] is an area of machine learning in which one or several agents take actions in an environment to maximize their cumulative reward. In other words, reinforcement learning involves learning to make decisions through interactions with the environment (see Fig. 1). When multiple agents cooperate with each other to learn the environment together, the approach is called Multi-Agent Reinforcement Learning (MARL) [36]. This field of machine learning has numerous practical and industrial applications [20].

Reinforcement learning is similar to how animals, such as dogs, are trained. Suppose we want to train a dog. Food and hunger can be used as positive and negative rewards, respectively, to teach it to perform particular actions in specific situations. As a result, the dog is trained through reinforcement. In reinforcement learning, the machine is like a dog trained to perform suitable actions in various situations within the environment.

This chapter introduces RL and deep RL. Section 2 explains the elements of RL. The Markov decision process is introduced in Sect. 3. The Bellman equation is covered in Sect. 4. Section 5 elaborates on value iteration and policy iteration for solving the Markov decision process. The difference between RL and the Markov decision process is explained in Sect. 6. Temporal difference evaluation is explained in Sect. 7. Section 8 covers Q-learning, epsilon-greedy policy, gradient Q-learning, and Deep Q Network (DQN). Policy gradient and the REINFORCE algorithm are explained in Sect. 9. Finally, AlphaGo is introduced in Sect. 10 as an example of deep RL applied to playing the game of Go.

B. Ghojogh and A. Ghodsi, *Elements of Deep Learning*,
https://doi.org/10.1007/978-3-032-10738-1_18

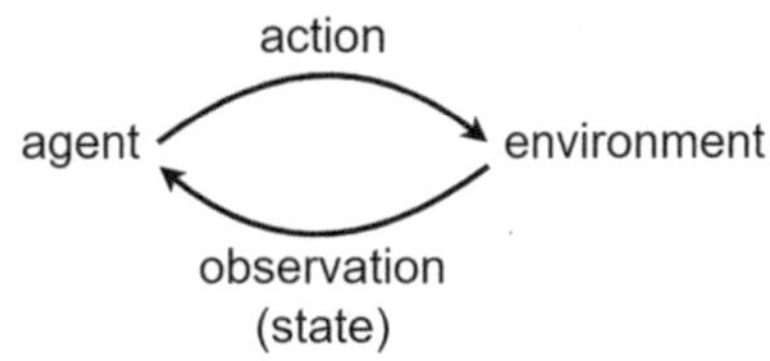

Fig. 1 Interaction with the environment in reinforcement learning

2 Elements of Reinforcement Learning

Reinforcement learning consists of the following elements:

- Environment: the environment in which the agents operate.
- Action: the action taken by the agents.
- State: a state or realization of the environment. The environment has a state at every particular time.
- Reward: the reward that the agents receive when they take an action. Better actions yield higher rewards.

The state s can be a vector or matrix representing the state of the environment. For example, in tic-tac-toe, the state is a nine-dimensional vector where each element can take one of three values, i.e., a value for cross, a value for circle, or a value for an empty square. In the inverted pendulum problem, the state is a four-dimensional vector containing the values for position, velocity, angle, and angular velocity. In Atari games, it can be the frame (RGB image) of the game at the current time. In the game of Go or chess, it can be a matrix or a reshaped vector containing values for the squares of the board.

Some examples of the elements of reinforcement learning are illustrated in Fig. 2 for the game of tic-tac-toe. In this game, the environment is the nine squares of the board, the action is placing a cross or circle in one of the empty squares, the state is a nine-dimensional vector, and the reward is obtained when three crosses are placed together horizontally or vertically.

While deep reinforcement learning is often illustrated with simple environments like tic-tac-toe or the inverted pendulum, by 2025 it is also applied in large-scale real domains such as robotics, recommendation systems, resource management, and complex games.

3 Markov Decision Process

3.1 Markov Process

Consider a time series of random variables $s_1, s_2, \ldots, s_n$. In general, the joint probability of these random variables can be written as:

$$\mathbb{P}(s_1, s_2, \ldots, s_n) = \mathbb{P}(s_1)\,\mathbb{P}(s_2 \mid s_1)\,\mathbb{P}(s_3 \mid s_2, s_1) \cdots \mathbb{P}(s_n \mid s_{n-1}, \ldots, s_2, s_1), \tag{1}$$

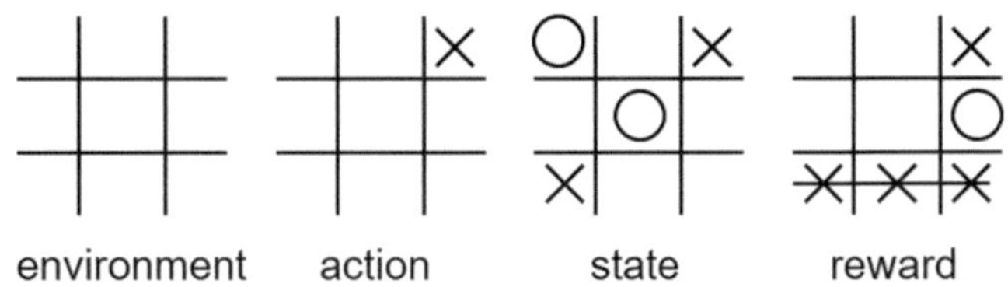

Fig. 2 Elements of reinforcement learning for the game of tic-tac-toe

according to the chain (or multiplication) rule in probability. The (first-order) Markov property is an assumption stating that in a time series of random variables $s_1, s_2, \ldots, s_n$, each random variable depends only on the immediately preceding variable and not on the others [10]. In other words:

$$\mathbb{P}(s_i \mid s_{i-1}, s_{i-2}, \ldots, s_2, s_1) = \mathbb{P}(s_i \mid s_{i-1}). \tag{2}$$

Hence, with the Markov property, the chain rule is simplified to:

$$\mathbb{P}(s_1, s_2, \ldots, s_n) = \mathbb{P}(s_1)\,\mathbb{P}(s_2 \mid s_1)\,\mathbb{P}(s_3 \mid s_2)\ldots\mathbb{P}(s_n \mid s_{n-1}). \tag{3}$$

The Markov property can be of any order. For example, in a second-order Markov property, a random variable depends on both the latest and the one-before-latest variables. Usually, the default Markov property is of order one. A stochastic process that has the Markov property is called a Markovian process or a Markov process.

3.2 Markov Decision Process

A Markov Decision Process (MDP) is a decision process that is also a Markov process. In other words, in an MDP, every state is conditioned only on the previous state, with all past information assumed to be encapsulated in that previous state.

An MDP is defined as a tuple $\{S, A, R, P, \gamma\}$ where:

- States: $s \in S$
- Actions: $a \in A$
- Rewards: $r \in R$, reward model: $\mathbb{P}(r_t|s_t, a_t)$
- Transition model: $\mathbb{P}(s_{t+1}|s_t, a_t)$
- Discount factor: $\gamma \in [0, 1]$
  - discounted: $\gamma < 1$, undiscounted: $\gamma = 1$
  - discounted rewards: $r_0 + \gamma r_1 + \gamma^2 r_2 + \cdots$

At every time slot $t \in \{0, 1, 2, \ldots\}$, the agent is in one of the possible states $s_t \in S$ and takes an action $a_t \in A$. By taking this action, it moves to a new state s_{t+1} according to the transition model, $s_{t+1} \sim \mathbb{P}(s_{t+1}|s_t, a_t)$. By performing this action, the agent receives a reward r_t, according to the reward model, $r_t \sim \mathbb{P}(r_t|s_t, a_t)$.

An MDP usually considers rewards not only in the current time slot but also over a future time period, called the horizon, h. In other words, in addition to the short-term reward, the total future reward is also important. Sometimes, it is better to sacrifice an immediate reward to gain a larger long-term reward. For example, in chess, a player may sacrifice their queen to win the game several moves later.

When considering the total reward over a horizon, the importance of the rewards is not treated equally. Therefore, there is a discount factor $\gamma \in [0, 1]$. The agent receives the total discounted rewards $r_0 + \gamma r_1 + \gamma^2 r_2 + \cdots$ over the horizon, where r_0 is the reward at the current time and r_i is the reward at the ith subsequent time slot. The rewards are discounted because short-term rewards (such as r_0 and r_1) should be valued more than long-term rewards (such as r_{1000}). Clearly, the rewards are discounted when $\gamma < 1$ and undiscounted when $\gamma = 1$.

3.3 Policy

A policy π is a mapping from states to actions:

$$\pi : S \to A, \quad \pi : s \mapsto a. \tag{4}$$

In other words, it determines which action should be performed in each state of the environment. The policy can be either deterministic:

$$a_t = \pi(s_t), \tag{5}$$

or stochastic:

$$a_t = \pi(a_t|s_t). \tag{6}$$

Note that the policy π essentially represents the action, as it is the mapping from state to action. The goal of an MDP is to find a policy π^* that maximizes the long-term reward:

$$\pi^* = \arg\max_{\pi} \sum_{t=0}^{h} \gamma^t \mathbb{E}_\pi[r_t], \tag{7}$$

where t is the time slot index, h is the horizon, and $\mathbb{E}_\pi[.]$ denotes the expectation with respect to the policy.

4 Bellman Equation

Value is defined as the maximum or best reward that the agent can obtain in a given state. Therefore, value is essentially a measure of reward. The value at state s_t and time t is denoted by $v(s_t)$ or $v_t(s)$.

Consider a time period, also called the horizon, h. The last time slot of this horizon is at time t, and its value is

$$v(s_t) = \max_{a_t} r(s_t, a_t).$$

The value at the second-to-last time slot of the horizon is

$$v(s_{t-1}) = \max_{a_{t-1}} r(s_{t-1}, a_{t-1}) + \gamma\mathbb{P}(s_t|s_{t-1}, a_{t-1})\, v(s_t).$$

The value at the third-to-last time slot of the horizon is

$$v(s_{t-2}) = \max_{a_{t-2}} r(s_{t-2}, a_{t-1}) + \gamma\mathbb{P}(s_{t-1}|s_{t-2}, a_{t-2})\, v(s_{t-1}),$$

and so on. This series can be generalized into the following equation, known as the Bellman equation [3]:

$$v(s_t) = \max_{a_t} r(s_t, a_t) + \gamma \sum_{s_{t+1}} \mathbb{P}(s_{t+1}|s_t, a_t)\, v(s_{t+1}). \tag{8}$$

Note that the Bellman equation can be considered the most important equation in Markov decision processes and reinforcement learning, as many algorithms and equations are derived from it.

In the literature of reinforcement learning, the variable at the next time slot $(t+1)$ may also be denoted with a prime, while the variable at the current time slot t is denoted without a prime. For example, the states s_t and s_{t+1} may also be denoted by s and s', respectively. The value may also have the superscript π, i.e., $v^{\pi}(s)$, indicating that it is associated with policy π. Therefore, another notation for the Bellman equation is

$$v(s) = \max_{a} r(s,a) + \gamma \sum_{s'} \mathbb{P}(s'|s,a)\, v(s'). \tag{9}$$

It is also sometimes expressed using expectation:

$$v(s) = \mathbb{E}[r|s, \pi(s)] + \gamma \sum_{s'} \mathbb{P}(s'|s, \pi(s))\, v(s'), \tag{10}$$

where $\pi(s) = a$.

5 Solving Markov Decision Process

Recall that the goal of an MDP is to find the optimal policy (see Sect. 3.2). There are two main approaches to solve the MDP problem, i.e., to find the optimal policy. These two approaches are value iteration and policy iteration.

5.1 Value Iteration

Let h denote the horizon and t denote the time slot within the horizon. The value iteration algorithm calculates the optimal values backward through the horizon using the Bellman equation:

$$\begin{aligned} t = 0 &: v_0^* \leftarrow \max_{a}\ r(s,a),\ \forall s, \\ t > 0 &: v_t^*(s) \leftarrow \max_{a}\ r(s,a) + \gamma \sum_{s'} \mathbb{P}(s'|s,a)\, v_{t-1}^*(s'),\ \forall s, t. \end{aligned} \tag{11}$$

The value iteration algorithm is presented in Algorithm 1.

After finding the optimal value using value iteration, the optimal policy can be determined for the last time slot and other time slots within the horizon as the actions that maximize the values:

```
1 v_0^* ← max_a r(s, a), ∀s
2 for t from 1 to h do
3     v_t^*(s) ← max_a r(s, a) + γ Σ_{s'} P(s'|s, a) v_{t-1}^*(s'), ∀s
4 Return v_h^*
```

Algorithm 1: Value iteration

$$t = 0 : \pi_0^*(s) \leftarrow \arg\max_a r(s, a), \ \forall s,$$
$$t > 0 : \pi_t^*(s) \leftarrow \arg\max_a \ r(s, a) + \gamma \sum_{s'} \mathbb{P}(s'|s, a)\, v_{t-1}^*(s'), \ \forall s, \tag{12}$$

according to the Bellman equation. As is evident from the above equations, the optimal policy π^* is non-stationary, i.e., it is time-dependent.

5.2 Policy Iteration

Another approach to finding the optimal policy in an MDP is policy iteration. Using an alternating optimization approach [9], policy iteration alternates between two steps iteratively until convergence. These two steps are policy evaluation and policy improvement. Policy evaluation assesses how good the current policy is by calculating its value, i.e., the expected reward:

$$v^\pi(s) \leftarrow r(s, \pi(s)) + \gamma \sum_{s'} \mathbb{P}(s'|s, \pi(s))\, v^\pi(s'), \ \forall s. \tag{13}$$

Policy improvement enhances the policy by selecting the action that maximizes the value:

$$\pi(s) \leftarrow \arg\max_a \ r(s, a) + \gamma \sum_{s'} \mathbb{P}(s'|s, a)\, v^\pi(s'), \ \forall s. \tag{14}$$

Note that the equations for policy evaluation and policy improvement are both derived from the Bellman equation, where, in policy evaluation, the action a is replaced with the policy $\pi(s)$, which inherently represents the action. The policy iteration algorithm, which includes both policy evaluation and policy improvement, is presented in Algorithm 2.

5.3 Modified Policy Iteration

In value iteration, the policy evaluation step is repeated over the horizon to maximize the value $v^\pi(s)$, and then the policy improvement step is performed only once to find the optimal policy. In policy iteration, however, policy evaluation and policy improvement are repeated alternately until convergence. The time complexities of each iteration in value iteration and policy iteration are $\mathcal{O}(|S|^2|A|)$ and $\mathcal{O}(|S|^3 + |S|^2|A|)$, respectively, where $|S|$ is the number of possible states and $|A|$ is the number of possible actions. The number of iterations for value iteration and policy iteration correspond to linear and quadratic convergence, respectively. As a result, compared to policy iteration, value iteration requires less computation per iteration but needs a larger number of iterations.

```
Initialize π(s) with a random action for every state s.
while not converged do
    v^π(s) ← r(s, π(s)) + γ Σ_{s'} P(s'|s, π(s)) v^π(s'), ∀s
    π(s) ← argmax_a r(s, a) + γ Σ_{s'} P(s'|s, a) v^π(s'), ∀s
Return v^π(s) and π(s)
```

Algorithm 2: Policy iteration

Initialize $\pi(s)$ with a random action for every state s.
while *not converged* **do**
  for *k times* **do**
    $v^{\pi}(s) \leftarrow r(s, \pi(s)) + \gamma \sum_{s'} \mathbb{P}(s'|s, \pi(s))\, v^{\pi}(s'),\ \forall s$
  $\pi(s) \leftarrow \arg\max_{a}\ r(s, a) + \gamma \sum_{s'} \mathbb{P}(s'|s, a)\, v^{\pi}(s'),\ \forall s$
Return $v^{\pi}(s)$ and $\pi(s)$

Algorithm 3: Modified policy iteration

Therefore, there is a method called modified policy iteration [22], which is an intermediate approach between value iteration and policy iteration, combining the advantages of both. Modified policy iteration is more practical and usually converges faster than either value iteration or policy iteration. This algorithm is similar to the policy iteration algorithm, except that it performs its policy evaluation step multiple times, inspired by value iteration. The algorithm for modified policy iteration is presented in Algorithm 3.

It is noteworthy that there exist approximate variants of both the policy iteration [4] and modified policy iteration [23] algorithms.

6 Reinforcement Learning

Reinforcement Learning (RL) is similar to an MDP except that the transition model is not known. Not having access to the transition model is more realistic, as it is usually unavailable in practice. As a result, reinforcement learning is a more challenging and, of course, more useful task compared to MDP.

Reinforcement learning is defined as a tuple $\{S, A, R, P, \gamma\}$:

- States: $s \in S$
- Actions: $a \in A$
- Rewards: $r \in R$, reward model: $\mathbb{P}(r_t|s_t, a_t)$
- Discount factor: $\gamma \in [0, 1]$
  - discounted: $\gamma < 1$, undiscounted: $\gamma = 1$
  - discounted rewards: $r_0 + \gamma r_1 + \gamma^2 r_2 + \cdots$

The goal of reinforcement learning, as in MDP, is to find the optimal policy.

There are two main approaches for finding the best policy in reinforcement learning:

- Model-based reinforcement learning [18]: This approach first finds the model and then solves the problem. Specifically, it learns the transition model and the reward model before solving the task. Some robotic applications use this type of reinforcement learning [13, 21], for example by moving the robot's arm manually to different states and tuning the model.
- Model-free reinforcement learning [6]: This approach finds the policy without explicitly knowing the transition model and the reward model. In model-free reinforcement learning, the model is not explicitly available but is implicitly learned to solve the problem.

This chapter focuses on model-free reinforcement learning algorithms.

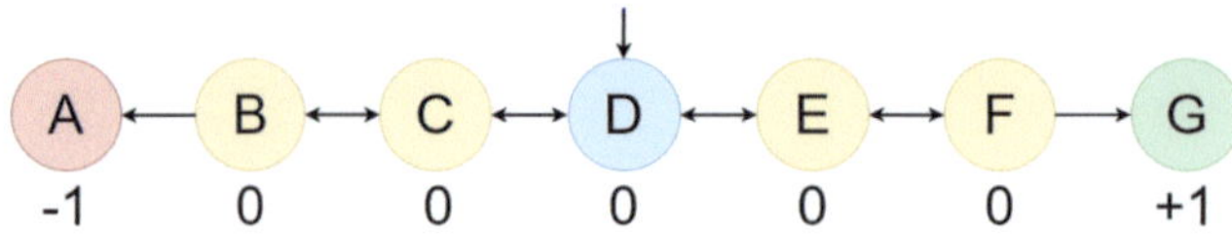

Fig. 3 The chain game starting at node D with actions of moving left or right. The game ends at nodes A (lose) or G (win). Therefore, the rewards of nodes A and G are -1 and $+1$, respectively, and the rest have zero rewards

7 Temporal Difference Evaluation

7.1 Monte-Carlo Evaluation

In model-free reinforcement learning, the value $v^{\pi}(s)$ should be estimated, given a policy π, without using any transition model. Recall that value is the expected total rewards with respect to the policy:

$$v^{\pi}(s) = \mathbb{E}_{\pi}\Big[\sum_{t=0}^{h} \gamma^t r_t\Big] \overset{(a)}{\approx} \frac{1}{n(s)} \sum_{k=1}^{n(s)} \Big[\sum_{t=0}^{h} \gamma^t r_t^{(k)}\Big], \tag{15}$$

where (a) is due to the Monte-Carlo approximation (evaluation) of expectation[1] [11], h is the horizon, $n(s)$ is the number of times the state s occurs, and $r_t^{(k)}$ is the reward at time t in the horizon for the kth episode. The larger the $n(s)$, the more accurate the approximation becomes. For this Monte-Carlo evaluation of the value, the agent should play in the environment k times. An episode refers to each complete run of the agent in the environment.

For example, in the game illustrated in Fig. 3, the possible actions at every middle node are moving left or right stochastically. Let the rewards of nodes A and G be -1 and $+1$, respectively, with the other nodes having zero rewards. The game starts from node D. The policy involves stochastic left or right movements with different probabilities at each node. For a particular policy (specific probabilities of actions at nodes), the game is played for k episodes. Each episode is represented as a trajectory ending at either node A or node G. The Monte-Carlo evaluation is the average of total rewards across these trajectories.

7.2 Temporal Difference Evaluation

The problem with Monte-Carlo evaluation is that it requires many iterations, i.e., a large $n(s)$, to achieve accuracy. To address this problem, temporal difference evaluation is introduced as follows. Let G_k be a one-trajectory Monte-Carlo evaluation of the value:

$$G_k := \sum_{t=0}^{h} \gamma^t r_t^{(k)}, \tag{16}$$

[1] The Monte-Carlo approximation estimates the mean (i.e., expectation) with the average. This approximation is justified because, according to the central limit theorem [8], the distribution of the sum of multiple random variables converges to the normal distribution. In real life (physics), many variables are summed together. For example, light beams are combined. Therefore, the most commonly observed continuous distribution in the world is the normal distribution (hence the name normal). The maximum likelihood estimation of the mean of a normal distribution is the average. Thus, it makes sense to approximate the mean with the average. This also explains why people often use the terms mean and average interchangeably.

which represents the total discounted reward. According to Eq. (15), the value is

$$
\begin{aligned}
v^{\pi}_{n(s)}(s) &\approx \frac{1}{n(s)} \sum_{k=1}^{n(s)} G_k \\
&= \frac{1}{n(s)} \Big(G_{n(s)} + \sum_{k=1}^{n(s)-1} G_k \Big) \\
&\overset{(17)}{=} \frac{1}{n(s)} \Big(G_{n(s)} + \big(n(s) - 1\big) v^{\pi}_{n(s)-1}(s) \Big) \\
&= v^{\pi}_{n(s)-1}(s) + \frac{1}{n(s)} \big(G_{n(s)} - v^{\pi}_{n(s)-1}(s) \big).
\end{aligned} \tag{17}
$$

Therefore, the Monte-Carlo evaluation can be expressed as an incremental update:

$$
v^{\pi}_{n}(s) = v^{\pi}_{n-1}(s) + \alpha_n \big(G_n - v^{\pi}_{n-1}(s) \big), \tag{18}
$$

where n is the simplified notation for $n(s)$ and $\alpha_n := 1/n(s)$.

The temporal difference evaluation is similar to the incremental update except that it replaces G_n with $r + \gamma v^{\pi}_{n-1}(s')$, where s' is the next state at time $t+1$. This replacement is justified because, according to Eq. (16), G_n is the value (or total rewards) of a trajectory. In other words, recall the Bellman equation in Eq. (10); G_n is the Monte-Carlo evaluation of the value using only one trajectory, which is not a very accurate approximation:

$$
\begin{aligned}
v(s) &= \mathbb{E}[r|s, \pi(s)] + \gamma \sum_{s'} \mathbb{P}(s'|s, \pi(s))\, v(s') \\
&\approx r + \gamma v^{\pi}(s'),
\end{aligned} \tag{19}
$$

where that one trajectory includes only a single s'. Therefore, the temporal difference evaluation is formulated as:

$$
v^{\pi}_{n}(s) = v^{\pi}_{n-1}(s) + \alpha_n \big(r + \gamma v^{\pi}_{n-1}(s') - v^{\pi}_{n-1}(s) \big). \tag{20}
$$

Theorem 1 *If α_n is appropriately decreased with the number of times a state is visited, then $v^{\pi}_{n}(s)$ converges to the correct value. Sufficient conditions for α_n are*

$$
\lim_{k \to \infty} \sum_{n=1}^{k} \alpha_n = \infty, \tag{21}
$$

$$
\lim_{k \to \infty} \sum_{n=1}^{k} (\alpha_n)^2 < \infty. \tag{22}
$$

α_n is called the learning rate, and it can be any function of n that satisfies the above sufficient conditions. Often, $\alpha_n = 1/n(s)$ is used, where $n(s)$ is the number of times the state s is visited.

Instead of attempting to calculate the total future reward, temporal difference simply tries to predict the combination of the immediate reward and its own reward prediction at the next moment in time. When the next moment arrives, the new prediction is compared against what was expected. In other

Input: policy π
while v^{π} *is not converged* **do**
  Execute $\pi(s)$
  Observe s and s'
  $n(s) \leftarrow n(s) + 1$
  $\alpha \leftarrow 1/n(s)$
  $v^{\pi}(s) \leftarrow v^{\pi}(s) + \alpha\big(r + \gamma v^{\pi}(s') - v^{\pi}(s)\big)$
  $s \leftarrow s'$
Return v^{π}

Algorithm 4: Temporal difference evaluation of value

words, the temporal difference is calculated. This temporal difference is then used to adjust the old prediction toward the new prediction.

An example of using temporal difference is as follows. Consider weather prediction where the temperatures of all days in the coming week are forecasted. When one of the days arrives, the predicted temperature for that day can be compared with the actual temperature. If the difference between them is small, it means that the prediction was accurate; if the difference is large, the predicted temperatures for later days should be adjusted based on the calculated error. Assume the predicted temperature for Tuesday was 19 and the actual temperatures for Monday and Tuesday are 20 and 23. The temporal difference evaluation suggests that the predicted temperature for Tuesday should be adjusted to $20 + \alpha(23 - 19)$, where $\alpha > 0$ is a small multiplier.

The algorithm for temporal difference evaluation of value is presented in Algorithm 4. As temporal difference evaluation makes use of incremental updates, it requires much less computation compared to Monte-Carlo evaluation. Monte-Carlo evaluation requires calculating total trajectories, whereas temporal difference calculates only the differences in values between successive states along the trajectory.

8 Q-Learning

8.1 Q Function

Instead of evaluating the state value $v^{\pi}(s)$, which is a function of state, it is possible to evaluate the state-action value $Q^{\pi}(s, a)$, which is a function of both state and action. The state-action value is called the Q function. The state value indicates how good it is to be in a particular state s, whereas the Q function determines how good it is to perform action a, following the policy π, in state s.

Note that the Q function, like the value function, essentially represents reward. Therefore, many of the equations and algorithms for the value function also hold for the Q function with some small modifications. These equations are explained in the following.

8.1.1 Bellman Equation for the Q Function

Recall the Bellman equation for the state value $v(s)$, introduced in Eq. (10):

$$v(s) = \mathbb{E}[r|s, a] + \gamma \sum_{s'} \mathbb{P}(s'|s, a)\, v(s').$$

Similarly, the Bellman equation for the state-action value (Q function) is

$$Q(s, a) = \mathbb{E}[r|s, a] + \gamma \sum_{s'} \mathbb{P}(s'|s, a) \max_{a'} Q(s', a'), \tag{23}$$

where a' is the next action in the next time slot. As a result, according to the Bellman equation, the optimal state value and the optimal Q function are

$$v^*(s) = \mathbb{E}[r|s, a] + \gamma \sum_{s'} \mathbb{P}(s'|s, a)\, v^*(s'), \tag{24}$$

$$Q^*(s, a) = \mathbb{E}[r|s, a] + \gamma \sum_{s'} \mathbb{P}(s'|s, a) \max_{a'} Q^*(s', a'), \tag{25}$$

respectively.

If the Q function is used, then the optimal value $v^*(s)$ and the optimal policy π^* are found by maximizing the Q function:

$$v^*(s) = \max_{a} Q^*(s, a), \tag{26}$$

$$\pi^*(s) = \arg\max_{a} Q^*(s, a). \tag{27}$$

Maximizing the Q function to find the optimal policy is roughly in the form of Eq. (7):

$$\pi^* = \arg\max_{\pi} \sum_{t=0}^{h} \gamma^t \mathbb{E}_\pi[r_t],$$

which finds the policy by maximizing the value or the total rewards.

8.1.2 Policy Iteration Using the Q Function

Recall Eq. (18), which was the incremental update for the state value:

$$v_n^\pi(s) = v_{n-1}^\pi(s) + \alpha_n\big(G_n - v_{n-1}^\pi(s)\big).$$

Likewise, it is possible to have an incremental update for the Q function:

$$Q_n^\pi(s, a) = Q_{n-1}^\pi(s, a) + \alpha_n\big(G_n - Q_{n-1}^\pi(s, a)\big), \tag{28}$$

where G_n is a one-trajectory Monte-Carlo evaluation of the value, defined in Eq. (16):

$$G_n := \sum_{t=0}^{h} \gamma^t r_t^{(n)}.$$

As a result, the policy iteration algorithm using the incremental update of the Q function is shown in Algorithm 5, where the algorithm alternates between policy evaluation:

$$Q^\pi(s, a) \leftarrow Q^\pi(s, a) + \alpha_n\big(G_n - Q^\pi(s, a)\big), \ \forall s, \tag{29}$$

1 Initialize $Q^{\pi}(s, a)$ with a random value for every state s and action a.
2 **while** *not converged* **do**
3 $Q^{\pi}(s, a) \leftarrow Q^{\pi}(s, a) + \alpha_n\big(G_n - Q^{\pi}(s, a)\big), \forall s$
4 $\pi(s) \leftarrow \arg\max_a Q^{\pi}(s, a), \forall s$
5 Return $Q^{\pi}(s, a)$ and $\pi(s)$

Algorithm 5: Policy iteration using the Q function

and policy improvement:

$$\pi(s) \leftarrow \arg\max_{a} Q^{\pi}(s, a), \forall s. \tag{30}$$

8.2 Q-Learning

Recall Eq. (19), which approximated the Monte-Carlo evaluation of the value using only one trajectory:

$$\begin{aligned} v(s) &= \mathbb{E}[r|s, \pi(s)] + \gamma \sum_{s'} \mathbb{P}(s'|s, \pi(s))\, v(s') \\ &\approx r + \gamma v^{\pi}(s'), \end{aligned}$$

Likewise, this approximation for the Q function is

$$\begin{aligned} Q(s, a) &\overset{(23)}{=} \mathbb{E}[r|s, a] + \gamma \sum_{s'} \mathbb{P}(s'|s, a) \max_{a'} Q(s', a') \\ &\approx r + \gamma \max_{a'} Q(s', a'). \end{aligned} \tag{31}$$

As a result, the temporal difference evaluation of the Q function is

$$Q_n^{*}(s, a) = Q_{n-1}^{*}(s, a) + \alpha_n\big(r + \gamma \max_{a'} Q_{n-1}^{*}(s', a') - Q_{n-1}^{*}(s, a)\big). \tag{32}$$

This is similar to the temporal difference update for the state value in Eq. (20). It is noteworthy that the Q function is updated merely by observing s, s', a, and r based on temporal difference.[2] This leads to the Q-learning algorithm [33, 34] presented below.

It is also worth comparing the updates in Q-learning and value iteration (recall Eq. (11)):

$$v_n^{*}(s) \leftarrow \max_{a} r(s, a) + \gamma \sum_{s'} \mathbb{P}(s'|s, a)\, v_{n-1}^{*}(s').$$

Value iteration is for MDPs, whereas Q-learning is for reinforcement learning. In MDPs, the transition model is known, but in reinforcement learning, the transition model is not available; therefore, reinforcement learning updates the Q function, which depends on both state and action.

[2] This is the essence of reinforcement learning, where the Q function (which represents reward) is updated solely based on the current state, next state, action, and current reward. In other words, it learns what is best to do by merely interacting with the environment.

Initialize $Q^*(s, a),\ \forall s, a$
Initialize $n(s, a) \leftarrow 0,\ \forall s, a$
while *Q^* is not converged* **do**
    Select and execute action a
    Observe s' and r
    $n(s, a) \leftarrow n(s, a) + 1$
    $\alpha \leftarrow 1/n(s, a)$
    $Q^*_n(s, a) \leftarrow Q^*_{n-1}(s, a) + \alpha_n \big(r + \gamma \max_{a'} Q^*_{n-1}(s', a') - Q^*_{n-1}(s, a)\big)$
    $s \leftarrow s'$
Return $Q^*(s, a)$

Algorithm 6: Q-learning

The Q-learning algorithm is presented in Algorithm 6. In this algorithm, the Q functions are initialized either randomly or by setting them to zero. This algorithm uses the temporal difference evaluation of the Q function, i.e., Eq. (32). Here, $n(s, a)$ is the number of times the state s and action a are observed together. The learning rate $\alpha > 0$ can be any function of $n(s, a)$ that satisfies the sufficient conditions in Theorem 1. A common choice for α is $1/n(s, a)$.

8.3 Epsilon-Greedy Policy

There are two approaches for selecting the action a in Q-learning: (1) choose action a at state s randomly, or (2) choose the action a that maximizes the Q function. The former approach promotes exploration, which takes longer to converge but explores more of the environment and possible actions. The latter approach promotes exploitation, which converges faster but may get stuck in local optima without finding the global best solution. Therefore, there is a trade-off between exploration and exploitation [31], and it is better to combine both approaches to benefit from their advantages. The epsilon-greedy policy [34] combines these two approaches by selecting the action as:

$$a \leftarrow \begin{cases} \arg\max_a Q(s, a) & \text{with probability } 1 - \epsilon \\ \text{a random action} & \text{with probability } \epsilon, \end{cases} \tag{33}$$

where $\epsilon \in (0, 1)$. The name of this algorithm arises because selecting a random action is a greedy strategy for exploring the environment. More theories on the epsilon-greedy policy can be found in [7].

In addition to this policy, it is possible to select random actions for exploration in the initial iterations and then use the best action for exploitation in later iterations. This is because, initially, the algorithm should explore the possible actions, but later, it is expected to be closer to the optimal solution and should not oscillate excessively around it.

8.4 Example for Q-Learning

Consider the game depicted in Fig. 3. This game is illustrated again in Fig. 4 with more details. In this game, the possible actions at every middle node are moving left or right stochastically. The rewards of nodes A and G are -1 and $+1$, respectively, while the other nodes have zero rewards. The game starts

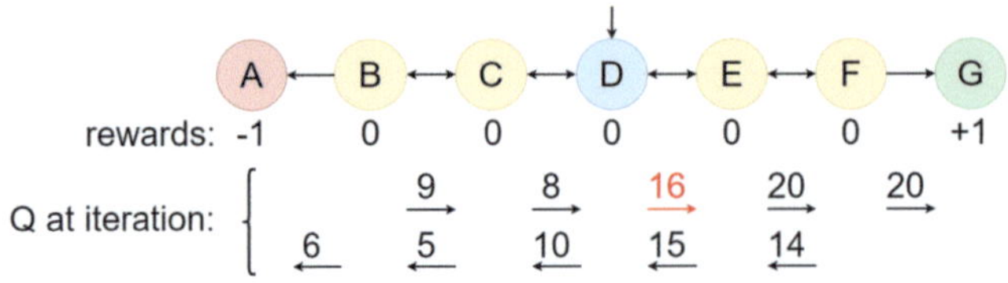

Fig. 4 The chain game starting at node D with actions of moving left or right. The game ends at nodes A (lose) or G (win). Therefore, the rewards of nodes A and G are -1 and $+1$, respectively, and the rest have zero rewards. Assume the Q values at the current iteration of Q-learning are as mentioned in this figure

from node D. The game is won if it reaches node G and lost if it reaches node A. The policy involves stochastic left or right movements with different probabilities at each node.

Suppose the iterations of Q-learning are being performed and we are in one of the iterations. Assume the current Q function at this iteration for the states and actions is as follows:

$$Q(\text{D}, \text{right}) = 16, \quad Q(\text{D}, \text{left}) = 10,$$
$$Q(\text{E}, \text{right}) = 20, \quad Q(\text{E}, \text{left}) = 15.$$

Assume that at the current iteration of Q-learning, we are at state D and choose the action "moving to right." Thus, the next state is E and the reward is 0 for state D:

$$s = \text{D}, \; a = \text{right}, \; s' = \text{E}, \; r = 0.$$

Assume $n(\text{D}, \text{right}) = 2$ so far in the iterations, i.e., we have moved to the right at state D twice. Let the discount factor be $\gamma = 0.9$ at this iteration. According to Eq. (32), by performing this action in this state, $Q(\text{D}, \text{right})$ is updated from 16 to 17 as:

$$\begin{aligned} Q_n^*(\text{D}, \text{right}) &= Q_{n-1}^*(\text{D}, \text{right}) + \alpha_n\big(r + \gamma \max_{a'} Q^*(\text{E}, a') - Q^*(\text{D}, \text{right})\big) \\ &= Q_{n-1}^*(\text{D}, \text{right}) \\ &\quad + \alpha_n\big(r + \gamma \max_{\{\text{left}, \text{right}\}} \{Q^*(\text{E}, \text{left}), Q^*(\text{E}, \text{right})\} - Q^*(\text{D}, \text{right})\big) \\ &= 16 + (1/2)(0 + (0.9 \times \max\{15, 20\}) - 16) = 17. \end{aligned}$$

Increasing the Q function (which essentially represents reward) for moving right at state D makes sense because it is better to move right to hopefully end up in state G eventually and win the game.

8.5 Q Function Approximation

The Q function can be approximated by any linear or nonlinear function. For example, it can be approximated by a neural network, which is a nonlinear function. Let $Q_{\boldsymbol{w}}(s, a)$ denote the approximated Q function with parameters $\boldsymbol{w}$; for instance, it can be a neural network with weights $\boldsymbol{w}$.

Recall Eq. (32), in which the temporal difference is the difference between $r + \gamma \max_{a'} Q_{n-1}^*(s', a')$ and $Q_{n-1}^*(s, a)$. Therefore, $r + \gamma \max_{a'} Q_{\boldsymbol{w}}(s', a')$ can serve as the target, and the Q function estimate, $Q_{\boldsymbol{w}}(s, a)$, needs to approach this target. As a result, the loss function for the neural network can be defined as the mean squared error between these two terms:

$$\ell(\boldsymbol{w}) = \frac{1}{2}\big(Q_{\boldsymbol{w}}(s,a) - r - \gamma \max_{a'} Q_{\boldsymbol{w}}(s',a')\big)^2. \tag{34}$$

This loss function has an issue referred to as the moving target problem. Both $Q_{\boldsymbol{w}}(s,a)$ and $r + \gamma \max_{a'} Q_{\boldsymbol{w}}(s',a')$ depend on the weights $\boldsymbol{w}$; therefore, changing the weights $\boldsymbol{w}$ modifies both the function estimate and the target. In this sense, the target becomes a moving target. In other words, when $Q_{\boldsymbol{w}}(s,a)$ is pushed toward the target $r + \gamma \max_{a'} Q_{\boldsymbol{w}}(s',a')$, the target also moves away. It is as if $Q_{\boldsymbol{w}}(s,a)$ chases the target while the target escapes, preventing convergence.

It has been empirically observed that if the Q function is linear, minimizing this loss function converges; however, if it is nonlinear, it may not converge. There are two approaches to resolve the moving target problem, namely gradient Q-learning and experience replay. Both approaches usually work in practice, but they do not have any theoretical guarantee of convergence. These approaches are explained in the following.

8.5.1 Gradient Q-Learning

The first approach for resolving the moving target problem is gradient Q-learning. In this approach, the Q function in the target $r + \gamma \max_{a'} Q_{\boldsymbol{w}}(s',a')$ is fixed and updated every k iterations. However, the estimated Q function $Q_{\boldsymbol{w}}(s,a)$ is updated by optimization in every iteration. In practice, this approach can be implemented using two neural networks – the Q-value estimate network and the target network – rather than a single network. Therefore, the target is updated with some delay compared to the Q-value estimate, allowing the estimate to eventually reach the target.

Let the weights of the Q-value estimate network and the target network be denoted by $\boldsymbol{w}$ and $\bar{\boldsymbol{w}}$, respectively. The loss function becomes

$$\ell(\boldsymbol{w}) = \frac{1}{2}\big(Q_{\boldsymbol{w}}(s,a) - r - \gamma \max_{a'} Q_{\bar{\boldsymbol{w}}}(s',a')\big)^2. \tag{35}$$

The gradient of this loss function, used in backpropagation, is

$$\frac{\partial \ell(\boldsymbol{w})}{\partial \boldsymbol{w}} = \big(Q_{\boldsymbol{w}}(s,a) - r - \gamma \max_{a'} Q_{\bar{\boldsymbol{w}}}(s',a')\big)\frac{\partial Q_{\boldsymbol{w}}(s,a)}{\partial \boldsymbol{w}}. \tag{36}$$

The algorithm for gradient Q-learning is presented in Algorithm 7. In this algorithm, $\alpha > 0$ is the learning rate for optimization. Updating $\bar{\boldsymbol{w}}$ can be performed either by optimizing $\bar{\boldsymbol{w}}$ in the target neural network or by copying the weights of the estimate network to the target network ($\bar{\boldsymbol{w}} \leftarrow \boldsymbol{w}$), in case the two networks have the same structure.

```
Initialize w and w̄
Observe the current state s
while not converged do
    Select and execute action a
    Observe the new state s'
    ∂ℓ(w)/∂w = (Q_w(s,a) − r − γ max_{a'} Q_w(s',a')) ∂Q_w(s,a)/∂w
    w ← w − α ∂ℓ(w)/∂w
    s ← s'
    if it is every k iterations then
        Update w̄
Return Q_w(s,a)
```

Algorithm 7: Gradient Q-learning

8.5.2 Experience Replay

Another approach for resolving the moving target problem is experience replay [1]. In this approach, previous experiences $\{s, a, s', r\}$ are stored in a buffer, and at every iteration of gradient Q-learning, a mini-batch of these previous experiences is sampled. This approach uses the same weights and a single network for both the Q-value estimate function and the target. By using previous experiences, the target function being optimized does not change dramatically. In other words, this reduces the effect of the moving target.

8.6 Deep Q Network

Deep reinforcement learning [2, 14, 26] refers to reinforcement learning where the Q function, value function, or policy function is estimated using a neural network as the estimator function. Therefore, deep reinforcement learning is essentially the same as reinforcement learning, except that its function estimator is a neural network.

The Deep Q Network (DQN) was the first breakthrough in deep reinforcement learning, showing that neural networks can approximate Q-functions and achieve human-level performance in Atari games [16, 17]. DQN combines Q-learning with convolutional networks, experience replay, and target networks.

DQN can play Atari games at a human level, where the trained network can even outperform humans [17]. As shown in Fig. 5, the input to DQN is the frame of the Atari game at each time slot. DQN uses a convolutional neural network to approximate the Q function. The output neurons of the network correspond to the possible actions performable using the Atari joystick. A softmax activation function is used in the last layer to output the probabilities of actions, and the action with the highest probability determines the action at the current time slot.

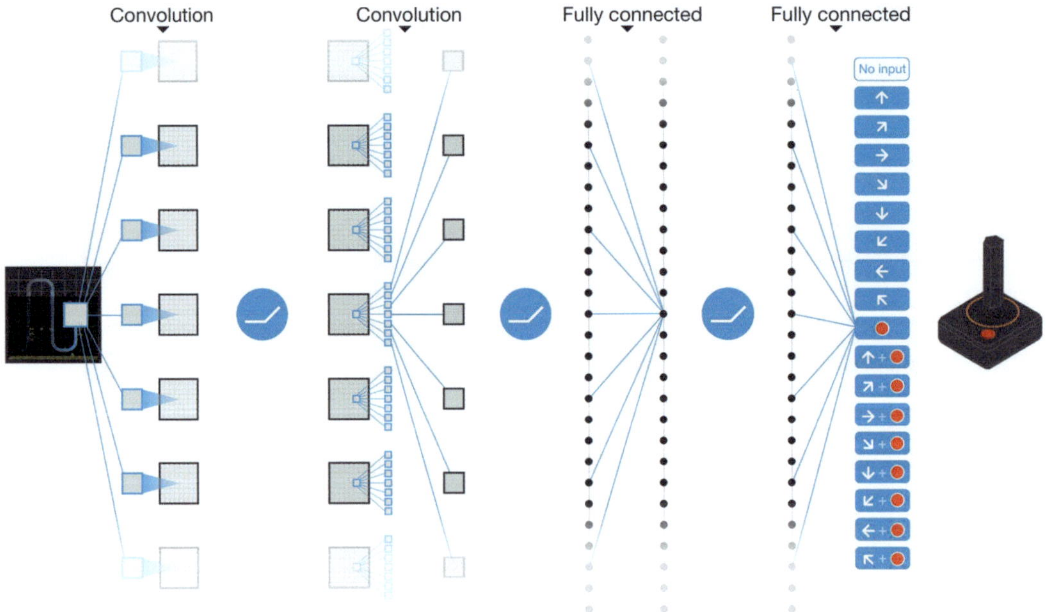

Fig. 5 Playing Atari games with DQN. *Image credit* [16]

9 Policy Gradient

Q-learning is a model-free value-based method. In Q-learning, the value function or Q function is explicitly learned while the policy function is implicit. Policy gradient [32], however, is a model-free policy-based algorithm. Policy gradient methods, such as REINFORCE [35], explicitly optimize the policy by following the gradient of expected return. Both Q-learning and policy gradient pursue the goal of reinforcement learning, which is to find the best mapping of policy $\pi : s \mapsto a$. The policy gradient approach attempts to learn the policy function explicitly.

In practice, modern algorithms like Actor-Critic [30, 32], Asynchronous Advantage Actor-Critic (A3C) [15], Proximal Policy Optimization (PPO) [25], and Soft Actor-Critic (SAC) [12] extend this idea to achieve stability and scalability in continuous and large action spaces.

9.1 Policy Gradient

The policy function can be a deterministic (non-random) function $\pi(s) = a$ or a stochastic (random) function defined as a conditional probability $\pi(a|s)$. Each of these policy functions has its pros and cons. The deterministic function is usually discrete, whereas the stochastic policy is typically a smooth and continuous probabilistic density function, which is more suitable for learning. The stochastic policy function is also more useful when the action space is continuous. In general, the stochastic policy function can be used for both discrete and continuous actions.

For discrete and continuous actions, the softmax function and Gaussian distribution can be useful, respectively:

$$\pi_{\boldsymbol{w}}(a|s) = \frac{\exp(f(s, a; \boldsymbol{w}))}{\sum_{\bar{a}} \exp(f(s, \bar{a}; \boldsymbol{w}))}, \tag{37}$$

$$\pi_{\boldsymbol{w}}(a|s) = \mathcal{N}\big(a \mid \mu(s; \boldsymbol{w}), \Sigma(s; \boldsymbol{w})\big), \tag{38}$$

where $f(s, a; \boldsymbol{w})$ is the function estimator of the policy, with parameter $\boldsymbol{w}$, before applying the softmax function, and $\mu(s; \boldsymbol{w})$ and $\Sigma(s; \boldsymbol{w})$ are the Gaussian mean and covariance functions of the states with parameter $\boldsymbol{w}$.

In supervised learning, the goal is to learn $\pi_{\boldsymbol{w}}(a|s)$ where the training data are state-action pairs, $\{(s_1, a_1^*), (s_2, a_2^*), \dots\}$, with the best actions as labels for the states. The cross-entropy of the softmax function can be minimized or the log-likelihood of the Gaussian distribution can be maximized. For example, maximization of the log-likelihood in the forms of one-shot optimization and temporal difference are

$$\begin{aligned} \boldsymbol{w}^* &= \arg\max_{\boldsymbol{w}} \sum_n \log \pi_{\boldsymbol{w}}(a_n^*|s_n), \\ \boldsymbol{w}_{n+1} &\leftarrow \boldsymbol{w}_n + \alpha_n \nabla_{\boldsymbol{w}} \log \pi_{\boldsymbol{w}}(a_n^*|s_n), \end{aligned} \tag{39}$$

respectively, where $\nabla_{\boldsymbol{w}}(.)$ is the derivative operator with respect to $\boldsymbol{w}$. Each of these equations updates the policy in supervised learning.

In reinforcement learning, however, the goal is to learn $\pi_{\boldsymbol{w}}(a|s)$ where the training data are state-action reward triplets, $\{(s_1, a_1, r_1), (s_2, a_2, r_2), \dots\}$. In each triplet, the reward provides a hint about how good the action is in that state. Policy gradient in reinforcement learning maximizes the discounted summation of rewards in the forms of one-shot optimization and temporal difference [32]:

Initialize $\boldsymbol{w}$
Observe the current state s
for *each episode* **do**
  Generate (play) the episode: $\{(s_0, a_0, r_0), (s_1, a_1, r_1), \dots, (s_T, a_T, r_T)\}$
  for *n from* 0 *to* T **do**
    $G_n \leftarrow \sum_{t=0}^{T-n} \gamma^t r_{n+t}$
    $\boldsymbol{w} \leftarrow \boldsymbol{w} + \alpha \gamma^n G_n \nabla_{\boldsymbol{w}} \log \pi_{\boldsymbol{w}}(a_n | s_n)$
Return $\pi_{\boldsymbol{w}}(a|s)$

Algorithm 8: REINFORCE algorithm

$$\begin{aligned} \boldsymbol{w}^* &= \arg\max_{\boldsymbol{w}} \sum_n \gamma^n \, \mathbb{E}[r_n | s_n, a_n], \\ \boldsymbol{w}_{n+1} &\leftarrow \boldsymbol{w}_n + \alpha_n \gamma^n G_n \nabla_{\boldsymbol{w}} \log \pi_{\boldsymbol{w}}(a_n | s_n), \end{aligned} \tag{40}$$

respectively, where G_n is the total reward defined in Eq. (16), i.e., $G_n = \sum_{t=0}^{\infty} \gamma^t r_{n+t}$. Each of these equations updates the policy in reinforcement learning.

Comparing Eqs. (39) and (40):

$$\begin{aligned} \boldsymbol{w}_{n+1} &\leftarrow \boldsymbol{w}_n + \alpha_n \nabla_{\boldsymbol{w}} \log \pi_{\boldsymbol{w}}(a_n^* | s_n), \\ \boldsymbol{w}_{n+1} &\leftarrow \boldsymbol{w}_n + \alpha_n \gamma^n G_n \nabla_{\boldsymbol{w}} \log \pi_{\boldsymbol{w}}(a_n | s_n). \end{aligned}$$

shows that, compared to supervised learning, reinforcement learning scales the gradient by the reward $\gamma^n G_n$. If the reward is large, that direction of the gradient is amplified, and the algorithm moves in that direction. However, if the reward is small, that direction of the gradient is ignored, and the algorithm does not move in that direction.

It is noteworthy that it is possible to combine the algorithms of Q-learning and policy gradient [19].

9.2 REINFORCE Algorithm

The REINFORCE algorithm is one of the earliest algorithms in reinforcement learning, proposed in 1992 [35]. This algorithm, shown in Algorithm 8, is a policy gradient algorithm and it uses the temporal update in Eq. (40). It plays the game or interacts with the environment multiple times. Each complete run of the game is called an episode. Every episode has T steps, where T may differ across episodes. In this algorithm, $\alpha > 0$ is the learning rate.

10 AlphaGo: Game of Go by Reinforcement Learning

The game of Go is an Asian-originated game in which two players take turns placing a stone on a vacant (empty) intersection (dot). One player uses white stones and the other uses black stones (see Fig. 6a). Connected stones that have no adjacent vacant intersections are removed. The player who controls the largest number of intersections at the end of the game is the winner. This game is difficult for computers because the game board is 19×19 (i.e., 361 intersections), and each intersection can be vacant, have a white stone, or have a black stone. Therefore, its state space is extremely large, with 3^{361} possibilities.

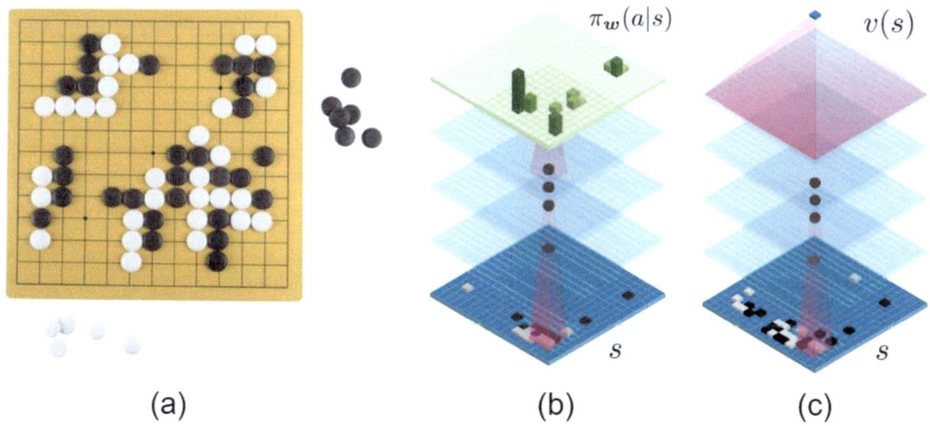

Fig. 6 **a** The game of Go, **b** the policy network, and **c** the value network in AlphaGo

AlphaGo [29] demonstrated the power of deep reinforcement learning by combining supervised learning, policy gradient fine-tuning, value networks, and Monte Carlo Tree Search. This approach was later generalized into AlphaZero [27] and MuZero [24], which achieved superhuman performance across Go, chess, and shogi using a single unified framework.

10.1 Policy Network in AlphaGo

10.1.1 Supervised Learning in AlphaGo

AlphaGo has a policy network, shown in Fig. 6b, for modeling $\pi(a|s)$, where the input to the network is the state s, representing the configuration of the board. The output of the policy network is the distribution of actions a. The intersection with the highest output probability determines where the stone should be placed. AlphaGo utilizes supervised learning for training the policy network, using 30 million board configurations played by professional players. The network maximizes the log-likelihood of the policy:

$$\underset{\boldsymbol{w}}{\text{maximize}} \quad \log \pi_{\boldsymbol{w}}(a|s),$$
$$\boldsymbol{w} \leftarrow \boldsymbol{w} + \alpha \frac{\partial \log \pi_{\boldsymbol{w}}(a|s)}{\partial \boldsymbol{w}},$$

where the second equation, which is gradient ascent, is performed by backpropagation (gradient descent) with the gradient multiplied by -1: $\boldsymbol{w} \leftarrow \boldsymbol{w} - \alpha \partial \log \pi_{\boldsymbol{w}}(a|s)/\partial \boldsymbol{w}$.

10.1.2 Policy Gradient in AlphaGo

After supervised learning, a policy function implemented as the policy network is obtained. This policy function can be further improved using policy gradient. In policy gradient training for the policy network, the network plays the game of Go in episodes against its previous version, which was fine-tuned in the previous episode. In every episode, the policy network is further trained and fine-tuned. For each episode, the total reward is defined as:

$$G_n = \begin{cases} 1 & \text{if it wins} \\ -1 & \text{if it loses.} \end{cases} \tag{41}$$

Equation (40) is used for fine-tuning the policy network in episodes via policy gradient:

$$\boldsymbol{w}_{n+1} \leftarrow \boldsymbol{w}_n + \alpha_n \gamma^n G_n \nabla_{\boldsymbol{w}} \log \pi_{\boldsymbol{w}}(a_n|s_n).$$

10.2 Value Network in AlphaGo

AlphaGo also has a value network, shown in Fig. 6c, to predict the value $v(s)$, i.e., to predict who will eventually win the game. This network is useful for AlphaGo to evaluate its status during the game and decide whether it should resign or continue playing. The input to the value network is the state s, which represents the configuration of the board. The output of the value network is the expected discounted summation of rewards, i.e., $v(s)$.

The value network is trained by gradient value learning. The training data are the (s, G) pairs, where G is defined as in Eq. (41):

$$G = \begin{cases} 1 & \text{if it wins} \\ -1 & \text{if it loses.} \end{cases}$$

The cost function minimized by the network is

$$\underset{\boldsymbol{w}}{\text{minimize}} \quad \frac{1}{2}(v_{\boldsymbol{w}}(s) - G)^2,$$

and the optimization in backpropagation is

$$\boldsymbol{w} \leftarrow \boldsymbol{w} - \alpha(v_{\boldsymbol{w}}(s) - G)\frac{\partial v_{\boldsymbol{w}}(s)}{\partial \boldsymbol{w}}.$$

10.3 Monte Carlo Tree Search (MCTS)

AlphaGo combines the policy network and the value network into a Monte Carlo Tree Search (MCTS) algorithm [5]. In MCTS, every node of the tree represents a state s, and every edge represents an action. Every possible episode of the game corresponds to one of the trajectories in the tree, each with its own reward. As this tree becomes very large, it can cause memory overflow, and searching within the tree becomes difficult and time-consuming. Therefore, MCTS grows the tree only along the trajectories of the most promising states. Note that many artificial intelligence algorithms for solving board games, such as AlphaGo for the game of Go [29] and AlphaZero for chess and shogi [24, 27, 28], use MCTS.

11 Conclusion

Deep reinforcement learning unifies reinforcement learning and deep learning, enabling agents to learn directly from raw high-dimensional inputs such as images. Starting from DQN and AlphaGo, the field has rapidly progressed to scalable algorithms like PPO, SAC, and distributed methods such as A3C. DRL powers real applications in robotics, recommendation systems, games, and large-scale decision-making. While efficiency, stability, and safety remain open challenges, deep reinforcement learning has become a cornerstone of modern AI research and practice.

12 Exercises

1. (Level: Moderate) Consider the following deterministic Markov Decision Process (MDP) with two states and two actions:

   - States: $S = \{s_1, s_2\}$
   - Actions: $A = \{a_1, a_2\}$
   - Discount factor: $\gamma = 0.9$

   The transition and reward structure is as follows:

From State	Action	To State	Reward
s_1	a_1	s_1	2
s_1	a_2	s_2	0
s_2	a_1	s_1	1
s_2	a_2	s_2	3

   (a) Value Iteration: Perform one iteration of value iteration with initial values $V_0(s_1) = 0$, $V_0(s_2) = 0$.
   (b) Policy Iteration (Policy evaluation step): Suppose we have a fixed policy π defined as:

   - $\pi(s_1) = a_2$
   - $\pi(s_2) = a_2$

   Compute the value function V^{π} using one iteration of policy evaluation starting from $V_0(s_1) = 0$, $V_0(s_2) = 0$.

2. (Level: Easy) Consider a reinforcement learning agent navigating a grid world where the agent can move left or right. Two policies are defined:

   - Policy A (Deterministic): Always move right.
   - Policy B (Stochastic): Move right with probability 0.7 and left with probability 0.3.

   Compute the expected return if the agent starts at s_0 and receives a reward of +1 for reaching s_1 and −1 for reaching s_2.
3. (Level: Moderate) Consider the following grid game:

A	B	C
D	E	F
G	H	I

   The game starts at square A. The actions are ℓ, r, u, d denoting moving left, right, up, and down, respectively, depending on square in the grid. For example, the actions for square A are r and d and the actions for square E are ℓ, r, u, and d.
   We want to perform Q-learning on this game. Let the rewards of the squares be:

0	1	2
1	3	4
2	4	5

Let the initial random Q values be:

$$\begin{aligned}
&Q(A, r) = 2,\ Q(A, d) = 3,\\
&Q(B, r) = 2,\ Q(B, \ell) = 0,\ Q(B, d) = 3,\\
&Q(c, \ell) = 1,\ Q(C, d) = 2\\
&Q(D, r) = 2,\ Q(D, u) = -1,\ Q(D, d) = 1,\\
&Q(E, r) = 1,\ Q(E, \ell) = 0,\ Q(E, u) = 0,\ Q(E, d) = 2,\\
&Q(F, \ell) = 1,\ Q(F, d) = 0,\ Q(F, u) = 2,\\
&Q(G, r) = 1,\ Q(G, u) = 2,\\
&Q(H, r) = 2,\ Q(H, \ell) = 1,\ Q(H, u) = 0,\\
&Q(I, \ell) = 1,\ Q(I, u) = 0.
\end{aligned}$$

(a) Starting at square A, perform two iterations of training in Q-learning.
Note: initialize all $n(s, a)$ to zero; let the discount factor be $\gamma = 0.9$; and choose the best action for selecting action a.

(b) After the training with two iterations (step a), perform two iterations in the test phase, starting at square A. Report the action a and the next state s' for every iteration.

4. (Level: Easy) Assuming you are training an intelligent learning assistant, where the assistant presents a question to a student and rewards the student based on their answer. This reward will be used to induct assistant in future learning and optimizing processes. The data you have is as follows:

- A correct answer gets a reward: $r_w = 1$
- A wrong answer gets a reward: $r_l = 0$

Assume the learning assistant predicts the student's answer with a score of $r_w = 0.8$, meaning the predicted probability of a correct answer is 0.8, and the predicted probability of a wrong answer is 0.2.

(a) Calculate the loss value for this example. Let the loss function be:

$$\mathcal{L}(\theta) = \mathbb{E}\big[\log(\sigma(r_w - r_l))\big],$$

where $\sigma(.)$ is the sigmoid function.

(b) Based on the loss, compute the gradient and update the model parameter θ. Let the learning rate be 0.1 and the current parameter value be $\theta = 0.7$.

5. (Level: Moderate) Consider a stochastic policy $\pi_\theta(a|s)$ defined over a single-state, two-action environment. The policy is parameterized using a softmax over preferences:

$$\pi_\theta(a_i) = \frac{e^{\theta_i}}{e^{\theta_1} + e^{\theta_2}}, \quad i \in \{1, 2\}.$$

At each time step:

- The agent receives a reward $R = 1$ if action a_1 is chosen,
- And $R = 0$ if a_2 is chosen.

Let the current policy parameters be $\theta = (\theta_1, \theta_2) = (0, 0)$.

(a) Compute the policy probabilities $\pi_\theta(a_1)$ and $\pi_\theta(a_2)$.
(b) Suppose the agent samples a_1 and receives reward $R = 1$. Compute the policy gradient $\nabla_\theta J(\theta)$ using the REINFORCE estimator:

$$\nabla_\theta J(\theta) = \mathbb{E}\left[R \cdot \nabla_\theta \log \pi_\theta(a)\right].$$

References

1. Marcin Andrychowicz et al. "Hindsight experience replay". In: *Advances in neural information processing systems* 30 (2017).
2. Kai Arulkumaran, Marc Peter Deisenroth, Miles Brundage, and Anil Anthony Bharath. "Deep reinforcement learning: A brief survey". In: *IEEE Signal Processing Magazine* 34.6 (2017), pp. 26–38.
3. Richard Bellman. "On the theory of dynamic programming". In: *Proceedings of the national Academy of Sciences* 38.8 (1952), pp. 716–719.
4. Dimitri P Bertsekas. "Approximate policy iteration: A survey and some new methods". In: *Journal of Control Theory and Applications* 9.3 (2011), pp. 310–335.
5. Cameron B Browne et al. "A survey of Monte Carlo tree search methods". In: *IEEE Transactions on Computational Intelligence and AI in games* 4.1 (2012), pp. 1–43.
6. Sinan Çalışır and Meltem Kurt Pehlivanoğlu. "Model-free reinforcement learning algorithms: A survey". In: *2019 27th signal processing and communications applications conference (SIU)*. IEEE. 2019, pp. 1–4.
7. Chris Dann, Yishay Mansour, Mehryar Mohri, Ayush Sekhari, and Karthik Sridharan. "Guarantees for epsilon-greedy reinforcement learning with function approximation". In: *International conference on machine learning*. PMLR. 2022, pp. 4666–4689.
8. Hans Fischer. *A history of the central limit theorem: from classical to modern probability theory*. Vol. 4. Springer, 2011.
9. Benyamin Ghojogh, Mark Crowley, Fakhri Karray, and Ali Ghodsi. "Background on Optimization". In: *Elements of Dimensionality Reduction and Manifold Learning*. Springer, 2023, pp. 75–120.
10. Benyamin Ghojogh, Fakhri Karray, and Mark Crowley. "Hidden Markov Model: Tutorial". In: *Engineering Archive* (2019).
11. Benyamin Ghojogh, Hadi Nekoei, Aydin Ghojogh, Fakhri Karray, and Mark Crowley. "Sampling algorithms, from survey sampling to Monte Carlo methods: Tutorial and literature review". In: *arXiv preprint* arXiv:2011.00901 (2020).
12. Tuomas Haarnoja, Aurick Zhou, Pieter Abbeel, and Sergey Levine. "Soft actor-critic: Off-policy maximum entropy deep reinforcement learning with a stochastic actor". In: *International conference on machine learning*. PMLR. 2018, pp. 1861–1870.
13. Xiang Li, Weiwei Shang, and Shuang Cong. "Model-based reinforcement learning for robot control". In: *2020 5th International Conference on Advanced Robotics and Mechatronics (ICARM)*. IEEE. 2020, pp. 300–305.
14. Yuxi Li. "Deep reinforcement learning: An overview". In: *arXiv preprint* arXiv:1701.07274 (2017).
15. Volodymyr Mnih et al. "Asynchronous methods for deep reinforcement learning". In: *International conference on machine learning*. PMLR. 2016, pp. 1928–1937.
16. Volodymyr Mnih et al. "Human-level control through deep reinforcement learning". In: *Nature* 518.7540 (2015), pp. 529–533.
17. Volodymyr Mnih et al. "Playing Atari with deep reinforcement learning". In: *arXiv preprint* arXiv:1312.5602 (2013).
18. Thomas M Moerland, Joost Broekens, Aske Plaat, Catholijn M Jonker, et al. "Model-based reinforcement learning: A survey". In: *Foundations and Trends® in Machine Learning* 16.1 (2023), pp. 1–118.
19. Brendan O'Donoghue, Remi Munos, Koray Kavukcuoglu, and Volodymyr Mnih. "Combining policy gradient and Q-learning". In: *International Conference on Learning Representations (ICLR)*. 2017.
20. Philip Osborne, Kajal Singh, and Matthew E Taylor. *Applying Reinforcement Learning on Real-World Data with Practical Examples in Python*. Springer Nature, 2022.

21. Athanasios S Polydoros and Lazaros Nalpantidis. “Survey of model-based reinforcement learning: Applications on robotics”. In: *Journal of Intelligent & Robotic Systems* 86.2 (2017), pp. 153–173.
22. Martin L Puterman and Moon Chirl Shin. “Modified policy iteration algorithms for discounted Markov decision problems”. In: *Management Science* 24.11 (1978), pp. 1127–1137.
23. Bruno Scherrer, Victor Gabillon, Mohammad Ghavamzadeh, and Matthieu Geist. “Approximate modified policy iteration”. In: *International conference on machine learning*. PMLR. 2012.
24. Julian Schrittwieser et al. “Mastering Atari, Go, chess and shogi by planning with a learned model”. In: *Nature* 588.7839 (2020), pp. 604–609.
25. John Schulman, Filip Wolski, Prafulla Dhariwal, Alec Radford, and Oleg Klimov. “Proximal policy optimization algorithms”. In: *arXiv preprint* arXiv:1707.06347 (2017).
26. Mohit Sewak. *Deep reinforcement learning*. Springer, 2019.
27. David Silver et al. “A general reinforcement learning algorithm that masters chess, shogi, and Go through self-play”. In: *Science* 362.6419 (2018), pp. 1140–1144.
28. David Silver et al. “Mastering chess and shogi by self-play with a general reinforcement learning algorithm”. In: *arXiv preprint* arXiv:1712.01815 (2017).
29. David Silver et al. “Mastering the game of Go with deep neural networks and tree search”. In: *Nature* 529.7587 (2016), pp. 484–489.
30. Richard S Sutton. “Learning to predict by the methods of temporal differences”. In: *Machine learning* 3.1 (1988), pp. 9–44.
31. Richard S Sutton and Andrew G Barto. *Reinforcement learning: An introduction*. MIT press, 2018.
32. Richard S Sutton, David McAllester, Satinder Singh, and Yishay Mansour. “Policy gradient methods for reinforcement learning with function approximation”. In: *Advances in neural information processing systems* 12 (1999).
33. Christopher JCH Watkins and Peter Dayan. “Q-learning”. In: *Machine learning* 8 (1992), pp. 279–292.
34. Christopher John Cornish Hellaby Watkins. “Learning from delayed rewards”. PhD thesis. King’s College, Cambridge United Kingdom, 1989.
35. Ronald J Williams. “Simple statistical gradient-following algorithms for connectionist reinforcement learning”. In: *Machine learning* 8 (1992), pp. 229–256.
36. Kaiqing Zhang, Zhuoran Yang, and Tamer Başar. “Multi-agent reinforcement learning: A selective overview of theories and algorithms”. In: *Handbook of reinforcement learning and control* (2021), pp. 321–384.

19 Few-Shot Learning and Meta-Learning

Abstract

Few-shot learning enables machine learning models to rapidly adapt to new tasks using only a small number of training examples, addressing challenges where data is scarce or rapid generalization is required. This chapter provides a comprehensive overview of few-shot learning methodologies, encompassing both episodic and non-episodic approaches, with an emphasis on meta-learning as a prominent framework for episodic few-shot learning. Key concepts such as episodes, tasks, and benchmark datasets are introduced to ground the discussion. Additionally, the chapter explores the challenges posed by domain shifts between training and testing data distributions, highlighting transfer learning, domain adaptation, and domain generalization as strategies to enhance model robustness. The role of meta-learning in enabling domain generalization is examined through algorithms such as Model-Agnostic Meta-Learning (MAML) and Model-Agnostic Learning of Semantic Features (MASF). This chapter aims to equip readers with both foundational knowledge and advanced techniques for few-shot learning and its applications in dynamic, low-data environments.

1 Introduction

Few-shot learning is the study of algorithms that can generalize to new tasks given only a handful of labeled examples. Its importance comes from applications where data collection is expensive, and it often combines ideas from transfer learning, meta-learning, and representation learning. It is useful in cases where there is little training data available or where rapid adaptation to new data is required. Several surveys on few-shot learning exist [1–3]. Note that there are even one-shot learning algorithms capable of learning a domain from a single data instance [4].

Few-shot learning algorithms can be categorized as either episodic or non-episodic. Episodic training is performed in episodes consisting of tasks. Most episodic few-shot learning methods are variants of meta-learning. Meta-learning, which refers to learning to learn, is a family of episodic few-shot learning algorithms. Various surveys on meta-learning are available [5–9]. This chapter introduces important algorithms in both non-episodic and episodic few-shot learning, as well as meta-learning.

It is noteworthy that in some applications, the test data may exhibit a domain shift from the training data. In other words, the distribution of the test data differs noticeably from that of the training data. This often occurs when the nature of the data changes over time or due to variations in data distributions. In such cases, transfer learning [10] can be used to fine-tune the machine learning model for the new test data. Two other main approaches for handling new domains or distributions of data are domain

B. Ghojogh and A. Ghodsi, *Elements of Deep Learning*,
https://doi.org/10.1007/978-3-032-10738-1_19

adaptation [11] and domain generalization [12]. Domain adaptation incorporates all possible domains (distributions) during training so that the model learns different possible scenarios and can handle test data from those domains. Domain generalization, however, is more challenging as it does not include all domains (distributions) in the training phase, allowing the model to generalize to unseen domains. Domain generalization and zero-shot learning are related but not identical. Domain generalization assumes training and test domains differ but share label space, while zero-shot learning assumes disjoint label spaces between training and test [13]. One of the use cases for few-shot learning is domain generalization [14, 15]. It is noteworthy that domain generalization from only a few samples of the new domain is also observed in humans. According to [16], children can infer a new category from even a single instance.

This chapter introduces few-shot learning and meta-learning. Section 2 defines episode, task, support set, query set, and well-known few-shot datasets. Non-episodic few-shot learning is introduced in Sect. 3. Meta-learning is introduced in Sect. 4, and Model-Agnostic Meta-Learning (MAML) is explained in Sect. 5. Finally, Sect. 6 introduces Model-Agnostic Learning of Semantic Features (MASF) and explains how meta-learning can be used for domain generalization.

2 Data in Few-Shot Learning

2.1 Episode and Task

In episodic few-shot learning, training is performed in episodes. Episodes are iterations of training where each episode involves learning one or multiple tasks. Each task in few-shot learning is a machine learning task such as classification, regression, or reinforcement learning. A task in episodic few-shot learning usually covers a subset of classes or cases in the training data. For example, in classification, if the training data contain c classes, each task may contain only N of the c classes, where $N \ll c$, especially when c is large. In other words, each task trains the machine learning model to classify among N classes, which are randomly sampled from the c classes. Performing this process across different tasks gradually introduces all c classes to the model, enabling it to learn to discriminate among all classes.

In regression, each task deals with a subset of possible values of the label. Going over different tasks introduces the full range of the label to the model. In reinforcement learning, each task covers different actions and scenarios in the environment.

2.2 Support and Query Sets

In few-shot learning, the training dataset is usually divided into two sets called the support set, denoted by $\mathcal{S}$, and the query set, denoted by $\mathcal{Q}$. Some papers refer to the support and query sets as the training and test sets, while others refer to them as the training and validation sets, respectively. Here, the terminologies "support set" and "query set" are used. The phrase "validation set" is reserved for data used to tune hyperparameters, and "test set" refers to data used to evaluate the model after training.

In each episode of training, one or multiple tasks $\{\mathcal{T}_i\}_i^t$ are selected. Each task contains a support set, which trains the model with weights θ to obtain the updated weights θ'. Every task also includes a query set, which fine-tunes the model with weights θ' to produce the final fine-tuned weights θ. In summary, every task contains a support set and a query set for updating and fine-tuning the weights of the neural network, respectively. This procedure will be explained in more detail in Sect. 5.

For few-shot classification, every support set is chosen such that it contains N classes (out of all c classes), with K data instances per class. Therefore, the support set is $\mathcal{S} = \{(\boldsymbol{x}_j, y_j)\}_{j=1}^{N \times K}$, where every y_j is one of the labels among the N classes.[1] Note that the number of instances per class may differ, but it is common practice to use the same number of instances per class. When the support set contains N classes with K instances per class, the few-shot learning setting is referred to as N-way K-shot.

The query set also contains the same N classes and a few data instances per class. It is important that the classes in the support and query sets are the same. If the number of instances per class in the query set is K_q, the query set is $\mathcal{Q} = \{(\boldsymbol{x}_j, y_j)\}_{j=1}^{N \times K_q}$, where every y_j is one of the labels among the N classes. Note that the class labels in the support and query sets must be the same. It is possible to have the same number of instances per class in both the support and query sets, i.e., $K = K_q$.

2.3 Well-Known Benchmark Datasets for Few-Shot Learning

2.3.1 Single Benchmarks

There are several well-known benchmark datasets for few-shot learning. One of them is Omniglot [17], which contains 1623 images of handwritten characters from 50 different alphabets. The images were drawn online by 20 people using Amazon's Mechanical Turk. The "Quick, Draw!" dataset (quickdraw.withgoogle.com) [18] contains over 50 million sketches across 345 categories and has become a standard benchmark for testing few-shot classification algorithms in sketch recognition. This dataset was collected by asking participants to draw a picture representing a given word within 20 s.

Another widely used dataset is ImageNet [19], which contains more than 14 million labeled images spanning over 21,000 classes. This dataset was originally proposed for the ImageNet Large-Scale Visual Recognition Challenge (ILSVRC), which uses 1000 classes. For the ILSVRC, the dataset has been reduced to 1.2 million training instances with 1000 classes and resized to 256×256 pixels. Because ImageNet is large scale, smaller subsets have been created, such as Mini-ImageNet [20], which contains 100 classes out of the 1000 classes of ILSVRC, with 600 images per class. It has been downsized to 84×84 pixels. There is also Tiny ImageNet, which contains 200 classes out of the 1000 classes of ILSVRC and approximately 110,000 images in total. The images in Tiny ImageNet have been downsized to 64×64 pixels. This dataset was proposed in the CS231n course (Convolutional Neural Networks for Visual Recognition) at Stanford University.

2.3.2 Meta-Datasets

Some benchmarks combine multiple few-shot datasets to be used for few-shot learning. These datasets are called meta-datasets because meta-learning, introduced later in this chapter, is an important family of algorithms in few-shot learning. A well-known meta-dataset is [21], which combines ImageNet, Omniglot, "Quick, Draw!", and several other few-shot datasets. Meta-album [22] is another meta-dataset for few-shot learning. There is also a meta-dataset [23] that uses decision trees induced from datasets for meta-learning.

[1] The labels in each task are renumbered from one to N because each task is treated as an N-class classification problem.

3 Non-episodic Few-Shot Learning

Non-episodic few-shot learning trains a backbone model on the entire dataset with standard supervised learning. At test time, the final classification head is replaced or adapted using only a few samples from new classes, making it closer to transfer learning. The last layer of neural network has c neurons to classify the data using the cross-entropy loss function, while the penultimate layer serves as the embedding layer, providing embeddings of the data.

It is possible to fine-tune the trained neural network to classify a new training dataset with only a few data instances. To achieve this, the output layer is replaced with a new output layer on top of the embedding layer, and the entire network is fine-tuned with the new training data. Since the initial weights for this fine-tuning have already been trained on the previous training data, the network can be fine-tuned effectively with just a few instances of the new data. This approach is known as the Baseline++ algorithm [24], which is inspired by [25, 26].

4 Meta-Learning for Few-Shot Learning

In contrast to non-episodic few-shot learning, episodic few-shot learning is trained end to end. Episodic few-shot learning is also referred to as meta-learning. Meta-learning, or "learning to learn" [27–29], has roots in the 1980s [28], but its modern neural formulations, such as matching networks [20] and MAML [30], have made it central to few-shot learning research.

Regular machine learning focuses on learning tasks, whereas meta-learning focuses on learning how to learn tasks. In fact, it is conceptually possible to have meta-meta-meta-... learning, referring to learning to learn to learn, and so on [28]. Additionally, some works have been proposed for learning the hyperparameters of the learner itself [31–33].

Meta-learning is useful for few-shot learning because, in each episode, it learns the weights θ' using the support set and then fine-tunes (or further learns) the weights to θ for the few available data instances. Therefore, it learns to learn.

There are several well-known episodic few-shot learning, or meta-learning, algorithms, including matching networks, prototypical networks, relation networks, and model-agnostic meta-learning. These networks are explained in the following sections. In the descriptions that follow, assume $f(\boldsymbol{x})$ denotes the embedding of $\boldsymbol{x}$ at the embedding layer, i.e., at the penultimate layer.

4.1 Matching Network

In every episode, the matching network [20] labels each query instance $\boldsymbol{x}^*$ as follows:

$$\mathbb{P}(y^* = \ell \mid \boldsymbol{x}^*, \mathcal{S}) = \sum_{i=1}^{|\mathcal{S}|} \frac{\exp\big(\widehat{f}(\boldsymbol{x}_i)^\top \widehat{f}(\boldsymbol{x}^*)\big)}{\sum_{j=1}^{|\mathcal{S}|} \exp\big(\widehat{f}(\boldsymbol{x}_j)^\top \widehat{f}(\boldsymbol{x}^*)\big)} \mathbb{I}(y_i = \ell), \tag{1}$$

where $\widehat{f}(\boldsymbol{x})$ is the normalized $f(\boldsymbol{x})$ to have unit length, y^* is the predicted label for the query $\boldsymbol{x}^*$, $|\mathcal{S}|$ is the cardinality of the support set, and $\mathbb{I}(.)$ denotes the indicator function, which is one if its condition is satisfied and zero otherwise.

Equation (1) provides the probability that the predicted label of $\boldsymbol{x}^*$ is ℓ, given the support set. This probability is computed as a softmax over the cosine similarities between the query instance and the support instances having that label. The loss function of the network maximizes Eq. (1) for the target labels in the query set.

4.2 Prototypical Network

4.2.1 Supervised Prototypical Network

The supervised prototypical network [34] labels each query instance $\boldsymbol{x}^*$ as follows:

$$\mathbb{P}(y^* = \ell \mid \boldsymbol{x}^*, \mathcal{S}) = \frac{\exp(-\|f(\boldsymbol{x}^*) - \widetilde{f_\ell}\|_2^2)}{\sum_{\ell'=1}^{N} \exp(-\|f(\boldsymbol{x}^*) - \widetilde{f_{\ell'}}\|_2^2)}, \tag{2}$$

where $\|.\|_2$ denotes the ℓ_2 norm, and $\widetilde{f_\ell}$ denotes the prototype of the embeddings of support instances in class ℓ. This prototype is a summary statistic, such as the average, of the embeddings of support instances with label ℓ:

$$\widetilde{f_\ell} := \frac{1}{K} \sum_{i=1}^{|\mathcal{S}|} f(\boldsymbol{x}_i)\, \mathbb{I}(y_i = \ell). \tag{3}$$

Here, K is the number of instances with label ℓ in the N-way K-shot setting.

Equation (2) provides the probability that the predicted label of $\boldsymbol{x}^*$ is ℓ, given the support set. This probability is computed as a softmax over the squared Euclidean distances between the query instance and the prototypes of the support instances with that label. The loss function of the network maximizes Eq. (1) for the target labels in the query set.

4.2.2 Semi-supervised Prototypical Network

Prototypical networks can be trained using partially labeled data [35], as described below. In a semi-supervised setting, part of the dataset is unlabeled while the rest is labeled. Let the labeled and unlabeled support sets be denoted by $\mathcal{S} = \{\boldsymbol{x}_i, y_i\}_{i=1}^{|\mathcal{S}|}$ and $\mathcal{R} = \{\boldsymbol{x}'_i, y'_i\}_{i=1}^{|\mathcal{R}|}$, respectively.

Each unlabeled data instance can be labeled using a clustering algorithm, such as K-means, applied to the labeled support data. However, because K-means assigns hard cluster memberships, it is not differentiable for optimization. Therefore, soft clustering is used, where first, the probability of partial assignment to the prototypes of classes is calculated:

$$z'_{i,\ell} := \frac{\exp(-\|f(\boldsymbol{x}'_i) - \widetilde{f_\ell}\|_2^2)}{\sum_{\ell'=1}^{N} \exp(-\|f(\boldsymbol{x}'_i) - \widetilde{f_{\ell'}}\|_2^2)}, \tag{4}$$

which is the assignment score of the unlabeled $\boldsymbol{x}'_i$ to the prototype of class ℓ.

Then, the prototype of each class is refined using the newly labeled data $\mathcal{R}$:

$$\widetilde{f_\ell} := \frac{\sum_{i=1}^{|\mathcal{S}|} f(\boldsymbol{x}_i)\mathbb{I}(y_i = \ell) + \sum_{i=1}^{|\mathcal{R}|} f(\boldsymbol{x}'_i) z'_{i,\ell}}{\sum_{i=1}^{|\mathcal{S}|} \mathbb{I}(y_i = \ell) + \sum_{i=1}^{|\mathcal{R}|} z'_{i,\ell}}. \tag{5}$$

One problem with this approach is that it assumes each unlabeled data instance belongs to one of the existing class labels; however, some instances may not belong to any of them. Therefore, an additional distractor class label can be defined. For simplicity, the prototype of the distractor class can be assumed to be the origin, i.e., a zero vector. Then, Eqs. (4) and (5) can be used with all class labels plus the distractor class, resulting in $N + 1$ class prototypes.

This approach still has a limitation in that it considers only one distractor class for all outlier instances, while multiple distractor classes may exist. To resolve this issue, soft masking can be used to partially mask the unlabeled instances. In this case, Eq. (5) is revised to:

$$\widetilde{f_\ell} := \frac{\sum_{i=1}^{|\mathcal{S}|} f(\boldsymbol{x}_i)\mathbb{I}(y_i = \ell) + \sum_{i=1}^{|\mathcal{R}|} f(\boldsymbol{x}'_i) z'_{i,\ell} m'_{i,\ell}}{\sum_{i=1}^{|\mathcal{S}|} \mathbb{I}(y_i = \ell) + \sum_{i=1}^{|\mathcal{R}|} z'_{i,\ell} m'_{i,\ell}}, \tag{6}$$

where $m'_{i,\ell}$ is the mask for the unlabeled $\boldsymbol{x}'_i$ in class ℓ, defined using the following sigmoid function:

$$m'_{i,\ell} := \frac{1}{1 + \exp\big(\gamma_\ell (d'_{i,\ell} - \beta_\ell)\big)}, \tag{7}$$

in which γ_ℓ and β_ℓ are parameters learned by a separate small neural network, and $d'_{i,\ell}$ is defined as the normalized distance of $\boldsymbol{x}'_i$ from the prototype of class ℓ:

$$d'_{i,\ell} := \frac{\|\boldsymbol{x}'_i - \widetilde{f_\ell}\|_2^2}{\frac{1}{|\mathcal{R}|} \sum_{j=1}^{|\mathcal{R}|} \|\boldsymbol{x}'_j - \widetilde{f_\ell}\|_2^2}. \tag{8}$$

4.3 Relation Network

The relation network [36] obtains the embeddings of the support set and calculates the prototypes as in Eq. (3). Then, each prototype is concatenated with every query instance in $\{\boldsymbol{x}_j\}_{j=1}^{|\mathcal{Q}|}$. Assume $[\widetilde{f_\ell}^\top, \boldsymbol{x}_j^\top]^\top$ denotes the concatenation of the ℓth prototype and the jth query instance. The embedding of this concatenation is $f([\widetilde{f_\ell}^\top, \boldsymbol{x}_j^\top]^\top)$. The last layer of the neural network has one neuron with a sigmoid activation function, $\sigma(.)$, to output a score in the range [0, 1]. The loss function minimizes the mean squared error between this score and the target label of the query instance:

$$\text{minimize} \sum_{\ell=1}^{N} \sum_{j=1}^{|\mathcal{Q}|} \Big(f([\widetilde{f_\ell}^\top, \boldsymbol{x}_j^\top]^\top) - \mathbb{I}(y_i = \ell) \Big). \tag{9}$$

This encourages the predicted label of a query instance concatenated with the correct prototype to be one, while pushing the label of a query instance concatenated with other prototypes toward zero.

5 Model-Agnostic Meta-Learning (MAML)

The most well-known episodic few-shot learning method is Model-Agnostic Meta-Learning (MAML). Gradient-based meta-learning usually refers to meta-learning methods inspired by MAML. Several different variants of MAML are introduced in the following. Refer to [37–41] for convergence analyses of MAML.

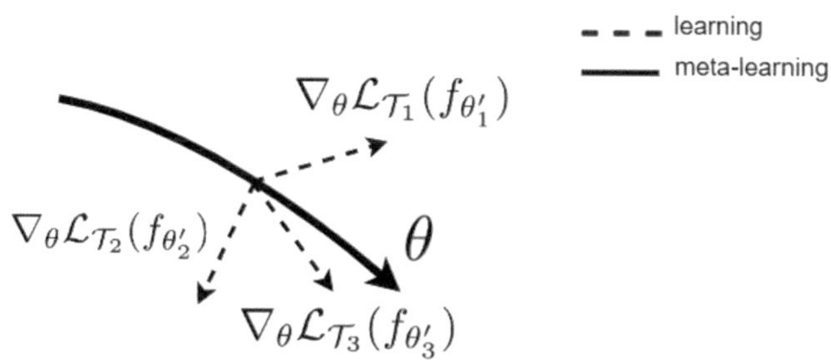

Fig. 1 The intuition of weight updates in MAML. The learning step updates the weights to θ'_1, θ'_2, and θ'_3. Then, the meta-learning step aggregates the derivatives $\nabla_\theta \mathcal{L}_{\mathcal{T}_1}(f_{\theta'_1})$, $\nabla_\theta \mathcal{L}_{\mathcal{T}_2}(f_{\theta'_2})$, and $\nabla_\theta \mathcal{L}_{\mathcal{T}_3}(f_{\theta'_3})$ to update the weights to θ

5.1 Vanilla MAML

The first MAML algorithm was proposed in 2017 [30]. As explained in Sect. 2.2, MAML uses tasks composed of support and query sets in every episode of training. Each task contains a support set and a query set with the same class labels. MAML samples a batch of tasks $\{\mathcal{T}_i\}_{i=1}^t$ from the support and query sets, where t denotes the number of sampled tasks per episode. The weights of the neural network, denoted by θ, are adapted (or updated) to θ' using the support data instances of the tasks. As there is a batch of tasks, there will be a set of adapted weights $\{\theta'_i\}_{i=1}^t$, each obtained by backpropagation using one of the sampled tasks from the support set. In other words, the parameters of the network are adapted to the tasks in the support set. It can be considered that there are t neural networks with the same structure but different weights $\{\theta'_i\}_{i-1}^t$.

After that, each neural network with weights θ'_i, denoted by $f_{\theta'_i}$, is evaluated on the query set in task $\mathcal{T}_i$. An aggregation of gradients from these evaluations provides the final gradient for the update using the query set. If a network has a high loss value, it will have a greater impact on the gradient direction. The intuition of weight updates in MAML is illustrated in Fig. 1.

The algorithm for vanilla MAML is provided in Algorithm 1. The weights of the neural network are initialized randomly to θ. In every episode or epoch of training, a batch of tasks $\{\mathcal{T}_i\}_{i=1}^t$ is sampled from the support and query sets. For each task $\mathcal{T}_i$, the gradient of the loss function $\mathcal{L}$ is evaluated on the K samples in the task. Let $\mathcal{L}_{\mathcal{T}_i}(f_\theta)$ denote the loss function of the network f_θ evaluated using the K support instances in task $\mathcal{T}_i$. A step of backpropagation is performed:

$$\theta'_i = \theta - \alpha \nabla_\theta \mathcal{L}_{\mathcal{T}_i}(f_\theta), \tag{10}$$

where $\alpha > 0$ is the learning rate and ∇_θ denotes the derivative with respect to θ. This results in a set of t adapted weights $\{\theta'_i\}_{i=1}^t$.

Each network $f_{\theta'_i}$ is evaluated on the query instances in task $\mathcal{T}_i$ to compute the loss function $\mathcal{L}_{\mathcal{T}_i}(f_{\theta'_i})$. These loss functions are summed together to form the meta-objective function $\sum_{i=1}^t \nabla_\theta \mathcal{L}_{\mathcal{T}_i}(f_{\theta'_i})$. A step of backpropagation, or meta-optimization, is then performed:

$$\theta = \theta - \beta \sum_{i-1}^{t} \nabla_\theta \mathcal{L}_{\mathcal{T}_i}(f_{\theta'_i}), \tag{11}$$

where $\beta > 0$ is the meta-learning rate. The weight θ is used as the starting point for the next epoch or episode. Note that the loss function $\mathcal{L}$ used in Eqs. (10) and (11) can be any loss function depending on the machine learning task, such as cross-entropy loss for classification, mean squared error for regression, or a negative value for reinforcement learning.

Initialize the weights θ
while *not converged (episodes)* **do**
  Sample a batch of tasks $\{\mathcal{T}_i\}_{i=1}^{t}$
  for *each task* $\mathcal{T}_i$ **do**
    Evaluate $\nabla_\theta \mathcal{L}_{\mathcal{T}_i}(f_\theta)$ on the K instances
    $\theta'_i = \theta - \alpha \nabla_\theta \mathcal{L}_{\mathcal{T}_i}(f_\theta)$
  $\theta = \theta - \beta \sum_{i=1}^{t} \nabla_\theta \mathcal{L}_{\mathcal{T}_i}(f_{\theta'_i})$

Algorithm 1: Vanilla MAML training algorithm

Having the two backpropagation updates in Eqs. (10) and (11) results in gradients through gradients or, in other words, second-order gradients (Hessians). Experiments have shown that it is possible to approximate the second-order gradients using first-order gradients only [30]. In other words, the flow of gradients through gradients can be stopped for simplicity. In addition to vanilla MAML, there are other Hessian-free MAML variants, such as [39, 42, 43], which use first-order approximations of gradients.

5.2 Unsupervised MAML (UMTRA)

MAML is supervised in the sense that it uses class labels to form the tasks containing support and query sets. The support and query sets have the same class labels in every task. UMTRA [44], short for Unsupervised Meta-learning with Tasks constructed by Random sampling and Augmentation, is an unsupervised (or self-supervised) version of MAML. This method can be used when the data are not labeled. It randomly samples N data instances from the dataset in every task and assumes that these instances belong to different classes or patterns. More discussion on the validity of this assumption is provided later. UMTRA then augments these instances using augmentation functions to generate the support and query sets.

5.2.1 Augmentations

Let the randomly selected data instances in a task $\mathcal{T}_i$ be denoted by $\{\boldsymbol{x}_j\}_{j=1}^{N}$. The support and query sets are obtained as follows:

$$\mathcal{S} = \bigcup_{j=1}^{N} \{\mathcal{A}_\ell(\boldsymbol{x}_j)\}_{\ell=1}^{K}, \tag{12}$$

$$\mathcal{Q} = \bigcup_{j=1}^{N} \{\mathcal{A}_\ell(\boldsymbol{x}_j)\}_{\ell=1}^{K}, \tag{13}$$

where $\cup$ denotes the union operation and $\mathcal{A}_\ell(.)$ is one of the possible augmentation functions, such as rotation, translation, flipping, and gray-scaling. In other words, each instance is augmented K times to form the K samples in the support set, and the same procedure is used to form the K samples per instance in the query set. In this way, there will be N hypothetical classes, each with K samples in the support set and K corresponding samples in the query set.

In the special case where $K = 1$ (i.e., N-way one-shot), Eqs. (12) and (13) simplify to:

$$\mathcal{S} = \{\boldsymbol{x}_j\}_{j=1}^{N}, \tag{14}$$

$$\mathcal{Q} = \{\mathcal{A}(\boldsymbol{x}_j)\}_{j=1}^{N}, \tag{15}$$

where $\mathcal{A}(.)$ is a random augmentation function.

5.2.2 Discussions on the Assumption of UMTRA

UMTRA uses the augmentation technique to construct the support and query sets and then uses the MAML algorithm to train the neural network. UMTRA relies on the assumption that the randomly sampled examples likely belong to different classes when the dataset has many balanced categories. While practical in datasets like ImageNet, this assumption can break down in unbalanced or small-class settings. In other words, this assumption is practical especially if $N \ll c$, where N is the number of sampled examples and c denotes the actual number of classes in the entire dataset. Assuming that the classes are balanced, each having about m instances, the probability of the N drawn instances belonging to different classes is [44]:

$$\mathbb{P} = \frac{c!\, m^N (mc - N)!}{(c - N)!\, (mc)!}. \tag{16}$$

This probability is high in large datasets with many classes. For example, it is 99% in the ImageNet dataset [20]. Similar assumptions have been used in some self-supervised learning algorithms [45, 46].

5.3 MAML++

MAML++ [47] addresses several issues in MAML. The issues it addresses and resolves are as follows:

- **Multi-step MAML**: Vanilla MAML uses one step of backpropagation for the support sets in each task. However, it is possible to have multiple steps for the updates using the support set. In other words, the inner loop over tasks in Algorithm 1 can itself include a nested loop for s steps. MAML++ considers s steps, rather than just one, for the updates using support sets in each task. Theoretical analysis of multi-step MAML can be found in [48].
- **Gradient instability**: In every episode, MAML updates the network weights using the query set after it has performed updates with all support sets of the tasks. It then updates the weights using a weighted summation of the losses:

$$\theta - \theta - \beta \sum_{i=1}^{t} \sum_{j=1}^{s} w_j \nabla_\theta \mathcal{L}_{\mathcal{T}_i}(f_{\theta'_i}), \tag{17}$$

  where s is the number of steps for updates using the support set and w_j is the weight associated with the jth step. The weights $\{w_j\}_{j=1}^{s}$ are initially set to equal values, but in later episodes, larger weights are assigned to later steps because the most recent steps should have more impact on the loss value. It has been shown empirically that this improves the gradient instability in MAML.[2]

[2] Note that MAML can be considered as a dynamical system transitioning from one state to another [49]. Therefore, the gradients in this dynamical system may become unstable under certain conditions.

- **First-order gradient approximation**: As discussed in Sect. 5.1, MAML approximates the second-order gradient, i.e., the gradient through gradients, by first-order gradients. Performing this approximation can cause errors, while not performing it results in higher computational costs and slower training. MAML++ aims to achieve the best of both approaches by using the first-order approximation of gradients during the first 50 epochs (episodes) and avoiding the approximation afterward. Therefore, its training is initially faster to find a good solution, and later it uses Hessians to improve the accuracy of the solution.
- **Batch normalization**: MAML uses batch normalization [50] by computing the mean and variance of the entire batch across all s steps. MAML++ calculates the mean and variance in each batch per step and performs batch normalization per step. Moreover, in contrast to MAML, MAML++ uses different biases for neural network layers in different steps. These modifications improve the generalization of meta-learning to unseen data during training.
- **Learning rate**: While MAML uses a fixed learning rate across different steps, MAML++ uses a different learning rate α for each step. Additionally, MAML++ employs a cosine annealing schedule [51] to adapt the meta-learning rate β.

5.4 Adaptive Learning Rate in MAML

Vanilla MAML uses fixed values for the learning rate α and the meta-learning rate β. It is possible to adapt the optimal values of these learning rates during the training of MAML [52]. In other words, the learning rates are also updated using backpropagation:

$$\alpha^{(\nu+1)} = \alpha^{(\nu)} - \alpha' \sum_{i=1}^{t} \frac{\partial \mathcal{L}_{\mathcal{T}_i}(f_{\theta_i'^{(\nu)}})}{\partial \alpha}, \tag{18}$$

$$\beta^{(\nu)} = \beta^{(\nu-1)} - \beta' \sum_{i=1}^{t} \frac{\partial \mathcal{L}_{\mathcal{T}_i}(f_{\theta_i'^{(\nu)}})}{\partial \beta}, \tag{19}$$

where ν is the iteration index of the MAML algorithm and α' and β' are the learning rates for updating α and β, respectively.

The gradient for updating the learning rate α is

$$\begin{aligned}
\frac{\partial \mathcal{L}_{\mathcal{T}_i}(f_{\theta_i'^{(\nu)}})}{\partial \alpha} &\overset{(a)}{=} \frac{\partial \mathcal{L}_{\mathcal{T}_i}(f_{\theta_i'^{(\nu)}})}{\partial \theta_i'^{(\nu)}} \times \frac{\partial \theta_i'^{(\nu)}}{\partial \alpha} \\
&\overset{(b)}{=} \nabla_{\theta_i'} \mathcal{L}_{\mathcal{T}_i}(f_{\theta_i'^{(\nu)}}) \frac{\partial\big(\theta^{(\nu-1)} - \alpha^{(\nu-1)} \nabla_\theta \mathcal{L}_{\mathcal{T}_i}(f_{\theta^{(\nu-1)}})\big)}{\partial \alpha} \\
&= \nabla_{\theta_i'} \mathcal{L}_{\mathcal{T}_i}(f_{\theta_i'^{(\nu)}})\big(-\nabla_\theta \mathcal{L}_{\mathcal{T}_i}(f_{\theta^{(\nu-1)}})\big),
\end{aligned}$$

where (a) is due to the chain rule in derivatives, and (b) follows from the update formula for $\theta_i'^{(\nu)}$ by gradient descent.

Likewise, the gradient for updating the meta-learning rate β is

Initialize the weights θ
$\nu = 1$
while *not converged (episodes)* **do**
  Sample a batch of tasks $\{\mathcal{T}_i\}_{i=1}^t$
  for *each task* $\mathcal{T}_i$ **do**
    Evaluate $\nabla_\theta \mathcal{L}_{\mathcal{T}_i}(f_\theta)$ on the K instances
    $\theta_i'^{(\nu)} = \theta^{(\nu)} - \alpha^{(\nu)} \nabla_\theta \mathcal{L}_{\mathcal{T}_i}(f_{\theta^{(\nu)}})$
  $\alpha^{(\nu+1)} = \alpha^{(\nu)} + \alpha' \sum_{i=1}^t \nabla_{\theta_i'} \mathcal{L}_{\mathcal{T}_i}(f_{\theta_i'^{(\nu)}}) \nabla_\theta \mathcal{L}_{\mathcal{T}_i}(f_{\theta^{(\nu-1)}})$
  $\beta^{(\nu)} = \beta^{(\nu-1)} + \beta' \sum_{i=1}^t \nabla_\theta \mathcal{L}_{\mathcal{T}_i}(f_{\theta_i'^{(\nu)}}) \nabla_\theta \mathcal{L}_{\mathcal{T}_i}(f_{\theta_i'^{(\nu)}})$
  $\theta^{(\nu)} = \theta^{(\nu-1)} - \beta^{(\nu)} \sum_{i=1}^t \nabla_\theta \mathcal{L}_{\mathcal{T}_i}(f_{\theta_i'^{(\nu)}})$
  $\nu = \nu + 1$

Algorithm 2: Adaptive learning rates in MAML

$$
\begin{aligned}
\frac{\partial \mathcal{L}_{\mathcal{T}_i}(f_{\theta_i'^{(\nu)}})}{\partial \beta} &\overset{(a)}{=} \frac{\partial \mathcal{L}_{\mathcal{T}_i}(f_{\theta_i'^{(\nu)}})}{\partial \theta^{(\nu)}} \times \frac{\partial \theta^{(\nu)}}{\partial \beta} \\
&\overset{(b)}{=} \nabla_\theta \mathcal{L}_{\mathcal{T}_i}(f_{\theta_i'^{(\nu)}}) \frac{\partial\big(\theta^{(\nu-1)} - \beta^{(\nu)} \nabla_\theta \mathcal{L}_{\mathcal{T}_i}(f_{\theta_i'^{(\nu)}})\big)}{\partial \beta} \\
&= \nabla_\theta \mathcal{L}_{\mathcal{T}_i}(f_{\theta_i'^{(\nu)}}) \big(-\nabla_\theta \mathcal{L}_{\mathcal{T}_i}(f_{\theta_i'^{(\nu)}})\big),
\end{aligned}
$$

where (a) is due to the chain rule, and (b) follows from the update formula for $\theta^{(\nu)}$ by gradient descent.

Substituting these derivatives back into Eqs. (18) and (19) gives

$$
\alpha^{(\nu+1)} = \alpha^{(\nu)} + \alpha' \sum_{i=1}^{t} \nabla_{\theta_i'} \mathcal{L}_{\mathcal{T}_i}(f_{\theta_i'^{(\nu)}}) \nabla_\theta \mathcal{L}_{\mathcal{T}_i}(f_{\theta^{(\nu-1)}}), \tag{20}
$$

$$
\beta^{(\nu)} = \beta^{(\nu-1)} + \beta' \sum_{i=1}^{t} \nabla_\theta \mathcal{L}_{\mathcal{T}_i}(f_{\theta_i'^{(\nu)}}) \nabla_\theta \mathcal{L}_{\mathcal{T}_i}(f_{\theta_i'^{(\nu)}}). \tag{21}
$$

The algorithm for MAML with adaptive learning rates is shown in Algorithm 2.

5.5 Multi-modal MAML

Vanilla MAML assumes tasks are drawn from a single distribution. In practice, tasks can be multi-modal (e.g., visual domains versus linguistic domains), and extensions such as multi-modal MAML [53] introduce task embeddings and modulation networks to handle this. One naive way to account for multiple modes in the task distribution is to learn a separate MAML for each task mode. However, since the modes may be correlated, this approach is not efficient.

Multi-modal MAML [53] learns a multi-modal distribution for tasks. This model is composed of two meta-learners: a model-based meta-learner [54–56] and a gradient-based meta-learner [30].

In the first stage, which is the model-based meta-learner, multi-modal MAML passes K samples to a task embedding network f to obtain the embedding vector $\boldsymbol{v}$. This embedding vector is then fed to a modulation network g to obtain ℓ modulation parameters $\{\tau_i\}_{i=1}^{\ell}$, where ℓ is the number of layers in the MAML (gradient-based meta-learning) network. In the second stage, the weights of the

MAML network $\{\theta_i\}_{i=1}^{\ell}$ are modulated by the modulation parameters to yield $\{\theta_i \odot \tau_i\}_{i=1}^{\ell}$, where $\odot$ is a modulation operator such as attention (see Chap. 9). The MAML network is then trained as in standard MAML training. In every episode of training, the weights of the task embedding network, the modulation network, and the MAML network are all updated.

5.6 Variational MAML

There are several variants of Bayesian [57, 58], variational [59], or probabilistic [60] MAML. Here, variational MAML [59] is introduced as an example of a Bayesian approach for MAML. Let the support set and query set be denoted by $\mathcal{S} = \{\boldsymbol{x}_i^{(s)}, y_i^{(s)}\}_{i=1}^{N\times K}$ and $\mathcal{Q} = \{\boldsymbol{x}_i^{(q)}, y_i^{(q)}\}_{i=1}^{N\times K}$, respectively. As in MAML, the weights are updated to $\{\theta_i'\}_{i=1}^{t}$ using the support sets in the inner loop, and then the weight θ is updated using the query sets in the outer loop (see Algorithm 1).

On the one hand, variational MAML approximates the task-specific posterior distribution $\mathbb{P}(\theta_i' \mid y_i^{(q)}, \theta)$ by the variational distribution $q(\theta_i'|\lambda_i)$, which is parameterized by the random variable λ_i. Therefore, it minimizes the Kullback–Leibler (KL) divergence between these two distributions:

$$
\begin{aligned}
\lambda_i^* &= \arg\min_{\lambda_i} \text{KL}\big[q(\theta_i'|\lambda_i) \,\|\, \mathbb{P}(\theta_i' \mid y_i^{(q)}, \theta)\big] \\
&\overset{(a)}{=} \arg\min_{\lambda_i} \int q(\theta_i'|\lambda_i) \ln\Big(\frac{q(\theta_i'|\lambda_i)}{\mathbb{P}(\theta_i' \mid y_i^{(q)}, \theta)}\Big) d\theta_i' \\
&\overset{(b)}{=} \arg\min_{\lambda_i} \int q(\theta_i'|\lambda_i) \ln\Big(\frac{q(\theta_i'|\lambda_i)\mathbb{P}(y_i^{(q)} \mid \theta)}{\mathbb{P}(y_i^{(q)} \mid \theta_i')\mathbb{P}(\theta_i' \mid \theta)}\Big) d\theta_i' \\
&= \arg\min_{\lambda_i} \int \Big[q(\theta_i'|\lambda_i) \ln\Big(\frac{q(\theta_i'|\lambda_i)}{\mathbb{P}(\theta_i' \mid \theta)}\Big) \\
&\qquad - \ln(\mathbb{P}(y_i^{(q)} \mid \theta_i')) + \ln(\mathbb{P}(y_i^{(q)} \mid \theta))\Big] d\theta_i' \\
&\overset{(c)}{=} \text{KL}\big[q(\theta_i'|\lambda_i) \,\|\, \mathbb{P}(\theta_i' \mid \theta)\big] - \mathbb{E}_{q(\theta_i'|\lambda_i)}[\ln(\mathbb{P}(y_i^{(q)} \mid \theta_i'))] + \underbrace{\ln(\mathbb{P}(y_i^{(q)} \mid \theta))}_{\text{constant w.r.t. } \lambda_i},
\end{aligned}
$$

where (a) follows from the definition of KL divergence, (b) is because $\mathbb{P}(\theta_i' \mid y_i^{(q)}, \theta) = \mathbb{P}(y_i^{(q)} \mid \theta_i')\mathbb{P}(\theta_i' \mid \theta)/\mathbb{P}(y_i^{(q)} \mid \theta)$, and (c) follows from the definitions of KL divergence and expectation. The last term is constant with respect to λ_i and can be dropped. The remaining terms are defined as follows:

$$
\mathcal{L}_i(\lambda_i, \theta) := \text{KL}\big[q(\theta_i'|\lambda_i) \,\|\, \mathbb{P}(\theta_i' \mid \theta)\big] - \mathbb{E}_{q(\theta_i'|\lambda_i)}[\ln(\mathbb{P}(y_i^{(q)} \mid \theta_i'))]. \tag{22}
$$

l_1 data instances are sampled from the distribution $q(\theta_i'|\lambda_i)$, and Eq. (22) is calculated. The parameter λ_i is updated by backpropagation:

$$
\lambda_i = \lambda_i - \frac{\alpha}{l_1} \nabla_{\lambda_i} \mathcal{L}_i(\lambda_i, \theta), \tag{23}
$$

where α is the learning rate, and λ_i is initialized by θ.

On the other hand, consider the log-likelihood of the query set given the support set:

$$\ln p(y_i^{(q)} \mid y_i^{(s)}, \theta) \overset{(a)}{=} \ln \mathbb{E}_{\mathbb{P}(\theta_i' \mid y_i^{(q)}, \theta)}\big[p(y_i^{(q)} \mid \theta_i')\big]$$
$$\overset{(b)}{\geq} \mathbb{E}_{\mathbb{P}(\theta_i' \mid y_i^{(q)}, \theta)}\big[\ln p(y_i^{(q)} \mid \theta_i')\big],$$

where (a) follows from the law of total expectation, and (b) follows from Jensen's inequality. We define the last term as follows:

$$\mathcal{L}_i(\theta) := \mathbb{E}_{\mathbb{P}(\theta_i' \mid y_i^{(q)}, \theta)}\big[\ln p(y_i^{(q)} \mid \theta_i')\big], \tag{24}$$

which is the lower bound on the log-likelihood. This lower bound can also be approximated using Monte Carlo approximation with l_2 samples drawn from the distribution $q(\theta_i'|\lambda_i)$:

$$\mathcal{L}_i(\theta) := \frac{1}{l_2} \sum_{j=1}^{l_2} \ln p(y_i^{(q)} \mid \theta_{i,j}'), \tag{25}$$

where $\theta_{i,j}'$ denotes the jth sample for θ_i'.

Variational inference maximizes the log-likelihood or, equivalently, minimizes the negative log-likelihood:

$$\theta^* = \arg\min_{\theta} -\frac{1}{t} \sum_{i=1}^{t} \ln p(y_i^{(q)} \mid y_i^{(s)}, \theta), \tag{26}$$

where t denotes the number of tasks drawn in each episode of training. Instead of maximizing the negative likelihood, it is possible to maximize its lower bound or, equivalently, to minimize the negative lower bound:

$$\theta^* = \arg\min_{\theta} -\frac{1}{t} \sum_{i=1}^{t} \mathcal{L}_i(\theta), \tag{27}$$

which is updated by backpropagation as follows:

$$\theta = \theta + \frac{\beta}{t} \sum_{i=1}^{t} \nabla_\theta \mathcal{L}_i(\theta), \tag{28}$$

where β is the meta-learning rate. The training algorithm for variational MAML is shown in Algorithm 3.

6 Meta-Learning for Domain Generalization

Different datasets may have different distributions or domains. There are two broad approaches to handle different domains in the data. One approach is called "domain adaptation" [11], in which all different domains are introduced to the model during the training phase. Therefore, the model can adapt to different domains by being trained on various domains. Another approach is "domain generalization" [14], where the target domain, to which we intend to generalize, has not been observed in the training phase; hence, the embedding space is learned using other source domains. In other words, the model

Initialize the weights θ
while *not converged (episodes)* **do**
  Sample a batch of tasks $\{\mathcal{T}_i\}_{i=1}^{t}$
  for *each task* $\mathcal{T}_i$ **do**
    $\lambda_i = \theta$
    Draw l_1 samples $\theta_i' \sim q(\theta_i' | \lambda_i)$ and calculate $\mathcal{L}_i(\lambda_i, \theta)$
    $\lambda_i = \lambda_i - \frac{\alpha}{l_1} \nabla_{\lambda_i} \mathcal{L}_i(\lambda_i, \theta)$
    Draw l_2 samples $\theta_i' \sim q(\theta_i' | \lambda_i)$ and calculate $\mathcal{L}_i(\theta)$
  $\theta = \theta + \frac{\beta}{t} \sum_{i=1}^{t} \nabla_\theta \mathcal{L}_i(\theta)$

Algorithm 3: Variational MAML training algorithm

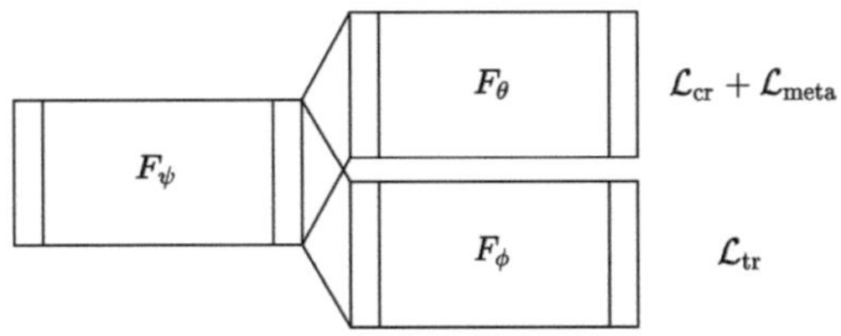

Fig. 2 The sub-networks and loss functions used in MASF

is trained on multiple source domains and is tested on unseen target domain(s). Domain generalization is a more challenging task than domain adaptation because the target domain is not introduced to the model.

As explained in Sect. 1, meta-learning can be used for domain generalization. Model-Agnostic Learning of Semantic Features (MASF) [14] uses meta-learning for domain generalization. It has been applied in biological applications for magnification and hospital generalization of histopathology data [61, 62].

Consider the K training source domains, $\{\mathcal{D}_k\}_{k=1}^{K}$. In every iteration of the MASF method, the source domains are split into meta-train and meta-test domains, denoted by $\mathcal{D}_{\text{tr}}$ and $\mathcal{D}_{\text{te}}$, respectively. MASF uses a combination of three different loss functions to learn the embedding space. These loss functions are detailed in the following. The MASF method contains three sub-networks, depicted in Fig. 2.

6.1 Loss for Supervised Embedding

The first loss is the cross-entropy loss for supervised embedding by classification:

$$\mathcal{L}_{\text{cr}}(\mathcal{D}_{\text{tr}}; \psi, \theta) := \frac{-1}{|\mathcal{D}_{\text{tr}}|} \sum_{\mathcal{D} \in \mathcal{D}_{\text{tr}}} \frac{1}{|\mathcal{D}|} \sum_{(\boldsymbol{x}, y) \in \mathcal{D}} \sum_{c=1}^{C} \mathbb{I}(y = c) \log \mathbb{I}(\widehat{y} = c), \tag{29}$$

where C is the number of classes, y is the ground-truth label for $\boldsymbol{x}$, $\widehat{y}$ is the predicted label for $\boldsymbol{x}$, and $\mathbb{I}(\cdot)$ is the indicator function, which is one when its condition is satisfied and zero otherwise. In practice, Eq. (29) is used in batches. The updated ψ and ϕ through the cross-entropy loss are denoted by ψ' and ϕ', respectively.

6.2 Loss for Domain Generalization

The next loss is the generalization loss, which accounts for domain generalization. Consider the feature extraction embeddings of samples from domain k. If the samples from this domain with class label c are denoted by $\{\boldsymbol{x}_{c,i}^{(k)}\}_{i=1}^{n_c}$ and their feature extraction embeddings are $\{F_{\psi'}(\boldsymbol{x}_{c,i}^{(k)})\}_{i=1}^{n_c}$, the Monte Carlo approximation for the mean embedding of class c in domain k is given as follows:

$$\boldsymbol{z}_c^{(k)} := \frac{1}{n_c} \sum_{i=1}^{n_c} F_{\psi'}(\boldsymbol{x}_{c,i}^{(k)}). \tag{30}$$

This mean is passed through the layers for supervised embedding, and its softmax with temperature $\tau > 1$ is

$$\boldsymbol{s}_c^{(k)} := \text{softmax}(F_{\theta'}(\boldsymbol{z}_c^{(k)})/\tau), \tag{31}$$

which provides a soft confusion matrix.

For two domains $\mathcal{D}_i$ and $\mathcal{D}_j$, the global loss, which is the symmetrized Kullback–Leibler (KL) divergence averaged over all C classes, is calculated as follows:

$$\ell_{\text{gen}}(\mathcal{D}_i, \mathcal{D}_j; \psi', \theta') := \frac{1}{C} \sum_{c=1}^{C} \frac{1}{2}\Big[\text{KL}(\boldsymbol{s}_c^{(i)} \| \boldsymbol{s}_c^{(j)}) + \text{KL}(\boldsymbol{s}_c^{(j)} \| \boldsymbol{s}_c^{(i)})\Big], \tag{32}$$

where KL(.‖.) denotes the KL divergence.

This loss is computed over all the meta-train and meta-test domains as follows:

$$\mathcal{L}_{\text{gen}}(\mathcal{D}_{\text{tr}}, \mathcal{D}_{\text{te}}; \psi', \theta') := \frac{1}{|\mathcal{D}_{\text{tr}}||\mathcal{D}_{\text{te}}|} \sum_{\mathcal{D}_i \in \mathcal{D}_{\text{tr}}} \sum_{\mathcal{D}_j \in \mathcal{D}_{\text{te}}} \ell_{\text{gen}}(\mathcal{D}_i, \mathcal{D}_j; \psi', \theta'), \tag{33}$$

where $|\cdot|$ denotes the cardinality of a set. In practice, Eqs. (30), (32), and (33) are applied on batches.

6.3 Loss for Metric Learning

Finally, a triplet loss [63] is used for metric learning. Consider a triplet consisting of anchor, positive, and negative instances denoted by $\boldsymbol{x}_a$, $\boldsymbol{x}_p$, and $\boldsymbol{x}_n$, respectively. The triplet loss aims to increase the inter-class variance while decreasing the intra-class variance of the data.

If R triplets, $\mathcal{T} := \{\boldsymbol{x}_a^r, \boldsymbol{x}_p^r, \boldsymbol{x}_n^r\}_{r=1}^{R}$, are randomly sampled from all the source domains $\{\mathcal{D}_k\}_{k=1}^{K}$, the average triplet loss is

$$\begin{aligned}\mathcal{L}_{\text{tri}}(\mathcal{T}; \psi', \phi') := \frac{1}{R} \sum_{r=1}^{R} \Big[&\| F_{\phi'}(F_{\psi'}(\boldsymbol{x}_a^r)) - F_{\phi'}(F_{\psi'}(\boldsymbol{x}_p^r))\|_2^2 \\ &+ \| F_{\phi'}(F_{\psi'}(\boldsymbol{x}_a^r)) - F_{\phi'}(F_{\psi'}(\boldsymbol{x}_n^r))\|_2^2 + \zeta\Big]_+,\end{aligned} \tag{34}$$

where $\|\cdot\|_2$ is the ℓ_2 norm, ζ is a margin, and $[\cdot]_+ := \max(\cdot, 0)$ denotes the standard hinge loss.

6.4 Updating Parameters

This subsection explains how the weights of the sub-networks shown in Fig. 2 are updated iteratively. A step of gradient descent is used to update the weights ψ and θ, given by

$$(\psi', \theta') := (\psi, \theta) - \alpha \nabla_{\psi,\theta} \mathcal{L}_{\text{cr}}(\mathcal{M}_{\text{tr}}; \psi, \theta), \tag{35}$$

where α is the learning rate and $\nabla_{\psi,\theta}$ denotes the gradient with respect to ψ and θ.

A linear combination of the generalization and triplet losses is used for the meta-loss:

$$\mathcal{L}_{\text{meta}}(\mathcal{M}_{\text{tr}}, \mathcal{M}_{\text{te}}, \mathcal{T}; \psi', \theta', \phi') := \beta_1 \mathcal{L}_{\text{gen}}(\mathcal{M}_{\text{tr}}, \mathcal{M}_{\text{te}}; \psi', \theta') + \beta_2 \mathcal{L}_{\text{tri}}(\mathcal{T}; \psi', \phi'), \tag{36}$$

where $\beta_1, \beta_2 > 0$.

After Eq. (35), the weights are updated using two additional gradient descent steps:

$$(\psi, \theta) := (\psi, \theta) - \eta \nabla_{\psi,\theta}(\mathcal{L}_{\text{cr}} + \mathcal{L}_{\text{meta}}), \tag{37}$$

$$\phi := \phi - \gamma \nabla_{\phi} \mathcal{L}_{\text{tri}}(\mathcal{T}; \psi', \phi'), \tag{38}$$

where η and γ are the learning rates.

Equations (35), (37), and (38) are repeated iteratively until convergence. These updates are performed using the backpropagation algorithm. Figure 2 illustrates these loss functions.

7 Conclusion

Few-shot learning and meta-learning have become crucial tools for enabling AI systems to generalize from limited data. From early metric-based methods to gradient-based frameworks like MAML and beyond, the field has matured with applications ranging from computer vision to reinforcement learning. However, challenges remain in handling domain shifts, scaling to real-world data imbalance, and bridging the gap between synthetic benchmarks and practical applications. The integration of meta-learning with modern representation learning (e.g., transformers and self-supervised pretraining) is a promising direction for the future.

8 Exercises

1. (Level: Easy) A few-shot learning model is trained using episodes. Each episode corresponds to a classification task. Suppose you are working with five-way one-shot learning. This means each task involves five classes and one support example per class. In an episode, you are given:

   - A support set $S = \{(\boldsymbol{x}_i, y_i)\}_{i=1}^{5}$, where each $y_i \in \{1, 2, 3, 4, 5\}$.
   - A query set $Q = \{(\boldsymbol{x}'_j, y'_j)\}_{j=1}^{15}$, with three query samples per class.

   Suppose the model assigns probabilities $p(y \mid \boldsymbol{x}')$ to each class $y \in \{1, 2, 3, 4, 5\}$ for every query point $\boldsymbol{x}'$. You compute the accuracy for this episode as the fraction of correct predictions on the query set. Given the model's predictions for the query set yield the following correct predictions per class:

- Class 1: 2 correct predictions out of 3
- Class 2: 3 correct predictions out of 3
- Class 3: 1 correct prediction out of 3
- Class 4: 2 correct predictions out of 3
- Class 5: 2 correct predictions out of 3

What is the accuracy for this episode?

2. (Level: Moderate) In Model-Agnostic Meta-Learning (MAML), we aim to find a model parameter initialization θ such that it can quickly adapt to a new task with a few gradient steps. Consider a task $\mathcal{T}_i$ with loss function $\mathcal{L}_{\mathcal{T}_i}(\theta)$. The adaptation step updates the parameters using gradient descent:

$$\theta_i' = \theta - \alpha \nabla_\theta \mathcal{L}_{\mathcal{T}_i}^{\text{support}}(\theta).$$

Then the meta-objective is to minimize the loss over the adapted parameters:

$$\min_\theta \sum_i \mathcal{L}_{\mathcal{T}_i}^{\text{query}}(\theta_i').$$

Suppose you are given a single task $\mathcal{T}_1$, with:

$$\mathcal{L}_{\mathcal{T}_1}^{\text{support}}(\theta) = (\theta - 2)^2,$$
$$\mathcal{L}_{\mathcal{T}_1}^{\text{query}}(\theta) = (\theta - 5)^2,$$

and learning rate $\alpha = 0.5$. Compute the value of the meta-loss $\mathcal{L}_{\mathcal{T}_1}^{\text{query}}(\theta_1')$.

3. (Level: Difficult) Consider a gradient-based meta-learning setup where for each task $\mathcal{T}_i$, you are given:

- A support set $\mathcal{D}_i^{\text{support}} = \{(x_j, y_j)\}_{j=1}^N$.
- A query set $\mathcal{D}_i^{\text{query}} = \{(x_k', y_k')\}_{k=1}^M$.

Let the model be a scalar function $f_\theta(x) = \theta x$, and the loss function be the squared loss:

$$\mathcal{L}(\theta; x, y) = (f_\theta(x) - y)^2 = (\theta x - y)^2.$$

Suppose:

- $\mathcal{D}_i^{\text{support}} = \{(x_1, y_1)\} = \{(2, 4)\}$.
- $\mathcal{D}_i^{\text{query}} = \{(x_1', y_1')\} = \{(1, 1)\}$.
- Adaptation learning rate $\alpha = 1$.

The adaptation step performs one gradient update:

$$\theta_i' = \theta - \alpha \nabla_\theta \mathcal{L}(\theta; x_1, y_1).$$

Compute:

(a) The meta-loss: $\mathcal{L}(\theta_i'; x_1', y_1')$.

(b) The meta-gradient: $\frac{d}{d\theta}\mathcal{L}(\theta_i'; x_1', y_1')$.

4. (Level: Difficult) In a domain generalization meta-learning setting, domain shifts are simulated by splitting data from a task (domain) $\mathcal{T}$ into:

- A support set $\mathcal{D}^{\text{support}} = \{(x_s, y_s)\} = \{(2, 4)\}$.
- A query set $\mathcal{D}^{\text{query}} = \{(x_q, y_q)\} = \{(1, -1)\}$.

Suppose the followings are defined:

$$f_\theta(x) = \theta x, \quad \mathcal{L}(\theta; x, y) = (\theta x - y)^2, \quad \alpha = 1, \quad \beta = 1.$$

The meta-objective that encourages generalization across domains is

$$\mathcal{L}_{\text{meta}}(\theta) = \mathcal{L}(\theta; x_s, y_s) + \beta \nabla_\theta \mathcal{L}(\theta; x_s, y_s)^\top \nabla_\theta \mathcal{L}(\theta'; x_q, y_q),$$

where

$$\theta' = \theta - \alpha \nabla_\theta \mathcal{L}(\theta; x_s, y_s).$$

Compute the meta-objective $\mathcal{L}_{\text{meta}}(\theta)$ explicitly as a function of θ. Explain how this encourages the model parameters to generalize well to the query set.

References

1. Archit Parnami and Minwoo Lee. “Learning from few examples: A summary of approaches to few-shot learning”. In: *arXiv preprint* arXiv:2203.04291 (2022).
2. Yisheng Song, Ting Wang, Puyu Cai, Subrota K Mondal, and Jyoti Prakash Sahoo. “A comprehensive survey of few-shot learning: Evolution, applications, challenges, and opportunities”. In: *ACM Computing Surveys* 55.13s (2023), pp. 1–40.
3. Yaqing Wang, Quanming Yao, James T Kwok, and Lionel M Ni. “Generalizing from a few examples: A survey on few-shot learning”. In: *ACM computing surveys (csur)* 53.3 (2020), pp. 1–34.
4. Gregory Koch, Richard Zemel, Ruslan Salakhutdinov, et al. “Siamese neural networks for one-shot image recognition”. In: *ICML deep learning workshop*. Vol. 2. 1. Lille. 2015, pp. 1–30.
5. Timothy Hospedales, Antreas Antoniou, Paul Micaelli, and Amos Storkey. “Meta-learning in neural networks: A survey”. In: *IEEE transactions on pattern analysis and machine intelligence* 44.9 (2021), pp. 5149–5169.
6. Mike Huisman, Jan N Van Rijn, and Aske Plaat. “A survey of deep meta-learning”. In: *Artificial Intelligence Review* 54.6 (2021), pp. 4483–4541.
7. Huimin Peng. “A comprehensive overview and survey of recent advances in meta-learning”. In: *arXiv preprint* arXiv:2004.11149 (2020).
8. Yingjie Tian, Xiaoxi Zhao, and Wei Huang. “Meta-learning approaches for learning-to-learn in deep learning: A survey”. In: *Neurocomputing* 494 (2022), pp. 203–223.
9. Ricardo Vilalta and Youssef Drissi. “A perspective view and survey of meta-learning”. In: *Artificial intelligence review* 18 (2002), pp. 77–95.
10. Karl Weiss, Taghi M Khoshgoftaar, and DingDing Wang. “A survey of transfer learning”. In: *Journal of Big data* 3 (2016), pp. 1–40.
11. Shai Ben-David et al. “A theory of learning from different domains”. In: *Machine Learning* 79.1-2 (2010), pp. 151–175.
12. Jindong Wang et al. “Generalizing to unseen domains: A survey on domain generalization”. In: *IEEE transactions on knowledge and data engineering* 35.8 (2022), pp. 8052–8072.
13. Wei Wang, Vincent W Zheng, Han Yu, and Chunyan Miao. “A survey of zero-shot learning: Settings, methods, and applications”. In: *ACM Transactions on Intelligent Systems and Technology (TIST)* 10.2 (2019), pp. 1–37.

14. Qi Dou, Daniel Coelho de Castro, Konstantinos Kamnitsas, and Ben Glocker. "Domain generalization via model-agnostic learning of semantic features". In: *Advances in neural information processing systems* 32 (2019).
15. Da Li, Yongxin Yang, Yi-Zhe Song, and Timothy Hospedales. "Learning to generalize: Meta-learning for domain generalization". In: *Proceedings of the AAAI conference on artificial intelligence*. Vol. 32. 1. 2018.
16. Linda B Smith and Lauren K Slone. "A developmental approach to machine learning?" In: *Frontiers in psychology* 8 (2017), p. 296143.
17. Brenden M Lake, Ruslan Salakhutdinov, and Joshua B Tenenbaum. "Human-level concept learning through probabilistic program induction". In: *Science* 350.6266 (2015), pp. 1332–1338.
18. Jonas Jongejan, Henry Rowley, Takashi Kawashima, Jongmin Kim, and Nick Fox-Gieg. *The Quick, Draw!—AI experiment*. quickdraw.withgoogle.com. 2016.
19. Olga Russakovsky et al. "Imagenet large scale visual recognition challenge". In: *International journal of computer vision* 115 (2015), pp. 211–252.
20. Oriol Vinyals, Charles Blundell, Timothy Lillicrap, Daan Wierstra, et al. "Matching networks for one shot learning". In: *Advances in neural information processing systems* 29 (2016).
21. Eleni Triantafillou et al. "Meta-Dataset: A Dataset of Datasets for Learning to Learn from Few Examples". In: *International Conference on Learning Representations (ICLR)*. 2020.
22. Ihsan Ullah et al. "Meta-album: Multi-domain meta-dataset for few-shot image classification". In: *Advances in Neural Information Processing Systems* 35 (2022), pp. 3232–3247.
23. Yonghong Peng, Peter A Flach, Carlos Soares, and Pavel Brazdil. "Improved dataset characterisation for meta-learning". In: *Discovery Science: 5th International Conference, DS 2002 L ü beck, Germany, November 24–26, 2002 Proceedings 5*. Springer. 2002, pp. 141–152.
24. Wei-Yu Chen, Yen-Cheng Liu, Zsolt Kira, Yu-Chiang Frank Wang, and Jia-Bin Huang. "A Closer Look at Few-shot Classification". In: *International Conference on Learning Representations (ICLR)*. 2019.
25. Spyros Gidaris and Nikos Komodakis. "Dynamic few-shot visual learning without forgetting". In: *Proceedings of the IEEE conference on computer vision and pattern recognition*. 2018, pp. 4367–4375.
26. Hang Qi, Matthew Brown, and David G Lowe. "Low-shot learning with imprinted weights". In: *Proceedings of the IEEE conference on computer vision and pattern recognition*. 2018, pp. 5822–5830.
27. Devang K Naik and Richard J Mammone. "Meta-neural networks that learn by learning". In: *International Joint Conference on Neural Networks (IJCNN)*. Vol. 1. IEEE. 1992, pp. 437–442.
28. Jürgen Schmidhuber. "Evolutionary principles in self-referential learning, or on learning how to learn: the meta-meta-... hook". PhD thesis. Technische Universität München, 1987.
29. Sebastian Thrun and Lorien Pratt. "Learning to learn: Introduction and overview". In: *Learning to learn*. Springer, 1998, pp. 3–17.
30. Chelsea Finn, Pieter Abbeel, and Sergey Levine. "Model-agnostic meta-learning for fast adaptation of deep networks". In: *International conference on machine learning*. PMLR. 2017, pp. 1126–1135.
31. Samy Bengio, Yoshua Bengio, Jocelyn Cloutier, and Jan Gecsei. "On the optimization of a synaptic learning rule". In: *Optimality in Artificial and Biological Neural Networks* (1992).
32. Yoshua Bengio, Samy Bengio, and Jocelyn Cloutier. *Learning a synaptic learning rule*. Citeseer, 1990.
33. Jürgen Schmidhuber. "Learning to control fast-weight memories: An alternative to dynamic recurrent networks". In: *Neural Computation* 4.1 (1992), pp. 131–139.
34. Jake Snell, Kevin Swersky, and Richard Zemel. "Prototypical networks for few-shot learning". In: *Advances in neural information processing systems* 30 (2017).
35. Mengye Ren et al. "Meta-Learning for Semi-Supervised Few-Shot Classification". In: *International Conference on Learning Representations (ICLR)*. 2018.
36. Flood Sung et al. "Learning to compare: Relation network for few-shot learning". In: *Proceedings of the IEEE conference on computer vision and pattern recognition*. 2018, pp. 1199–1208.
37. Maria-Florina Balcan, Mikhail Khodak, and Ameet Talwalkar. "Provable guarantees for gradient-based meta-learning". In: *International Conference on Machine Learning*. PMLR. 2019, pp. 424–433.
38. Lisha Chen, Songtao Lu, and Tianyi Chen. "Understanding benign overfitting in gradient-based meta learning". In: *Advances in neural information processing systems* 35 (2022), pp. 19887–19899.
39. Alireza Fallah, Aryan Mokhtari, and Asuman Ozdaglar. "On the convergence theory of gradient-based model-agnostic meta-learning algorithms". In: *International Conference on Artificial Intelligence and Statistics*. PMLR. 2020, pp. 1082–1092.
40. Simon Guiroy, Vikas Verma, and Christopher Pal. "Towards understanding generalization in gradient based meta learning". In: *arXiv preprint* arXiv:1907.07287 (2019).
41. Haoxiang Wang, Ruoyu Sun, and Bo Li. "Global convergence and generalization bound of gradient-based meta-learning with deep neural nets". In: *arXiv preprint* arXiv:2006.14606 (2020).
42. Alex Nichol and John Schulman. "Reptile: a scalable metalearning algorithm". In: *arXiv preprint* arXiv:1803.02999 2.3 (2018), p. 4.
43. Xingyou Song et al. "ES-MAML: Simple Hessian-free meta learning". In: *arXiv preprint* arXiv:1910.01215 (2019).

44. Siavash Khodadadeh, Ladislau Boloni, and Mubarak Shah. "Unsupervised meta-learning for few-shot image classification". In: *Advances in neural information processing systems* 32 (2019).
45. Aaron van den Oord, Yazhe Li, and Oriol Vinyals. "Representation learning with contrastive predictive coding". In: *arXiv preprint* arXiv:1807.03748 (2018).
46. M. Hadi Sepanj, Benyamin Ghojogh, and Paul Fieguth. "Self-supervised Learning Using Nonlinear Dependence". In: *arXiv preprint* arXiv2501.18875 (2025).
47. Antreas Antoniou, Harrison Edwards, and Amos Storkey. "How to train your MAML". In: *International conference on learning representations (ICLR)*. 2018.
48. Kaiyi Ji, Junjie Yang, and Yingbin Liang. "Theoretical convergence of multi-step model-agnostic meta-learning". In: *Journal of machine learning research* 23.29 (2022), pp. 1–41.
49. Jin-Hwa Kim, Junyoung Park, and Yongseok Choi. "Multi-step estimation for gradient-based meta-learning". In: *arXiv preprint* arXiv:2006.04298 (2020).
50. Sergey Ioffe. "Batch normalization: Accelerating deep network training by reducing internal covariate shift". In: *arXiv preprint* arXiv:1502.03167 (2015).
51. Ilya Loshchilov and Frank Hutter. "SGDR: Stochastic gradient descent with warm restarts". In: *International Conference on Learning Representations (ICLR)*. 2017.
52. H Behl, A Baydin, and P Torr. "Alpha MAML: Adaptive model-agnostic meta-learning". In: *6th ICML Workshop on Automated Machine Learning, Thirty-Sixth International Conference on Machine Learning (ICML)*. 2019.
53. Risto Vuorio, Shao-Hua Sun, Hexiang Hu, and Joseph J Lim. "Toward multimodal model-agnostic meta-learning". In: *Workshop on Meta-Learning at NeurIPS 2018*. 2018.
54. Tsendsuren Munkhdalai and Hong Yu. "Meta networks". In: *International conference on machine learning*. PMLR. 2017, pp. 2554–2563.
55. Sachin Ravi and Hugo Larochelle. "Optimization as a model for few-shot learning". In: *International conference on learning representations*. 2017.
56. Jane X Wang et al. "Learning to reinforcement learn". In: *arXiv preprint* arXiv:1611.05763 (2016).
57. Erin Grant, Chelsea Finn, Sergey Levine, Trevor Darrell, and Thomas Griffiths. "Recasting gradient-based meta-learning as hierarchical Bayes". In: *International Conference on Learning Representations (ICLR)*. 2018.
58. Jaesik Yoon et al. "Bayesian model-agnostic meta-learning". In: *Advances in neural information processing systems* 31 (2018).
59. Cuong Nguyen, Thanh-Toan Do, and Gustavo Carneiro. "Uncertainty in model-agnostic meta-learning using variational inference". In: *Proceedings of the IEEE/CVF Winter Conference on Applications of Computer Vision*. 2020, pp. 3090–3100.
60. Chelsea Finn, Kelvin Xu, and Sergey Levine. "Probabilistic model-agnostic meta-learning". In: *Advances in neural information processing systems* 31 (2018).
61. Milad Sikaroudi, Benyamin Ghojogh, Fakhri Karray, Mark Crowley, and Hamid R Tizhoosh. "Magnification generalization for histopathology image embedding". In: *2021 IEEE 18th International Symposium on Biomedical Imaging (ISBI)*. IEEE. 2021, pp. 1864–1868.
62. Milad Sikaroudi, Shahryar Rahnamayan, and Hamid R Tizhoosh. "Hospital-agnostic image representation learning in digital pathology". In: *2022 44th Annual International Conference of the IEEE Engineering in Medicine & Biology Society (EMBC)*. IEEE. 2022, pp. 3055–3058.
63. Florian Schroff, Dmitry Kalenichenko, and James Philbin. "Facenet: A unified embedding for face recognition and clustering". In: *Proceedings of the IEEE Conference on Computer Vision and Pattern Recognition*. 2015, pp. 815–823.

Neural Network Compression and Knowledge Distillation

20

Abstract

Running large neural networks on phones, IoT devices, and embedded systems is difficult because of limited memory, storage, compute, and battery. Neural network compression methods aim to shrink models while keeping accuracy acceptable. This chapter surveys key methods for compressing neural networks, including pruning, weight sharing, matrix decomposition, and quantization, with a particular focus on knowledge distillation as a powerful approach for achieving efficient model compression. By presenting foundational concepts and recent advances, this chapter aims to provide a comprehensive understanding of how neural networks can be effectively compressed for deployment in resource-limited environments.

1 Introduction

Many large neural networks cannot be easily implemented on edge devices and embedded systems due to resource limitations such as restricted storage memory and battery capacity [80]. An example is the use of large language models in cell phones to enable user-friendly interactions. Therefore, a large neural network needs to be compressed to become smaller while maintaining its performance as much as possible.

Neural network compression methods can be categorized into multiple algorithmic approaches. These categories include neural network pruning [11, 53, 78], weight sharing, matrix decomposition, quantization [30, 53, 99], and knowledge distillation [28]. This chapter introduces these methods for neural network compression with an emphasis on knowledge distillation.

This chapter introduces neural network compression and knowledge distillation. Section 2 presents various methods for neural network compression, including neural network pruning, weight sharing, matrix decomposition, and quantization. Knowledge distillation, which can also be used for neural network compression, is explained in Sect. 3. There are several good surveys on neural network compression, some of which were mentioned in the above references. Additional literature reviews on neural network compression include [1, 12, 13, 61, 68].

B. Ghojogh and A. Ghodsi, *Elements of Deep Learning*,
https://doi.org/10.1007/978-3-032-10738-1_20

2 Neural Network Compression

This section introduces neural network pruning, weight sharing, matrix decomposition, and quantization techniques for neural network compression. Knowledge distillation for network compression will be explained in Sect. 3.

2.1 Neural Network Pruning

Neural network pruning reduces model size by removing weights, neurons, or even entire channels that contribute little to predictions [11, 53, 78]. The goal is to simplify the model structure while retaining accuracy.

This pruning is not random; rather, it removes the least important weights for the classification or regression task [78]. Note that random pruning differs from dropout: pruning is a permanent removal of connections for compression, while dropout [85] is a temporary training-time regularizer (see Chap. 5). It is also noteworthy that weight pruning has a biological analogy, as the human brain also gradually removes unimportant information from past experiences [94].

2.1.1 Weight Pruning by Thresholding

It is possible to use a threshold per layer or for the entire neural network and remove the weights with magnitudes less than this threshold [31].

2.1.2 Weight Pruning by Regularization

Some methods, such as [97], regularize the loss function of the neural network with ℓ_1 or ℓ_2 norms to encourage the weights to become smaller in magnitude. This approach is referred to as weight decay [14, 48] (see Chap. 5). The weights with small magnitudes can then be removed from the network.

Some algorithms, such as [63, 98], use ℓ_1-norm (or LASSO) regularization [90] to impose sparsity on the neural network. This leads to network compression because many of the weights become zero when sparsity is enforced.

2.1.3 Weight Pruning by Sensitivity of Loss to Weights

The sensitivity of the loss function to the weights can be computed to determine how important each weight is and how much it contributes to achieving a small loss value [78]. In this approach, only the important weights that form the structure of the neural network are retained, while the rest are removed. This method is referred to as skeletonization [65] because it preserves only the skeleton of the neural network.

One can calculate the sensitivity of the loss to the weights using the derivative of the loss with respect to the weights and then drop the weights with low sensitivity [46]. Another approach is to remove weights whose variance in sensitivity is low, meaning their presence does not make a significant difference [23]. While some methods use first-order derivatives to calculate sensitivity [63], it is also possible to use second-order derivatives, i.e., the Hessian, for this calculation. Optimal brain damage [50] and optimal brain surgeon [33] are examples that use second-order derivatives for sensitivity analysis.

2.1.4 Search-Based Weight Pruning

Important weights to be retained can also be identified using intelligent search techniques such as metaheuristic optimization [89]. For example, genetic algorithms [36] have been used to prune weights in fully connected [100] and Convolutional Neural Networks (CNNs) [38]. Other intelligent search algorithms, such as simulated annealing [71], particle swarm optimization [92], particle filters [2], and reinforcement learning-based search [34], have also been employed for neural network weight pruning.

2.1.5 Pruning for Sub-network Architecture

Some pruning methods focus on pruning the neural network to find the optimal sub-network architecture(s). These methods are based on the lottery ticket hypothesis [25], which states that in every network, there exist sparse sub-networks that can achieve the same performance as the full network when trained from scratch with the same weight initialization. Some of these pruning methods include weight-agnostic neural networks [26], deep rewiring [5], sparse evolutionary training [62], dynamic sparse reparameterization [64], and edge-popup [75].

2.2 Neural Network Compression by Weight Sharing

Weight sharing is a neural network compression algorithm that shares weights between layers of a neural network. By doing so, the shared weights are treated as identical, which reduces the size of the network. For example, in Convolutional Neural Networks (CNNs), the kernel filters can be shared between layers. One can fit a Gaussian mixture model [27] to the weights of a neural network to cluster them into multiple Gaussian distributions [70]. Another approach is to group the weight connections into hash buckets using hash functions [10].

Another method of weight sharing is to reuse the weight layers recursively so that a smaller number of layers can function as if there were many more layers. In this approach, the output of a layer is also fed back to it as its input, creating recursion. This technique has been used to compress Convolutional Neural Networks [22].

2.3 Neural Network Compression by Matrix Decomposition

One approach to compress neural networks is to decompose or factorize the weight matrices or tensors of neural networks into the multiplication of multiple lower-rank matrices. This process removes redundant parameters by replacing the original weight matrices with matrices of lower rank, thereby reducing the storage capacity required for the weight matrices. Various matrix decomposition algorithms can be used for this purpose. For example, singular value decomposition [69, 81, 103, 104], Kronecker decomposition [88], block-term tensor decomposition [17, 106], and low-rank decomposition [40, 107, 108] have been applied to neural network compression using matrix factorization.

2.4 Neural Network Compression by Quantization

A neural network can be compressed by quantizing its weight values to occupy a reduced number of bits in storage [30, 53, 99]. For example, if the weights are converted from 64-bit float values to 8-bit integer values, storing the weights in memory requires significantly less storage. Although this compression may result in some distortions [4], some studies have shown that quantization not only compresses the neural network but may even improve its performance in certain cases [54].

2.4.1 Quantization Ranges

Quantization ranges do not necessarily need to be fixed; they should be determined based on the data and the neural network. This is because the weight and activation values are not uniformly distributed [72], so the denser regions of values should be quantized with finer ranges. For example, quantization ranges can be obtained using residual quantization [8], least squares quantization [55], and product quantization [42].

It has been empirically observed that the first and last convolutional layers in Convolutional Neural Networks (CNNs) are more sensitive to quantization [16, 110], so it may be preferable not to quantize them. However, Jungwook Cho et al. [16] has shown that these layers can be quantized if the quantization ranges are not too coarse.

2.4.2 Loss-Aware Quantization

Quantization can also take the loss function of the neural network into account. For example, Explicit Loss-Aware Quantization (ELQ) [109] considers the loss function during quantization to minimize perturbations in the loss value caused by quantization. Some works, such as [21, 84], have also used second-order derivatives, i.e., the Hessian, for quantization.

2.4.3 Backpropagation with Quantization

It should be noted that when the weights of a neural network are quantized, backpropagation (see Chap. 3) encounters difficulties because the values become discrete at the quantization level. Several methods have been developed to enable backpropagation through quantized weights in quantized neural networks [39, 110].

2.4.4 Quantization Used With Other Compression Methods

Some studies have combined quantization with other network compression methods such as weight pruning [32] and knowledge distillation [60].

3 Knowledge Distillation

3.1 Vanilla Knowledge Distillation

Knowledge Distillation (KD) is one of the methods used for neural network compression [28]. It was first proposed by Buciluă [7] and later improved by Hinton [35]. In knowledge distillation (KD), a large pretrained model (called teacher) guides the training of a smaller model (called student). The student learns both from ground-truth labels and from the teacher's softer probability outputs. Knowledge is distilled from the teacher network to the student network, resulting in the student network becoming a compressed version of the teacher network.

Assume the task is the classification of data into c classes. The teacher and student neural networks have c neurons and use the softmax activation function in their last layers for this classification. Let $f_t(\boldsymbol{x}) \in \mathbb{R}^c$ and $f_s(\boldsymbol{x}) \in \mathbb{R}^c$ denote the logits (outputs) of the teacher and student networks before applying the softmax activation.

The loss function for training the student network in KD is composed of two parts, i.e., Cross-Entropy (CE) and Knowledge Distillation (KD):

$$\mathcal{L} := \frac{1}{b} \sum_{i=1}^{b} \big((1-\lambda)\mathcal{L}_{\text{CE}}(\boldsymbol{x}_i) + \lambda \mathcal{L}_{\text{KD}}(\boldsymbol{x}_i)\big), \tag{1}$$

where b is the batch size, $\boldsymbol{x}_i$ is the ith data instance in the batch, and $\lambda \in (0, 1)$ is the hyperparameter weighting the KD part relative to the CE part in the loss value. Let $\boldsymbol{y}_i \in \{0, 1\}^c$ denote the one-hot encoding of the class label of data instance $\boldsymbol{x}_i$. The CE and KD parts of the loss are

$$\mathcal{L}_{\text{CE}}(\boldsymbol{x}_i) := -\sum_{l=1}^{c} (\boldsymbol{y}_i)_l \log\big(f_s(\boldsymbol{x}_i)_l\big), \tag{2}$$

$$\mathcal{L}_{\text{KD}}(\boldsymbol{x}_i) := \tau^2 \text{KL}\Big(\sigma\big(\frac{f_t(\boldsymbol{x}_i)}{\tau}\big) \Big\| \sigma\big(\frac{f_s(\boldsymbol{x}_i)}{\tau}\big)\Big), \tag{3}$$

where $(\boldsymbol{y}_i)_l$ and $f_s(\boldsymbol{x}_i)_l$ are the lth elements of $\boldsymbol{y}_i$ and $f_s(\boldsymbol{x}_i)$, respectively, $\sigma(.)$ denotes the softmax activation function, $\text{KL}(.\|.)$ denotes the Kullback-Leibler (KL) divergence [49], and $\tau > 0$ is the temperature hyperparameter controlling the softness of the output probability distributions. The larger the τ, the softer the output probabilities are.

In regular KD, the teacher network is assumed to be already trained. The student network is trained by minimizing the loss function (1). The training dataset can be the same as the one used to train the teacher network, or it can be another dataset from the same domain (without domain shift). The CE part of the loss, i.e., Eq. (2), uses the ground-truth labels in the training dataset, which are called the hard labels. The KD part of the loss, i.e., Eq. (3), uses the output predictions of the teacher network, which are called the soft labels.

Minimizing the CE part of the loss, i.e., Eq. (2), trains the student network to classify the training dataset according to its ground-truth labels. Minimizing the KD part of the loss, i.e., Eq. (3), trains the student network to mimic the teacher network by minimizing the difference in output distributions between the student and teacher networks for the same input data. Minimizing both the CE and KD parts in the loss function (1) enables the student network to classify the training dataset correctly while mimicking the teacher network. Therefore, the student network becomes a compressed version of the teacher network.

3.2 The Capacity Gap Problem

3.2.1 PAC Learning in Knowledge Distillation

KD has been analyzed within the framework of PAC learning [93] (see Chap. 6) by the work in [57]. According to [57], KD is effective if the following inequality holds:

$$\mathcal{O}\Big(\frac{|\mathcal{F}_s|_c + |\mathcal{F}_t|_c}{n^\alpha}\Big) + \epsilon_t + \epsilon_l \leq \mathcal{O}\Big(\frac{|\mathcal{F}_s|_c}{\sqrt{n}}\Big) + \epsilon_s, \tag{4}$$

where $\mathcal{F}_s$ and $\mathcal{F}_t$ are the function classes corresponding to the student and teacher networks, respectively, $|.|_c$ denotes the function class capacity, $\mathcal{O}(.)$ represents the estimation error, ϵ_s is the approximation error of the best student estimator function in $\mathcal{F}_s$ with respect to the underlying function, ϵ_t is the approximation error for the teacher function in $\mathcal{F}_t$ with respect to the underlying function, ϵ_l is the approximation error of the best student estimator function in $\mathcal{F}_s$ with respect to the teacher function, and n is the number of training data instances. The parameter $\alpha \in [0.5, 1]$ is related to the difficulty of the problem; smaller values of α correspond to harder and slower training of the student, while larger values indicate easier and faster training. Equation (4) has useful interpretations, one of which is discussed in the following section.

3.2.2 Statement of the Capacity Gap Problem

Consider a scenario where the capacity of the student network, i.e., $|\mathcal{F}_s|_c$, is much less than the capacity of the teacher network, i.e., $|\mathcal{F}_t|_c$:

$$|\mathcal{F}_s|_c \ll |\mathcal{F}_t|_c. \tag{5}$$

In such a scenario, Eq. (4) does not hold because the left-hand side of the inequality becomes larger than the right-hand side. Therefore, if the capacity of the student network is significantly less than that of the teacher network, KD becomes ineffective. In other words, as the gap between the student and teacher networks increases, training the student network using KD becomes more difficult. This phenomenon is known as the capacity gap problem in KD [57, 59].

If the student network is too small compared to the teacher network, it cannot capture the transferred knowledge well. This is called the capacity gap problem, and it motivates solutions like teacher assistants or staged distillation. The following sections discuss some of these techniques to address the capacity gap problem.

3.2.3 Knowledge Distillation with Teacher Assistants

When the size difference between the teacher and student networks is significant, i.e., when the capacity gap increases, it becomes difficult to distill knowledge directly from the large teacher network to the small student network. To address this, a middle-sized network called a teacher assistant can be introduced, whose size lies between those of the teacher and student networks [59]. As shown in Fig. 1b, knowledge is first distilled by KD from the teacher network to the teacher assistant network. Then, knowledge is distilled from the teacher assistant network to the student network. This staged compression resolves the capacity gap problem by gradually compressing the neural network.

If the capacity gap between the student and teacher networks is extremely large, it is possible to introduce a hierarchy of teacher assistant networks of different sizes, arranged like a pyramid [59] (see Fig. 1c). Each teacher assistant network is smaller than the teacher network but larger than the student network. Knowledge is distilled sequentially from the teacher to the largest teacher assistant, then to smaller teacher assistants, and finally from the smallest teacher assistant to the student.

3.2.4 Annealing Knowledge Distillation

Another way to resolve the capacity gap problem in KD is annealing [41], similar to temperature annealing in the simulated annealing algorithm [47]. The paper [41] performs KD training in two stages. In the first stage, it trains the student network to gradually mimic the teacher network using annealing, and in the second stage, it fine-tunes the student network by minimizing the cross-entropy loss function. The loss function to minimize is

$$\mathcal{L} := \begin{cases} \mathcal{L}_{\text{KD}}^{\text{annealing}}(\boldsymbol{x}_i; j) & \text{Stage 1: } 1 \leq \tau_j \leq \tau_{\max} \\ \mathcal{L}_{\text{CE}}(\boldsymbol{x}_i) & \text{Stage 2,} \end{cases} \tag{6}$$

where j denotes the epoch index in backpropagation, τ_j is the temperature parameter at the jth epoch, $\mathcal{L}_{\text{KD}}^{\text{annealing}}$ is the KD loss with annealing, and $\mathcal{L}_{\text{CE}}$ is the cross-entropy loss defined in Eq. (2).

The KD loss with annealing is defined as the mean squared error between the student's output and the softened version of the teacher's output [41]:

$$\mathcal{L}_{\text{KD}}^{\text{annealing}}(\boldsymbol{x}_i; j) := \|f_s(\boldsymbol{x}_i) - \phi(\tau_j) f_t(\boldsymbol{x}_i)\|_2^2, \tag{7}$$

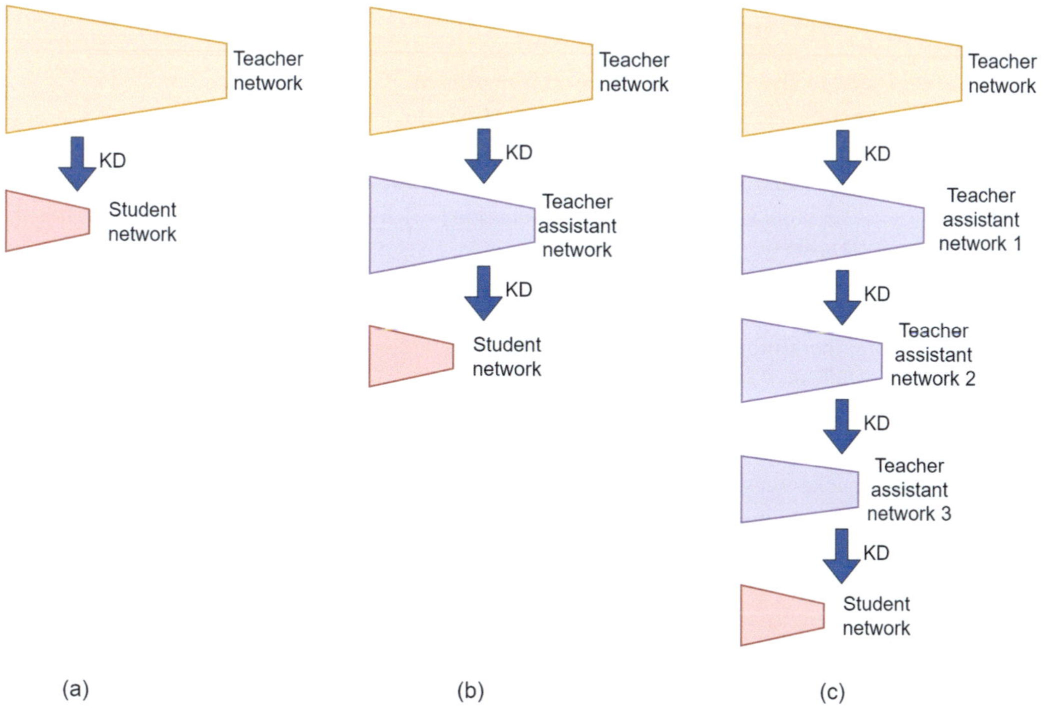

Fig. 1 **a** Distilling knowledge from a teacher network to a student network, **b** distilling knowledge from a teacher network to a teacher assistant network and then to a student network, and **c** knowledge distillation from a teacher network to a student network through a hierarchy of teacher assistant networks with different sizes

where $\phi(\tau_j)$ determines the degree to which the teacher's output is softened and is defined as:

$$\phi(\tau_j) := 1 - \frac{\tau_j - 1}{\tau_{\max}}, \quad \tau_j \in \{1, 2, \ldots, \tau_{\max}\}, \tag{8}$$

$$\tau_{j+1} := \tau_j - 1, \quad \tau_1 = \tau_{\max}. \tag{9}$$

This loss function causes the student's output to gradually mimic the teacher's output. Note that instead of Eq. (7), it is also possible to use the KL divergence between the student's output and the softened teacher's output; however, the paper [41] used mean squared error.

The first stage begins with $\tau_1 = \tau_{\max}$, so $\phi(\tau_1) = \frac{1}{\tau_{\max}}$, and the student's output learns to match the most softened variant of the teacher's output. Gradually, the temperature is annealed and cooled down. The temperature τ_j is reduced from $\tau_{\max}$ toward 1, i.e., $\tau_j \to 1$, so $\phi(\tau_j) \to 1$ (see Eq. (8)) and the student's output learns to match the least softened variant of the teacher's output. In other words, during the first stage, the student initially learns to mimic the teacher with leniency, and this requirement becomes stricter over time. This process smoothly transfers knowledge from the teacher to the student.

The first stage ends after a predetermined number of epochs, after which the second stage begins, where the student network is trained using hard labels. In fact, the second stage fine-tunes the student network using the hard labels in the training dataset.

3.3 The Checkpoint Search Problem

3.3.1 Statement of the Checkpoint Search Problem

When the teacher network is available for the evaluation phase, it is usually used at its best-performing checkpoint, which has the smallest validation error. However, this best-performing checkpoint is not necessarily the optimal choice for compression. In other words, a teacher network that achieves the lowest validation error is not always the best for distillation [15]. Sometimes earlier checkpoints transfer knowledge more effectively.

Therefore, it may be necessary to search for the best checkpoint of the teacher network suitable for KD. This issue is referred to as the checkpoint search problem in KD [79]. Obviously, this search is time-consuming; therefore, other techniques are often employed to address this problem. Some of these methods are introduced in the following.

3.3.2 Early Stopping in Training the Teacher Network

According to the checkpoint search problem, the fully trained teacher is not necessarily the best for KD to the student. One way to resolve this problem is to use early stopping[1] [74, 105] during the training of the teacher network before distilling knowledge to the student network [15]. In fact, some theories suggest that distilling knowledge from a teacher network to a student network can be interpreted as harvesting dark, potentially useless knowledge by applying early stopping during the training of the teacher network [20].

3.3.3 Co-training the Teacher and Student Networks

Another way to tackle the checkpoint search problem is to co-train the teacher and student networks together, in contrast to regular KD where the student network is trained only after the teacher network has been fully trained.

– **Pro-KD**: Several approaches exist for co-training the teacher and student networks. One such approach is Progressive Knowledge Distillation (Pro-KD) [79], which consists of two stages inspired by [41]. The teacher and student are co-trained in the first stage, and the student network is fine-tuned in the second stage.

Since the teacher is usually early-stopped and not trained for a long duration, the student is trained for more epochs than the teacher network. Therefore, for every epoch of teacher network training, the student network is trained for e epochs where $e \geq 1$. In the first stage, at every epoch k of the teacher network training, the teacher network minimizes the cross-entropy loss function:

$$\mathcal{L}_{\text{CE}}(\boldsymbol{x}_i) := -\sum_{l=1}^{c} (\boldsymbol{y}_i)_l \log\big(f_t(\boldsymbol{x}_i)_l\big). \tag{10}$$

During the first stage, at every epoch j of the student network training, the following annealing KD loss is minimized:

$$\mathcal{L}_{\text{KD}}^{\text{annealing}}(\boldsymbol{x}_i) := \left\| f_s(\boldsymbol{x}_i) - \frac{f_t(\boldsymbol{x}_i)}{\tau_k} \right\|_2^2, \tag{11}$$

$$\tau_{k+1} := \tau_k - 1, \quad \tau_1 = \tau_{\max}, \tag{12}$$

[1] For more information on early stopping, refer to Chap. 5.

where $\tau_k \in \{1, 2, \ldots, \tau_{\max}\}$ and k is the epoch index in teacher network training. This annealing KD loss transfers knowledge smoothly and gradually from the teacher to the student.

In the second stage of Pro-KD, training of the teacher network is stopped (early stopping), and the student network is trained by minimizing the cross-entropy loss function defined in Eq. (2). Since the first stage co-trains the teacher and student networks and employs early stopping for the teacher network, it effectively addresses the checkpoint search problem in KD.

– **Co-Distill**: Other methods also co-train the teacher and student networks. For example, Co-Distill [6] performs co-training and additionally refines noisy labels in KD. It refines training labels based on loss values observed in early epochs of training. Empirically, data instances with non-noisy (clean) labels usually exhibit small training loss values at early epochs [3, 18, 95]. Therefore, by evaluating early epoch loss values or training a separate classifier, labels can be identified as noisy or clean. If a label is noisy, the teacher network re-labels it during the co-training of teacher and student networks. It is noteworthy that other works have addressed KD with noisy labels, such as [52, 83, 102].

3.4 Intermediate Layer Matching in Knowledge Distillation

3.4.1 Intermediate Layer Matching and the Problem of Different Architectures

Standard KD usually uses only the teacher's final outputs (soft labels). However, more advanced methods also align intermediate representations between teacher and student; in other words, they transfer knowledge from intermediate layers of the teacher network to the corresponding intermediate layers of the student network.

However, the number of intermediate layers differs between the teacher and student networks; in other words, the teacher and student networks have different architectures because the student network is smaller than the teacher network. Therefore, simply copying knowledge from a teacher layer to a student layer is not possible. Instead, knowledge should be distilled from multiple layers of the teacher network to each layer of the student network [9].

3.4.2 Attention-Based Knowledge Distillation

One way to distill knowledge from the intermediate layers of the teacher network to the intermediate layers of the student network is by using the attention mechanism (see Chap. 9) in KD. Attention functions as a form of weighted averaging, allowing a weighted average of intermediate layers in the teacher network to be computed for each intermediate layer in the student network. Several methods have applied attention in KD, such as "Show, Attend, and Distill" [43], ALP-KD [73], and Universal-KD [101].

It is noteworthy that other approaches exist for KD from intermediate layers of the teacher network to those of the student network. One example is Patient-KD [86], which patiently distills knowledge from multiple layers of the teacher network to the student network.

Here, Universal-KD [101] is described as an example of attention-based KD methods. In contrast to Patient-KD [86], which matches the hidden representations of the intermediate layers in the teacher and student networks, Universal-KD matches the output representations of the intermediate layers. Let $\boldsymbol{h}_i^t$ and $\boldsymbol{h}_j^s$ denote the hidden representations of the ith layer in the teacher network and the jth layer in the student network, respectively. Let $f_i^t(\boldsymbol{h}_i^t) = \sigma(\boldsymbol{W}_i^t \boldsymbol{h}_i^t)$ denote the output of the ith teacher layer after applying the weight matrix $\boldsymbol{W}_i^t$ and the activation function $\sigma(.)$. Similarly, $f_j^s(\boldsymbol{h}_j^s) = \sigma(\boldsymbol{W}_j^s \boldsymbol{h}_j^s)$ is defined for the jth layer in the student network.

The attention weight of the jth student layer to the ith teacher layer is defined as:

$$\alpha_{ij} = \frac{\exp\big(f_i^t(\boldsymbol{h}_i^t)^\top f_j^s(\boldsymbol{h}_j^s)\big)}{\sum_{i=1}^{\ell_t} \exp\big(f_i^t(\boldsymbol{h}_i^t)^\top f_j^s(\boldsymbol{h}_j^s)\big)}, \tag{13}$$

where ℓ_t is the number of layers in the teacher network.

The weighted average of the intermediate layers in the teacher network, for the jth student layer, is

$$\mathcal{F}^t(j) := \sum_{i=1}^{\ell_t} \alpha_{ij} f_i^t(\boldsymbol{h}_i^t). \tag{14}$$

The Universal-KD loss function for the jth student layer is

$$\mathcal{L}_j^s := \mathrm{KL}\big(\mathcal{F}^t(j) \| f_j^s(\boldsymbol{h}_j^s)\big). \tag{15}$$

This loss is minimized for each student layer. Minimizing this KL divergence aligns the distribution of the jth student layer with the weighted average of all layers in the teacher network, where the weights are given by the attention scores.

The total Universal-KD loss function sums the losses over all student layers:

$$\mathcal{L}_{\text{universal}} := \sum_{j=1}^{\ell_s} \mathcal{L}_j^s, \tag{16}$$

where ℓ_s is the number of layers in the student network. Finally, the Universal-KD loss is combined with the KD loss in Eq. (1) to form the overall loss function:

$$\mathcal{L} := \frac{1}{b} \sum_{i=1}^{b} \big(\lambda_1 \mathcal{L}_{\text{CE}}(\boldsymbol{x}_i) + \lambda_2 \mathcal{L}_{\text{KD}}(\boldsymbol{x}_i) + \lambda_3 \mathcal{L}_{\text{universal}}(\boldsymbol{x}_i)\big), \tag{17}$$

where $\lambda_1 > 0$, $\lambda_2 > 0$, and $\lambda_3 > 0$ are hyperparameters that weight the relative importance of the three components of the loss function.

3.5 Finding Optimal Weights in the Loss of Knowledge Distillation

In the regular KD loss function, i.e., Eq. (1), the weights of the CE and KD parts of the loss function are $(1 - \lambda)$ and λ, respectively. The question arises: what is the optimal value of the hyperparameter λ for each data instance $\boldsymbol{x}_i$? It is possible to learn the optimal weights of the CE and KD parts of the loss for each training data instance using meta-learning [58]. Meta-learning [37] refers to learning how to learn and can be used to learn the optimal values of hyperparameters as well. One of the most well-known meta-learning algorithms is Model-Agnostic Meta-Learning (MAML) [24].

Consider the loss function (1) with weights treated as parameters to be learned:

$$\mathcal{L} := \frac{1}{b} \sum_{i=1}^{b} \big(\lambda_i^{\text{CE}} \mathcal{L}_{\text{CE}}(\boldsymbol{x}_i) + \lambda_i^{\text{KD}} \mathcal{L}_{\text{KD}}(\boldsymbol{x}_i)\big), \tag{18}$$

where λ_i^{CE} and λ_i^{KD} are the weights of the CE and KD parts of the loss, respectively, for the ith training data instance in the batch. They satisfy the following condition:

$$\lambda_i^{\text{CE}} + \lambda_i^{\text{KD}} = 1. \tag{19}$$

Following the MAML algorithm [24], meta-learning training is episodic, where each training iteration, or episode, contains a support set and a query set. In other words, the training dataset is divided into mini-batches, each containing a support set and a query set [91]. During each episode, the network weights are updated by backpropagating error using the support set. These updated weights are not finalized yet. The query set is then fed to the network with the updated weights, and the error on the query set is backpropagated. This second backpropagation updates the network weights finally at the end of the episode. In other words, the query set is used to evaluate the quality of the update made by the support set.

In applying meta-learning to KD [58], in each backpropagation iteration, a meta-student network $\widehat{S}_\theta$ is initialized with the weights of the student network S_θ from the previous iteration. Then, the support mini-batch $\{\boldsymbol{x}_i^s\}_{i=1}^b$ is fed to the meta-student network $\widehat{S}_\theta$ and the teacher network to obtain predictions $f_{\widehat{s}}(\boldsymbol{x}_i^s)$ and $f_t(\boldsymbol{x}_i^s)$, respectively. Let $\{\epsilon_i^{\text{CE}}\}_{i=1}^b$ and $\{\epsilon_i^{\text{KD}}\}_{i=1}^b$ denote the perturbations of the weights $\{\lambda_i^{\text{CE}}\}_{i=1}^b$ and $\{\lambda_i^{\text{KD}}\}_{i=1}^b$ from the previous iteration. The perturbed loss, denoted by $\widehat{\mathcal{L}}$, is calculated as:

$$\widehat{\mathcal{L}} := \frac{1}{b}\sum_{i=1}^{b}\big(\epsilon_i^{\text{CE}}\mathcal{L}_{\text{CE}}(\boldsymbol{x}_i^s) + \epsilon_i^{\text{KD}}\mathcal{L}_{\text{KD}}(\boldsymbol{x}_i^s)\big). \tag{20}$$

One pass of backpropagation (see Chap. 3), minimizing Eq. (20), is performed on the student network with weights θ to obtain updated weights $\widehat{\theta}$:

$$\widehat{\theta} := \theta - \eta\nabla_\theta\widehat{\mathcal{L}}, \tag{21}$$

where η is the learning rate and $\nabla_\theta\widehat{\mathcal{L}}$ is the gradient of $\widehat{\mathcal{L}}$ with respect to θ.

Next, a query mini-batch $\{\boldsymbol{x}_i^q\}_{i=1}^b$ is fed to the meta-student network $\widehat{S}_{\widehat{\theta}}$ with weights $\widehat{\theta}$ and the teacher network to obtain outputs $f_{\widehat{s}}(\boldsymbol{x}_i^q)$ and $f_t(\boldsymbol{x}_i^q)$, respectively. The meta-loss is calculated as:

$$\mathcal{L}_{\text{meta}} := \frac{1}{b}\sum_{i=1}^{b}\big(\mathcal{L}_{\text{CE}}(\boldsymbol{x}_i^s) + \mathcal{L}_{\text{KD}}(\boldsymbol{x}_i^s)\big). \tag{22}$$

The optimal weights can be computed as the gradients of this meta-loss function with respect to ϵ_i^{CE} and ϵ_i^{KD}:

$$\widetilde{\lambda}_i^{\text{CE}} := \frac{\partial\mathcal{L}_{\text{meta}}}{\partial\epsilon_i^{\text{CE}}}, \quad \widetilde{\lambda}_i^{\text{KD}} := \frac{\partial\mathcal{L}_{\text{meta}}}{\partial\epsilon_i^{\text{KD}}}. \tag{23}$$

This is justified because a larger gradient indicates greater sensitivity and thus greater importance of the weight. Finally, as the weights $\widetilde{\lambda}_i^{\text{CE}}$ and $\widetilde{\lambda}_i^{\text{KD}}$ do not necessarily satisfy Eq. (19), they are normalized:

$$\lambda_i^{\text{CE}} = \frac{\max(\widetilde{\lambda}_i^{\text{CE}}, \delta)}{\max(\widetilde{\lambda}_i^{\text{CE}}, \delta) + \max(\widetilde{\lambda}_i^{\text{KD}}, \delta)}, \tag{24}$$

$$\lambda_i^{\text{KD}} = \frac{\max(\widetilde{\lambda}_i^{\text{KD}}, \delta)}{\max(\widetilde{\lambda}_i^{\text{CE}}, \delta) + \max(\widetilde{\lambda}_i^{\text{KD}}, \delta)}, \tag{25}$$

where it is ensured that the weights are not less than $\delta > 0$ for stability. This procedure is repeated so that the student network is trained while simultaneously learning the optimal weights of the CE and KD parts in its loss function through meta-learning.

3.6 Data-Free Knowledge Distillation

In some scenarios, there is no access to the original training data used to train the teacher network. This may be due to data privacy concerns or the data simply no longer being available. In such cases, data-free knowledge distillation should be employed.

One approach involves capturing summary statistics, such as the mean and covariance of data at each hidden layer. Then, Gaussian noise is fed as input to the trained teacher network, and the summary statistics of the resulting activations are compared to those of the original training data. By minimizing the mean squared error of these summary statistics at each layer, the input is iteratively modified to reconstruct the original training data. The reconstructed data can then be used to train the student network via KD. This methodology underlies many data-free knowledge distillation algorithms such as [56, 96].

Data-free (zero-shot) KD distills knowledge without access to the original training data, often by synthesizing pseudo-data that matches the teacher's internal statistics [67, 76]. If limited access to a few samples of the teacher's training data is available for training the student network, few-sample or few-shot KD [51] can be used.

3.7 Knowledge Distillation in BERT Models

Knowledge distillation has been used to compress many large language models into smaller models that consume less storage, power, and inference time. This compression is especially useful for edge devices such as cell phones, which seek to provide user-friendly experiences with language models. Using large language models on cell phones is generally impractical due to insufficient memory and high battery consumption. In such cases, large language models like BERT [19] should be compressed using neural network compression techniques. For example, DistilBERT [82], TinyBERT [44], and MobileBERT [87] are smaller student models distilled from BERT that are faster and lighter for deployment.

3.8 Other Notes on Knowledge Distillation

Data augmentation techniques [66] have also been employed for KD. For example, Mate-KD [77] and MiniMax-KNN KD [45] perform data augmentation by generating augmented data based on the maximum divergence loss between the teacher and student networks.

It is also possible to collaboratively train multiple student networks with different capacities using KD [29]. This approach enables multiple compressions at different levels. Another method to achieve multiple levels of compression is to use a hierarchy of teacher assistant networks [59], as discussed in Sect. 3.2.3.

4 Conclusion

4.1 Conclusion of Chapter

Neural network compression is critical for deploying deep learning in the real world, especially on resource-constrained devices. Techniques such as pruning, quantization, matrix decomposition, and weight sharing simplify models structurally, while knowledge distillation leverages information transfer from large to small models. By 2025, compression is especially important for large language and vision models, where distillation methods like DistilBERT and MobileBERT are widely used. Together, these approaches enable practical deployment of neural networks without sacrificing too much accuracy. Compression remains an active research area, particularly as models continue to grow in size and real-time applications demand efficiency. In short: compression shrinks models, distillation transfers knowledge, and both together make deep learning usable in practice.

4.2 Conclusion of the Book Part "Emerging Topics in Deep Learning"

Several emerging topics are shaping the frontiers of deep learning research, including self-supervised learning, deep metric learning, graph neural networks, deep reinforcement learning, few-shot learning, meta-learning, neural network compression, and knowledge distillation. Each of these areas represents an active and rapidly evolving field.

Deep metric learning and self-supervised learning focus on learning semantic embedding spaces that enable discrimination between classes or generation of meaningful representations. Graph neural networks are designed to process graph-structured data, such as social networks and recommendation systems. Deep reinforcement learning aims to learn policies for selecting optimal actions within an environment. Few-shot learning enables models to generalize effectively from a limited number of training examples, with meta-learning (or "learning to learn") serving as a prominent approach within this paradigm.

Neural network compression techniques reduce model size and computational requirements, enabling deployment on edge devices and efficient memory usage. Knowledge distillation, a key compression strategy, transfers knowledge from a larger model to a smaller one while maintaining performance.

Collectively, these emerging topics are expanding the capabilities of deep learning and are expected to play a central role in advancing future artificial intelligence applications across a wide range of domains.

5 Exercises

1. (Level: Difficult) A fully connected neural network layer has 1 million parameters. After pruning, 70% of the weights are set to zero and removed (i.e., only 30% of weights remain). Then, weight sharing is applied to the remaining weights by clustering them into 256 clusters. Each cluster shares

a single representative weight value. Assume that to store the cluster assignments (indices) for each weight, you use 8 bits per weight (since $2^8 = 256$ clusters). Each representative weight value is stored as a 32-bit float.

(a) Calculate the total number of bits needed to store the pruned and quantized weights using weight sharing.
(b) What is the overall compression ratio compared to the original 32-bit full model?

2. (Level: Moderate) A fully connected neural network layer has an input size of 1000 and an output size of 500. The weight matrix W is therefore of size 500×1000. Suppose you apply low-rank matrix decomposition to compress this weight matrix by approximating it as:

$$\boldsymbol{W} \approx \boldsymbol{U}\boldsymbol{V}^T,$$

where $\boldsymbol{U} \in \mathbb{R}^{500\times r}$ and $\boldsymbol{V} \in \mathbb{R}^{1000\times r}$ with $r < 500$ being the rank of the decomposition.

(a) How many parameters are saved if you use a rank $r = 50$ approximation instead of the full 500×1000 weight matrix?
(b) What is the compression ratio?

3. (Level: Moderate) A neural network layer has a weight matrix with 1 million parameters. Currently, each parameter is stored as a 32-bit floating point number. You decide to compress the model by uniformly quantizing the weights to 8-bit integers.

(a) What is the percentage reduction in memory size after quantization?
(b) If you use 4-bit quantization instead, what would be the compression ratio compared to the original 32-bit representation?

4. (Level: Moderate) A teacher neural network outputs a probability distribution over 4 classes for a given input:

$$\boldsymbol{p}^{(T)} = [0.1, 0.6, 0.2, 0.1].$$

A student network outputs its own predicted distribution:

$$\boldsymbol{p}^{(S)} = [0.25, 0.5, 0.15, 0.1].$$

Knowledge distillation uses a softened softmax with temperature $T = 2$. The softened probabilities are computed by:

$$q_i = \frac{\exp(z_i/T)}{\sum_j \exp(z_j/T)},$$

where z_i are the logits before softmax. Assuming the teacher's softened output probabilities at temperature $T = 2$ are

$$\boldsymbol{q}^{(T)} = [0.15, 0.45, 0.25, 0.15],$$

and the student's softened output probabilities at temperature $T = 2$ are

$$\boldsymbol{q}^{(S)} = [0.20, 0.40, 0.25, 0.15],$$

calculate the Kullback-Leibler (KL) divergence from teacher to student at temperature $T = 2$:

$$\mathrm{KL}(\boldsymbol{q}^{(T)} \| \boldsymbol{q}^{(S)}) = \sum_i q_i^{(T)} \log \frac{q_i^{(T)}}{q_i^{(S)}}.$$

Use natural logarithms.

5. (Level: Difficult) A large teacher network T and a smaller student network S are trained for classification with $C = 5$ classes. The teacher outputs logits:

$$\boldsymbol{z}^{(T)} = [2.0,\ 1.0,\ 0.5,\ 0.1,\ -1.0],$$

and the student outputs logits:

$$\boldsymbol{z}^{(S)} = [1.5,\ 1.2,\ 0.7,\ 0.0,\ -0.5].$$

The knowledge distillation loss uses softened softmax with temperature $T = 3$. The softened probabilities are defined as:

$$q_i^{(T)} = \frac{\exp\left(\frac{z_i^{(T)}}{T}\right)}{\sum_{j=1}^{C} \exp\left(\frac{z_j^{(T)}}{T}\right)}, \quad q_i^{(S)} = \frac{\exp\left(\frac{z_i^{(S)}}{T}\right)}{\sum_{j=1}^{C} \exp\left(\frac{z_j^{(S)}}{T}\right)}.$$

The knowledge distillation loss is the scaled Kullback-Leibler divergence:

$$\mathcal{L}_{\mathrm{KD}} = T^2 \sum_{i=1}^{C} q_i^{(T)} \log \frac{q_i^{(T)}}{q_i^{(S)}}.$$

(a) Calculate the softened probability vectors $\boldsymbol{q}^{(T)}$ and $\boldsymbol{q}^{(S)}$.
(b) Calculate the knowledge distillation loss $\mathcal{L}_{\mathrm{KD}}$.

References

1. Ali Alqahtani, Xianghua Xie, and Mark W Jones. "Literature review of deep network compression". In: *Informatics*. Vol. 8. 4. MDPI. 2021.
2. Sajid Anwar, Kyuyeon Hwang, and Wonyong Sung. "Structured pruning of deep convolutional neural networks". In: *ACM Journal on Emerging Technologies in Computing Systems (JETC)* 13.3 (2017), pp. 1–18.
3. Devansh Arpit et al. "A closer look at memorization in deep networks". In: *International conference on machine learning*. PMLR. 2017, pp. 233–242.
4. Ron Banner, Yury Nahshan, Elad Hoffer, and Daniel Soudry. "ACIQ: Analytical clipping for integer quantization of neural networks". In: *openreview.net* (2018).
5. Guillaume Bellec, David Kappel, Wolfgang Maass, and Robert Legenstein. "Deep Rewiring: Training very sparse deep networks". In: *International Conference on Learning Representations*. 2018.
6. Shivendra Bhardwaj et al. "Knowledge Distillation with Noisy Labels for Natural Language Understanding". In: *Proceedings of the Seventh Workshop on Noisy User-generated Text (W-NUT 2021)*. 2021, pp. 297–303.
7. Cristian BuciluÇŽ, Rich Caruana, and Alexandru Niculescu-Mizil. "Model compression". In: *Proceedings of the 12th ACM SIGKDD international conference on Knowledge discovery and data mining*. 2006, pp. 535–541.
8. Andrés Buzo, A Gray, R Gray, and John Markel. "Speech coding based upon vector quantization". In: *IEEE Transactions on Acoustics, Speech, and Signal Processing* 28.5 (1980), pp. 562–574.
9. Pengguang Chen, Shu Liu, Hengshuang Zhao, and Jiaya Jia. "Distilling knowledge via knowledge review". In: *Proceedings of the IEEE/CVF conference on computer vision and pattern recognition*. 2021, pp. 5008–5017.

10. Wenlin Chen, James Wilson, Stephen Tyree, Kilian Weinberger, and Yixin Chen. “Compressing neural networks with the hashing trick”. In: *International conference on machine learning*. PMLR. 2015, pp. 2285–2294.
11. Hongrong Cheng, Miao Zhang, and Javen Qinfeng Shi. “A survey on deep neural network pruning: Taxonomy, comparison, analysis, and recommendations”. In: *IEEE Transactions on Pattern Analysis and Machine Intelligence* (2024).
12. Yu Cheng, Duo Wang, Pan Zhou, and Tao Zhang. “A survey of model compression and acceleration for deep neural networks”. In: *arXiv preprint* arXiv:1710.09282 (2017).
13. Yu Cheng, Duo Wang, Pan Zhou, and Tao Zhang. “Model compression and acceleration for deep neural networks: The principles, progress, and challenges”. In: *IEEE Signal Processing Magazine* 35.1 (2018), pp. 126–136.
14. Ching-Tai Chiu, Kishan Mehrotra, Chilukuri K Mohan, and Sanjay Ranka. “Modifying training algorithms for improved fault tolerance”. In: *Proceedings of 1994 IEEE International Conference on Neural Networks (ICNN’94)*. Vol. 1. IEEE. 1994, pp. 333–338.
15. Jang Hyun Cho and Bharath Hariharan. “On the efficacy of knowledge distillation”. In: *Proceedings of the IEEE/CVF international conference on computer vision*. 2019, pp. 4794–4802.
16. Jungwook Choi et al. “PACT: Parameterized clipping activation for quantized neural networks”. In: *arXiv preprint* arXiv:1805.06085 (2018).
17. Lieven De Lathauwer. “Decompositions of a higher-order tensor in block terms—Part II: Definitions and uniqueness”. In: *SIAM Journal on Matrix Analysis and Applications* 30.3 (2008), pp. 1033–1066.
18. M Dehghani, A Mehrjou, S Gouws, J Kamps, and B Schölkopf. “Fidelity-Weighted Learning”. In: *6th International Conference on Learning Representations (ICLR 2018)*. 2018.
19. Jacob Devlin, Ming-Wei Chang, Kenton Lee, and Kristina Toutanova. “BERT: Pre-training of deep bidirectional transformers for language understanding”. In: *Proceedings of the 2019 Conference of the North American Chapter of the Association for Computational Linguistics: Human Language Technologies*. 2019.
20. Bin Dong, Jikai Hou, Yiping Lu, and Zhihua Zhang. “Distillation $\approx$ Early Stopping? Harvesting Dark Knowledge Utilizing Anisotropic Information Retrieval For Overparameterized Neural Network”. In: *NeurIPS 2019 Workshop on Machine Learning with Guarantees*. 2019.
21. Zhen Dong, Zhewei Yao, Amir Gholami, Michael W Mahoney, and Kurt Keutzer. “HAWQ: Hessian aware quantization of neural networks with mixed-precision”. In: *Proceedings of the IEEE/CVF international conference on computer vision*. 2019, pp. 293–302.
22. David Eigen, Jason Rolfe, Rob Fergus, and Yann LeCun. “Understanding deep architectures using a recursive convolutional network”. In: *arXiv preprint* arXiv:1312.1847 (2013).
23. Andries P Engelbrecht. “A new pruning heuristic based on variance analysis of sensitivity information”. In: *IEEE transactions on Neural Networks* 12.6 (2001), pp. 1386–1399.
24. Chelsea Finn, Pieter Abbeel, and Sergey Levine. “Model-agnostic meta-learning for fast adaptation of deep networks”. In: *International conference on machine learning*. PMLR. 2017, pp. 1126–1135.
25. Jonathan Frankle and Michael Carbin. “The Lottery Ticket Hypothesis: Finding Sparse, Trainable Neural Networks”. In: *International Conference on Learning Representations*. 2019.
26. Adam Gaier and David Ha. “Weight agnostic neural networks”. In: *Advances in neural information processing systems* 32 (2019).
27. Benyamin Ghojogh, Aydin Ghojogh, Mark Crowley, and Fakhri Karray. “Fitting a mixture distribution to data: tutorial”. In: *arXiv preprint* arXiv:1901.06708 (2019).
28. Jianping Gou, Baosheng Yu, Stephen J Maybank, and Dacheng Tao. “Knowledge distillation: A survey”. In: *International Journal of Computer Vision* 129.6 (2021), pp. 1789–1819.
29. Qiushan Guo et al. “Online knowledge distillation via collaborative learning”. In: *Proceedings of the IEEE/CVF Conference on Computer Vision and Pattern Recognition*. 2020, pp. 11020–11029.
30. Yunhui Guo. “A survey on methods and theories of quantized neural networks”. In: *arXiv preprint* arXiv:1808.04752 (2018).
31. Masafumi Hagiwara. “Removal of hidden units and weights for back propagation networks”. In: *Proceedings of 1993 International Conference on Neural Networks (IJCNN-93-Nagoya, Japan)*. Vol. 1. IEEE. 1993, pp. 351–354.
32. Song Han, Huizi Mao, and William J Dally. “Deep compression: Compressing deep neural networks with pruning, trained quantization and Huffman coding”. In: *arXiv preprint* arXiv:1510.00149 (2015).
33. Babak Hassibi, David Stork, and Gregory Wolff. “Optimal brain surgeon: Extensions and performance comparisons”. In: *Advances in neural information processing systems* 6 (1993).
34. Yihui He et al. “AMC: AutoML for model compression and acceleration on mobile devices”. In: *Proceedings of the European conference on computer vision (ECCV)*. 2018, pp. 784–800.
35. Geoffrey Hinton, Oriol Vinyals, and Jeff Dean. “Distilling the Knowledge in a Neural Network”. In: *NIPS 2014 Deep Learning Workshop*. 2015.
36. John H Holland. “Genetic algorithms”. In: *Scientific american* 267.1 (1992), pp. 66–73.
37. Timothy Hospedales, Antreas Antoniou, Paul Micaelli, and Amos Storkey. “Meta-learning in neural networks: A survey”. In: *IEEE transactions on pattern analysis and machine intelligence* 44.9 (2021), pp. 5149–5169.

38. Yiming Hu, Siyang Sun, Jianquan Li, Xingang Wang, and Qingyi Gu. "A novel channel pruning method for deep neural network compression". In: *arXiv preprint* arXiv:1805.11394 (2018).
39. Benoit Jacob et al. "Quantization and training of neural networks for efficient integer-arithmetic-only inference". In: *Proceedings of the IEEE conference on computer vision and pattern recognition*. 2018, pp. 2704–2713.
40. Max Jaderberg, Andrea Vedaldi, and Andrew Zisserman. "Speeding up convolutional neural networks with low rank expansions". In: *arXiv preprint* arXiv:1405.3866 (2014).
41. Aref Jafari, Mehdi Rezagholizadeh, Pranav Sharma, and Ali Ghodsi. "Annealing Knowledge Distillation". In: *Proceedings of the 16th Conference of the European Chapter of the Association for Computational Linguistics: Main Volume*. 2021, pp. 2493–2504.
42. Herve Jegou, Matthijs Douze, and Cordelia Schmid. "Product quantization for nearest neighbor search". In: *IEEE transactions on pattern analysis and machine intelligence* 33.1 (2010), pp. 117–128.
43. Mingi Ji, Byeongho Heo, and Sungrae Park. "Show, attend and distill: Knowledge distillation via attention-based feature matching". In: *Proceedings of the AAAI Conference on Artificial Intelligence*. Vol. 35. 9. 2021, pp. 7945–7952.
44. Xiaoqi Jiao et al. "TinyBERT: Distilling BERT for Natural Language Understanding". In: *Findings of the Association for Computational Linguistics: EMNLP 2020*. 2020, pp. 4163–4174.
45. Ehsan Kamalloo, Mehdi Rezagholizadeh, Peyman Passban, and Ali Ghodsi. "Not Far Away, Not So Close: Sample Efficient Nearest Neighbour Data Augmentation via MiniMax". In: *Findings of the Association for Computational Linguistics: ACL-IJCNLP 2021*. 2021, pp. 3522–3533.
46. Ehud D Karnin. "A simple procedure for pruning back-propagation trained neural networks". In: *IEEE transactions on neural networks* 1.2 (1990), pp. 239–242.
47. Scott Kirkpatrick, C Daniel Gelatt Jr, and Mario P Vecchi. "Optimization by simulated annealing". In: *Science* 220.4598 (1983), pp. 671–680.
48. Anders Krogh and John A Hertz. "A simple weight decay can improve generalization". In: *Advances in neural information processing systems*. 1992, pp. 950–957.
49. Solomon Kullback and Richard A Leibler. "On information and sufficiency". In: *The annals of mathematical statistics* 22.1 (1951), pp. 79–86.
50. Yann LeCun, John Denker, and Sara Solla. "Optimal brain damage". In: *Advances in neural information processing systems* 2 (1989).
51. Tianhong Li, Jianguo Li, Zhuang Liu, and Changshui Zhang. "Few sample knowledge distillation for efficient network compression". In: *Proceedings of the IEEE/CVF conference on computer vision and pattern recognition*. 2020, pp. 14639–14647.
52. Yuncheng Li et al. "Learning from noisy labels with distillation". In: *Proceedings of the IEEE international conference on computer vision*. 2017, pp. 1910–1918.
53. Tailin Liang, John Glossner, Lei Wang, Shaobo Shi, and Xiaotong Zhang. "Pruning and quantization for deep neural network acceleration: A survey". In: *Neurocomputing* 461 (2021), pp. 370–403.
54. Darryl Lin, Sachin Talathi, and Sreekanth Annapureddy. "Fixed point quantization of deep convolutional networks". In: *International conference on machine learning*. PMLR. 2016, pp. 2849–2858.
55. Stuart Lloyd. "Least squares quantization in PCM". In: *IEEE transactions on information theory* 28.2 (1982), pp. 129–137.
56. Raphael Gontijo Lopes, Stefano Fenu, and Thad Starner. "Data-free knowledge distillation for deep neural networks". In: *NIPS 2017 Workshop on Learning with Limited Data*. 2017.
57. David Lopez-Paz, Léon Bottou, Bernhard Schölkopf, and Vladimir Vapnik. "Unifying distillation and privileged information". In: *International Conference on Learning Representations (ICLR)*. 2016.
58. Peng Lu et al. "RW-KD: Sample-wise loss terms re-weighting for knowledge distillation". In: *Findings of the Association for Computational Linguistics: EMNLP 2021*. 2021, pp. 3145–3152.
59. Seyed Iman Mirzadeh et al. "Improved knowledge distillation via teacher assistant". In: *Proceedings of the AAAI conference on artificial intelligence*. Vol. 34. 04. 2020, pp. 5191–5198.
60. Asit Mishra and Debbie Marr. "Apprentice: Using Knowledge Distillation Techniques To Improve Low-Precision Network Accuracy". In: *International Conference on Learning Representations*. 2018.
61. Rahul Mishra, Hari Prabhat Gupta, and Tanima Dutta. "A survey on deep neural network compression: Challenges, overview, and solutions". In: *arXiv preprint* arXiv:2010.03954 (2020).
62. Decebal Constantin Mocanu et al. "Scalable training of artificial neural networks with adaptive sparse connectivity inspired by network science". In: *Nature communications* 9.1 (2018), p. 2383.
63. Dmitry Molchanov, Arsenii Ashukha, and Dmitry Vetrov. "Variational dropout sparsifies deep neural networks". In: *International conference on machine learning*. PMLR. 2017, pp. 2498–2507.
64. Hesham Mostafa and Xin Wang. "Parameter efficient training of deep convolutional neural networks by dynamic sparse reparameterization". In: *International Conference on Machine Learning*. PMLR. 2019, pp. 4646–4655.
65. Michael C Mozer and Paul Smolensky. "Skeletonization: A technique for trimming the fat from a network via relevance assessment". In: *Advances in neural information processing systems* 1 (1988).

66. Alhassan Mumuni and Fuseini Mumuni. "Data augmentation: A comprehensive survey of modern approaches". In: *Array* 16 (2022), p. 100258.
67. Gaurav Kumar Nayak, Konda Reddy Mopuri, Vaisakh Shaj, Venkatesh Babu Radhakrishnan, and Anirban Chakraborty. "Zero-shot knowledge distillation in deep networks". In: *International Conference on Machine Learning*. PMLR. 2019, pp. 4743–4751.
68. James O' Neill. "An overview of neural network compression". In: *arXiv preprint* arXiv:2006.03669 (2020).
69. Alexander Novikov, Dmitrii Podoprikhin, Anton Osokin, and Dmitry P Vetrov. "Tensorizing neural networks". In: *Advances in neural information processing systems* 28 (2015).
70. Steven J Nowlan and Geoffrey E Hinton. "Simplifying neural networks by soft weight sharing". In: *Neural computation* 4.4 (1992), pp. 473–493.
71. Asaf Noy et al. "ASAP: Architecture search, anneal and prune". In: *International conference on artificial intelligence and statistics*. PMLR. 2020, pp. 493–503.
72. Eunhyeok Park, Sungjoo Yoo, and Peter Vajda. "Value-aware quantization for training and inference of neural networks". In: *Proceedings of the European Conference on Computer Vision (ECCV)*. 2018, pp. 580–595.
73. Peyman Passban, Yimeng Wu, Mehdi Rezagholizadeh, and Qun Liu. "ALP-KD: Attention-based layer projection for knowledge distillation". In: *Proceedings of the AAAI Conference on artificial intelligence*. Vol. 35. 15. 2021, pp. 13657–13665.
74. Lutz Prechelt. "Early stopping-but when?" In: *Neural Networks: Tricks of the trade*. Springer, 1998, pp. 55–69.
75. Vivek Ramanujan, Mitchell Wortsman, Aniruddha Kembhavi, Ali Farhadi, and Mohammad Rastegari. "What's hidden in a randomly weighted neural network?" In: *Proceedings of the IEEE/CVF conference on computer vision and pattern recognition*. 2020, pp. 11893–11902.
76. Ahmad Rashid, Vasileios Lioutas, Abbas Ghaddar, and Mehdi Rezagholizadeh. "Towards Zero-Shot Knowledge Distillation for Natural Language Processing". In: *Proceedings of the 2021 Conference on Empirical Methods in Natural Language Processing*. 2021, pp. 6551–6561.
77. Ahmad Rashid, Vasileios Lioutas, and Mehdi Rezagholizadeh. "MATE-KD: Masked Adversarial TExt, a Companion to Knowledge Distillation". In: *Proceedings of the 59th Annual Meeting of the Association for Computational Linguistics and the 11th International Joint Conference on Natural Language Processing (Volume 1: Long Papers)*. 2021, pp. 1062–1071.
78. Russell Reed. "Pruning algorithms-a survey". In: *IEEE transactions on Neural Networks* 4.5 (1993), pp. 740–747.
79. Mehdi Rezagholizadeh et al. "Pro-KD: Progressive Distillation by Following the Footsteps of the Teacher". In: *Proceedings of the 29th International Conference on Computational Linguistics*. 2022, pp. 4714–4727.
80. Wolfgang Roth et al. "Resource-efficient neural networks for embedded systems". In: *Journal of Machine Learning Research* 25 (2024), pp. 1–51.
81. Tara N Sainath, Brian Kingsbury, Vikas Sindhwani, Ebru Arisoy, and Bhuvana Ramabhadran. "Low-rank matrix factorization for deep neural network training with high-dimensional output targets". In: *2013 IEEE international conference on acoustics, speech and signal processing*. IEEE. 2013, pp. 6655–6659.
82. V Sanh. "DistilBERT, A Distilled Version of BERT: Smaller, Faster, Cheaper and Lighter". In: *The 5th Workshop on Energy Efficient Machine Learning and Cognitive Computing - NeurIPS 2019*. 2019.
83. Bharat Bhusan Sau and Vineeth N Balasubramanian. "Deep model compression: Distilling knowledge from noisy teachers". In: *arXiv preprint* arXiv:1610.09650 (2016).
84. Sheng Shen et al. "Q-BERT: Hessian based ultra low precision quantization of BERT". In: *Proceedings of the AAAI Conference on Artificial Intelligence*. Vol. 34. 05. 2020, pp. 8815–8821.
85. Nitish Srivastava, Geoffrey Hinton, Alex Krizhevsky, Ilya Sutskever, and Ruslan Salakhutdinov. "Dropout: a simple way to prevent neural networks from overfitting". In: *The journal of machine learning research* 15.1 (2014), pp. 1929–1958.
86. Siqi Sun, Yu Cheng, Zhe Gan, and Jingjing Liu. "Patient Knowledge Distillation for BERT Model Compression". In: *Proceedings of the 2019 Conference on Empirical Methods in Natural Language Processing and the 9th International Joint Conference on Natural Language Processing (EMNLP-IJCNLP)*. Association for Computational Linguistics. 2019.
87. Zhiqing Sun et al. "MobileBERT: a Compact Task-Agnostic BERT for Resource-Limited Devices". In: *Proceedings of the 58th Annual Meeting of the Association for Computational Linguistics*. 2020, pp. 2158–2170.
88. Marzieh Tahaei, Ella Charlaix, Vahid Nia, Ali Ghodsi, and Mehdi Rezagholizadeh. "KroneckerBERT: Significant compression of pre-trained language models through Kronecker decomposition and knowledge distillation". In: *Proceedings of the 2022 Conference of the North American Chapter of the Association for Computational Linguistics: Human Language Technologies*. 2022, pp. 2116–2127.
89. EG Talbi. "Metaheuristics: From Design to Implementation". In: *John Wiley & Sons google schola* 2 (2009), pp. 268–308.
90. Robert Tibshirani. "Regression shrinkage and selection via the lasso". In: *Journal of the Royal Statistical Society Series B: Statistical Methodology* 58.1 (1996), pp. 267–288.
91. Eleni Triantafillou et al. "Meta-dataset: A dataset of datasets for learning to learn from few examples". In: *International Conference on Learning Representations*. 2020.

92. Juanjuan Tu, Yongzhao Zhan, and Fei Han. "A neural network pruning method optimized with PSO algorithm". In: *2010 Second International Conference on Computer Modeling and Simulation*. Vol. 3. IEEE. 2010, pp. 257–259.
93. Vladimir Vapnik. *The nature of statistical learning theory*. Springer science & business media, 2000.
94. Christopher A Walsh. "Peter Huttenlocher (1931–2013)". In: *Nature* 502.7470 (2013), pp. 172–172.
95. Yisen Wang et al. "Symmetric cross entropy for robust learning with noisy labels". In: *Proceedings of the IEEE/CVF international conference on computer vision*. 2019, pp. 322–330.
96. Zi Wang. "Data-free knowledge distillation with soft targeted transfer set synthesis". In: *Proceedings of the AAAI conference on artificial intelligence*. Vol. 35. 11. 2021, pp. 10245–10253.
97. Andreas Weigend, David Rumelhart, and Bernardo Huberman. "Generalization by weight-elimination with application to forecasting". In: *Advances in neural information processing systems* 3 (1990).
98. Wei Wen, Chunpeng Wu, Yandan Wang, Yiran Chen, and Hai Li. "Learning structured sparsity in deep neural networks". In: *Advances in neural information processing systems* 29 (2016).
99. Olivia Weng. "Neural network quantization for efficient inference: A survey". In: *arXiv preprint* arXiv:2112.06126 (2021).
100. Darrell Whitley, Timothy Starkweather, and Christopher Bogart. "Genetic algorithms and neural networks: Optimizing connections and connectivity". In: *Parallel computing* 14.3 (1990), pp. 347–361.
101. Yimeng Wu, Mehdi Rezagholizadeh, Abbas Ghaddar, Md Akmal Haidar, and Ali Ghodsi. "Universal-KD: Attention-based output-grounded intermediate layer knowledge distillation". In: *Proceedings of the 2021 Conference on Empirical Methods in Natural Language Processing*. 2021, pp. 7649–7661.
102. Guodong Xu, Ziwei Liu, Xiaoxiao Li, and Chen Change Loy. "Knowledge distillation meets self-supervision". In: *European conference on computer vision*. Springer. 2020, pp. 588–604.
103. Jian Xue, Jinyu Li, and Yifan Gong. "Restructuring of deep neural network acoustic models with singular value decomposition". In: *Interspeech*. 2013, pp. 2365–2369.
104. Jian Xue, Jinyu Li, Dong Yu, Mike Seltzer, and Yifan Gong. "Singular value decomposition based low-footprint speaker adaptation and personalization for deep neural network". In: *2014 IEEE International Conference on Acoustics, Speech and Signal Processing (ICASSP)*. IEEE. 2014, pp. 6359–6363.
105. Yuan Yao, Lorenzo Rosasco, and Andrea Caponnetto. "On early stopping in gradient descent learning". In: *Constructive Approximation* 26.2 (2007), pp. 289–315.
106. Jinmian Ye et al. "Learning compact recurrent neural networks with block-term tensor decomposition". In: *Proceedings of the IEEE conference on computer vision and pattern recognition*. 2018, pp. 9378–9387.
107. Xiyu Yu, Tongliang Liu, Xinchao Wang, and Dacheng Tao. "On compressing deep models by low rank and sparse decomposition". In: *Proceedings of the IEEE conference on computer vision and pattern recognition*. 2017, pp. 7370–7379.
108. Meng Zhang, Fei Liu, and Dongpeng Weng. "Speeding-up and compression convolutional neural networks by low-rank decomposition without fine-tuning". In: *Journal of Real-Time Image Processing* 20.4 (2023), p. 64.
109. Aojun Zhou, Anbang Yao, Kuan Wang, and Yurong Chen. "Explicit loss-error-aware quantization for low-bit deep neural networks". In: *Proceedings of the IEEE conference on computer vision and pattern recognition*. 2018, pp. 9426–9435.
110. Shuchang Zhou et al. "DoReFa-Net: Training low bitwidth convolutional neural networks with low bitwidth gradients". In: *arXiv preprint* arXiv:1606.06160 (2016).

Part V

Deep Learning in Practice

21 Practical Tips in Deep Learning

Abstract

Deep learning has become an essential tool in both research and industrial applications. While a solid theoretical understanding is important for developing and analyzing deep learning algorithms, successful real-world deployment often depends on practical insights and heuristics. This chapter focuses on the effective application of deep learning in practice, with particular emphasis on industry use cases. Key topics include strategies for data collection and splitting into training, validation, test, and development sets; understanding a range of performance metrics—from human-level benchmarks to deployment outcomes, and providing essential rules of thumb for efficient neural network training. Practical guidance is offered on handling equal-sized and variable-sized data, designing data pipelines, selecting network architectures, choosing hyperparameters—including activation functions, number of layers, optimizers, learning rates, and mini-batch sizes—and optimizing hardware and memory efficiency. This chapter aims to bridge the gap between deep learning theory and its practical application in real-world environments.

1 Introduction

Deep learning is useful not only in academic projects but also in industrial and practical applications. Knowing the theory is essential for designing new algorithms and understanding how they work. However, applying these algorithms in practice requires practical tips and rules of thumb. This chapter explains how to use deep learning effectively in practice, with a particular focus on industry applications. It is recommended to study Chap. 5 which covers overfitting and regularization, before reading this chapter.

Data collection and splitting—including training, validation, test, and dev sets—is explained in Sect. 2. The breakdown of performance, encompassing human-level performance, training performance, test performance, dev performance, and deployment performance, is discussed in Sect. 3. Finally, Sect. 4 provides several rules of thumb for training neural networks.

B. Ghojogh and A. Ghodsi, *Elements of Deep Learning*,
https://doi.org/10.1007/978-3-032-10738-1_21

2 Data Collection and Split

2.1 Training, Validation, and Test Sets

As discussed in Chap. 5, any dataset used in machine learning should be split into training, validation, and test sets. However, in practical deep learning, engineers usually collect and prepare datasets in a slightly different way. In industry, there is often a dataset, available either publicly or privately, that is useful for the company's task. This dataset is then split into training, validation, and test sets.

Unlike non-deep machine learning, which usually uses the validation set in cross-validation for reporting generalization accuracy or tuning hyperparameters, in deep learning the validation set is primarily used for early stopping. After each epoch of training, both the training error and validation error are typically computed to check whether the validation error has become significantly worse than the training error; if so, training is stopped. Moreover, the best validation performance is usually recorded as the best checkpoint of weights to be used in the inference (test) phase.

Cross-validation—introduced in Chap. 5—is generally not used in deep learning. This is because cross-validation is often infeasible in deep learning since each complete training of deep networks is very time-consuming; it may even take days to complete. In some industry settings, engineers split data into training and test sets and use the test set as both validation and test. While this may save compute, it risks overestimating generalization (see Chap. 5); best practice remains to keep a dedicated validation set separate from the final test set.

After the training of the neural network is finished using the training set for backpropagation (see Chap. 3) and the validation set for early stopping and checkpoint selection, the trained network should be tested on the test set. The test set is unseen during training; thus, the performance of the neural network on the test set is a reasonably good estimation of its generalization.

2.2 Dev Set

In some cases, the dataset—which was split into training, validation, and test sets—is a public dataset. In other cases, the labels of the dataset are noisy or auto-labeled because manual labeling is a time- and budget-consuming task. In such situations, a dev set (development set) should be prepared in the company as an in-house private dataset. This dev set should be collected from the task being worked on to resemble as closely as possible the data that the neural network will encounter after deployment.

The dev set should also be manually and accurately labeled to ensure it does not contain errors or noise. For example, if the task involves image processing of pictures from a specific camera, the dev set can be created by capturing some pictures with that specific camera and manually labeling their contents. Because the dev set is similar to the data encountered after deployment, performance on the dev set provides a better estimate of the model's generalization to unseen data compared to the test set.

3 Breakdown of Performance

3.1 Performances

As shown in Fig. 1, there can be a hierarchy of performance as follows [1]:

1. Human-level performance: This is the performance of humans on the task; for example, classifying images. This performance is usually very high because it leverages human intelligence. Ignor-

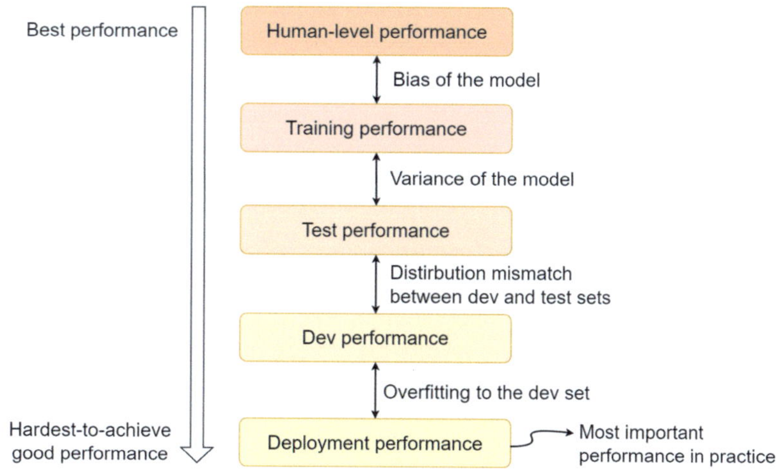

Fig. 1 The breakdown of performance as human-level performance, training performance, test performance, dev performance, and deployment performance

ing some cases where artificial intelligence surpasses human capabilities, this performance level determines the irreducible error.

2. Training performance: This is the performance of the trained neural network on the training set. This performance is typically high because it evaluates the network on its own training data, meaning the network has learned its patterns during training.
3. Test performance: This is the performance of the trained neural network on the test set. It is often inferior to the training performance because the test set was unseen during training. It is rare but possible to see test accuracy slightly higher than training accuracy, often due to strong regularization or stochasticity (e.g., dropout [2]). Large differences usually indicate a biased or unrepresentative split. To avoid this, the dataset should be reshuffled and split again.
   Note that some engineers use the test performance for tuning hyperparameters of neural networks, such as learning rate, mini-batch size, and the number of layers. This is not a recommended approach because, as explained in Chap. 5, it is considered cheating. You may be pleased for a few days thinking the model is performing well, but after deployment, you will find that it does not work effectively on deployment data.
4. Dev performance (inference performance): This is the performance of the trained neural network on the dev set. As the dev set represents the data encountered after deployment, its performance is a good measure of how the model will function in deployment. Therefore, dev performance is a more realistic indicator than test performance.
5. Deployment performance: This is the performance of the trained neural network after it is deployed in the real-world application. For example, it is the performance after the neural network is installed in a product that is sold and used by customers.

The main goal is to improve deployment performance. Even improving the test and dev performance ultimately serves to improve deployment performance because performance after deployment holds real value. The other performance measures are theoretical; test performance is the most theoretical, while dev performance has some level of practicality.

3.2 Interpretation of Gaps Between Performances

The performances generally follow the order from best to worst as human-level performance, training performance, test performance, dev performance, and deployment performance. The gaps between these performances can be interpreted as shown in Fig. 1. Moreover, there are solutions to reduce these gaps. These interpretations and solutions are introduced below.

3.2.1 Human-Level Performance Versus Training Performance

If the gap between human-level performance and training performance is large, it indicates bias in the model (see Chap. 5 for more information on bias). This is because it shows that the model does not have sufficient flexibility to learn the training data. To reduce this gap, the flexibility of the model should be increased. For example, additional features can be added to the data, the nonlinearity of the model can be increased (e.g., by increasing the depth of the neural network), or regularization can be reduced (see Chap. 5).

3.2.2 Training Performance Versus Test Performance

If the gap between training performance and test performance is large, it is a sign of high variance in the model (see Chap. 5 for more information on variance). This indicates that the model has excessive variance or complexity in fitting the training data. In other words, the model has been overfitted to the training data (see Chap. 5 for the concept of overfitting). To reduce this gap, the variance or complexity of the model should be decreased; in other words, a simpler model should be used. For example, fewer features from the data can be used, the nonlinearity of the model can be decreased (e.g., by reducing the depth of the neural network), or regularization can be increased or added (see Chap. 5).

Moreover, collecting and using more training data can help reduce the gap between training and test performance. Training with more data reduces overfitting. For example, this is what large language models do by training on huge Internet corpora to avoid overfitting (see Chap. 11).

3.2.3 Test Performance Versus Dev Performance

If the gap between test performance and dev performance is large, it indicates a distribution mismatch between the dev set and the training/test set. The training and test sets are split from the same dataset, so they usually have similar domains and distributions. However, the distribution of the dev set is more realistic and more similar to the data encountered after deployment.

To reduce this gap, either the distribution of the dev set should be made more similar to the training and test sets, or vice versa. If the distribution of the dev set is made similar to the training and test sets, it is merely fooling ourselves. Therefore, the distribution of the training and test sets should be made similar to the dev set to reduce the gap. To achieve this, one can take out a part of the dev set and include it in the test set. In this way, the test set becomes more realistic by mimicking the distribution of the dev set.

3.2.4 Dev Performance Versus Deployment Performance

If the gap between dev performance and deployment performance is large, it implies overfitting to the dev set, meaning that the dev set is not diverse or large enough to represent realistic data after deployment (see Chap. 5 for more information on overfitting). To resolve this issue, the dev set should be enriched by collecting more data with greater diversity across different scenarios. In this way, the dev set becomes more representative of real-world data after deployment, making it harder to overfit.

Moreover, including various scenarios in the dev set makes it possible to evaluate the performance of the trained neural network when facing different cases and situations.

4 Rules of Thumb for Training Neural Networks

The following sections present several practical guidelines for training neural networks. These rules of thumb address handling equal-sized and variable-sized data, designing data pipelines, selecting network architectures, choosing hyperparameters—including activation functions, number of layers, optimizers, learning rates, and mini-batch sizes—and optimizing hardware and memory efficiency.

4.1 Equal-Sized Versus Varying-Sized Data

If the data instances have equal sizes, i.e., an equal number of features, then regular machine learning and deep learning methods can be used. Examples of this type of data include tabular data and image datasets with equal image sizes.

In some cases, data instances have varying sizes, i.e., different numbers of features. This usually occurs with sequence-type data. Some examples of sequence data include text (natural language processing), speech (speech processing), actions (action recognition from successive body poses), and amino acid sequences (in proteins and antibodies). Text data can contain sentences with different numbers of words. Speech data can have parts with various lengths, or the same part of speech can differ in pacing. Action data can include actions with different numbers of frames, such as different actions or the same action performed at different speeds. Proteins and antibodies, which are amino acid sequences, also have different numbers of amino acids.

Two approaches can be used for handling varying-sized data in machine learning and deep learning:

- One approach is to use machine learning algorithms that can handle different lengths. Some examples of this approach are hidden Markov models [3], recurrent neural networks and long short-term memory networks (see Chap. 8), and Transformers (see Chaps. 9 and 11).
- Another approach is to make the sizes of the data instances the same. This can be performed in two ways:
  - One way is to crop the data instances or zero-pad them (see Chap. 4).
  - Another way is to warp the data instances to become equal-sized. For example, Dynamic Time Warping (DTW) [4] can be used for warping body poses in action recognition [5]. As another example in biology applications, warping of amino acid sequences can be performed either by numbering, such as IMGT numbering [6], or by dynamic-programming sequence alignment, such as the Needleman–Wunsch algorithm [7].

4.2 Data Pipeline

A data pipeline is a sequence of steps designed to efficiently feed data into a machine learning model during training or inference. Typical pipelines include preprocessing, data augmentation, caching, shuffling, batching, prefetching, and device transfer, each aimed at maximizing throughput and GPU utilization. A well-designed pipeline can significantly reduce training time, particularly for large datasets.

Data loading efficiently retrieves raw data from disk without creating a bottleneck. Techniques include memory-mapped files, which allow access to large datasets without fully loading them into RAM, and custom dataset classes that wrap data in iterable objects.

Data preprocessing transforms raw inputs into a format suitable for the model. Common steps include normalization (scaling inputs to zero mean and unit variance), tokenization (converting text into integer sequences or embeddings), and resizing or cropping (standardizing input dimensions).

Data augmentation [8, 9] enhances dataset diversity to improve generalization and reduce overfitting. Examples include image rotations, flipping, color jittering, and random cropping; text augmentations such as synonym replacement, back-translation, and random masking; and time-series augmentations like time warping, jittering, and scaling.

Caching avoids redundant computations of expensive preprocessing or augmentations by storing data in RAM or on disk, trading memory usage for faster access. Batching groups samples to enable efficient GPU computation, while shuffling randomizes the data of each epoch to prevent the model from learning spurious order patterns. Prefetching [10] overlaps data loading and preprocessing with GPU computation to reduce idle time. Finally, device transfer and memory pinning ensure batches are efficiently moved to GPU or TPU memory.

4.3 Deciding the Architecture of Neural Network

When deciding on the architecture of a neural network, the type of data should be considered. If the data are tabular or vector-based, a fully connected neural network should be used (see Chap. 2). If the data are images or image-like data (e.g., MRI scans) with one or multiple channels, a Convolutional Neural Network (CNN) should be used (see Chap. 4). If the data are sequences, sequence models such as Recurrent Neural Networks (RNNs) and Long Short-Term Memory (LSTM) networks (see Chap. 8), as well as Transformers (see Chaps. 9 and 11), should be used.

4.4 Hyperparameters of Neural Networks

The important hyperparameters of neural networks are listed below. Hints for choosing these hyperparameters are provided for each of them in the following sections. Note that there are also additional practical hints provided by Google on the hyperparameters of neural networks in [11]; interested readers can refer to this source for tips beyond those presented here.

4.4.1 Choice of Activation Functions

For discussions on activation functions, refer to Sect. 2.8 in Chap. 2. ReLU [12] remains the default activation function in most hidden layers, though variants such as GELU [13] and Swish [14] are now common in modern architectures like Transformers.

The activation function of the last layer of a neural network depends on the application:

- If the task is binary classification or multi-class classification, the sigmoid (logistic) function and the softmax function are used, respectively. One output neuron is used for binary classification, while for multi-class classification, the number of classes determines the number of output neurons.
- If the task is unbounded regression, the linear activation function, or no activation function, is used in the last layer.
- If the task is bounded regression, the sigmoid or hyperbolic tangent functions are used for bounds $[0, 1]$ and $[-1, 1]$, respectively.
- If the task is embedding learning, metric learning, or manifold learning (see Chap. 16), no activation function is usually used in the last layer.

4.4.2 Choice of Number of Layers or Depth of Neural Network

Increasing the number of layers, i.e., the depth of the network, increases its complexity and flexibility. This allows the neural network to learn more complex and nonlinear data. However, the size of the training data should also increase along with the depth of the network; otherwise, the network will become overfitted to the training data (see Chap. 5).

4.4.3 Choice of Optimizer

Refer to Chap. 3 for information about different optimizers in deep learning. Most frameworks, such as PyTorch [15] and TensorFlow [10], provide many optimizers. In practice, AdamW [16] has become the default for many tasks, while SGD with momentum remains important in computer vision. Other recent optimizers such as Lion [17], Sophia [18], Adafactor [19], RAdam [20], and AdaBelief [21] are also widely used.

AdamW improves generalization, particularly in large-scale deep learning tasks such as NLP transformers, and has become the default optimizer in frameworks like Hugging Face and PyTorch. Lion replaces variance tracking with a sign-based momentum update, making it more memory efficient by eliminating the need for second-moment accumulators. Sophia is a second-order optimizer that efficiently approximates the Hessian diagonal using curvature information collected during training, enabling faster convergence than AdamW in large-scale models. Adafactor is a memory-efficient variant of Adam designed for very large transformer models, while RAdam (Rectified Adam) stabilizes early training by correcting variance in adaptive steps. Finally, AdaBelief modifies Adam by estimating variance from the deviation between gradients and their exponential moving averages.

4.4.4 Choice of Learning Rate

The optimal learning rate is highly task-dependent, typically ranging between 1e-4 (i.e., 10^{-4}) and 1e-2 (i.e., 10^{-2}) for most architectures. Values like 1e-6 (i.e., 10^{-6}) are usually too small except in fine-tuning very large models. The optimal learning rate also depends on the dataset and the architecture of the neural network. It should not be too large, as this can cause oscillation in the gradient descent and loss values. Conversely, it should not be too small, as this will result in little or no progress in reducing the loss function.

It is also possible to use an adaptive learning rate for training. An adaptive learning rate starts with a larger learning rate to encourage exploration in the optimization landscape of the loss function, assuming the current solution is far from optimal. Then, the learning rate is gradually reduced during the training process to decrease exploration and increase exploitation. As the solution approaches the optimum, it does not need extensive exploration; rather, it should search locally around the best solution found so far in the hope of finding an even better solution.

The adaptive learning rate is often referred to as a learning rate schedule, and several scheduling strategies have been developed. In step decay [22], the learning rate is reduced by a fixed factor after a set number of epochs: the network initially explores with a higher learning rate and later stabilizes with a lower one. In exponential decay [23], the learning rate decreases smoothly and continuously rather than in discrete steps. Cosine annealing (SGDR) [24] schedules the learning rate according to a cosine curve, starting high and gradually decreasing toward zero, with a particularly gentle decline near the end. This soft landing often improves generalization and is widely used in modern architectures such as transformers and ResNets. The one-cycle policy [25] consists of two phases: first increasing the learning rate from a low to a high value to escape poor local minima, then decreasing it—typically with a cosine schedule. This strategy has proven effective in both computer vision and NLP. Finally, warm restarts [24] reset the learning rate to a high value after each cosine annealing cycle, mimicking a restart of training before gradually decaying again.

4.4.5 Choice of Mini-batch Size

The (mini-)batch size, also called the batch size, is usually a power of two, such as 32 or 64, by convention. It is generally chosen to be neither too small nor too large. In practice, smaller batches may sometimes generalize better, while large batches improve hardware throughput. Very large batch sizes increase memory demands and often require careful learning rate scaling and warmup to maintain stability.

4.5 Hardware and Memory Efficiency

Several techniques have been developed to train models efficiently on real hardware, including mixed precision training, gradient checkpointing, and distributed training.

Mixed precision training [26] uses half-precision (FP16 or BF16) for most computations while retaining critical operations, such as loss scaling and weight updates, in single precision (FP32). This approach reduces memory usage and accelerates computation on modern GPUs and TPUs with specialized hardware, enabling the training of larger models while maintaining numerical stability.

Gradient checkpointing (or activation checkpointing) [27] stores only a subset of activations during the forward pass, recomputing the others as needed during backpropagation. This significantly reduces memory consumption, allowing very deep networks or long-sequence models (e.g., Transformers) to be trained on limited GPU memory.

Distributed training distributes computation across multiple GPUs or nodes to increase memory and processing capacity, facilitating the training of trillion-parameter models and extremely large datasets. Key strategies include data parallelism [28], where each worker processes different batches and gradients are averaged; model parallelism [29], which partitions the model across devices; and the Zero Redundancy Optimizer (ZeRO) [30], which partitions optimizer states, gradients, and parameters to minimize redundancy.

Recent advances in compiler-driven optimization further improve hardware efficiency. PyTorch 2.x introduces "torch.compile," a high-level API that automatically converts PyTorch programs into optimized, compiled versions [31]. By performing graph capture and backend-specific optimizations, "torch.compile" accelerates execution and reduces memory overhead without requiring manual code changes. Similarly, TensorFlow leverages XLA (Accelerated Linear Algebra) [32] to compile computation graphs into hardware-optimized instructions. XLA performs operation fusion, memory layout optimization, and kernel specialization, enabling faster training and inference on CPUs, GPUs, and TPUs. Together, these tools exemplify how modern frameworks combine dynamic programmability with compiler-level optimization to maximize hardware utilization.

5 Conclusion

Proper data splitting into training, validation, and test sets is essential for reliable model evaluation. In real-world scenarios, a development (dev) set may also be used to assess model performance more realistically. Performance can be evaluated at multiple levels, including human-level, training, test, dev, and deployment metrics, with the gaps between these levels offering valuable insights, as discussed in this chapter.

This chapter also presented practical tips and rules of thumb for effective deep learning. Topics included handling equal-sized versus variable-sized data, designing data pipelines, selecting neural network architectures, and choosing hyperparameters such as activation functions, number of layers, optimizers, learning rates, and mini-batch sizes. Additionally, strategies for optimizing hardware and memory efficiency were introduced. These practical guidelines are particularly useful for applying deep learning in real-world and industrial contexts.

6 Exercises

1. (Level: Easy) Explain the role of the dev set in the practical deployment of deep learning models. How does it differ from the test set in terms of purpose and data distribution, and why is it considered a better estimate of generalization performance after deployment?
2. (Level: Easy) Which of the following correctly matches a performance gap with its most likely cause?

   (a) A large gap between human-level performance and training performance $\rightarrow$ Overfitting due to excessive model complexity.
   (b) A large gap between training performance and test performance $\rightarrow$ Model bias due to underfitting.
   (c) A large gap between test performance and dev performance $\rightarrow$ Distribution mismatch between training/test data and dev data.
   (d) A large gap between dev performance and deployment performance $\rightarrow$ Training data is too similar to test data.

3. (Level: Easy) Which of the following statements is correct according to the rules of thumb for training neural networks?

   (a) A convolutional neural network is recommended for processing sequence data such as text or speech.
   (b) ReLU is typically used in the last layer for binary classification problems.
   (c) AdamW optimizer is the default optimizer for many tasks.
   (d) Using a very large mini-batch size usually improves generalization.

References

1. Anand Avati. *Stanford CS229: Machine Learning—Summer 2019—Lecture 22 - Practical Tips and Course Recap*. https://www.youtube.com/watch?v=OftHZYlrK-E. 2021.
2. Nitish Srivastava, Geoffrey Hinton, Alex Krizhevsky, Ilya Sutskever, and Ruslan Salakhutdinov. "Dropout: a simple way to prevent neural networks from overfitting". In: *The journal of machine learning research* 15.1 (2014), pp. 1929–1958.
3. Benyamin Ghojogh, Fakhri Karray, and Mark Crowley. "Hidden Markov model: tutorial". In: *Engineering Archive* (2019).
4. Meinard Müller. "Dynamic time warping". In: *Information retrieval for music and motion* (2007), pp. 69–84.
5. Hoda Mohammadzade, Soheil Hosseini, Mohammad Reza Rezaei-Dastjerdehei, and Mohsen Tabejamaat. "Dynamic time warping-based features with class-specific joint importance maps for action recognition using Kinect depth sensor". In: *IEEE Sensors Journal* 21.7 (2021), pp. 9300–9313.
6. Marie-Paule Lefranc et al. "IMGT unique numbering for immunoglobulin and T cell receptor variable domains and Ig superfamily V-like domains". In: *Developmental & Comparative Immunology* 27.1 (2003), pp. 55–77.
7. Saul B Needleman and Christian D Wunsch. "A general method applicable to the search for similarities in the amino acid sequence of two proteins". In: *Journal of molecular biology* 48.3 (1970), pp. 443–453.
8. Luis Perez and Jason Wang. "The effectiveness of data augmentation in image classification using deep learning". In: *arXiv preprint* arXiv:1712.04621 (2017).
9. Connor Shorten and Taghi M Khoshgoftaar. "A survey on image data augmentation for deep learning". In: *Journal of big data* 6.1 (2019), pp. 1–48.

10. Martín Abadi et al. "Tensorflow: Large-scale machine learning on heterogeneous distributed systems". In: *arXiv preprint* arXiv:1603.04467 (2016).
11. Varun Godbole, George E. Dahl, Justin Gilmer, Christopher J. Shallue, and Zachary Nado. *Deep Learning Tuning Playbook*. https://github.com/google-research/tuning_playbook. 2023.
12. Vinod Nair and Geoffrey E Hinton. "Rectified linear units improve restricted boltzmann machines". In: *Proceedings of the 27th international conference on machine learning (ICML-10)*. 2010, pp. 807–814.
13. Dan Hendrycks and Kevin Gimpel. "Gaussian error linear units (GELUs)". In: *arXiv preprint* arXiv:1606.08415 (2016).
14. Prajit Ramachandran, Barret Zoph, and Quoc V Le. "Searching for activation functions". In: *arXiv preprint* arXiv:1710.05941 (2017).
15. Adam Paszke et al. "PyTorch: An imperative style, high-performance deep learning library". In: *Advances in neural information processing systems* 32 (2019).
16. Ilya Loshchilov, Frank Hutter, et al. "Fixing weight decay regularization in Adam". In: *International Conference on Learning Representations*. 2018.
17. Lizhang Chen, Bo Liu, Kaizhao Liang, and Qiang Liu. "Lion secretly solves constrained optimization: As lyapunov predicts". In: *International Conference on Learning Representations*. 2024.
18. Hong Liu, Zhiyuan Li, David Hall, Percy Liang, and Tengyu Ma. "Sophia: A scalable stochastic second-order optimizer for language model pre-training". In: *International Conference on Learning Representations*. 2024.
19. Noam Shazeer and Mitchell Stern. "Adafactor: Adaptive learning rates with sublinear memory cost". In: *International Conference on Machine Learning*. PMLR. 2018, pp. 4596–4604.
20. Liyuan Liu et al. "On the variance of the adaptive learning rate and beyond". In: *International Conference on Machine Learning*. 2020.
21. Juntang Zhuang et al. "Adabelief optimizer: Adapting stepsizes by the belief in observed gradients". In: *Advances in neural information processing systems* 33 (2020), pp. 18795–18806.
22. Alex Krizhevsky, Ilya Sutskever, and Geoffrey E Hinton. "Imagenet classification with deep convolutional neural networks". In: *Advances in neural information processing systems* 25 (2012).
23. Geoffrey E Hinton, Nitish Srivastava, Alex Krizhevsky, Ilya Sutskever, and Ruslan R Salakhutdinov. "Improving neural networks by preventing co-adaptation of feature detectors". In: *arXiv preprint* arXiv:1207.0580 (2012).
24. Ilya Loshchilov and Frank Hutter. "SGDR: Stochastic gradient descent with warm restarts". In: *International Conference on Machine Learning*. 2017.
25. Leslie N Smith. "A disciplined approach to neural network hyper-parameters: Part 1–learning rate, batch size, momentum, and weight decay". In: *arXiv preprint* arXiv:1803.09820 (2018).
26. Paulius Micikevicius et al. "Mixed precision training". In: *International Conference on Machine Learning*. 2018.
27. Tianqi Chen, Bing Xu, Chiyuan Zhang, and Carlos Guestrin. "Training deep nets with sublinear memory cost". In: *arXiv preprint* arXiv:1604.06174 (2016).
28. Jeffrey Dean et al. "Large scale distributed deep networks". In: *Advances in neural information processing systems* 25 (2012).
29. Mohammad Shoeybi et al. "Megatron-LM: Training multi-billion parameter language models using model parallelism". In: *arXiv preprint* arXiv:1909.08053 (2019).
30. Samyam Rajbhandari, Jeff Rasley, Olatunji Ruwase, and Yuxiong He. "Zero: Memory optimizations toward training trillion parameter models". In: *SC20: International Conference for High Performance Computing, Networking, Storage and Analysis*. IEEE. 2020, pp. 1–16.
31. Jason Ansel et al. "Pytorch 2: Faster machine learning through dynamic python bytecode transformation and graph compilation". In: *Proceedings of the 29th ACM International Conference on Architectural Support for Programming Languages and Operating Systems, Volume 2*. 2024, pp. 929–947.
32. Amit Sabne. *XLA: Compiling machine learning for peak performance*. https://research.google/pubs/xla-compiling-machine-learning-for-peak-performance/. 2020.

Coding Neural Networks for Deep Learning

22

Abstract

This chapter provides an overview of essential libraries for coding neural networks, as well as tools for managing experiment logging and visualization in deep learning. It introduces popular neural network coding frameworks and tools for tracking and visualizing training progress. Additionally, the chapter presents practical sample PyTorch code implementations for common neural network architectures and machine learning tasks, including feedforward networks for regression and Convolutional Neural Networks (CNNs) for classification. These examples serve as practical guides to complement the theoretical concepts discussed in earlier chapters.

1 Introduction

There are various libraries available for coding neural networks in deep learning. This chapter introduces neural network coding libraries, as well as logging and visualization libraries. It also provides sample PyTorch codes for different neural network architectures and machine learning tasks. It is recommended to study the previous chapters of the book to understand the theoretical foundations of the modules used in the codes presented in this chapter.

Section 2 introduces deep learning libraries, with neural network coding libraries covered in Sect. 2.1 and logging and visualization libraries discussed in Sect. 2.2. Sample PyTorch codes for neural networks are provided in Sect. 3. A sample PyTorch code for a feedforward neural network and a regression task is presented in Sect. 3.2, while Sect. 3.3 provides a sample PyTorch code for a Convolutional Neural Network (CNN) and a classification task.

2 Deep Learning Libraries

2.1 Neural Network Coding Libraries

Many coding libraries have been developed in different programming languages, especially Python, for deep learning. These libraries are introduced below. Note that among these libraries, PyTorch [28] and TensorFlow [1] are the most widely used, although JAX [10], FLAX [13], and PyTorch Lightning [22] are also gaining attention. TensorFlow was developed by Google, and PyTorch was developed

B. Ghojogh and A. Ghodsi, *Elements of Deep Learning*,
https://doi.org/10.1007/978-3-032-10738-1_22

by Facebook. Recently, PyTorch has become more widely used due to its simplicity and NumPy-style interface. Today, most deep learning code is developed using PyTorch.

2.1.1 Theano

One of the very first deep learning libraries was Theano, developed by Yoshua Bengio and his students at the University of Montreal from 2007 to 2010 [4]. After its initial proposal, it was further developed by a large team over the years [29] until its support ended in 2017. This library could define, optimize, and evaluate mathematical expressions involving multi-dimensional vectors and matrices. However, its complex syntax limited its adoption, and it lost ground to more user-friendly libraries.

2.1.2 Caffe

Between 2013 and 2014, the Caffe library was developed at the Berkeley Vision and Learning Center [17]. Its high speed and availability of pre-trained models made it widely used, especially in computer vision applications. However, it offered limited flexibility for non-convolutional neural networks and dynamic architectures. Consequently, researchers gradually lost interest in this library.

2.1.3 Lasagne

Lasagne, released in 2014 [20], was a lightweight library built on top of Theano for neural networks. It provided a more user-friendly wrapper over Theano and was modular and easy to use for defining neural networks. However, because it depended on Theano, it inherited its static graph limitation. It is now deprecated and no longer maintained following Theano's discontinuation.

2.1.4 TensorFlow

TensorFlow was developed by Google Brain in 2015 [1]. It was designed for both flexibility and large-scale production deployment. One of its early contributions, TensorBoard for visualization, was widely praised. TensorFlow 1 used a static computation graph, which made model debugging complex until TensorFlow 2 introduced dynamic graph features. TensorFlow 2 improved usability with eager execution and Keras integration. TensorFlow remains popular in both academia and industry.

2.1.5 Keras

Keras, first created in 2015 [7], was originally designed to make building neural networks easy, but early versions were limited and less flexible for advanced research. Today, Keras is fully integrated into TensorFlow (available as "tf.keras"). This integration allows both simple "plug-and-play" use and advanced customization. For example, researchers can now write their own training loops step by step, instead of relying only on Keras' automatic training functions.

2.1.6 MXNet

MXNet, released in 2015 [6], was developed by Apache and later heavily backed by Amazon. It is known for its scalability and speed, especially for distributed computing. MXNet is also the deep learning backend of choice for Amazon's AWS services. MXNet was once strongly supported by Amazon, but since 2022 its development has slowed. Today, it has a much smaller community and is rarely chosen over TensorFlow, PyTorch, or JAX.

2.1.7 CNTK (Microsoft Cognitive Toolkit)

CNTK was developed by Microsoft in 2016 [27]. Its main advantage was its high-performance training engine. It was used internally at Microsoft for speech and image tasks. However, its API was difficult to work with, and it lacked community support. Its development ceased in 2020 and was superseded by PyTorch in Microsoft projects.

2.1.8 Chainer

Chainer, a Japanese library released in 2015 [32], was the first major library to introduce define-by-run (dynamic graph) computation, which inspired PyTorch. It offered flexibility for research, making it popular among researchers. However, Chainer was slower than static frameworks in some use cases. Its development discontinued in favor of PyTorch, with its last major release in 2019.

2.1.9 Deeplearning4j (DL4J)

Deeplearning4j (DL4J), released in 2016 [8], was developed in the Java programming language with Python bindings. It was designed for JVM[1] -based enterprise use cases and integrated with Hadoop and Spark. However, it is not native to the Python community and has a smaller ecosystem compared to TensorFlow and PyTorch. It is still maintained, but its use is limited. Most deep learning work today happens in Python-based frameworks, so Deeplearning4j has a limited but focused application.

2.1.10 PaddlePaddle

PaddlePaddle, developed by Baidu in 2016 [26], was the first large-scale open-source deep learning platform from China. It focuses on production readiness and large-scale applications. Globally, it is less popular, and historically, its documentation has been less accessible to non-Chinese users. It remains actively developed, especially in China.

2.1.11 Torch and PyTorch

Torch, released in 2015, was a deep learning framework based on Lua that was initially popular in academic research. PyTorch [28], a Python adaptation of Torch by Facebook AI Research, was released in 2016. PyTorch quickly gained popularity due to its intuitive dynamic computation graph, ease of use, and Python-like and NumPy-like syntax. It enabled more interactive model experimentation and became highly favored in both academia and industry.

PyTorch has dynamic computation graphs, an intuitive API, and strong research adoption. Its earlier versions lacked production tools, an issue now addressed with TorchScript and TorchServe. It is the most widely used deep learning library today and is often preferred over TensorFlow.

2.1.12 Gluon

Gluon was built on MXNet in 2017 [11] and allowed an easy interface for building models. It provided a user-friendly imperative programming model similar to PyTorch. However, as MXNet itself declined, Gluon also lost support. By 2025, it is effectively no longer used in practice.

2.1.13 ONNX (Open Neural Network Exchange)

ONNX (Open Neural Network Exchange) was developed by Facebook and Microsoft in 2017 [9]. It provides a standard format that allows models to be transferred between frameworks (e.g., PyTorch to TensorFlow). ONNX simplifies deployment, especially to edge devices. It is supported by major deep learning frameworks and is widely used in industry for interoperability.

[1] Java Virtual Machine.

2.1.14 FastAI

Built on top of PyTorch, FastAI, released in 2018 [15], aimed to make deep learning more accessible with high-level abstractions and a user-friendly interface. As a high-level library, it is less flexible for low-level model customization. FastAI remains popular for educational purposes and rapid prototyping.

2.1.15 JAX

JAX, developed by Google Research in 2018 [10], combines a NumPy-style API with auto-grad and XLA for fast differentiable computing. JAX provides powerful tools for high-performance computing and differentiable programming, making it particularly useful for research in optimization and dynamic neural network architectures. It is popular in cutting-edge machine learning research and is used by Google DeepMind and OpenAI. Its popularity is rapidly growing, especially in research.

2.1.16 FLAX

FLAX, developed in 2020 [13], was built on JAX. It is a high-level neural network library offering flexibility for JAX users. It requires familiarity with JAX and functional programming. FLAX is used in many state-of-the-art research projects, such as vision Transformers (see Chap. 9) and diffusion models (see Chap. 15). It is actively developed and widely used in the research community.

2.1.17 PyTorch Lightning

PyTorch Lightning, released in 2021 [22], was built on PyTorch. It simplifies training code by managing routine components like loops and logging. It offers a clean structure for reproducibility and rapid prototyping. PyTorch Lightning became popular, but newer PyTorch features (such as "torch.compile" in PyTorch 2.0 [2]) and tools like Hugging Face Accelerate [12] now provide some of the same benefits. Lightning is still useful, but it is not as central as it was a few years ago.

2.2 Logging and Visualization Libraries

There are various tools for logging and visualizing plots, graphs, and metrics (e.g., loss values and accuracies) during neural network training.

Many tools have been developed; TensorBoard [1] was the standard for years, but today many teams prefer tools like Weights & Biases (wandb.ai) [33] or MLflow (mlflow.org) [34]. Another open-source option that is growing is Aim (aimstack.io) [3]. PyTorch Lightning Logs can use both TensorBoard and W&B for logging.

In addition to these widely used logging and visualization tools, there are a few others such as Comet.ml (2017), Sacred + Omniboard (2017), Neptune.ai (2018), and ClearML (2019). The following subsections introduce TensorBoard, W&B, PyTorch Lightning Logs, and MLflow.

2.2.1 TensorBoard

TensorBoard, released in 2015 [1], was developed for monitoring TensorFlow experiments. It is one of the earliest and most popular tools for visualizing deep learning metrics and graphs. TensorBoard tracks metrics such as loss and accuracy, visualizes model architectures, and displays histograms, distributions, and embeddings. It also supports hyperparameter tuning and profiling.

Although TensorBoard is often used by TensorFlow and Keras users, it is also usable with PyTorch through integrations, such as via the "torch.utils.tensorboard" module. While it is less native for PyTorch users, integration is still possible.

2.2.2 W&B (wandb)

Weights & Biases (W&B), also called wandb, was released in 2018 [33]. It is a powerful experiment tracking and visualization tool for deep learning. W&B provides real-time logging of metrics, hyperparameter tuning, and artifact management. It offers a comprehensive web-based dashboard for visualizing metrics, tracking model versions, and sharing experiment results.

It is particularly useful for cross-framework tracking and is valuable for large teams working with multiple experiments. Its limitation is that it is primarily a cloud-based solution with limited features on the free tier.

2.2.3 PyTorch Lightning Logs

PyTorch Lightning Logs, released in 2019, is an integration of TensorBoard and W&B. The logging module of PyTorch Lightning integrates seamlessly with TensorBoard, W&B, and other tools to log deep learning experiment metrics. It can log metrics, images, and models directly within the PyTorch Lightning framework with minimal configuration. It is best suited for PyTorch users who prefer using PyTorch Lightning to handle boilerplate code and model training. Its limitation is that it is dependent on PyTorch Lightning, so it is not a standalone tool.

2.2.4 MLflow

MLflow, introduced in 2018 by Databricks [34], is an open-source platform designed to manage the complete machine learning lifecycle. Unlike TensorBoard or W&B, which are primarily focused on logging and visualization, MLflow provides a more comprehensive framework covering four key components: experiment tracking, model packaging, model registry, and deployment.

Its tracking module allows logging and visualizing metrics, parameters, and artifacts across multiple runs, similar to TensorBoard and W&B. In addition, MLflow supports framework-agnostic model packaging, enabling trained models to be saved and deployed in standardized formats. The model registry facilitates version control, stage transitions (e.g., staging, production), and collaborative model management within teams. MLflow is particularly useful in industry settings where experiment tracking, reproducibility, and deployment pipelines are critical. Its main limitation is that while it provides broad functionality, its visualization and logging capabilities are less specialized than those of dedicated tools like TensorBoard or W&B.

3 Sample PyTorch Codes of Neural Networks

As explained in Sect. 2.1, PyTorch is the most widely used library for coding neural networks. In the following, a few sample PyTorch codes are provided for implementing feedforward and Convolutional Neural Networks (CNNs).

3.1 Package Imports and Code Initials

3.1.1 Package Imports

The required packages (libraries), for both Sects. 3.2 and 3.3, can be installed using pip:

```
pip install torch
pip install torchvision
pip install numpy
pip install matplotlib
pip install tqdm
pip install tensorboard
```

```
pip install wandb
pip install mlflow
```

The necessary packages to be imported are

```
# importing packages (libraries):
import torch
import torch.nn as nn
from torch.utils.data import DataLoader, Dataset
import torchvision
import torchvision.transforms as transforms
import matplotlib.pyplot as plt
import numpy as np
from tqdm import tqdm
import os
from torch.utils.tensorboard import SummaryWriter  # import only if using
    TensorBoard
import wandb  # import only if using W\&B (wandb)
import mlflow  # import only if using MLFlow
```

The imported packages are

- The "torch.nn" package is the neural network module within PyTorch.
- The "torch.utils.data" package includes classes and functions for loading data and preparing it into mini-batches.
- The "torchvision" package provides multiple tools for working with image data in PyTorch. For example, "torchvision.datasets" contains widely used datasets such as MNIST [21].
- The "matplotlib" package is a collection of tools for plotting graphs.
- The "numpy" package provides tools for working with arrays, vectors, and matrices.
- The "tqdm" package is used for displaying progress bars.
- The "os" (operating system) package is used for path-related functions such as checking if a file exists in the computer.
- The "torch.utils.tensorboard," "wandb," and "mlflow" are the libraries for TensorBoard [1], W&B (wandb) [33], and MLFlow [34] loggers.

3.1.2 Determining the Device

The device on which the code is running should be determined as either CPU or GPU (CUDA):

```
device = torch.device("cuda:0" if torch.cuda.is_available() else "cpu")
```

If a GPU is available on the machine, code execution becomes faster. In this case, operations will be performed on the GPU, but results should be brought back to the CPU for reporting.

3.2 Sample PyTorch Code of Feedforward Neural Network for Regression

A sample PyTorch code for a feedforward neural network (see Chap. 2) is provided below. This code is designed for a regression task (see Chap. 1).

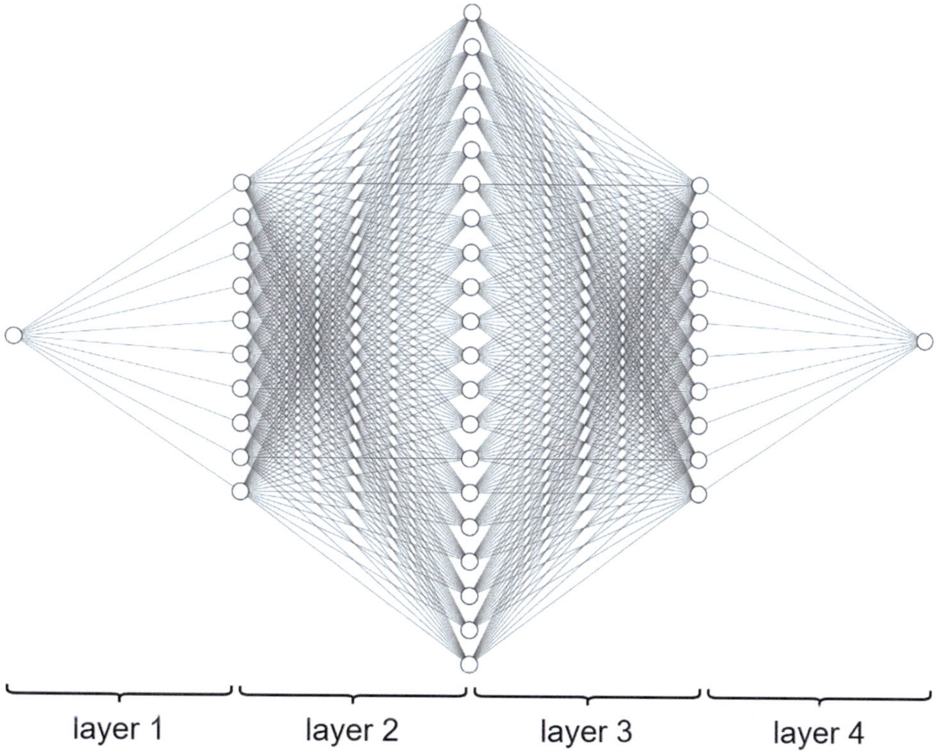

Fig. 1 The architecture of the feedforward neural network in the sample PyTorch code

3.2.1 Defining the Neural Network

To define the architecture of the neural network, a class is created that inherits from the "nn.Module" class. The layers (with their sizes) and activation functions are defined in the constructor of the class. The "nn.Linear(a, b)" class represents a linear or fully connected layer connecting a input neurons to b output neurons. The "nn.ReLU" class represents the ReLU activation function (see Sect. 2.8 in Chap. 2.

The following code creates a feedforward neural network with one input neuron connected to 10 neurons, then to 20 neurons, then to 10 neurons, and finally to one output neuron. The hidden layers use the ReLU activation function, while the last layer does not use any activation function. The architecture of this neural network is illustrated in Fig. 1.

The "forward()" function takes x as input, passes it through the first layer and its activation function, then through the second layer and its activation function, followed by the third layer and its activation function, and finally through the fourth layer.

```
# defining the architecture of neural network:
class NeuralNetwork(nn.Module):
    def __init__(self):
        super(NeuralNetwork, self).__init__()
        self.layer1 = nn.Linear(1, 10)
        self.layer2 = nn.Linear(10, 20)
        self.layer3 = nn.Linear(20, 10)
        self.layer4 = nn.Linear(10, 1)
        self.relu1 = nn.ReLU()
        self.relu2 = nn.ReLU()
        self.relu3 = nn.ReLU()
```

```
    def forward(self, x):
        x = self.relu1(self.layer1(x))
        x = self.relu2(self.layer2(x))
        x = self.relu3(self.layer3(x))
        x = self.layer4(x)
        return x
```

The class of the neural network is instantiated to have an object named "net." This object is brought to the device.

```
# instantiate the class of neural network:
net = NeuralNetwork().to(device)
print(net)
```

Printing this network displays the architecture of the neural network, where the bias refers to the biases of the layers, which capture the shift of signals (see Chap. 2):

```
NeuralNetwork(
  (layer1): Linear(in_features=1, out_features=10, bias=True)
  (layer2): Linear(in_features=10, out_features=20, bias=True)
  (layer3): Linear(in_features=20, out_features=10, bias=True)
  (layer4): Linear(in_features=10, out_features=1, bias=True)
  (relu1): ReLU()
  (relu2): ReLU()
  (relu3): ReLU()
)
```

3.2.2 Optimizer and Loss Function

As explained in Chaps. 3 and 21, the most widely used optimizers today are AdamW [24], which is the default in many tasks, and SGD (Stochastic Gradient Descent) with momentum, which is still popular for computer vision. Adam (Adaptive Moment Estimation) [18] is also an option for the optimizer. Newer optimizers such as Lion [5] and Sophia [23] are also being explored in research. The SGD, Adam, and AdamW optimizers are provided by PyTorch as "torch.optim.SGD," "torch.optim.Adam," and "torch.optim.AdamW," respectively.

The weights of the neural network, i.e., "net.parameters()," are passed to the optimizer. The "lr" parameter in the optimizer specifies the learning rate.

```
# define optimizer:
optimizer_type = 'Adam'
if optimizer_type == 'SGD':
    optimizer = torch.optim.SGD(net.parameters(), lr=0.02)
elif optimizer_type == 'Adam':
    optimizer = torch.optim.Adam(net.parameters(), lr=0.02)
elif optimizer_type == 'AdamW':
    optimizer = torch.optim.AdamW(net.parameters(), lr=0.02)
```

The loss functions are available in "torch.nn." For regression tasks, the loss function can be Mean Squared Error (MSE), as explained in Chap. 2.

```
# define the loss function:
loss_func = torch.nn.MSELoss()
```

3.2.3 Data Loader

A class should be defined that inherits from the "torch.utils.data.Dataset" class. This class is used to define the dataset for PyTorch. The constructor receives the data x and labels y as NumPy arrays and casts them to Torch tensors. If the dataset is small enough to fit in RAM, it is also possible to cache it. The "__len__()" function defines the length of the dataset, while the "__getitem__()" function defines how the ith data point and its corresponding label should be fetched for "item=i."

```
class Data(Dataset):
    def __init__(self, x, y, cache=False):
        self.data = torch.from_numpy(x.reshape(-1,1)).float()
        self.label = torch.from_numpy(y.reshape(-1,1)).float()
        if cache:
            self.data = self.data.clone().pin_memory()  # preload in RAM
            self.label = self.label.clone().pin_memory()

    def __len__(self):
        return len(self.data)

    def __getitem__(self, idx):
        return self.data[idx], self.label[idx]
```

The "batch_size" determines the (mini-)batch size. As mentioned in Chap. 21, this size is usually a power of two. The trade-offs and adjustments for batch size are discussed in Chap. 21. The function "load_dataset()" receives the data features x and labels y for both the training and test datasets and outputs data loaders for training and testing. To do this, it instantiates the "torch.utils.data.DataLoader" class twice to create two objects: "train_loader" and "test_loader."

The training data should be shuffled while the test data do not need to be shuffled. The number of workers is the number of subprocesses for loading data. The option "num_workers = 0" loads data in the main process—which is safe but slower—but "num_workers > 0" loads in parallel which is faster but needs tuning depending on CPU cores. The option "pin_memory=True" keeps batches in page-locked memory for faster GPU transfer. It is usually helpful when training on CUDA. Option "prefetch_factor" determines how many batches each worker preloads in advance. A larger value for this yields a smoother training but with more memory use. Finally, the option "persistent_workers" keeps workers alive across epochs; it is faster if dataset is large.

```
batch_size = 16
def load_dataset(x_train, y_train, x_test, y_test, cache=False,
    num_workers=0, prefetch_factor=2, pin_memory=False,
    persistent_workers=False):
    # data loader for training data:
    train_ds = Data(x_train, y_train, cache=cache)
    train_loader = DataLoader(train_ds, batch_size=batch_size, shuffle=True,
                              num_workers=num_workers, pin_memory=pin_memory,
                              prefetch_factor=prefetch_factor if num_workers
                                  > 0 else None,
                              persistent_workers=persistent_workers if
                                  num_workers > 0 else False)

    # data loader for test data:
    test_ds = Data(x_test, y_test, cache=cache)
    test_loader = DataLoader(test_ds, batch_size=batch_size, shuffle=False,
                             num_workers=num_workers, pin_memory=pin_memory,
                             prefetch_factor=prefetch_factor if num_workers
                                 > 0 else None,
                             persistent_workers=persistent_workers if
                                 num_workers > 0 else False)

    return train_loader, test_loader
```

3.2.4 Dataset

A nonlinear random dataset is created using a sinusoidal function. The samples are reshaped to be row-wise because PyTorch requires data instances to be stacked row-wise, where rows represent instances and columns represent features.

```
# dataset:
x_train = np.random.rand(100) * 10
y_train = np.sin(x_train) + np.random.rand(100)*0.8
x_test = np.random.rand(100) * 10
y_test = np.sin(x_test) + np.random.rand(100)*0.8

# reshape to have samples in rows:
x_train = x_train.reshape((-1, 1))
x_test = x_test.reshape((-1, 1))
```

The created dataset can be visualized. The visualization of the training and test data is shown in Fig. 2. In this plot, the x-axis represents the scalar data points, and the y-axis represents the labels for these data points.

```
# visualize data:
plt.scatter(x_train, y_train, c='r', marker='o', label='training')
plt.scatter(x_test, y_test, c='b', marker='s', label='test')
plt.xlabel('x')
plt.ylabel('y')
plt.legend()
plt.show()
```

The created dataset is passed to the "load_dataset()" function to obtain the training and test data loaders.

```
# load dataset:
train_loader, test_loader = load_dataset(x_train, y_train, x_test, y_test)
```

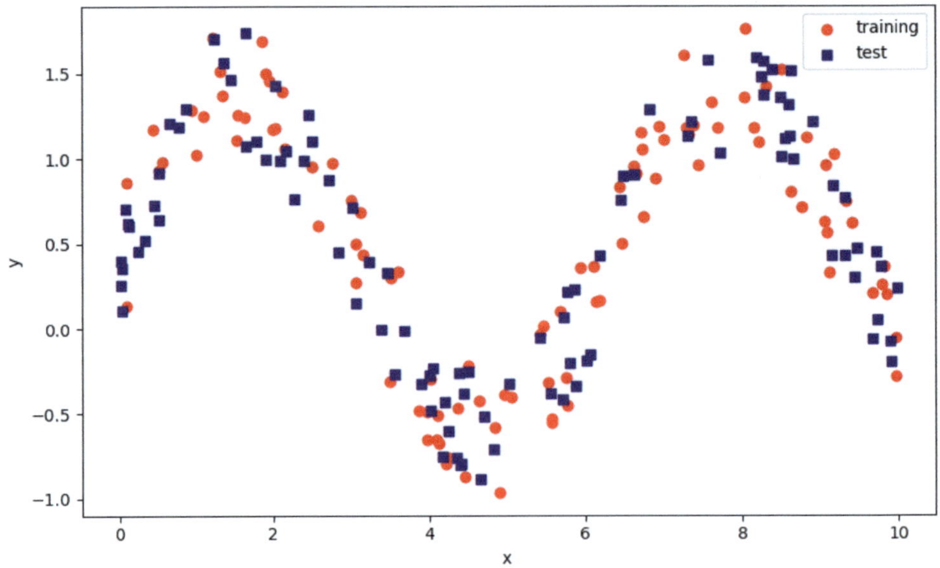

Fig. 2 The created toy dataset containing training and test sets

3.2.5 Learning Rate Scheduler

It is possible to optionally have a learning rate scheduler. Common learning rate schedules include step decay [19], cosine annealing (SGDR) [25], one-cycle policy [30], and warm restarts [25]. Cosine annealing and one-cycle are especially popular in large-scale training. Exponential decay [14] is less commonly used now.

```
# define the learning rate scheduler:
scheduler_type = None  # options: 'step', 'exponential', 'cosine', 'onecycle',
    'warm_restarts', None
if scheduler_type == 'step':
    scheduler = torch.optim.lr_scheduler.StepLR(optimizer, step_size=50, gamma=0.5)
elif scheduler_type == 'exponential':
    scheduler = torch.optim.lr_scheduler.ExponentialLR(optimizer, gamma=0.95)
elif scheduler_type == 'cosine':  # Cosine Annealing
    scheduler = torch.optim.lr_scheduler.CosineAnnealingLR(optimizer, T_max=100)
elif scheduler_type == 'onecycle':
    scheduler = torch.optim.lr_scheduler.OneCycleLR(
        optimizer,
        max_lr=0.1,
        steps_per_epoch=len(train_loader),
        epochs=1000
    )
elif scheduler_type == 'warm_restarts':  # Cosine Annealing with Restarts
    scheduler =
        torch.optim.lr_scheduler.CosineAnnealingWarmRestarts(optimizer, T_0=100, T_mult=2)
elif scheduler_type is None:  # no scheduler
    scheduler = None
```

3.2.6 Saving and Loading Checkpoints

Checkpointing makes training resumable and allows to later reload the best model. It is possible to save the best model—having the best validation performance—and later use it either for inference or for resuming training. The following function saves the checkpoint of the model at a given epoch. The saved checkpoint contains the information of the neural network model, optimizer, learning rate scheduler, epoch, and loss values.

```
def save_checkpoint(model, optimizer, scheduler, epoch, loss, path="checkpoint.pth"):
    checkpoint = {
        "epoch": epoch,
        "model_state": model.state_dict(),
        "optimizer_state": optimizer.state_dict(),
        "scheduler_state": scheduler.state_dict() if scheduler is not None else None,
        "loss": loss,
    }
    torch.save(checkpoint, path)
    print(f"Checkpoint saved at epoch {epoch} to {path}")
```

The following function loads the checkpoint at a given path for the sake of inference or resuming training.

```
def load_checkpoint(model, optimizer=None, scheduler=None, path="checkpoint.pth", device="cpu"):
```

```
    checkpoint = torch.load(path, map_location=device, weights_only=False)
    model.load_state_dict(checkpoint["model_state"])
    if optimizer is not None:
        optimizer.load_state_dict(checkpoint["optimizer_state"])
    if scheduler is not None and checkpoint["scheduler_state"] is not None:
        scheduler.load_state_dict(checkpoint["scheduler_state"])
    start_epoch = checkpoint["epoch"] + 1
    loss = checkpoint["loss"]
    print(f"Checkpoint loaded: epoch={checkpoint['epoch']},
        loss={loss:.4f}")
    return model, optimizer, scheduler, start_epoch
```

It is possible to resume training from a previously saved checkpoint. For example, the following code resumes training from the checkpoint of the hundredth epoch if it exists.

```
resume = True
if resume and os.path.exists("checkpoint_epoch100.pth"):
    net, optimizer, scheduler, start_epoch = load_checkpoint(
        net, optimizer, scheduler, path="checkpoint_epoch100.pth",
            device=device
    )
else:
    start_epoch = 0
```

3.2.7 Logging and Monitoring

There are various logging and monitoring libraries as discussed before. For example, we can use TensorBoard [1], W&B (wandb) [33], or MLFlow [34]. The defined logger can be used during training for monitoring the loss values.

```
# logger:
logger_type = 'tensorboard'   # options: 'tensorboard', 'wandb', 'mlflow',
    None
logger = None
if logger_type == 'tensorboard':
    logger = SummaryWriter(log_dir="runs/experiment1")
elif logger_type == 'wandb':
    wandb.init(project="my_project", name="run1")
    logger = wandb
elif logger_type == 'mlflow':
    mlflow.start_run(run_name="run1")
```

While or after training, we should do the following instructions to view the loss plots:

- TensorBoard:
  Run the following in the terminal:

```
tensorboard --logdir=runs
```

  Then, load the page `http://localhost:6006/` in a web browser.
- W&B (wandb):
  For the first time—before runnning the code—run the following in the terminal:

```
wandb login
```

It will ask for an API key from https://wandb.ai/authorize. Go to this URL link and log in (or create account) and it will give you a code to be entered in the terminal.
Then, run the code and load the page https://wandb.ai/ in a web browser.

- MLflow:
  Run the following in the terminal:

```
mlflow ui --port 5000
```

Then, load the page `http://localhost:5000/` in a web browser where 5000 is the port number.

3.2.8 Training the Neural Network

The neural network is trained in epochs, where each epoch consists of using all the training data once (see Chap. 3). In every epoch, mini-batches of data are obtained using the training data loader. In other words, each epoch is divided into non-overlapping mini-batches of data.

Each mini-batch is passed to the network object to obtain predictions. The predictions and the target labels of the mini-batch are then passed to the loss function to calculate the loss. This loss value is appended to the list "loss_list_in_epoch," where the loss is brought to the CPU, detached from the device, and its value is accessed using "item()."

The optimizer is reset using "zero_grad()" to clear the gradients calculated in the previous mini-batch. The two commands "loss.backward()" and "optimizer.step()" together perform one pass of backpropagation, i.e., one step of gradient descent (see Chap. 3). The "loss_list" collects the mean (average) values of the loss function across mini-batches. These average loss values can be plotted against the number of epochs to observe the reduction pattern of the loss function over time.

If a learning rate scheduler is used, the command "scheduler.step()" performs one step of the scheduler. The one-cycle policy scheduler [30] needs step per (mini-)batch but the other schedulers need step per epoch. The loggers also record the loss values, either per step or per epoch, to report them in the loss plots.

The following code saves the checkpoint every 100 epochs. Alternatively, it is possible to have a more sophisticated approach where the validation loss values are tracked and if there is an improvement in the best-so-far validation performance, a checkpoint is saved. This is almost equivalent to early stopping (see Chap. 5) where training is stopped when the validation loss goes up.

```
n_epochs = 1000
loss_list = []
for epoch in tqdm(range(start_epoch, n_epochs), desc='epochs'):
    loss_list_in_epoch = []
    for step, (data_batch, label_batch) in enumerate(train_loader):
        data_batch, label_batch = data_batch.to(device),
            label_batch.to(device)
        prediction = net(data_batch)
        loss = loss_func(prediction, label_batch)
        loss_list_in_epoch.append(loss.cpu().detach().item())
        optimizer.zero_grad()
        loss.backward()
        optimizer.step()

        # log per step (optional)
        if logger_type == 'tensorboard':
            logger.add_scalar("batch_loss", loss.item(), epoch *
                len(train_loader) + step)
        elif logger_type == 'wandb':
```

```
                wandb.log({"batch_loss": loss.item()})
            elif logger_type == 'mlflow':
                mlflow.log_metric("batch_loss", loss.item(), step=epoch *
                    len(train_loader) + step)

            # one-cycle needs step per batch
            if (scheduler is not None) and (scheduler_type == 'onecycle'):
                scheduler.step()

        # other schedulers step per epoch
        if (scheduler is not None) and (scheduler_type != 'onecycle'):
            scheduler.step()

        epoch_loss = np.mean(loss_list_in_epoch)
        loss_list.append(epoch_loss)

        # save checkpoint every 100 epochs
        if (epoch + 1) % 100 == 0:
            save_checkpoint(net, optimizer, scheduler, epoch, epoch_loss,
                path=f"checkpoint_epoch{epoch+1}.pth")

        # log per epoch
        if logger_type == 'tensorboard':
            logger.add_scalar("epoch_loss", epoch_loss, epoch)
        elif logger_type == 'wandb':
            wandb.log({"epoch_loss": epoch_loss, "epoch": epoch})
        elif logger_type == 'mlflow':
            mlflow.log_metric("epoch_loss", epoch_loss, step=epoch)
```

3.2.9 Test (Inference) Phase

After the network is trained, it can be used for inference. In the test (inference) phase, mini-batches of test data features and labels are loaded using the test loader. Each mini-batch is brought to the device and passed to the trained neural network object. The predictions are appended to a list to store them.

Note that the test phase should be enclosed within the "torch.no_grad()" block because gradients should not be backpropagated during this phase.

```
prediction_list = []
with torch.no_grad():
    for step, (data_batch, label_batch) in enumerate(test_loader):
        prediction = net(data_batch.to(device))
        prediction_list.extend(prediction)
```

Figure 3 illustrates the predictions of the labels for the test data along with the true (target) labels of the test data. The following code plots this graph. In this plot, the x-axis represents the scalar data points, and the y-axis represents the labels for these data points. As depicted in the figure, the predictions are roughly accurate, following the sinusoidal pattern of the labels.

```
# visualize the predicted and actual data:
plt.scatter(x_test, y_test, c='b', marker='s', label='test')
plt.scatter(x_test, torch.stack(prediction_list).cpu().numpy(), c='g',
    marker='D', label='prediction')
plt.xlabel('x')
plt.ylabel('y')
plt.legend()
plt.show()
```

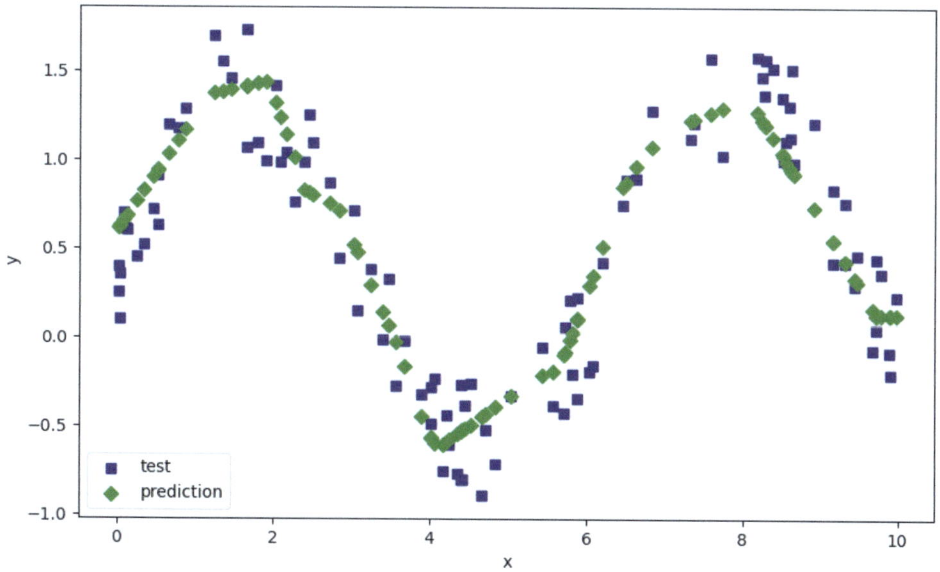

Fig. 3 Predictions of the labels for test points by the feedforward neural network. This plot shows both the target labels (denoted by "test") and the predicted labels (denoted by "prediction")

3.3 Sample PyTorch Code of Convolutional Neural Network for Classification

A sample PyTorch code for a Convolutional Neural Network (CNN) (see Chap. 4) is provided below. This code is designed for a classification task (see Chap. 1).

3.3.1 Defining the Neural Network

To define the architecture of the Convolutional Neural Network, a class is created that inherits from the "nn.Module" class. The layers (with their sizes) and activation functions are defined in the constructor of the class. The "nn.Conv2d(a, b, kernel_size=c, padding=d)" class is a convolutional layer with a input channels, b output channels, kernel size (width) c, and d paddings (see Chap. 4). The "nn.ReLU" class is the ReLU activation function (see Sect. 2.8 in Chap. 2). The "nn.MaxPool2d(a, b)" class represents max pooling with width a and height b (see Chap. 4). The "nn.Linear(a, b)" class is a linear or fully connected layer connecting a input neurons to b output neurons.

The following code creates a Convolutional Neural Network with the following layer configurations. The "forward()" function takes x as input and passes it through these layers:

- **The first layer (convolutional layer)**:
  - The input (MNIST) data has a size of [batch_size, 1, 28, 28].
  - The convolution layer "conv1" has 16 output channels, a 5×5 kernel, and "padding=2" (note that without "padding=2," the output of the convolution will have smaller width and height than its input).
  - The output of this convolution, before pooling, has a size of [batch_size, 16, 28, 28].
  - The output of the convolution passes through the ReLU activation function "relu1" (see Sect. 2.8 in Chap. 2) and is then fed to the max pooling layer "pool1."
  - The max pooling layer "pool1" uses a 2×2 kernel with "stride=2."
  - The output of this layer, after pooling, has a size of [batch_size, 16, 14, 14].

- **The second layer (convolutional layer)**:
  - The input to this layer has a size of [batch_size, 16, 14, 14], which is the output size of the first convolutional layer.
  - The convolution layer "conv2" has 32 output channels, a 5×5 kernel, and "padding=2" (note that without "padding=2," the output of the convolution will have smaller width and height than its input).
  - The output of this convolution, before pooling, has a size of [batch_size, 32, 14, 14].
  - The output of the convolution passes through the ReLU activation function "relu2" and is then fed to the max pooling layer "pool2."
  - The max pooling layer "pool2" uses a 2×2 kernel with "stride=2."
  - The output of this layer, after pooling, has a size of [batch_size, 32, 7, 7].
- **The third layer (fully connected)**:
  - The command "x = x.view(x.size(0), −1)" flattens the 3D output of the second layer, with size [batch_size, 32, 7, 7], to [batch_size, 1568], where $1568 = 32 \times 7 \times 7$.
  - The flattened output is then passed as input to a fully connected layer connecting 1568 neurons to 100 neurons. The input and output of this layer have sizes [batch_size, 1568] and [batch_size, 100], respectively.
  - A ReLU activation function is applied to the output of the third layer.
- **The fourth layer (fully connected)**:
  - The output of the third layer, with size [batch_size, 100], is passed to the fourth layer, whose output has a size of [batch_size, 10], corresponding to the ten classes of digits in the MNIST dataset.
  - No activation function is used for this last layer because the "torch.nn.CrossEntropyLoss" loss function used in this network implicitly includes a softmax activation function.

The architecture of this neural network is illustrated in Fig. 4.

```
# defining the architecture of neural network:
class NeuralNetwork(nn.Module):
    def __init__(self):
        super(NeuralNetwork, self).__init__()
        self.conv1 = nn.Conv2d(1, 16, kernel_size=5, padding=2)
        self.relu1 = nn.ReLU()
        self.pool1 = nn.MaxPool2d(2, 2)

        self.conv2 = nn.Conv2d(16, 32, kernel_size=5, padding=2)
        self.relu2 = nn.ReLU()
        self.pool2 = nn.MaxPool2d(2, 2)

        self.fc1 = nn.Linear(32 * 7 * 7, 100)
        self.relu3 = nn.ReLU()
        self.fc2 = nn.Linear(100, 10)

    def forward(self, x):
        x = self.pool1(self.relu1(self.conv1(x)))
        x = self.pool2(self.relu2(self.conv2(x)))
        x = x.view(x.size(0), -1)
        x = self.relu3(self.fc1(x))
        x = self.fc2(x)
        return x
```

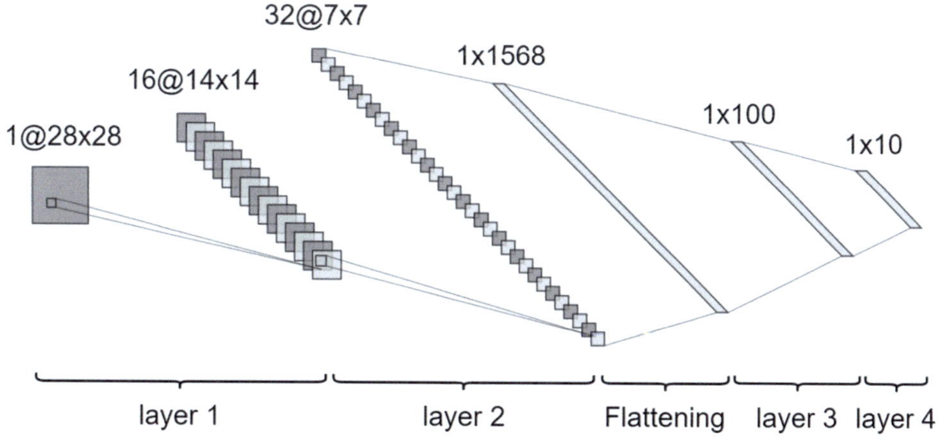

Fig. 4 The architecture of the convolutional neural network in the sample PyTorch code

The class of the neural network is instantiated to create an object named "net." This object is then moved to the device.

```
# instantiate the class of neural network:
net = NeuralNetwork().to(device)
print(net)
```

Printing this network displays the architecture of the neural network:

```
NeuralNetwork(
  (conv1): Conv2d(1, 16, kernel_size=(5, 5), stride=(1, 1), padding=(2, 2))
  (relu1): ReLU()
  (pool1): MaxPool2d(kernel_size=2, stride=2, padding=0, dilation=1,
      ceil_mode=False)
  (conv2): Conv2d(16, 32, kernel_size=(5, 5), stride=(1, 1), padding=(2, 2))
  (relu2): ReLU()
  (pool2): MaxPool2d(kernel_size=2, stride=2, padding=0, dilation=1,
      ceil_mode=False)
  (fc1): Linear(in_features=1568, out_features=100, bias=True)
  (relu3): ReLU()
  (fc2): Linear(in_features=100, out_features=10, bias=True)
)
```

3.3.2 Optimizer and Loss Function

The explanation of the optimizer is the same as that for the fully connected network.

```
# define optimizer:
optimizer_type = 'Adam'
if optimizer_type == 'SGD':
    optimizer = torch.optim.SGD(net.parameters(), lr=0.02)
elif optimizer_type == 'Adam':
    optimizer = torch.optim.Adam(net.parameters(), lr=0.001)
elif optimizer_type == 'AdamW':
    optimizer = torch.optim.AdamW(net.parameters(), lr=0.001)
```

The loss functions are available in "torch.nn." For classification tasks, the loss function can be cross-entropy, as explained in Chap. 2. Note that the "torch.nn.CrossEntropyLoss" loss function implicitly includes a softmax activation function.

```
# define the loss function:
loss_func = torch.nn.CrossEntropyLoss()
```

3.3.3 Data Loader

A function is defined to download the MNIST dataset (an image dataset of handwritten digits) [21] from the "torchvision.datasets.MNIST" class and save it in the path "./data/." The saved files in this folder are binary files of the dataset. The images are passed to "torchvision.transforms" to first convert them to Torch tensors and then normalize them. The "torchvision.transforms.Compose()" class composes multiple transformations sequentially.

The following function shuffles the training dataset to avoid any bias. However, shuffling is not required for the test data. The training and test data are passed to "torch.utils.data.DataLoader" to create the train and test data loaders, each with a (mini-)batch size of 64.

```
batch_size = 64
def load_dataset():
    transform = transforms.Compose([
        transforms.ToTensor(),
        transforms.Normalize((0.5,), (0.5,))
    ])

    # data loader for training data:
    train_set = torchvision.datasets.MNIST(root='./data', train=True,
        download=True, transform=transform)
    train_loader = DataLoader(train_set, batch_size=batch_size,
        shuffle=True)

    # data loader for test data:
    test_set = torchvision.datasets.MNIST(root='./data', train=False,
        download=True, transform=transform)
    test_loader = DataLoader(test_set, batch_size=batch_size, shuffle=False)

    return train_loader, test_loader
```

3.3.4 Dataset

The defined function is called to obtain the train and test loaders.

```
# load dataset:
train_loader, test_loader = load_dataset()
```

A few examples of the training images from the MNIST dataset are visualized, as shown in Fig. 5.

```
# visualize some training data:
examples = next(iter(train_loader))
images, labels = examples
for i in range(6):
    plt.subplot(2, 3, i+1)
    plt.imshow(images[i][0], cmap='gray')
    plt.title(f"Label: {labels[i].item()}")
    plt.axis('off')
plt.tight_layout()
plt.show()
```

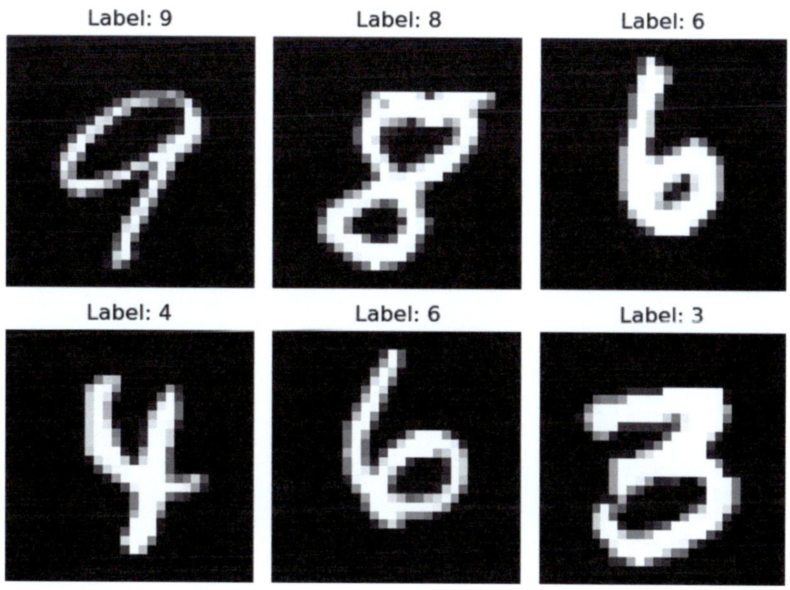

Fig. 5 A few examples of training images from the MNIST dataset

3.3.5 Training the Neural Network

The explanations for training this neural network are the same as those for training the feedforward neural network explained earlier. For this example, the network is trained for five epochs on MNIST, which is a simple dataset. In practical tasks, models are usually trained for longer and often use techniques like dropout [31] or batch normalization [16] for better generalization (see Chap. 5).

```
n_epochs = 5
loss_list = []
for epoch in tqdm(range(n_epochs), desc='epochs'):
    loss_list_in_epoch = []
    for step, (data_batch, label_batch) in enumerate(train_loader):
        data_batch, label_batch = data_batch.to(device),
            label_batch.to(device)
        prediction = net(data_batch)
        loss = loss_func(prediction, label_batch)
        loss_list_in_epoch.append(loss.cpu().detach().item())
        optimizer.zero_grad()
        loss.backward()
        optimizer.step()
    loss_list.append(np.mean(loss_list_in_epoch))
```

3.3.6 Test (Inference) Phase

In the test (inference) phase, mini-batches of test images and labels are loaded using the test loader. Each mini-batch is brought to the device and passed to the trained neural network object. The predictions are the probability distributions over the ten classes for each image. The class with the maximum probability in each prediction determines the predicted class for that image. The correctly predicted labels are recorded to calculate the accuracy.

This code achieved an accuracy of 98.89% for label prediction on the test images. Note that the test phase should be enclosed within the "torch.no_grad()" block because gradients should not be backpropagated during this phase.

```
correct = 0
total = 0
predictions_all = []
images_all = []
with torch.no_grad():
    for step, (data_batch, label_batch) in enumerate(test_loader):
        data_batch, label_batch = data_batch.to(device),
            label_batch.to(device)
        outputs = net(data_batch)
        _, predicted = torch.max(outputs.data, 1)
        total += label_batch.size(0)
        correct += (predicted == label_batch).sum().item()

        # save some images and predictions for visualization
        if step == 0:
            predictions_all.extend(predicted[:6].cpu().numpy())
            images_all.extend(data_batch[:6].cpu())

# print test accuracy:
print(f"Accuracy of the network on the 10000 test images: {100 * correct /
    total:.2f} %")
```

It is possible to visualize a few examples of the test images along with their predicted labels by the trained network. As shown in Fig. 6, the network has correctly predicted the labels for these examples.

```
# visualize predictions:
for i in range(6):
    plt.subplot(2, 3, i+1)
    plt.imshow(images_all[i][0], cmap='gray')
    plt.title(f"Predicted: {predictions_all[i]}")
    plt.axis('off')
plt.tight_layout()
plt.show()
```

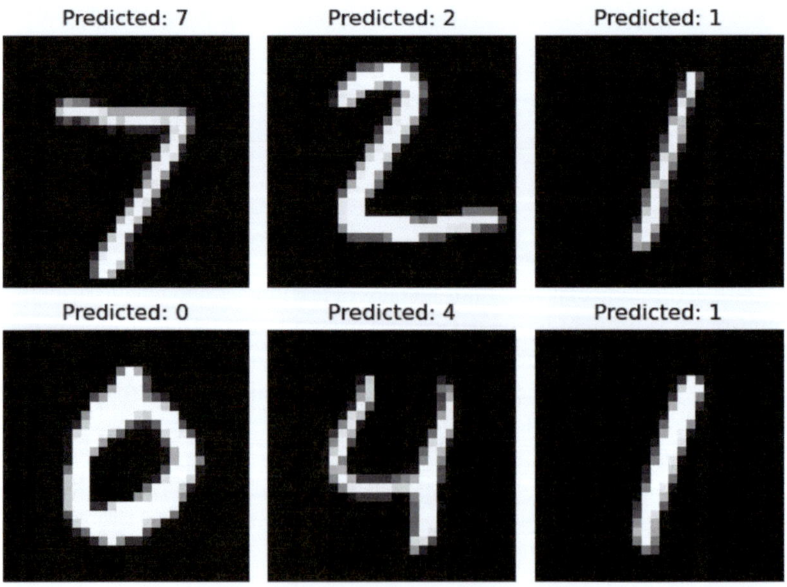

Fig. 6 A few examples of test images and their predicted labels by the Convolutional Neural Network

4 Conclusion

4.1 Conclusion of Chapter

A variety of libraries have been developed for implementing deep learning models, with TensorFlow and PyTorch being two of the most widely used. More recent frameworks, such as JAX and PyTorch Lightning, have also gained popularity for their flexibility and efficiency. In addition, several logging and visualization tools, including TensorBoard, Weights & Biases (W&B), and MLflow, support monitoring and debugging of deep learning experiments. This chapter provided sample PyTorch code for both feedforward and convolutional neural networks, offering readers practical examples to gain hands-on experience with neural network implementation.

4.2 Conclusion of the Book Part "Deep Learning in Practice"

Practical deep learning requires several key considerations. Data should be properly split into training, validation, and test sets, with an optional development (dev) set for more realistic evaluation in industrial settings. Model performance can be assessed at multiple levels, including human-level, training, test, dev, and deployment metrics, with the gaps between these levels providing valuable insights.

Practical guidelines include handling equal-sized versus variable-sized data, designing efficient data pipelines, selecting appropriate neural network architectures, and choosing hyperparameters such as activation functions, number of layers, optimizers, learning rates, and mini-batch sizes. Techniques for optimizing hardware and memory efficiency are also crucial for real-world deployment.

A variety of software libraries support practical deep learning implementation. TensorFlow and PyTorch are the most widely used frameworks, while logging and visualization tools such as TensorBoard, Weights & Biases (W&B), and MLflow facilitate monitoring of training progress and model performance. These tools bridge the gap between theoretical understanding and practical implementation, enabling efficient development and experimentation with deep neural networks.

By mastering these practical considerations and tools, practitioners can effectively apply deep learning in real-world and industrial contexts, driving innovation and advancing artificial intelligence applications.

4.3 Conclusion of the Entire Book

This book has provided a comprehensive overview of deep learning, from foundational concepts to practical applications and emerging research directions. Early neural network models, including the Perceptron, ADALINE, and CNNs, laid the groundwork for modern architectures, while techniques such as backpropagation, regularization, and optimization enable effective training and generalization. Sequence models, particularly Transformers, state space models, and large language models, have revolutionized natural language processing, with alignment methods like RLHF and DPO ensuring outputs meet human standards. Emerging topics, including self-supervised learning, graph neural networks, few-shot learning, meta-learning, and neural network compression, continue to expand the capabilities and applications of deep learning. Practical considerations–such as data management, model selection, hyperparameter tuning, hardware optimization, and the use of libraries and visualization tools–are essential for deploying deep learning effectively in real-world and industrial contexts. By integrating theory, emerging methods, and practical guidance, this book equips readers with the knowledge and tools to advance both research and applications in deep learning, fostering innovation and driving the continued evolution of artificial intelligence.

5 Exercises

1. (Level: Easy) Which of the following statements about deep learning libraries is false?
   (a) Theano was one of the first deep learning libraries but lost popularity due to its complex syntax and was discontinued in 2017.
   (b) PyTorch introduced dynamic computation graphs inspired by Chainer and is now more widely used than TensorFlow.
   (c) Keras was initially built on top of MXNet and is known for its high flexibility in complex research applications.
   (d) ONNX provides a standard format to allow interoperability between different deep learning frameworks.
2. (Level: Easy) In PyTorch, which of the following commands is used to disable gradient calculation for all tensors inside a block, often used during model evaluation?
   (a) `torch.no_grad()`
   (b) `torch.disable_grad()`
   (c) `torch.no_grad_mode()`
   (d) `torch.set_grad(False)`
3. (Level: Easy) In PyTorch, which of the following classes is typically used to define a feedforward neural network layer that applies a linear transformation to the incoming data?
   (a) `nn.Conv2d`
   (b) `nn.Linear`
   (c) `nn.ReLU`
   (d) `nn.BatchNorm1d`
4. (Level: Easy) Which of the following PyTorch loss functions is most appropriate for a regression task, and which one is commonly used for a multi-class classification task?
   (a) Regression: `nn.CrossEntropyLoss`, Classification: `nn.MSELoss`
   (b) Regression: `nn.MSELoss`, Classification: `nn.CrossEntropyLoss`
   (c) Regression: `nn.BCELoss`, Classification: `nn.L1Loss`
   (d) Regression: `nn.NLLLoss`, Classification: `nn.BCELoss`

References

1. Martín Abadi et al. "Tensorflow: Large-scale machine learning on heterogeneous distributed systems". In: *arXiv preprint* arXiv:1603.04467 (2016).
2. Jason Ansel et al. "Pytorch 2: Faster machine learning through dynamic python bytecode transformation and graph compilation". In: *Proceedings of the 29th ACM International Conference on Architectural Support for Programming Languages and Operating Systems, Volume 2*. 2024, pp. 929–947.
3. Gor Arakelyan, Gev Soghomonyan, et al. *Aim: An open-source, self-hosted AI metadata and experiment tracking tool*. https://aimstack.io. 2022.
4. James Bergstra et al. "Theano: A CPU and GPU Math Compiler in Python". In: *Proceedings of the 9th Python in Science Conference*. 2010, pp. 18–24. DOI: https://doi.org/10.25080/Majora-92bf1922-003.
5. Lizhang Chen, Bo Liu, Kaizhao Liang, and Qiang Liu. "Lion secretly solves constrained optimization: As lyapunov predicts". In: *International Conference on Learning Representations*. 2024.
6. Tianqi Chen et al. "MXNet: A flexible and efficient machine learning library for heterogeneous distributed systems". In: *arXiv preprint* arXiv:1512.01274 (2015).

7. François Chollet et al. *Keras*. https://keras.io. 2015.
8. Deeplearning4j authors. *Deeplearning4j*. https://github.com/deeplearning4j/deeplearning4j. 2016.
9. Open Neural Network Exchange. *Open Neural Network Exchange*. https://onnx.ai/ 2017.
10. Roy Frostig, Matthew James Johnson, and Chris Leary. "Compiling machine learning programs via high-level tracing". In: *Systems for Machine Learning* 4.9 (2018).
11. Gluon authors. *Gluon API*. https://github.com/gluon-api/gluon-api/. 2017.
12. Sylvain Gugger et al. *Accelerate: Training and inference at scale made simple, efficient and adaptable*. https://github.com/huggingface/accelerate. 2022.
13. Jonathan Heek et al. *Flax: A neural network library and ecosystem for JAX*. Version 0.10.6. 2024. URL: http://github.com/google/flax.
14. Geoffrey E Hinton, Nitish Srivastava, Alex Krizhevsky, Ilya Sutskever, and Ruslan R Salakhutdinov. "Improving neural networks by preventing co-adaptation of feature detectors". In: *arXiv preprint* arXiv:1207.0580 (2012).
15. Jeremy Howard and Sylvain Gugger. "Fastai: a layered API for deep learning". In: *Information* 11.2 (2020), p. 108.
16. Sergey Ioffe and Christian Szegedy. "Batch normalization: Accelerating deep network training by reducing internal covariate shift". In: *International conference on machine learning*. pmlr. 2015, pp. 448–456.
17. Yangqing Jia et al. "Caffe: Convolutional architecture for fast feature embedding". In: *Proceedings of the 22nd ACM international conference on Multimedia*. 2014, pp. 675–678.
18. Diederik P Kingma and Jimmy Ba. "Adam: A method for stochastic optimization". In: *arXiv preprint* arXiv:1412.6980 (2014).
19. Alex Krizhevsky, Ilya Sutskever, and Geoffrey E Hinton. "Imagenet classification with deep convolutional neural networks". In: *Advances in neural information processing systems* 25 (2012).
20. Lasagne authors. *Lasagne*. https://github.com/Lasagne/Lasagne. 2014.
21. Yann LeCun, Léon Bottou, Yoshua Bengio, and Patrick Haffner. "Gradient-based learning applied to document recognition". In: *Proceedings of the IEEE* 86.11 (2002), pp. 2278–2324.
22. Lightning-AI. *Pytorch Lightning*. https://github.com/Lightning-AI/pytorch-lightning. 2021.
23. Hong Liu, Zhiyuan Li, David Hall, Percy Liang, and Tengyu Ma. "Sophia: A scalable stochastic second-order optimizer for language model pre-training". In: *International Conference on Learning Representations*. 2024.
24. Ilya Loshchilov, Frank Hutter, et al. "Fixing weight decay regularization in Adam". In: *International Conference on Learning Representations*. 2018.
25. Ilya Loshchilov and Frank Hutter. "SGDR: Stochastic gradient descent with warm restarts". In: *International Conference on Machine Learning*. 2017.
26. Yanjun Ma, Dianhai Yu, Tian Wu, and Haifeng Wang. "PaddlePaddle: An open-source deep learning platform from industrial practice". In: *Frontiers of Data and Domputing* 1.1 (2019), pp. 105–115.
27. Microsoft. *Microsoft Cognitive Toolkit (CNTK)*. https://github.com/microsoft/CNTK. 2016.
28. Adam Paszke et al. "PyTorch: An imperative style, high-performance deep learning library". In: *Advances in neural information processing systems* 32 (2019).
29. Rami Al-Rfou, Guillaume Alain, Amjad Almahairi, et al. "Theano: A Python framework for fast computation of mathematical expressions". In: *arXiv preprint* arXiv:1605.02688 (2016).
30. Leslie N Smith. "A disciplined approach to neural network hyper-parameters: Part 1–learning rate, batch size, momentum, and weight decay". In: *arXiv preprint* arXiv:1803.09820 (2018).
31. Nitish Srivastava, Geoffrey Hinton, Alex Krizhevsky, Ilya Sutskever, and Ruslan Salakhutdinov. "Dropout: a simple way to prevent neural networks from overfitting". In: *The journal of machine learning research* 15.1 (2014), pp. 1929–1958.
32. Seiya Tokui, Kenta Oono, Shohei Hido, and Justin Clayton. "Chainer: a next-generation open source framework for deep learning". In: *Proceedings of workshop on machine learning systems (LearningSys) in the twenty-ninth annual conference on neural information processing systems (NIPS)* Vol. 5. 2015, pp. 1–6.
33. Weights and Biases, Inc. *Wandb*. https://github.com/wandb/wandb. 2018.
34. Matei Zaharia et al. "Accelerating the machine learning lifecycle with MLflow". In: *IEEE Data Eng. Bull.* 41.4 (2018), pp. 39–45.

Index

B. Ghojogh and A. Ghodsi, *Elements of Deep Learning*,
https://doi.org/10.1007/978-3-032-10738-1

M

N

O

P

T

FSC
www.fsc.org

MIX
Papier aus verantwortungsvollen Quellen
Paper from responsible sources
FSC® C105338

If you have any concerns about our products,
you can contact us on
ProductSafety@springernature.com

In case Publisher is established outside the EU,
the EU authorized representative is:
Springer Nature Customer Service Center GmbH
Europaplatz 3, 69115 Heidelberg, Germany

Printed by Libri Plureos GmbH
in Hamburg, Germany